SHARKS

OF THE WORLD

SHARKS
OF THE WORLD

LEONARD COMPAGNO
MARC DANDO
SARAH FOWLER

Princeton University Press
Princeton and Oxford

Princeton Field Guides

Rooted in field experience and scientific study, Princeton's guides to animals and plants are the authority for professional scientists and amateur naturalists alike. **Princeton Field Guides** present this information in a compact format carefully designed for easy use in the field. The guides illustrate every species in color and provide detailed information on identification, distribution, and biology.

Recent Titles

Mammals of North America, by Roland W. Kays and Don E. Wilson
Reptiles and Amphibians of Europe, by E. Nicholas Arnold
Marine Mammals of the North Atlantic, by Carl Christian Kinze
Birds of Chile, by Alvaro Jaramillo
Birds of the West Indies, by Herbert Raffaele, James Wiley, Orlando Garrido, Allan Keith, and Janis Raffaele
Birds of Northern India, by Richard Grimmett and Tim Inskipp
Birds of Africa South of the Sahara, by Ian Sinclair and Peter Ryan
Reptiles of Australia, by Steve Wilson and Gerry Swan
Birds of Australia, Seventh Edition, by Ken Simpson and Nicolas Day
Birds of the Middle East, by R. F. Porter, S. Christensen, and P. Schiermacker-Hansen
Sharks of the World, by Leonard Compagno, Marc Dando, and Sarah Fowler

Published in the United States, Canada and the Philippine Islands
by Princeton University Press,
41 William Street, Princeton, New Jersey 08540

Published by arrangement with HarperCollins*Publishers* Ltd.

LIbrary of Congress Control Number 2004111901

ISBN 0-691-12071-4 (cloth)
0-691-12072-2 (paper)

nathist.princeton.edu

Designed and produced by Fluke Art
Printed and bound in Thailand by Imago

1 3 5 7 9 10 8 6 4 2

Contents

Foreword

I remember the first time I understood the literal meaning of the word breathtaking. I had just entered Venice and was so dazzled by the beauty of the city that I could not breathe. The second time I had this reaction was when I was preparing for a board meeting of the UK conservation charity the Shark Trust with my fellow trustee Sarah Fowler, who is a co-author of *Collins Field Guide to Sharks of the World*. Sarah showed me some of Marc Dando's artwork for this tome and I was so impressed by the breathtaking quality of the illustrations that I forgot about the board meeting and had to be dragged back into mundane reality.

Sharks are notoriously difficult to depict realistically in artwork. Some species are easily identified while others are only subtly distinguishable from each other. The latter makes artistic depiction all the more challenging. The ability to capture the essence of a species is a very special skill and Marc Dando most certainly has it. Identifying a shark is a bit like meeting someone you knew thirty years ago and recognising them again: each shark species has that sort of persisting individuality shining forth through the bumps and bruises of its individuals. But few can convey this in their artwork – and also give the animals such life!

With Leonard Compagno, the world's leading shark systematist, Sarah Fowler has done a grand job in presenting to the public in admirably clear form, the natural history and breathtaking variety of species of this charismatic group of animals. This concise volume is in fact a monumental achievement: three and a half years of work compressed into a single, authoritative tome, drawing on the authors' combined total of over 60 years' work in shark taxonomy, marine nature conservation and marine illustration. Before, popular shark identification books inevitably had errors in them (copied from other volumes) that generated on-going confusion. Now there are no excuses: *Sharks of the World*, written by the real experts, is at the leading edge of the subject and destined to be, the moment it hits the bookshops, a classic.

Readers interested in the remarkably sophisticated natural history of sharks can dip in to study a specific issue while those of more courageous temperament can attempt to master the bewildering variety of animals included herein: a heroic task given that 453 shark species are illustrated.

Sharks are being exterminated by commercial fishermen at a rate at which the figures seem almost ungraspable: over 100,000,000 a year. I hope this outstanding book will make people aware of how varied, intriguing and sophisticated (not to mention magnificent) sharks truly are and therefore ignite concern for their conservation. They say that fact is stranger than fiction and for sharks it most certainly is: turn off the Jaws video, read this volume and discover for yourself the true world of sharks.

Jeremy Stafford-Deitsch

Acknowledgements

It would be impossible here to list all those individuals who have directly or indirectly contributed to this field guide, both over the past decades and during the last couple of years of writing. Many members of the IUCN Shark Specialist Group have been extremely generous with their time and knowledge, enthusiastically passing on scientific information and illustrations and spreading their dedication to shark research, conservation and management not just among the authors but also, through the authors, to the users of this book. We are fortunate to be able to work with these talented and generous individuals. Particular thanks go to Clinton Duffy, Dave Ebert, Ian Fergusson, Phil Heemstra, Michelle Heupel, Brett Human, Peter Kyne, Julio Lamilla, Peter Last, Kazuhiro Nakaya, Andy Payne, Barrie Rose, Bernard Seret, Malcolm Smale, John Stevens, and Kazunari Yano.

Closer to home, our families have all suffered from the birth pangs of this book. SF would like to thank Jonathan, Matthew and Rebecca Spencer who have given up much family time and donated uncounted hours to this work. LJVC could contribute to this work as part of his job, but he would especially like to thank his wife, Martina Roeleveld, for her support and patience during its development and production, particularly during the final and rather hectic phase of its preparation. MD would not have been able to start, let alone complete, this exhaustive book without the support of his wife Julie. He would also like to thank Ryan and Megan for their understanding in the hours given up to this book.

LJVC's work was supported by the South African National Research Foundation, the South African Museum, and the Food and Agriculture Organisation of the United Nation.

Final thanks goes to all those at Collins, especially Myles Archibald who had the foresight to sign up such a massive project and Helen Brocklehurst who has steered the project through.

Introduction

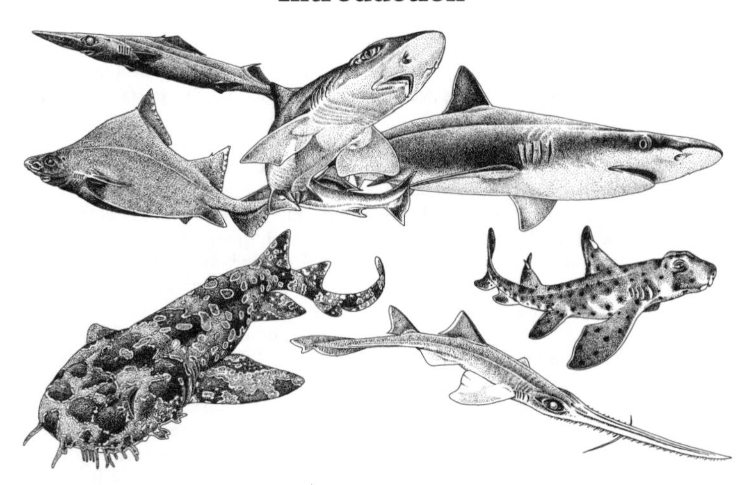

This book is the first single-volume illustrated field guide to all of the known sharks of the world, from freshwater and the intertidal to the deep ocean.

The difficulties involved in collating and matching many sources of information (including written descriptions, photographs, illustrations and preserved museum specimens) when writing this guide and producing accurate colour illustrations of all species as they appear in life have sometimes been almost overwhelming! Although the authors have drawn heavily on the technical *FAO Species Catalogue Sharks of the World*, the original two-volume publication (Compagno 1984) is now rather dated. The major revision of the FAO catalogue is also still in progress, with only one of the three new volumes so far published (Compagno 2001) and a second one nearly complete. Indeed, during the writing of this field guide, several completely new species have been discovered and described.

Scientists and shark enthusiasts will both notice that a few well-known species names are missing from this publication, because recent research has demonstrated that they are actually the same as other species with different names that were described earlier, or because the species name was actually used first for other taxa. This means that this small volume currently (if only briefly) represents the most up-to-date review of some groups of shark species. Unfortunately, we recognise that by the time it appears in the shops we may already know about other new species not included here. We also suspect that some individual shark species that are described from different regions or oceans under the same name are actually several quite distinct, isolated species with a similar appearance (sibling species) or are regional subspecies that are genetically isolated and further research is required to prove this. But, one of the exciting results of producing a field guide of this sort is that it will encourage and stimulate more people to go out to study sharks – whether by observation or photography or by collecting specimens for museum collections – and to discover new information that contradicts or amplifies the descriptions and illustrations in this volume. Indeed, we actively encourage readers to do so, because our knowledge of these amazing animals in many parts of the world is seriously lacking and there is so much to learn about shark biodiversity and general biology before the latest projected phase of marine extinctions makes this impossible.

How to use this book

The structure of this book will be familiar to anyone who has used other species field guides. See Glossary on pages 327 to 338 for an explanation of technical terms used in the text, and the illustrations on pages 12 to 14 to identify the parts of the animal named.

The brief introductory chapters are followed on page 54 by an illustrated dichotomous key to the orders and families of sharks, to help readers get to the right sections of the book for a shark that they have seen. We recommend using this key when trying to identify any shark. Start at the beginning and follow the numbers through the pages until you identify the order or family within which the shark occurs. You can then turn to the plates and more detailed text at the back of the book to identify the species.

The key is followed by the coloured plates, which are organised in taxonomic sequence, commencing with the most primitive species (the frilled sharks) and ending with the most recently evolved hammerhead sharks. These illustrate the species as they appear in life (sometimes obtained from very sparse information indeed). The text alongside each plate provides a small amount of information on the species illustrated. One of the challenges faced when preparing this book is the very large number of shark species that are known from just one or two specimens. Often these examples have been lost from collections and in some cases there are no illustrations (and certainly no photographs or colour illustrations of the fresh animals) available to work from.

More information about each species is given in the following part of the book. The species descriptions are also arranged in taxonomic sequence, organised by order and subdivided further by family. Each section starts by briefly describing the characteristics of each order, then the family, including the most important external diagnostic identification characters of the species that they contain (e.g. the number and shape of fins, fin spines, gills, position of mouth etc.), their biology, and their status. We recommend reading these basic descriptions first, to ensure that they match the animal you are trying to identify, before trying to identify any species.

If all else fails (it may be very difficult indeed to distinguish between some species, e.g. the lantern sharks or demon catsharks), determined readers should consult the most recently published volumes of the *FAO Species Catalogue Sharks of the World* or FAO's regional species catalogues. The updated *Catalogue of Sharks of the World* is being posted on the FAO website (www.fao.org) as each volume is published.

The introduction to each family is followed by an entry for each of the individual species within it, identified both by its most widely used common name and by its scientific name. Species are listed in the alphabetical order of their scientific names within each family. Every species that has been formally described by a scientist has two scientific names that are always used together and italicised. The first name indicates the species' genus, the second the individual species name. Thus, order Lamniformes, the mackerel sharks, includes seven families. One of these is family Lamnidae, which includes three genera: *Carcharodon*, *Isurus* and *Lamna*. There are two species of *Lamna*, the Salmon Shark *Lamna ditropis* from the North Pacific, and the Porbeagle Shark *Lamna nasus* from the North Atlantic and southern oceans. You will, therefore, find all species in family Lamnidae listed in alphabetical order, first by genus, then by species, with the White Shark *Carcharodon carcharias* first on the list and Porbeagle Shark *Lamna nasus* at the end.

Each species description is accompanied by a simple line drawing of the side view of the animal. The underside of the head and other diagnostic features (e.g. teeth) are also illustrated, where sufficient information is available to enable this to be done.

The species description starts with measurements, where known (this information is incomplete for many species). All measurements are given as total length in centimetres, measured as the 'point to point' distance (i.e. not over the curve of the body) from the tip of the snout to the tip of the caudal fin (tail). Where available, these include total length at 'birth', at 'maturity' and the 'maximum' size. For egg laying species, we also try to provide information on size of eggcase and length of young when they 'hatch'. 'Mature' indicates the size at which this species first reaches sexual maturity (this is often different for males (♂) and for females (♀); the latter are generally larger). 'Immature', 'Adolescent' or 'Adult' lengths may be given if information on size at maturity is not available. The 'Maximum' length is often the largest recorded in the scientific literature, not the maximum that a species is capable of reaching. An apology or explanation is due for anglers, who like to know the weight rather than or as well as the length of the sharks that they catch. Weights are not often recorded in the scientific literature, because these are very variable, depending upon the time of year, state of pregnancy, recent meals and so on. Length, on the other hand, is a more constant, reliable and useful measure of size (and age) of animal.

The identification text must be read in conjunction with the section summarising the identification characters for the order and family. It includes each species' characteristic coloration, shape and other distinctive features, some of which are illustrated. Please consult the drawings on pages 12 to 14 to identify the parts of the animal referred to in the text.

Distribution lists the oceans and (for sharks with a limited geographic range) the countries where the species has been recorded. A distribution map (as known at the time of writing) is also provided. Readers should note that many distribution maps are likely incomplete, because there has been so little monitoring and shark research in many parts of the world. It is quite possible that many species have a wider distribution than illustrated, but it is unlikely that a Pacific Ocean species will turn up in, for example, the Mediterranean or Caribbean Seas. On the other hand, it is also possible that certain species recorded from widely separated locations (particularly poor swimmers that live on or near the seabed) may in due course be discovered to be different species with different biological characteristics, possibly even occurring in different habitats.

Habitat describes where the animal most commonly lives (e.g. bottom or pelagic, intertidal or deepwater), including the depths below sea-level in metres where it has been reported.

Behaviour may include movements and migrations and characteristic feeding activities, where known.

Biology includes what is known about the reproduction of the species (see also pages 38–41) and its prey.

Status very briefly describes the conservation, management and fisheries status of the animal and notes when it is known to have caused injury to people . This starts with the *2003 IUCN Red List Threatened Species Assessment*, where available. See page 51 for more information and please always check www.redlist.org for the most recent status of each species; more species are continually being assessed and old assessments reviewed as more information becomes available.

The glossary of terms is provided on pages 327 to 335, a checklist of species on pages 357 to 368, and a list of further reading on page 342.

Topography

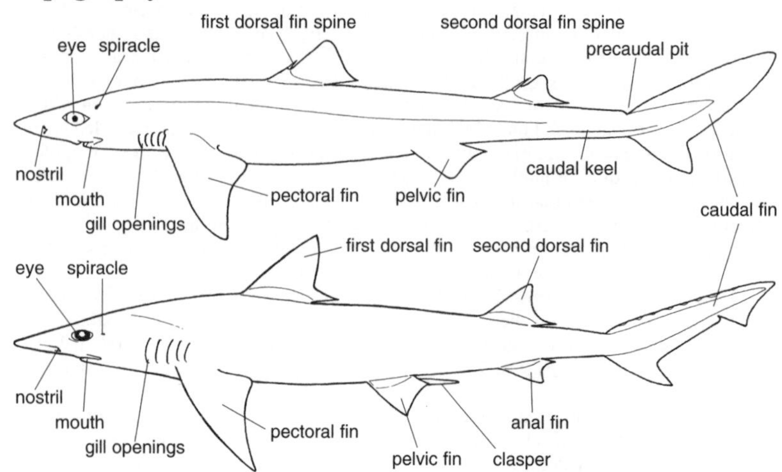

Figure 1: Lateral views – (top) female squalid shark; (bottom) male hemigaleid shark

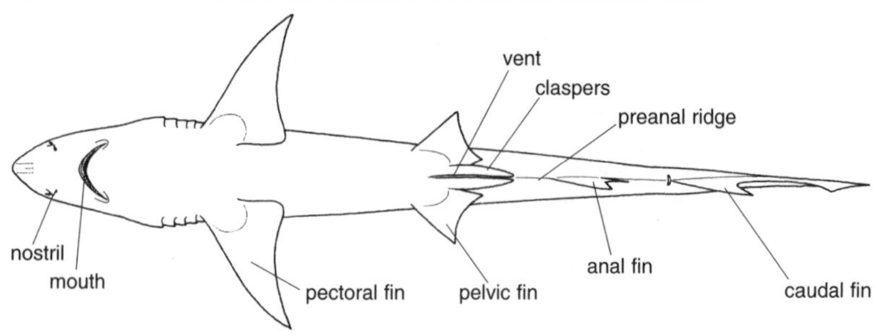

Figure 2: Ventral view – male hemigaleid shark

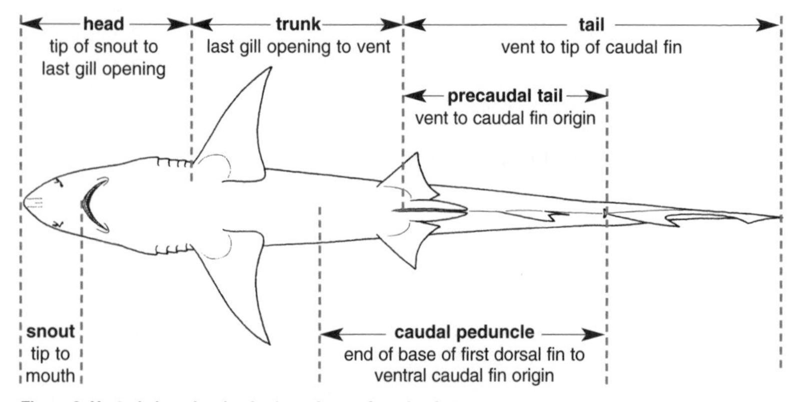

Figure 3: Ventral view showing body regions referred to in text

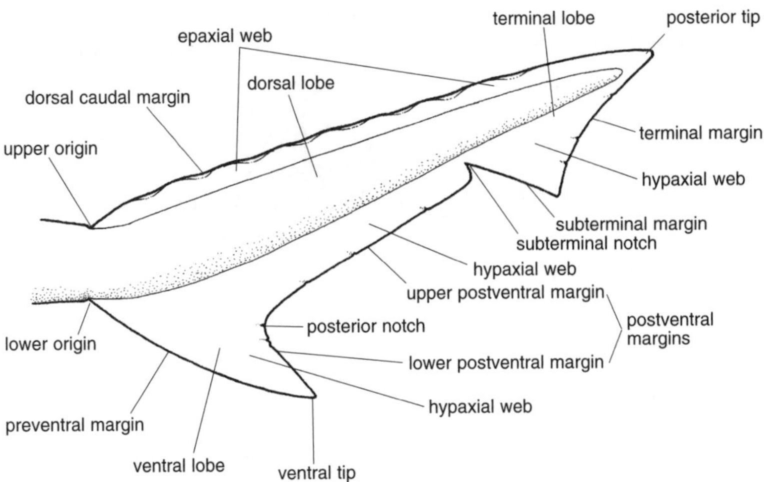

Figure 4: Caudal fin topography

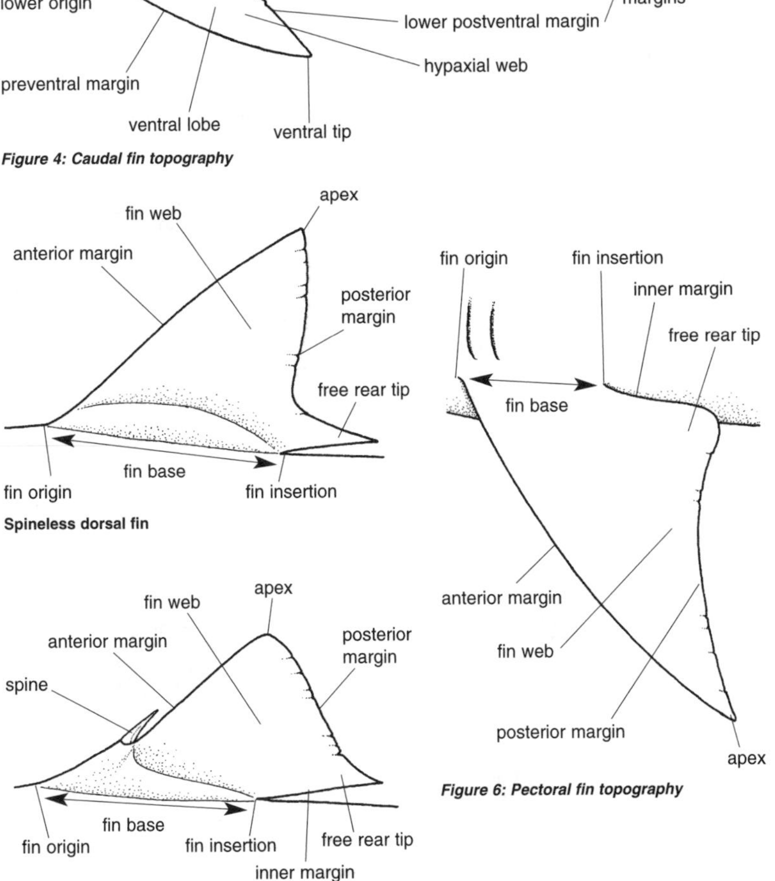

Figure 5: Dorsal fin topography

Figure 6: Pectoral fin topography

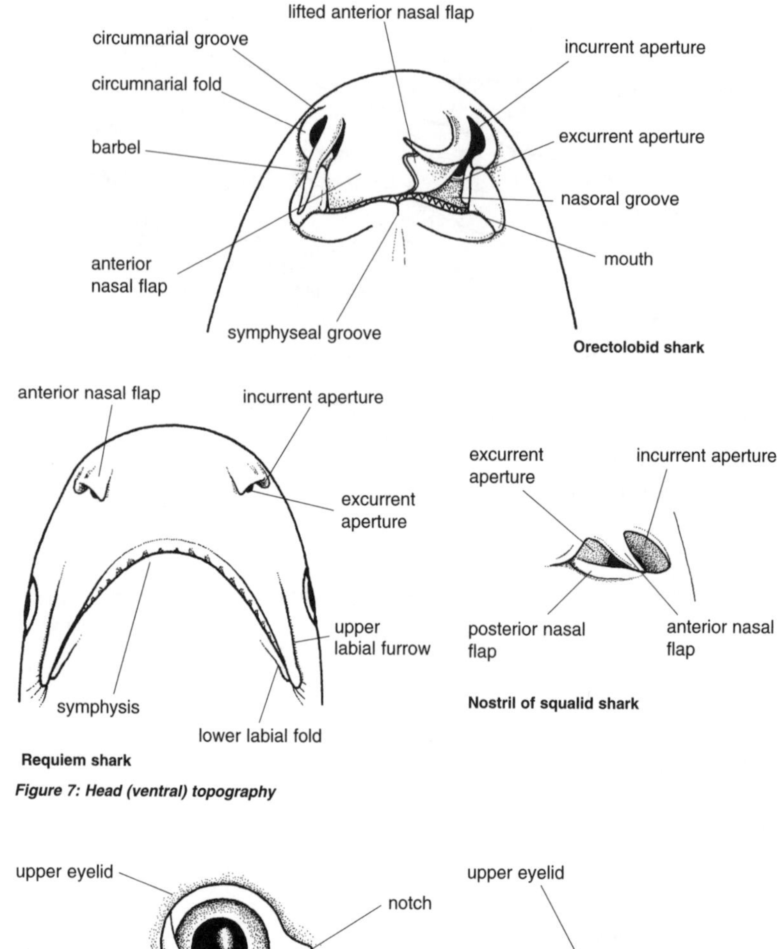

Figure 7: Head (ventral) topography

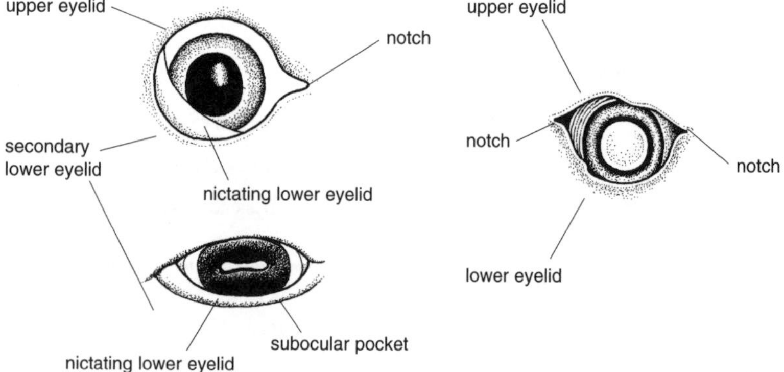

Figure 8: Eyes of (top left) requiem, (right) kitefin and (bottom) catsharks

What is a shark?

Sharks belong to the taxonomic class Chondrichthyes, or cartilaginous fishes. As the name suggests, all of these fishes have a simple internal skeleton formed from flexible cartilage. There is no true bone present in their skeleton, fins or scales. They also have a true upper and lower jaw (unlike the primitive jawless lampreys, which also have a cartilaginous skeleton) and nostrils below their head.

There are two main groups of chondrichthyan fishes. The largest of these is the subclass Elasmobranchii ('elasmo' means plate and 'branchii' means gills), which includes the sharks and rays. The elasmobranchs are easily recognised by the multiple (five to seven) paired gill openings on the sides of their head. Subclass Holocephali ('holo': whole, 'cephali': head) contains the chimaeras, which are a very much smaller group of living animals, although they were more abundant in the fossil record. A soft gill cover with just a single opening on each side of the head protects the four pairs of gill openings in holocephalans.

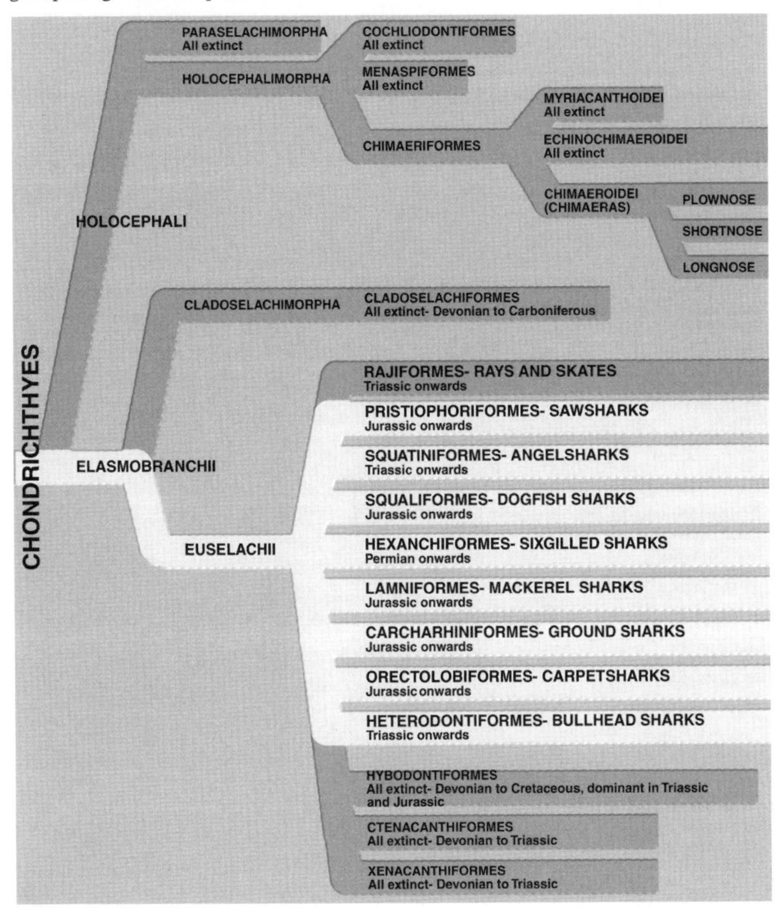

Figure 9: Family tree of Chondrichthyes.

The sharks comprise about 500 (and rising) known species of elasmobranchs. They are usually cylindrical, sometimes flattened, and characterised by five to seven paired gill openings on the side of their head and pectoral fins that are not attached to the head above the gill openings. They also have a large caudal fin (tail) and one or two dorsal fins (sometimes with spines). Some species of shark are flattened like the majority of rays, but can still be distinguished by the diagnostic location of their gill openings and shape of their pectoral fins, as described above.

Rays, or batoid fishes, which are not included in this book, are flattened sharks, sometimes known affectionately as 'pancake sharks'. There are over 600 species of ray, characterised by the expanded and flattened pectoral fins that are fused to the sides of the head above the gill openings, rather like wings, on a generally short, flat body. Some species of ray, however, look more like flattened sharks than the more familiar disc-shaped stingrays and skates. Examples of these are the guitarfish and sawfishes, which look very like the angel sharks and sawsharks.

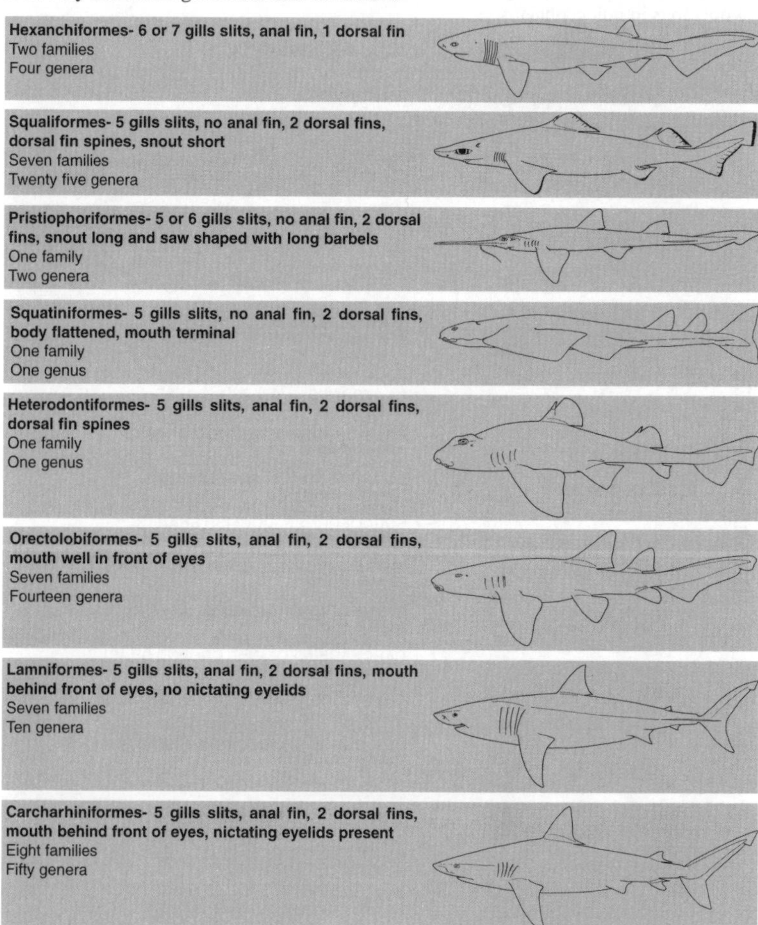

Figure 10: There are ten extant orders of Chondrichthyes, here are the eight which are sharks.

The evolution of fishes

All vertebrates (mammals, birds, reptiles, amphibians and fishes) are descended from small, simple, superficially fish-like creatures that occurred in the world's oceans about 500 million years ago (the Cambrian period). These animals, our common ancestors, were fairly long and flattened from side to side (rather leaf-shaped), with blocks of muscle attached to a rod running along the length of the body and a simple gut. They had no eyes, fins or bones and were obviously rather poor swimmers. The *Branchiostoma* or lancelet (which may be found burrowing into sand in shallow water) is the closest example of this form of animal still in existence today.

Figure 11: Branchiostoma lying at surface and burrowing in sand.

Over the next few tens of millions of years, however, these simple animals evolved into creatures that are recognisably fish-like in the fossil record. The original simple rod developed into a cartilaginous backbone, simple eyes and unpaired fins appeared, and the head and upper body became protected by bony plates. The absence of jaws presumably meant that these animals could only feed by scooping small food items or detritus off the seabed. Later still, biting jaws evolved, probably adapted from the first of the series of cartilaginous arches that support the gills. Jaws, together with improved swimming ability as a result of the evolution of paired fins, enabled fishes to feed upon a much wider range of prey living in a wider range of habitats and to evolve into much more varied forms including some that could breathe air and crawl out on land. Many of these ancestral fishes and their descendents have since become extinct.

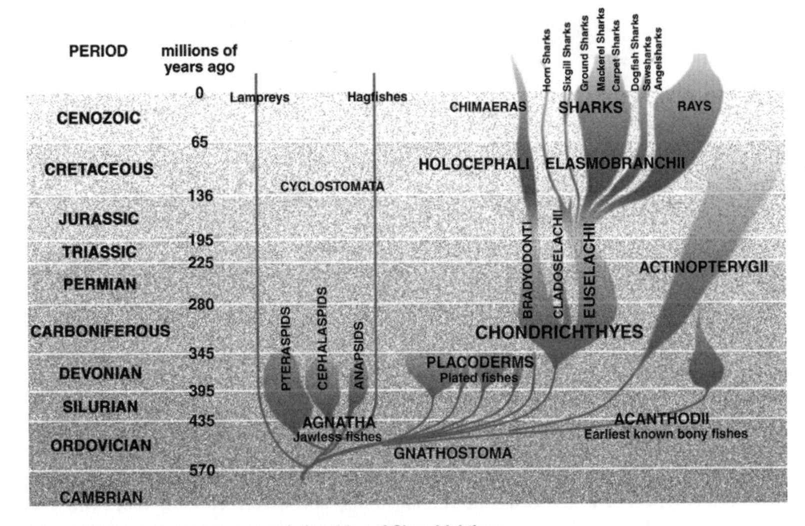

Figure 12: Possible evolutionary relationships of Chondrichthyes.

Most scientists agree that by around 400 million years ago (200 million years before the dinosaurs walked the earth) the evolution of the ancestral fishes had evolved and diverged into the two main groups that survive today, as well as into other groups that exist only in the fossil record. One of these two groups, the class Chondrichthyes, kept the cartilaginous skeleton (hence the name derived from the Greek 'chondros': cartilage and 'ichthos': fish) and lack internal bones and external flattened bony scales. The other successful evolutionary line replaced cartilage with bone and is known as the bony fishes, class Osteichthyes ('osteos': bone, 'ichthos': fish), including the modern ray-finned bony fishes or teleosts and their primitive relatives including paddlefish, sturgeon, garpike, bowfin, and reedfish. The lobefin fishes, Sarcopterygii ('sarcos', flesh, and 'pteryx', fin or wing) which include our ultimate ancestors and the living coelacanth and lungfishes, are often included with the ray-finned bony fishes in the Osteichthyes or placed in a class of their own. Bony fishes are by far the most abundant of vertebrates, with an estimated 25,000 or so living species. Sharks and their relatives were much more abundant and diverse in the fossil record, which includes over 3000 fossil species (many more must have become extinct without trace). Some fossil species that lived 150 million years ago are almost identical to some modern sharks, rays and chimaeras. Today, we know of about 1100 living species, but this number is rising as more species new to science are continually being discovered and named.

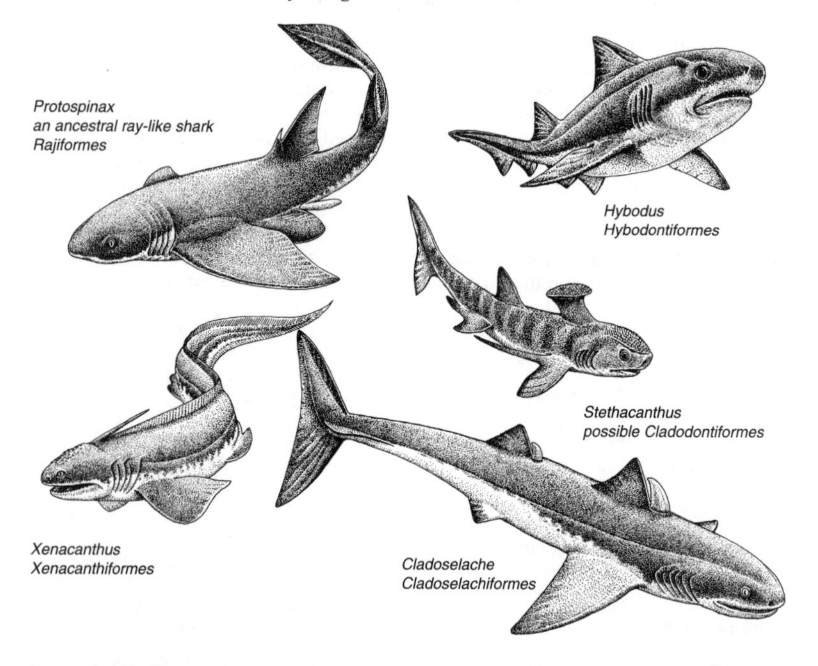

Figure 13: Prehistoric elasmobranchs, four of the orders, Hybodontiformes, Cladodontiformes, Xenacanthiformes and Cladoselachiformes have no living representative.

This field guide covers some 453 species of sharks, including all described and valid species and well-known undescribed species (particularly from Australia, where sharks have been particularly well-studied in recent years). There are still more, poorly known, undescribed species, which could not be illustrated and described here, that bring the number up to about 500.

Biology

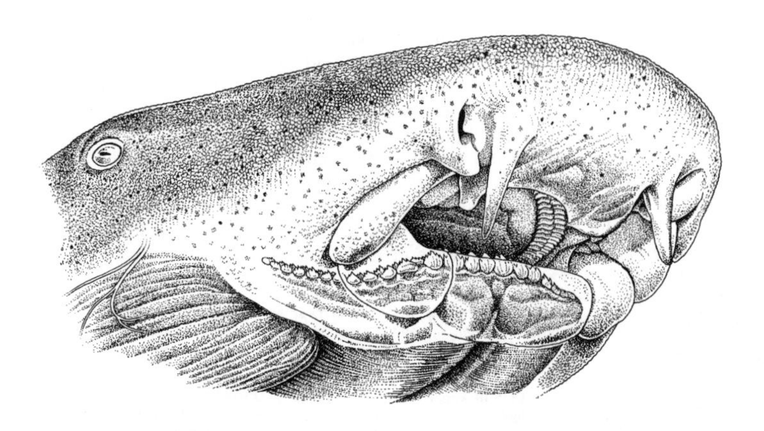

The extremely short descriptions of the biology and life history of all sharks that are summarised species by species in just a few words on pages 20 to 42 cannot possibly do justice to the highly sophisticated and varied characteristics of these species, developed over the past few million years of their existence. These introductory pages therefore provide a slightly more detailed introduction to the biology of this extraordinary group of animals. We hope that this will interest our readers and perhaps lead you to undertake some more detailed reading elsewhere.

Starting from the outside, the following pages describe shark body structure from scales, skin and fins, through to skeleton, teeth, musculature and movement, blood and circulation. We describe briefly the necessary physiological adaptations associated with being a successful predator in a cold salty environment, as well as the awe-inspiring special senses of these large-brained animals and a selection of their wide range of highly successful feeding strategies. Perhaps the most varied biological characters of the sharks are their breeding strategies, ranging from egg-laying, to a wide range of strategies for giving birth to live young (including egg-eating or cannibalism within the uterus and highly specialised placentas).

The specialised biology of these animals is naturally matched by very complex behaviours that have only recently become the subject of scientific study. Some behavioural traits unfortunately bring sharks into uncomfortably close contact with man, almost always to the detriment of the shark. We look briefly at the ancient and modern myths associated with sharks and the current threats to their survival, in comparison with the very small threat that they pose to us.

Body structure

The basic body structure of sharks has remained virtually unchanged for hundreds of millions of years. It consists of a head (from snout to gill region), body or trunk (from pectoral girdle to vent), and tail. The head is comprised of the snout (known as a rostrum in sawsharks) in front of the eyes and mouth, the orbital region (including the eyes and mouth), and the branchial region, with gills and spiracles (the latter have been lost in many groups of sharks). The body extends from the paired pectoral fins to the paired pelvic fins (with claspers in males) and vent. The tail is subdivided into the precaudal tail (which may have an anal fin) and the caudal fin. The first dorsal fin is usually located on the body, the second, if present, on the precaudal tail. While these characters are constant, there is huge variation in size and shape. The largest sharks are gigantic, reaching a length of up to 20 m in the Whale Shark. Most sharks are around one metre in length, but some (such as some Lanternsharks) are tiny and mature at less than 30 cm.

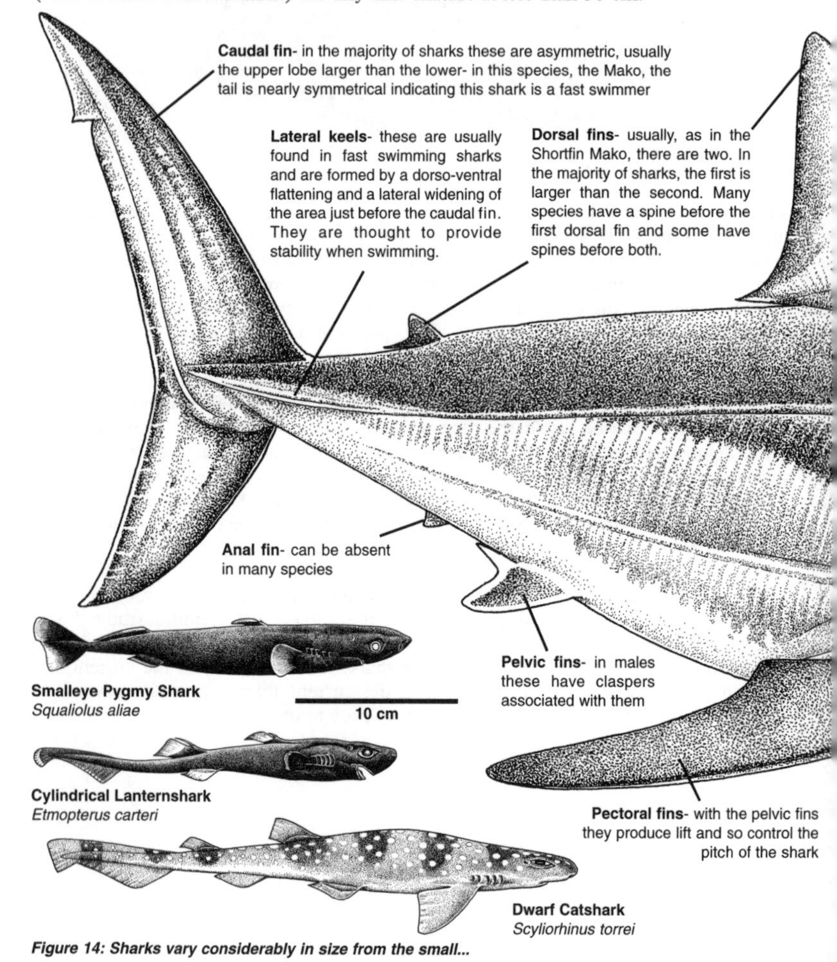

Figure 14: Sharks vary considerably in size from the small...

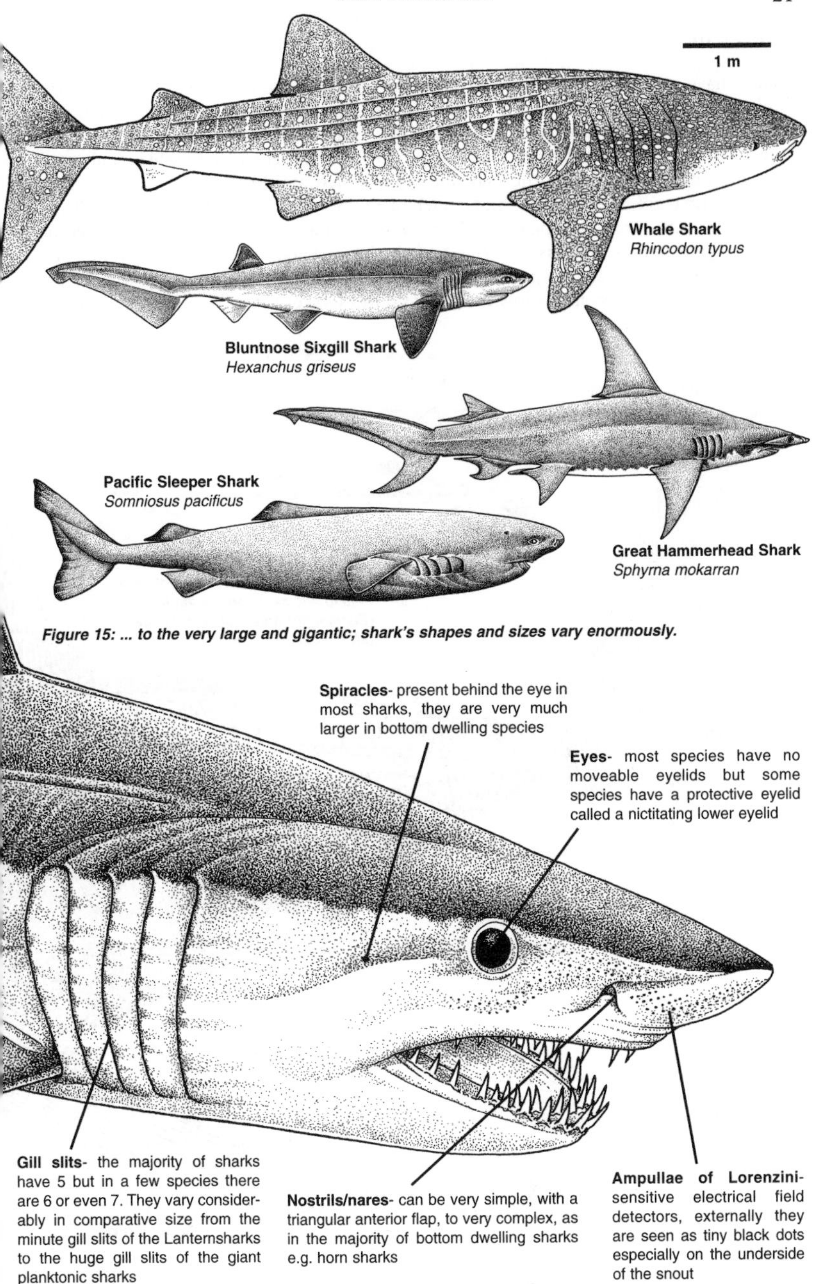

Figure 15: ... to the very large and gigantic; shark's shapes and sizes vary enormously.

Figure 16: External features of the Shortfin Mako, *Isurus oxyrhinchus.*

Skeleton

Sharks have a very simple skeleton compared with bony fishes. One of the most important characteristics is that it is made of cartilage, not bone, and is very light and flexible like the cartilage in human ears or noses. (The cartilage in older larger sharks may be partly calcified, harder and more bone-like.) There is a box-like skull with associated bones supporting the gills, the jaws (which are thought to have developed from the first gill arch and are not attached to the skull), a long vertebral column running from the skull into the tail in the form of a string of hourglass-shaped vertebrae beneath an arch that protects the spinal cord, and cartilaginous structures supporting the fins (and in males, claspers). The lack of connections between most of these structures makes sharks incredibly flexible with most species capable of turning rapidly in a very tight circle.

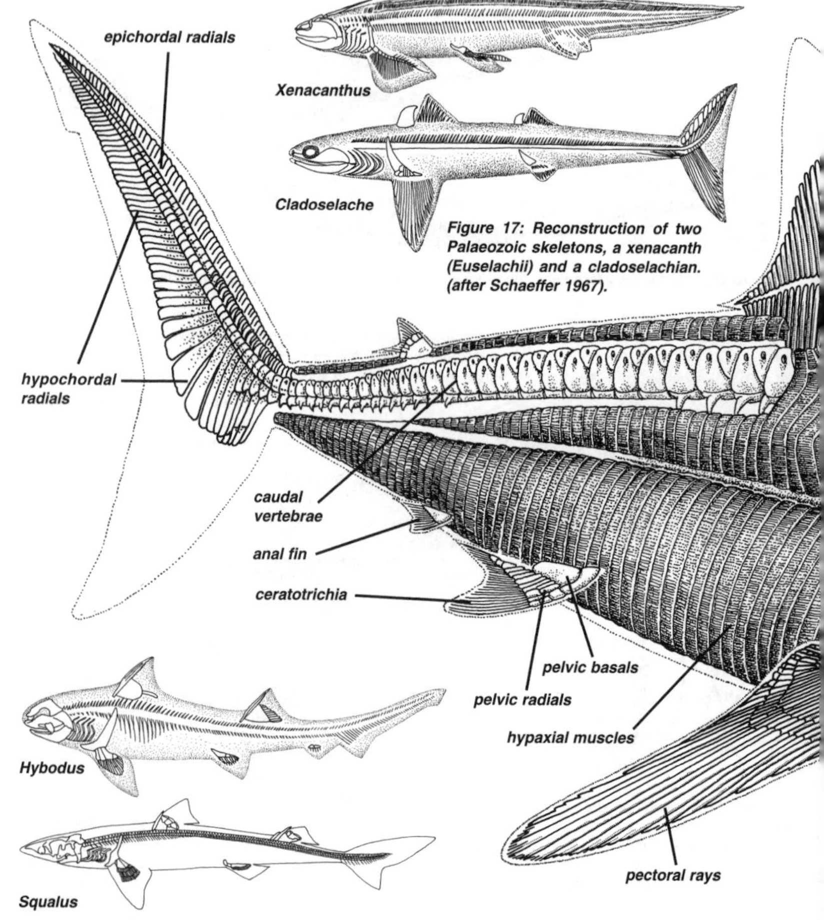

Figure 17: Reconstruction of two Palaeozoic skeletons, a xenacanth (Euselachii) and a cladoselachian. (after Schaeffer 1967).

Figure 18: Recontruction of an hybodont and modern living shark skeletons (after Schaeffer 1967).

Ironically, it is the structure of the shark skeleton, which made it one of the ocean's most advanced and efficient animals, that now poses one of the greatest threats to its future survival. Shark fins, which produce essential 'lift' when swimming, are supported at their base by cartilaginous basal and radial rods and by the caudal vertebrae in the tail. Most of the fin web, however, is supported and stiffened by fine elongated fibres of collagenous tissue known as ceratotrichia. These long fibres are the most important ingredient of shark fin soup. Although flavourless, they are highly valued for their texture and are one of the most valuable seafood products that enters international trade (far more valuable than shark meat). Because shark fins consist primarily of cartilage, skin and ceratotrichia, they are lightweight and easy to dry or freeze and store for long periods on board ship or on land until they can be traded.

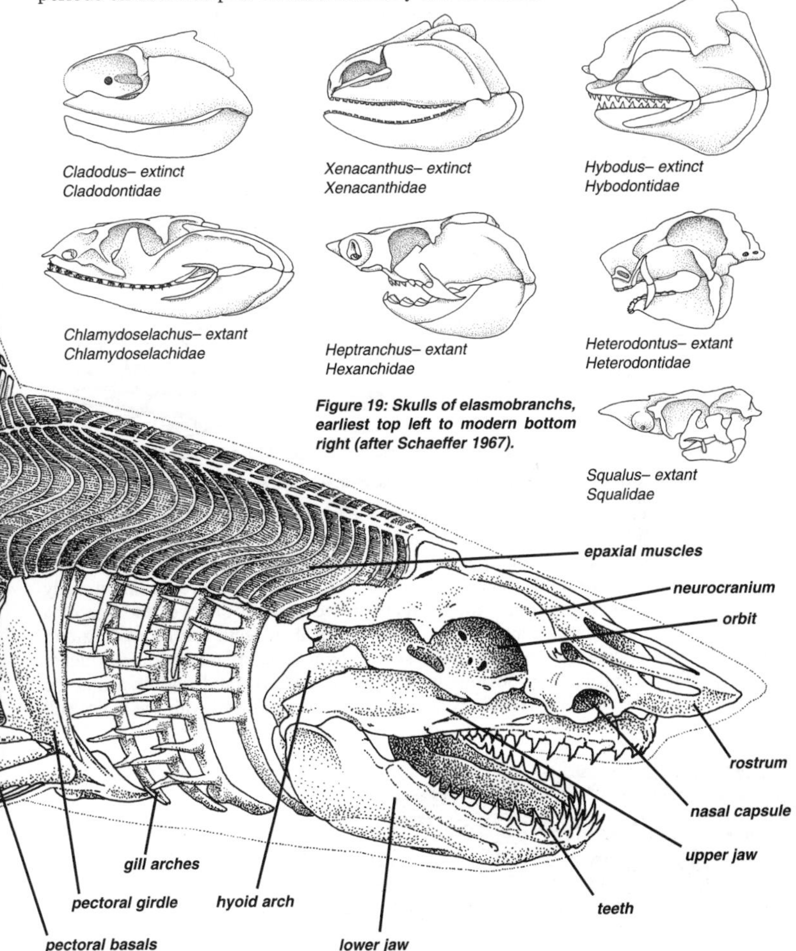

Figure 19: Skulls of elasmobranchs, earliest top left to modern bottom right (after Schaeffer 1967).

Figure 20: A skeleton and muscle cutaway of the Shortfin Mako, *Isurus oxyrhinchus.*

Skin and scales

Sharks are typically protected by a very tough skin, usually covered with small sharp tooth-like placoid scales, also known as dermal denticles, although some species have none on their ventral surface (belly). These denticles are very similar in structure to shark teeth, with crowns covered in hard enamel anchored into the skin by dentine bases. Their shapes are incredibly varied, both between species and on different parts of the body. They include low flattened plates or shields, complex ridged and pointed crowns, and long hooked structures. The shape can be an important diagnostic for distinguishing between some very similar species, but it has not been possible to go into this much detail in a short field guide. One of the most important functions of the dermal denticles, apart from physical protection, is to provide a surface that minimises surface drag and maximises swimming efficiency by producing a laminar flow along the tiny gullies and ridges on the surface of the skin. Industry is now trying to mimic this function of shark skin to produce energy efficient surfaces for swimming costumes and underwater vehicles.

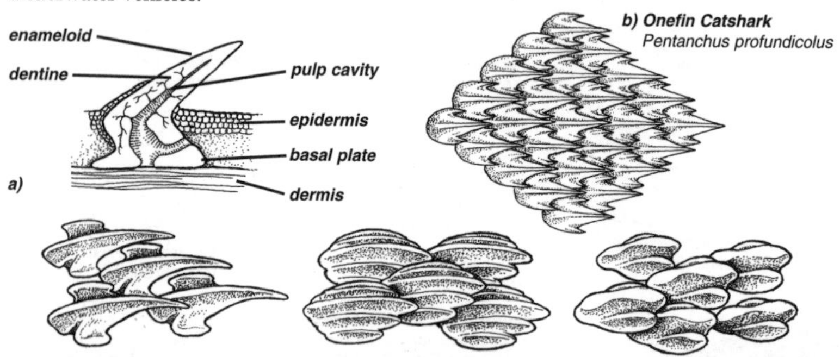

b) Whale Shark *Rhincodon typus* **b) requiem shark** *Carcharhinus sp.* **b) Tiger Shark** *Galeocerdo cuvier*

Figure 21: a) cutaway of a dermal denticle; b) examples of dermal denticles in four species of shark (from Sealife 2000)

Dermal denticles fall out continuously during the life of the shark, with replacements growing out through the skin. In some species the dermal denticles are enlarged to form specialised structures such as fin spines, rostral teeth in Sawsharks and the enlarged thorns or bucklers of Bramble Sharks; these do not fall out but grow continuously (annual growth rings in fin spines may sometimes be used to discover the age of a shark). Because male sharks may hold the female with their teeth during mating and inflict fairly serious bites (see page 39), the skin of females is often very much thicker than males. This is particularly obvious in Blue Sharks. Shark (and ray) skin may be used to produce thick durable leather. When the leather is produced with the scales still embedded, providing a tough non-slip surface, this is known as shagreen.

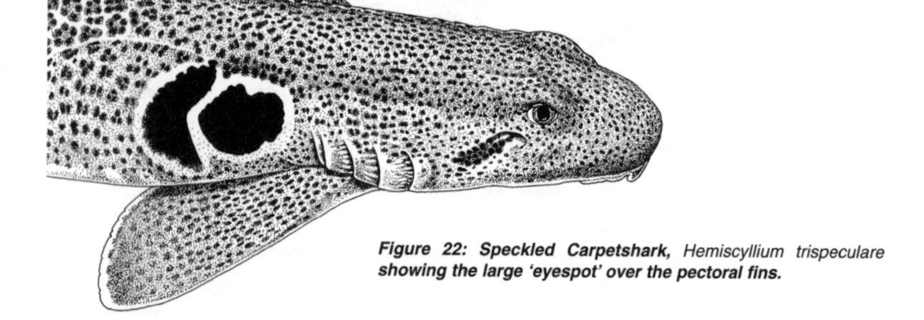

Figure 22: Speckled Carpetshark, *Hemiscyllium trispeculare* ***showing the large 'eyespot' over the pectoral fins.***

Sharks exhibit a huge range of colours, produced by pigment cells in the skin. Most are fairly drab shades of greys and browns, but some are spectacularly marked. Even apparently bright colour markings provide incredibly good camouflage against the seabed. It is possible that the large eyespot markings on some small species like epaulette sharks may help to protect them by deluding possible predators into thinking that they belong to a much larger and more dangerous animal.

Teeth and jaws

All sharks have multiple rows of teeth along the edges of their upper and lower jaws. New teeth form in a deep groove just inside the mouth, behind the rows of teeth. These replacement teeth move forward from inside the mouth on a sort of conveyor belt formed by the skin in which they are anchored, with new teeth continuously replacing the oldest front teeth as these become worn and fall out. Estimates of discard and replacement rates vary between species and at different times of year. They may fall out in just 8-10 days per row, or not for several months. That adds up to a lot of teeth in a lifetime in sharks that may easily live for well over 40 years and which start shedding teeth before they have even been born.

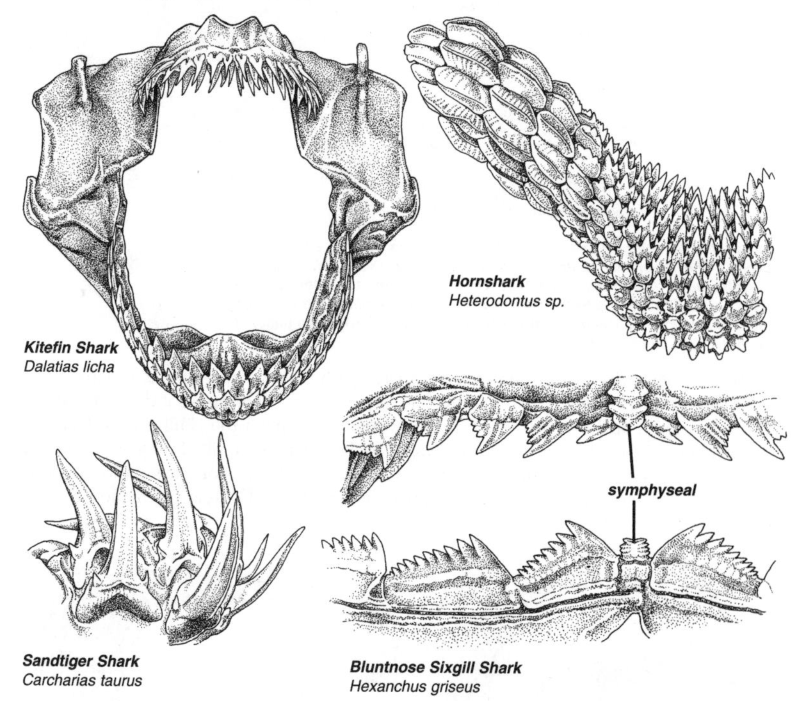

Figure 23: A few of the various shark teeth types, the shape of which often indicate its feeding habits. The Kitefin Shark has many clutching teeth in the upper jaw and a single working row of tearing teeth in the lower; hornsharks have similar upper/lower heterodont clutching (front)-grinding (back) teeth. The Sandtiger's large dagger-like piercing-tearing teeth are tricuspidate; Bluntnose Sixgill Sharks have cutting teeth, the lower jaws are wide and multicuspidate and the uppers are high with slightly less cusps. Crushing-type dentition is found in the houndsharks, with their large numbers of small slightly raised paving slab-like teeth (not illustrated).

Because shark teeth are so hard, formed of enamel over a dentine root, they are incredibly persistent and numerous in the fossil record. It is possible to find apparently pristine, razor-sharp teeth, hundreds of millions of years old, in sedimentary rocks and eroded out of rocks into rivers and seabeds. The huge teeth of the gigantic Megatooth Shark, *Carcharodon megalodon*, can still appear relatively fresh; many, however, are heavily mineralised, even though this species became extinct about four million years ago. It is commonplace to see recently shed teeth of living sharks lying on the bottom of aquaria.

Shark teeth take many forms and sizes. They range from thin spear-shaped or fang-like teeth used for seizing slippery fishes and squid, to very large flat crushing teeth (sometimes arranged like a pavement) specialised for breaking up shellfish, and flat saw-edged teeth for cutting bites out of large prey. Many teeth are so tiny that a microscope is needed to examine them. Not only do the shapes of teeth vary depending upon their position in the jaw, they may also change with age in sharks that start out with spear-like teeth for feeding on small fishes, then become flatter and adapted for cutting chunks out of much larger prey.

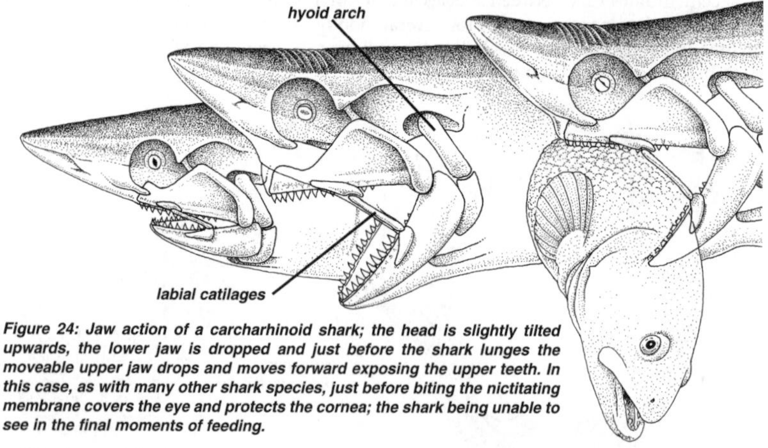

Figure 24: Jaw action of a carcharhinoid shark; the head is slightly tilted upwards, the lower jaw is dropped and just before the shark lunges the moveable upper jaw drops and moves forward exposing the upper teeth. In this case, as with many other shark species, just before biting the nictitating membrane covers the eye and protects the cornea; the shark being unable to see in the final moments of feeding.

The jaws of sharks are most unusual compared with those of mammals and bony fishes as the upper jaw is not part of the skull. The upper jaw can then move independently, so both the upper and lower jaws may be protruded away from the skull during feeding, which enables the teeth to be rotated outwards so that a larger chunk of prey can be bitten off, or providing for more effective suction-feeding. For this reason jaws removed from sharks and put on display are often mounted upside down! Correct orientation can be seen in the figure below.

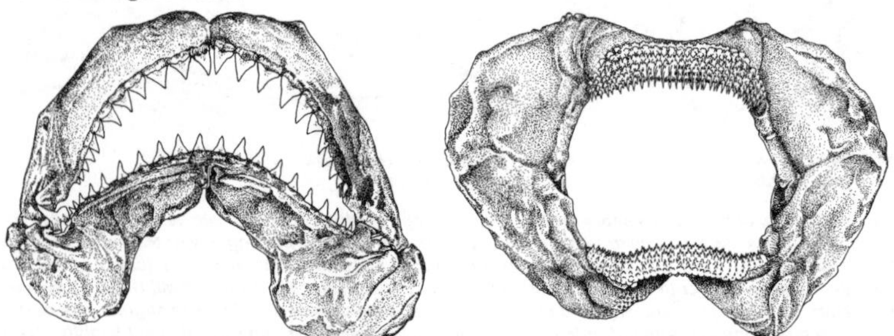

Figure 25: Jaws of (left) the Speartooth Shark *Glyphis glyphis* ***and (right) Zebra Shark*** *Stegasoma fasciatum.*

Movement and respiration

The characteristic zigzag segmented muscle fibres of sharks can easily be recognised in the skinned carcasses of dogfish seen on the fishmonger's slab, while shark steaks, cut across the body, show the long bundles of muscle fibres that run from head to tail. Forward movement is produced when these muscle fibres contract first on one side of the body, then on the other, pulling against the central vertebral column to produce a series of undulations along the body and a sinuous swimming movement powered by the caudal fin. While forward thrust and acceleration is produced by these muscle contractions, the more that a shark's body bends, the less efficient is its use of energy.

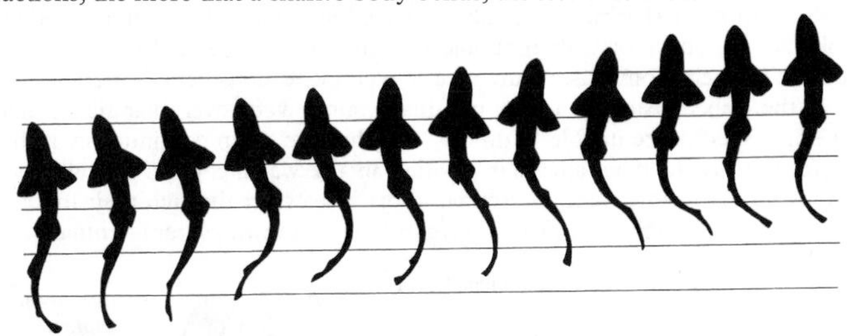

***Figure 26:** Ventral view showing swimming sequence in a catshark, the sinusoidal movement begins at the head and flows down the body becoming more pronounced at the caudal fin (from Sealife 2000).*

Efficient swimmers, therefore, initially accelerate with a series of waves along the body then stiffen their body, producing less pronounced undulations and using less energy when cruising. The fastest sharks (the mackerel shark family) tend to be teardrop in shape, with crescent-shaped tails very similar to those of tunas, while the slow-swimming frilled and cow sharks are longer and thinner with a very long upper tail lobe. (The extremely long upper tail lobe of thresher sharks appears to be an adaptation for herding and striking prey, rather than for propulsion.) The large paired pectoral fins act rather like wings to counter the downward movement produced (along with the forward thrust) by the tail. Fast oceanic swimmers like makos and other mackerel sharks also have two broad keels in front of the tail that provide them with extra stability.

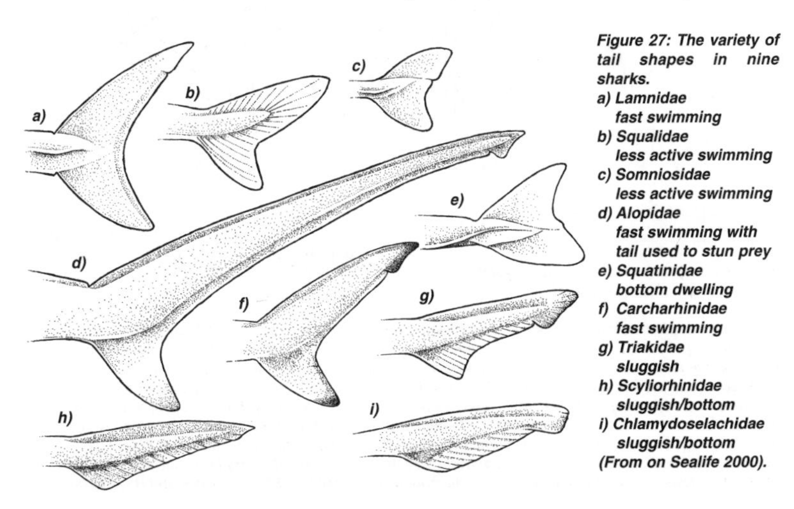

***Figure 27:** The variety of tail shapes in nine sharks.*
a) Lamnidae fast swimming
b) Squalidae less active swimming
c) Somniosidae less active swimming
d) Alopidae fast swimming with tail used to stun prey
e) Squatinidae bottom dwelling
f) Carcharhinidae fast swimming
g) Triakidae sluggish
h) Scyliorhinidae sluggish/bottom
i) Chlamydoselachidae sluggish/bottom
(From on Sealife 2000).

A surprisingly large number of sharks, including some of the bullhead sharks, probably spend as much time walking around on the seabed as they do swimming. They have developed well-muscled mobile paired fins to enable them to do this. Some live in the intertidal and some bamboosharks are even reported to scuttle around out of the water from rock pool to rock pool.

Unlike bony fishes, sharks have no swim bladder to provide them with neutral buoyancy and prevent sinking, but still manage to be almost neutrally buoyant in seawater. While their relatively light cartilaginous skeleton is an advantage, most buoyancy is provided by low density oils stored in the large liver (which may represent up to 25% of the total weight of some deepwater sharks, compared with around 5% in mammals). Sandtiger sharks, commonly displayed in aquaria, have an interesting trick to enable them to hover virtually motionless in mid-water – they gulp air at the surface and store it in their stomachs. Many sharks will spend considerable periods of time resting on the seabed, using their large mouths to pump water over their gills. Other very active pelagic species are unable to do this; they have to swim continuously if they are to survive, not just to maintain their position in the water column but also because continuous forward movement is essential to drive seawater through their mouths and over their gills so that they can extract oxygen. This is known as 'ram-ventilation'.

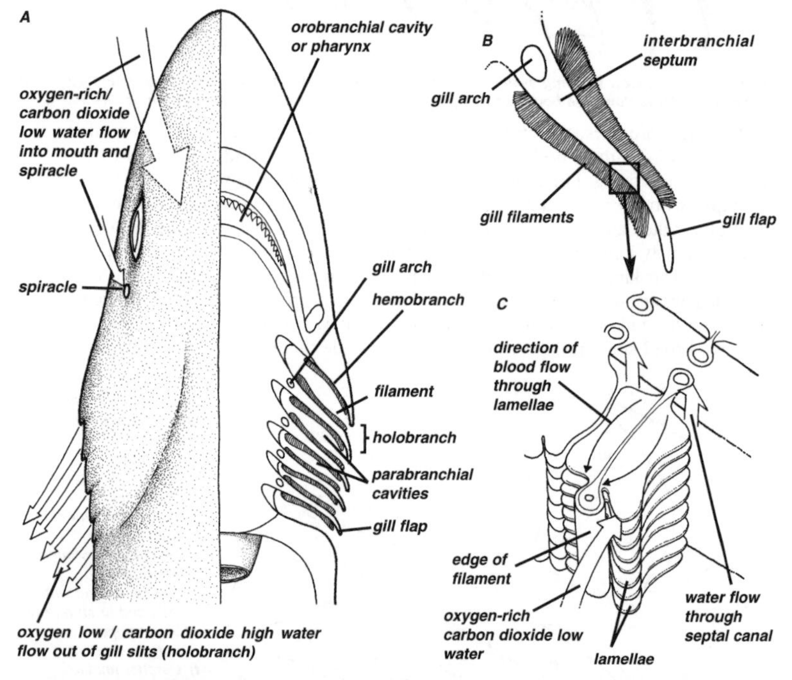

Figure 28: Respiration in a typical shark; a: cutaway and external view of water flow through the mouth and out through the gill slits; b: enlarged lateral cutaway view of one gill. c: diagram of gill septum with attached lamellae; the blood flow through the lamellae is in the opposite direction to the flow of water. Blood heavily depleted of oxygen and high in carbon dioxide is therefore in close proximity to water with the lowest concentration of oxygen and highest in carbon dioxide; the net gaseous exchange is that of oxygen in and carbon dioxide out. At the other end of the lamellae, the now more oxygen rich blood with lower carbon dioxide is surrounded with high oxygen and low carbon dioxide water, so there is still a net flow of oxygen into the blood and carbon dioxide out. This contraflow system, along with the very large surface area of the gill filaments, allows sharks to respire even in turbid conditions. (Based on Sharks, Skates and Rays 1999).

Warm blooded sharks

Actively swimming sharks need a lot of oxygen, so have large hearts (posterior to the gills and protected by the pectoral girdle) that pump a lot of blood around the body and through their large gills. They also produce plenty of heat in their muscles. Most species of shark lose this heat almost immediately when the heated blood from their muscles passes through the narrow-walled blood vessels in the gills in order to pick up oxygen from the colder seawater, then return cold oxygenated blood to the body. Some, however, including the Porbeagle and White Shark, can retain this warmth in their bodies so that, like mammals, they keep a constant body heat even when swimming in seawater several tens of degrees colder than their core temperature. This makes them more efficient predators and faster growing than similar sized cold-blooded sharks. It is difficult enough for marine mammals to maintain body temperature in the cold ocean with the help of thick layers of fat under the skin and lungs well hidden inside the warm body. With sharks, the most important challenge is to stop heat loss through the gills, which continually expose all the blood in the shark's body to seawater temperatures and provide a huge surface area for heat exchange. They have met this challenge by developing a network of tiny blood vessels (capillaries) packed tightly together to act as a heat exchange system. This is known to scientists as a 'rete mirabile' or 'marvellous net'. Within this structure, thin walled blood vessels coming from the gills and carrying cold oxygenated blood runs alongside but in the opposite direction to thin capillaries carrying warm deoxygenated blood from the muscles on its way back to the gills. These two sets of vessels exchange heat so effectively as they run alongside each other that the warmth from the body ends up returning to the muscles with the oxygenated blood instead of being dissipated into the sea. When the shark enters warmer water it does not need to keep so much of the warmth generated by its muscles, so some of the blood is directed through other pathways, around instead of through the rete, allowing the excess heat to be lost into the sea.

Osmoregulation

All marine fishes have to cope with the challenge of being immersed in a salty aquatic environment that draws body water out across skin and gill membranes into the sea (high concentrations of salt always absorb water). Bony fishes cope by drinking lots of seawater and excreting the excess salts through their gills or gut. Sharks have a completely different strategy. They do not drink seawater. Instead they retain high concentrations of 'waste' chemicals or 'salts' in their body in order to change the direction in which water tends travels – out of the sea and into their more concentrated body fluids. One of the most important chemicals retained in sharks is urea, which is excreted rapidly in the urine of mammals and most other animals. Sharks also excrete excess salt across their gills, like bony fishes, and have a special gland that extracts salt out of the blood for excretion through the gut. This adaptation means that most sharks are confined to fully marine environments, being unable to change their management of body fluids to cope with less salty water. The really difficult test is experienced by species such as the Bull Shark, which move freely from the sea to estuaries and freshwater rivers and lakes. These animals have to cope with huge quantities of water flooding into their bodies when they enter freshwater. They do so by completely changing kidney function to excrete large quantities of urea in watery urine and by reversing the direction of movement of salts across the gills, from excretion to absorption from the environment.

Senses

Perhaps the single most important point to note is that sharks are highly evolved complex predatory animals with relatively large brains. Indeed, those of the galeomorph sharks (bullhead, carpet, mackerel and ground sharks) and stingrays are entirely comparable in size and complexity to the brains of birds and mammals, although many squalomorph sharks have smaller brains. The largest and most complex of all are found in the hammerhead sharks and the devil rays. The sections of the brain that deal with muscle control, learning, memory and all the different senses used by sharks (sight, pressure and hearing, electro-senses, smell and taste) have been identified, but scientists still do not understand the functions of many parts of these large brains. Although shark senses have been the subject of many documentaries, but a short summary can be provided here.

Smell and taste are of huge importance to sharks. Not only are they used to detect potential prey or other food items at long distance, but they are likely also important when it comes to detecting shark pheromones in order to find mates, and possibly even for location during their long transoceanic migrations between feeding, mating and pupping grounds.

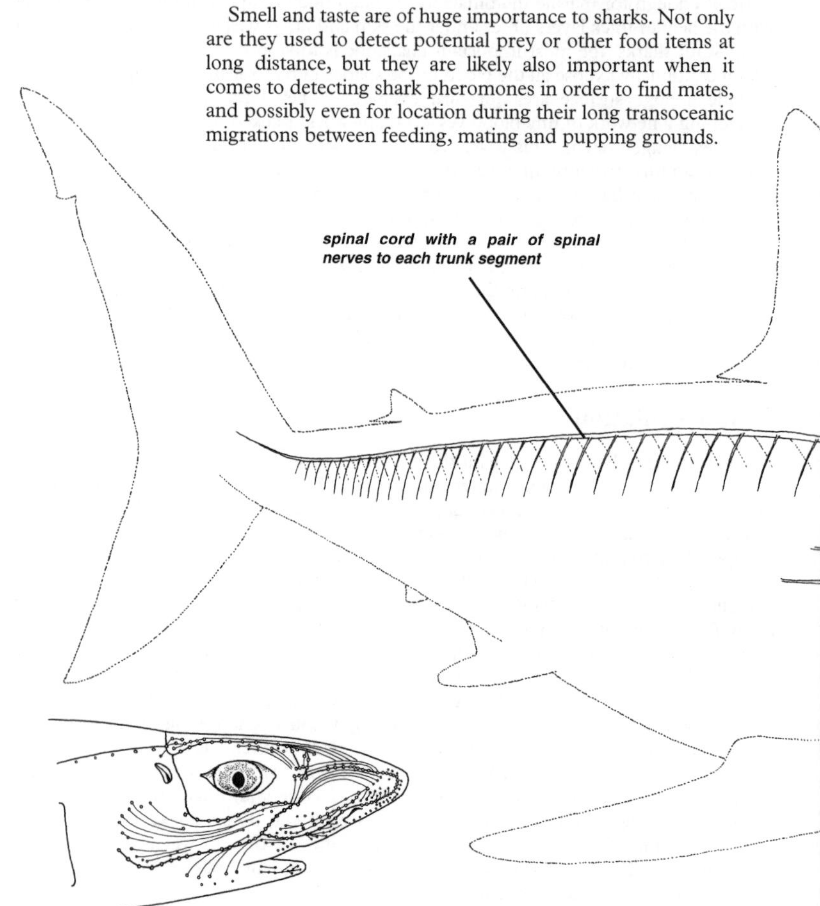

Figure 29: Close-up diagram of a squaloid shark showing the arrangement of the tiny pores of the ampullae of Lorenzini and the arrangement of the lateral line canals in the head region.

The complex system of nasal flaps in front of the mouth (on the underside of the head) directs the incoming water flow over highly sensitive membranes where even a tiny trace of amino acids in the water can be picked up. Sharks zero in on the source of a scent by swimming up current, often using a sinuous track and constantly changing direction to ensure that the signal received by each nostril is as strong as that received by the other. In some species the nose and mouth (which contains taste buds) are connected, in others they are completely separate.

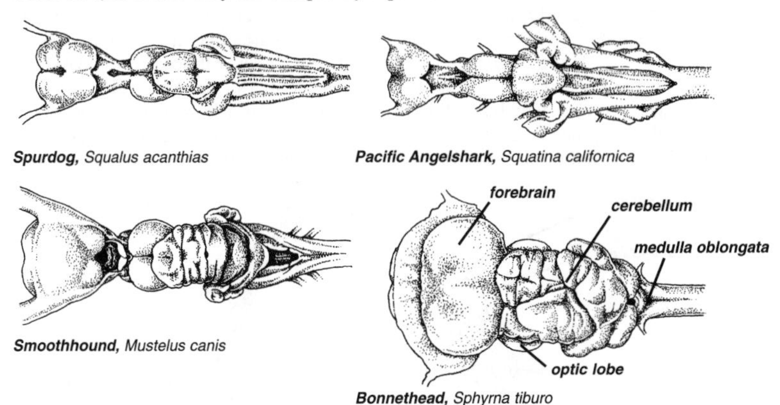

Figure 30: Dorsal views of four shark brains showing the comparative sizes of forebrain and cerebellum, it can especially be seen that the increase of the hammerhead's sensory organs of the head has brought about a considerable increase in size the of the forebrain, (based on Sharks, Skates and Rays 1999).

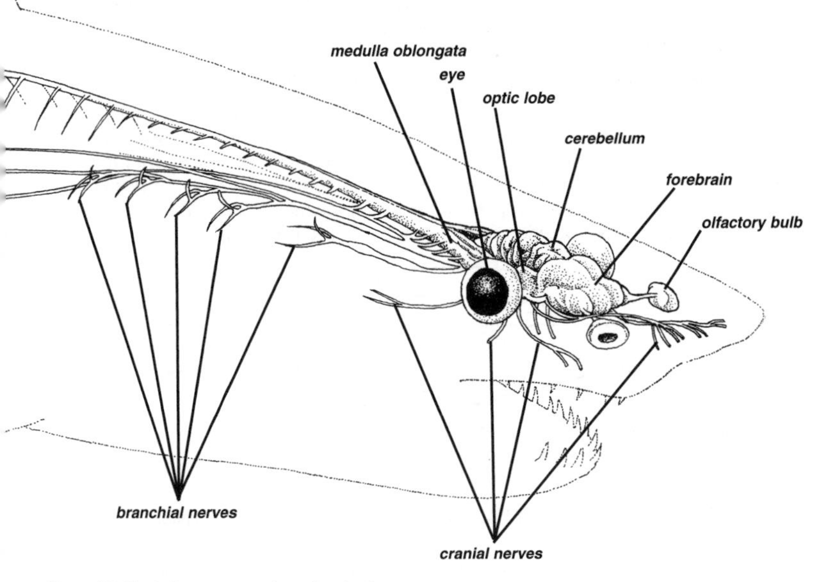

Figure 31: Central nervous system of a shark.

Vision in sharks is probably equally important, taking over from taste and scent at close quarters. Shark eyes (particularly those of top predators like the White Shark) are very sophisticated and extremely similar to those of mammals. The iris surrounds a pupil that can be opened wide to let in a lot of light or contracted to a pinhole. Behind this is a crystalline lens used to focus images onto the retina, which (just as in mammals) contains structures known as 'cones' for good vision (colour vision in many species), and rods for high sensitivity to low levels of light. Like many other animals, a reflective mirror or 'tapetum' behind the retina allows light to bounce around inside the eye in order to improve vision in very low levels of light, but sharks that are active in shallow water by day go one better by using a layer of pigment to cover the mirror by day (cats and dogs can not do that!). Deepwater and nocturnal sharks are remarkable for their huge glowing green eyes, which are designed to capture as much light as possible in the darkness of the deep ocean. Sharks can not close their eyelids to protect their eyes, but some groups have developed a third eyelid called the nictitating lower eyelid to do this.

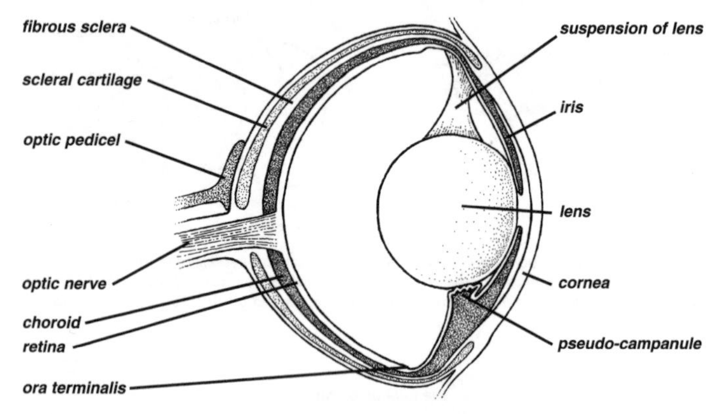

Figure 32: Lateral section through the eye of a shark (from Sealife 2000).

Another shark sense that is not just similar to that in mammals, but rather more advanced, is the detection of changes in pressure. Sound is simply changes in pressure causing vibrations in air or water that are picked up by the inner ear; shark inner ears (they have no outer ears) are very similar in structure to those present in mammals and are incredibly sensitive to sound, particularly low-frequency signals. Many species can even identify the precise direction from which the sound is coming.

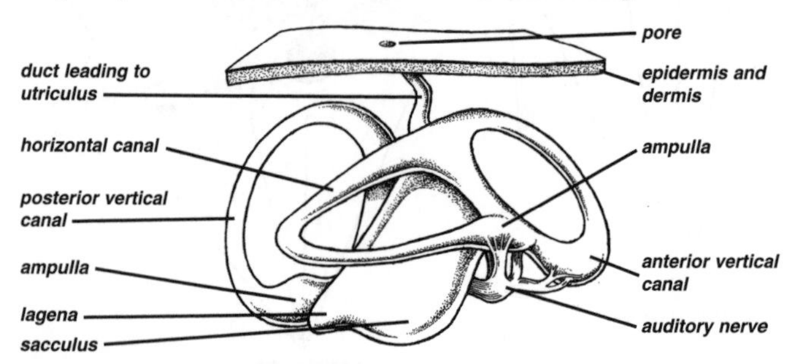

Figure 33: Ear of a shark comprising of three sac-like structures and three canals. Each sac has a sensitive spot that contains the ear stones and sensory cells which detect orientation and speed change; the three ampulla at the end of each canal detect movement in all panes, up, down and sideways (from Sealife 2000).

Touch is simply the detection of changes in pressure applied to the skin, but sharks have an additional remote sense that has no equivalent in mammals. This is the special hydrodynamic or mechanosensory system, which operates using the lateral line (a row of small pores leading into a fluid-filled canal system beneath the skin) that runs along each side of the body of the shark and additional rows of sense organs around the head and mouth and sometimes scattered elsewhere along the body. Pressure changes in the water outside the animal are transferred into these sensory organs, causes tiny hairs inside them to move and send nerve impulses to the brain, which can interpret them as nearby prey, predators or other sharks, even if other senses cannot be used.

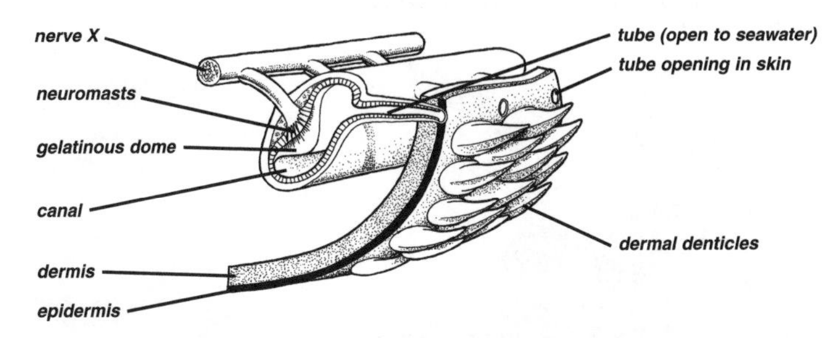

Figure 34: Close-up cutaway of the lateral line system in a shark (from Sealife 2000).

The most extraordinary special shark sense detects the minute electric fields given off by living animals and inanimate objects, and water moving through the earth's magnetic field. Sharks' electrosensory system relies upon unique receptors within special jelly-filled sensory organs known as ampullae of Lorenzini that are scattered around the head and mouth of sharks. They can use these organs to detect prey at close quarters (up to 50 cm or so) even when completely buried in the seabed. Hammerheads, with their broad wing-like heads, are particularly skilled at this, presumably because they have more ampullae than any other group of sharks and can precisely triangulate the location of the prey using those located at the extremities of their head. Deepwater sharks have been known to bite on transatlantic cables because of the electric fields that they produce. Electro-reception is also used to orient sharks within the earth's magnetic field, a very important sense when undertaking long cross-ocean migrations.

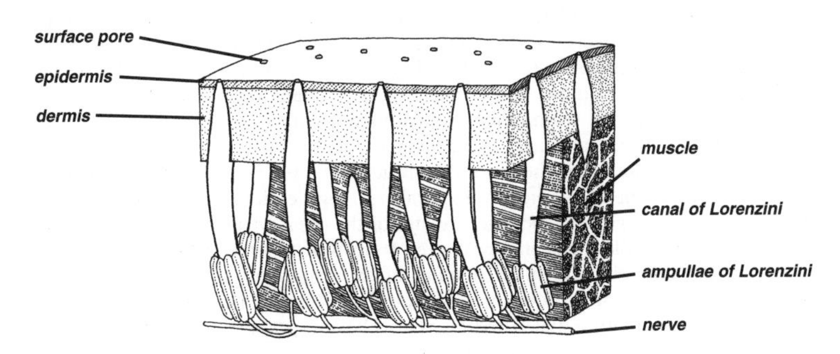

Figure 35: Close-up cutaway of the ampullae of Lorenzini.

Feeding and digestion

Sharks have a largely undeserved reputation for eating just about anything that they may come across that will fit into their mouths, regardless of origin. This is true for a few species (Tiger Sharks, for example, have been found with an extraordinary range of inedible objects in their stomachs) and many sharks will take a wide range of prey items. In contrast, a great many are rather 'picky' about what they eat and are specialised to feed on a relatively small range of prey items. None though are herbivorous.

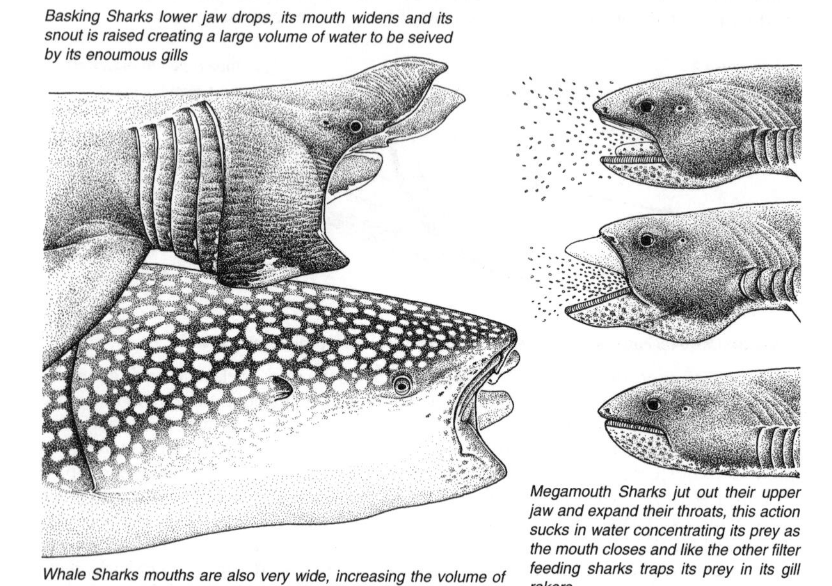

Basking Sharks lower jaw drops, its mouth widens and its snout is raised creating a large volume of water to be seived by its enoumous gills

Megamouth Sharks jut out their upper jaw and expand their throats, this action sucks in water concentrating its prey as the mouth closes and like the other filter feeding sharks traps its prey in its gill rakers

Whale Sharks mouths are also very wide, increasing the volume of water to be filtered

***Figure 36: Mouth positions of the Basking** Cetorhinus maximus **and Whale Sharks** Rhincodon typus **when feeding with mouth closed shown behind and suctorial method of feeding in Megamouth Sharks** Megachasma pelagios.*

The huge plankton feeders: the Whale, Basking and Megamouth Sharks, are prime examples of such specialisation, although they have, interestingly, developed into plankton feeders completely independently and use different strategies. Whale Sharks feed using suction to take in large concentrations of plankton and small fishes. Basking Sharks are ram-feeders, swimming steadily, huge mouth wide open, through plankton blooms. Megamouth Sharks may make their suction feeding extra efficient with the use of luminescent tissue inside the mouth to attract prey in the deep ocean. The key to this form of feeding is the development of gill rakers, long slender filaments that form a very efficient sieve, analogous to the baleen plates of the great whales. Their planktonic prey is trapped in these filaments and swallowed from time to time in huge mouthfuls. The teeth of these huge giants are very small compared with the size of the animal, because they do not need them for feeding, although they may still be important in courtship and mating (large mature male Basking Sharks have very worn teeth, presumably from holding onto the rough skin of their mates).

Other highly specialist feeders include the cookiecutter sharks, which feed on very large circular chunks of flesh sliced out of much larger pelagic fishes and marine mammals. In contrast to the huge plankton feeders, with their tiny teeth, these little sharks have enormous, saw-like cutting teeth in their lower jaw. Although they have never been observed feeding, they are assumed to latch onto their prey (large fishes and marine mammals), use their thick lips and suction to seal off the area being bitten, and then twist their bodies in order to cut out a neat, circular plug of flesh. Cookiecutter sharks tend to be rather poorly muscled with small fins. It is hard to imagine that they could swim as fast as the big oceanic animals on which they feed (although they can certainly damage large fishes caught in fishing gear, which are 'sitting targets'). The presence of light organs in these animals has led scientists to speculate that cookiecutters, possibly swimming in schools, use light as a decoy to attract their enormous prey to a possible meal, and then ambush them painfully. Whatever the method they use to get to their prey, it also seems to work for submarines; these little sharks are also infamous as the perpetrators of the mysterious bites taken out of the rubber covers on submarine hydrophones!

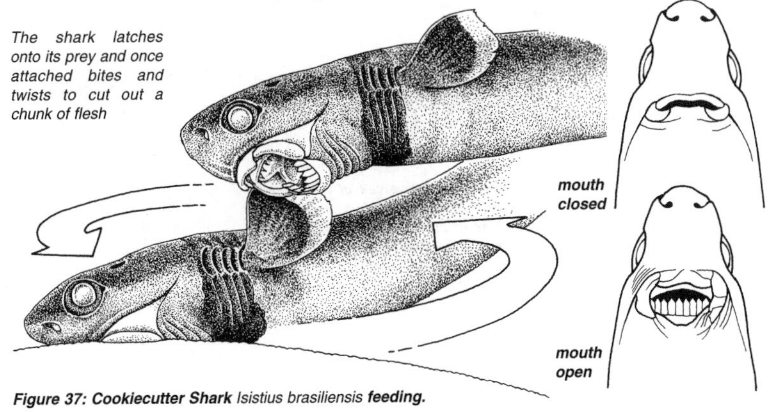

Figure 37: Cookiecutter Shark** Isistius brasiliensis **feeding.

Some species that live on the seabed are highly effective ambush predators. The flattened angelsharks and wobbegongs are perfectly camouflaged for lying in wait in order to suck passing prey into their cavernous mouths. Many bottom-dwelling species feed only on hard-shelled crustaceans and molluscs, which they grind up with their large flat molariform teeth.

Other sharks are specialist feeders on fishes or squid, which are swallowed whole. The extraordinary fangs of the Viper Shark can be pointed outwards to strike at and capture prey that is then swallowed intact. The White Shark and other large predators can either swallow small prey whole or take huge bites out of large animals (see fig. 24). Thresher sharks use their long tails to stun the shoaling fishes on which they feed, and sawsharks may either stir prey up from the seabed or slash at swimming prey with their tooth-studded rostra.

Many sharks, including some squaloid dogfish and the Whitetip Reef Shark, are cooperative feeders. They hunt in packs in order to herd and capture elusive prey and can dismember prey animals very much larger than themselves. These social sharks are often highly migratory, travelling huge distances around ocean basins in large schools. These migrations may partly be necessary to continually find new sources of food.

Once it has been captured and swallowed, digestion of the prey can be a long business in sharks, particularly in the cold-blooded species. The food moves from the mouth into the stomach, which is 'J' shaped and used for storing and some initial digestion of the food. Unwanted items may never get any further than the stomach and are, in due course, 'coughed' back up again. Many sharks have the ability to turn their stomach inside out and evert it out of their mouths in order to get rid of any unwanted contents. This is fairly commonly seen in hooked sharks. Although it looks horrible and can be mistaken for a fatal injury, sharks regularly do this voluntarily and an everted stomach is no reason to destroy a shark that would otherwise have been returned to the water alive. The next part of the gut is the intestine. One of the biggest differences between sharks and mammals is its extremely short length. This is achieved by use of a 'spiral valve' with multiple turns within a single short section of gut instead of a very long tube-like intestine. The spiral valve takes three main forms: a true spiral winding around a central column, a series of interconnecting cones or funnels, or a cylindrical scroll valve attached to the wall of the intestine but free in the centre. Whatever the type of valve, it provides a very long surface area for the digestion of the food, requiring it to pass around and around inside the apparently short gut until fully digested, when remaining waste products pass into the cloaca and vent. The length of the valve and the number of turns it contains depends upon the type of food typically eaten, and may range from just a few turns to many dozen.

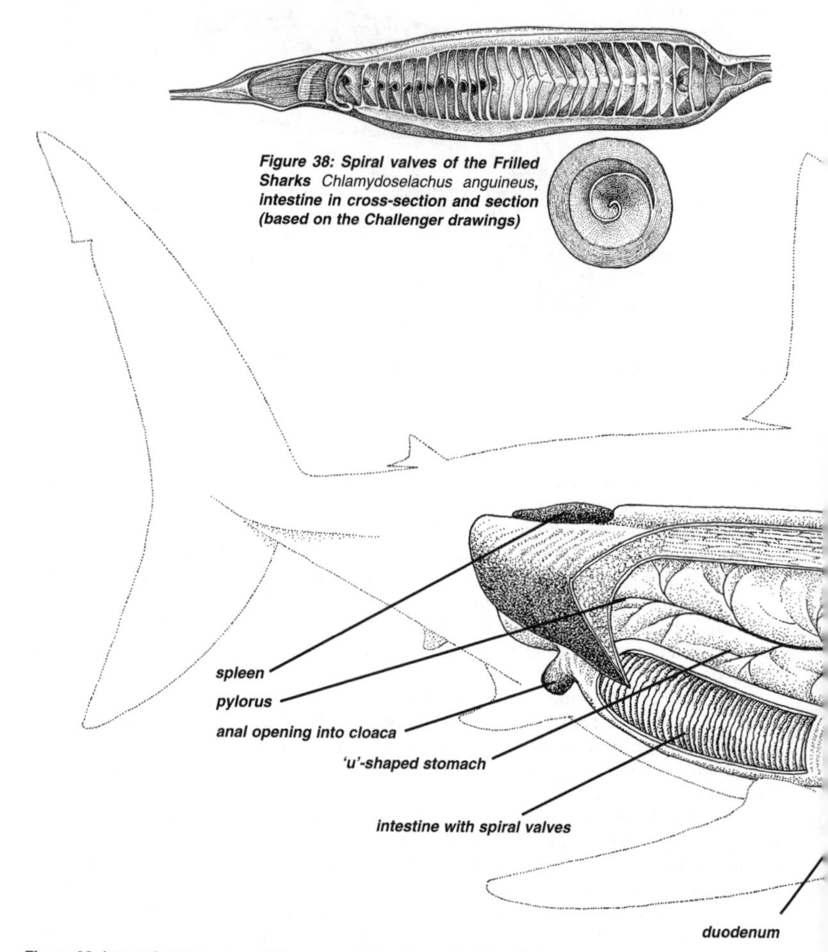

Figure 38: Spiral valves of the Frilled Sharks *Chlamydoselachus anguineus,* **intestine in cross-section and section (based on the Challenger drawings)**

Figure 39: Internal organs, except the urogenital system, of a Shortfin Mako, *Isurus oxyrhinchus.*

One the many fascinating discoveries by scientists studying shark intestines are the internal parasites that live inside them. All animals have internal parasites, but the interesting thing about those in the spiral valve is that a different species of intestinal worm occurs in every few turns of the valve, specialised to feed on that particular stage of the partly digested food.

The most obvious internal organ in sharks is not the gut, but the huge liver, which often fills most of the body cavity. In the Basking Shark, the liver makes up about a quarter of the body weight and may weigh up to a tonne; this was the major product from traditional Basking Shark fisheries because it contained up to 80% in weight of very high quality light squalene oil, important for industrial, cosmetic and pharmaceutical use. Deepwater shark fisheries, targeting sharks with large livers and a high squalene content, supply these oil markets today. The liver has two main purposes: buoyancy (remove the liver from a dead shark in the water and it will rapidly sink), and storage of energy (particularly in species that feed relatively infrequently). Storage of energy or fuel is particularly important in mature female sharks, which use up a huge amount of energy during their long pregnancies. The size and weight of a female shark liver at the beginning of a pregnancy is several times greater than it is when she gives birth. No wonder that many female sharks need at least one year 'off' between pregnancies to rebuild these vital energy stores!

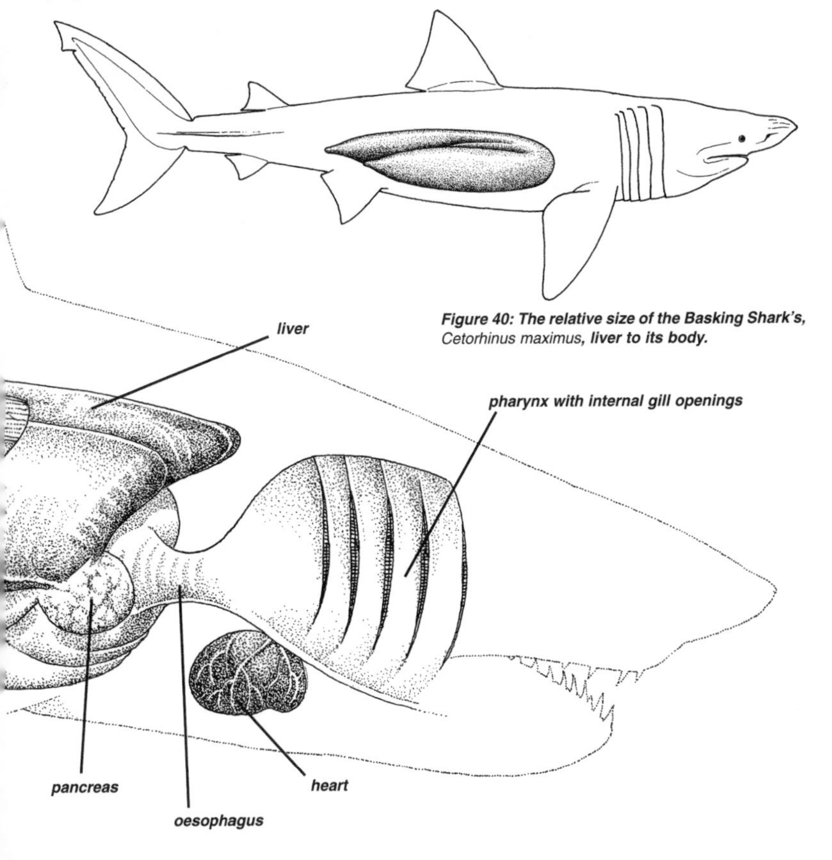

Figure 40: The relative size of the Basking Shark's, *Cetorhinus maximus*, ***liver to its body.***

Life history

Much about shark life history and reproduction is similar to those in birds and mammals, but far more varied and, in many cases, more advanced. They have, after all, had hundreds of millions of years to develop internal fertilisation, live birth and even placental reproduction long before mammals came on the scene. Furthermore, so far as reproduction and life history strategies are concerned, sharks and bony fishes have as little in common as do bony fishes and birds or mammals. Sharks produce only a small number of relatively large young that have a relatively good chance of survival to adulthood (compared with the tiny percentage of bony fishes that develop successfully from huge numbers of very small eggs produced during spawning).

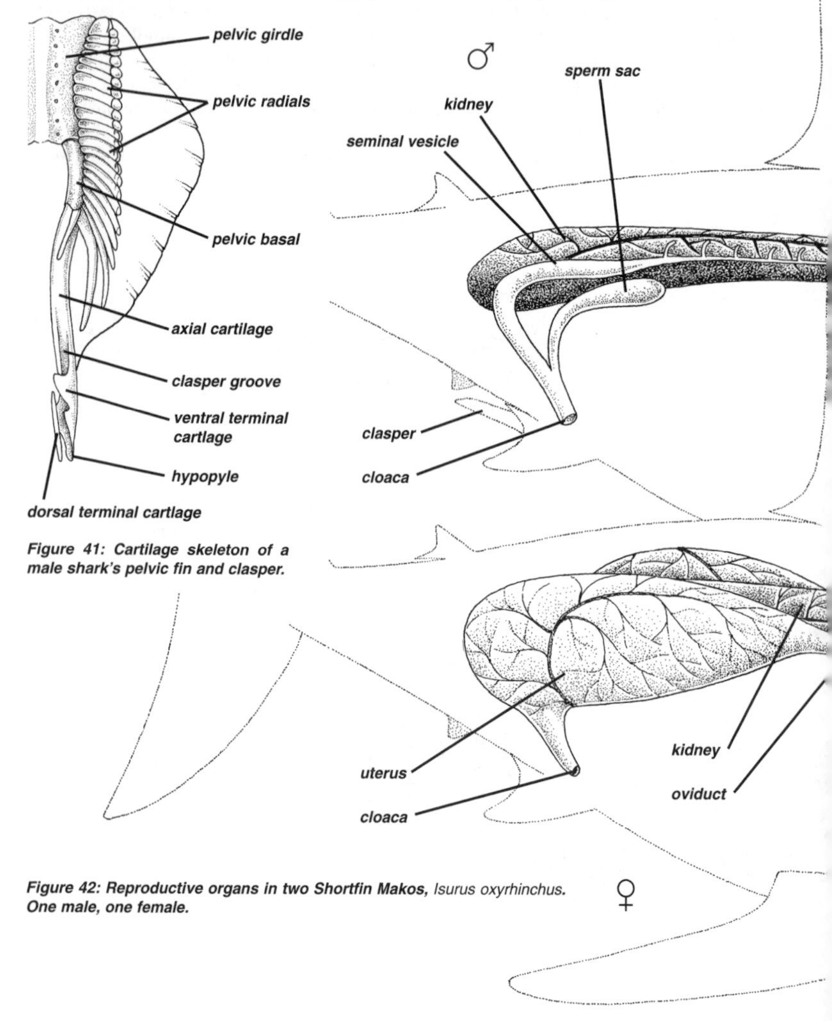

Figure 41: Cartilage skeleton of a male shark's pelvic fin and clasper.

Figure 42: Reproductive organs in two Shortfin Makos, *Isurus oxyrhinchus.* ***One male, one female.***

Reproduction takes many forms in sharks, but all require internal fertilisation, usually following a period of courtship to ensure the cooperation of females. Mating has only been observed in a few species. In larger sharks, it involves the male biting the female to hold her alongside while using one or two of its paired claspers (grooved extensions of the pelvic fins) to insert packages of sperm into the female reproductive tract. Smaller species twine around each other. After fertilisation of the large yolky eggs has taken place, a tough protective covering is laid down.

The eggs may be laid (often in pairs – one from each ovary) at daily or weekly intervals almost immediately after fertilisation and deposition of the protective egg capsule. They are carefully anchored onto the seabed so that they can develop safely in the nursery ground, sometimes for over a year, until the young have absorbed the yolk sac and are ready to hatch as miniature versions of the adults.

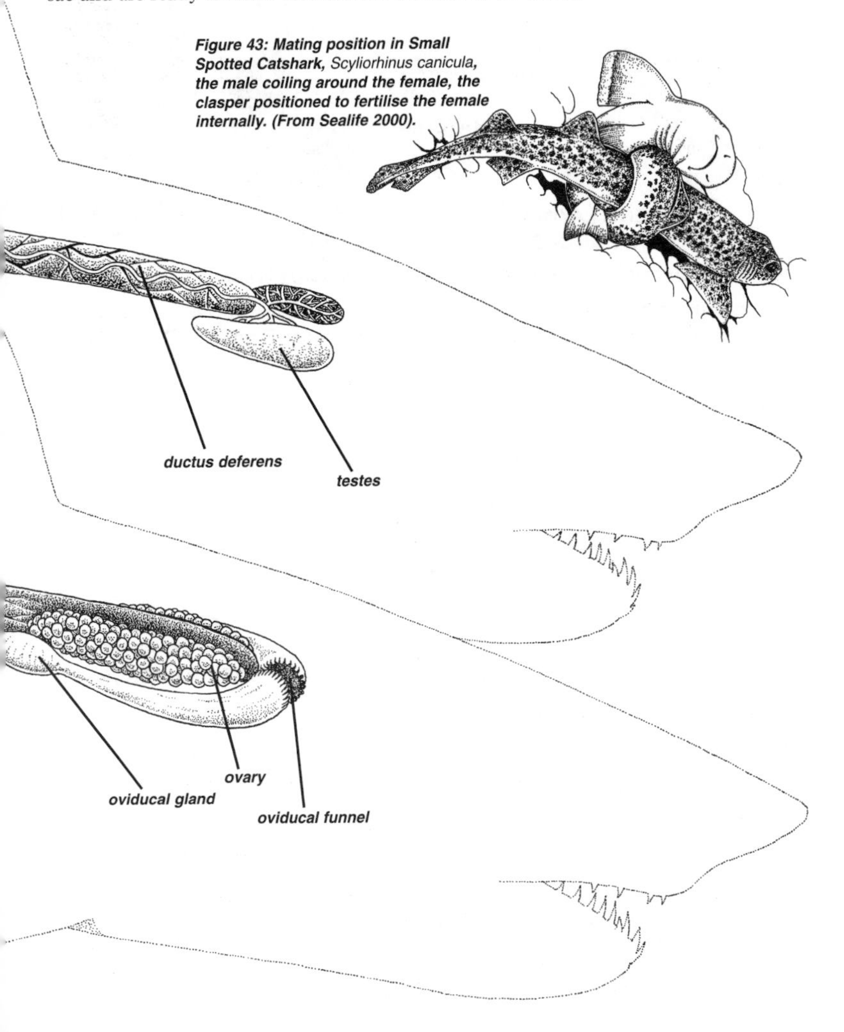

Figure 43: Mating position in Small Spotted Catshark, *Scyliorhinus canicula,* ***the male coiling around the female, the clasper positioned to fertilise the female internally. (From Sealife 2000).***

To reduce the length of time in which the eggs are exposed to seabed predators, many sharks retain their eggs for many months before laying them. In these cases, the final stages of development outside the female may take just a few weeks. Laying eggs that hatch outside the female is known as oviparity. About 40% of all sharks are oviparous. The rest have developed a range of reproductive strategies for giving birth to live young, known as viviparity.

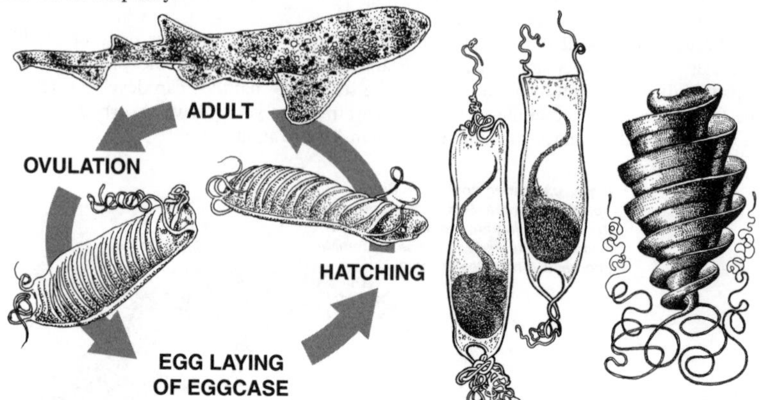

Figure 44: a) Oviparity in a swellshark, ; b) eggcases of a catshark and a hornshark.

A simplest extension of oviparity is ovoviviparity (or aplacental yolk sac viviparity), which is used by about 25% of shark species, including Spiny Dogfish and Whale Shark. The female retains the eggs until they have absorbed the egg yolk, completed their development and hatched safely inside the female. She then gives birth to fully developed young.

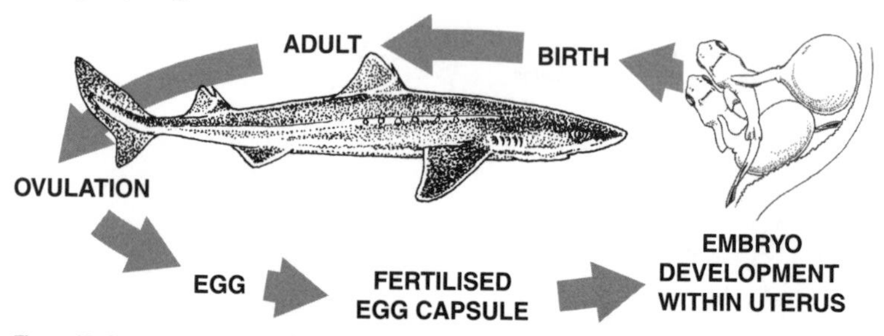

Figure 45: Ovoviviparity in a spurdog.

Instead of nourishing their young with a single yolk sac, many species have developed ways to increase the amount of food available to the young inside the female, so that they are larger and better developed at birth, thus presumably more likely to survive the first dangerous weeks of life in the ocean. Some of the lamnoid sharks produce a great many infertile eggs that are steadily released to feed the growing young inside the uterus, a process known as oophagy. Some species produce only one fertile egg from each ovary; all the others being infertile and destined to feed the pups. Others produce several fertilised eggs and can give birth to fairly large litters. The Sandtiger Shark is notorious for 'intrauterine cannibalism'. Its pups not only eat infertile eggs while still inside the female, they also eat the other developing pups (their brothers and sisters) until just one survives. An unborn Sandtiger pup once bit a scientist who unwisely put his hand inside the uterus.

The most advanced form of reproduction, used by about 10% of living sharks (including Blue Sharks and hammerheads), is placental viviparity. In these species the yolk sac develops into a placenta, becoming attached to the wall of the uterus with the yolk stalk connection between the yolk sac and the embryo turning into a placental cord to transfer nutrients from the mother to the pup. Large litters can be nourished in this way. This strategy is very similar to mammalian reproduction, but developed in sharks long before mammals evolved. The main difference between sharks and mammals is the absence of parental care of young after birth.

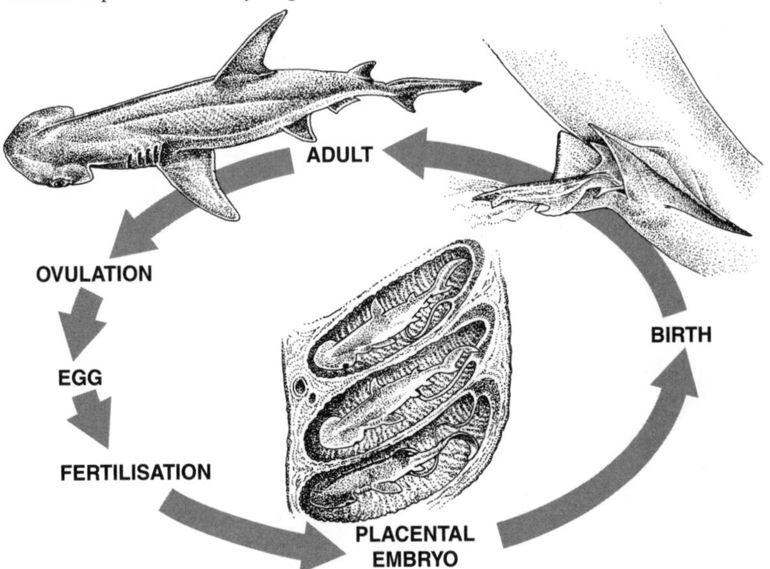

Figure 46: Viviparity in a hammerhead shark.

There are exceptions, but sharks generally have to be fairly large in order to be able to produce large young. They may take many years to grow slowly until large enough and old enough to reproduce (particularly if they live in the cold, low energy deep ocean environment where food is scarce). The large sharks that occupy the very top of the marine food web have very few natural predators and have adapted to produce only the very few young needed to maintain a stable population. Producing such large young also takes time. In general, therefore, sharks are long-lived, slow-growing animals that produce only a relatively small number of pups during their lifetime. Readers will quickly notice, however, just how little information is available on the life history and reproduction of most species in the following sections of this guide. Studies of age and growth are made even more problematical because there are few hard structures in the body that lay down growth rings (vertebrae and fin spines do not always produce clear rings). Where growth rings can be found, it is hard to determine whether rings are laid down annually, seasonally, or more or less frequently than that. The fastest growing, rapidly maturing and fecund species of small coastal sharks can begin to reproduce at as little as two or three years old. The difficulties of studying the larger species, particularly those that live in the deep ocean, means that we do not yet know how old some of the slowest growing sharks are before they begin to reproduce or how long they can live. It would not be surprising, however, to discover that some deepsea sharks may be immature until they reach 40 years old or more, and could live for well over 100 years.

Behaviour

The more scientists learn about sharks, the more we recognise just how complex is the behaviour of these intelligent animals. More and more species are being discovered to be social, living in large groups and hunting prey in packs. Even solitary sharks meet for breeding or on rich hunting grounds and have developed a complex system of signals and behaviours that minimise conflict and structure social interactions. Many, including Blue Sharks and Spiny Dogfish, spend most of their lives in single sex groups that are also segregated by age. Once they reach maturity, they may meet with members of the opposite sex just once a year to mate. Pups are generally found in nursery grounds, well separated from the adults that might otherwise eat them, juveniles congregate into groups, adult males are found together, and mature females can be separated into pregnant and non-pregnant groups.

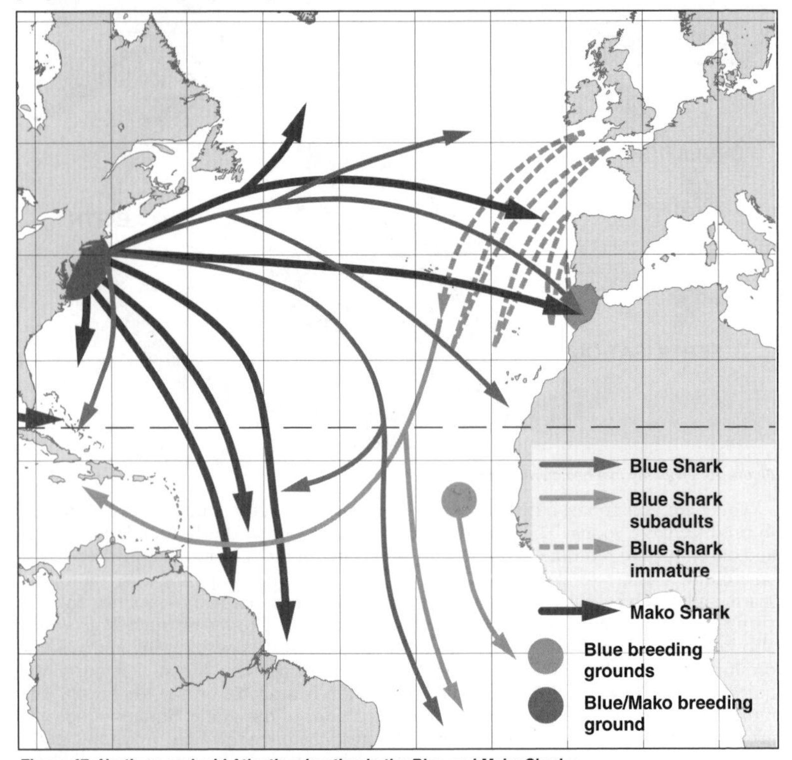

Figure 47: Northern and mid Atlantic migration in the Blue and Mako Shark.

Migration patterns are even more complex than in birds, with different journeys carried out by mature and immature animals, males and females. Migrations may regularly cross ocean basins and, what is even more interesting, are not always annual. Tagging studies have demonstrated that many species of shark with a two year reproductive cycle also exhibit a two year migration pattern, only visiting their pupping grounds every other year in order to give birth. Perhaps they are feeding or mating in other parts of the ocean during alternate years. Females usually only visit shallow nursery grounds very briefly to give birth and do not feed during this period.

Sharks and people

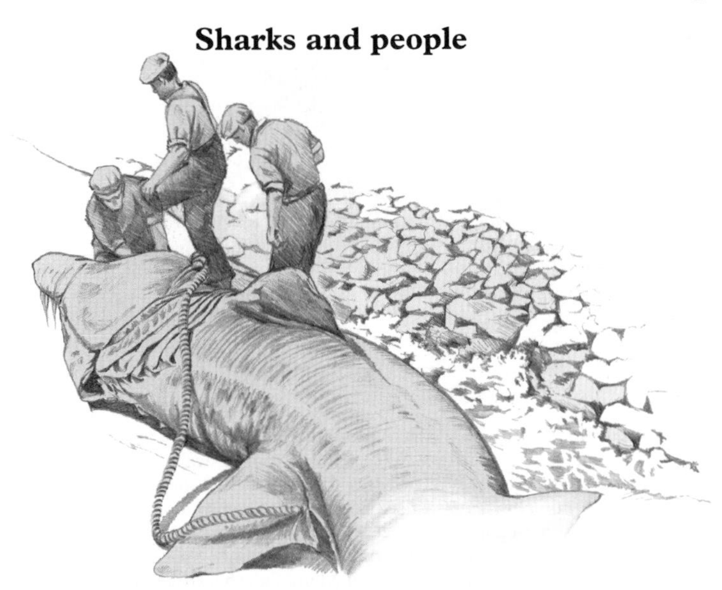

The classic horror film 'Jaws' and subsequent less successful imitations have had a huge influence on perceptions, worldwide, of the relationship between people and sharks. The result of exposure to this fictional representation of one or two species means that most people think of sharks as being a serious and ever-present threat to those individuals who dare to venture into their environment. While it is certainly true that a few shark species are potentially dangerous to people and a relatively small number of shark attacks every year receive huge publicity, the whole picture is very different. Sharks are in crisis, worldwide, as a result of man's activities.

These top marine predators have adapted over several hundred million years to a position at or near the top of the marine food web. They do not have many natural enemies and have developed life history strategies that protect their most vulnerable life stages from predation, for example by giving birth to very large well-developed pups that can fend for themselves and use of nursery grounds unvisited by large predators. There have certainly been some massive extinction events in the distant past, primarily associated with major environmental changes in world oceans (far more shark species have become extinct than are living today), but mankind has completely changed the picture. We are now heading for a new extinction event in world oceans, this time driven not by long-term climate change or other massive natural catastrophic events (volcanic eruptions or comet strikes), but by our unsustainable exploitation of marine resources. What is most extraordinary is the very recent nature of this impact. While most extinctions of large land animals took place between tens of thousands to hundreds of years ago, as mankind spread across the world's land masses, our impact upon the large marine animals that spend their entire lives at sea is very recent. Coastal fishing pressure increased sharply in the 1950s, when cheap artificial fibre monofilament nets became widely available, and significant depletion and perhaps disappearance of some coastal species probably began around this time (although information on identity and numbers of species caught is sadly lacking in most parts of the world). Most large-scale target shark fisheries only really began to expand in the mid 1980s as demand increased for their fins, meat and cartilage. The dawn of latest extinction crisis was less than 20 years ago.

Truth, legend and fiction

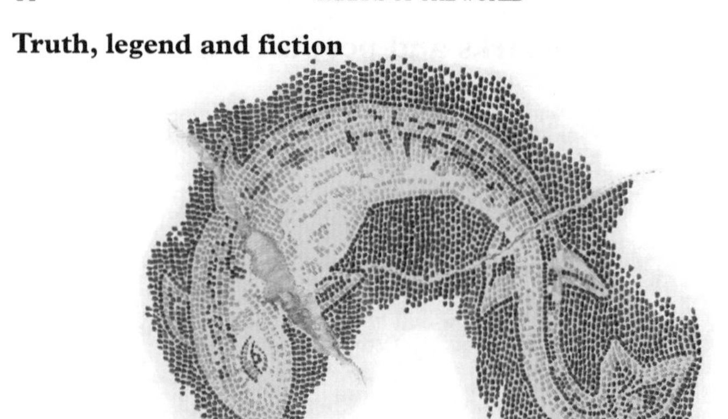

Where sharks are concerned, the truth is often harder to locate than the fabulous stories that appear in legend and fiction. What biblical animal swallowed Jonah: a shark or a whale? There are records of White Sharks discovered with animals completely intact in their stomachs – dolphins, a 50 kg sea lion and, apparently, an intact and fully clothed sailor (the latter reported in the 16th century by French naturalist Guillaume Rondelet) and sharks can certainly disgorge their stomach contents. So perhaps a White Shark is a more likely culprit than the whale described in modern translations of the Bible. There are also records of Mediterranean sharks from the early Greek writer Herodotus in the fifth century BC, who describes shark attacks on shipwrecked sailors, and Aristotle (third century BC) who reported accurately several aspects of shark biology, despite misunderstanding their feeding strategies (he suggested that sharks have to turn upside down to take food – a constraint placed upon them by Nature in order to give their prey an opportunity to escape). Later still, the Roman Pliny the Elder warned in his *Natural History Encyclopaedia* (around 77 AD) of small sharks biting the exposed white parts of divers' bodies unless deliberately frightened away (he recognised the timid nature of these animals), anticipating modern records of sharks biting the flashing white hands or feet of swimmers and surfers presumably misidentified as small fishes.

European views of sharks as ravening beasts seem to differ quite considerably from those of other peoples who live in much closer association with the sea and have a greater awareness of the true nature of its inhabitants. Pacific and Indian Ocean peoples, particularly islanders and members of coastal fishing communities, have rather different traditional views of sharks. Many attribute sharks with supernatural powers, protecting rather than threatening fishermen, and shark worship was once fairly common, perhaps sometimes including human sacrifice in a few cultures. Some tribes claim sharks as ancestors. Shark calling is still practiced by some Pacific Islanders and sharks are very important in the culture and folklore of several races in Melanesia, West Africa, Australasia, and the Amazon basin. The remains of many large sharks have been discovered entombed beneath the ruins of the Aztec Great Temple beneath central Mexico City together with human and crocodile bones, perhaps representing the great sea monster whose body, ripped apart by the gods in Aztec mythology, first formed the earth, or sacrifices to placate the civilisation's blood-thirsty gods.

It is somewhat depressing to discover that the modern myths of sharks and shark attacks portrayed in the media have today reached even the most remote of coastal fishing communities in developing countries. Lack of reliable electricity supplies has not prevented widespread viewing of modern shark horror films and, despite no known local reports of shark attack, resulted in a widespread fear of sharks that is not supported by centuries of experience at sea.

Fisheries

Sharks, and their relatives the rays and chimaeras, provide about 1% of world fisheries landings reported to FAO, or some 700,000 to 800,000 tonnes; perhaps 70–100 million animals. About 60% of these are sharks, around 40% skates and rays. These figures are undoubtedly an underestimate, because many countries under-report their catches and ignore bycatch, discards and recreational fisheries. Actual catches may be twice that reported. Reported catch levels have remained relatively constant over the past several decades, but only because depleted fisheries in some regions have been replaced by yields from new fisheries elsewhere. In many countries, recent steep increases in fishing effort have resulted in short-term increases in shark landings, which are now being followed by a marked decline in catch rates and a reduction in the quantity and diversity of shark species being landed. It generally takes around ten years or less for a shark fishery to go through a classic 'boom and bust' pattern of initial high landings followed by a rapid population crash. In most cases the target fishery then ceases (unless the sharks are of very high value) but the remnants of the population continue to be taken as bycatch in other fisheries for more abundant and resilient species of bony fish and may potentially be driven to extinction by this 'incidental fishing pressure'.

Sharks are probably taken in largest quantities worldwide in coastal multi-species or 'catch all' fisheries, which harvest and utilise a wide range of marine fishes, or may discard the least valuable portion of the catch, including some sharks. These fisheries are largely unmonitored and unmanaged and their landing sites provide a rich hunting ground for shark enthusiasts seeking new species. In many cases, retained shark 'bycatch' may be a very valuable source of income for fisheries nominally targeting declining stocks of large pelagic tunas and billfishes, which may only still be economically viable because of the value of shark fin. Target shark fisheries are relatively uncommon by comparison (not least because they tend to be relatively short-lived), but nonetheless, are or have been, important in some regions. Examples are the Norwegian Porbeagle fishery, European Spiny Dogfish fishery, Australian Southern School and Gummy Shark fishery, coastal fisheries for a variety of large sharks in Australia and the USA, and Reef Shark fisheries in the Red Sea and Indian Ocean.

The most recent fisheries frontier is in the deep sea, off the edge of the continental shelf and around oceanic islands, where there is a remarkable diversity of shark species adapted to relatively narrow depth bands and a very stable, low energy environment. These species tend to grow and reproduce even more slowly than sharks that occur in coastal or oceanic waters, making them even more vulnerable to depletion by fisheries than their shallow water cousins. Deep sea fisheries are not only exploiting unnamed and unknown shark species never seen by scientists, but could be driving species to extinction before they can be described.

Recreational shark fisheries are increasing in economic importance compared with commercial fisheries in many parts of the world. Anglers value sharks, particularly large sharks, very highly as game fish. As fish stocks decline, recreational fishing can yield greater economic benefits than commercial fisheries to many coastal communities as a result of income from boat charters, sales of fishing gear, accommodation and food. An environmental benefit of sports fishing over commercial fishing is that the former may easily be undertaken on a 'catch and release' basis, which obviously has far less impact upon the target population than do traditional consumptive commercial fisheries. On the other hand, some unregulated sports fisheries are reported to have killed more sharks than died in commercial fisheries in the same waters, before angling took up a conservation ethic.

Other shark fisheries supply the aquarium trade (both for public display and hobbies), although the numbers taken are tiny compared with the tens of millions removed by commercial fisheries. Shark meshing and drumline fishing programmes also target large sharks in Australia, South Africa and New Zealand. These are intended to reduce population size in order to minimise risk of encounters between swimmers and sharks off tourist beaches. They do not physically exclude sharks from beaches. These bather safety programmes are responsible for the deaths a few thousand sharks annually, worldwide.

Utilisation

Shark meat has been an important source of food for thousands of years, not just for coastal peoples but also for those who never saw the sea but traded for the dried salted meat available along the great continental trading routes – indeed shark and ray meat is still prepared and traded in this form in many parts of the world as well as in the more familiar fresh and frozen form now shipped around the globe. The high levels of urea in the blood and flesh of sharks necessary for osmoregulation (the control of body fluid content) in the sea can make eating shark meat an unpleasant experience unless it has been properly handled and stored after capture. This is because enzymes produced by bacteria convert the urea to ammonia after the shark has died. Only a tiny quantity of urea (1.8%) and even less ammonia (0.03%) is enough to taint shark meat and make it very unpleasant to eat.

Shark fin may have been one of the most valuable seafood products in the world for over two thousand years – it used to be so rare and precious and so difficult to prepare well that it was originally only eaten at the banquets of Chinese emperors. Until very recently, it was a dish only for the most important of special occasions.

Other important traditional shark products include the oil that can be easily extracted from the livers (simply by chopping it up and leaving it to decay in containers in the sun). This has been used for hundreds of years to waterproof boats, for lighting and for a variety of medical purposes. The high levels of vitamin A in shark liver stimulated some important shark fisheries in the early 20th century until synthetic supplies became available and liver oil is also used for a variety of industrial purposes.

Shark skin makes a very durable thick leather (far stronger than cow hide) when correctly treated and is commonly sold as belts and wallets. Dried but untreated skin, with the denticles still embedded, is called 'shagreen' and makes a very effective sandpaper or a non-slip binding for sword hilts. Polished shark leather with denticles is known as 'boroso' leather and used for a wide variety of very expensive products. Shark and ray leather has recently been hitting the headlines again as a new, exciting and highly priced product.

Many shark products are low value, low quality, with the exception of shark fin, which is one of the most valuable fisheries products in the world, or traded in only tiny volumes compared with most other fisheries products. It is not surprising, therefore, that these animals are a low research and management priority compared with, for example, large scale and highly valuable fisheries for herring, tunas, jacks, cod, or lobster. Increased awareness of the highly vulnerable nature of these animals over the past ten years has so far done very little to change this. It is still very difficult, if not impossible for the responsible consumer to identify shark products to species level, or to find products that are derived from sustainably managed fisheries.

One of the highest value and most sustainable shark 'products', however, is derived from sharks swimming freely in their natural environment. Dive ecotourism is a booming industry in many parts of the world. It was calculated, over ten years ago, that a single reef shark was worth tens of thousands of dollars as a generator of tourist revenue at popular dive sites as far apart as the Maldives and the Bahamas. Today, sharks at important dive sites off the Australian coast may be worth hundreds of thousands of dollars each. Newly developed dive tourism industries focused on Whale Sharks (which operate for only a few months of the year) are certainly worth several million, possibly tens of millions of dollars annually to local communities. This does not mean that shark ecotourism has no impact on shark populations. There are serious concerns over its impact upon shark behaviour, habitats and ecology. Some governments are now beginning to require shark ecotourism to be licensed to ensure that it does not damage the resource upon which it relies. Shark watchers should always ensure that they select ecotourism operators whose staff have been properly trained, have a good reputation and the necessary licenses.

Figure 48: Encouraged shark encounters may be a thrill of a lifetime but it can cause serious behavioural changes in local shark populations. Properly managed shark encounters can benefit both the diver and the shark and could help secure the future of many shark species.

Research

There is a small but active core of shark researchers working on shark science, including taxonomy, biology, ecology, behaviour and fisheries and making exciting new discoveries every year that are disseminated at annual research conferences. A list of scientific organisations is given on page 343. Shark research is, however, often viewed as the 'Cinderella' of fisheries research, largely because of the generally low value and low volume of sharks in fisheries, and is almost non-existent in many countries. Indeed, much early research was driven by fear of shark attack and the need to understand more about the animals responsible for such highly publicised encounters between man and shark.

Experts largely agree that the greatest shark research priorities are to improve understanding of their biology and ecology; critical habitats; population structure; age and growth; reproduction and life history, and trends in catches per unit effort. Rare, vulnerable and threatened species require special attention, but rarely receive this. In many cases, the biodiversity and distribution of shark species is very inadequately known, because of a lack of basic surveys or of fisheries monitoring at species level; this book may help to redress this need.

In addition to monitoring landings of sharks from commercial and sports fisheries (information on species identification, size structure of fished population, size at birth and maturity and number of pups can all be obtained from these sources), tag and release programmes form an important source of information on shark distribution, life cycles, migrations, age and growth. Recreational anglers, commercial fishermen and divers are increasingly becoming involved in such research programmes and, if conscientious with their observations, can be a hugely valuable asset to researchers and research programmes. Even if not involved in tagging sharks, observers should still look out for tagged sharks and record details, taking care to follow the instructions (if any) provided on shark tags. At the very least, they should describe the tag, record its number, the length of the shark tagged, the date and position of capture and send this information off to their national fisheries laboratory so that it can be forwarded to the correct research group. In some cases, when tags are found on dead sharks, researchers will want to be able to study the whole animal, or at least parts of it (if so, these instructions may be provided on the tag).

Use of satellite, acoustic and pop-up archival tags is becoming increasingly common. These tags are often very expensive and difficult to attach to sharks and programmed to become detached automatically after a set period or to transmit information while still attached. They should never, therefore, be removed prematurely from live sharks, because this can potentially result in the loss of valuable and expensive information. Instead all tag details should be recorded before the shark is released.

Conservation and Management

Most shark fisheries around the world are virtually unmonitored and completely unmanaged. The value of and demand for shark products is rising, resulting in increasing fishing pressure, while stocks decline and collapse because of the life history constraints that make it difficult for shark populations to increase sufficiently rapidly to replace those removed by fisheries. Several scientific papers published in the last few years have identified huge declines in formerly wide spread shark stocks, some of which have been depleted by over 90% as a result of target and/or bycatch fisheries, while population declines of 70% over the past 20-30 years are the norm. The urgency of introducing effective shark conservation and management programmes has been widely acknowledged by United Nations (UN) fisheries and environment agencies and national governments over the past decade, but little progress has been made in improving the situation. As noted above, lack of action is the result both of the traditional low economic value of shark fisheries and small volumes of products produced, and the poor public image of animals viewed primarily as dangerous marine predators. Shark eradication programmes have, in the past, been better resourced than shark conservation and management programmes!

Other major threats to sharks include habitat alteration, damage and loss from coastal developments, pollution, and the impacts of fisheries on the seabed and food species.

Today, the case for shark conservation and management is more compelling than ever before. We now understand the importance of ensuring that shark fisheries are sustainably managed in order to yield long-term benefits to coastal communities, whether reliant upon commercial fisheries, sports fishing or marine ecotourism. We also understand that sharks play a vital role in maintaining the biodiversity, structure and stability of the marine ecosystem. Removing sharks does not necessarily mean that their prey animals will increase and result in improved yields of other fish species in fisheries; rather their loss may result in reduced populations of other species lower down the food chain because sharks also reduce other fish predator numbers. So, the loss of top predators may result in an unpredictable range of undesirable side effects for other fishery resources and may damage marine ecosystem functions.

Figure 49: The sight, like this Whitetip Reef Shark, of finned dead and dying sharks should now be consigned to the past.

Important shark conservation initiatives recognising the importance and urgency of shark management include the adoption of the UN Food and Agriculture Organization's (FAO) International Plan of Action for the Conservation and Management of Sharks (IPOA–Sharks) in 1999, which urged shark fishing States to produce Shark Assessment reports and adopt Shark Management Plans by 2001. At the last count, roughly half of the 113 States that report shark fisheries landings to FAO had reported some progress with implementation of the wholly voluntary IPOA–Sharks, but most of this progress was unspecified and had most likely not included any conservation or management action. Only 16 shark fishing States had reported that they had actually produced Shark Assessment Reports or Management Plans. This provides an over-optimistic view of improvements in shark fisheries management worldwide; so far only a few these documents have been translated into improvements in shark fisheries management.

The IPOA–Sharks largely affects fisheries on the continental shelf, inside States' Exclusive Economic Zones. It does not cover shark fisheries on the high seas, outside the jurisdiction of coastal states. Many pelagic shark species have, however, been listed on the United Nations Convention on the Law of the Sea (UNCLOS), which is supposed to be implemented by the UN Agreement on Straddling and Highly Migratory Fish Stocks. Although this Agreement was officially ratified several years ago it is still not implemented. Most of the Regional Fisheries Management Organisations established to manage fisheries for high seas and shared fish stocks are not monitoring shark catches and none of them are managing shark fisheries.

It is widely recognised that the need for introducing shark fisheries management is now very urgent and that a precautionary approach to their management is essential. A truly precautionary approach to shark fisheries would mean that fisheries are only conducted when detailed stock assessments are available, and there is detailed and effective monitoring and management of fisheries. This is quite impossible in the majority of shark fishing States, where fisheries management may be ineffective and the data needed to form a basis for precautionary management are seriously lacking. At the very least, however, lack of data and shortcomings in management capacity should not be used as an excuse for not managing sharks and shark fisheries. We do know enough about their biology and shark fisheries dynamics to begin to implement some basic precautionary fisheries management, and international agreements require that shark fisheries management begins to be implemented at once.

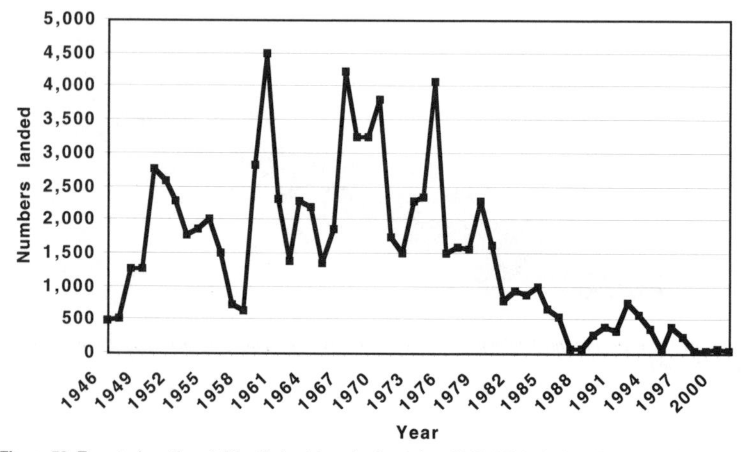

Figure 50: Targeted northeast Atlantic basking shark catches 1946–2001, the last 6 years all in Norway.

Figure 51: Removed shark fins drying in the sun. The majority of these sharks would have been thrown back alive left to die on the bottom. In this way only the highly prized part of the shark is brought back to shore and the 'uneconomic' rest of the shark is dumped.

Recognising the highly vulnerable nature of sharks, the unsustainable fisheries that threaten to drive some species to extinction, and the widespread failure of fisheries management bodies to address threats to sharks, a number of wildlife Agreements are now focusing on shark conservation. The Convention on International Trade in Endangered Species (CITES) was among the first to raise the alarm over unsustainable shark fisheries driven by international trade demand for shark fin, in 1994, and has been urging improved conservation and management through Resolutions on Shark Conservation and Management ever since. The Conference of Parties to CITES (representing over 160 nations) agreed in 2002 to list Whale Shark and Basking Shark on Appendix II. This is not a trade ban, but requires all international trade in their products to be accompanied by permits verifying that the harvest of these products had not been detrimental to the wild population fished. More shark listing proposals may be expected in 2004 and at subsequent meetings of Parties. The Convention on Migratory Species has also listed White Sharks and Whale Sharks on its appendices, recognising the need for collaborative management of these vulnerable species. Some regional environmental Agreements are now also beginning to add shark species to lists of species requiring protection or management. Several species are now legally fully protected under wildlife or fisheries legislation in several States.

The IUCN (World Conservation Union) Shark Specialist Group (SSG) provides an important source of information for the decisions made by such conventions. The SSG is currently undertaking a review of the threatened status of all species of sharks, rays and chimaeras for publication in the *IUCN Red List of Threatened Species* (www.redlist.org). Assessments published at the time of writing are included in the species descriptions in the following pages. In 2003, 21% of the species that had been assessed (less than a third of the total) were identified as Threatened with extinction (Critically Endangered, Endangered or Vulnerable) and 24% as Near Threatened. Another 24% were Data Deficient, meaning that inadequate information was available to make an assessment of their threatened status. The remainder were Least Concern, meaning that they are not currently considered to be at risk of extinction. Species at high extinction risk include deepwater sharks, sharks restricted to freshwater and coastal endemics whose entire range is subjected to high levels of fishing effort.

Shark encounters

Sharks belong in an environment that is alien to mankind. The sea is deep and dark, mysterious, dangerous and, until recently, largely unknown. Early mariners in European waters hated to sail out of sight of land in case a sudden storm led to the loss of their vessel and all its crew and passengers. The sea is full of shipwrecks and it is impossible to imagine how many human lives have been lost through drowning. Even today, hundreds of people drown accidentally every year (while many more are killed in traffic accidents on the way to the beach). For some reason, however, it is not the sea itself that still scares most people, but (perhaps because of their scarcity and unknown nature) one of the most elusive and rarely seen groups of animals that it contains. The relatively tiny risk of injury from shark attack (perhaps 75-100 per year out of the many tens of millions of individuals using the sea for work or recreation) is completely out of proportion to the fear that arises from the mere potential for such attacks occurring. The writings of the Greek Herodotus contain the first known records of shark attacks, these on shipwrecked Persian sailors in around 500 BC. Pliny the Elder recorded that Mediterranean sponge divers were having uncomfortably close encounters with sharks (presumably white sharks) well over 2000 years ago. Interestingly, people who were far more at home in the ocean, the Polynesian ocean navigators who routinely undertook extraordinary long distance sea journeys between the tiny atolls and volcanic sparsely scattered around the immense Pacific Ocean, had and still have a far more balanced view of sharks.

The truth about shark attack is very different. Injury rates are tiny compared with the huge numbers of individuals at risk. The shock and horror that arise when rare events of serious injury from shark bites, sometimes leading to death, are understandable; these are tragedies – not least because the chances of them occurring at all are so very small.

The huge majority of shark attacks are caused by mistaken identity – a shark identifies a sequence of noise, vibration or movement as coming from a potential prey item and takes a single bite at the pale flashing hand or foot of a swimmer or surfer that it has mistaken for a fish. The shark then retreats instantly when it realises its mistake. Most times, the injury is so small that no medical attention is needed – indeed, the attacker is very often much smaller than the victim. Other examples of 'hit and run' behaviour occur when large sharks bump and bite the victim, or carry out a sneak attack 'out of nowhere' with a single bite. A single bite may be sufficient to immobilise the victim, even if the attack is not followed up, and may result in death if the victim is not immediately helped to shore and given medical attention. Fortunately people are far more likely than seals to receive the help necessary for them to survive such incidents (although there are marine mammal rescue operations that do treat shark-injured seals). Examples of multiple bites and sharks actually feeding on human prey are very rare.

Shark attacks can occur anywhere that sharks and people may meet in the water and in almost any water depth (provided that there is enough to allow the shark to swim). Incidents are naturally more likely to occur when very large numbers of people and sharks are using the same waters, but the incidence of shark attacks has not increased in line with the rising numbers of people using the water for recreational purposes. The pattern of these sharks incidents seems to be a continuing series of sporadic, erratic, and almost accidental encounters and many of these being avoidable. Studies of the circumstances under which shark bites occur have been studied in detail by the scientists who maintain shark attack databases; most incidents fit into one of several classic patterns. Many attacks occur near dawn or dusk, when sharks are hunting actively. Many occur in murky water, for example near river mouths, after high rainfall, or during onshore winds. These conditions of poor visibility seem more likely to lead to cases of mistaken identity resulting in bites, while large predatory sharks can also more closely approach their prey without being observed under these circumstances. Many attacks seem to occur because the bite victim was swimming erratically (thrashing around) or swimming with dolphins (which are often accompanied by sharks). Swimming alone, rather than in a group, can also increase chances of injury from a shark, as can urinating in the water, feeding sharks, spearfishing, swimming near fishing boats that are discarding bait or slurry, or wearing strongly contrasting coloured clothing. Swimming right up to sharks, following them around closely, picking them up or pulling their tails are also almost fail-safe ways to attract a shark bite. Avoiding the above behaviour patterns and treating sharks with respect will greatly decrease your already minute chances of a damaging encounter with a shark. Despite much misinformation about 'rogue sharks' in the media, 'serial maneating' by individual sharks does not seem to happen, unlike the occasional Tiger, Lion and Leopard.

The authors hope that readers of this book will realise that, despite the hyperbole, the true story of man and sharks is not one of man-eating sharks, but of the hugely more common shark-eating man. Sharks, not mankind, are the species in danger, the animals now being driven rapidly towards extinction. We all need sharks in our oceans. We need them for food, particularly in countries where protein is scarce. We need them for the recreational benefits they bring, for the maintenance of the entire marine ecosystem, and for the simple joy of knowing that they are still somewhere out there – some of the most beautiful, highly evolved, streamlined and quite simply 'coolest' animals in the world.

Help us to help them by supporting shark conservation organisations and campaigns and acting responsibly.

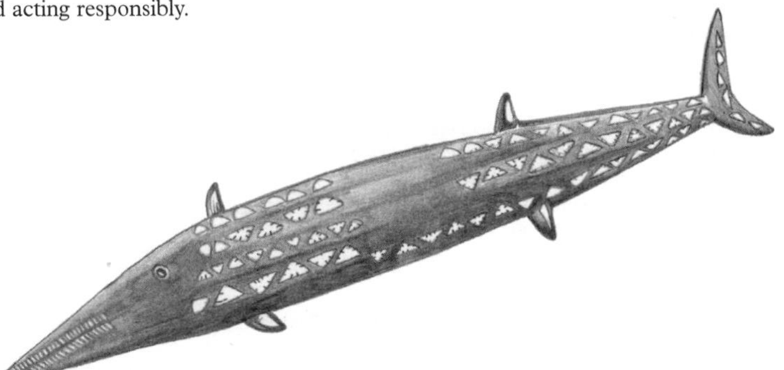

Figure 52: Carved wooden shark effigy from the Solomon Islands where the shark is more revered than maligned.

Key to orders and families of living sharks

1a Anal fin absent, *drawing 1.* ➔**2**

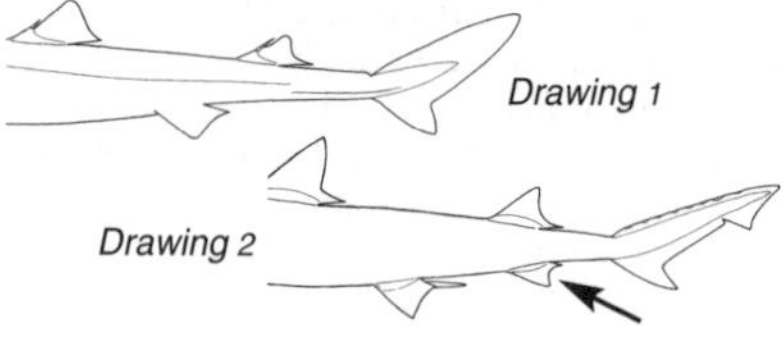

Drawing 1

Drawing 2

1b Anal fin present, *drawing 2.* ➔**10**

2a [1a] Mouth at end of head; body flat and ray-like; very large pectoral fins with anterior triangular lobes that overlap gill slits; caudal fin with base slanted ventrally (hypocercal) *see figure 3.*

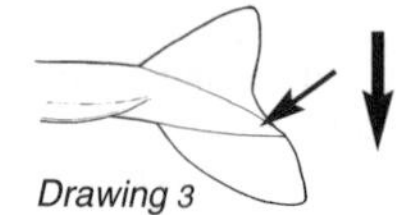

Drawing 3

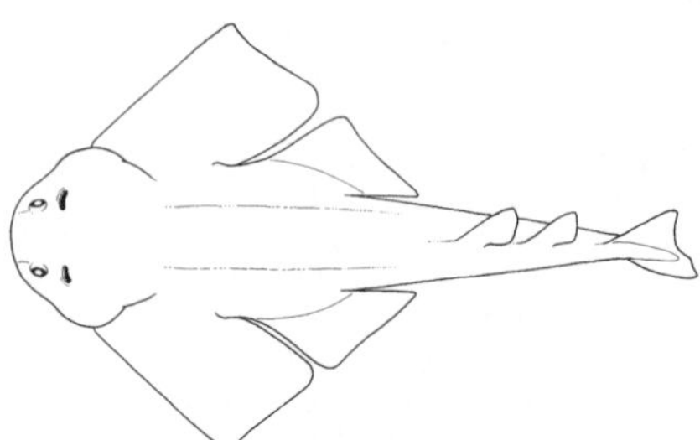

➔ **Order Squatiniformes, Family Squatinidae. Angel Sharks: plates 17-19**

2b [1a] Mouth beneath head; body cylindrical, compressed, or slightly flattened, not ray-like; pectoral fins smaller, without anterior lobes; caudal fin with base slanted dorsally (heterocercal) *see drawing 4* or horizontal (diphycercal). ➔**3**

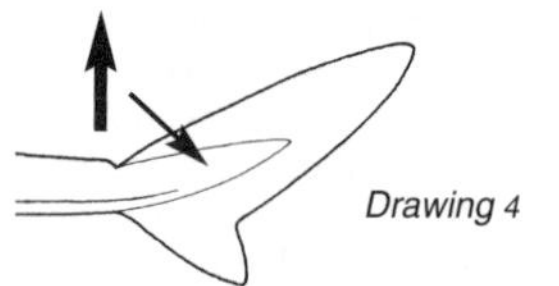

Drawing 4

3a [2b] Very long flattened sawlike snout, with rows of large and small sharp-pointed tooth-like denticles on sides and underside; a pair of long, tape-like barbels on lower surface of snout in front of nostrils.

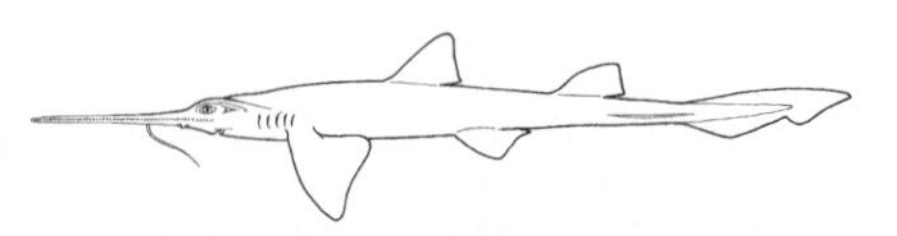

➔ **Order Pristiophoriformes, Family Pristiophoridae. Sawsharks: plates 15-16**

3b [2b] Snout normal, not sawlike.
Order Squaliformes. Dogfish sharks. ➔**4**

4a [3b] Spiracles small and well behind eyes; fifth gill slits much longer than first four; body covered with moderately large and close set, thornlike denticles or sparse, large, platelike denticles; pelvic fins much larger than second dorsal fin; first dorsal fin origin over or behind pelvic fin origins.

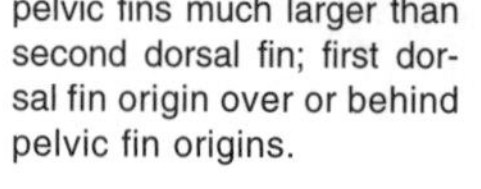

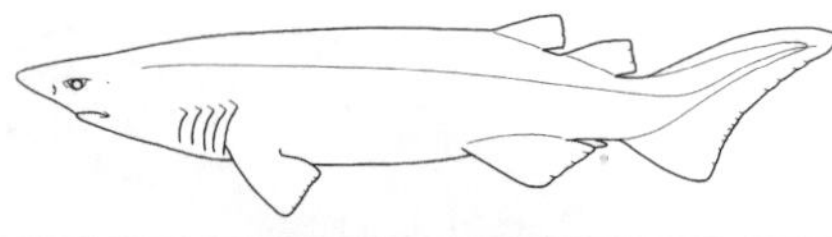

➔ **Family Echinorhinidae. Bramble and Prickly Sharks: plate 5**

4b [3b] Spiracles larger and close behind eyes; fifth gill slits not abruptly larger than first to fourth; denticles variable in shape but small to moderately large; pelvic fins usually as large as second dorsal fin or smaller; first dorsal fin origin well in front of pelvic origins. ➔**5**

5a [4b] Body very high and compressed, triangular in cross-section, with heavy lateral keels between pectoral and pelvic bases; dorsal fins extremely high.

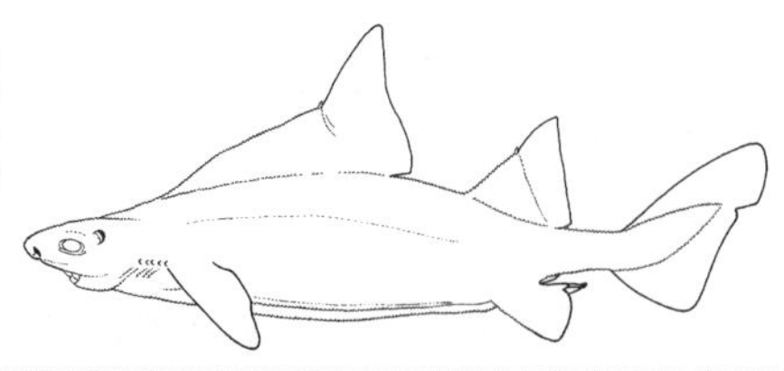

➔ **Family Oxynotidae. Rough Sharks: plate13**

5b [4b] Body low and more cylindrical in cross-section, with a low lateral keels between pectoral and pelvic bases or no keels; dorsal fins low. ➔**6**

6a [5b] Teeth similar and bladelike in both jaws, with a deflected horizontal cusp, a low blade, and no cusplets *see drawing 5*; caudal peduncle usually with an upper precaudal pit (weak or absent in *Cirrhigaleus*); strong lateral keels present on caudal peduncle *see drawing 6*; dorsal fin spines without grooves; subterminal notch absent from caudal fin.

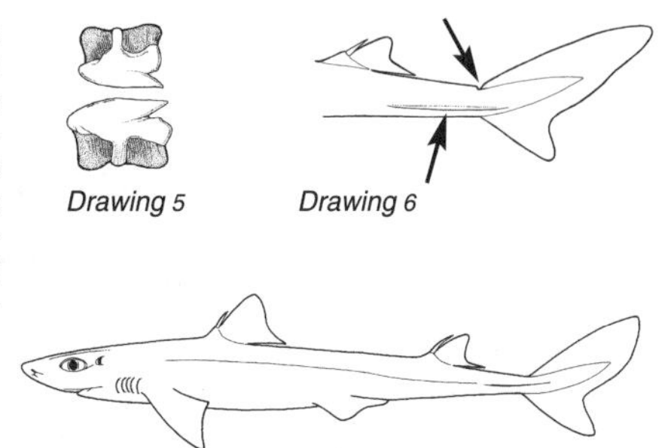

Drawing 5 *Drawing 6*

➔ **Family Squalidae. Spiny Dogfishes, Spurdogs, and Mandarin Sharks: plates 2-3**

6b [5b] Teeth variable but not bladelike and with deflected horizontal cusps and low blades in both jaws, *see drawing 7*; caudal peduncle without precaudal pits; lateral keels usually absent on caudal peduncle, *see drawing 8* (weak ones present in some members of Family Dalatiidae); dorsal fin spines, when present, with lateral grooves; subterminal notch usually present and well-developed on caudal fin. ➔**7**

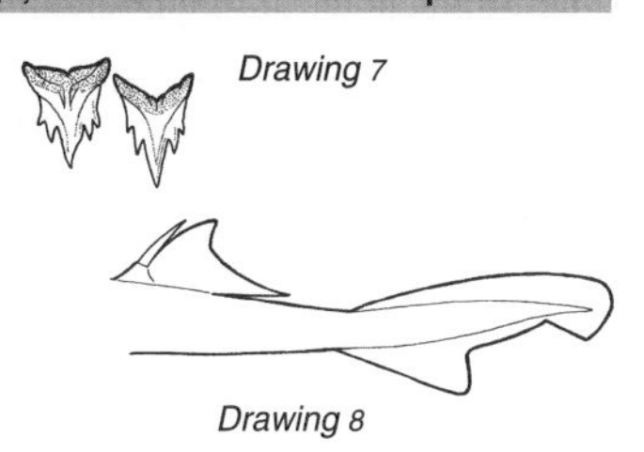

Drawing 7

Drawing 8

7a [6b] Teeth either hooklike or with cusps and cusplets in both jaws, or upper teeth with cusps and cusplets and lower teeth compressed, bladelike and more or less overlapping; underside of body, flanks, and tail usually with more or less conspicuous dense black markings (photomarks) with light organs (photophores).

 **Family Etmopteridae. Lantern Sharks: plates 6-9**

7b [6b] Upper teeth with strong cusps but without cusplets; lower teeth laterally expanded, compressed, bladelike, and overlapping, much larger than uppers *see drawing 9*; underside of body, flanks, and tail without conspicuous dense black markings that have light organs, though light-producing organs may be present elsewhere. ➔**8**

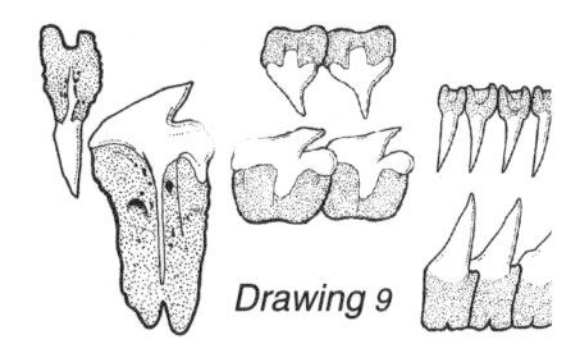

Drawing 9

8a [7b] Upper teeth relatively broad and bladelike, lower teeth low and wide and bladelike, *see drawing 10* below.

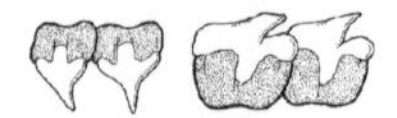

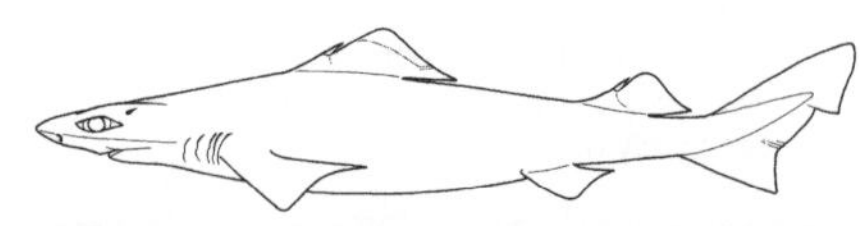

➔ **Family Centrophoridae. Gulper Sharks and Birdbeak Dogfishes: plates 4-5**

8b [7b] Upper teeth relatively narrow and not blade-like, lower teeth high and wide and bladelike. ➔**9**

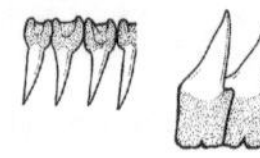

Drawing 11

9a [8b] Head moderately broad and somewhat flattened or conical; snout flat and narrowly rounded to elongate-rounded in dorsoventral view; abdomen usually with lateral keels; both dorsal fins either with or without (*Somniosus*, *Scymnodalatias*) fin spines.

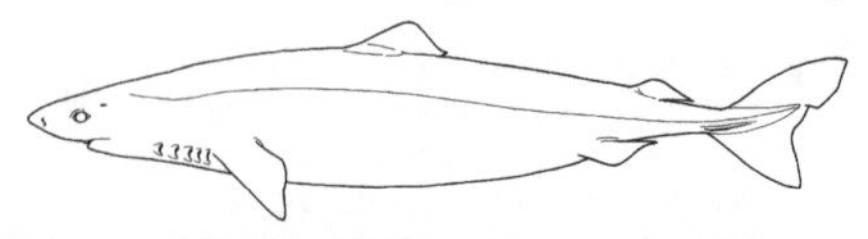

➔ **Family Somniosidae. Sleeper Sharks: plate 10-12**

9b [8b] Head narrow and rounded-conical; snout conical and narrowly rounded to elongate-rounded in dorsoventral view; abdomen without lateral ridges; most genera lacking dorsal fin spines (*Squaliolus* with a small first dorsal spine only).

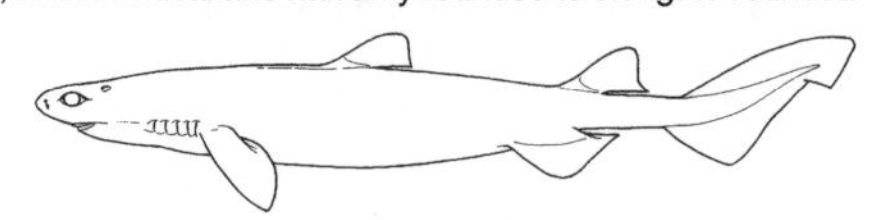

➔ **Family Dalatiidae. Kitefin Sharks: plate 14**

10a [1b] 6 or 7 gill slits on each side of head; one dorsal fin present, this set far back. Order Hexanchiformes; cow and frilled sharks. ➔**11**

10b [1b] Two dorsal fins (except the scyliorhinid *Pentanchus profundicolus* with one dorsal fin); 5 gill slits on each side of head. ➔**12**

11a [10a] Mouth terminal on head; teeth tricuspidate, *see drawing 12*, and similar in both jaws; 6 pairs of gill slits, first pair connected across the underside of the throat; body elongated and eel-like.

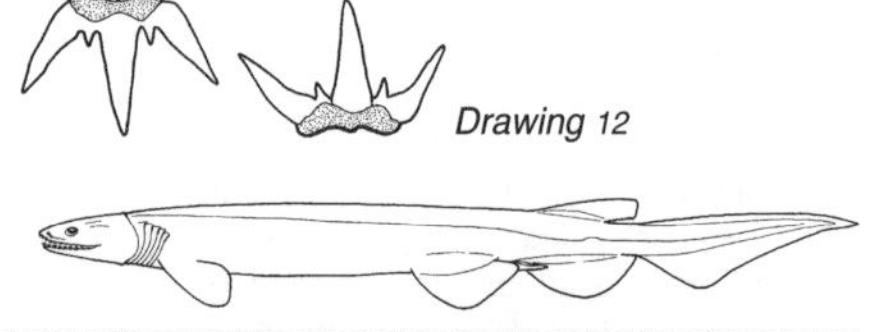

Drawing 12

➔ **Family Chlamydoselachidae. Frilled Sharks: plate 1**

11b [10a] Mouth subterminal on head; front teeth unicuspidate in upper jaw and comb-shaped and bladelike in lower jaw, *see drawing 13*; 6 or 7 pairs of gill slits, first not connected across underside of throat; body fairly stocky, not eel-like.

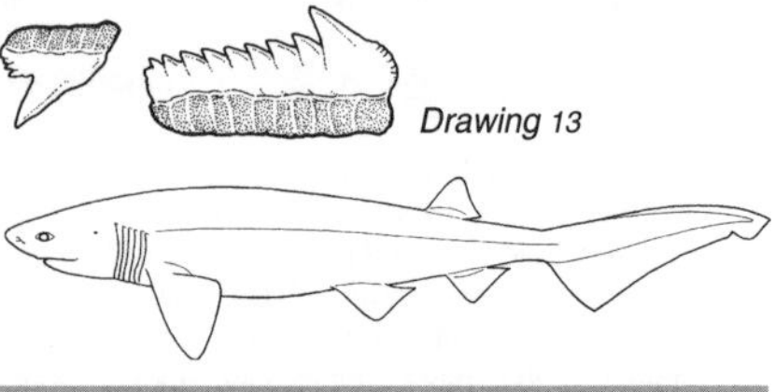

Drawing 13

➔ **Family Hexanchidae. Cow Sharks: plate 1**

12a [10b] A strong spine on each dorsal fin.

➔ **Order Heterodontiformes, Family Heterodontidae. Bullhead and Horn Sharks: plates 20-21**

12b [10b] Dorsal fins without spines. ➔**13**

13a [12b] Eyes behind mouth; deep nasoral grooves connecting nostrils and mouth; a pair of barbels just medial to incurrent apertures of nostrils, *see drawing 14* (rudimentary in Family Rhincodontidae). Order Orectolobiformes. Carpet sharks. ➔**14**

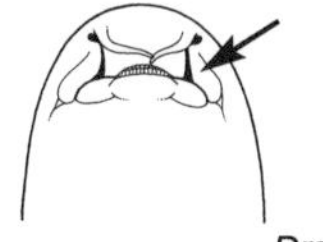

Drawing 14

13b [12b] Eyes partly or entirely over mouth; nasoral grooves usually absent, when present (a few members of the Family Scyliorhinidae) broad and shallow; barbels, when present, developed from anterior nasal flaps of nostrils, not separate from them, *see drawing15.* ➔**20**

Drawing 15

14a [13a] Mouth huge and nearly at end of head; external gill slits very large; caudal peduncle with strong lateral keels; caudal fin with a strong ventral lobe, but with a vestigial terminal lobe and subterminal notch.

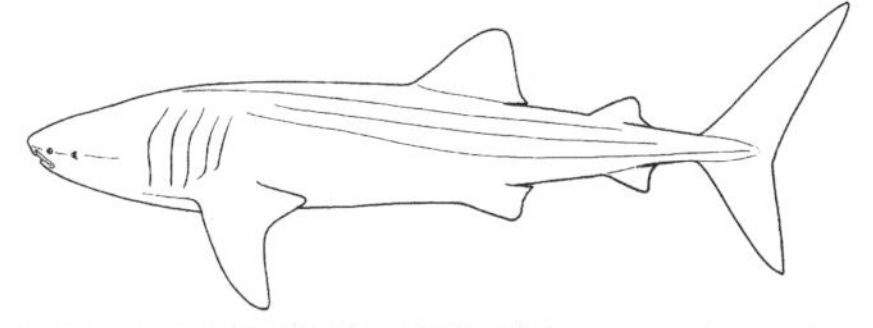

➔ **Family Rhincodontidae. Whale Sharks: plate 28**

14b [13a] Mouth smaller and subterminal; external gill slits small; caudal peduncle without strong lateral keels; caudal fin with a weak ventral lobe or none, but with a strong terminal lobe and subterminal notch, *see drawing 16.* ➔**15**

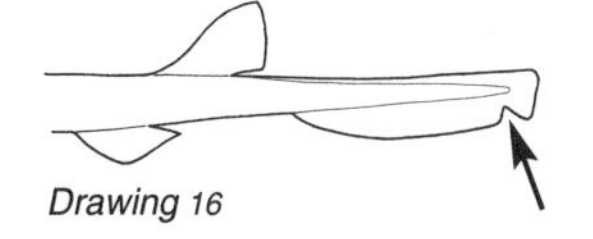

Drawing 16

15a [14b] Caudal fin about as long as rest of shark.

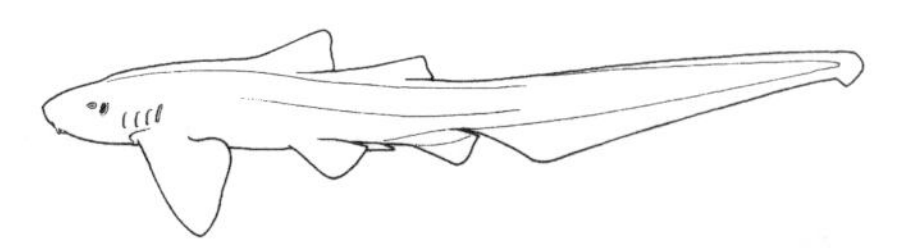

➔ **Family Stegostomatidae. Zebra Sharks: plate 26**

15b [14b] Caudal fin much shorter than rest of shark. ➔**16**

16a [15b] Head and body greatly flattened, head with skin flaps on sides; two rows of large, fanglike teeth at symphysis of upper jaw and 3 in lower jaw.

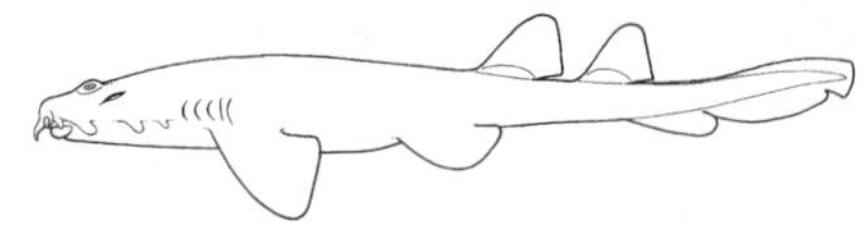

➔ **Family Orectolobidae. Wobbegongs: plate 24-24**

16b [15b] Head and body cylindrical or moderately flattened, head without skin flaps; teeth small. **➔17**

17a [16b] No circumnarial lobe and groove around outer edges of nostrils, *see drawing 17.*

Drawing 17

➔ **Family Ginglymostomatidae. Nurse Sharks: plate 26**

17b [16b] A circumnarial lobe and groove around outer edges of nostrils, *see drawing 18.* **➔18**

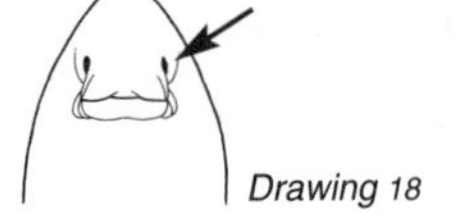

Drawing 18

18a [17b] Spiracles minute; origin of anal fin well in front of second dorsal origin, separated from lower caudal origin by space equal or greater than its base length.

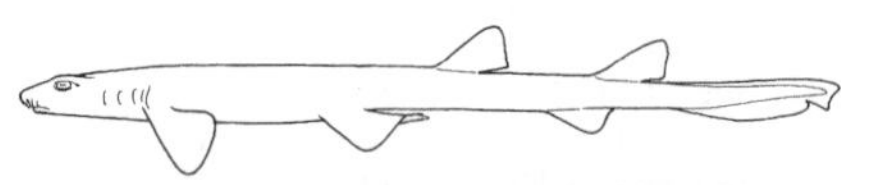

➔ **Family Parascylliidae. Collared and Barbelthroat Carpetsharks: plate 22**

18b [17b] Spiracles large; origin of anal fin well behind second dorsal origin, separated from lower caudal origin by space less than its base length. **➔19**

19a [18b] Nasal barbels very long; distance from vent to lower caudal-fin origin shorter than distance from snout to vent; anal fin high and angular.

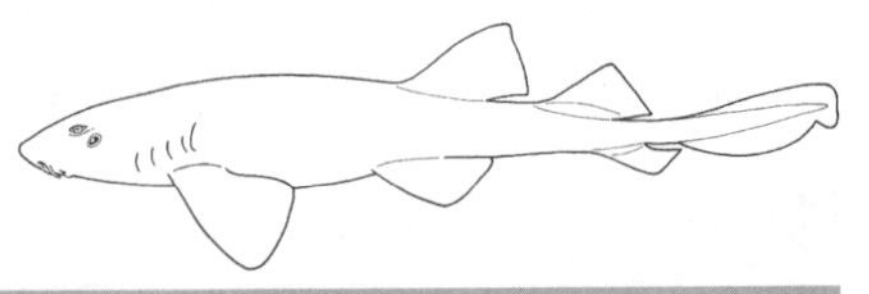

➔ **Family Brachaeluridae. Blind Carpetsharks: plate 22**

19b Nasal barbels short; distance from vent to lower caudal-fin origin longer than distance from snout to vent; anal fin low, rounded and keel-like.

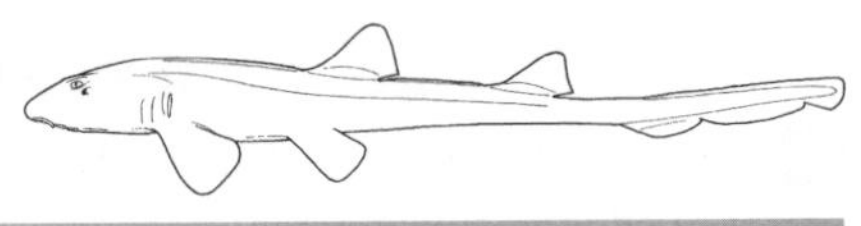

➔ **Family Hemiscyllidae. Bamboo and Epaulette Sharks: plate 25**

20a [13b] No nictitating eyelids; largest teeth in mouth usually are 2 or 3 rows of anteriors on either side of upper and lower jaw symphyses; upper anterior teeth separated from large lateral teeth at sides of jaw by a gap that may have one or more rows of small intermediate teeth, *drawing 19*; plankton-feeding species (Families Megachasmidae and Cetorhinidae) have reduced teeth with anterior, intermediate, and lateral teeth poorly differentiated; intestine with ring valve. Order Lamniformes. Mackerel sharks. ➔**21**

Drawing 19

20b [13b] Nictitating eyelids present; largest teeth in mouth are well lateral on dental band, not on either side of symphysis; no gap or intermediate teeth separating large anterior teeth from still larger teeth in upper jaw,*see drawing 20*; intestine usually with spiral or scroll valve. Order Carcharhiniformes. Ground sharks. ➔**27**

Drawing 20

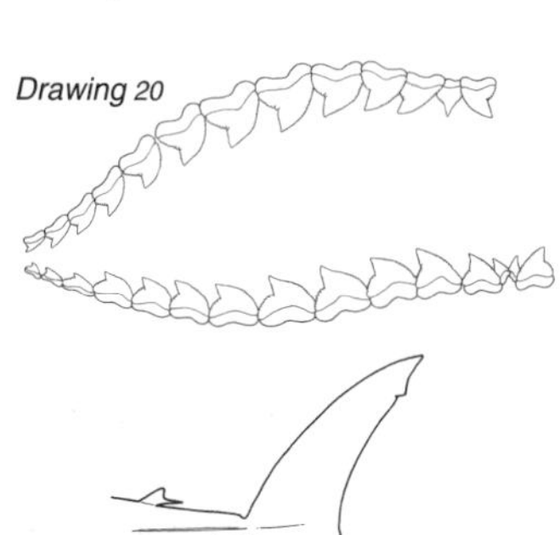

21a [20a] A strong keel present on each side of caudal peduncle; caudal fin crescentic and nearly symmetrical, with a long lower lobe,*see drawing 21*. ➔**22**

Drawing 21

21b [20a] No keels on caudal peduncle, or weak ones (Family Pseudocarchariidae); caudal fin asymmetrical, not crescentic, with ventral lobe relatively short but strong or absent. ➔**23**

22a [21a] Teeth large and few, sharp-edged; gill openings large but not extending onto upper surface of head; no gill rakers on internal gill arches.

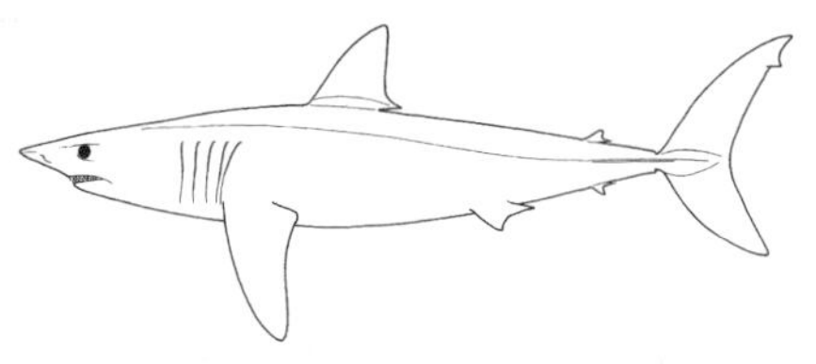

➔ **Family Lamnidae. Porbeagle, Salmon, White and Mako Sharks: plate 30**

22b [21a] Teeth minute and very numerous, not sharp-edged; gill openings huge, extending onto upper surface of head; gill rakers present on internal gill arches, sometimes absent after shedding.

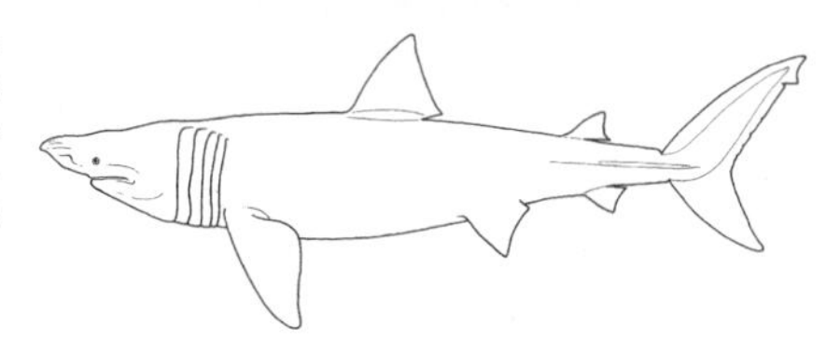

➔ **Family Cetorhinidae. Basking Sharks: plate 28**

23a [21b] Snout elongated and bladelike; anal fin much larger than dorsal fins; no precaudal pits; caudal fin without ventral lobe.

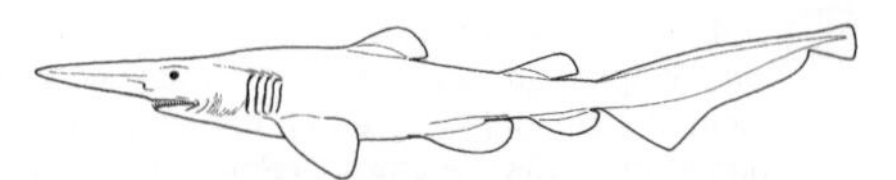

➔ **Family Mitsukurinidae. Goblin Sharks: plate 27**

23b [21b] Snout conical or flattened, short and not bladelike; anal fin subequal to dorsal fins in size or smaller than them; upper and sometimes lower precaudal pits present; caudal fin with strong ventral lobe. **➔24**

24a [23b] Caudal fin about as long as rest of shark.

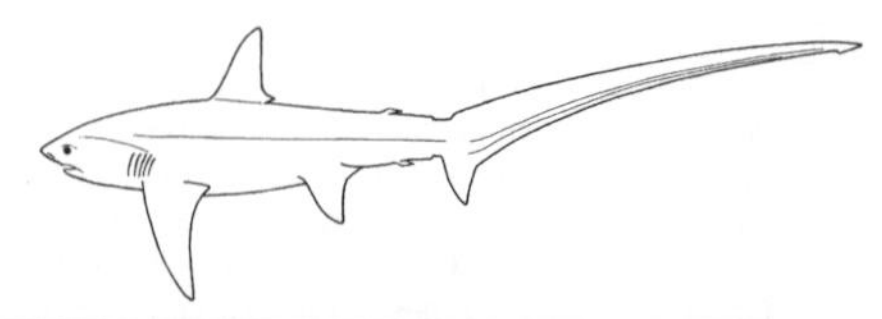

➔ **Family Alopiidae. Thresher Sharks: plate 29**

24b [23b] Caudal fin less than half the length of rest of shark. **➔25**

25a [24b] Mouth huge, terminal on head, level with snout; teeth small, very numerous, and hook-shaped; internal gill openings screened by numerous long papillose gill rakers.

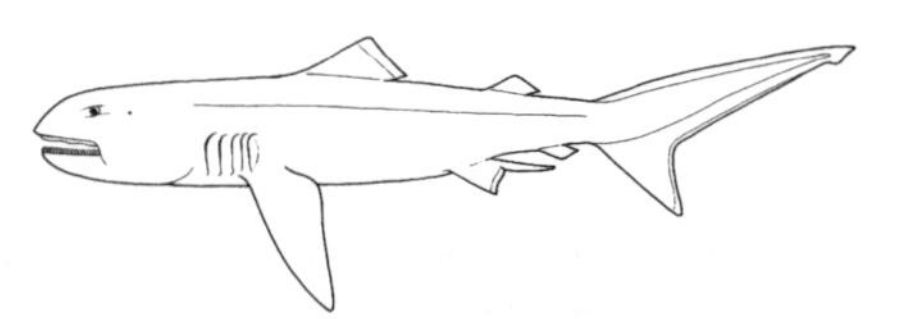

➔ **Family Megachasmidae. Megamouth Sharks: plate 28**

25b [24b] Mouth smaller, subterminal on head, behind snout tip; teeth bladelike with large anterior teeth, intermediate teeth, and lateral teeth in upper jaw; internal gill openings without gill rakers. **➔26**

26a [25b] Eyes very large; gill slits extending onto upper surface of head; both upper and lower precaudal pits present; a low keel on each side of caudal peduncle

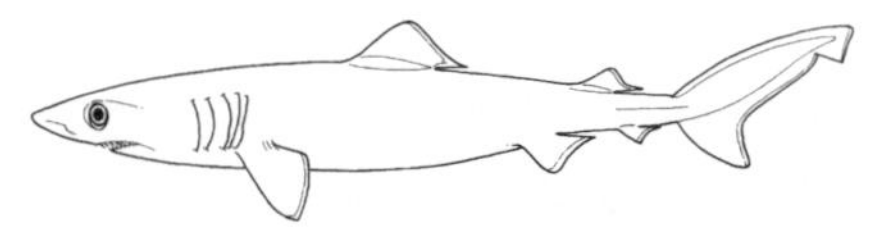

➔ **Family Pseudocarchariidae. Crocodile Sharks: plate 27**

26b [25b] Eyes smaller; gill slits not extending onto upper surface of head; lower precaudal pit absent; no keels on caudal peduncle.

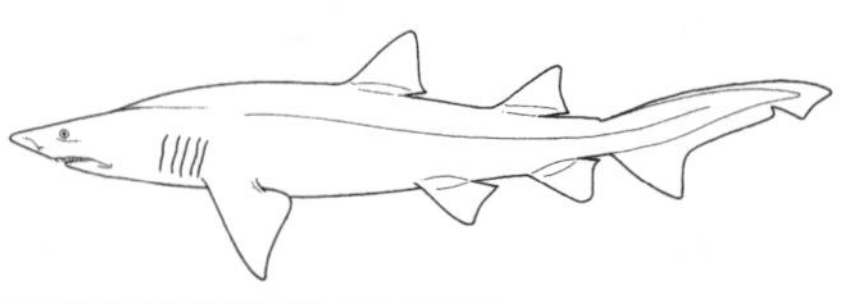

➔ **Family Odontaspididae. Sand Tiger or Raggedtooth Sharks: plate 27**

27a [20b] Head with lateral expansions or blades, like a double-edged axe.

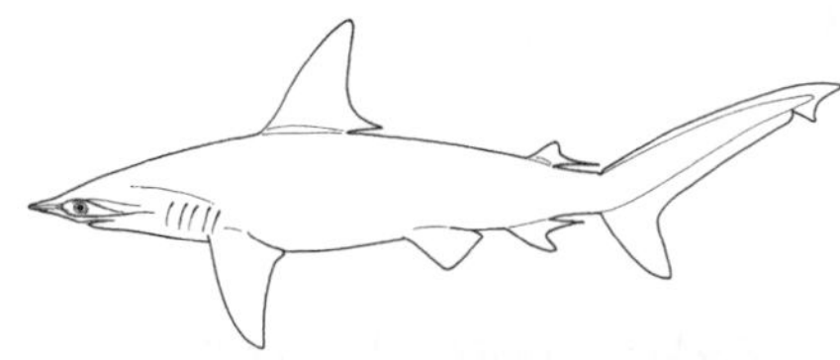

➔ **Family Sphyrnidae. Hammerhead, Bonnethead and Winghead Sharks: plates 63-64**

27b [20b] Head normal, not expanded laterally. **➔28**

28a [27b] Origin of first dorsal fin over or behind pelvic-fin bases.

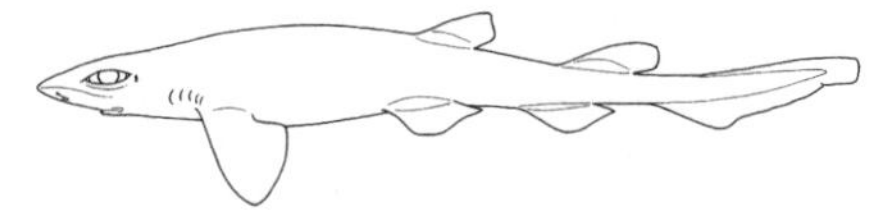

➔ **Family Scyliorhinidae. Catsharks: plates 31-42**

28b [27b] Origin of first dorsal fin well ahead of pelvic-fin bases. **➔29**

29a [28b] No precaudal pits, dorsal caudal fin margin smooth. **➔30**

29b [28b] Precaudal pits and rippled dorsal caudal margin present, *see drawing 22*, (ripples sometimes irregular in *Scoliodon* and *Triaenodon* of family Carcharhinidae). **➔33**

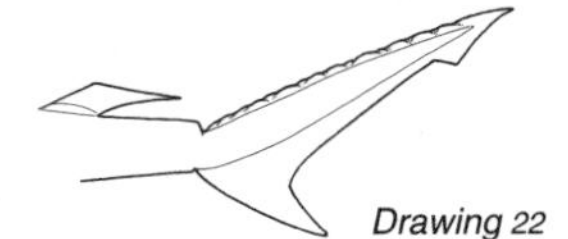

Drawing 22

30a [29a] Labial furrows very short or absent, when present confined to mouth corners; posterior teeth on dental bands comblike, *see drawing 23*. **➔31**

Drawing 23

30b [29a] Labial furrows relatively long with uppers extending partway or all the way anterior to level of symphysis, *see drawing 24*; posterior teeth on dental bands not comblike. **➔32**

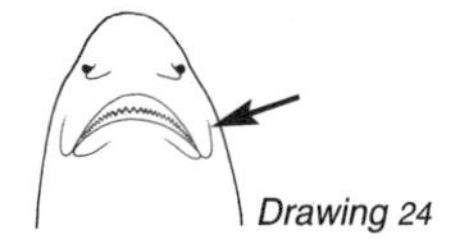

Drawing 24

31a [30a] Snout bell-shaped in dorsoventral profile, with a deep groove in front of eye, *see drawing 25*; internarial space over 1.5 times nostril width; inside of mouth and edges of gill arches without papillae; first dorsal fin more or less elongated, base closer to pectoral fins than pelvic fins.

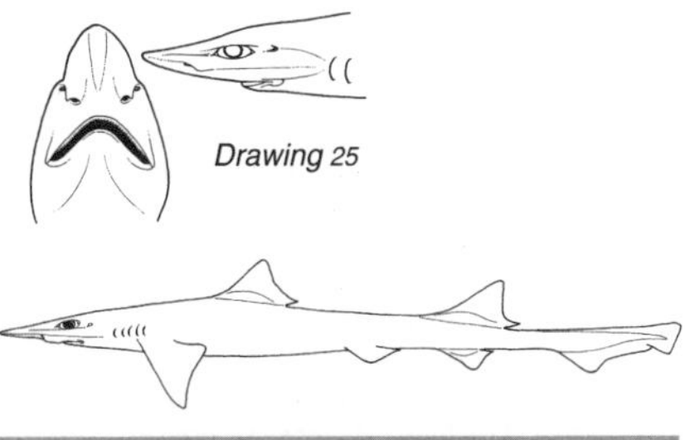

Drawing 25

➔ **Family Pseudotriakidae. False Catsharks and Gollumsharks: plate 43**

31b [30a] Snout rounded-parabolic or subangular in dorso-ventral profile, *see drawing 26*, without a deep groove in front of eye; internarial space less than 1.3 times nostril width; inside of mouth and edges of gill arches with papillae; first dorsal short, base closer to pelvic fins than pectoral fins.

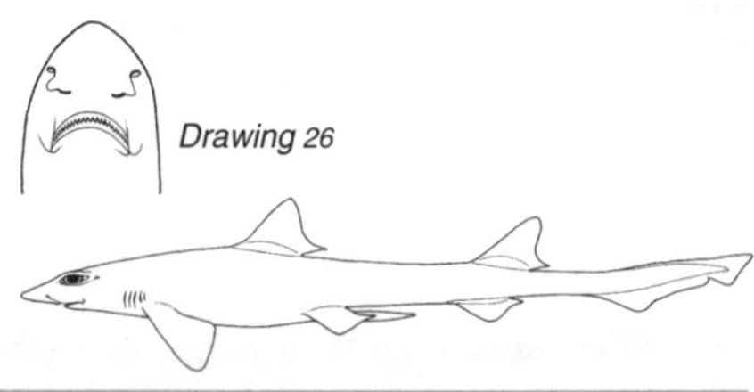

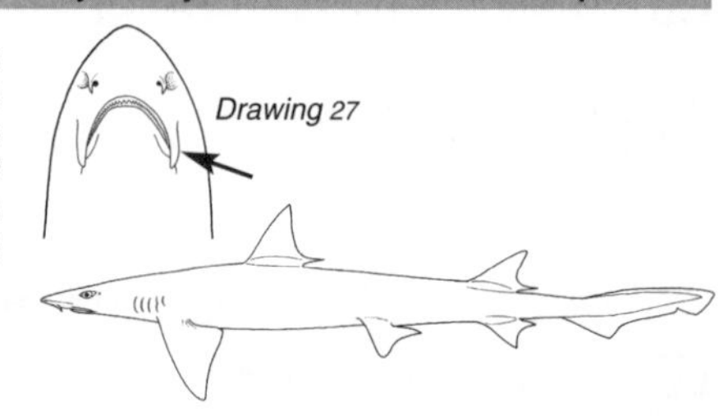

→ **Family Proscylliidae. Finback Catsharks: plate 43**

32a [30b] Anterior nasal flaps formed as slender barbels; upper labial furrows extremely long, nearly equal to internarial width and over half mouth width, *see drawing 27*; intestinal valve with 14 to 16 turns; no supraorbital crests on cranium.

→ **Family Leptochariidae. Barbeled Houndsharks: plate 43**

32b [30b] Anterior nasal flaps usually not barbel-like (except for *Furgaleus*); upper labial furrows shorter, considerably less than internarial and less than half of mouth width; intestinal valve with 4 to 10 turns; supraorbital crests present on cranium.

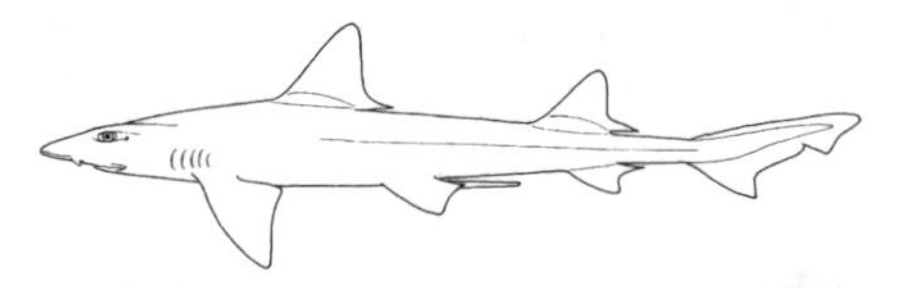

→ **Family Triakidae. Houndsharks: plates 44-48**

33a [29b] Posterior nasal flaps well-developed on rear edges of excurrent apertures of nostrils, *see drawing 28*; symphysial tooth rows well-developed in upper and lower jaws; second dorsal fin height about 0.4 to 0.7 times first dorsal height; intestine with a spiral valve having 4 to 6 turns.

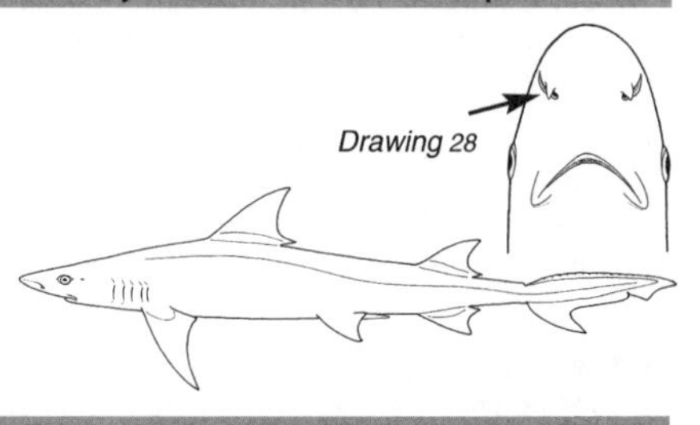

→ **Family Hemigaleidae. Weasel sharks: plate 49**

33b [29b] Posterior nasal flaps poorly developed on rear edges of excurrent apertures of nostrils; symphysial tooth rows usually poorly developed in upper and lower jaws; second dorsal fin height from 0.2 to 1.0 times first dorsal height, but most species with fin less than 0.4 times first dorsal height; intestine with a scroll valve.

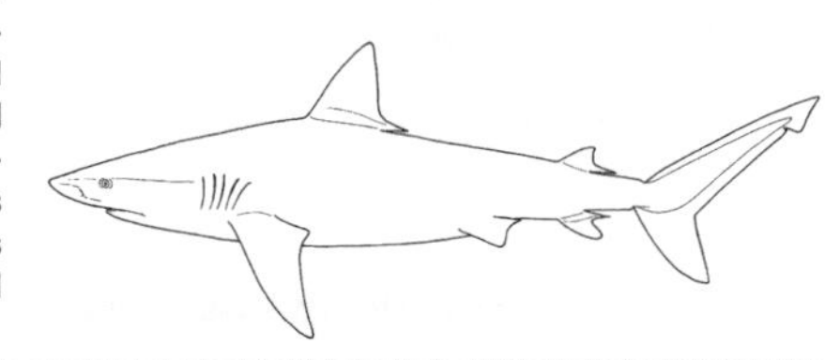

→ **Family Carcharhinidae. Requiem Sharks: plates 50-62**

Colour plates

Abbreviations used in plate text

Afr.	Africa
Ant.O	Antarctic Ocean
AO	Arctic Ocean
Atl.	Atlantic Ocean
Arg.	Argentina
Austr.	Australia
Carib.	Caribbean
IO	Indian Ocean
IP	Indo-Pacific
IWP	Indo West Pacific
Med.	Mediterranean Sea
New Caled.	New Caledonia
NZ	New Zealand
Pac.	Pacific Ocean
SO	Southern Ocean
WW	worldwide
N	north
S	south
E	east
W	west
abd.	abdomen
ad.	adult
A fin	anal fin
ant.	anterior
C	caudal
C fin	caudal fin
charac.	characteristic
cl.	cool
consp.	conspicuous
contin.	continental
C ped.	caudal peduncle
D fin	dorsal fin
D1 fin	first dorsal fin
D2 fin	second dorsal fin
dent.	denticle/denticles
dors.	dorsal
dk.	dark
dkr	darker
gill	gill opening
gills	gill openings
horiz.	horizontal
ht.	height
incosp.	inconspicuous
indent.	indentations
insert.	insertion
irr.	irregular
juv.	juvenile
lab. furr.	labial furrows
lat.	lateral
LL	lateral line
long.	longitudinal
lt.	light
ltr.	lighter
lwr.	lower
mod.	moderate
nr.	near
occ.	occasional
occly.	occasionally
on/nr.	on or near
P fin	pectoral fin
poss.	possibly
post.	posterior
prom.	prominent
pv.	post ventral
reg.	regular
rel.	relatively
sep.	separated
spec.	specimen/s
ST	subterminal
subC	subcaudal
surf.	surface
symm.	symmetrical
symph.	symphysis
T	terminal
temp.	temperate
trans.	transverse
triang.	triangular
trop.	tropical
Trop.	Tropics
upp.	upper
v.	very
V fin	(ventral) pelvic fin
vent.	ventral
vert.	vertical
wm.	warm
<	less than
>	more than
=	equal to
≈	approximately equal to
~	approximately
≤	less or equal to
≥	more or equal to

Plate 1 Hexanchidae & Chlamydoselachidae

1 spineless D fin, 1 A fin; 6 or 7 pairs of gill slits; large mouth; tiny spiracles.

Frilled Shark *Chlamydoselachus anguineus* p65

Patchy WW; benthic, epibenthic and pelagic, 50–1500 m. Large terminal mouth with widely spaced slender three-cusped needle sharp teeth; 6 pairs of gills; low D fin smaller than A fin, P fins < V fins.

Southern African Frilled Shark *Chlamydoselachus* sp. A p66

S Afr.; benthic, epibenthic and pelagic; 50–1500 m. Similar to above species, difficult to distinguish externally; head usually longer, body shorter than above.

Sharpnose Sevengill Shark *Heptranchias perlo* p66

Patchy trop. and temp., not NE Pac.; mainly deepwater 27–720 m, max. 1000 m. Slender; acutely pointed head with narrow mouth and large eyes; juv. black D fin apex fades with growth.

Bluntnose Sixgill Shark *Hexanchus griseus* p67

Patchy WW, absent polar regions? 500–1100 m, max. 1875 m or more, juv. inshore in cold water, ad. shallow water nr. submarine canyons. Large heavy body; broad head with wide mouth and small white-ringed eyes; soft supple fins; lt. coloured LL and post. fin edges.

Bigeye Sixgill Shark *Hexanchus nakamurai* p67

Patchy wm. temp. and trop., absent E Pac.?; on/nr. seabed, 90–621 m, occ. nr. surface or inshore. Slender body; narrow head with narrow mouth and large eyes; fins with white post. edges and tips, C fin ST notch deep.

Broadnose Sixgill Shark *Notorynchus cepedianus* p68

Patchy inshore cl. temp.; surf-line–50 m, large sharks to 136 m or more. Bluntly pointed broad head with wide mouth and small eyes; numerous small black spots (some plain or white spotted), newborn's black D fin apex fades with growth.

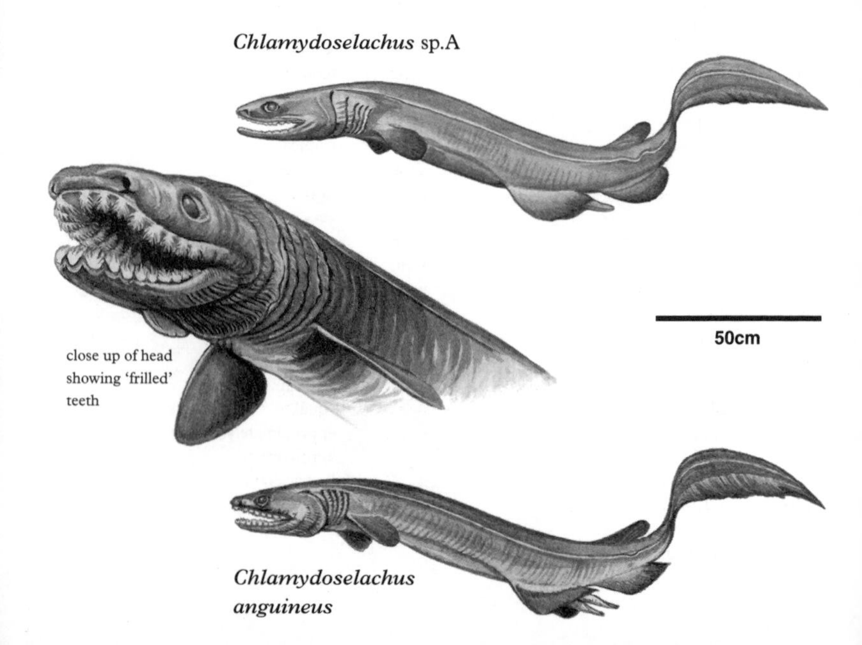

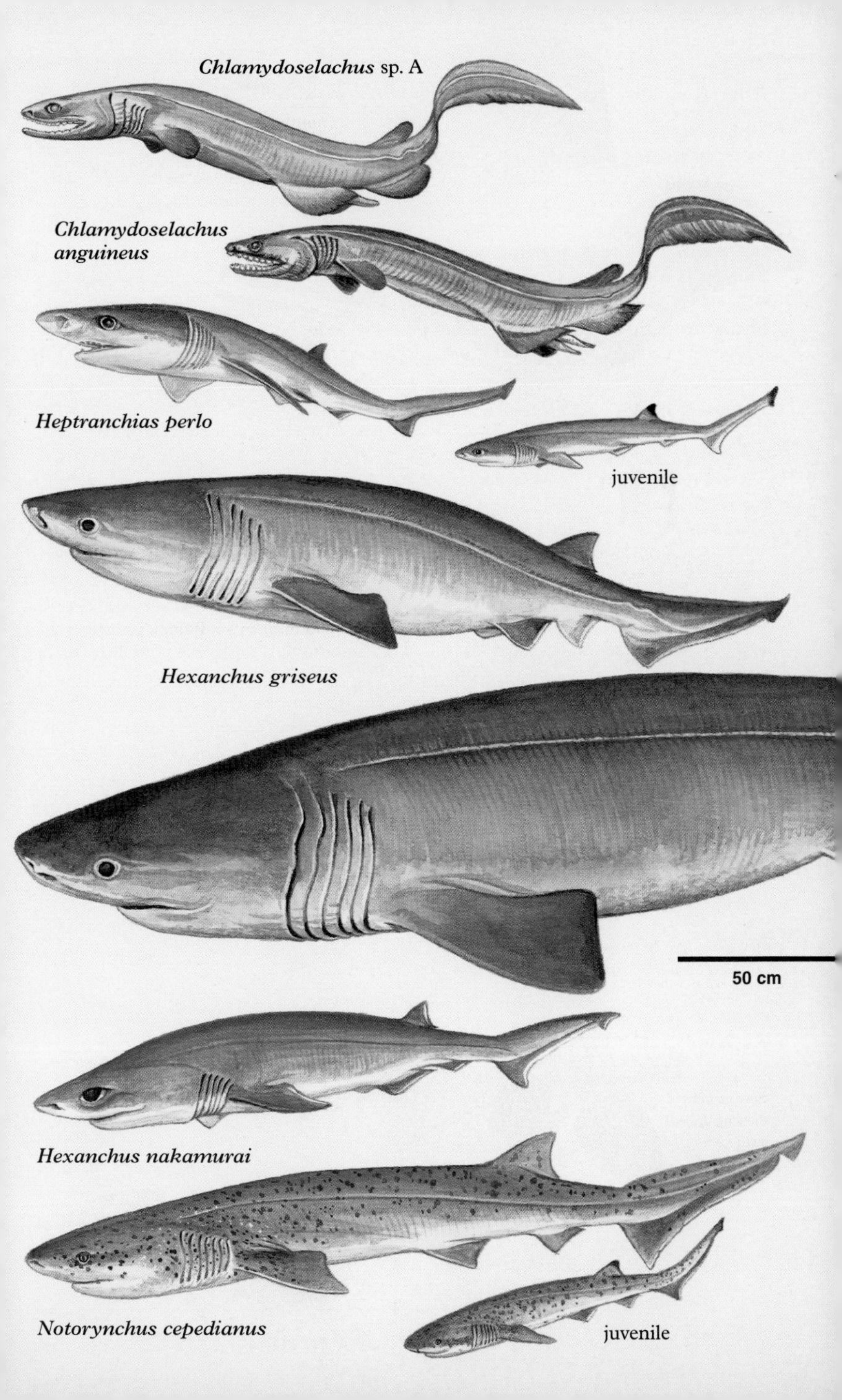

Chlamydoselachus sp. A
Chlamydoselachus anguineus
Heptranchias perlo
juvenile
Hexanchus griseus
50 cm
Hexanchus nakamurai
Notorynchus cepedianus
juvenile

Plate 2 Squalidae I

Head with shortish snout and large spiracle close behind eye; cylindrical body; 2 D fins with strong ungrooved spines, no A fin; many *Squalus* species difficult to distinguish.

Roughskin Spurdog *Cirrhigaleus asper* **p72**

Wm. temp. to trop. W Atl. to IO and central Pac.; on/nr. bottom 73-600 m, sometimes off bays and river mouths. Stocky rough-skinned body; broad flat head with short broadly rounded snout, large stubby barbels do not reach mouth; 2 high stout D spines; conspicuous white-edged fin margins.

Mandarin Dogfish *Cirrhigaleus barbifer* **p73**

Patchy S Jap. to NZ; on/nr. bottom 146–640 m. Similar to *C. asper* distinguished by long moustache-like barbels which reach mouth; conspicuous white-edged fin margins.

Bartail Spurdog *Squalus* **sp. A.** **p78**

NE Austr.; offshore 220–450 m. Fairly slender; broad head, short broad snout with large medial barbel; high short D1 fin with spine < apex, D2 fin smaller with spine> apex, P fins post. margin deeply concave with narrow tip; D fin apices dusky, C fin with dk. dors. margin and white post. margins, obvious dk. bar along C base.

Eastern Highfin Spurdog *Squalus* **sp. B.** **p79**

E Austr.; 240–450 m. Similar to *S.* sp.A distinguished by small barbel; shorter D1, thicker D2 spine, broad triangular P fins, C fin shorter; dusky D fin apices and edges, C fin with lt. post. margins and V lobe, no dk. marks on C fin.

Western Highfin Spurdog *Squalus* **sp. C.** **p79**

W Austr.; 220-510 m. Similar to *S.* sp.B distinguished by slightly longer snout; slender D2 spine slightly < fin apex; greyish D fins with pale apices; C fin dusky with narrow pale post. margins.

Fatspine Spurdog *Squalus* **sp. D.** **p80**

W Austr.;180-200 m. Slender; broad head with broad short snout and small medial barbels; D1 fin mod. high, D fin spines stout, D1 spine < fin apex, D2 spine = fin apex; pale fins, dusky D fin apices, C fin with dusky dorsal margin and almost white V lobe, post. margins and post. tip.

Western Longnose Spurdog *Squalus* **sp. E.** **p80**

W Austr.; 300-510 m. Slender; narrow head with long narrow snout and small medial barbels;D1 fin mod. high, slender D fin spines, D1 low, D2 high = fin apex; D fins with dusky apices and post. margins free tips white, C fin with dk. dors. margin and dk. blotch on UPV margin (fades in ad.).

Eastern Longnose Spurdog *Squalus* **sp. F.** **p81**

E Austr.; 220-500 m. Similar to *S.* sp.F distinguished by shorter snout; D1 fin lower, C fin larger; dk. bar along dors. and vent. C base, C fin white-tipped.

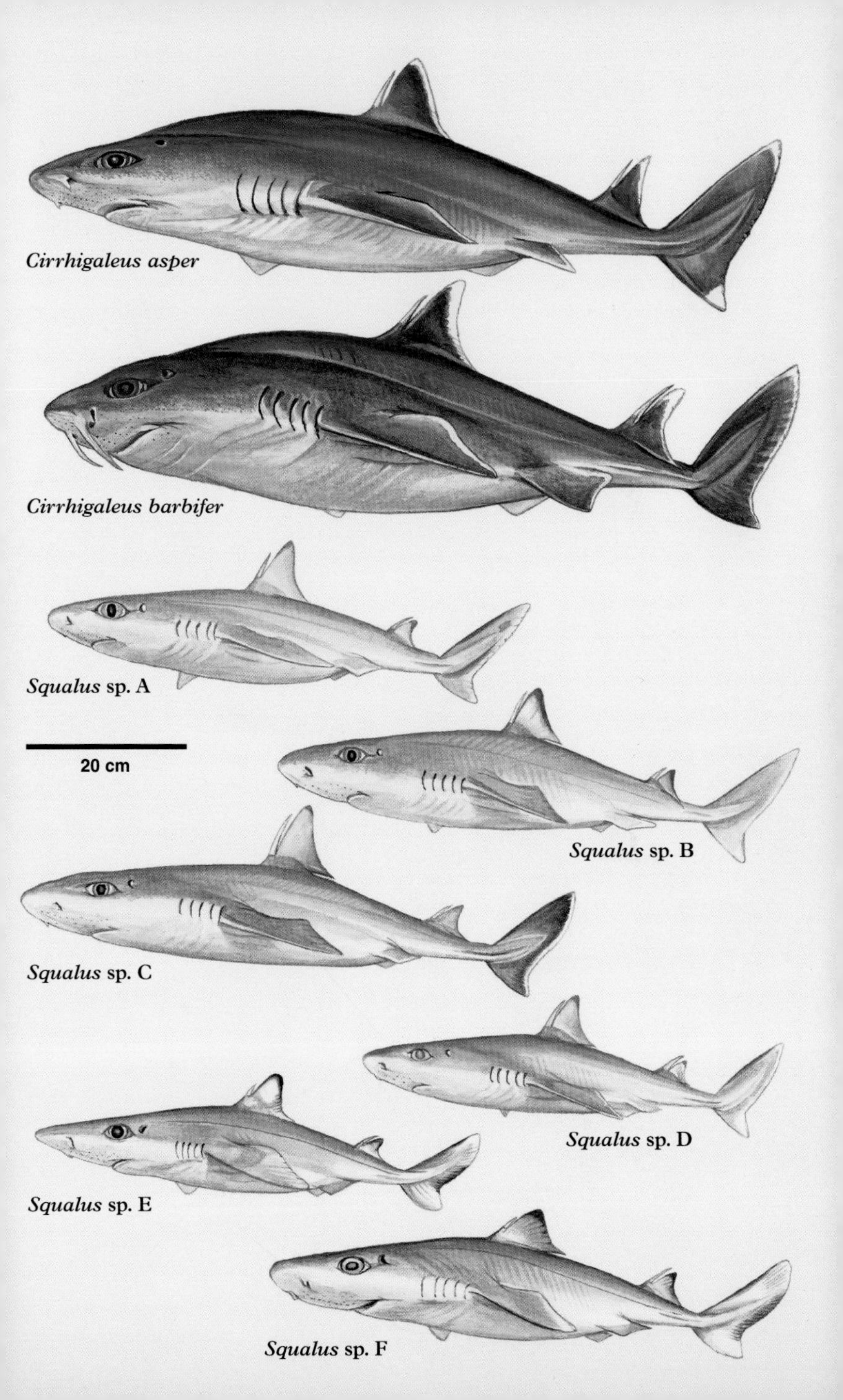

Cirrhigaleus asper
Cirrhigaleus barbifer
Squalus sp. A
20 cm
Squalus sp. B
Squalus sp. C
Squalus sp. D
Squalus sp. E
Squalus sp. F

Plate 3 Squalidae II

Piked Dogfish *Squalus acanthias* **p73**

WW except trop. and nr. poles; 0–600 m poss. deeper, epipelagic cold water 0-200 m. Slender; narrow head, long pointed snout, no barbel; D1 fin low, D1 spine slender and v. short; often with white spots on back and sides, black D fin tips in young, C fin with no blackish marks.

Longnose Spurdog *Squalus blainvillei* **p74**

Temp. to trop. E Atl.; on/nr. muddy bottom 16– >440 m. Heavy body; broad head, rel. short broad snout, large barbels; erect high D1 fin and heavy tall D1 spine; D fins white edged.

Cuban Dogfish *Squalus cubensis* **p75**

Wm. temp. to trop. W Atl.; on/nr. bottom 60–380 m. Slender; broad head, short rounded snout and small barbels; D1 mod. high, D spines slender and high, P fins deeply concave; P fins with white tip and post. margin, D fins black patch at tips, P and C fins with white post. margins.

Japanese Spurdog *Squalus japonicus* **p75**

N Pac.; on/nr. bottom 120–340 m. Fairly slender; long narrow tapering snout, small barbels; D1 fin mod. high, long slender spine; D fin apices dusky, dors. C margin dusky, P and C fins white post. margins.

Shortnose Spurdog *Squalus megalops* **p76**

E Atl. and IP; on/nr. bottom 0–732 m. Small, slender; broad head, short broad snout, small barbels; D1 fin mod. high, short slender spine; D fin apices dk., dors. C margin dk., P and C fins white post. margins.

Blacktail Spurdog *Squalus melanurus* **p77**

SW Pac.; 34-480 m. Slender; broad head, v. long broad snout, small barbels; D1 fin high, long slender spine; D fin apices black, dors. C margin partial black and V lobe with black patch.

Shortspine Spurdog *Squalus mitsukurii* **p77**

Apparently extensive; on/nr. bottom 4–954 m. Fairly slender; broad head, rel. long broad snout, large barbels; D1 fin spine stout and short; D fin apices dusky, P, V and C fins with white post. margins, post. C notch dusky.

Cyrano Spurdog *Squalus rancureli* **p78**

SW Pac., New Heb.; 320–400 m. Slender; broad head, v. long broad snout, small barbels; D1 fin high, long slender spine; D fin webs dusky, apices blackish, P and V fins with white post. margins, C fin web dusky, upp. pv. margin and vent. tip white.

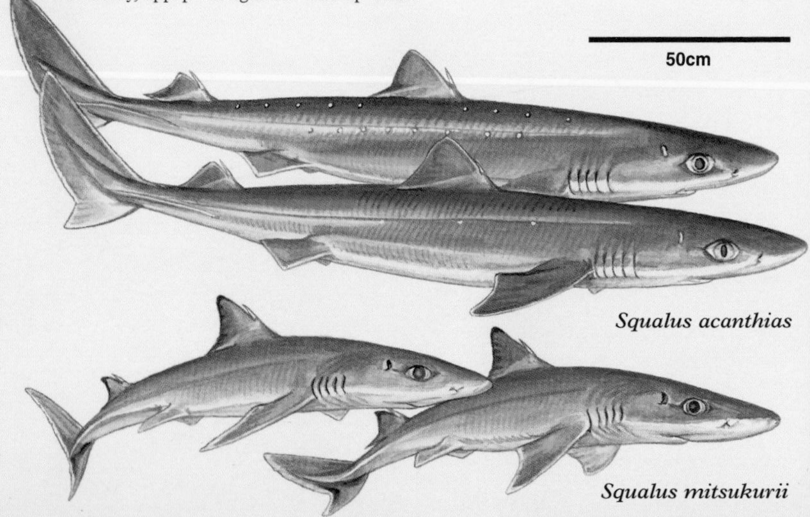

Squalus acanthias

Squalus mitsukurii

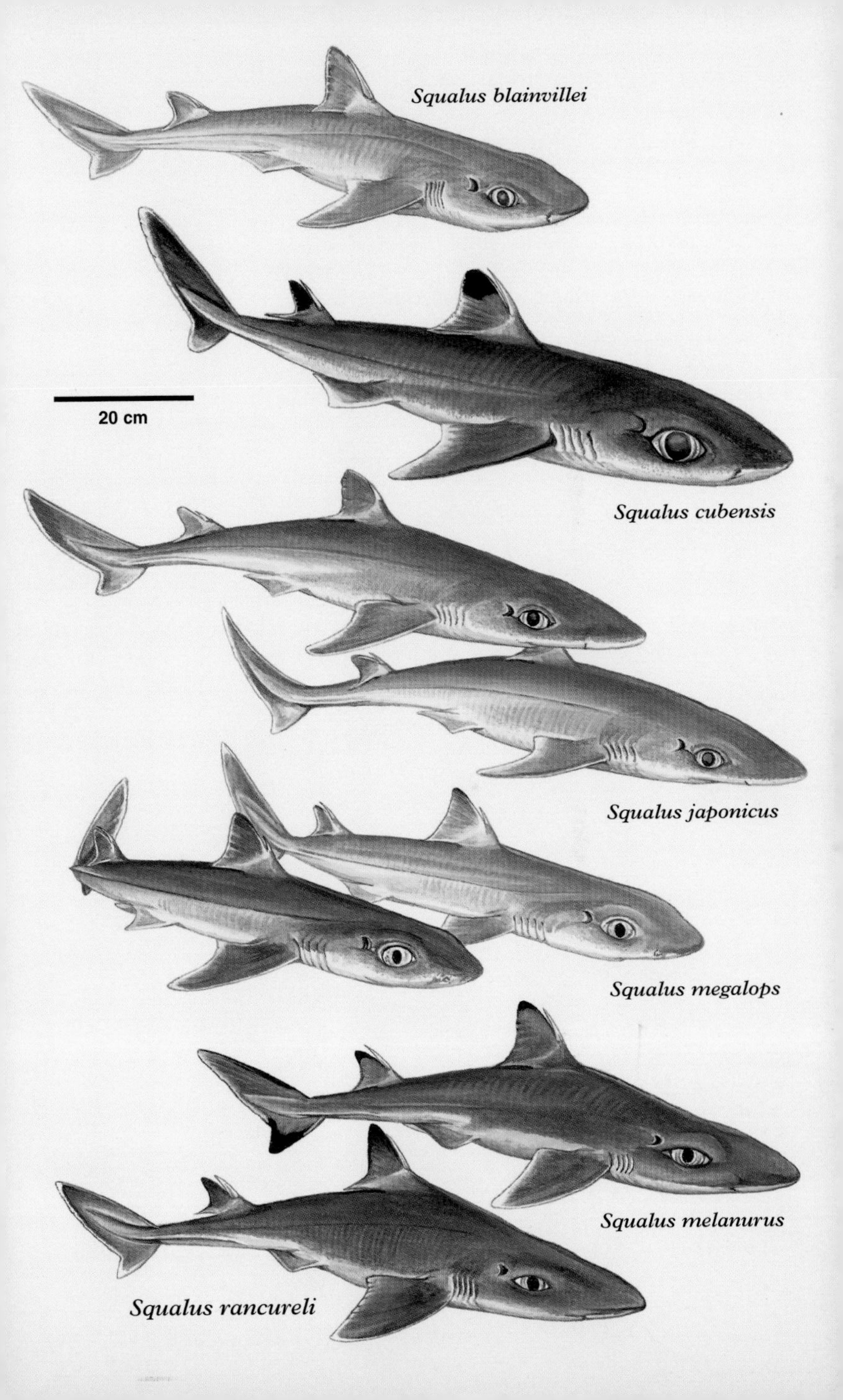
Squalus blainvillei
20 cm
Squalus cubensis
Squalus japonicus
Squalus megalops
Squalus melanurus
Squalus rancureli

Plate 4 Centrophoridae I

Cylindrical; huge green or yellowish eyes; 2 D fins with grooved spines, no anal fin.

Needle Dogfish *Centrophorus acus* **p82**

NW Pac.; 150–950 m, poss. 1786 m. Rough skin; fairly long flat snout; D1 fin low and long, D2 fin similar, P fin free rear tip long and angular; darker edges to post. fin margins.

Gulper Shark *Centrophorus granulosus* **p83**

Atl., W IO and W Pac.; on/nr. bottom 50–1440 m. Smooth skin; short thick snout; D2 fin shorter but nearly as high as D1, P fin free rear tip long and acute; fin webs dusky, dk. tips only in juvs.

Longnose Gulper Shark *Centrophorus harrissoni* **p84**

E Austr.; 250–790 m. Smooth skin; long flat narrow snout; D2 fin shorter and lower than D1, P fin free rear tip v. long and narrowly angular, notch in T margin; D fins with dk. oblique bar.

Lowfin Gulper Shark *Centrophorus lusitanicus* **p85**

E Atl., W IO and W Pac; 300–1400 m. Smooth skin; long flat snout; D1 fin v. long D2 1/3rd as long but ≥ high, P fin free rear tips long and narrowly angular; notch in T margin; dusky fin webs.

Smallfin Gulper Shark *Centrophorus moluccensis* **p85**

Scattered I-W Pac.; 125–820 m. Smooth skin; short thick snout; D2 fin < half height of D1, P fin free rear tips long and narrowly angular, C fin deep vent. lobe; D1 fin dk. blotch nr. apex in juv., C fin and sometimes P and V fins with narrow pale post. margins.

Taiwan Gulper Shark *Centrophorus niaukang* **p86**

Patchy Atl. and IP; 98–~1000 m. Fairly rough skin; short broad snout; D1 fin long, D2 fin nearly as large, P fin free rear tips short and angular, C fin post. margin straight with weak vent. lobe.

Leafscale Gulper Shark *Centrophorus squamosus* **p86**

Atl., W IO and Pac.; 230–2400 m. Rough skin; short thick slightly flattened snout; D1 fin long and low D2 shorter, higher, P fin free rear tip short, C fin post. margin weakly concave; dusky fins.

Mosaic Gulper Shark *Centrophorus tessellatus* **p87**

W and central Pac., Atl. and IO?; on/nr. bottom 260–730 m. Smooth skin; fairly long thick snout; D2 fin ≈ D1, P fin free rear tips long and narrowly angular; lt. margins to fins.

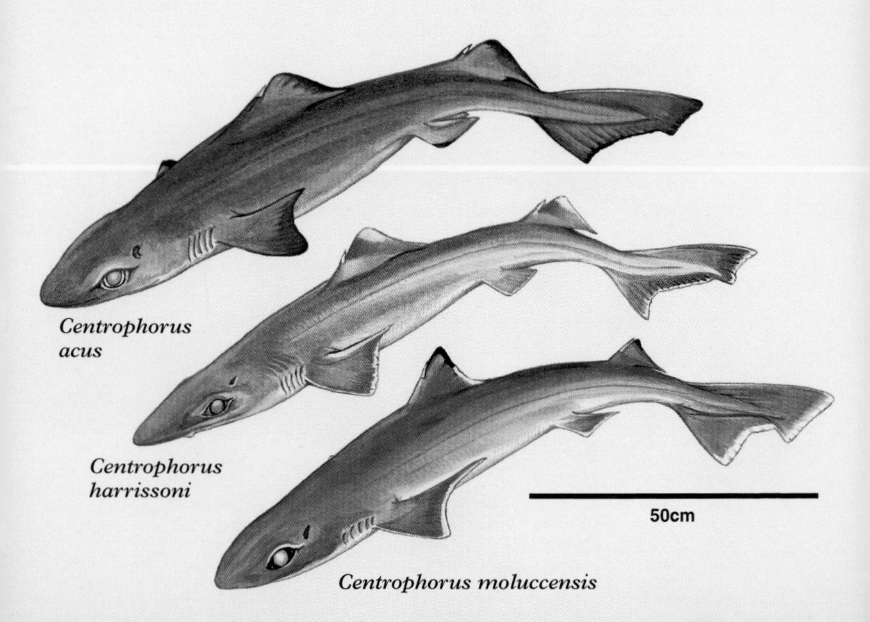

Centrophorus acus

Centrophorus harrissoni

Centrophorus moluccensis

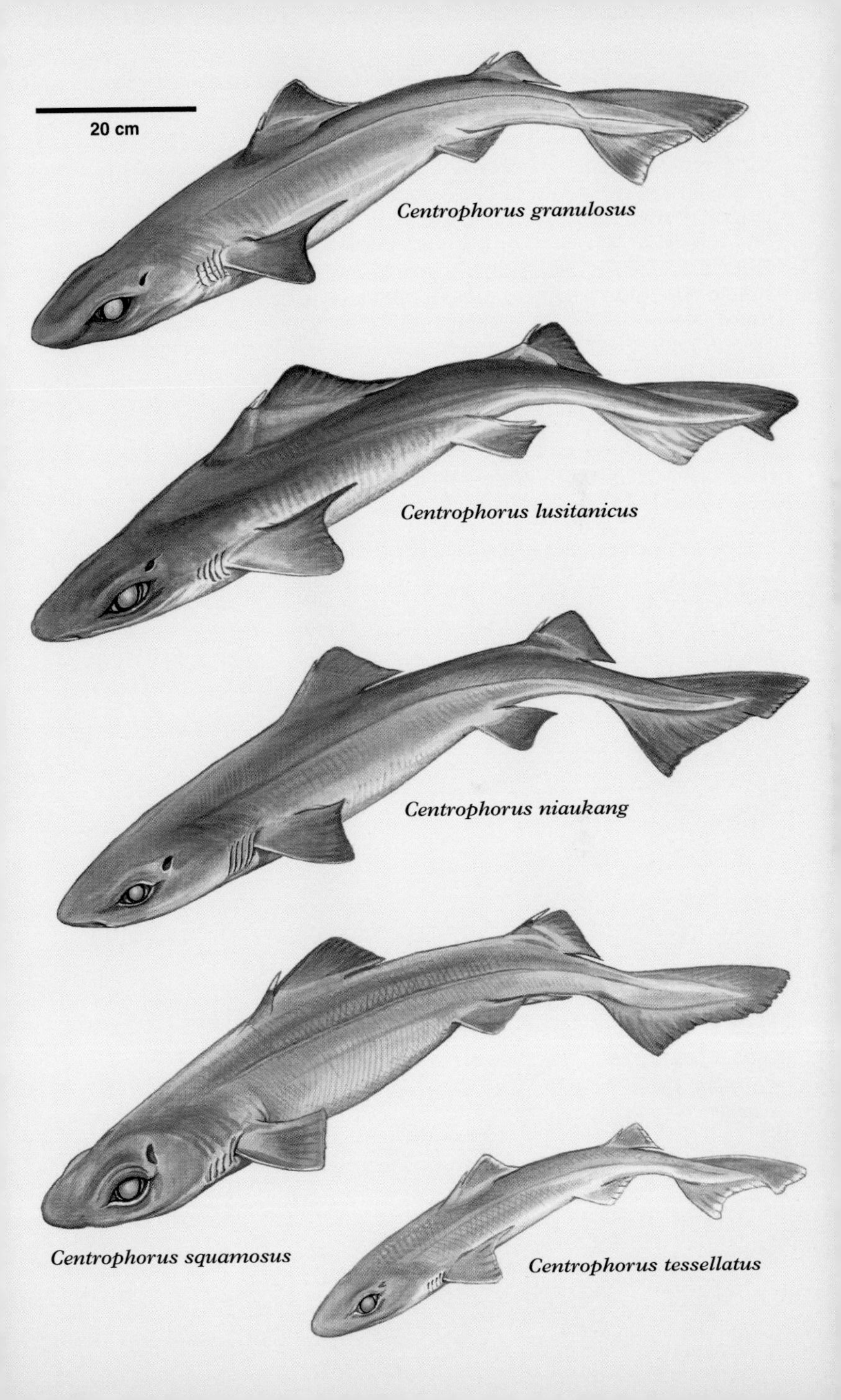
20 cm
Centrophorus granulosus
Centrophorus lusitanicus
Centrophorus niaukang
Centrophorus squamosus
Centrophorus tessellatus

Plate 5 Centrophoridae II & Echinorhinidae

Dwarf Gulper Shark *Centrophorus atromarginatus* **p82**
NW IO and W Pac.; 150–450 m. Smooth skin; fairly long flat snout; D1 fin low, D2 nearly as large, P fin free rear tips long and narrowly angular; dusky fin webs, dk. fin tips only in juv.

Blackfin Gulper Shark *Centrophorus isodon* **p84**
Scattered IO and W Pac.; 760–770 m. Smooth skin; long flat snout; D1 fin short, D2 lower but as long, P fin free rear tips long and narrowly angular; blackish fin webs esp. D2 fin and C fin.

Birdbeak Dogfish *Deania calcea* **p87**
E Atl. and Pac.; 70–1470 m. Rough skin; extremely long flat snout; no subC keel; D1 fin low and long, D2 shorter and higher with longer spine; juv. with blackish markings on fins, dusky on head.

Rough Longnose Dogfish *Deania hystricosum* **p88**
E Atl. and NW Pac.; 470–1300 m. Very rough skin; extremely long flat snout; no subC keel; D1 fin low, D2 shorter and higher; no obvious markings.

Arrowhead Dogfish *Deania profundorum* **p89**
E Atl., W Atl. and I-W Pac.; on/nr. bottom 275–1785 m. Smooth skin; extremely long flat snout; subC keel ; D1 fin rel. short and high, D2 similar but taller with much longer spine.

Longsnout Dogfish *Deania quadrispinosum* **p89**
S Afr., Austr. and NZ; 150–1360 m. Rough skin; extremely long flat snout; no subC keel; D1 fin rel. high, angular and short, D2 higher with longer spine; sometimes white-edged fins.

Echinorhinidae: Broad flat head with tiny spiracles; 2 similar-sized D fins set close together and well back, no A fin; large thorn-like skin denticles.

Bramble Shark *Echinorhinus brucus* **p70**
E Atl., Med. and Indo-W Pac.; on/nr. bottom, 200–900 m. Irr. scattered whitish consp. denticles which can fuse into plates; may have red/black spots/blotches on back and sides, fin edges blackish.

Prickly Shark *Echinorhinus cookei* **p71**
Pac.; close to bottom, 11-1100 m. Lt. colored inconsp. denticles numerous and regularly spaced, but few below snout; white around mouth and snout; post. fin edges blackish.

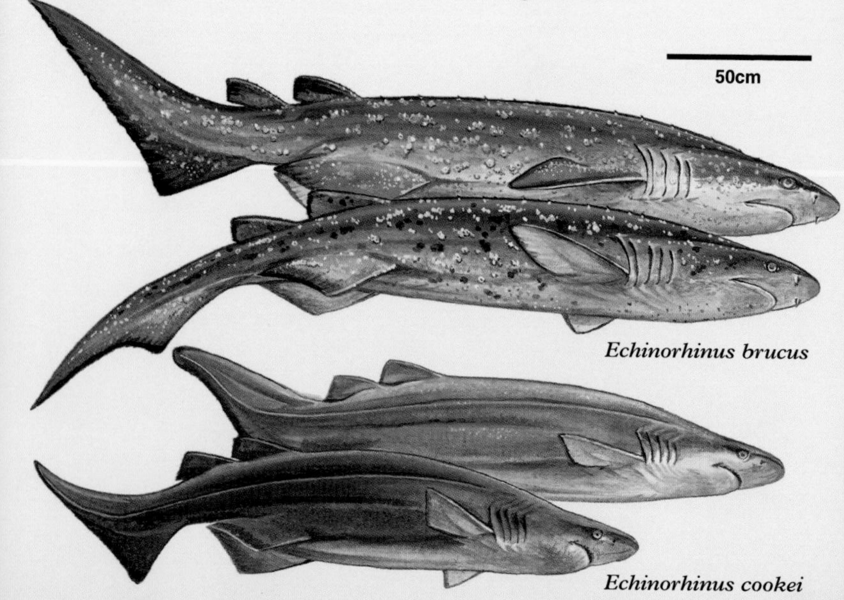

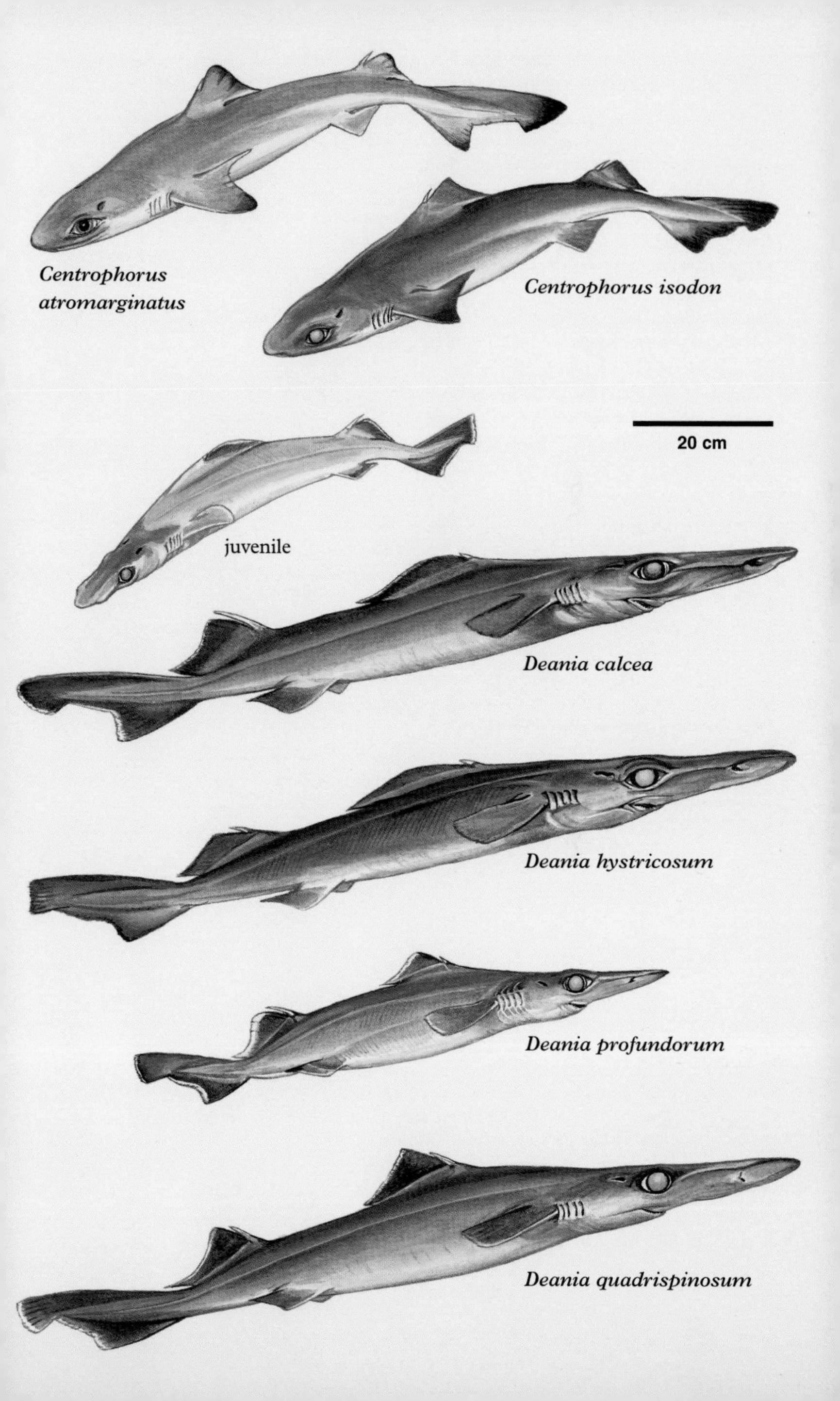

Centrophorus atromarginatus
Centrophorus isodon
20 cm
juvenile
Deania calcea
Deania hystricosum
Deania profundorum
Deania quadrispinosum

Plate 6 Etmopteridae I

Dwarf to medium-sized with photophores either inconspicuous or forming distinct black marks; 2 D fins with strong grooved spines, D2 fin and spine much larger than D1, no A fin and no keel; many species difficult to distinguish.

Hooktooth Dogfish *Aculeola nigra* **p90**

E Pac.; benthic and epibenthic 100–735 m.Stocky; broad blunt snout with broad long arched mouth; gills quite large; short D spines much < low D fins, P fins with rounded tips.

Highfin Dogfish *Centroscyllium excelsum* **p91**

NW Pac.; deep seamounts 800–1000 m. D1 fin high and rounded with short spine, D2 spine v. long reaching > apex, short C peduncle; intense black marks around mouth and beneath P fins; dorsal denticles sparse and irregular, ventral none.

Black Dogfish *Centroscyllium fabricii* **p91**

Widespread temp. Atl.; 180–1600 m. Fairly stout compressed long abd.; arched mouth; D1 fin low, short C peduncle; numerous close-set denticles.

Granular Dogfish *Centroscyllium granulatum* **p92**

S Amer.; 400–448m. Small slender cylindrical; mouth narrowly arched; long abdomen; D1 fin small much < D2, D2 spine v. large > apex; long C peduncle; closely covered by sharp denticles.

Bareskin Dogfish *Centroscyllium kamoharai* **p92**

W Pac.; bottom 500–1200 m.Stout and compressed; v. short broad ly arched mouth; D1 fin v. low and rounded, D2 fin slightly > spine ≈ apex; short C peduncle; skin smooth, almost naked.

Combtooth Dogfish *Centroscyllium nigrum* **p93**

Central and E Pac.; on/nr. bottom 400–1143 m. Fairly stout; short broadly arched mouth; D fins = size, D1 spine short D2 = fin apex; prominent white fin tips.

Ornate Dogfish *Centroscyllium ornatum* **p93**

N IO; nr. bottom 521–1262 m. Narrowly arched mouth; D1 fin low and rounded v. long spine ≈ D2 spine; denticles close-set and numerous.

Whitefin Dogfish *Centroscyllium ritteri* **p94**

NW Pac. 320–1100 m. Only *Centroscyllium* with obvious black markings on underside head, abdomen, P fins and stripe below C peduncle; mouth broadly arched; D1 fin low and rounded with v. short spine and ≈ D2 fin, D2 spine > apex; numerous close-set denticles; white post. fin margins.

Plate 7 Etmopteridae II

Southern Lanternshark *Etmopterus granulosus* **p100**

S Amer.; 220–637 m. Heavy bodied; big head with v. short gills; tail short and broad; D2 fin much > D1; consp. lines of large rough denticles on body not head; underside abruptly black, black marks above V fin and C fin base. See next plate.

Great Lanternshark *Etmopterus princeps* **p104**

N Atl.; on/nr. bottom, 350– 4,500 m. Stout; gills v. long; tail mod. long and broad; D2 fin much > D1 but< twice area of D1; denticles largely cover snout; no conspic. markings. See next plate.

Velvet Belly *Etmopterus spinax* **p107**

E Atl. and W Med.; nr. or well above bottom 70–2000 m. Long, fairly stout; gills v. short; long tail; D2 fin ~ twice area D1; denticles largely cover snout; underside abruptly black, black marks V fin area to tail. See next plate.

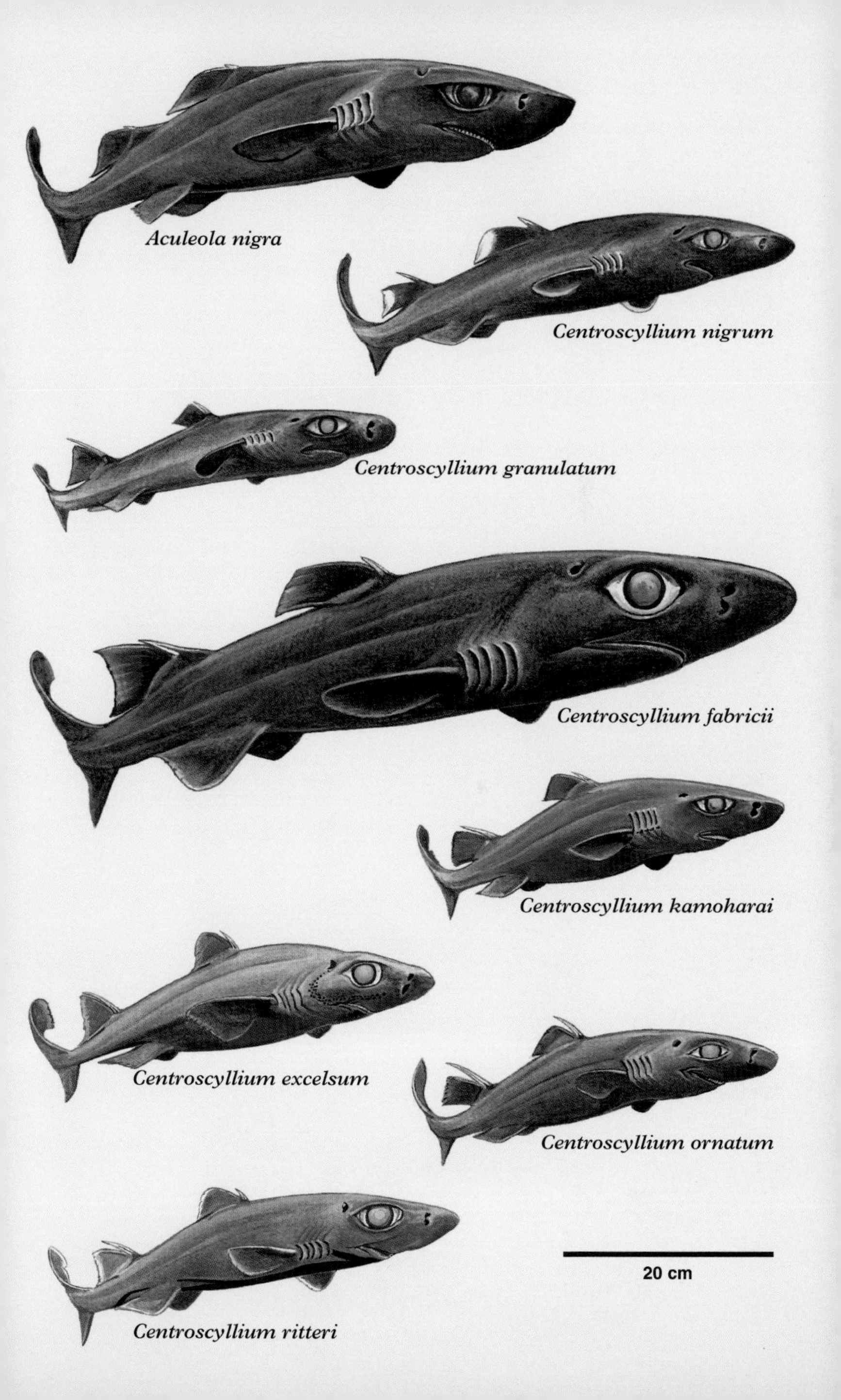

Aculeola nigra
Centroscyllium nigrum
Centroscyllium granulatum
Centroscyllium fabricii
Centroscyllium kamoharai
Centroscyllium excelsum
Centroscyllium ornatum
Centroscyllium ritteri
20 cm

Plate 7 Etmopteridae II

Lined Lanternshark *Etmopterus bullisi* **p96**

NW Atl.; on/nr. bottom 275–824 m. Slender; gills v. short; long tail; consp. long. row of denticles on sides and back; underside black, lt. band midline eye to D1 fin, black vent. and tail markings.

Broadband Lanternshark *Etmopterus gracilispinis* **p100**

W Atl. and W IO; on/nr. bottom 70–480 m, 2240 m off Arg. Stout; gills v. short; slender short tail; D2 fin ~ twice area D1; no reg. rows of denticles; underside grading black, inconsp. black vent. and tail markings.

Caribbean Lanternshark *Etmopterus hillianus* **p101**

NW Atl.; on/nr. bottom 311–695 m. Mod. stout; gills v. short; mod. long tail; D2 fin much > D1 but < twice area of D1; no reg. rows of denticles but largely cover snout; underside black, black vent. and tail markings.

African Lanternshark *Etmopterus polli* **p103**

E and W Atl.?; on/nr. bottom 300–1000 m. Fairly stout; gills short; fairly long tail; D2 fin ≥D1; denticles widespaced, no reg. rows, largely cover snout; blackish underside, black vent. and tail markings.

Smooth Lanternshark *Etmopterus pusillus* **p105**

Widespread Atl., I–W Pac.; on/nr. bottom 274–1000 m. Fairly slender; gills rather long; fairly short broad tail; D2 fin < twice area D1; denticles widespaced, not in rows, cover snout; obscure black vent. markings.

Fringefin Lanternshark *Etmopterus schultzi* **p106**

W Atl.; on/nr. bottom 220–915 m. Slender; gills v. short; mod. long tail; D2 fin ~ twice area D1; denticles widespaced, no rows, largely cover snout; fin margins naked; dk. vent. and tail markings.

Thorny Lanternshark *Etmopterus sentosus* **p107**

W IO; nr. bottom; 200–500 m? Slender; gills quite long; mod. long broad tail; D2 fin area > twice D1; 2–3 rows of denticles on flanks; fin margins largely naked; underside inconsp. black, black vent. and tail markings.

Green Lanternshark *Etmopterus virens* **p110**

NW Atl.; 196–915 m. Mod. slender; gills v. short; long narrow tail; D2 fin area > twice D1; denticles widespaced, no rows, largely cover snout; underside black, black vent. and tail markings.

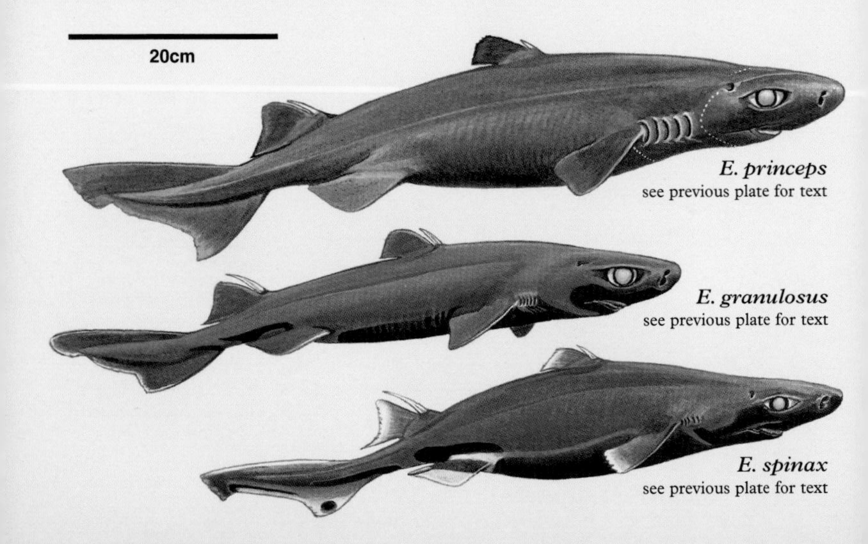

E. princeps
see previous plate for text

E. granulosus
see previous plate for text

E. spinax
see previous plate for text

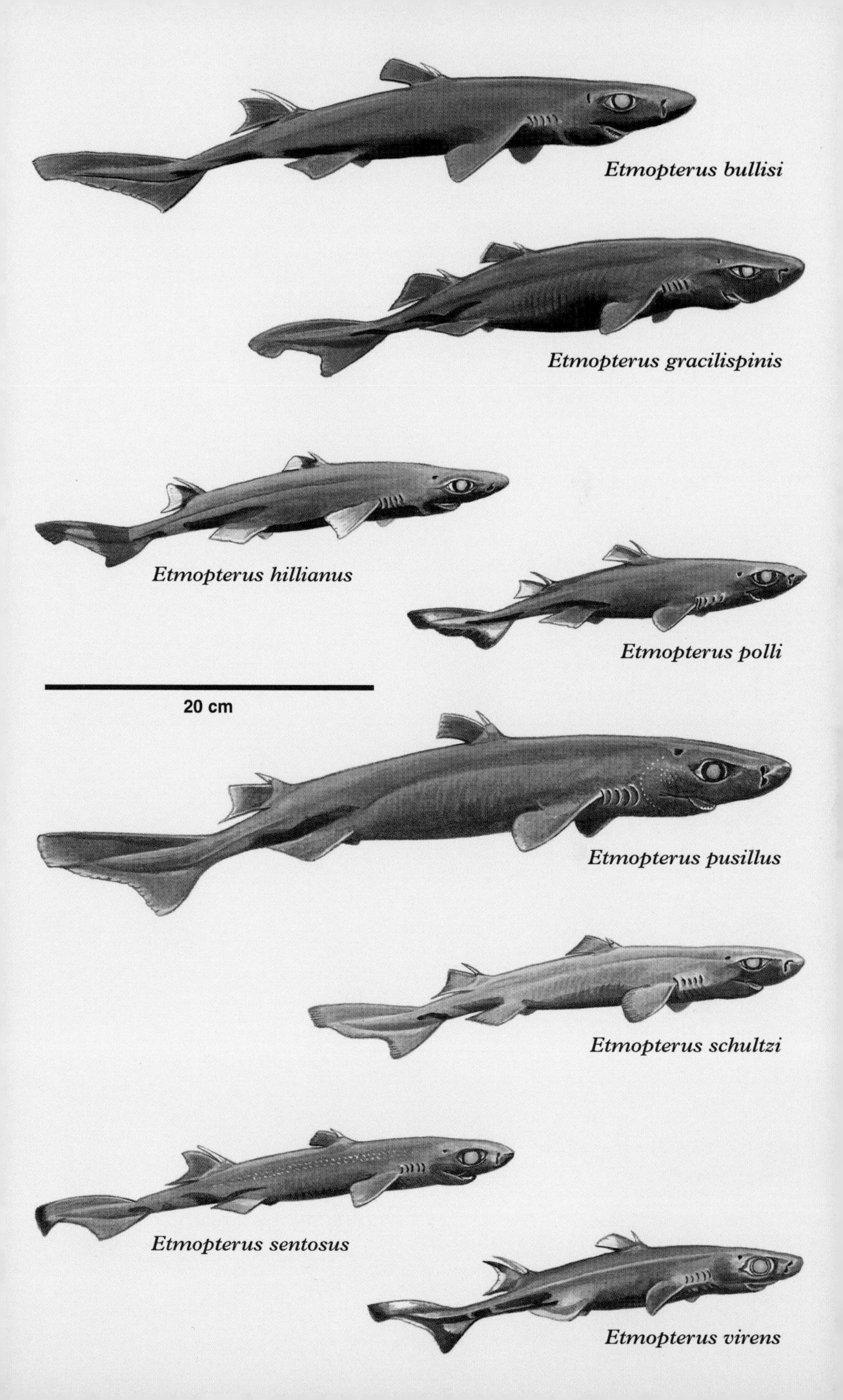
Etmopterus bullisi
Etmopterus gracilispinis
Etmopterus hillianus
Etmopterus polli
20 cm
Etmopterus pusillus
Etmopterus schultzi
Etmopterus sentosus
Etmopterus virens

Plate 8 Etmopteridae III

Giant Lanternshark *Etmopterus baxteri* **p94**
S Austr., NZ and S Afr.; 250–1500 m. Stout; short thick snout; short broad tail; D fins widely spaced, D2 fin twice area D1, D2 spine strongly curved; rough-textured; incosp. white blotch on head, black vent. and tail markings.

Blurred Smooth Lanternshark *Etmopterus bigelowi* **p95**
Central to S Atl., W IO and Pac.; 163– 1000 m+. Slender; broad head with long thick flat snout; long tail; D1 fin < D2; smooth skin; underside darker, white spot on head, lt. edges to fins, no consp. markings.

Shorttail Lanternshark *Etmopterus brachyurus* **p95**
W Pac.; nr. bottom 481m. Heavy bodied; broad head with short thick flat snout; short tail; D2 fin > D1, D2 spine strongly curved; consp. line of denticles; black vent. and tail markings.

Cylindrical Lanternshark *Etmopterus carteri* **p96**
NW Atl.; 283–356 m. Head semicylindrical, ~ deep as wide at eyes, snout v. short and rounded; gills broad; uniformly dark without concentrations of photophores, fins with pale webs.

Tailspot Lanternshark *Etmopterus caudistigmus* **p97**
New Caledonia; 638–793 m. Slender; narrow head with long thick narrow snout; long tail; D2 fin > D1; long. row of small close-set denticles on body and tail; obvious photophores on tail.

Combtooth Lanternshark *Etmopterus decacuspidatus* **p97**
NW Pac.; on/nr. bottom 512–692 m. Mod. slender; gills v. short; fairly long broad tail; D2 fin ~ twice area D1; no reg. rows of denticles; underside black, black vent. and tail markings.

Blackbelly Lanternshark *Etmopterus lucifer* **p102**
S Atl. and I–W P; on/nr.bottom 183–823 m. Stocky; gills mod. long; mod. long tail; D2 fin v. large; long. rows of denticles snout to tail; underside black, black vent. and tail markings.

Slendertail Lanternshark *Etmopterus molleri* **p102**
W Pac.; on/nr. bottom 238–655m. Slender; D2 fin much > D1; reg. long. rows of denticles snout to tail, except above P fins; underside abruptly black with black stripes on sides.

Dwarf Lanternshark *Etmopterus perryi* **p103**
NW Atl.; 283–375 m. One of smallest living sharks; long broad flat head; D2 fin > D1; underside black, consp. black vent. and tail markings.

West Indian Lanternshark *Etmopterus robinsi* **p106**
W Atl.; 412–787 m. Mod. stout; gills v. short; D2 fin much > D1 but < twice area of D1; denticles widespaced, no rows, largely cover snout; underside abruptly black, black vent. and tail markings.

Hawaiian Lanternshark *Etmopterus villosus* **p109**
Hawaiian Is.; on/nr. bottom 406–911 m. Stout; gills mod. long; short broad tail; D2 fin much > D1 but < twice area of D1; denticles widespaced, rows on trunk and tail, cover snout; underside slightly darker, indistinct black vent. markings.

Etmopterus baxteri

Etmopterus bigelowi

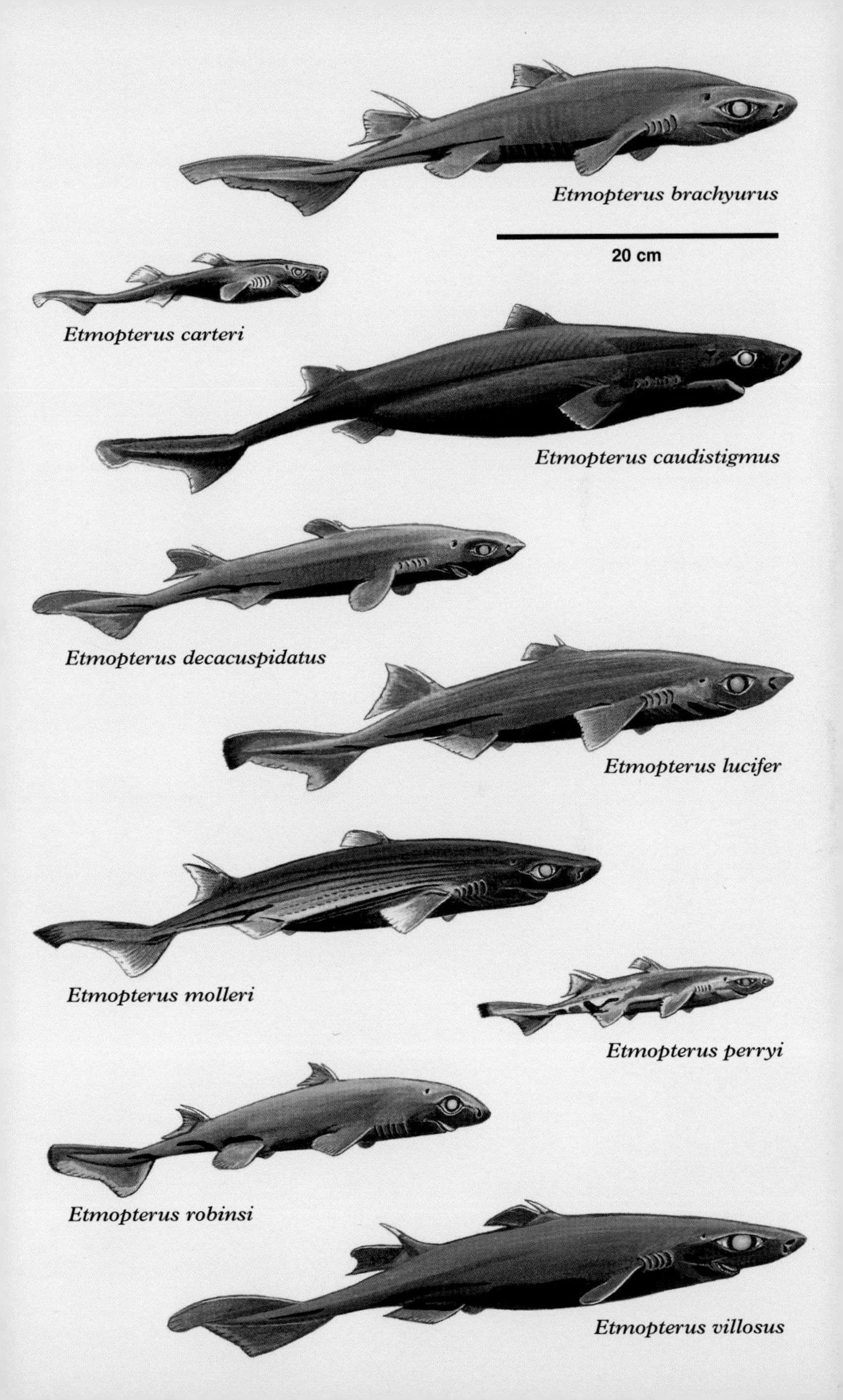
Etmopterus brachyurus
20 cm
Etmopterus carteri
Etmopterus caudistigmus
Etmopterus decacuspidatus
Etmopterus lucifer
Etmopterus molleri
Etmopterus perryi
Etmopterus robinsi
Etmopterus villosus

Plate 9 Etmopteridae IV

Pink Lanternshark *Etmopterus dianthus* **p98**
SW Pac.; 700–800 m. Stout; D1 fin small and low with short spine, D2 fin < twice D1 with D2 spine ≈ fin; bristle-like denticles not in rows; pinkish fresh, brownish-grey preserved, dusky to black below.

Lined Lanternshark *Etmopterus dislineatus* **p98**
NE Austr.; on/nr. bottom 590–700 m. Elongate attractive lanternshark; D1 fin small and low ~ half size D2 fin; bristle-like denticles not in rows; dk. broken lines on flanks with black tail markings.

Blackmouth Lanternshark *Etmopterus evansi* **p99**
Pac.; 430–550 m.Weakly defined rows of denticles on dors. midline and C ped. but not head; dk. borders around mouth, above eyes and sometimes gills, also black tail markings.

Pygmy Lanternshark *Etmopterus fusus* **p99**
E IO; 430–550 m. Cylindrical with long C ped.; D2 fin > twice height D1 fin; reg. rows of denticles on flanks and C ped. but not head; faint dk. marks on flanks and tail, fins pale with dk. markings.

Smalleye Lanternshark *Etmopterus litvinovi* **p101**
SE Pac.; 630–1100 m. Stout; large flat head, gills long; mod. long tail; D1 fin origin behind P fin rear tip, D2 fin slightly > D1 fin, interdorsal space short; no rows of denticles; no distinct markings.

False Lanternshark *Etmopterus pseudosqualiolus* **p104**
W Pac.; 1043–1102 m. Fusiform sim. shape to *E. carteri*; v. short deep snout with short round eyes; dk. brown to black, paler tail with inconsp. dk. marks, distal fin webs pale, term. C lobe dk.

Densescale Lanternshark *Etmopterus pycnolepis* **p105**
SE Pac.; 330–763 m. Slender; narrow head, gills long; mod. long tail; D1 fin origin ahead P fin rear tip, D2 fin > D1; v. small denticles in dense rows head to tail; black body and tail markings.

Splendid Lanternshark *Etmopterus splendidus* **p108**
NW Pac.; 210 m. Spindle-shaped sim. to *E. fusus*; back purplish-black underside bluish-black, pre-C fins pale red-brown webs and lighter patch on C fin.

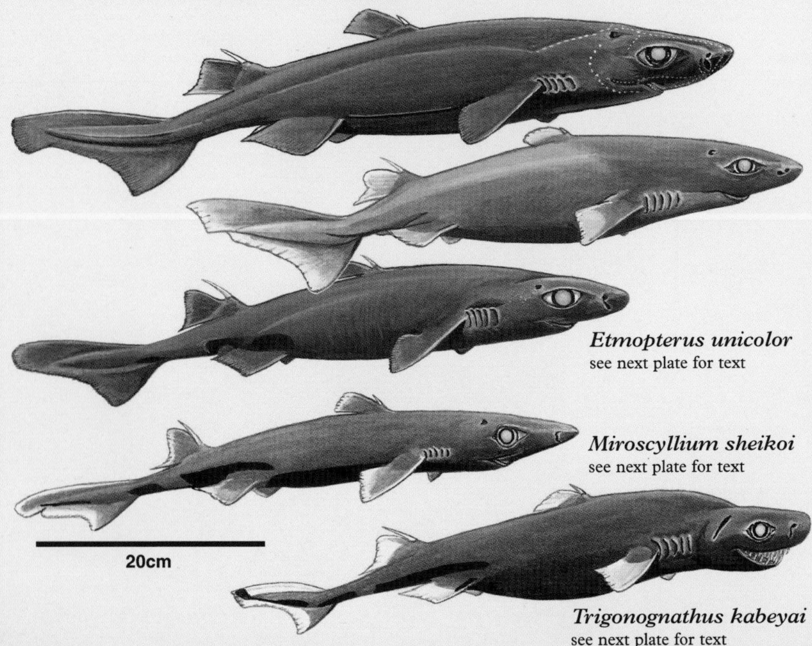

Etmopterus unicolor
see next plate for text

Miroscyllium sheikoi
see next plate for text

Trigonognathus kabeyai
see next plate for text

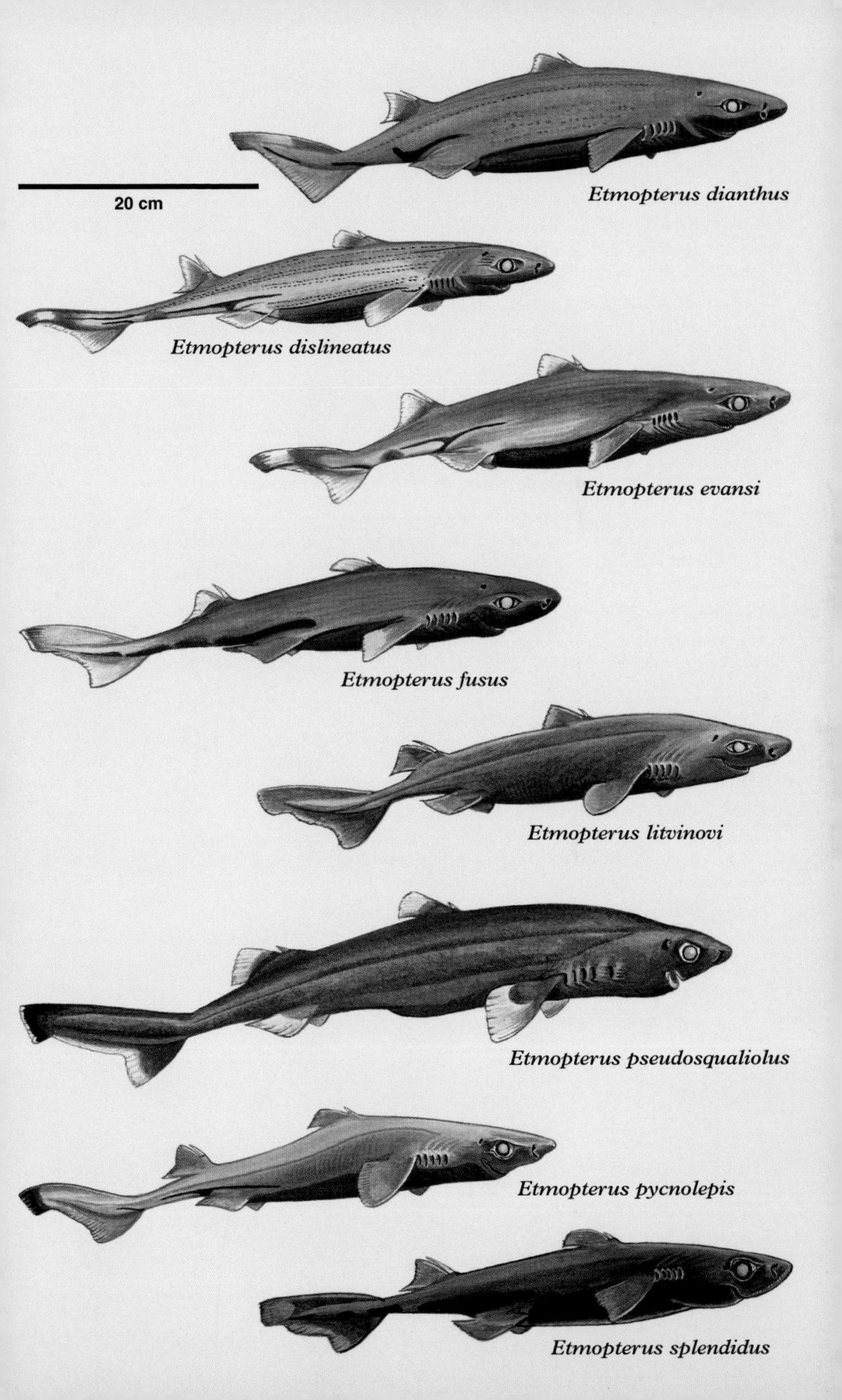
20 cm
Etmopterus dianthus
Etmopterus dislineatus
Etmopterus evansi
Etmopterus fusus
Etmopterus litvinovi
Etmopterus pseudosqualiolus
Etmopterus pycnolepis
Etmopterus splendidus

Plate 9 Etmopteridae IV contd.

Brown Lanternshark *Etmopterus unicolor* **p108**

?SE Atl. to W IO; 402–1380 m. Robust; fairly large gills; mod. long tail; long low D1 fin with v. short spine, D2 fin ~ twice height D1 with strong spine; denticles not in rows but cover snout; dk. brown to brownish-black with darker underside. **See previous plate.**

Rasptooth Dogfish *Miroscyllium sheikoi* **p110**

Japan; 340–370 m. Unmistakable long flat snout with short mouth; grooved D fin spines, D2 fin spine > D1; black underside, black tail markings. **See previous plate.**

Viper Dogfish *Trigonognathus kabeyai* **p111**

N and central Pac.; on bottom 330–360 m. V. long snake-like mouth with huge curved fang-like teeth, deep pockets on head in front of v. large elongated spiracles; grooved spines on both D fins; black underside, black tail markings. **See previous plate.**

Plate 10 Somniosidae I

Small to gigantic (40 cm to 6 m or more); fairly broad head and flat snout, spiracles large and close behind eyes; lateral ridges on abdomen; P fins low with short rounded free rear tips, 2 small D fins with spines which may be covered with skin, no A fin, C fin heterocercal with strong posterior notch.

Largespine Velvet Dogfish *Proscymnodon macracanthus* **p113**

SW Atl. and W Pac.; depth not recorded. Stocky, tapering strongly behind P fins; fairly long snout, thick fleshy lips and short upper labial furrows; P fins large, prominent stout D fin spines, D2 fin > D1; uniform dk. brown to blackish.

Plunket Shark *Proscymnodon plunketi* **p114**

IP; nr. bottom 219–1427 m. Stocky, tapering strongly behind P fins; v. short snout; D fin spines just visible, D fins = sized, D1 fin extends forward in prominent ridge, length D2 fin to upper C fin origin = D bases, D2 free rear tip well in front of upper C fin; uniform dk. brown to blackish.

Portuguese Dogfish *Centroscymnus coelolepis* **p112**

Atl. and IP; on/nr. bottom 128–3675 m. Stocky; short snout, short labial furrows; D fin spines usually just visible, small D fins = sized, D2 fin close to asymmetrical C fin; uniform golden-brown to blackish.

Roughskin Dogfish *Centroscymnus owstoni* **p112**

Atl. and Pac.; on/nr. bottom 426–1459 m. Similar to *C. coelolepis* but longer snout, longer lower D1 fin, taller triangular D2 fin with D fin spines barely exposed.

Longnose Velvet Dogfish *Centroselachus crepidater* **p113**

E Atl. and IP but not NE Pac.; on/nr. bottom 270–2080 m. Slender; v. long snout, small mouth with encircling v. long upper labial furrows; D fin spines v. small, D fins ≈ sized, D1 fin extends forward in prominent ridge origin over P bases, D2 free rear tip nearly reaches upper C origin.

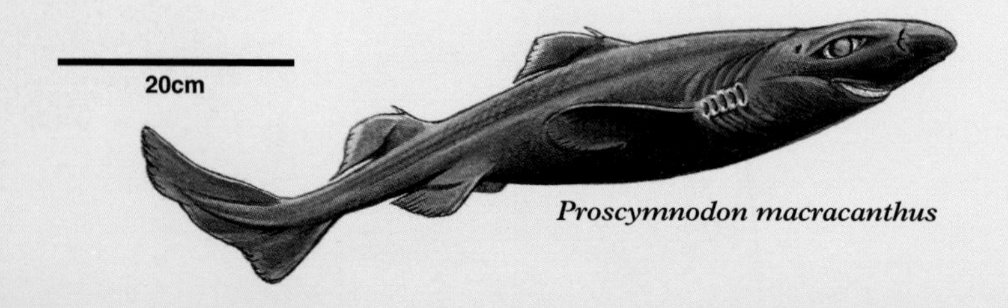

Proscymnodon macracanthus

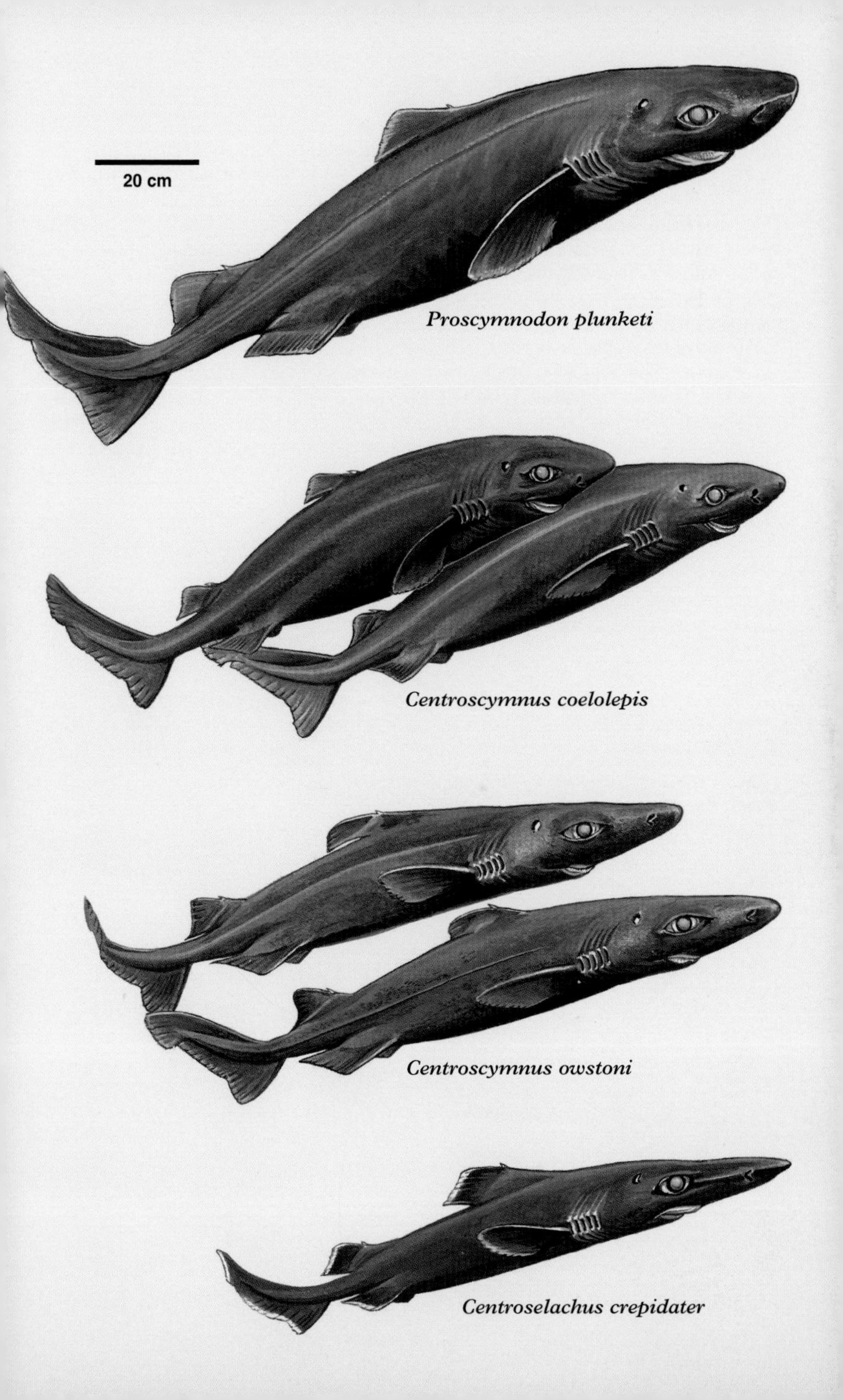
20 cm
Proscymnodon plunketi
Centroscymnus coelolepis
Centroscymnus owstoni
Centroselachus crepidater

Plate 11 Somniosidae II

Japanese Velvet Dogfish *Zameus ichiharai* **p121**

Japan; on/nr. bottom 450–830 m. Low flat head, mod. long snout, short narrow mouth, short gills ≤ eye length; caudal peduncle long; P fins narrow leaf-shaped, D fin spines small, V fins small ≈ D2 fin, C fin with strong ST notch and short vent. lobe; uniform black.

Velvet Dogfish *Zameus squamulosus* **p121**

Patchy WW not E Pac.; on/nr. bottom 550–1450 m. Slender; low flat head, fairly long narrow snout, short narrow mouth, nasoral grooves much > upper labial furrows; long tail; D fin spines small, D2 fin > D1 fin and ≈ V fins, C fin with strong ST notch and short vent. lobe; uniform black.

Knifetooth Dogfish *Scymnodon ringens* **p117**

E Atl. and W Pac.; on/nr. bottom 200–1600 m. Thick high head, short broad snout, v. large broad arched mouth, long gills ≥ eye length; D fin spines small, D2 fin slightly > D1 fin, C fin asymmetric with weak ST notch and no vent. lobe; uniform black.

Whitetail Dogfish *Scymnodalatias albicauda* **p114**

SO; 0–200+ m. Short broad round snout, long broad arched mouth, horiz. elongate eyes; P fins elongate, no D fin spines, D2 fin slightly > D1 fin and v. close to C fin; dk. brown or mottled grey, whitish-grey fin margins, C fin white blotches on dk. term. lobe.

Azores Dogfish *Scymnodalatias garricki* **p115**

N Atl. Ridge; caught 300 m. Small; long broad rounded snout, long broad arched mouth, horiz. elongate eyes; no D fin spines, D1 fin mid back; uniform dk. brown.

Sparsetooth Dogfish *Scymnodalatias oligodon* **p116**

SE Pac.; caught 0–200 m. Small; long broad pointed snout, long broad arched mouth, horiz. elongate eyes; no D fin spines, D1 fin mid back, C fin vent. lobe weak; uniform dk. brown.

Sherwood Dogfish *Scymnodalatias sherwoodi* **p116**

SW Pac.; 400–500 m. Mod. long flat pointed snout, long broad arched mouth, horiz. elongate eyes; P fins leaf-shaped, no D fin spines, D1 fin mid back, D2 fin origin above rear 1/3 V fin bases, C fin vent. lobe short strong; dk. brown above, lighter below, with lt. margins on gills and P fins.

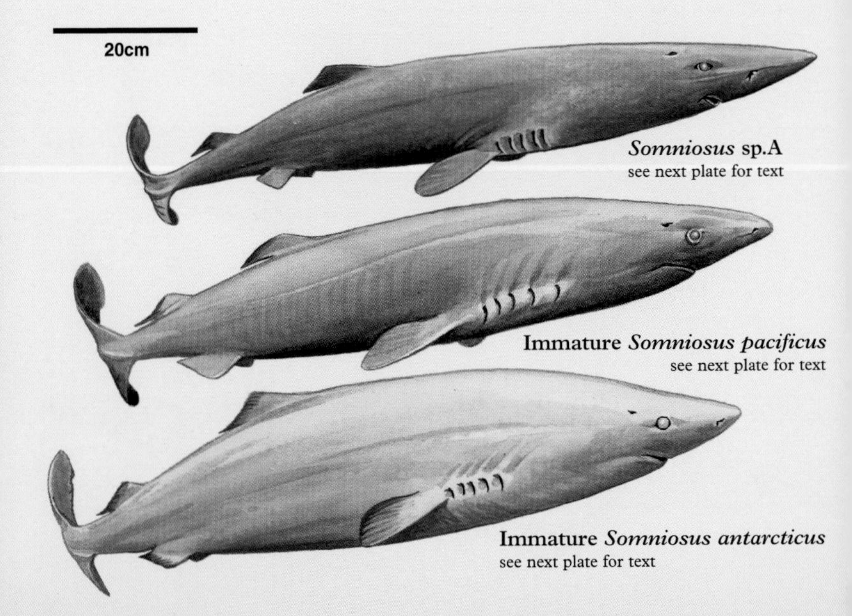

Somniosus sp.A
see next plate for text

Immature *Somniosus pacificus*
see next plate for text

Immature *Somniosus antarcticus*
see next plate for text

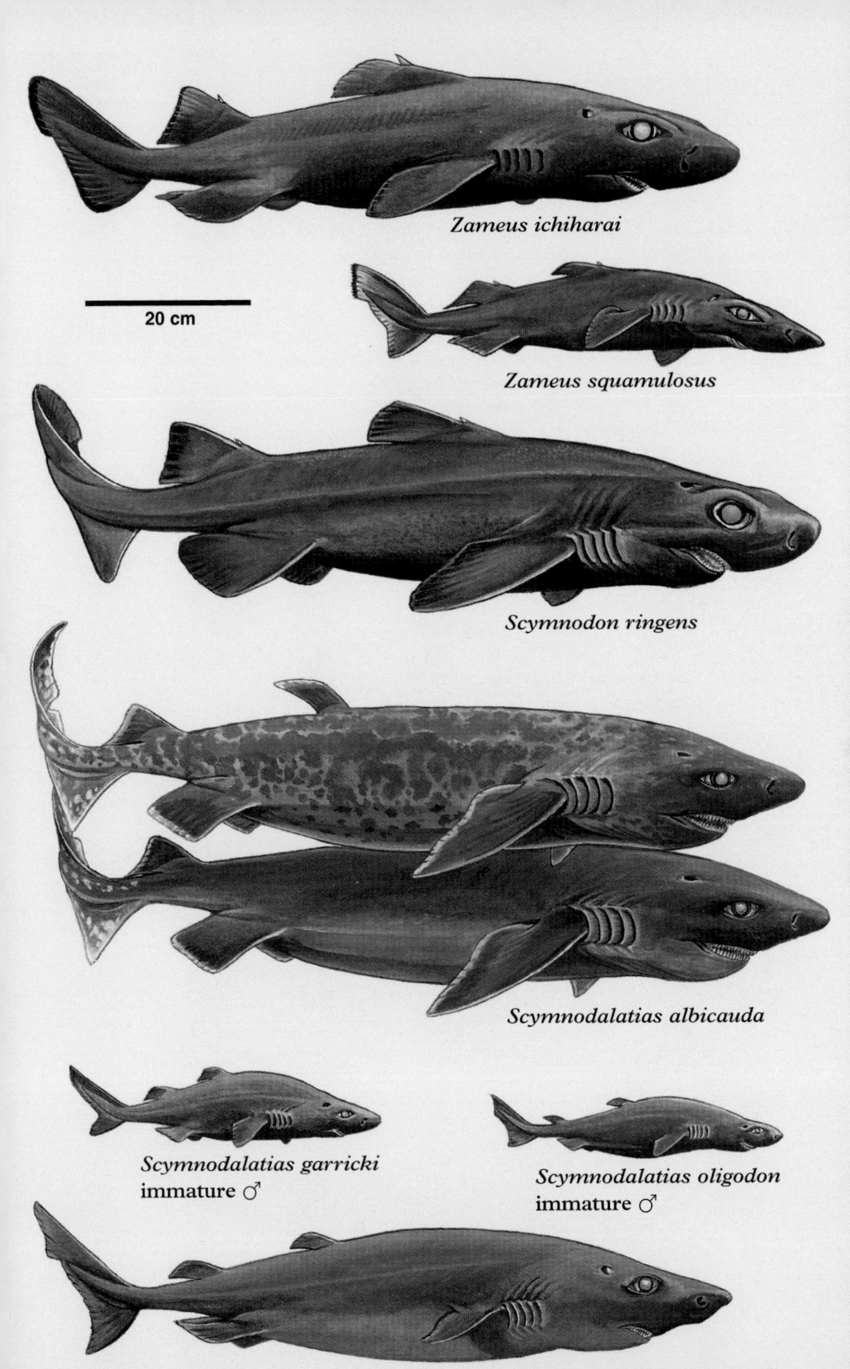
Zameus ichiharai
20 cm
Zameus squamulosus
Scymnodon ringens
Scymnodalatias albicauda
Scymnodalatias garricki
immature ♂
Scymnodalatias oligodon
immature ♂
Scymnodalatias sherwoodi

Plate 12 Somniosidae III

Frog Shark *Somniosus longus* **p118**

W and SE Pac.; on/nr. bottom 250–1160 m. Slender cylindrical body; short head with short rounded snout; short caudal peduncle, keels on base of C fin; no D fin spines, D1 fin ~ long as D2 fin and closer to P fins than V fins, C fin long vent. lobe, short dors.; smooth skin; uniformly blackish.

Little Sleeper Shark *Somniosus rostratus* **p120**

NE and NW Atl. and W Med.; on/nr. bottom 200–2200 m. Similar to *S. longus* but smaller eyes, shorter D2 fin and shorter gills, ~ eye length (*S. longus* > twice eye length).

Pacific Sleeper Shark *Somniosus pacificus* **p119**

N Pac.; 2000+ m. Gigantic, heavy (when mature) cylindrical body; short rounded snout; short caudal peduncle, keels absent or present on base of C fin; small precaudal fins, no D fin spines, low D fins = sized, D1 fin closer to V fins than P fins, distance between D fin bases ~ 70% snout to gill 1, C fin long vent. lobe, short dors.; skin rough and bristly; uniform greyish. **See previous plate for immature.**

Greenland Shark *Somniosus microcephalus* **p118**

N Atl. and AO; 1200+ m, inshore in winter. Gigantic, heavy body; short rounded snout; short caudal peduncle, keels on base of C fin; small precaudal fins, no D fin spines, low D fins = sized, D1 fin closer to P fins than V fins, distance between D fin bases ≈ snout to gill 1, C fin long vent. lobe, short dors.; skin rough; med. grey or brown, occly. transverse bands, small spots/blotches or light spots.

Southern Sleeper Shark *Somniosus antarcticus* **p117**

SO; 1200+ m. Gigantic, heavy cylindrical body; short rounded snout; short caudal peduncle, keels on base of C fin; v. small precaudal fins, no D fin spines, v. low D fins = sized, D1 fin closer to V fins than P fins, distance between D fin bases ~80% snout to gill 1, C fin long vent. lobe, short dors.; skin rough; uniform grey to blackish. **See previous plate for immature.**

Longnose Sleeper Shark *Somniosus sp.A* **p120**

Portugal; deepwater (only one specimen known). Mod. long head with elongated flat pointed snout; v. long caudal peduncle; no D fin spines, D fins = sized, D1 fin closer to P fins than V fins, C fin short but with long vent. lobe; uniform grey. **See previous plate.**

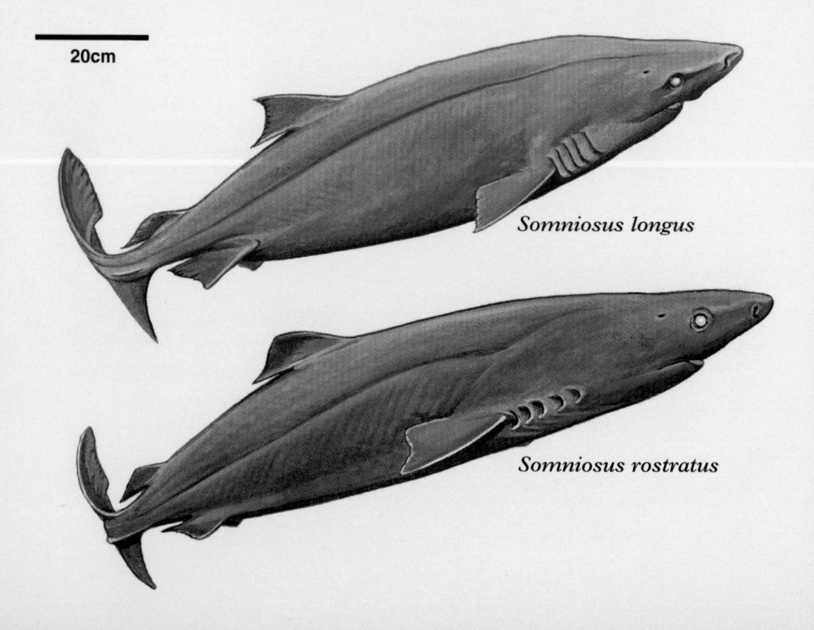

100 cm
Somniosus pacificus
Somniosus microcephalus
Somniosus antarcticus

Plate 13 Oxynotidae

Unmistakable small sharks. Compressed, triangular in cross-section, lateral ridges on abd.; flat blunt snout, small thick-lipped mouth, close-set large nostrils; 2 high spined D fins; rough skin.

Prickly Dogfish *Oxynotus bruniensis* **p122**

Temp. SW Pac.; 46–1067 m. Small circular spiracles; D fins with triangular tips, trailing edges straight/slightly concave, D1 fin spine leans forward; uniformly lt. grey-brown.

Caribbean Roughshark *Oxynotus caribbaeus* **p122**

NW Atl.; bottom 402–457 m. Small circular spiracles; D fins with narrow triangular tips, trailing edges concave, D1 fin spine leans forward; grey or brownish with dk. bands, blotches and small spots separated by prom. lt. areas over P and V fins.

Angular Roughshark *Oxynotus centrina* **p123**

E Atl. and Med.; bottom 50–660 m. Large expanded ridges over eyes covered with large denticles, v. large vert. elongate spiracles; D1 fin spine leans forward; grey or grey-brown with dk. blotches on head and sides (less prom. in adults) and lt. horiz. line over cheeks below eye.

Japanese Roughshark *Oxynotus japonicus* **p124**

Japan; 225–270 m. Large vert. oval spiracles; D fins with narrow triangular tips, trailing edges shallowly concave, D1 fin spine leans back; uniformly dk. brown except white lips, nasal flap margins, fin axils and inner clasper margins.

Sailfin Roughshark *Oxynotus paradoxus* **p124**

NE Atl.; 265–720 m. Small almost circular spiracles; D fins v. tall with narrow pointed tips, trailing edges v. concave, D1 fin spine leans back; uniformly dk. brown or blackish.

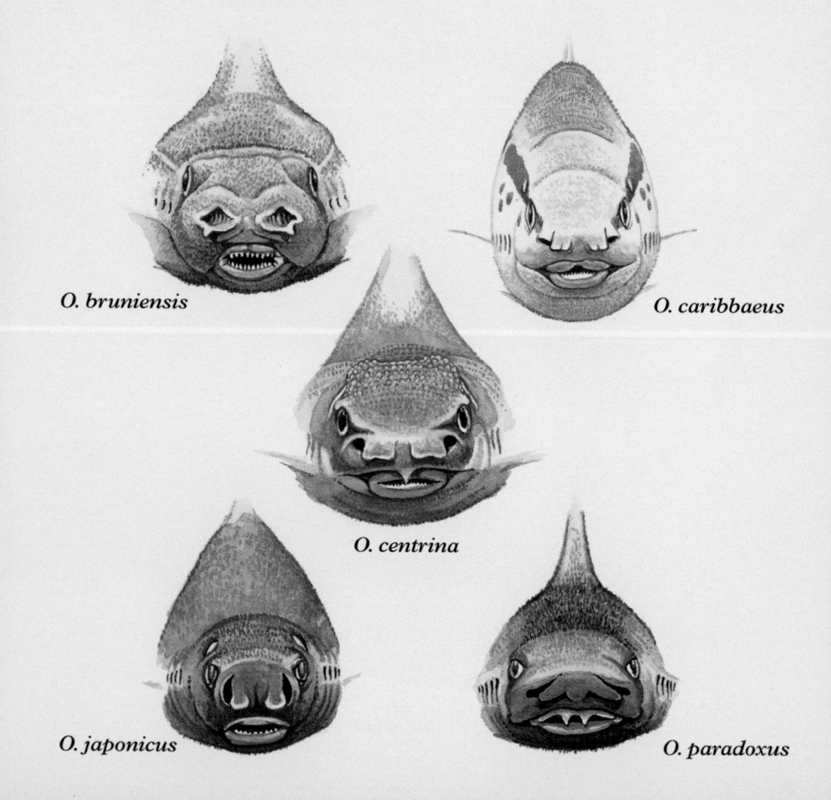

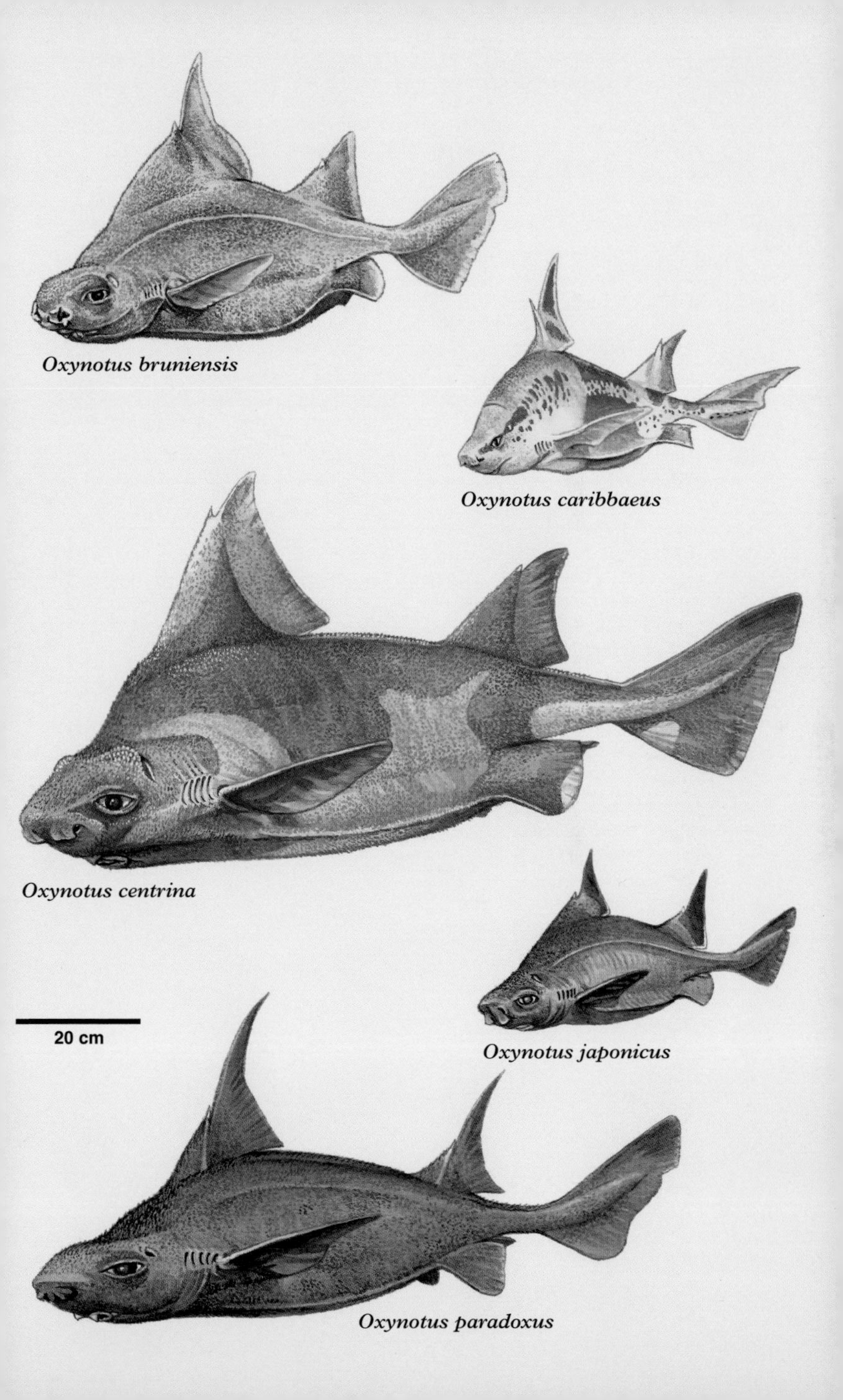
Oxynotus bruniensis
Oxynotus caribbaeus
Oxynotus centrina
Oxynotus japonicus
20 cm
Oxynotus paradoxus

Plate 14 Dalatiidae

Narrow conical head with short snout; P fins with short broad round free rear tips, 2 D fins without spines, no A fin, C fin with long dors. lobe and well-developed ST notch.

Kitefin Shark *Dalatias licha* p125

Atl., IO and PO; 37–1800 m. Med.-sized cylindrical; short blunt snout, thick fringed lips; D fins = sized, D1 fin base closer to P fins than V fins, weak C fin vent. lobe; uniform brown to blackish.

Taillight Shark *Euprotomicroides zantedeschia* p126

S Atl.; poss. surface–641 m. Tiny cylindrical; short blunt snout, thick fringed lips, v. long gill 5; luminous gland; D1 fin base closer to V than P fins, D2 fin origin in front of V fins, C fin paddle-shaped; all fins with prom. white margins.

Pygmy Shark *Euprotomicrus bispinatus* p126

S Atl., S IO and PO; 1500–9938 m. Tiny cylindrical; bulbous snout, large eyes, gills tiny; luminous gland; low lateral keels on C ped.; D1 fin tiny flag-like, D2 fin origin well behind V fin, C fin paddle-shaped nearly symm.; uniformly black.

Longnose Pygmy Shark *Heteroscymnoides marleyi* p127

SO; poss. surface–502 m. Dwarf cylindrical; v. long bulbous snout, thin unpleated lips, small gills; D1 fin origin over P fins, D2 fin slightly > D1, C fin paddle-shaped nearly symm. with strong v. lobe; uniformly dk. brown.

Cigar/Cookiecutter Shark *Isistius brasiliensis* p127

Atl., S IO and Pac.; surface–3500 m. Small cigar-shaped; v. short bulbous snout, suctorial lips; luminous organs lwr. surf. except collar and fins; D1 fin base over V fin origins, P fins > D fins, C fin paddle-shaped nearly symm.; med. grey to grey-brown, prom. dk. collar and lt. edges to fins.

South China Cookiecutter Shark *Isistius labialis* p128

W Pac.; 520 m. Similar to *I. brasiliensis* but eyes further forward, C fin less symm. with shorter v. lobe, dk. collar breaks up on vent.surf. and dk. markings on dors. surf.

Largetooth Cookicutter Shark *Isistius plutodus* p128

W and NE Atl. and W Pac.; 100–200 m. Similar to *I. brasiliensis* and *I. labialis*, but larger jaws and teeth, C fin smaller and no or faint collar markings.

Pocket Shark *Mollisquama parini* p129

SE Pac.; 330 m. Dwarf spindle-shaped; long bulbous conical snout, lateral papillae on upper lips; gland opening above P fin; D2 fin base > twice D1, C fin paddle-shaped; prom. white fin margins.

Dalatias licha

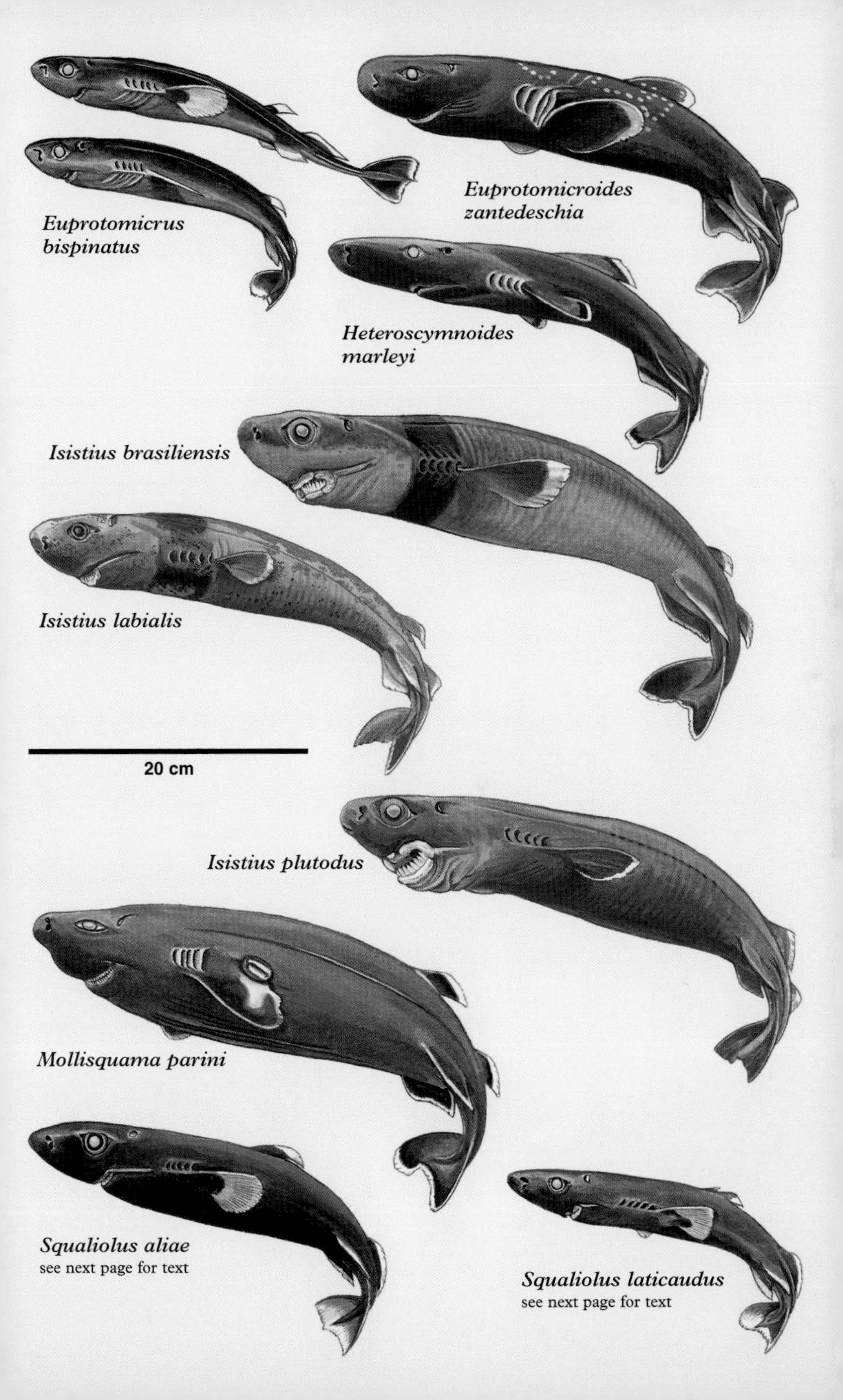

Euprotomicrus bispinatus
Euprotomicroides zantedeschia
Heteroscymnoides marleyi
Isistius brasiliensis
Isistius labialis
20 cm
Isistius plutodus
Mollisquama parini
Squaliolus aliae
see next page for text
Squaliolus laticaudus
see next page for text

Plate 14 Dalatiidae contd.

Smalleye Pygmy Shark *Squaliolus aliae* **p129**

SE IO and W. Pac.; 200–2000 m. Dwarf spindle-shaped; long bulbous conical snout, lateral papillae on upper lips, eyes smaller than *S. laticaudus*; photophores on lwr. surf.; only D1 fin with spine, D2 fin base > twice D1, C fin paddle-shaped; prom. lt. fin margins. **See previous plate.**

Spined Pygmy Shark *Squaliolus laticaudus* **p130**

Atl. and IP; epipelagic. V. similar to *S. aliae* but has larger eye with broad arched upper eyelid, lateral papillae absent or weak. **See previous plate.**

Plate 15 Pristiophoridae I

Slender distinctive sharks; flat heads, long flat saw-like snout with long vent. barbels and close-set rows of lateral and vent. sawteeth, large spiracles; thick lateral ridges on C ped; long C fin dors. lobe.

Sixgill Sawshark *Pliotrema warreni* **p131**

SE Atl. and SW IO; on/nr. bottom 37–500 m. Only sawshark with 6 pairs of gills; barbs on post. edges of larger rostral sawteeth, barbels closer to mouth than other species.

Longnose Sawshark *Pristiophorus cirratus* **p132**

S Austr.; on/nr. bottom to 311 m. Stocky sawshark; lat. teeth 9–11 in front of barbels, 9–10 behind; barbels much closer to rostral tip than mouth; pale yellow to greyish brown with sometimes faint blotches spots and bars, dk. brown rostral midline stripes and edges, tooth margins blackish.

Japanese Sawshark *Pristiophorus japonicus* **p132**

NW Pac.; on/nr. bottom to poss. 165 m. Stocky sawshark; lat. teeth 15–26 in front of barbels, 8–17+ behind, vent. teeth 9–14 in front of barbels, 8–9 behind; barbels closer to mouth than rostral tip; brown or reddish brown, dk. brown rostral midline stripes and edges.

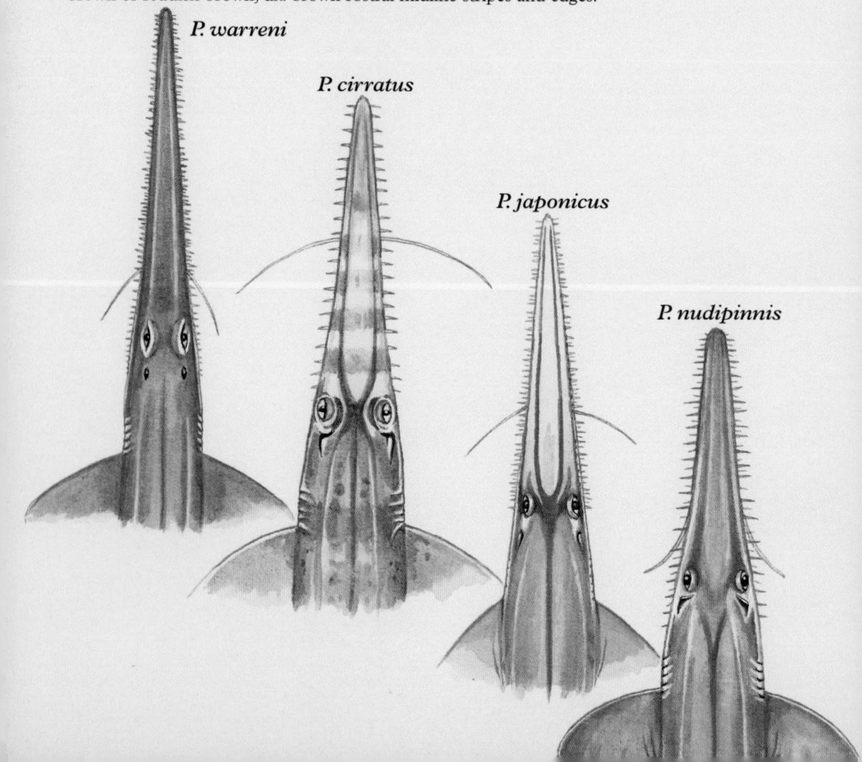

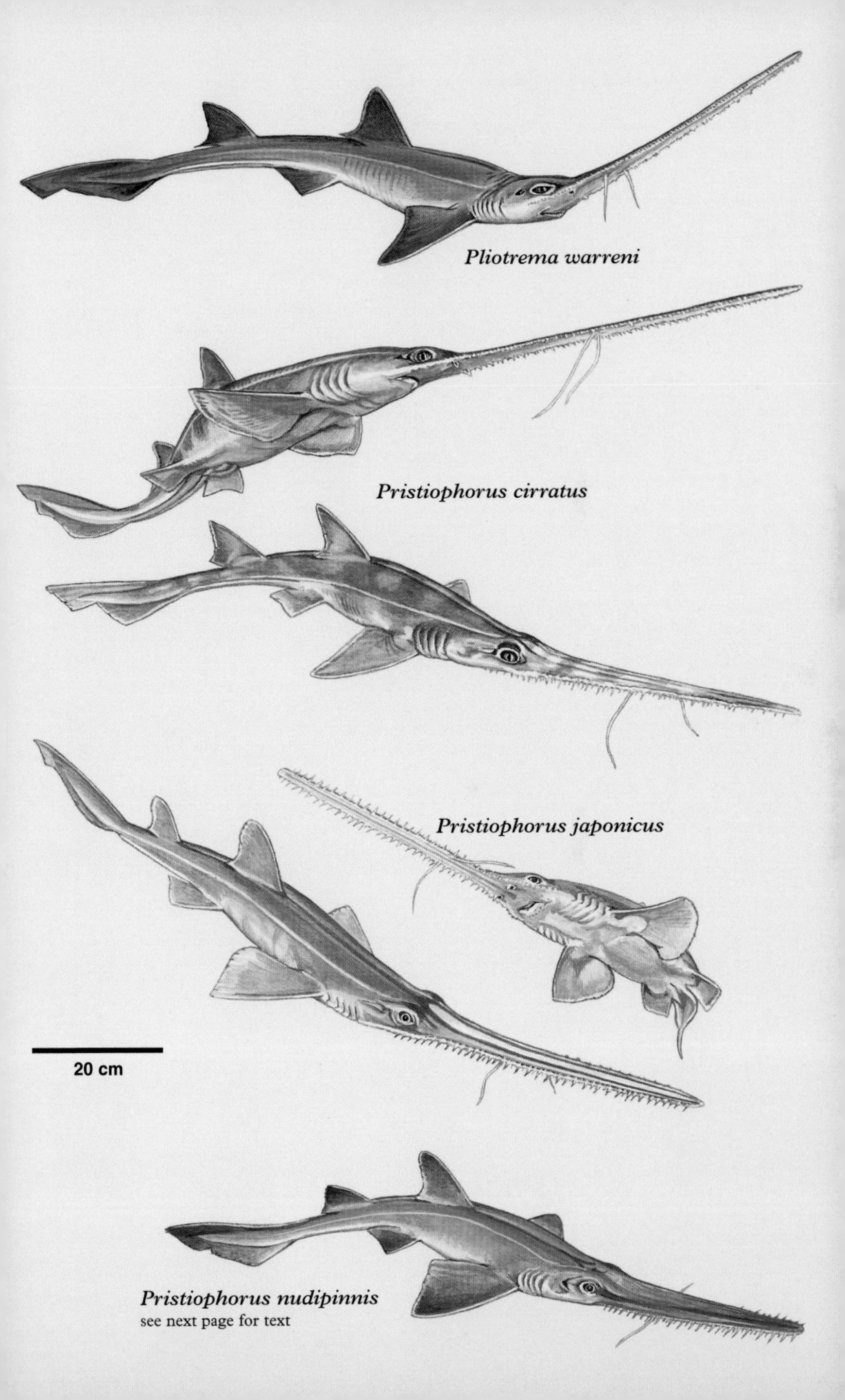

see next page for text

Plate 15 Pristiophoridae I contd.

Shortnose Sawshark *Pristiophorus nudipinnis* **p133**

S Austr.; on/nr. bottom to 70+ m. Stocky sawshark; lat. teeth 12–14 in front of barbels, 6–8 behind, vent. teeth 13–14 in front of barbels, 4 behind; barbels closer to mouth than rostral tip; uniformly slate grey above, indistinct dusky rostral midline stripes and edges. **See previous plate.**

Plate 16 Pristiophoridae II

Bahamas Sawshark *Pristiophorus schroederi* **p133**

NW Atl.; on/nr. bottom 438–952 m. Slender sawshark; lat. teeth 13 in front of barbels, 10 behind, juvs. with 1 smaller tooth between larger teeth; barbels midway between mouth and rostral tip; uniformly lt. grey above, dk. brown rostral midline stripes and edges, juvs. with dk. ant. edges to D fins.

Eastern Sawshark *Pristiophorus* sp. A **p134**

SE Austr.; 100–630 m. Stocky sawshark; lat. teeth 11–15 in front of barbels, 9–10 behind, juvs. with 1 smaller tooth between larger teeth; barbels closer to rostral tip than mouth; uniformly grey-brown above, dk. brown rostral midline and edges, juvs. with dk. ant. edges to P and D fins.

Tropical Sawshark *Pristiophorus* sp. B **p134**

E Austr.; 300–400 m. Slender sawshark; juvs. usually 2–3 smaller teeth between larger teeth; barbels slightly closer to mouth than rostral tip or =; uniformly pale yellow-brown above, no spots/bars.

Philippine Sawshark *Pristiophorus* sp. C **p135**

Philippinne Is.; on bottom 229–593 m. Slender sawshark; lat teeth 13–14 in front of barbels, 7–8 behind, juvs. 2–3 smaller teeth between larger teeth; barbels slightly closer to mouth than rostral tip; uniformly dk. brown above, no spot/bars.

Dwarf Sawshark *Pristiophorus* sp. D **p135**

W IO; 286–500 m. Small sawshark; 2 rows of 4–5 pits beneath rostrum, prom. ridges on large teeth bases; barbels much closer to mouth than rostral tip; broad triangular D1 fin; brown above with pale rostrum, dk. brown rostral midline and edges, dk. ant. edges to P and D fins (consp in juvs.).

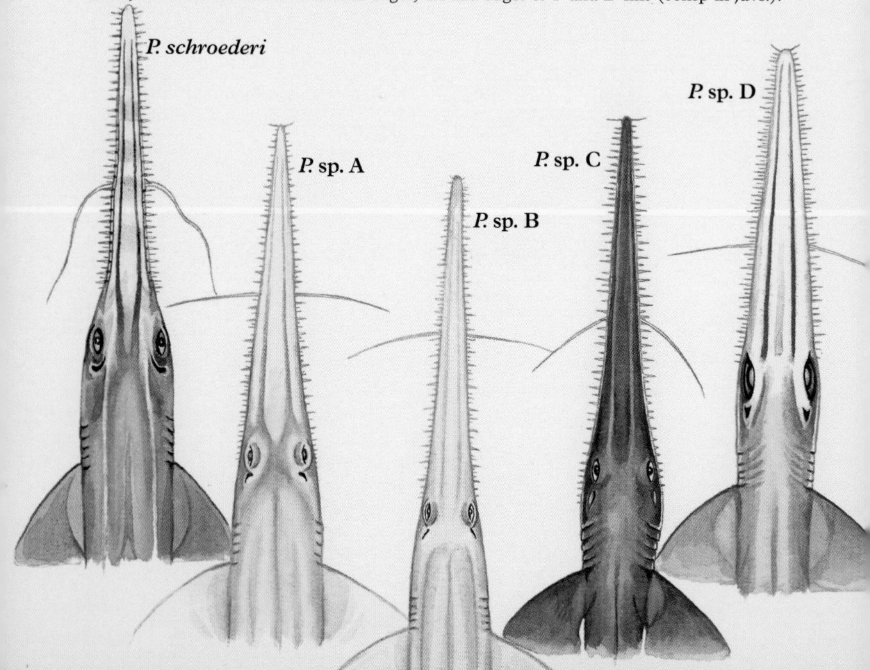

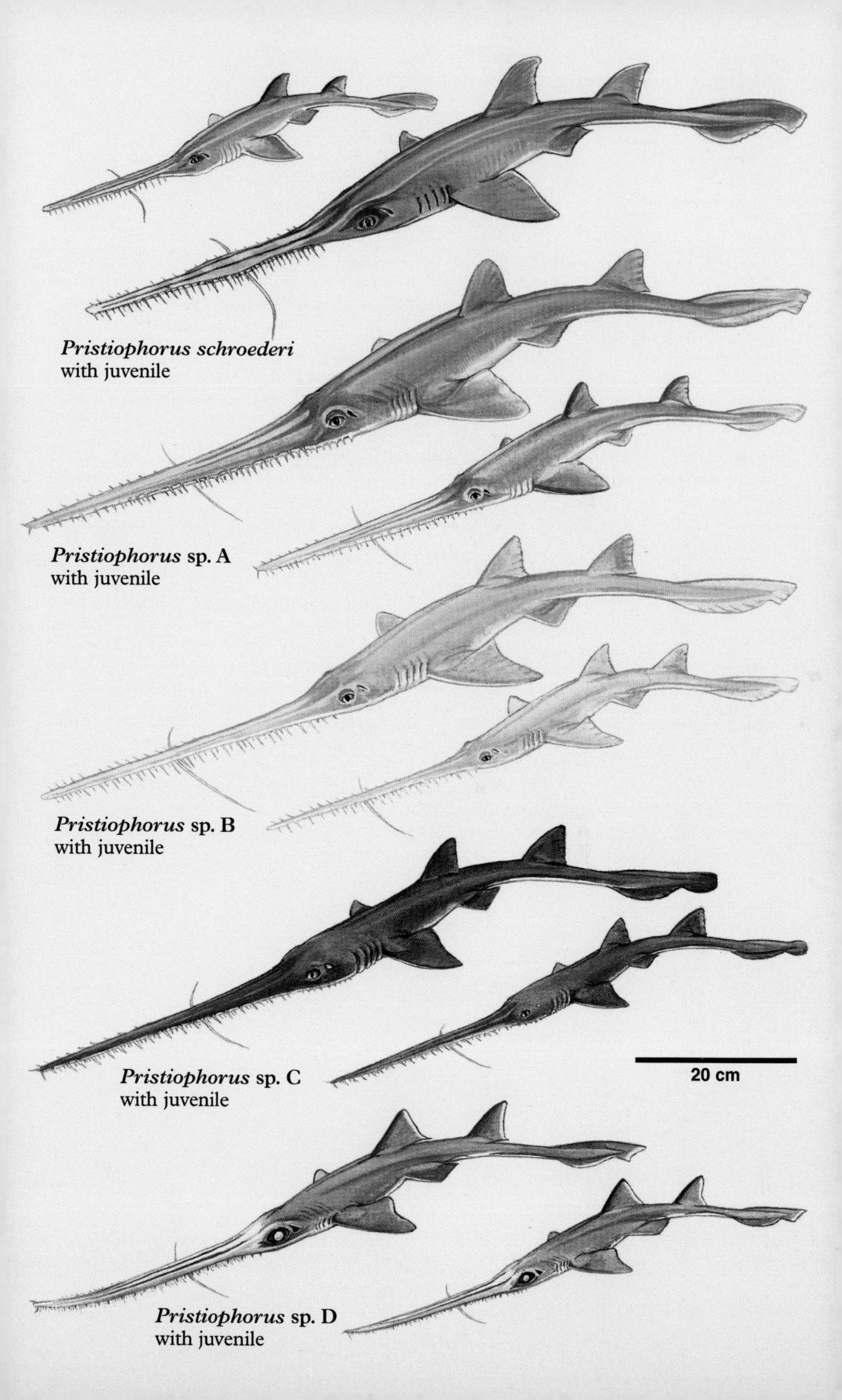

Pristiophorus schroederi
with juvenile
Pristiophorus sp. A
with juvenile
Pristiophorus sp. B
with juvenile
Pristiophorus sp. C
with juvenile
20 cm
Pristiophorus sp. D
with juvenile

Plate 17 Squatinidae I

Distinct broad flattened body; short snout with large mouth and nostrils, eyes on top of head close to and level with large spiracles, gills on side of head; short thick keel at base of C ped.; v. large P fins, 2 spineless D fins set back on C ped., D1 fin over or behind v. large V fins free rear tip, no A fin, C fin long dors. lobe and > vent. lobe; dors. surf. usually patterned, pale below. Some species are hard to distinguish.

Sawback Angelshark *Squatina aculeata* p137

E Atl.; mud 30–500 m. Concave between eyes; heavily fringed nasal barbels and ant. nasal flaps; large thorns on head and 1 row on along back; dull grey or lt. brown sparsely scattered with small irr. white spots and reg. small dk. brownish spots, large dk. blotches on dors. surf. and tail, no ocelli.

African Angelshark *Squatina africana* p138

E and S Afr.; sand and mud surf–494 m. Concave between eyes; simple flat nasal barbels with tapering or spatulate tips, ant. nasal flaps smooth or weakly fringed; large thorns on head not back; grey- or red-brown, many lt. and dk. spots with larger symm. dk. bands or saddles, blotches on P fins, dk. tail base with white margins, juvs. often wirh large ocelli.

Argentine Angelshark *Squatina argentina* p138

SW Atl.; 51–320 m. Concave between eyes; simple spatulate nasal barbels, weakly fringed or smooth ant. nasal flaps; large thorns on snout not back; convex ant. margin to P fins forming distinct 'shoulder'; purplish brown with many scattered dk. brown spots mostly in circular groups around darker spot, paler D fins, no ocelli.

Sand Devil *Squatina dumeril* p141

NW Atl.; on/nr.bottom inshore–1290 m. Strongly concave between eyes; simple tapering nasal barbels, weakly fringed or smooth ant. nasal flaps; discrete thorns on snout and between eyes and spiracles in young, more numerous and in patches in adult; fairly broad, post. angular P fins; uniform bluish to ash grey, irr. dusky or blackish spots may be present, young with white spots, D and C fins darker with lt. bases, D fins with lt. tips, underside white with red spots and reddish fin margins.

Angelshark *Squatina squatina* p146

NE Atl.; mud or sand 5–>150 m. Large, stocky; lateral head folds with single triangular lobe each side, simple nasal barbels with straight or spatulate tips, smooth or weakly fringed ant. nasal flaps; small thorns on mid back in young, adult just v. rough, patches small thorns on snout and between eyes; v. high broad P fins; grey to red- or green-brown with scattered small white spots and blackish dots and spots, nuchal spot maybe present, young often white reticulations and large dk. blotches, no ocelli.

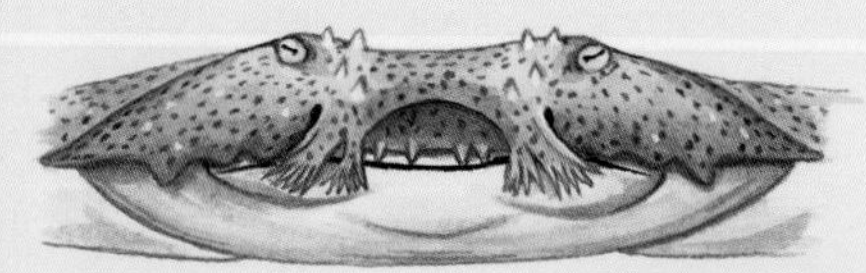

S. aculeata

S. dumeril

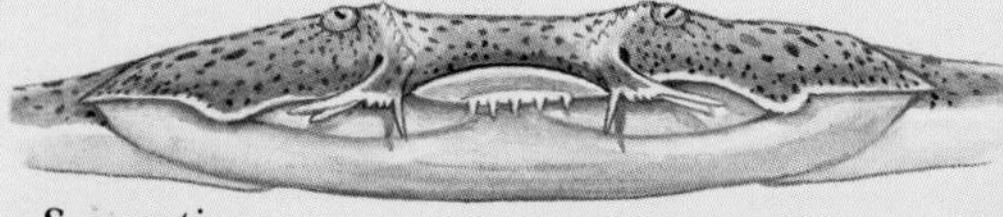

S. squatina

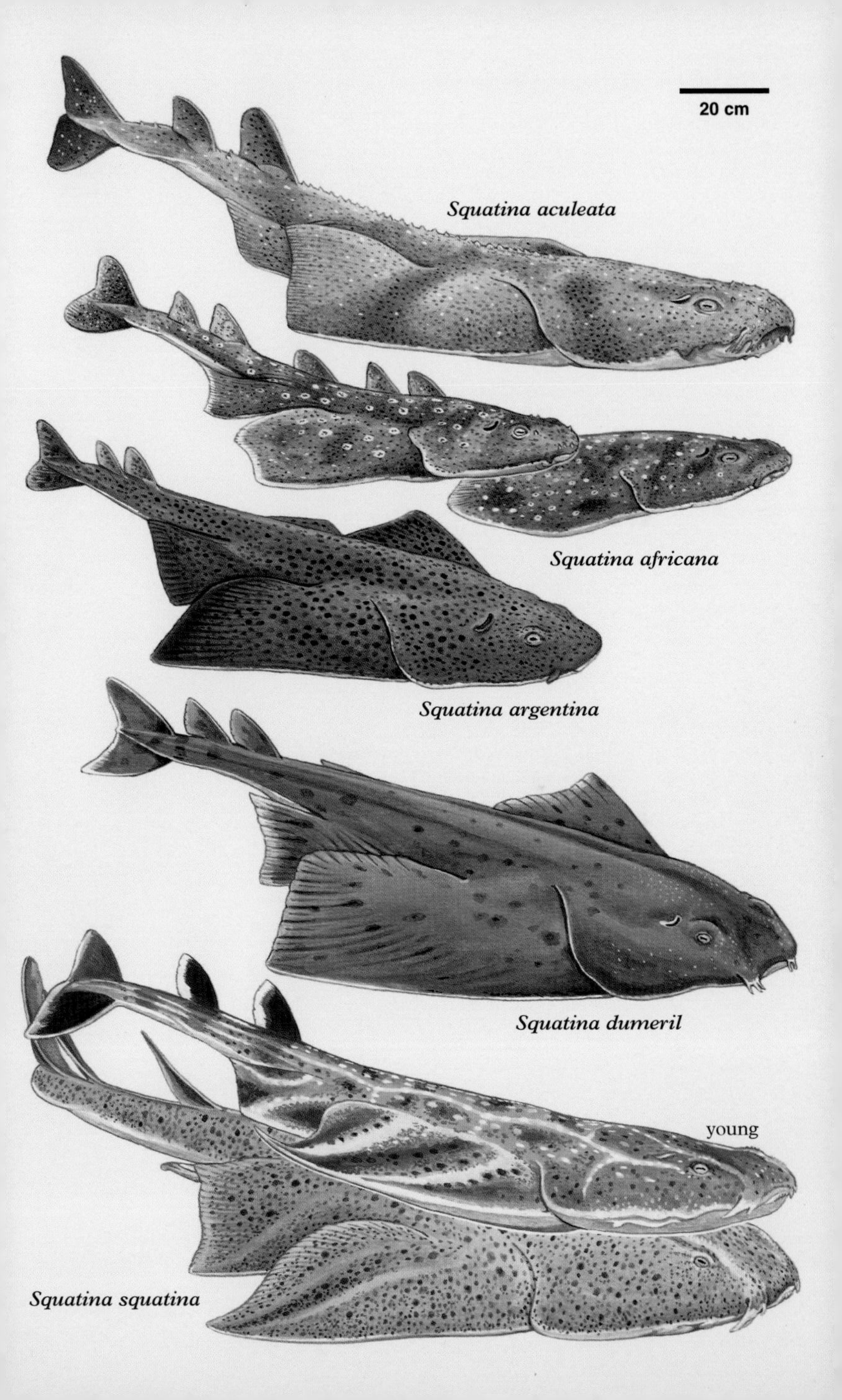
20 cm
Squatina aculeata
Squatina africana
Squatina argentina
Squatina dumeril
young
Squatina squatina

Plate 18 Squatinidae II

Chilean Angelshark *Squatina armata* **p139**

SE Pac.; depth not known. Narrow head; large thorns on snout, between eyes and large spiracles, double row on midline of back, between and behind D fins also row on ant. margin of P fins; grey to reddish-brown above, no described markings.

Australian Angelshark *Squatina australis* **p140**

S Austr.; sand, mud, sea grass surf–130 m. Flat or convex between eyes; heavily fringed nasal barbels and ant. nasal flaps, small spiracles; no large thorns in adults, young enlarged dent. on snout, head and pre-dors. rows on back; dull grey to brown with dense white spots and smaller dk. brown spots, no large ocelli.

Pacific Angelshark *Squatina californica* **p140**

NE Pac.; rocks and nr. kelp 1–200 m. Concave between large eyes; simple conical nasal barbels with spatulate tips, ant. nasal flaps weakly fringed; thorns in young, small or absent in adults; fairly broad long high P fins; red-brown to blackish, scattered lt. spots around dk. blotches in adults, white edged P and V fins, pale D fins with dk. blotches on base, dk. spot at base of pale C fin, ocelli in young.

Taiwan Angelshark *Squatina formosa* **p142**

NW Pac.; 183–385 m. Concave between large eyes; simple v. flat nasal barbels with round tips, ant. nasal flaps smooth or weakly fringed; patches of large dent. on snout and between eyes (rows on back in young); yellow-grey or brown, numerous small dk. brown spots and large irr. blotches, small lt. spots between head and D1 fin, saddle/band alongside D fins, small paired dk. ocelli.

Japanese Angelshark *Squatina japonica* **p143**

NW Pac.; depth poorly known. Concave between large eyes; cylindrical nasal barbels with slightly expanded tips, ant. nasal flaps smooth or weakly fringed; small thorns on snout, between eyes and spiracles, 1 row on midline of back, skin rough; fairly broad, high, rounded P fins; rusty or blackish-brown with dense dk. and small irr. white spots, large paired red-brown blotches from base of head to D fins, no ocelli.

Angular Angelshark *Squatina punctata* **p145**

SW Atl.; 10–80 m. Broadly concave between eyes; lateral head folds without triangular lobes, expanded slightly spatulate unfringed nasal barbels, ant. nasal flaps weakly fringed; short stout symm. thorns on snout and between eyes and a pair between spiracles also row on midline of back; P fins rel. small, high and angular with nearly straight ant. margin; dk. tan above with reg. pattern of several small to large blackish spots, small irr. dk. spots sometimes present, no ocelli.

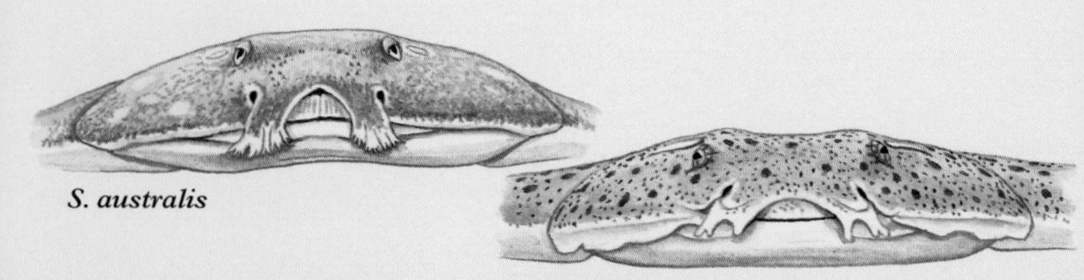

S. australis

S. californica

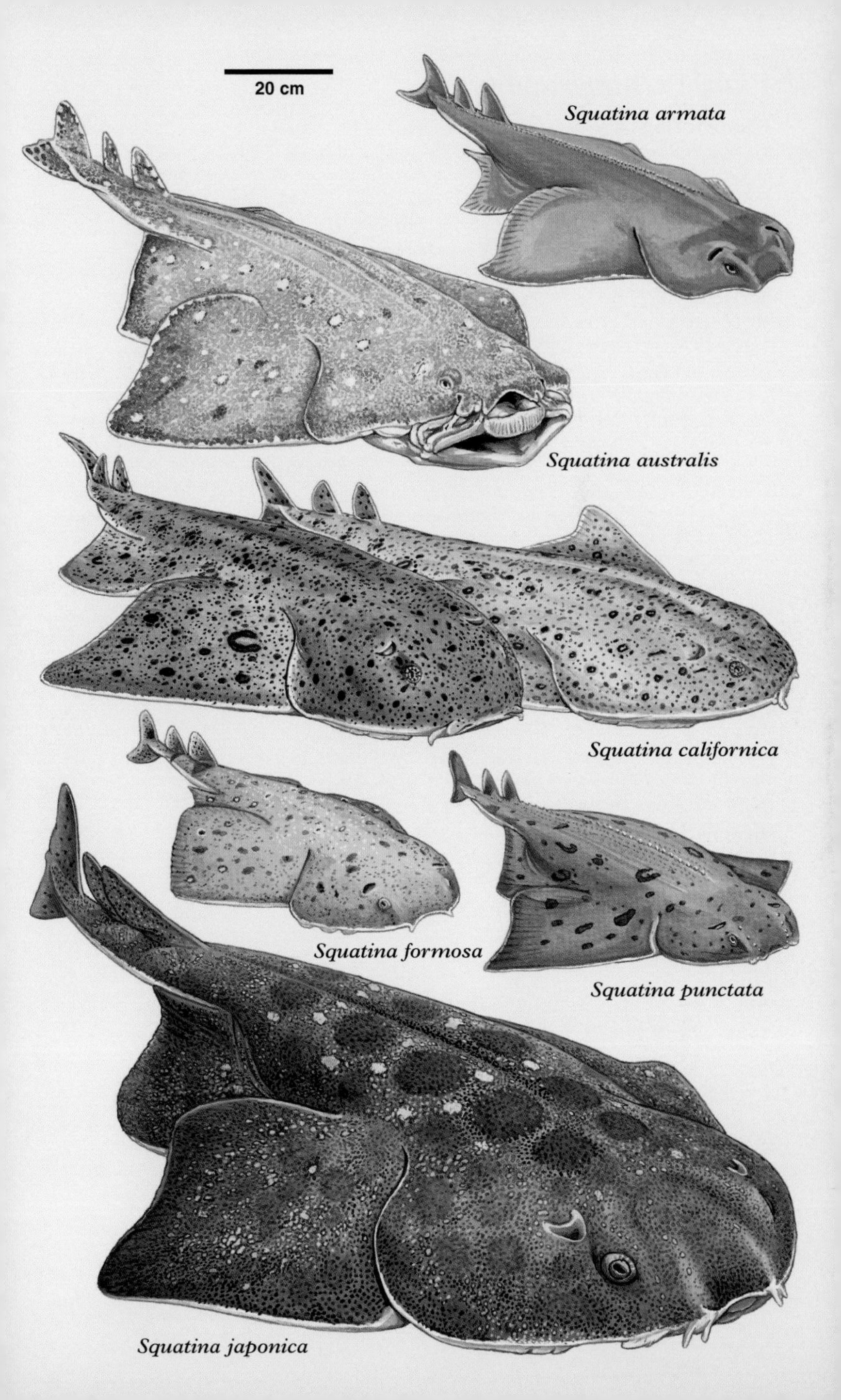
20 cm
Squatina armata
Squatina australis
Squatina californica
Squatina formosa
Squatina punctata
Squatina japonica

Plate 19 Squatinidae III

Hidden Angelshark *Squatina guggenheim* p142

SW Atl.; 35–115 m. Concave between eyes; cylindrical bases to nasal barbels with expanded unfringed tips, ant. nasal flaps v. weakly fringed; short stout symm. grouped thorns on snout and between eyes, pair between spiracles; P fins quite large, high and angular; dk. tan with numerous yellow spots and larger blackish marks, a few ocelli on P fins.

Clouded Angelshark *Squatina nebulosa* p144

NW Pac.; surf–330 m. Lateral head folds with 2 triangular lobes each side; simple tapering nasal barbels, ant. nasal flaps smooth or weakly fringed; no thorns; v. broad, low, obtuse, rounded P fins; brown to blue-brown with scattered lt. spots and numerous small back dots, dk. spot at base of P fin, dk. blotches below D fins, ocelli absent or small.

Smoothback Angelshark *Squatina oculata* p144

E Atl. and Med.; 20–500 m. Strongly concave between eyes; weakly bifurcate or lobed nasal barbels, ant. nasal flaps weakly fringed; large thorns on snout and above eyes; grey-brown with small white and blackish spots, large dk. blotches on base and rear tips of P fins, tail base and under D fins, D and C fin margins white, P and V fin margins dusky, white nuchal spot, sometimes symm. dk. ocelli.

Ornate Angelshark *Squatina tergocellata* p146

SW Austr.; on/nr. bottom 130–400 m. Strongly concave between large eyes; strongly fringed nasal barbels and ant. nasal flaps; no dorsal thorns; pale yellow-brown with grey-blue or white spots, 3 pairs of large ocelli with dk. rings surrounding thread-like patterned centres.

Ocellated Angelshark *Squatina tergocellatoides* p147

NW Pac.; depth unknown. Concave between eyes; strongly finely fringed nasal barbels, ant. nasal flaps strongly fringed; no large predors. thorns on back; light yellowish-brown with dense scattering of small round white spots, D fins with black base and ant. margins, 6 pairs of large ocelli of dk. rings around lt. centres on P and V fins and tail base.

Eastern Angelshark *Squatina* sp. A p148

E Austr.; 60–315 m. Low lateral head folds, v. short snout, concave between eyes; nasal barbels with expanded tips and lobate fringes; strong thorns above eyes, no thorns on back; yellow-brown to chocolate brown with dense pattern of symm. small white dk.-edged spots, many large brownish blotches and white nuchal spot, lt. unspotted D and C fins.

Western Angelshark *Squatina* sp. B p148

W Austr.; 130–310 m. V. short snout, concave between eyes; nasal barbels with expanded tips and lobate fringes; strong thorns above eyes, single row of thorns on back; med. to pale brownish or greyish with widely spaced blue spots and brown blotches, lt. unspotted D and C fins, single white nuchal spot, no ocelli.

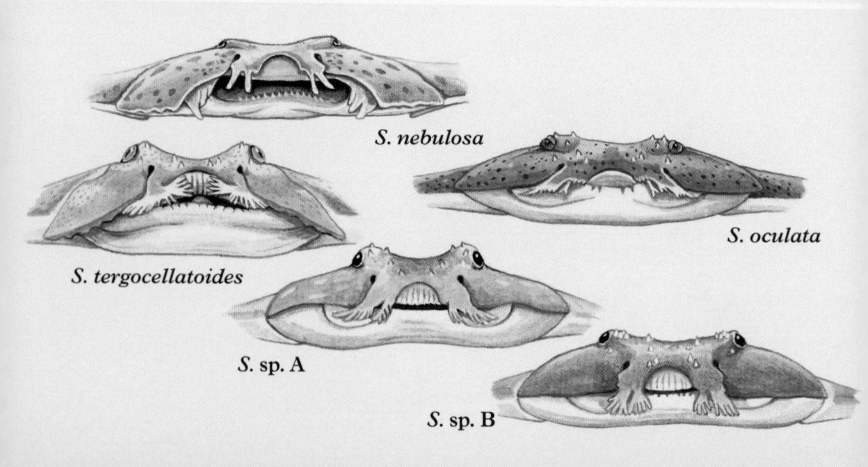

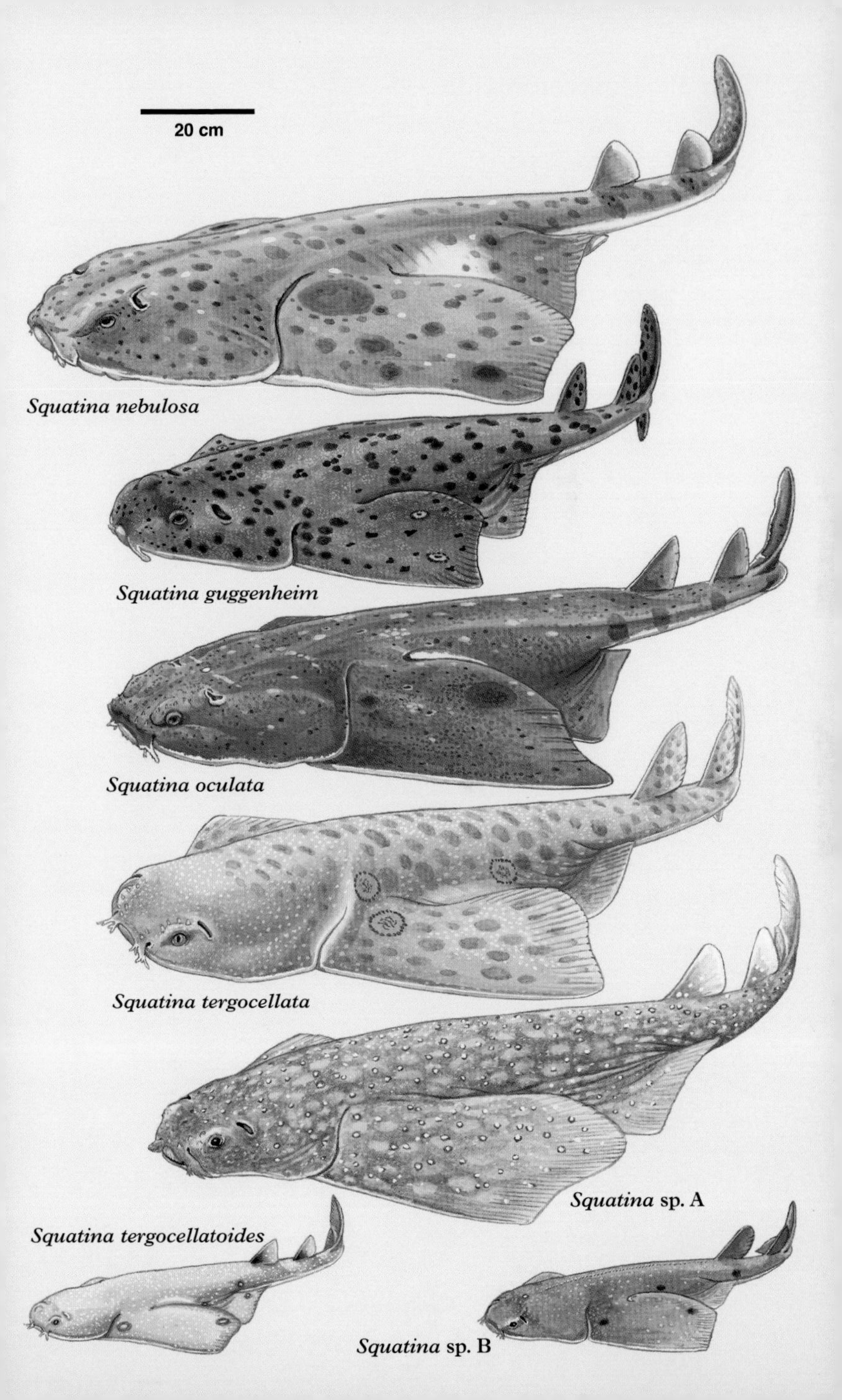
20 cm
Squatina nebulosa
Squatina guggenheim
Squatina oculata
Squatina tergocellata
Squatina sp. A
Squatina tergocellatoides
Squatina sp. B

Plate 20 Heterodontidae I

An ancient order with long fossil record, now only one living family of nine species. Stout bodied; blunt pig-like snout, small mouth, enlarged gill 1, prominent eye ridges; paddle-like paired fins, two spined D fins, no A fin; rough skin.

Horn Shark *Heterodontus francisci* p150

E Pac.; rock, kelp and sand, intertidal–150 m. D 1 fin origin over P fin bases; dk. to lt. brown usually with small (< 1/3 eye diam.) dk. spots, dusky patch under eyes, no lt. bar between eye ridges, no harness pattern, young more vivid with obvious dk. saddles.

Crested Bullhead Shark *Heterodontus galeatus* p151

E Austr.; rock, seaweed and sea grass, intertidal–93 m. Extremely high eye ridges; lt. brown or yellow-brown, no spots, broad dk. blotch under eyes, dk. bar between eye ridges, 5 dk. broad bands or saddles, young with more obvious dk. bands or saddles.

Japanese Bullhead Shark *Heterodontus japonicus* p151

NW Pac.; rock and kelp, intertidal–50 m. D 1 fin origin over P fin bases; tan to brown, no spots, dk. blotch under eyes less distinct in ad., ~12 irr. dk. saddles and bands, hatchlings more vivid.

Port Jackson Shark *Heterodontus portusjacksoni* p152

Austr. except N coast; temperate, intertidal–275 m. Grey to lt. brown or whitish, no spots, dk. band under eyes, dk. bar between eye ridges, unique dk. striped harness pattern.

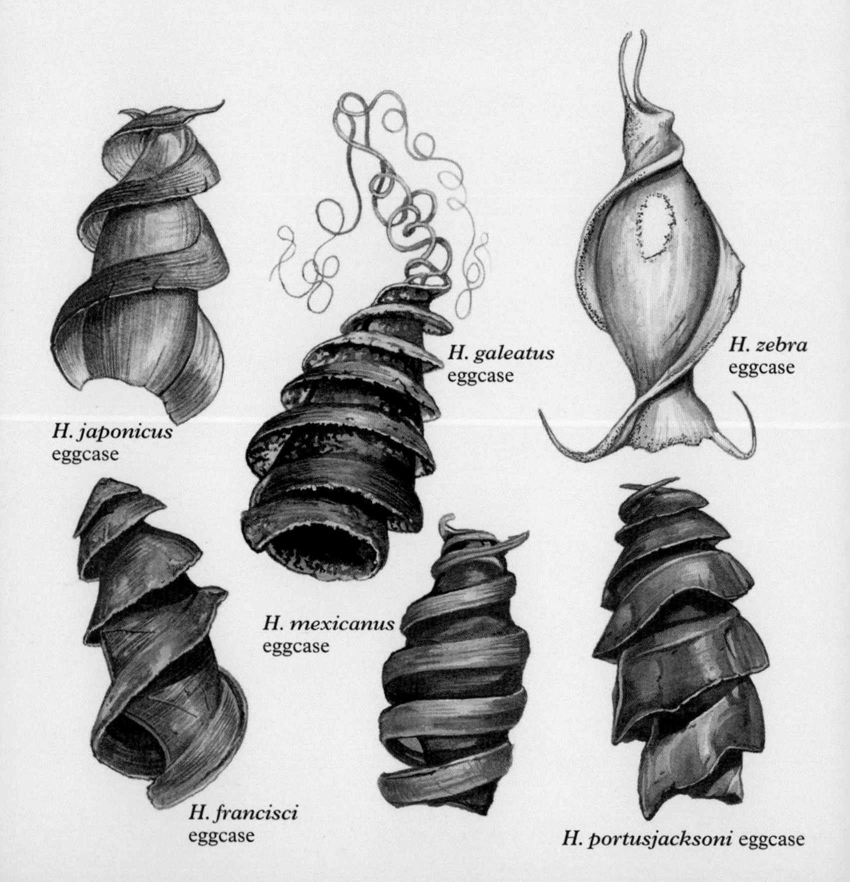

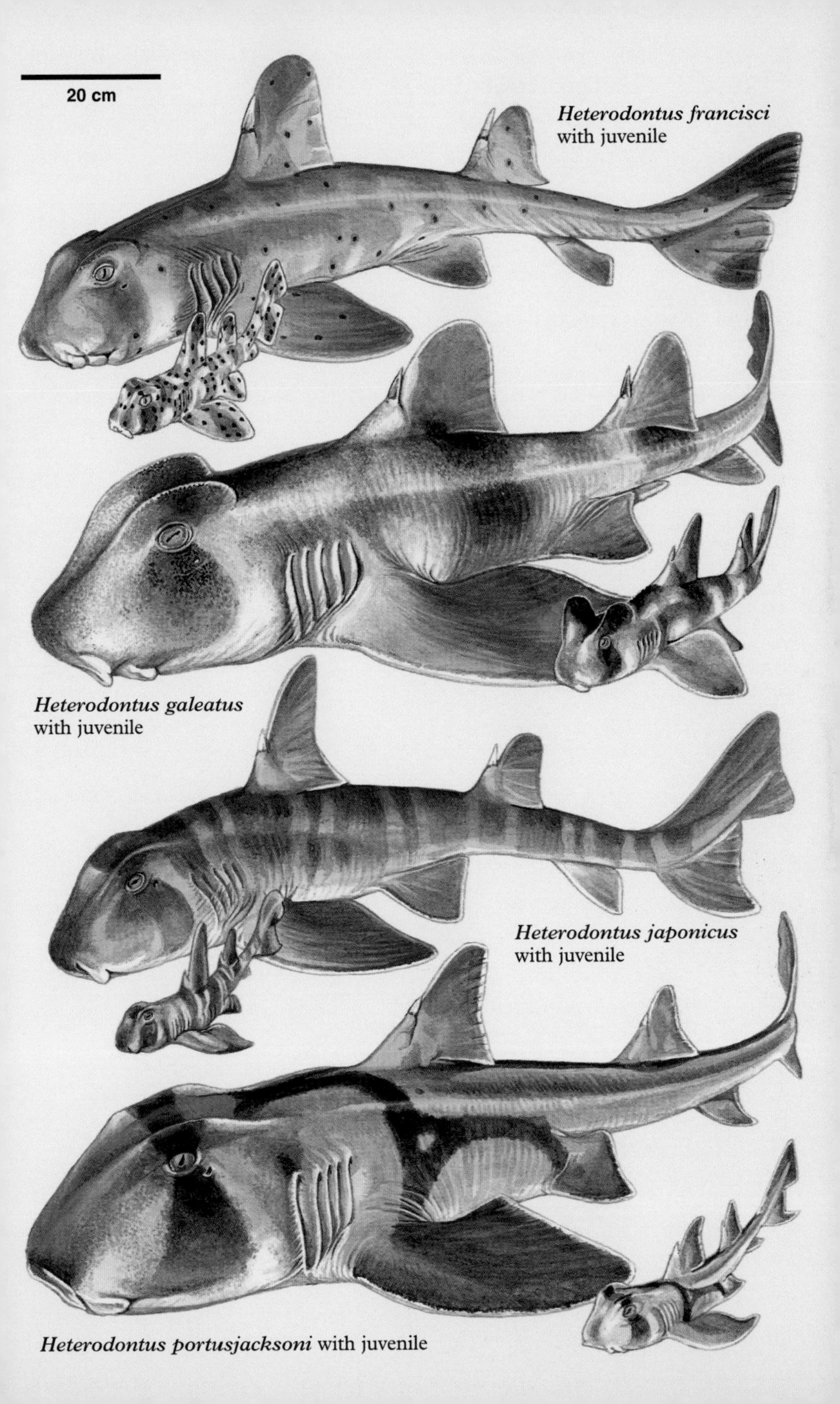
20 cm
Heterodontus francisci
with juvenile
Heterodontus galeatus
with juvenile
Heterodontus japonicus
with juvenile
Heterodontus portusjacksoni with juvenile

Plate 21 Heterodontidae II

Mexican Hornshark *Heterodontus mexicanus* **p152**

E Pac.; rock, coral reefs and sand, intertidal–50 m. D 1 fin origin over P fin bases; lt. grey-brown brown with large (≥ eye diam.) black spots, 1–2 indistinct blotches under eye, lt. bar between eye ridges, no harness pattern, young with more obvious dk. blotches.

Galapagos Bullhead Shark *Heterodontus quoyi* **p153**

E Pac.; rock and coral reefs, 16–30 m. D 1 fin origin over P fin inner margins; lt. grey or brown usually with large (< 1/2 eye diam.) black spots, mottled dk. spots or blotches under eye, no lt. bar between eye ridges, no harness pattern, young with less distinct markings and small spots.

Whitespotted Bullhead Shark *Heterodontus ramalheira* **p154**

N and w. IO; 40–275 m. Dk. red-brown usually with small white spots, no dusky patch under eye in adults, no lt. bar between eye ridges, no harness pattern, hatchlings with unique striking whorl pattern lost with age, parallel dk. lines between and under eyes change to dusky patch in large juv. and lost in adult.

Zebra Bullhead Shark *Heterodontus zebra* **p154**

W Pac.; rock, kelp and sand, intertidal–200 m. White or cream, no spots, no lt. bar between eye ridges, no harness pattern but striking black to dk. brown zebra-like narrow vertical saddles and bands, young similar but bands are red-brown.

Oman Bullhead Shark *Heterodontus* sp. **p155**

N IO; soft bottom? 80 m. D 1 fin origin over P fin inner margins; tan to brown, no spots, dk. blotch under eye, dk. bar between eye ridges, no harness pattern, 4–5 broad dk. brown saddles, dk. tipped fins with white spot on D fin tips, no information on hatchlings.

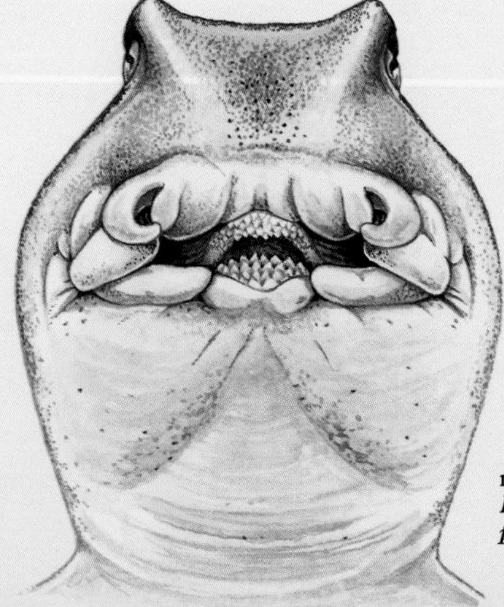

mouth of ***Heterodontus portusjacksoni***

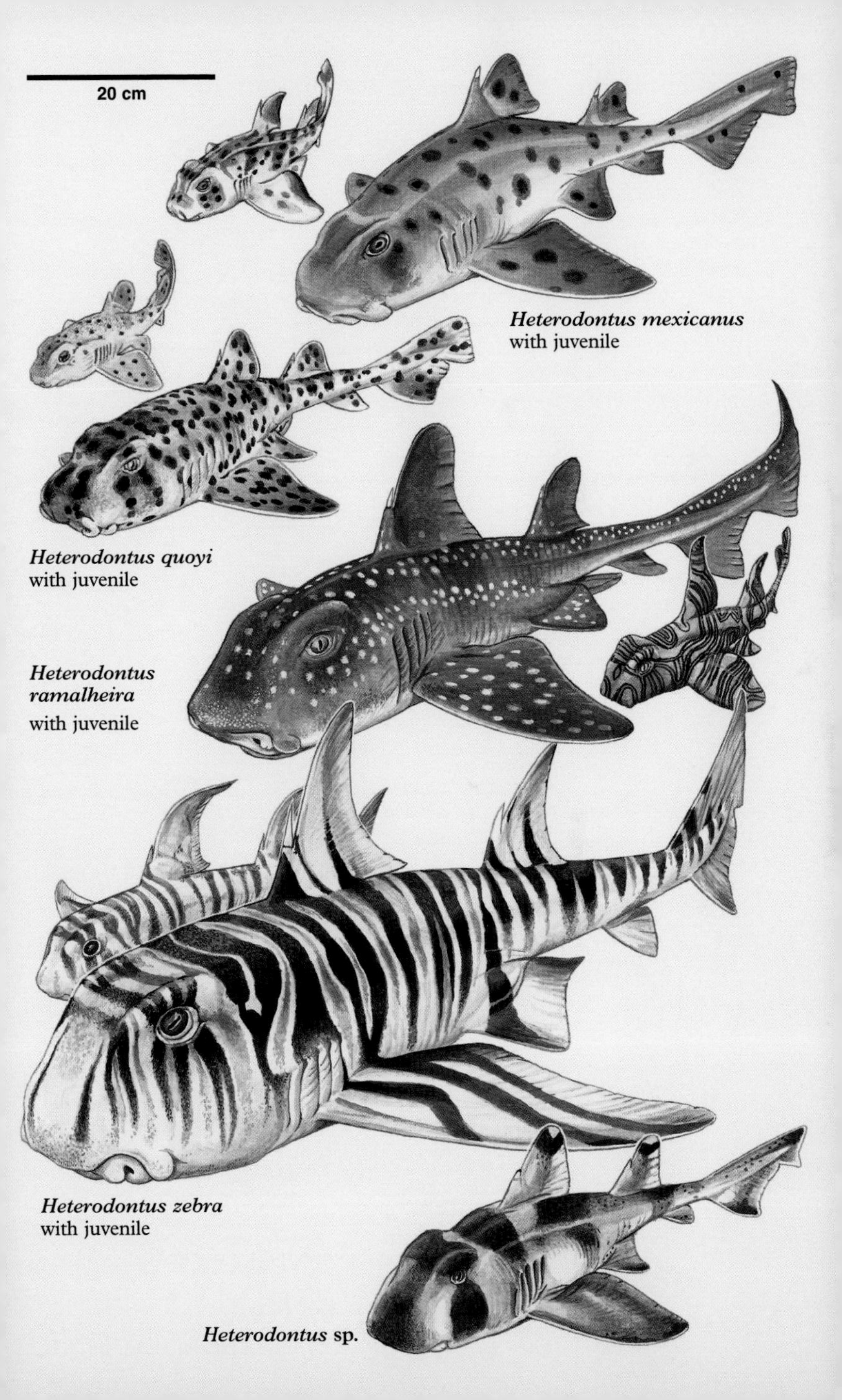
20 cm
Heterodontus mexicanus
with juvenile
Heterodontus quoyi
with juvenile
Heterodontus
ramalheira
with juvenile
Heterodontus zebra
with juvenile
Heterodontus sp.

Plate 22 Parascylliidae and Brachaeluridae

Parascylliidae: Slender; mouth in front of eyes, nostrils with barbels and naso-oral grooves, tiny spiracles; D1 fin originates behind V fin bases, D2 fin well behind A fin origin. *Cirrhoscyllium*: unique cartilage-cored pair of barbels on throat, saddles, no spots; *Parascyllium*: no barbels on throat, saddles and spots.

Barbelthroat Carpetshark *Cirrhoscyllium expolitum* **p156**

S China Sea; bottom 183–190 m. Head 3 times length D1 fin base; 6–10 diffuse saddles not 'C'-shaped, 1 saddle long and round between P and V fins extends over V fin bases.

Taiwan Saddled Carpetshark *Cirrhoscyllium formosanum* **p157**

NW Pac.; ~110 m. Head 2.3–2.6 times length D1 fin base; 6 diffuse saddles not 'C'-shaped, 1 saddle long and round between P and V fins extends over V fin bases.

Saddled Carpetshark *Cirrhoscyllium japonicum* **p157**

NW Pac.; 250–290 m. Long snout; 9 bold saddles 1 'C'-shaped between P and V fin bases.

Collared Carpetshark *Parascyllium collare* **p157**

E Austr.; rocky 20–160 m. Circumnarial grooves; lt. yellow- to red-brown with dk. unspotted sharp edged collar, 5 dusky saddles, large dk. spots except P fins <6 spots on sides of tail between D fins.

Rusty Carpetshark *Parascyllium ferrugineum* **p158**

S Austr.; on/nr. bottom 5–150 m. Circumnarial grooves; grey-brown with indistinct collar, 6–7 dusky saddles, dk. spots, >6 spots on sides of tail between D fins, more heavily spotted in Tasmania.

Necklace Carpetshark *Parascyllium variolatum* **p158**

S Austr.; bottom to 180 m. Dk. grey to chocolate brown with dk. white spotted collar, unmistakable highly variable pattern of dk. blotches and dense white spots, obvious black spots on fins.

Ginger Carpetshark *Parascyllium sparsimaculatum* **p159**

W Austr.; 245–435 m. Circumnarial grooves; pale brown or grey with indistinct unspotted collar, 5 indistinct dk. saddles, sparse large spots and blotches,< 6 spots on sides of tail between D fins.

Brachaeluridae: Stout; mouth in front of eyes, long barbels and nasooral and circumnarial grooves, large spiracles; D2 fin well infront of A fin.

Blind Shark *Brachaelurus waddi* **p159**

E Austr.; rock and seagrass to 140 m. D1 fin ≈ D2 fin, A and C fins almost touch; brown with small sparsely scattered white spots, dark saddles in young obsolete or indistinct in adults.

Bluegrey Carpetshark *Heteroscyllium colcloughi* **p160**

E Austr.; bottom to 6 m. Barbels with post. hooked flap; D1 fin > D2 fin; grey with no spots, young with distinct black markings fade with age.

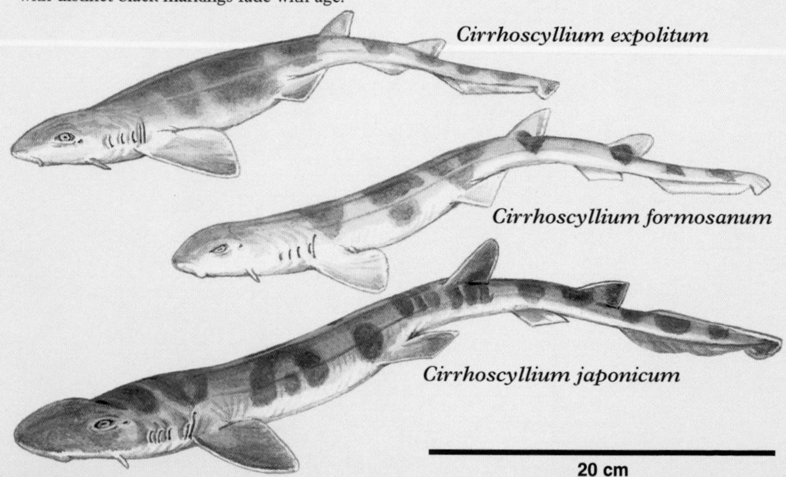

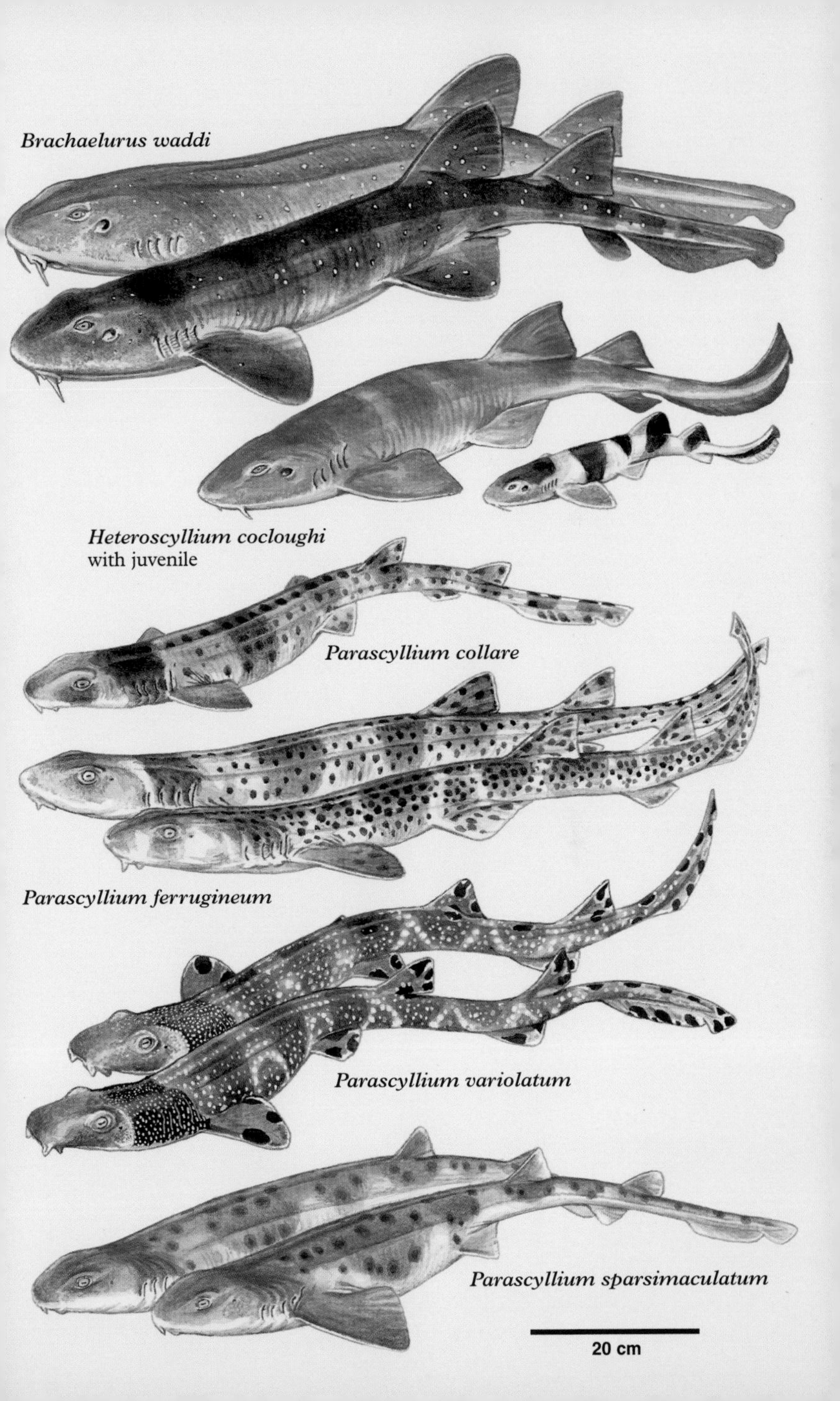

Brachaelurus waddi
Heteroscyllium colcloughi
with juvenile
Parascyllium collare
Parascyllium ferrugineum
Parascyllium variolatum
Parascyllium sparsimaculatum
20 cm

Plate 23 Orectolobidae I

Very distinctive, flattened well camouflaged; broad flat head, short mouth in front of eyes almost at short snout end, long barbels with nasoral circumnarial and symphysial grooves, dermal flaps on sides of head, spiracles larger than eyes; D1 fin origin over V fin bases, anal fin.

Tasselled Wobbegong *Eucrossorhinus dasypogon* **p161**

SW Pac.; inshore coral reefs. Numerous highly branched dermal lobes on head, 'beard' on chin; v. broad paired fins; indistinct saddles, variable reticulated pattern of narrow dark lines on lt. background, scattered enlarged dk. blotches on line junctions.

Northern Wobbegong *Orectolobus wardi* **p163**

Austr.; shallow reefs < 3m. Unbranched nasal barbels, 2 dermal lobes below and in front of eye, dermal lobes behind spiracle unbranched and broad; simple colour pattern, 3 large dark lt. edged rounded saddles, lt. background with indistinct dkr. mottling and a few dk. spots.

Japanese Wobbegong *Orectolobus japonicus* **p161**

NW Pac.; inshore rocky and coral reefs. Long branched nasal barbels, 5 dermal lobes below and in front of eye; obvious dk. dors. saddles with spots and blotches and dk. (not black), corrugated edges, lt. background with dk. broad reticular lines.

Cobbler Wobbegong *Sutorectus tentaculatus* **p164**

S and W Austr.; depth unrecorded. Rather slender with rows of warty dermal tubercules in rows on back and bases of v. low long D fins; simple unbranched nasal barbels, chin smooth, few slender short unbranched dermal lobes form isolated groups in 4–6 pairs; striking pattern of broad dk. dors. saddles with jagged corrugated edges, lighter background with irregular dk. spots.

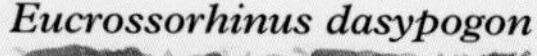

Eucrossorhinus dasypogon

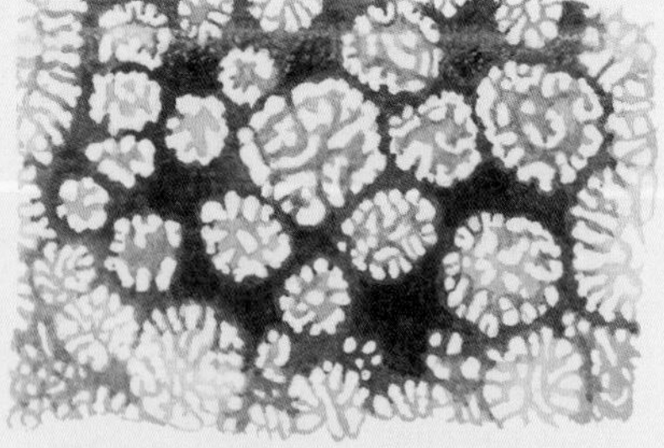

Orectolobus wardi

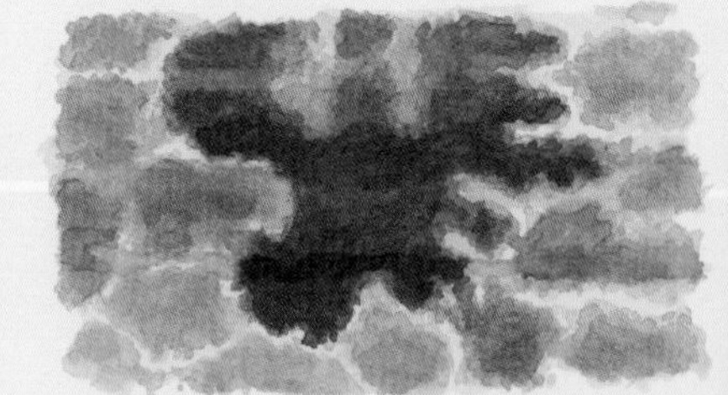

Orectolobus japonicus

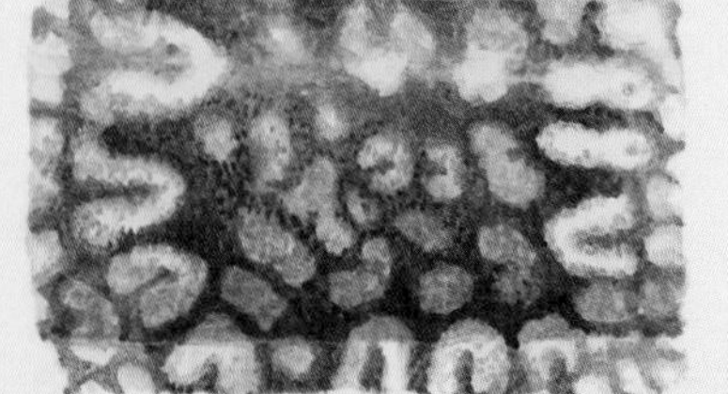

Sutorectus tentaculatus

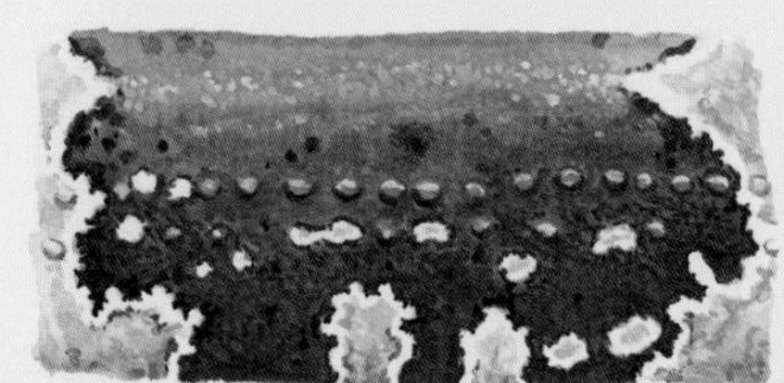

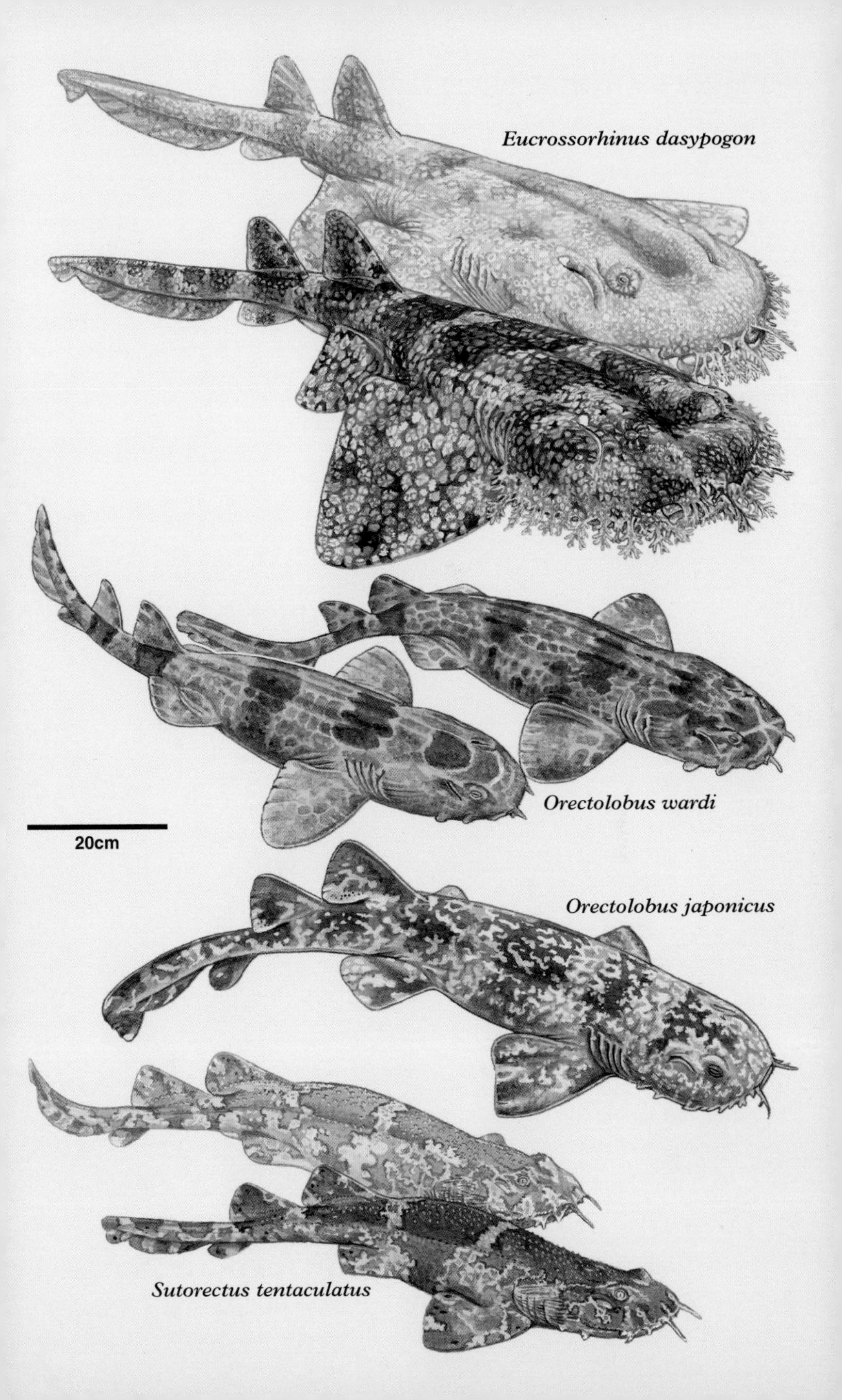

Eucrossorhinus dasypogon
Orectolobus wardi
20cm
Orectolobus japonicus
Sutorectus tentaculatus

Plate 24 Orectolobidae II

Ornate Wobbegong *Orectolobus ornatus* **p163**

Austr.; intertidal–>100 m. Nasal barbels with a few branches, 5 dermal lobes below and in front of eye, behind spiracles weakly or unbranched broad lobes; obvious broad dk. dors. saddles with lt. spots and consp. black corrugated borders, lt. background with consp. dk. lt. centred spots.

Spotted Wobbegong *Orectolobus maculatus* **p162**

Austr.; intertidal–>110 m. Long branched nasal barbels, 6–10 dermal lobes below and infront of eye; dk. back, broad dkr. dors. saddles with white O-shaped spots and blotches and corrugated edges, lt. background with dk. broad reticular lines.

Western Wobbegong *Orectolobus* sp. A **p164**

W Austr.; intertidal– <100 m. Long nasal barbels with 1 small branch, 4 dermal lobes below and in front of eye, behind spiracles weakly or unbranched and slender lobes; strongly contrasting broad dk. rectangular dorsal saddles with lt. spots and deeply corrugated, background ltr. with numerous broad dk. blotches without numerous lt. O-shaped rings.

Orectolobus ornatus

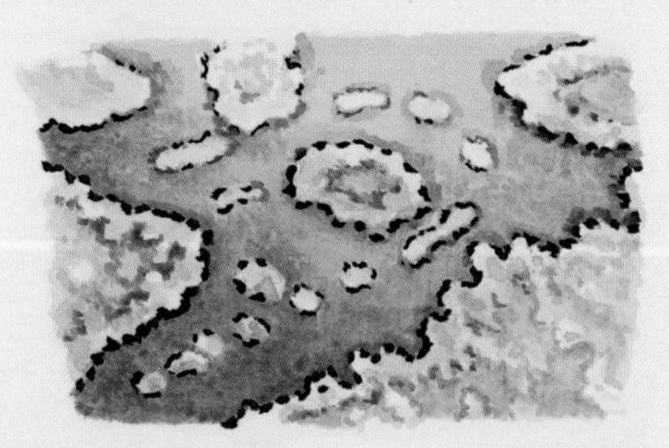

Orectolobus maculatus

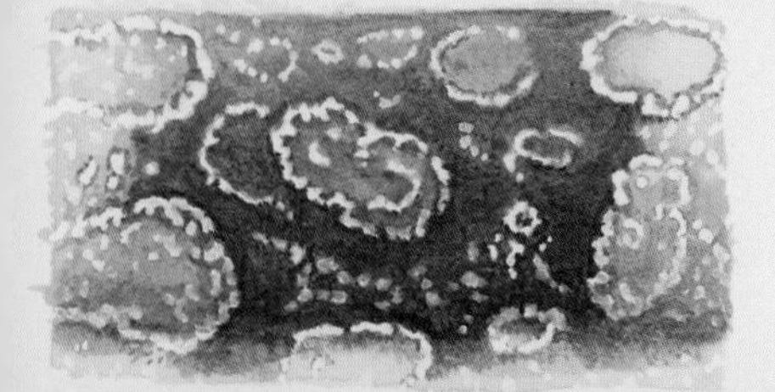

Orectolobus sp. A

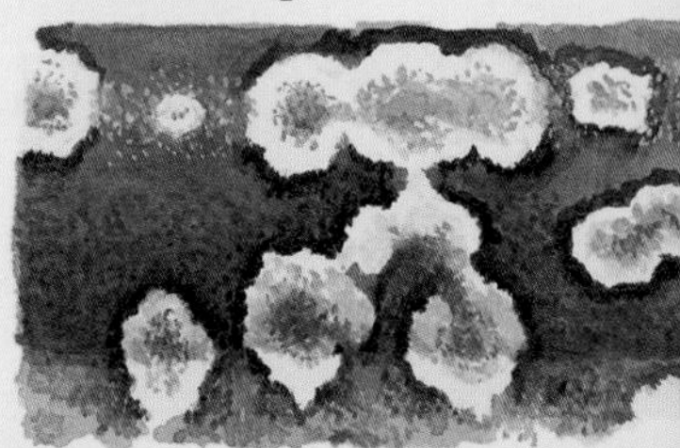

50 cm
Orectolobus ornatus
Orectolobus maculatus
Orectolobus sp. A

Plate 25 Hemiscylliidae

Slender, v. long tail; mouth well in front of eyes, short barbels, nasoral and circumnarial grooves, large spiracles below eyes; 2 = sized unspined D fins, D2 fin origin well ahead of A fin, D and A fins set far back, notch separates A and C fins; juvs. often different from adults.

Arabian Carpetshark *Chiloscyllium arabicum* **p165**

NW IO; bottom 3–100 m. Prom. ridges on back; thick tail; D fins straight post. margins, D1 fin origin opp. or just behind V fin insert., D2 base > D1; adult unpatterned, juvs. with lt. spots on fins.

Burmese Bambooshark *Chiloscyllium burmensis* **p166**

NE IO; depth unknown. No body ridges; thick tail; D fins straight/convex post. margins, D1 fin origin opp. V fin insert., long low A fin; adult with dk. fin webs, juv. colour unknown.

Grey Bambooshark *Chiloscyllium griseum* **p166**

I-W. Pac.; inshore 5–80 m. No body ridges; thick tail; D fins straight/convex post. margins, D1 fin origin over rear. V fin bases, long low A fin; adult unpatterned, juvs. with dk. saddles and transverse bands.

Indonesian Bambooshark *Chiloscyllium hasselti* **p166**

I-W Pac.; inshore–12 m. No body ridges; slender tail; D fins straight/convex post. margins, D1 fin origin over rear of V fin base; adult with dusky fins, juvs. with prom. saddles and blotches on fins.

Slender Bambooshark *Chiloscyllium indicum* **p167**

I-W Pac.; inshore. V. slender, lat. ridges; slender tail; D fins straight/convex post. margin, D1 fin origin opp. or just behind V fin insert.; adult with small dk. spots, bars and dashes on lt. brown background, juvs. with no dk. post. margins to fins.

Whitespotted Bambooshark *Chiloscyllium plagiosum* **p167**

I-W Pac.; bottom, inshore. Lat. ridges on trunk; thick tail; D fins straight/convex post. margins, D1 fin origin opp. or just behind V fin insert.; dk. bands, not consp. edged, with numerous lt. and dk. spots.

Brownbanded Bambooshark *Chiloscyllium punctatum* **p168**

I-W Pac.; reefs to >85 m. No body ridges; thick tail; D fins concave post. margins and long free tips, D1 fin origin over front of V fin base; adult unpatterned, juvs. with dk. bands, scattered small dk. spots.

Indonesian Speckled Carpetshark *Hemiscyllium freycineti* **p168**

New Guinea; coral reefs and seagrass. Thick tail; small dk. spots on snout, large and small on body, mod. large dk. epaulette spot without white ring or dk. blotches, juv. with dk. paired fins (scattered spots in adult) and broad dark bands under head and on tail.

Papuan Epaulette Carpetshark *Hemiscyllium hallstromi* **p169**

S PNG; bottom, inshore. Thick tail; no dk. spots on snout, large sparse dk. spots on body, mod. large black epaulette spot with white ring and black blotches, juv. with black webbed paired fins (dusky in adult) and dark bands on tail.

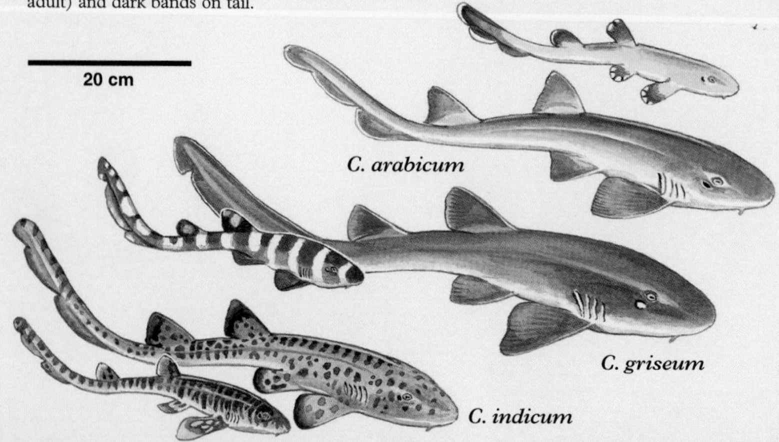

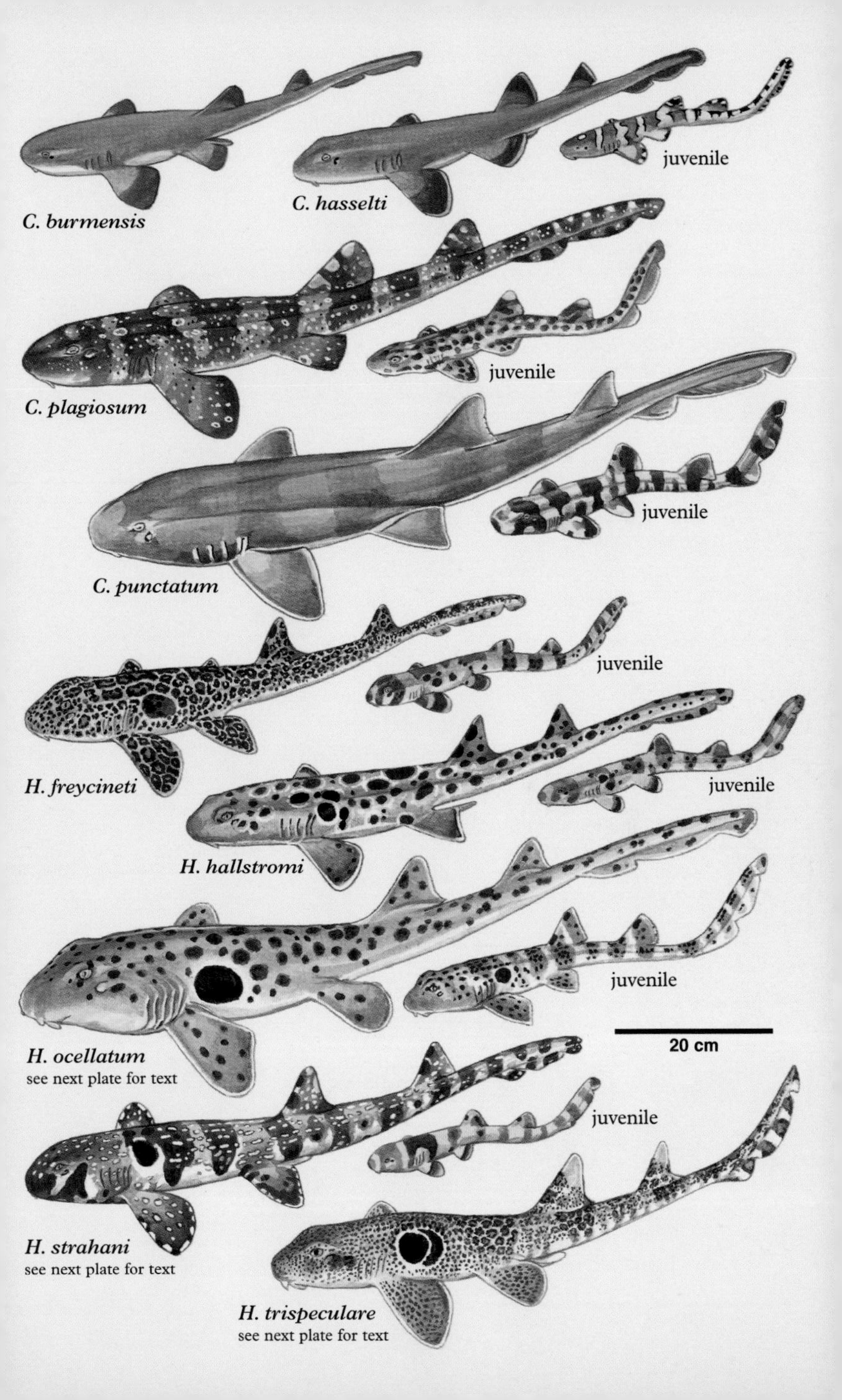

C. burmensis
C. hasselti
juvenile
C. plagiosum
juvenile
C. punctatum
juvenile
H. freycineti
juvenile
H. hallstromi
juvenile
H. ocellatum
see next plate for text
juvenile
20 cm
H. strahani
see next plate for text
juvenile
H. trispeculare
see next plate for text

Plate 25 Hemiscylliidae contd

Epaulette Shark *Hemiscyllium ocellatum* **p169**

SW Pac.; coral, shallow water. Thick tail; no dk. spots on snout, small dk. spots on body and unpaired fins, large black epaulette spot with white ring and black blotches, juv. with black webbed paired fins (dusky in adult) and dark bands on tail. See previous plate.

Hooded Carpetshark *Hemiscyllium strahani* **p170**

E PNG; coral 3–13 m. Thick tail; black mask on snout and head, white spots on body and fins over dark saddles and blotches, black epaulette spot partially merged with shoulder saddle, margins of paired fins white spots on black, dk. rings round tail, juvs. no white spots with less bold markings. See previous plate.

Speckled Carpetshark *Hemiscyllium trispeculare* **p170**

N Austr.; coral, shallow water. Thick tail; small dk. spots on snout, body and fins numerous large and small dk. spots separated by a reticular network, no white spots, large black epaulette spot with white ring and two curved black blotches, juv. coloration unknown. See previous plate.

Plate 26 Ginglymostomatidae & Stegostomatidae

Head broad and flat, no lateral skin flaps, subterminal mouth in front of eyes, barbels, long nasoral grooves, small spiracles, small gills 5 almost overlaps 4; precaudal tail much shorter than head and body; 2 spineless D fins, D2 same level and size as A fin, C fin with strong term. lobe and ST notch.

Nurse Shark *Ginglymostoma cirratum* **p171**

E Pac., W and E Atl.; usually to 12 m (40–130 m off Brazil). Long barbels, tiny spiracles; D fins broadly rounded, D1 fin > D2 and A fins, C fin >25% of total length; uniform yellow- to grey-brown, young with small dk. lt. ringed ocellar spots and obscure saddles.

Tawny Nurse Shark *Nebrius ferrugineus* **p172**

Trop. IP; to >70 m. Fairly long barbels, tiny spiracles; fins angular, D1 fin > D2 and A fins, D1 fin base over V fin bases, C fin >25% of total length; uniform shades of brown, depending on habitat.

Shorttail Nurse Shark *Pseudoginglymostoma brevicaudatum* **p173**

W IO; coral reefs depth unknown. Short barbels, tiny spiracles; all fins rounded, D1 fin ≈ D2 ≈ A fin, C fin <20% of total length; uniform dk. brown.

Zebra Shark *Stegostoma fasciatum* **p173**

I-W Pac.; to 62 m. Large, slender, ridged body; small transverse mouth in front of eyes, small barbels, large spiracles; D1 fin set forwards on back, D1 fin > D2, A fin close to C fin, broad C fin ~50% total length; distinct spotting in adults, juvs. with unique banding with intermediates to spotted adult.

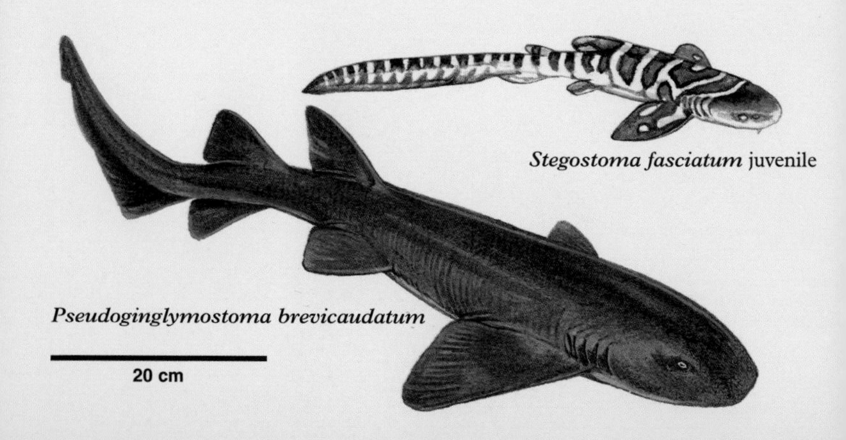

Stegostoma fasciatum juvenile

Pseudoginglymostoma brevicaudatum

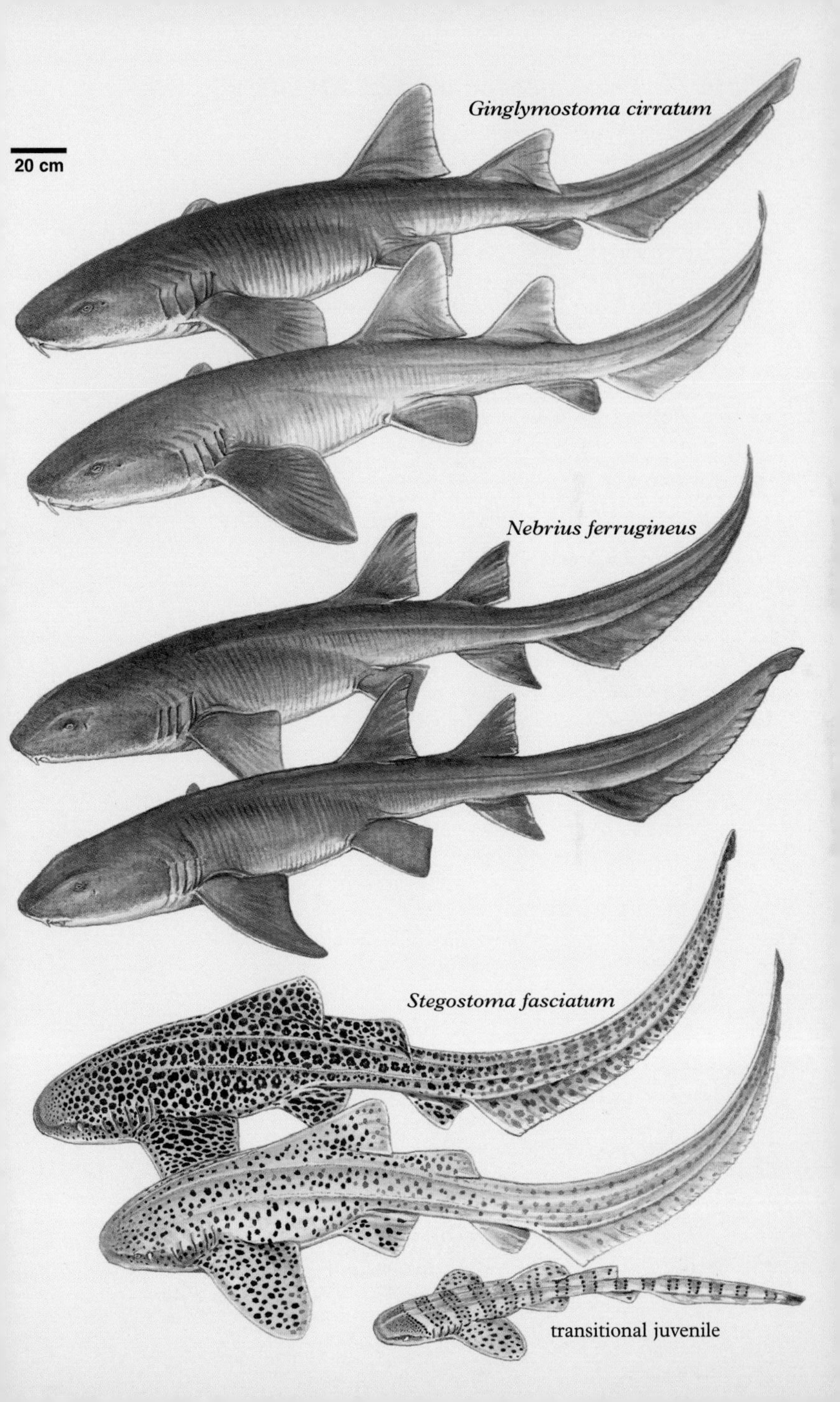
Ginglymostoma cirratum
20 cm
Nebrius ferrugineus
Stegostoma fasciatum
transitional juvenile

Plate 27 Pseudocarchariidae, Mitsukurinidae & Odontaspididae

Crocodile Shark *Pseudocarcharias kamoharai* **p177**

Trop. WW; well offshore to at least 550 m. Cylindrical, slender, v. distinctive; conical head, no barbels or grooves, nostrils free from mouth, huge eyes, large mouth with prom. long slender teeth and highly protrusible jaws, v. small spiracle, 5 broad gills; small fins, 2 spineless D fins, A fin, long dors. C fin lobe; grey or grey-brown, lt. below, lt.-edged fins.

Goblin Shark *Mitsukurina owstoni* **p178**

Patchy Atl., W IO and Pac.; deepwater 95–1300+ m. Unmistakable soft, flabby body; flat elongated snout, no barbels or grooves, nostrils free from mouth, large mouth with long-cusped slender teeth and protrusible jaws, v. small spiracle, 5 broad gills; 2 spineless D fins, A fin; pinkish white.

Three species of Odontaspididae. Large cylindrical heavy body; conical head, pointed snout, no barbels or grooves, nostrils free from mouth, large mouth extending behind eyes, v. small spiracles well behind eyes, long broad gills in front of P fins; upper precaudal pits present, no lwr.; 2 spineless D fins, A fin, C fin asymm. without keels.

Sandtiger Shark *Carcharias taurus* **p175**

Wm. temp. and trop. Atl., Med. and I-W Pac.; to at least 191 m. Flattened conical snout, large slender pointed teeth; large D fins and A fin of sim. size, D1 fin closer to V fin than P fin, C fin short vent. lobe; lt. brown, white below, often with scattered dk. spots.

Smalltooth Sandtiger *Odontaspis ferox* **p176**

WW wm. temp. and trop. deepwater; on/nr bottom. 13–420 m. Differs from *C. taurus* with long conical snout, fairly large eyes; D1 fin closer to P in than V fin, D1 fin > D2 and A fins; grey or grey-brown, lt. below, often with dk. spots.

Bigeye Sandtiger *Odontaspis noronhai* **p176**

Atl. and central Pac., WW?; midwater to nr. bottom 600–>1000 m. D1 fin > D2 and A fins; differs from other sandtigers with large eyes; uniform dk. red-brown to black above and below with white blotch on D1 fin.

Pseudocarcharias kamoharai

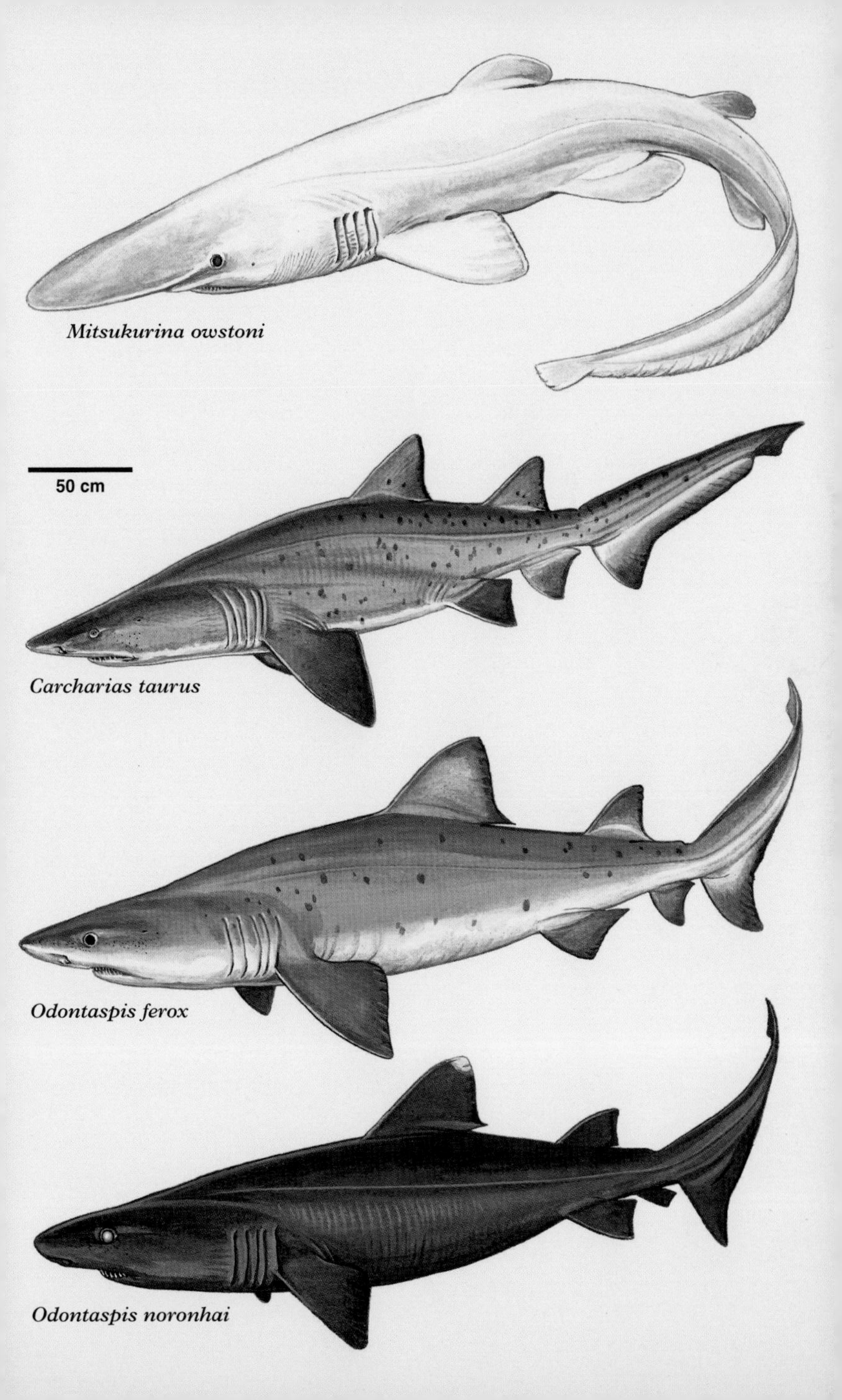
Mitsukurina owstoni
50 cm
Carcharias taurus
Odontaspis ferox
Odontaspis noronhai

Plate 28 Rhincodontidae, Megachasmidae & Cetorhinidae

Whale Shark *Rhincodon typus* p174

WW all trop. and wm. temp. seas except Med.; pelagic to >700 m. Unmistakable, huge with prom. ridges on body, lowest terminating in a keel on C ped; broad flat head, short snout, huge transverse mouth in front of eyes, comparatively small barbels, long nasoral grooves, huge gills, spiracles close to and > eyes; 2 spineless D fins, A fin, C fin lunate and unnotched; checkerboard pattern of yellow or white spots and blotches on a grey, blue-grey to green-brown background, white or yellow underside.

Basking Shark *Cetorhinus maximus* p181

WW, cold to wm. temp.; assoc. with coastal and oceanic fronts. Unmistakable, v. large cylindrical body, strong lat. keels on C ped; conical head, pointed snout, huge mouth with tiny teeth, nostrils free from mouth, no barbels or grooves, huge gills almost encircle head, v. small spiracles well behind eyes; 2 spineless D fins, A fin, lunate C fin; variable colour, dkr. above often with mottled pattern on back and sides, white blotches under head.

Megamouth Shark *Megachasma pelagios* p178

Probably WW in trop.; well offshore 5–166 m. Unmistakable, large cylindrical body; large long head, short round snout, huge term. mouth with numerous small hooked teeth, nostrils free from mouth, no barbels or grooves, large gills, v. small spiracles behind eyes; 2 spineless D fins, A fin; grey above, white below, lt. margins to blackish P and V fins, dk. spotting on lower jaw.

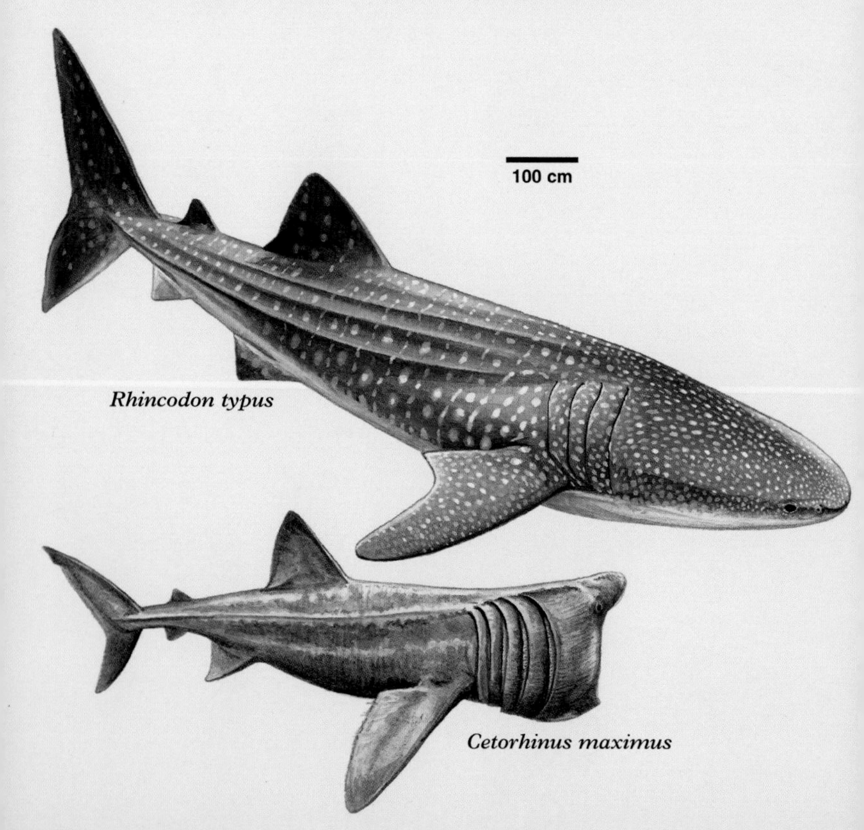

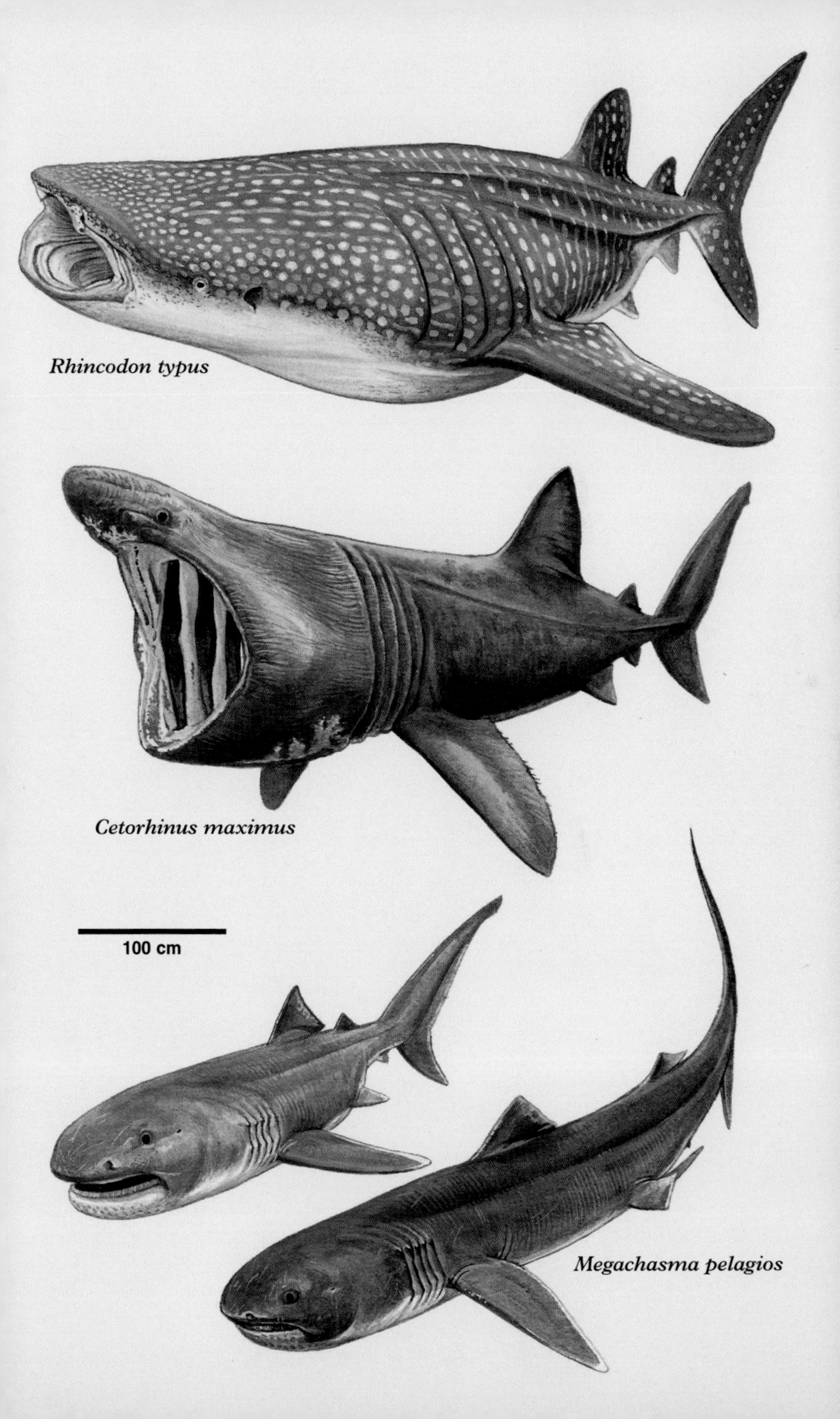
Rhincodon typus
Cetorhinus maximus
100 cm
Megachasma pelagios

Plate 29 Alopiidae

Three species. Cylindrical body with very long curving tail, conical head with mouth extending behind eyes, nostrils free from mouth, no barbels and no grooves, v. small spiracles behind eyes; 2 spineless D fins, anal fin present.

Pelagic Thresher *Alopias pelagicus* p179

IP; oceanic sometimes inshore 0–152 m. V. narrow head with straight forehead and arched profile, no labial furrows, fairly large eyes; long, straight broad tipped P fins, C fin dors. lobe nearly = to rest of body; deep blue above, white below but no white above P fins bases.

Bigeye Thresher *Alopias superciliosus* p180

WW trop. and temp. seas; oceanic to coastal to >500 m. Distinctive huge eyes extend onto flat topped head, deep horiz. groove above gills; large, v. long narrow P fins; purplish-grey or grey-brown above, lt. grey to white below which does not extend above P fin bases.

Thresher Shark *Alopias vulpinus* p180

WW trop. to cold temp. seas; nearshore to far offshore to at least 366 m. Fairly large eyes, labial furrows; long, curved narrow, pointed P fins, C fin dors. lobe nearly = to rest of body; blue-grey to dk. grey above, silvery or coppery sides, underside white extending in a patch above P fins, white dot on P fin tips.

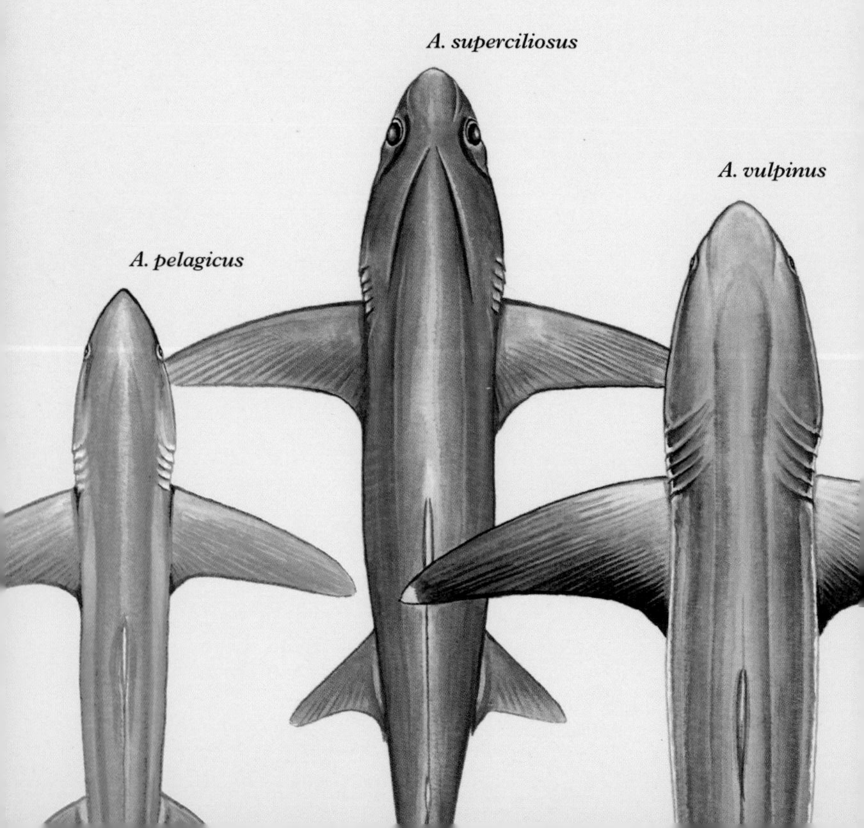

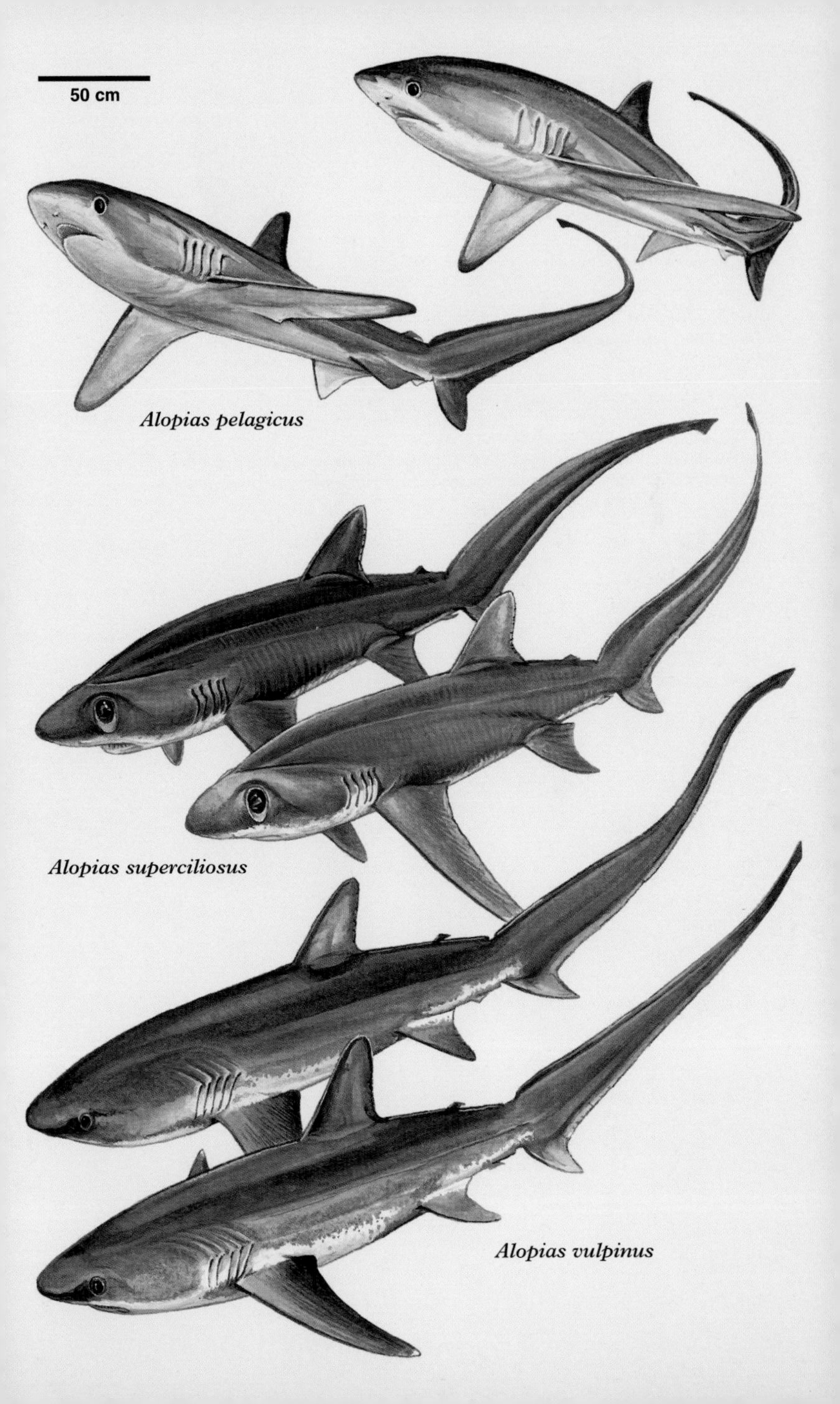
50 cm
Alopias pelagicus
Alopias superciliosus
Alopias vulpinus

Plate 30 Lamnidae

Five species. Spindle-shaped body; conical head, fairly short snout, nostrils free from mouth, no barbels or grooves, large mouth extending behind eyes, broad gills, v. small spiracles well behind eyes; keels present on C. ped; v. small D2 and A fins, nearly symm. lunate C fin.

White Shark *Carcharodon carcharias* **p181**

WW except polar seas; coastal and oceanic 0–1300 m. Heavy body; rel. long snout, v. black eyes, v. long gills; strong keels on C ped.; large D1 fin; grey above with sharp demarcation to white below, D1 fin with dk. free rear tip, black tip to underside of P fins, usually black spot at P fin insertion, old adults often become paler above and off-white below giving rise to a 'white' shark.

Shortfin Mako *Isurus oxyrinchus* **p182**

WW in temp. and trop. seas; coastal and oceanic 0–500 m. 'U'-shaped mouth, black eyes, v. long gills; strong keels on C ped.; back brilliant blue or purple, paler and more silvery on sides, white below, underside of snout and mouth white in adult (dusky in Azores species 'marrajo-criollo), ant. half V fins dk. rear and underside white.

Longfin Mako *Isurus paucus* **p183**

Prob. WW; poss. epipelagic in deepwater. Differs from *I. oxyrinchus* by a less pointed snout; P fins as long as head and rel. broad tipped; dusky underside to snout and mouth in adults.

Salmon Shark *Lamna ditropis* **p183**

N Pac.; cool coastal and oceanic to 225+ m. Heavy body; short snout, long gills; strong keels on C ped. with short secondary keels on C fin base; dk. grey or blackish above, white below with dusky blotches and dk. underside to snout in adults, white patch over P fin bases, D1 fin with dk. free rear tip.

Porbeagle Shark *Lamna nasus* **p184**

N Atl. and cool S hemisphere waters; inshore to offshore to 700 m. V. sim. to *L. ditropis* except for more pointed snout, no white patch above P fins, v. distinctive white free tip to D1 fin, N population have no dusky blotches, S population have distinctive dusky hood and dusky blotches sim. to *L. distropis*.

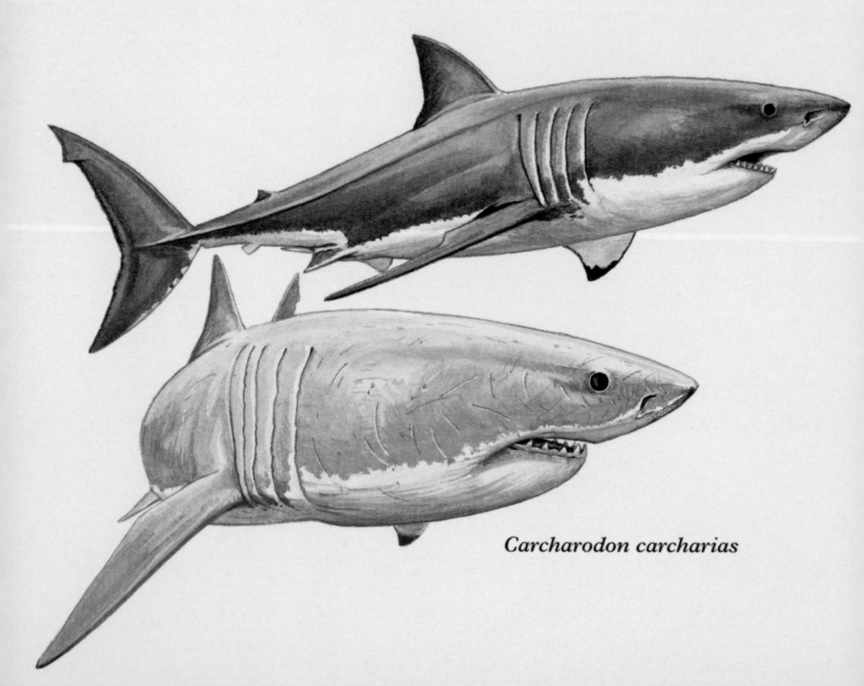

Carcharodon carcharias

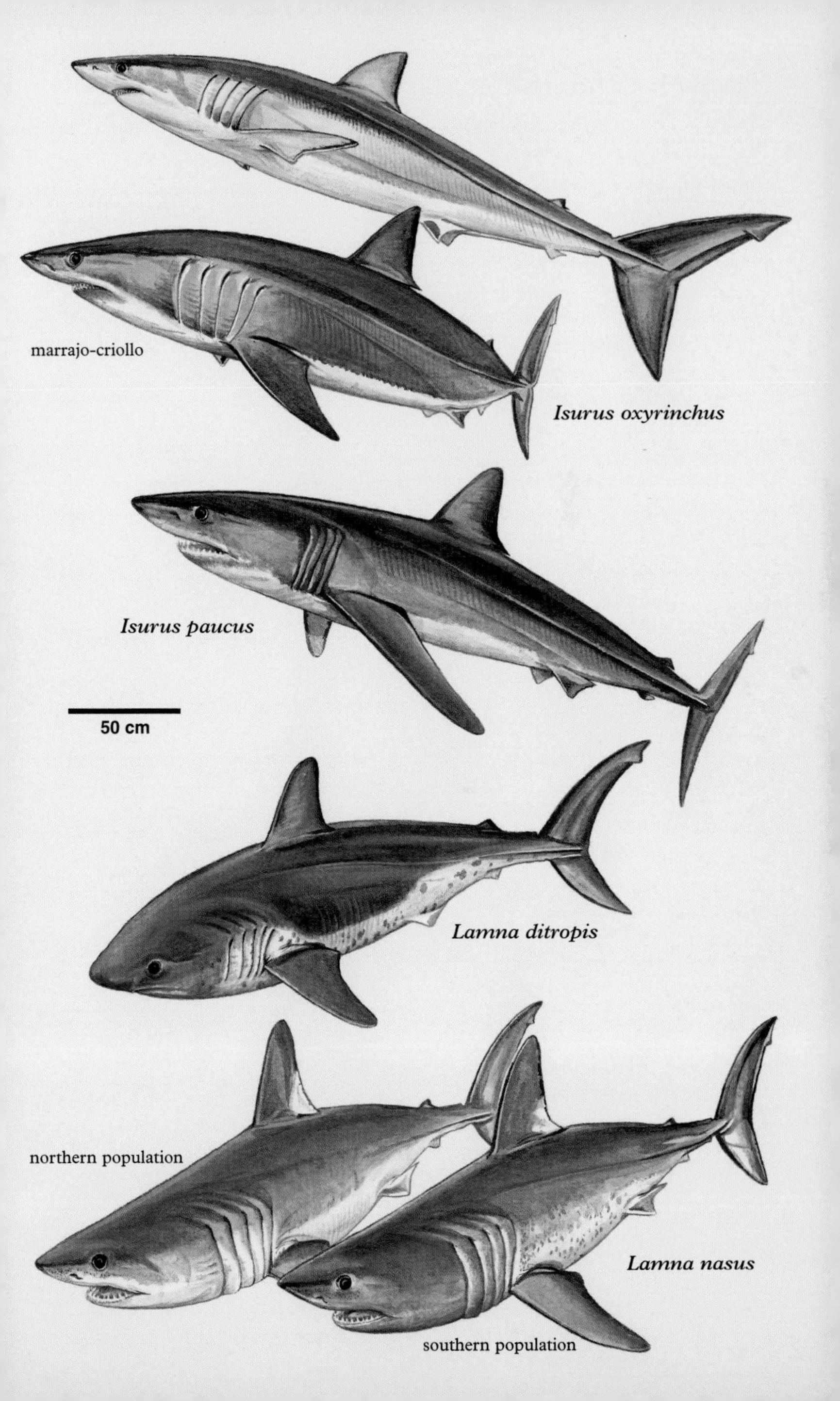
marrajo-criollo
Isurus oxyrinchus
Isurus paucus
50 cm
Lamna ditropis
northern population
Lamna nasus
southern population

Plate 31 Scyliorhinidae I- *Apristurus*

White-bodied Catshark *Apristurus albisoma* **p186**
New Caledonia; poss. soft bottom 935–1564 m. D1 fin slightly < D2, large short deep A fin.

White Ghost Catshark *Apristurus aphyodes* **p187**
NE Atl.; poss. soft bottom 1014–1800 m. D1 fin ≈ D2, large short mod. deep A fin.

Brown Catshark *Apristurus brunneus* **p187**
E Pac.; on/well above bottom 33–1298 m. D1 fin = D2, v. large long A fin.

Hoary Catshark *Apristurus canutus* **p188**
Carib.; insular slopes 573–1200 m. D1 fin < D2, v. large long A fin.

Flaccid Catshark *Apristurus exsanguis* **p189**
NZ; insular slopes 573–1200 m. D1 fin ≈ D2, v. long low A fin.

Smallbelly Catshark *Apristurus indicus* **p190**
W IO; deep water contin. slope bottom 1289–1840 m. D1 fin < D2 extends as long low ridge.

Iceland Catshark *Apristurus laurrussoni* **p193**
W Atl.; on/nr. bottom contin. slopes 560–2060 m. D1 fin slightly > D2, v. large mod. long A fin.

Ghost Catshark *Apristurus manis* **p195**
Atl.; contin. slopes 658–1740 m. D1 fin = D2, large broadly rounded A fin.

Smalleye Catshark *Apristurus microps* **p196**
Atl. and SW. IO; contin. slopes 700–2200 m. D1 fin ≈ D2, v. large long rounded–angular A fin.

Smallfin Catshark *Apristurus parvipinnis* **p197**
Carib.; on/nr. bottom contin. slope 635–1135 m. D1 fin v. small, large long A fin.

Deepwater Catshark *Apristurus profundorum* **p199**
N Atl.; contin. slope ~1500 m. D1 fin = D2, large long rounded–angular A fin.

Broadgill Catshark *Apristurus riveri* **p199**
G of M and Carib.; on/nr. bottom contin. slope 732–1461 m. D1 fin v. < D2, deep angular A fin.

Saldanha Catshark *Apristurus saldanha* **p200**
S Afr.; contin. slope 344–1009 m. D1 fin slightly < D2, long angular A fin.

20 cm

A. aphyodes

A. brunneus

A. albisoma

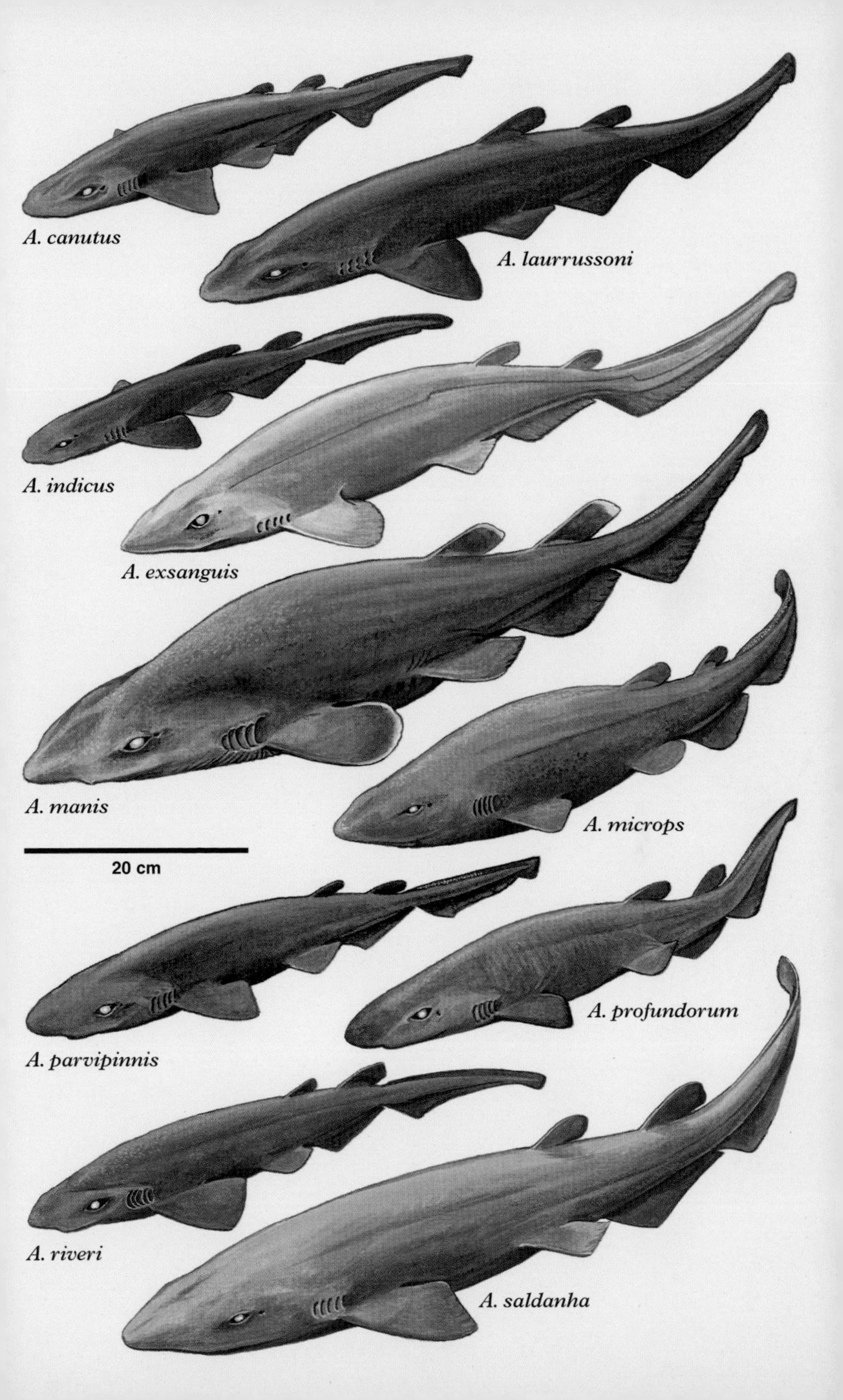

A. canutus
A. laurrussoni
A. indicus
A. exsanguis
A. manis
A. microps
20 cm
A. parvipinnis
A. profundorum
A. riveri
A. saldanha

Plate 32 Scyliorhinidae II- *Apristurus*

Longfin Catshark *Apristurus herklotsi* **p190**
S and E China Seas; bottom 520–910 m. D1 fin ~1/3 D2, v. long shallow angular A fin.

Shortnose Demon Catshark *Apristurus internatus* **p190**
E China Sea; contin. slope 670 m. D1 fin slightly < D2, v. large mod. long deep angular A fin.

Japanese Catshark *Apristurus japonicus* **p192**
E China Sea and Sea of Japan; nr. bottom 820–915 m. D1 fin = D2, v. large mod. long angular A fin.

Longnose Catshark *Apristurus kampae* **p192**
NE Pac.; contin. shelf and upper slope 180–1888 m. D1 fin ≈ D2, v. deep rounded A fin.

Longhead Catshark *Apristurus longicephalus* **p194**
W Pac. and W. IO; prob. nr. bottom 500–1140 m. D1 fin slightly < D2, large long angular A fin.

Flathead Catshark *Apristurus macrorhynchus* **p194**
E China Sea and S. Sea of Japan; bottom 220–1140 m. D1 fin 2/3 D2, v. large long angular A fin.

Largenose Catshark *Apristurus nasutus* **p197**
E Pac.; on/nr. bottom 400–925 m. D1 fin slightly < D2, large angular A fin, post. fin margins pale.

Spatulasnout Catshark *Apristurus platyrhynchus* **p198**
W Pac., S and E China Seas; cont. and insular slopes 594–985 m. D1 fin << D2, large long angular A fin.

Pale Catshark *Apristurus sibogae* **p200**
Indonesia; straits slope 655 m. D1 fin << D2, long angular A fin, v. large P fins.

South China Catshark *Apristurus sinensis* **p201**
S China Sea; cont. slope 537–1000 m. D1 fin ~1/2 D2, large angular A fin.

Spongehead Catshark *Apristurus spongiceps* **p201**
W Pac. and Hawaii; on/nr. bottom 572–1482 m. Pleated throat area, D1 fin ≈ D2, v. large A fin.

Panama Ghost Catshark *Apristurus stenseni* **p202**
E Pac.; cont. slope 915–975 m. D1 fin ≈ D2, mod. high rounded–angular A fin.

20 cm

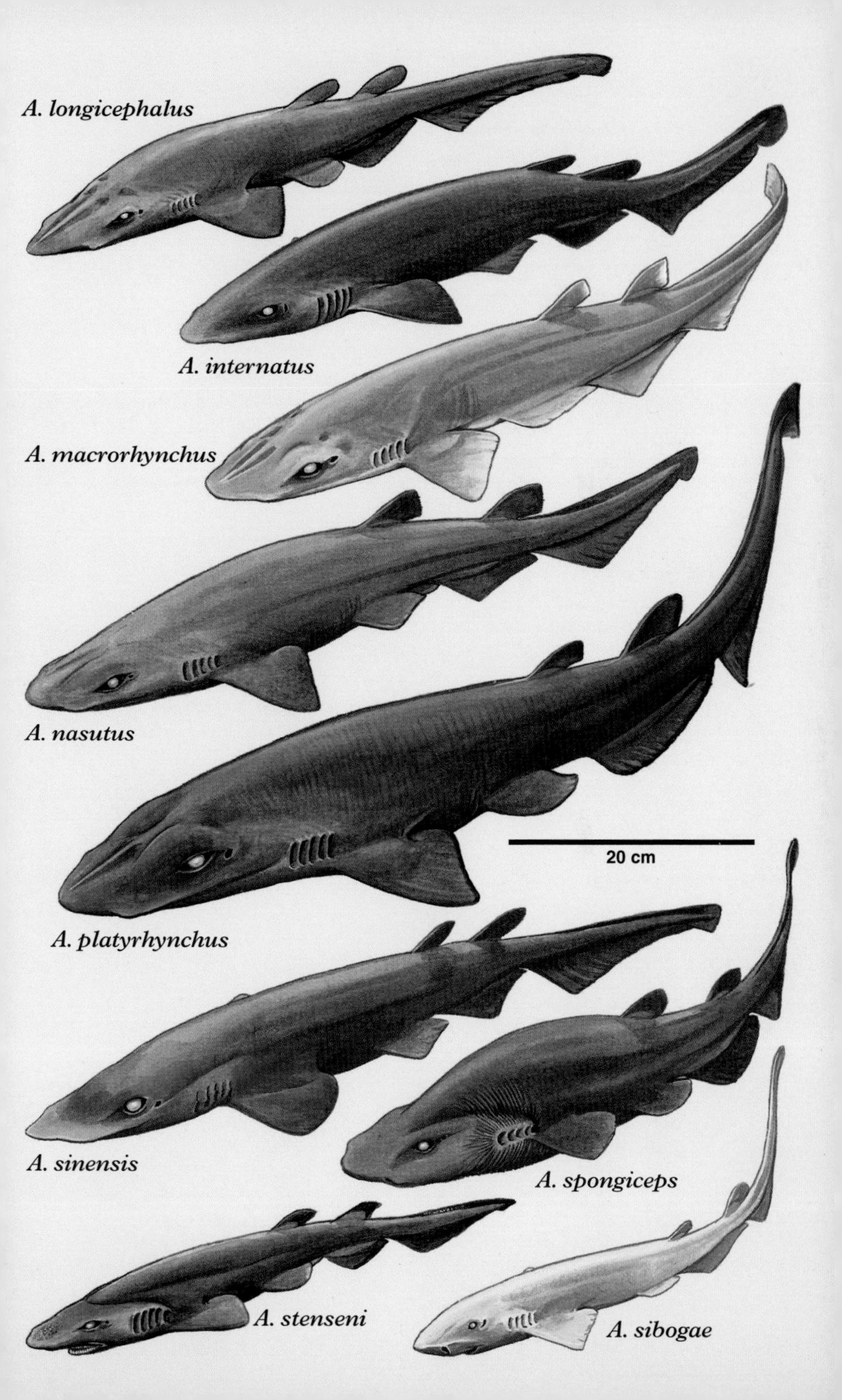
A. longicephalus
A. internatus
A. macrorhynchus
A. nasutus
20 cm
A. platyrhynchus
A. sinensis
A. spongiceps
A. stenseni
A. sibogae

Plate 33 Scyliorhinidae III- *Apristurus*

Stout Catshark ***Apristurus federovi*** **p189**
N Japan; 810–1430 m. D1 fin slightly < D2, v. large deep rounded A fin.

Humpback Catshark ***Apristurus gibbosus*** **p189**
S China Sea; contin. slope 913 m. D1 fin slightly < D2, v. large mod. long deep angular A fin.

Broadnose Catshark ***Apristurus investigatoris*** **p191**
Andaman Sea; bottom contin. slope 1040 m. D1 fin 2/3 area D2, v. large low rounded–angular A fin.

Broadmouth Catshark ***Apristurus macrostomus*** **p195**
S China Sea; captured 913 m. D1 fin < 1/2 D2, v. large long angular A fin.

Small Dorsal Catshark ***Apristurus micropterygeus*** **p196**
S China Sea; captured 913 m. D1 fin v. small ~1/9 D2, v. large long angular A fin.

Fat Catshark ***Apristurus pinguis*** **p198**
E China Sea; captured 200–1040 m. D1 fin ≈ D2, large long deep rounded–angular A fin.

Freckled Catshark ***Apristurus*** **sp. A.** **p203**
S Austr. and NZ?; 940–1290 m. D1 fin << D2, long shallow triangular A fin.

Bigfin Catshark ***Apristurus*** **sp. B.** **p203**
E and W Austr.; contin. slope 730–1000 m. D1 fin << D2, v. long deep angular A fin.

Fleshynose Catshark ***Apristurus*** **sp. C.** **p204**
S Austr. and NZ; contin. shelf 900–1150 m. D1 fin = D2, long deep angular A fin.

Roughskin Catshark ***Apristurus*** **sp. D.** **p204**
S Austr. and NZ; contin. shelf 830–1380 m. D1 fin slightly < D2, fairly long rounded A fin.

Bulldog Catshark ***Apristurus*** **sp. E.** **p204**
SE Austr.; contin. slope captured 1020–1500 m. D1 fin slightly < D2, distinctly short deep A fin.

Bighead Catshark ***Apristurus*** **sp. F.** **p205**
W Austr.; contin. slope 1030–1050 m. D1 fin lower than D2, large A fin.

Pinocchio Catshark ***Apristurus*** **sp. G.** **p205**
Austr.; contin. slope and seamounts 590–1000 m. V. long narrow pointed snout; D1 fin < D2, long v. shallow angular A fin.

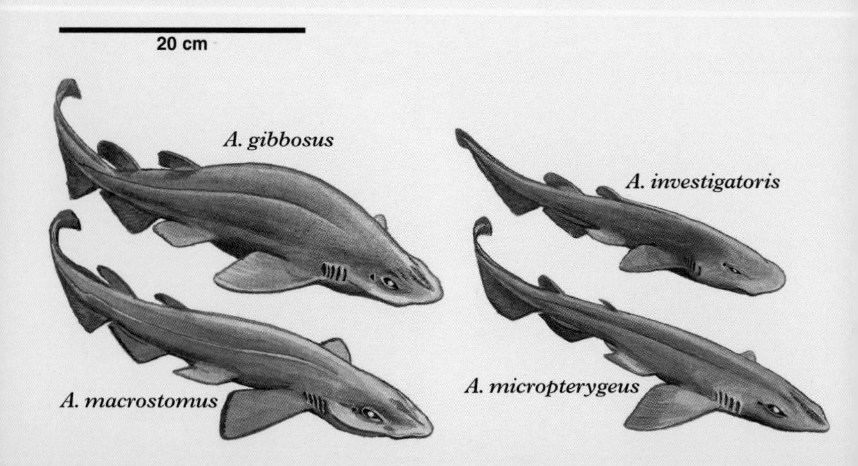

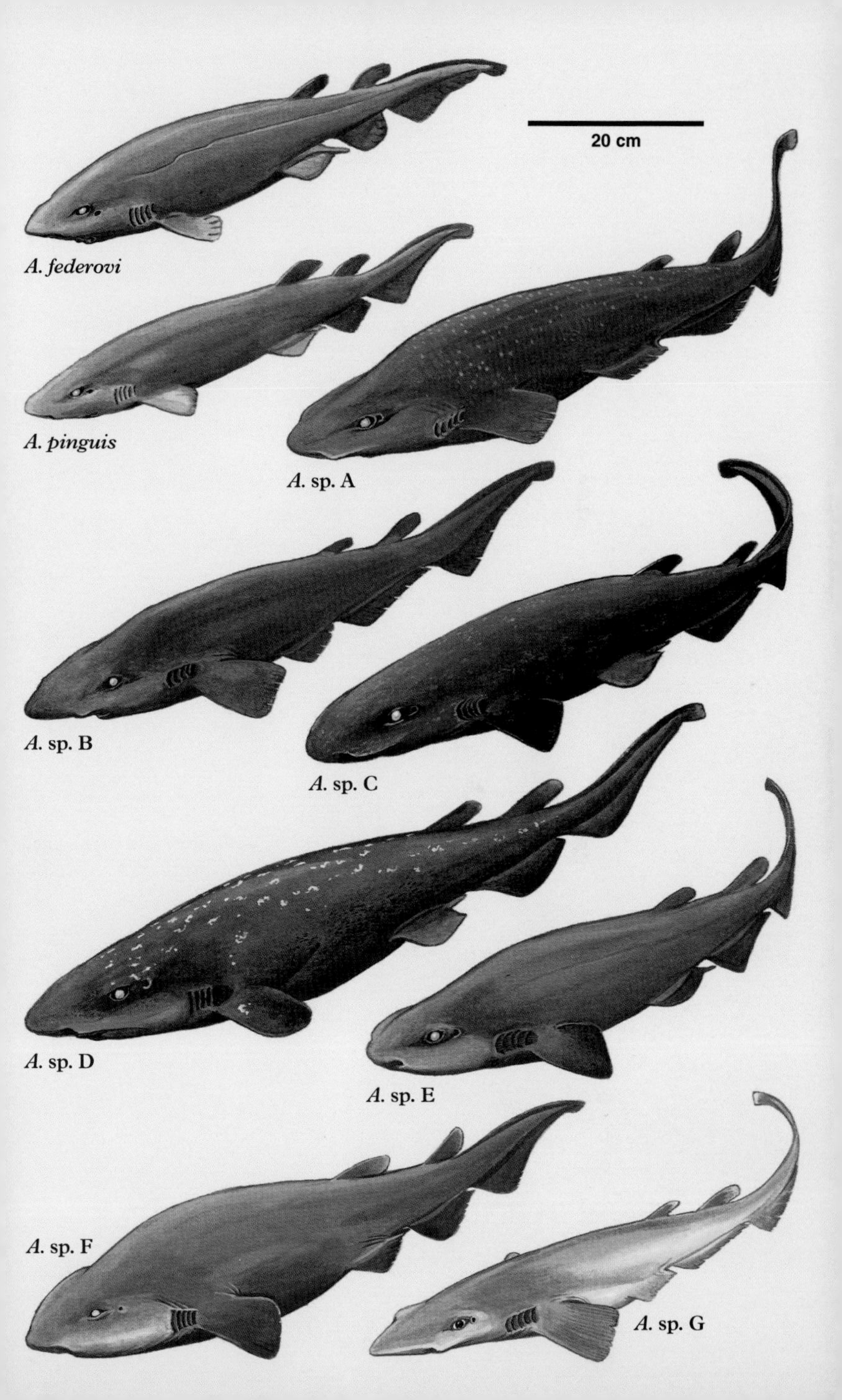
20 cm
A. federovi
A. pinguis
A. sp. A
A. sp. B
A. sp. C
A. sp. D
A. sp. E
A. sp. F
A. sp. G

Plate 34 Scyliorhinidae IV

Banded Sand Catshark *Atelomycterus fasciatus* **p209**
NW Austr.; bottom contin. shelf 27–122 m. Lt. coloured with brown saddles, small black spots, some white.

Australian Marbled Catshark *Atelomycterus macleayi* **p210**
N Austr.; sand and rock 0.5–3.5 m. Lt. grey to grey-brown with darker saddles, small black spots.

Coral Catshark *Atelomycterus marmoratus* **p210**
IWP; crevices/holes on reefs. Dk. with no clear saddles, white blotches over large black spots and bars.

Grey Spotted Catshark *Asymbolus analis* **p206**
E–SE Austr.; bottom contin. shelf 40–175 m. Greyish with obscure blotches, = sized dk. brown spots.

Blotched Catshark *Asymbolus funebris* **p206**
SW Austr.; outer contin. shelf 195 m. Brown with dk. brown blotches, 3 pre-dorsal saddles, no spots.

Western Spotted Catshark *Asymbolus occiduus* **p207**
SW Austr.; bottom outer contin. shelf 98–250 m. Yellow-green with 8–9 saddles, = sized dk. brown spots.

Pale Spotted Catshark *Asymbolus pallidus* **p207**
NE Austr.; bottom contin. shelf 270–400 m. Pale yellowish with no saddles, = sized dk. brown spots.

Dwarf Catshark *Asymbolus parvus* **p208**
Trop. NW Austr.; bottom contin. shelf 59–252 m. Pale brown with faint saddles, white spots and lines.

Orange Spotted Catshark *Asymbolus rubiginosus* **p208**
E Austr.; bottom contin. shelf 25–540 m. Pale brown with obscure dk. saddles, dk. brown spots.

Variegated Catshark *Asymbolus submaculatus* **p208**
SW Austr.; caves or ledges contin. shelf–150 m. Grey–brown with rusty brown blotches, small black spots.

Gulf Catshark *Asymbolus vincenti* **p209**
S Austr.; often in sea grass <100, 130–220 m in Bight. Grey-brown to chocolate with 7–8 saddles, small faint white spots.

Atelomycterus fasciatus

10 cm

Atelomycterus macleayi

Atelomycterus marmoratus

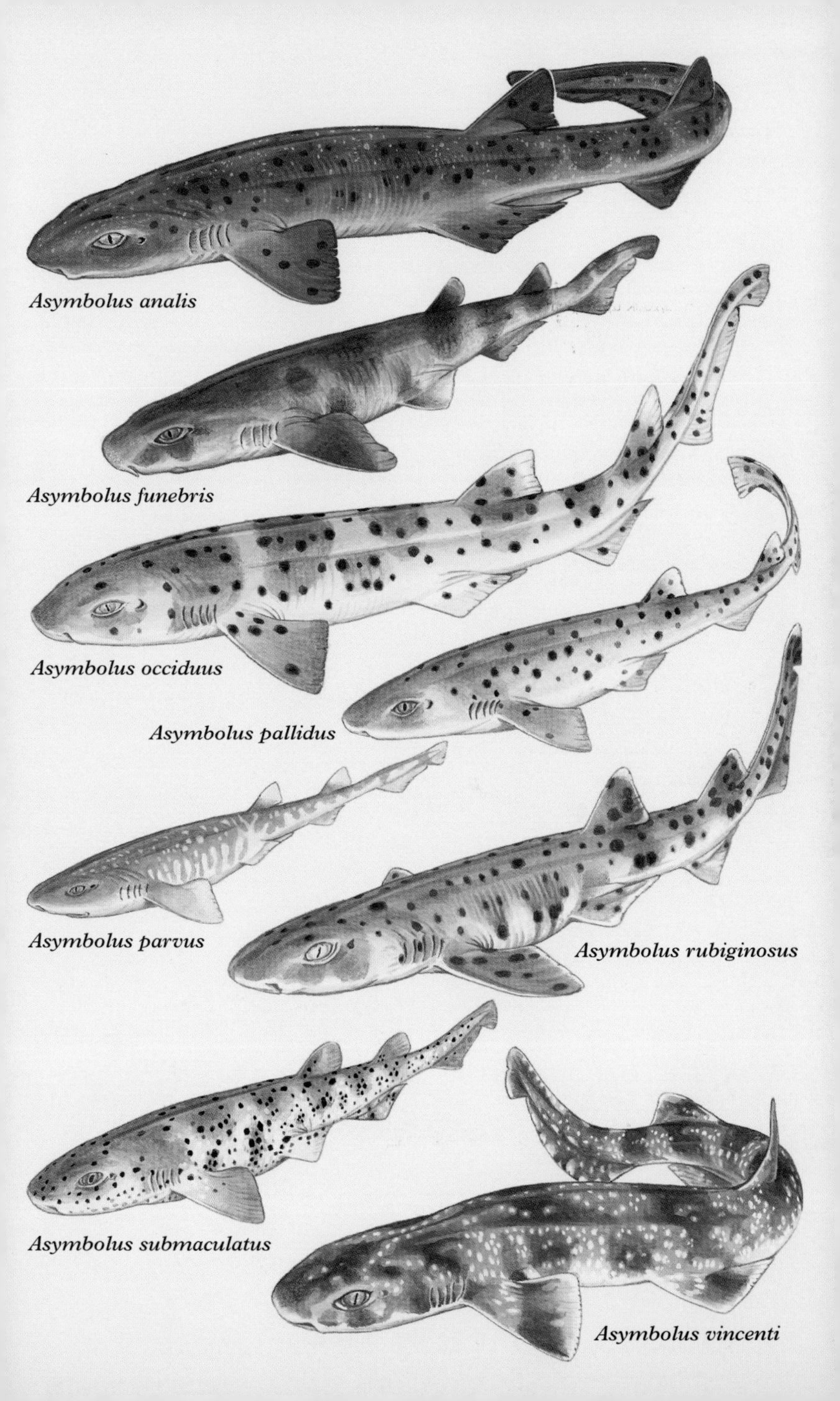
Asymbolus analis
Asymbolus funebris
Asymbolus occiduus
Asymbolus pallidus
Asymbolus parvus
Asymbolus rubiginosus
Asymbolus submaculatus
Asymbolus vincenti

Plate 35 Scyliorhinidae V

New Caledonia Catshark *Aulohalaelurus kanakorum* **p211**
New Caledonia; coral reef 49 m. Dk. grey with variegated large dk. and white blotches.

Blackspotted Catshark *Aulohalaelurus labiosus* **p211**
SW Austr.; reefs to 4 m+. Lt. grey to yellow-brown with dk. saddles, small to large black spots fine white spots and white D, C and A fin tips.

Dusky Catshark *Bythaelurus canescens* **p212**
SE Pac.; mud/rock upper contin. slope 250–700 m. Uniform dk. brown, young have white fin tips.

Broadhead Catshark *Bythaelurus clevai* **p213**
W IO; upper insular slopes 400–500m. Grey to white below, large consp. dk. brown blotches and spots.

New Zealand Catshark *Bythaelurus dawsoni* **p213**
NZ; on/nr. bottom upper slopes 371–420 m. Lt. brown or grey, paler below, white fin tips, dk. bands on C fin, line of white spots on sides of small catsharks.

Bristly Catshark *Bythaelurus hispidus* **p214**
N IO; bottom upper contin. slope 293–766 m. Pale brown to whitish, faint crossbands and spots.

Spotless Catshark *Bythaelurus immaculatus* **p214**
S China Sea; bottom contin. slope 534–1020 m. Yellow-brown similar to *B. dawsoni* but no white fin tips.

Mud Catshark *Bythaelurus lutarius* **p214**
E Afr.; on/nr. mud bottom 338–766 m. Grey-brown light below, sometimes dusky saddles. D1<D2.

Dusky Catshark *Bythaelurus* sp. A **p215**
NW Austr.; contin, slope 900 m. Bristly skin; dk. grey brown, few pale blotches below.

Galapagos catshark *Bythaelurus* sp. B **p215**
Galapagos; bottom insular slopes 400–600 m. Grey with variegated pattern lg. white blotches and spots.

Aulohalaelurus kanakorum

Aulohalaelurus labiosus

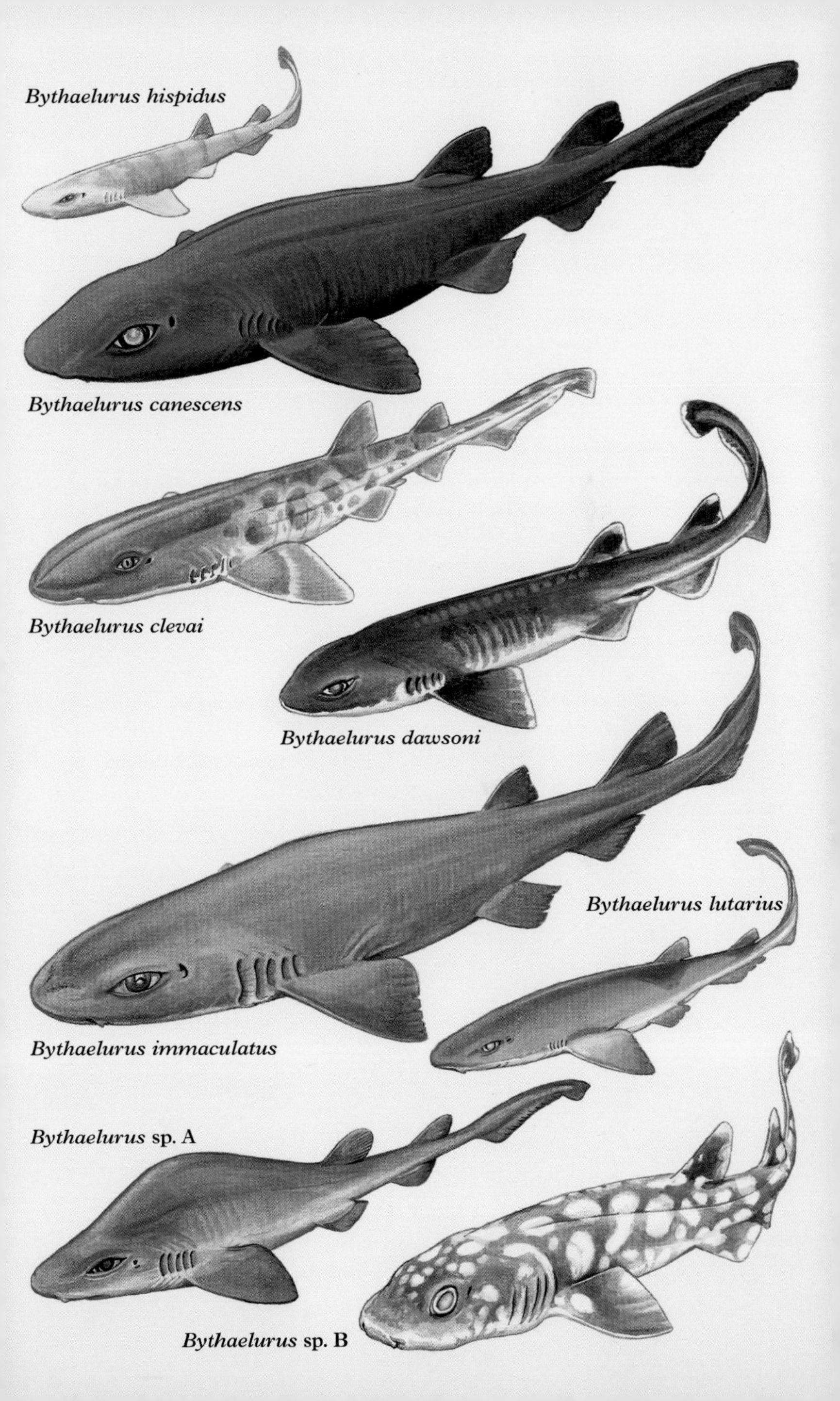
Bythaelurus hispidus
Bythaelurus canescens
Bythaelurus clevai
Bythaelurus dawsoni
Bythaelurus lutarius
Bythaelurus immaculatus
Bythaelurus sp. A
Bythaelurus sp. B

Plate 36 Scyliorhinidae VI

Reticulated Swellshark *Cephaloscyllium fasciatum* **p216**
IWP.; on/nr. mud bottom 219–450 m. Lt. grey with variegated pattern of dk. lines and spots.

Draughtsboard Shark *Cephaloscyllium isabellum* **p216**
NZ; rock/sandy bottom shore to 673 m. Up to 11 dk. brown saddles and alternating blotches.

Australian Swellshark *Cephaloscyllium laticeps* **p217**
S Austr.; contin. shelf to 60 m+. Lt. grey or chestnut with pattern of dk. brown to grey saddles and blotches.

Indian Swellshark *Cephaloscyllium silasi* **p217**
IO; bottom contin. shelf 300m. Lt. brown with 7 dk. brown saddles, dk. blotch over P fins.

Balloon Shark *Cephaloscyllium sufflans* **p218**
W IO; sand/mud bottom 40–440 m. Lt. grey with 7 lt. grey-brown obscure saddles, sometimes absent.

Japanese Swellshark *Cephaloscyllium umbratile* **p218**
W Pac.; prb. shallow water. Pale brown with dk. brown saddles and mottling sep. by lt. red-brown areas.

Swellshark *Cephaloscyllium ventriosum* **p219**
E Pac.; rocky bottom in kelp beds. Lt. yellow-brown with close-set dk. brown saddles and blotches.

Whitefin Swellshark *Cephaloscyllium* sp. A **p219**
SE Austr.; upper contin. slope 240–550 m. Med. brown or grey with dk. blotches and saddles.

Saddled Swellshark *Cephaloscyllium* sp. B **p220**
NE Austr.; contin. slope 380–590 m. Med. brown or grey with obv. dk. saddles, usually 5 pre-dors.

Northern Draughtboard Swellshark *Cephaloscyllium* sp. C **p220**
E Austr.; outer contin. shelf 90–140 m. Dk. grey-brown with obscure saddles and blotches, white flecks.

Narrowbar Swellshark *Cephaloscyllium* sp. D **p221**
NE Austr.; captured contin. slope 440 m. Dk. brown to cream with numerous close narrow dk. bars.

Speckled Swellshark *Cephaloscyllium* sp. E **p221**
N trop. Austr.; contin. slope 390–700 m. Pale grey with mottled dk. blotches and saddles, white spots.

Dwarf Oriental Swellshark *Cephaloscyllium* sp. **p222**
W Pac.; poss. inshore. Sim. *C. umbratile* simpler colour pattern with few dk. spots.

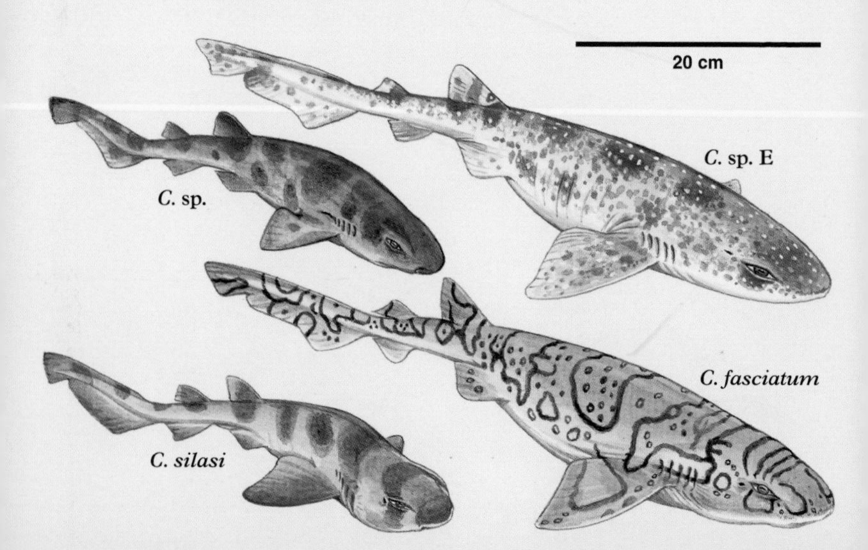

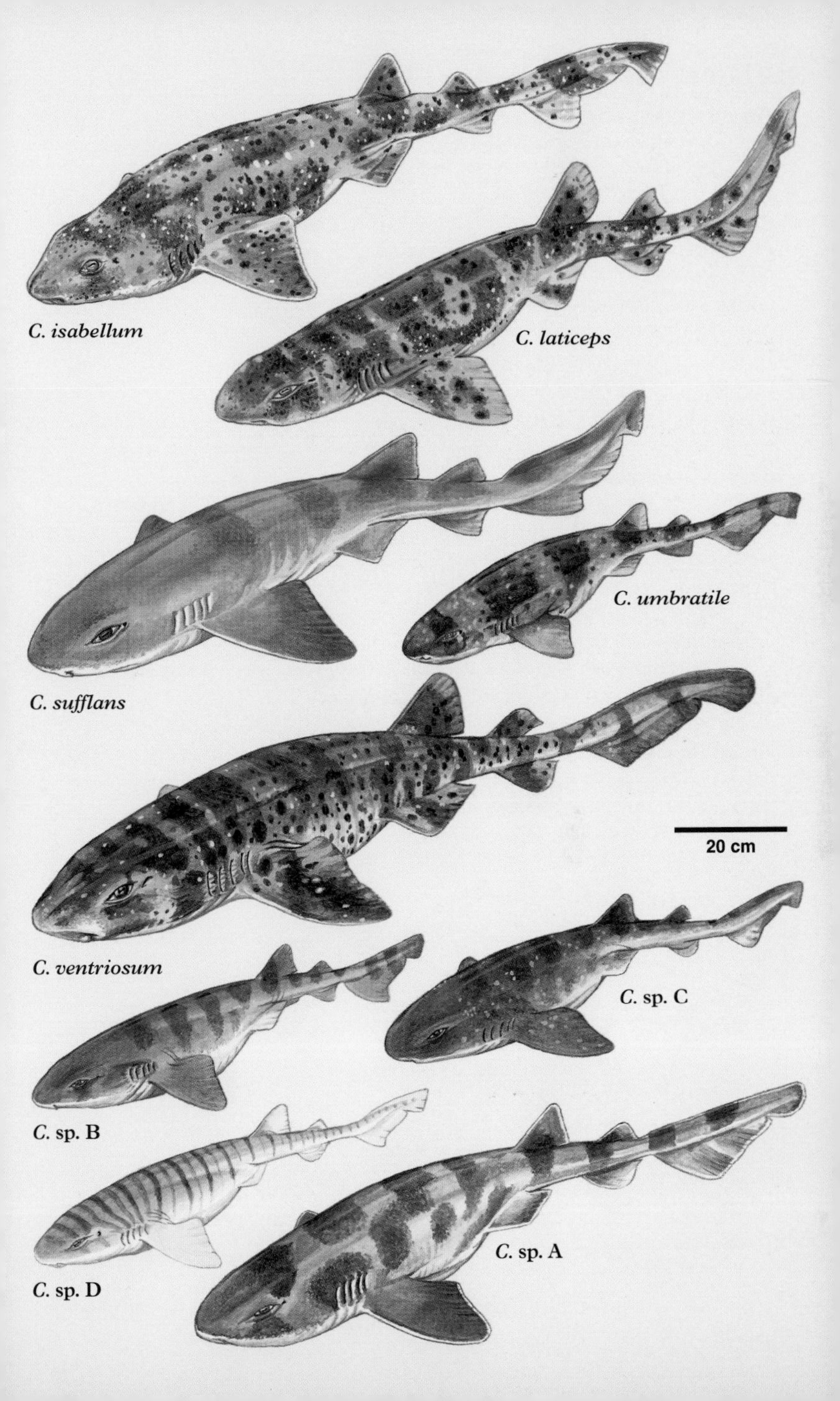
C. isabellum
C. laticeps
C. umbratile
C. sufflans
20 cm
C. ventriosum
C. sp. C
C. sp. B
C. sp. A
C. sp. D

Plate 37 Scyliorhinidae VII

Antilles Catshark *Galeus antillensis* **p223**
Straits Fl., Hispaniola to Martinique; on/nr. bottom 293–658 m. V. sim. to *G. arae* but larger; usually < 11 saddle blotches, dk. bands on tail, no black marks on D fin tips, C fin with light web.

Roughtail Catshark *Galeus arae* **p223**
N Carolina to Mex., Cuba and Belize to Costa Rica; on/nr. bottom 292–732 m. V. sim. to *G. antillensis.*

Atlantic Sawtail Catshark *Galeus atlanticus* **p224**
NE Atl. and Med.; contin. slope 400–600 m. Grey above white below with dk. grey blotches and saddles.

Australian Sawtail Catshark *Galeus boardmani* **p224**
Austr. (not N); on/nr. bottom 150–640 m. Dk. grey-brown saddles/bars, 3 pre- and 3 post-dors.

Blackmouth Catshark *Galeus melastomus* **p227**
NE Atl. and Med.; 55–1000 m. Striking pattern 15–18 dk. saddles, blotches and spots, white-edged fins.

Southern Sawtail Catshark *Galeus mincaronei* **p227**
S Brazil; deep reefs 430 m. Striking pattern 11 dk. oval saddles and spots outlined in white.

Mouse Catshark *Galeus murinus* **p228**
NE Atl.; contin. slope on/nr. bottom 380–1250 m. Brown above, paler below, no markings.

Broadfin Sawtail Catshark *Galeus nipponensis* **p228**
NW Pac.; bottom 362–540 m. Numerous obscure dusky saddles and blotches, white below.

Peppered Catshark *Galeus piperatus* **p229**
Mex. and N Gulf of Calif.; bottom 275–1326 m. Black dots all over body and tail, with/without saddles.

African Sawtail Catshark *Galeus polli* **p229**
E Atl.; 159–720 m. Usually 11 or < well-defined dk. saddle blotches outlined in white or uniform dk.

Blacktip Sawtail Catshark *Galeus sauteri* **p230**
Jap. to Philipp.; 60–90 m+. Not patterned, often D fins and C fin lobes with prom. black tips.

Dwarf Sawtail Catshark *Galeus schultzi* **p230**
Philipp.; mainly insular slope 329–431 m. Obscure dk. saddles below D fins and 2 bands on tail.

Springer's Sawtail Catshark *Galeus springeri* **p231**
Caribb. Sea; 457–699 m. Only *Galeus* with pre-dors. dk. longi. stripes outlined in white.

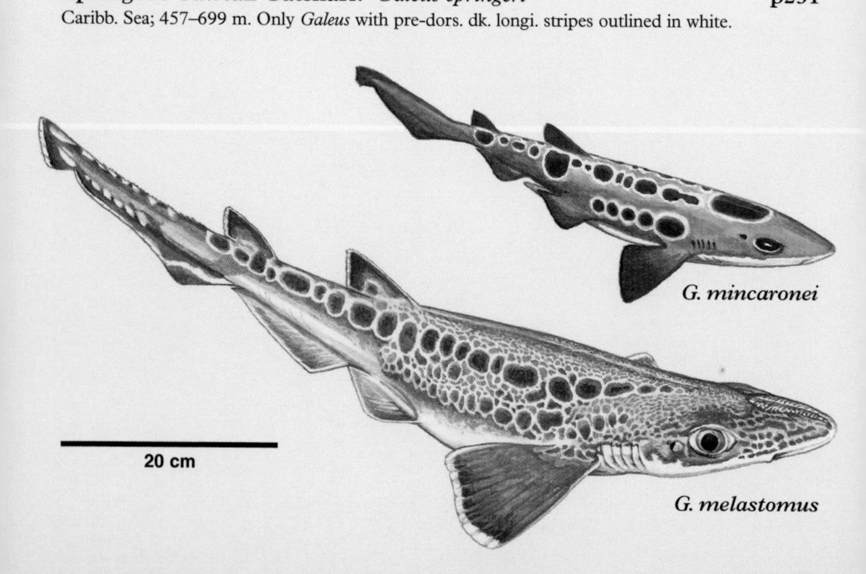

G. mincaronei

G. melastomus

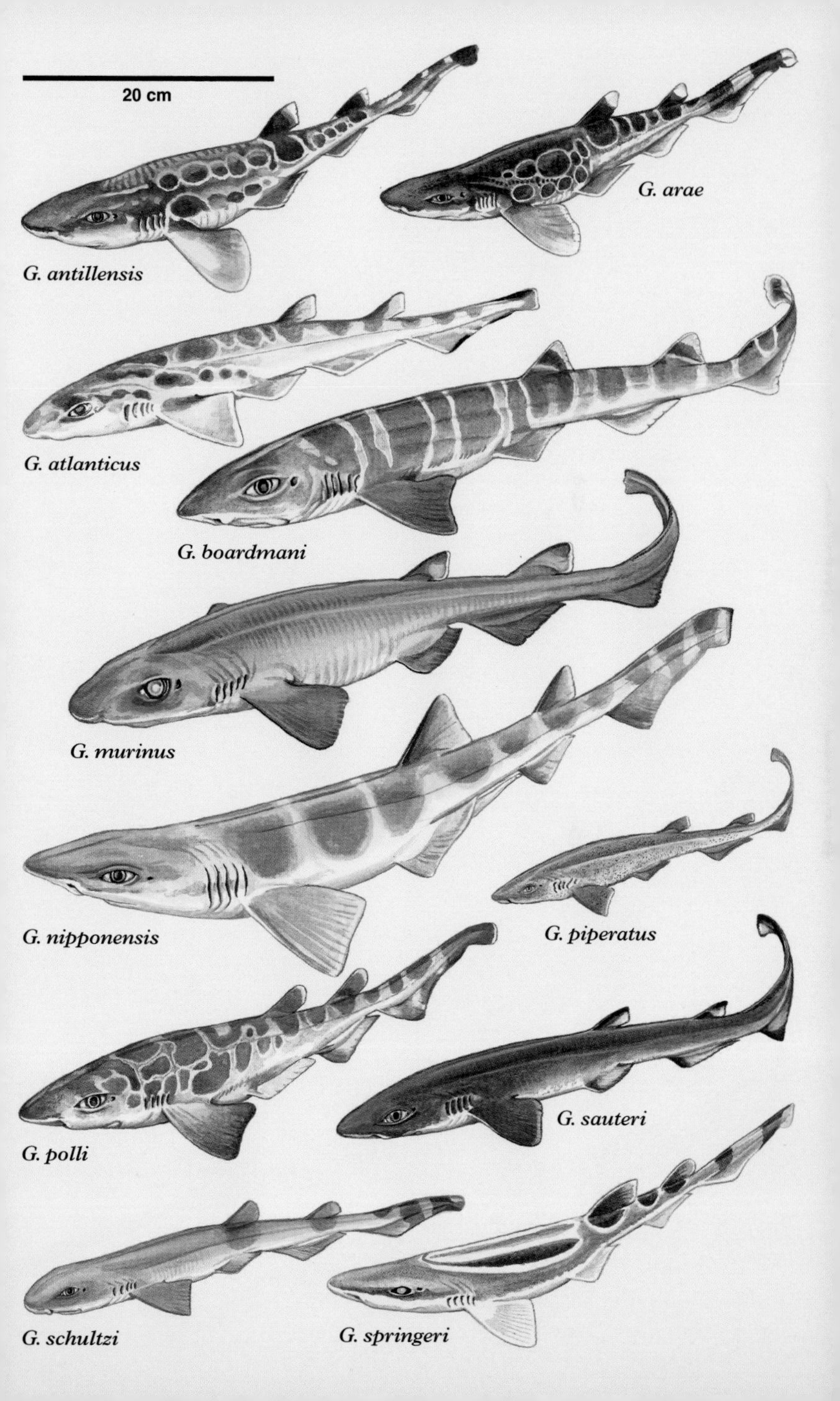

20 cm
G. antillensis
G. arae
G. atlanticus
G. boardmani
G. murinus
G. nipponensis
G. piperatus
G. polli
G. sauteri
G. schultzi
G. springeri

Plate 38 Scyliorhinidae VIII

Longfin Sawtail Catshark *Galeus cadenati* **p225**
SW Caribb.; 439–548 m. V. similar to *G. arae* and *G. antillensis* but with longer shallower A fin.

Gecko Catshark *Galeus eastmani* **p225**
Jap. and E China Sea; on/nr. bottom in deepwater. Obscure saddles, white-edged D and C fins.

Slender Sawtail Catshark *Galeus gracilis* **p226**
N Austr.; on/nr. bottom 290–470 m. 4 dusky saddles, 1 beneath each D fin and 2 on C fin.

Longnose Sawtail Catshark *Galeus longirostris* **p226**
Japan; on/nr. bottom 350–550 m. Grey above white below, young with obscure saddle blotches.

Northern Sawtail Catshark *Galeus sp.* B. **p231**
NE Austr.; bottom 310–420 m. 10–16 pre-dors. pale-edged saddles/bars, pale centred bars below D fins.

Campeche Catshark *Parmaturus campechiensis* **p239**
G of Mex.; on/nr. bottom 1097 m. D1 fin slightly < D2, D1 in front of V fin origin, D2 fin = A fin.

New Zealand Filetail *Parmaturus macmillani* **p239**
NZ and Madag.; 950–1003 m+. D1 fin ≈ D2, D1 origin opp./just behind V fin origin, D2 fin < A fin.

Blackgill Catshark *Parmaturus melanobranchius* **p240**
NW Pac.; mud bottom 540–835 m. D1 fin < D2, D1 origin well behind V fin origin, D2 fin ≈ A fin.

Salamander Catshark *Parmaturus pilosus* **p240**
NW Pac.; 358–895 m. D1 fin ≈ D2, D1 origin ~ opp. V fin origin, D2 fin << A fin.

Filetail Catshark *Parmaturus xaniurus* **p241**
NE Pac.; on/nr. bottom 91–1251 m. D1 fin = D2, D1 origin just behind V fin origin, D2 fin << A fin.

Shorttail Catshark *Parmaturus* sp. A. **p241**
NE Austr.; deepwater plateau 590 m. D1 fin < D2, D1 origin ~ V fin insert, D2 fin slightly < A fin.

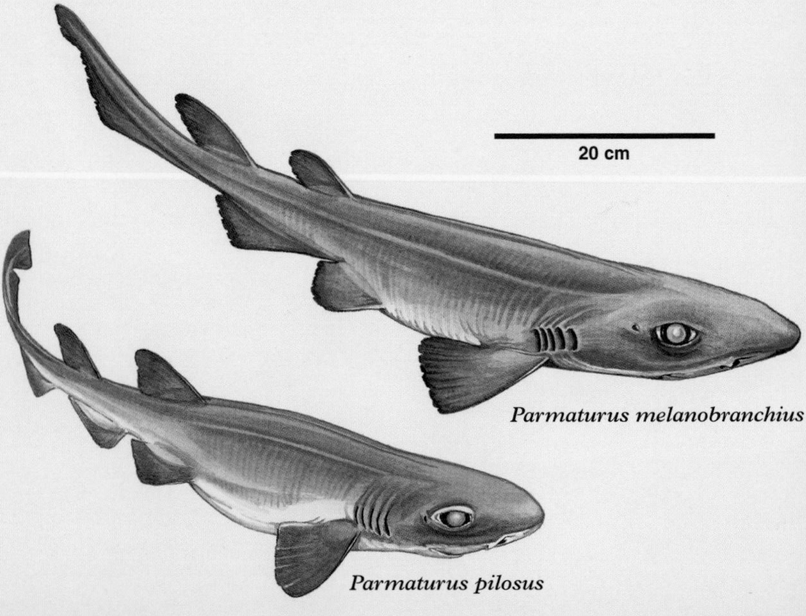

20 cm

Parmaturus melanobranchius

Parmaturus pilosus

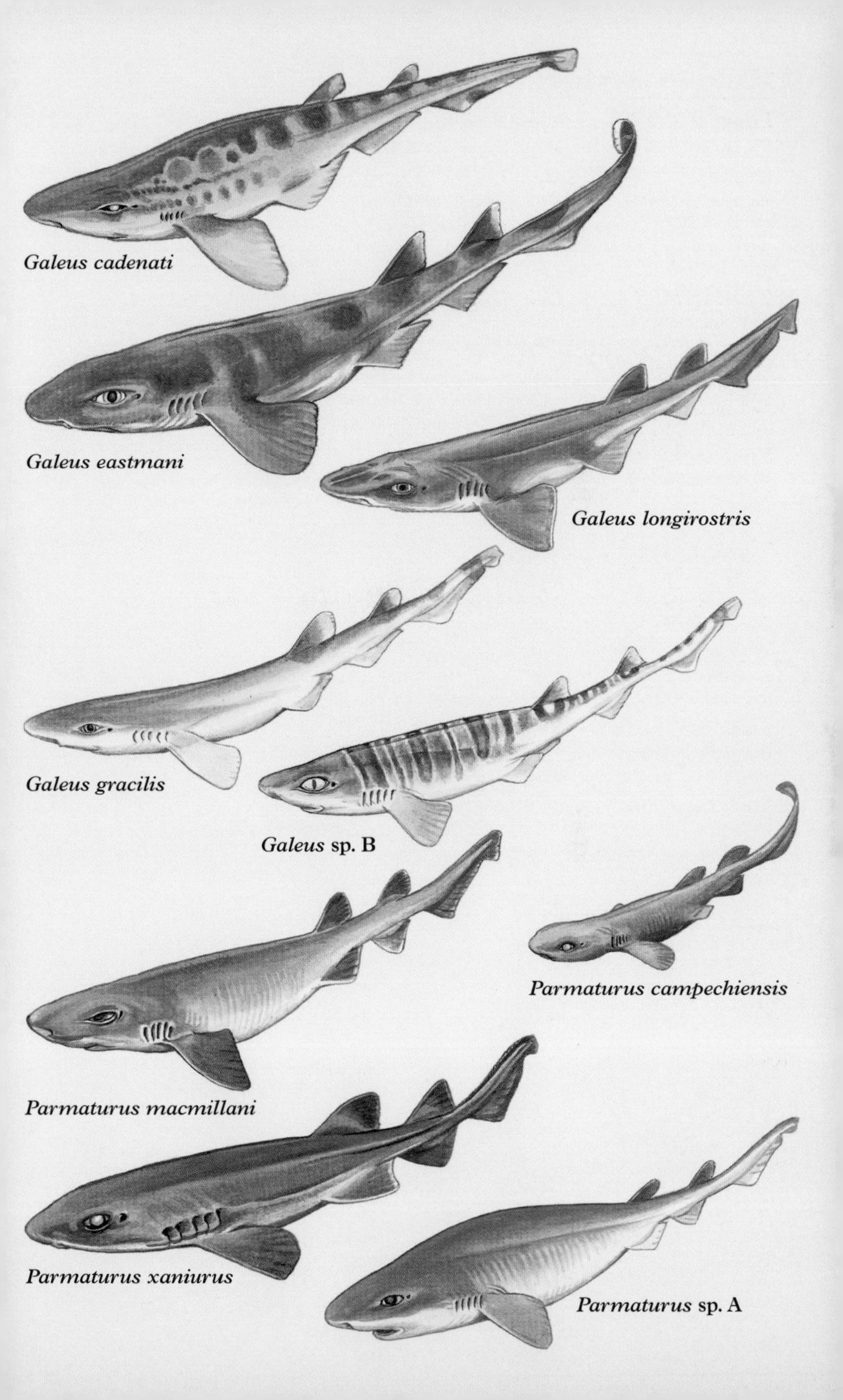

Galeus cadenati
Galeus eastmani
Galeus longirostris
Galeus gracilis
Galeus sp. B
Parmaturus campechiensis
Parmaturus macmillani
Parmaturus xaniurus
Parmaturus sp. A

Plate 39 Scyliorhinidae IX

Pyjama Shark or Striped Catshark *Poroderma africanum* **p242**
SE Atl. and W IO; surfline–282 m. Prominent short nasal barbels; D fins set far back, D1 >> D2; unmistakable colour pattern of striking longi. stripes, no spots.

Leopard Catshark *Poroderma pantherinum* **p243**
SE Atl.and W IO; surfline–256 m. Long nasal barbels reach mouth; D fins set far back, D1 >> D2; Striking colour pattern of leopard-like rosettes of dk. spots and lines around light centres, variations dense spots to v. large spots and partial longi. stripes.

Speckled Catshark *Halaelurus boesemani* **p232**
IWP; contin. and insular shelves 37–91 m. Pointed snout, eyes raised above head, gills on upper surface of head; 8 dk. saddles sep. by narrow bars, dk. blotches on D and C fins, numerous small spots.

Blackspotted or Darkspot Catshark *Halaelurus buergeri* **p232**
NW Pac.; contin. shelf 80–100 m. Pointed snout, eyes raised above head, gills on upper surface of head; variegated pattern of dusky bands outlined by large black spots on lt. background.

Lined Catshark *Halaelurus lineatus* **p233**
W IO; contin. shelf surfline–290 m. Narrow head with upturned knob on snout, eyes raised above head, gills on upper surface of head; ~13 pairs of narrow dk. brown stripes outlining dusky saddles, numerous spots and squiggles.

Tiger Catshark *Halaelurus natalensis* **p233**
SE Atl. and W IO; on/nr. bottom inshore–114 m. Broad head with pointed upturned snout tip, eyes raised above head, gills on upper surface of head; 10 pairs broad dk. brown bars around lighter reddish saddles, no spots.

Quagga Catshark *Halaelurus quagga* **p234**
IO; offshore on contin. shelf on/nr. bottom 54–186 m. Pointed snout, eyes raised above head, gills on upper surface of head; > 20 dk. brown narrow stripes, pairs of bars form saddles under D fins, no spots.

Lollipop Catshark *Cephalurus cephalus* **p222**
Baja Calif. and G of M; on/nr. bottom 155–927 m. Unmistakeable tadpole shaped; expanded flattened head and gill region; small slender body and tail; D fins small, D1 slightly ahead V fin origins.

Onefin Catshark *Pentanchus profundicolus* **p242**
Philipp.; insular slope on bottom 673–1070 m. Broad round snout with short mouth, short gills with incised gill septa; only species with 5 gills and 1 D fin, v. long shallow A fin; uniform brownish.

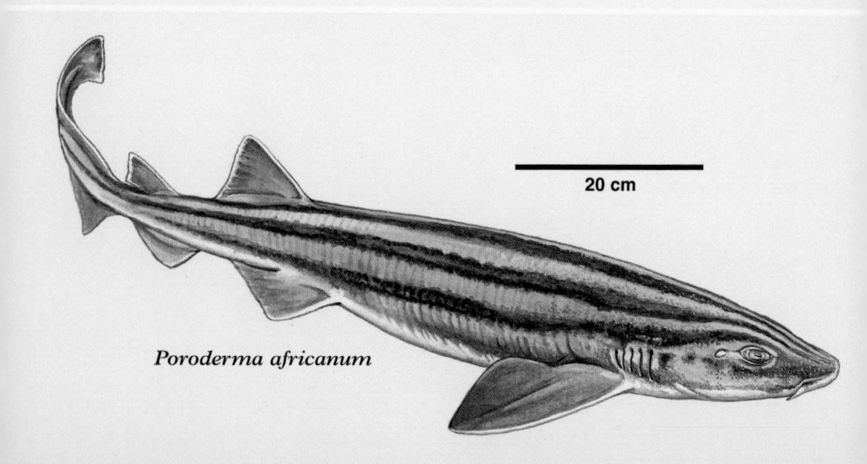

Poroderma africanum

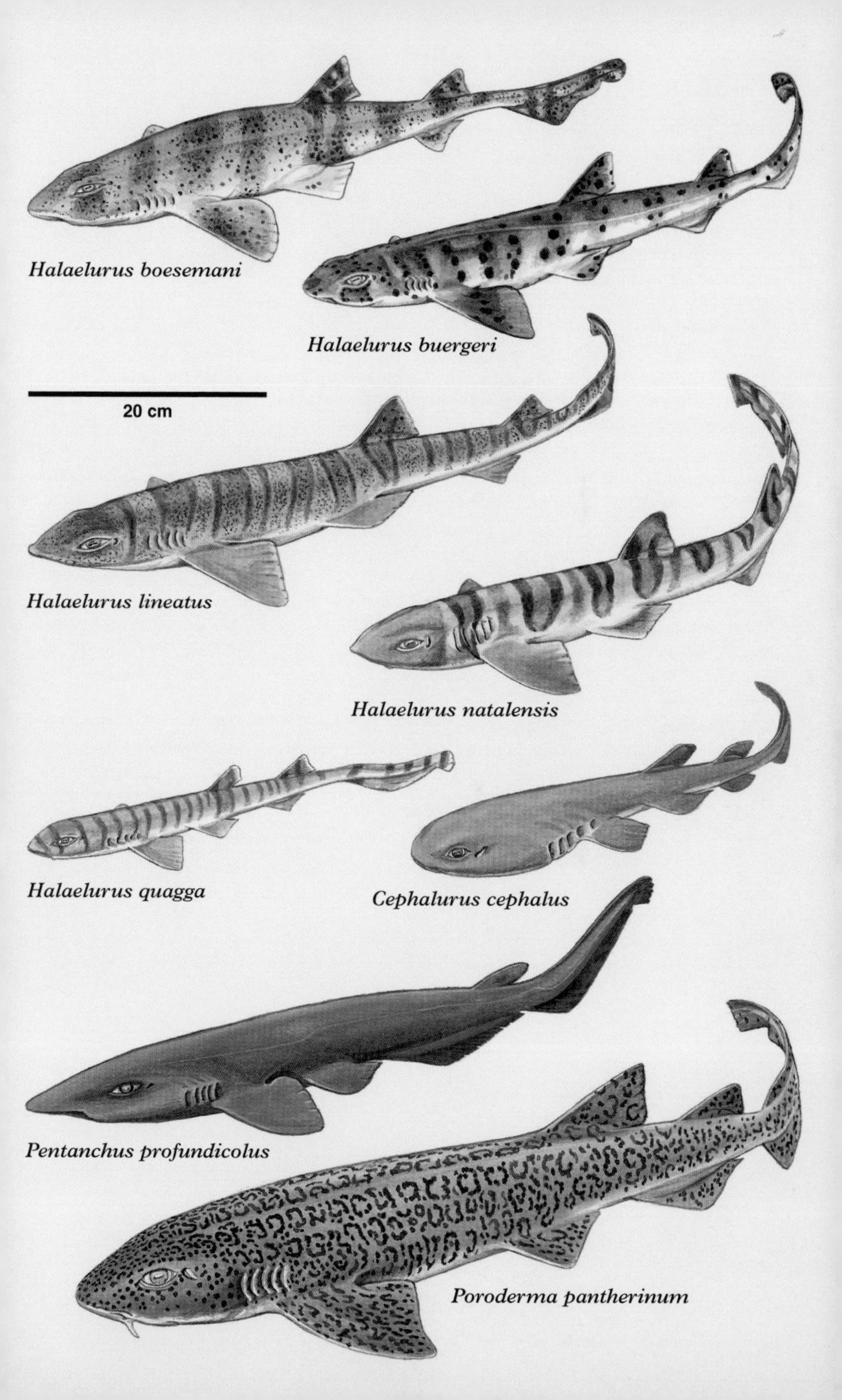
Halaelurus boesemani
Halaelurus buergeri
20 cm
Halaelurus lineatus
Halaelurus natalensis
Halaelurus quagga
Cephalurus cephalus
Pentanchus profundicolus
Poroderma pantherinum

Plate 40 Scyliorhinidae X

Puffadder Shyshark or Happy Eddie *Haploblepharus edwardsii* **p234**
S Afr.; contin. shelf on/nr. sand and rock bottom 0–288 m. Slender; stocky broad head, v. large nostrils with greatly expanded nasal flaps reach mouth, gills on upper body; prom. golden brown or reddish saddles with darker brown margins, numerous white spots, white below.

Brown Shyshark or Plain Happy *Haploblepharus fuscus* **p235**
S Afr.; inshore contin. shelf on/nr. sand and rock bottom 0–35 m. Stocky; stocky broad head, v. large nostrils with greatly expanded nasal flaps reach mouth, gills on upper body; sometimes obscure saddles or small white or black spots, white below.

Dark Shyshark or Pretty Happy *Haploblepharus pictus* **p235**
Namibia to S Afr.; contin. shelf kelp, sand and rock reefs close inshore–35m. Stocky; broad head, v. large nostrils with greatly expanded nasal flaps, gills on upper body; dors. saddles without obvious darker edges sparsely dotted with large white spots, white below.

Eastern or Natal Shyshark or Happy Chappie *Haploblepharus* sp. A. **p236**
S Afr.; contin. shelf close inshore. Slender; stocky broad head, v. large nostrils with greatly expanded nasal flaps, gills on upper body; body and tail with H-shaped saddles with conspic. dk. margins, numerous small white spots, body and fins with dk. mottling, white below.

Tropical or Crying Izak Catshark *Holohalaelurus melanostigma* **p236**
W IO; contin. slope 607–658 m. Short snout with long mouth; D fins short and angular; slender tail; numerous large dk. brown spots, some fused to form reticulations, blotches and stripes, horizontal tear-marks on snout, dk. lines and 'C'-shaped mark on bases and webs of D fins.

Whitespotted or African Spotted Catshark *Holohalaelurus punctatus* **p237**
W IO; contin. shelf and upper slope 220–420 m. Very broad head, short snout, long mouth; D fins short and angular; slender tail; upper surface densely covered with small dk. brown spots, few white spots scattered on back and D fin inserts., faint sometimes saddles, no reticulations, blotches or tear-marks, 'C' or'V'-shaped dk. marks on D fin webs, whitish below with tiny black dots underneath head.

Izak Catshark *Holohalaelurus regani* **p237**
Typical subspecies Namibia to S Afr., NE subspecies S Afr. and Mozambique; typical 40–910 m, NE 200–740 m. Short snout with long mouth; D fins short and angular; slender tail; upper surface covered with dk. brown reticulations bars and blotches, NE subspecies more checkerboard-like or spotted, no white spots or tear-marks, white below, young dark and slender with line of white spots on sides with black bars on fins and tail.

East African Spotted or Grinning Izak *Holohalaelurus* sp. A. **p238**
W IO; upper contin. slope 238–300 m. Very broad head with short snout and long v. wide mouth; D fins short and angular with highlighted narrow dk. bar on fin webs; slender tail; upper surfaces densely covered with small dk. brown spots, no dk. reticulations bars, blotches, stripes or tear-marks, few consp. large white spots above P fin inserts., whitish below with tiny black dots underneath head.

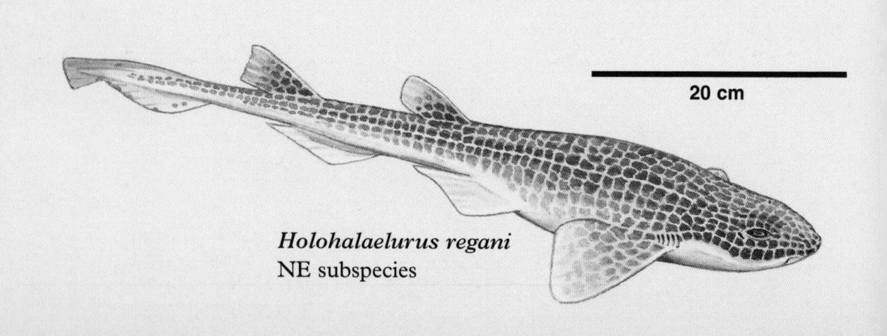

Holohalaelurus regani
NE subspecies

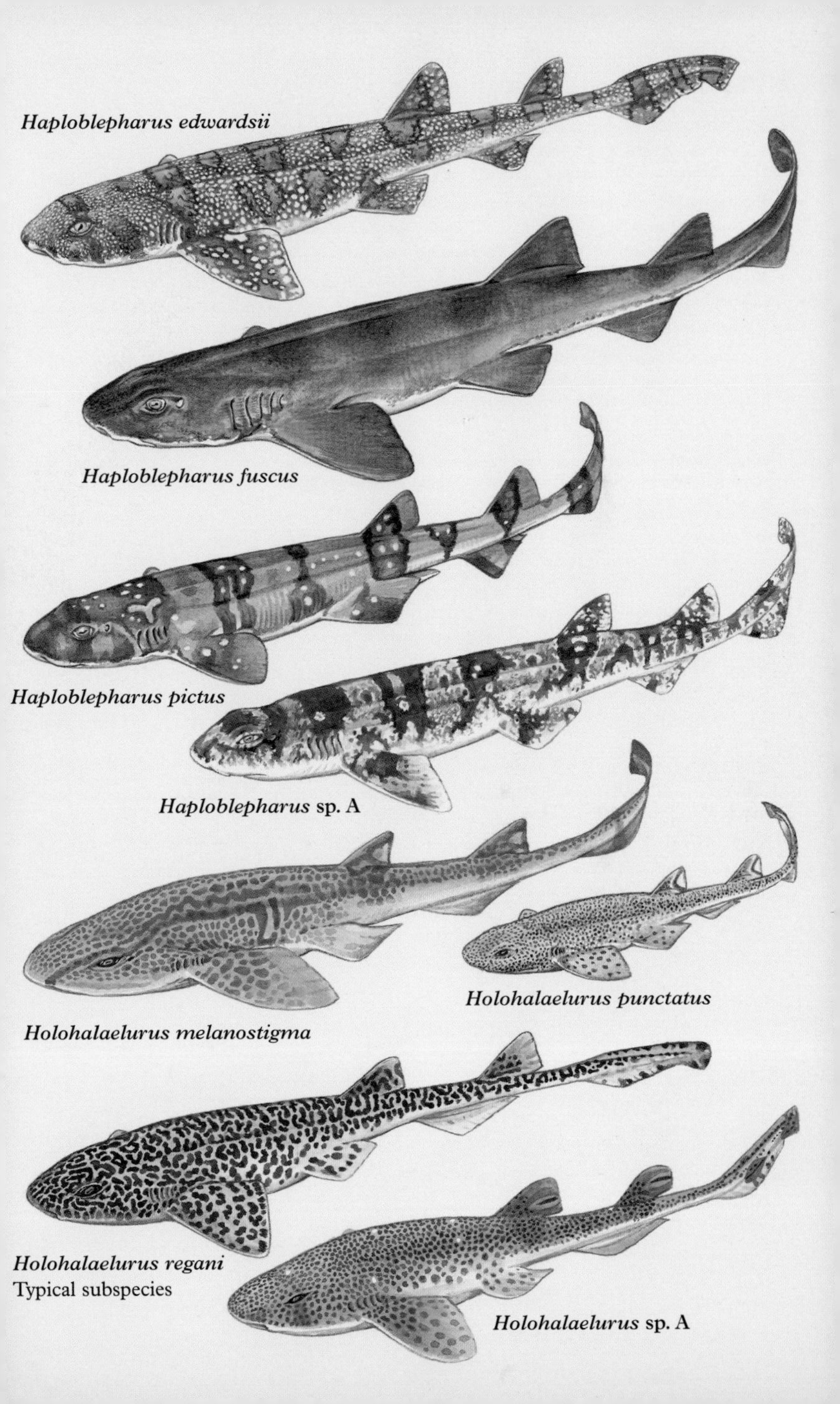
Haploblepharus edwardsii
Haploblepharus fuscus
Haploblepharus pictus
Haploblepharus sp. A
Holohalaelurus melanostigma
Holohalaelurus punctatus
Holohalaelurus regani
Typical subspecies
Holohalaelurus sp. A

Plate 41 Scyliorhinidae XI

Narrowmouth Catshark *Schroederichthys bivius* **p244**
S Chile to S Brazil; 14–359 m. Fairly slim; short narrow rounded snout, narrow, lobate ant. nasal flaps; short tail; 7–8 dk. brown saddles 2 consp. interdors., scattered large dk. and small white spots.

Redspotted Catshark *Schroederichthys chilensis* **p244**
Peru to Chile; 1–50 m. Mod. slim; short broad rounded snout, ant. nasal flaps broad and triangular; short tail; consp. dk. saddles 2 interdors., numerous dk. spots and few/ absent white spots.

Narrowtail Catshark *Schroederichthys maculatus* **p245**
Carrib. contin. shelf; 190–410 m. V. slender; rounded snout, long broad, triangular ant. nasal flaps; long tail; juv. 6–9 inconsp. brown saddles 3 interdors. absent in ad., juv. and ad. numerous small white spots.

Lizard Catshark *Schroederichthys saurisqualus* **p245**
S Brazil; deep reef, 122–435 m. Slender trunk and tail; rounded snout, mod. wide mouth, ant. nasal flaps long narrow and lobate; 10 consp. dusky saddles 4 interdors., scattered dk. and small white spots.

Slender Catshark *Schroederichthys tenuis* **p246**
Surinam to N Brazil; on/nr. bottom 72–450 m. V. slender; broad snout, narrow, lobate narrow mouth, ant. nasal flaps; 7–8 consp.dk. saddles 4 interdors., outlined by many small dk. spots, no white.

Smallspotted Catshark *Scyliorhinus canicula* **p247**
NE Atl. and Med.; nearshore–110 m. Slender; greatly expanded ant. nasal flaps reach mouth, lower lab. furr. only; D1 fin >D2; 8–9 dusky saddles maybe inconsp., numerous small dk. spots, occ. a few white.

Yellowspotted Catshark *Scyliorhinus capensis* **p248**
SE Atl. and W IO; bottom 26–695 m. Ant. nasal flaps small, no nasoral grooves, lower lab. furr. only; 8–9 irr. dk. grey saddles, numerous small bright yellow spots, no dk. spots.

West African Catshark *Scyliorhinus cervigoni* **p248**
Trop. W Afr.; rocky/mud bottom 45–500 m. V. stout; ant. nasal flaps just reach mouth, no nasoral grooves, lower lab. furr. only; 8–9 dk. saddles centred on midline dk. spots, few rel. large dk. spots, no white.

Comoro Catshark *Scyliorhinus comoroensis* **p249**
Comoro Is.; bottom 200–400 m. V. large ant. nasal flaps reach mouth, no nasoral grooves, lower lab. furr. only; bold defined dk. grey-brown saddles centred on midline dk. spots, numerous small white spots.

Nursehound *Scyliorhinus stellaris* **p252**
NE Atl. and Med.; rocky/seaweed bottom 1–125 m. Stocky; ant. nasal flaps do not reach mouth and widely sep., no nasoral grooves, lower lab. furr. only; saddles faint/absent, numerous large to small dk. spots.

Scyliorhinus capensis

Scyliorhinus stellaris

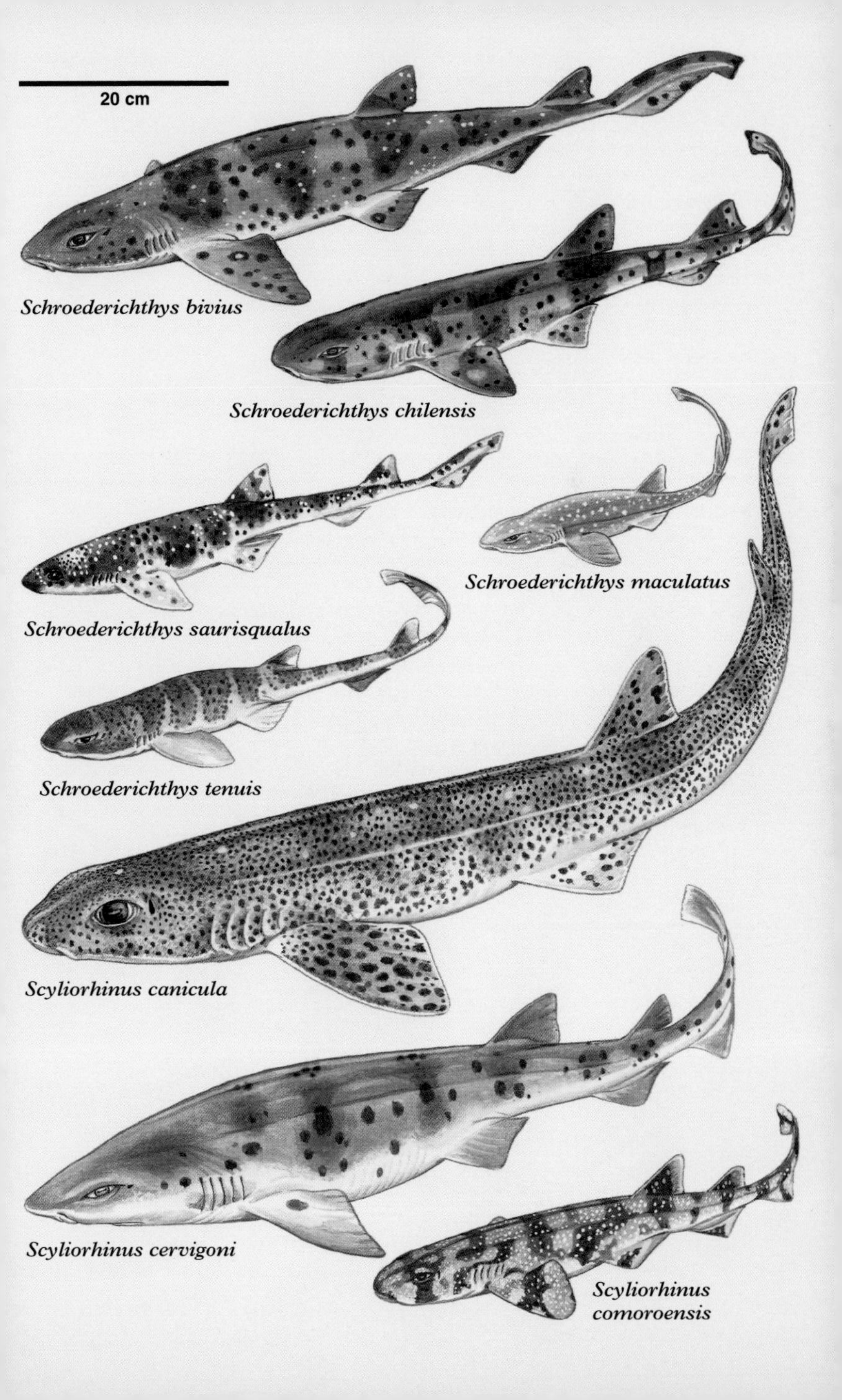

20 cm
Schroederichthys bivius
Schroederichthys chilensis
Schroederichthys maculatus
Schroederichthys saurisqualus
Schroederichthys tenuis
Scyliorhinus canicula
Scyliorhinus cervigoni
Scyliorhinus comoroensis

Plate 42 Scyliorhinidae XII

Polkadot Catshark *Scyliorhinus besnardi* **p246**
N Urug. to Brazil; bottom 140–190 m. Slender; ant. nasal flaps do not reach mouth, no nasoral grooves lower lab. furr. only; D1 fin >> D2; no prom. saddles, sparse almost round black spots, no white.

Boa Catshark *Scyliorhinus boa* **p247**
S Caribb. to Surinam; on/nr. bottom 229–676 m. Slender; ant. nasal flaps do not reach mouth, lower lab. furr. only; D1 fin >> D2; inconsp. saddles and blotches, outlined by black spots and lines, occ. a few white.

Brownspotted Catshark *Scyliorhinus garmani* **p249**
Poss. Philipp.; unknown depth. Stocky; ant. nasal flaps do not reach mouth, no nasoral grooves, lower lab. furr. only; D1 fin >> D2; 7 indistinct saddles, scattered large round brown spots, no white.

Freckled Catshark *Scyliorhinus haeckelli* **p250**
Venez. to Urug.; on/nr. bottom 37–439 m. Slender; ant. nasal flaps do not reach mouth, lower lab. furr. only; D1 fin >> D2; 7–8 dusky saddles maybe inconsp., dk. bar under eye, small black spots, no light spots.

Whitesaddled Catshark *Scyliorhinus hesperius* **p250**
Caribb. Hond. to Colomb.; on/nr. bottom 274–457 m. Slender; ant. nasal flaps nearly reach mouth, lower lab. furr. only; D1 fin >> D2; 7–8 well defined dk. saddles covered by large white spots.

Blotched Catshark *Scyliorhinus meadi* **p251**
N Carolina to Yucatan; on/nr. bottom 329–548 m. Stocky; broad head, ant. nasal flaps nearly reach mouth, lower lab. furr. only; D1 fin >> D2; 7–8 darker saddles maybe obscure, no spots.

Chain Catshark *Scyliorhinus retifer* **p251**
Mass. to Fl., G of M and Caribb.; on/nr. bottom 73–754 m. Ant. nasal flaps do not reach mouth, lower lab. furr. well developed; D1 fins set well back; black chain patterning outlines faint dusky saddles, no spots.

Izu Catshark *Scyliorhinus tokubee* **p252**
Japan; offshore 100 m. Narrow ant. nasal flaps nearly reach mouth, lower lab. furr. only; D1 fin >> D2; sharply defined red-brown saddles and large lat. blotches, v. numerous small yellow spots, no dk.

Cloudy Catshark *Scyliorhinus torazame* **p253**
Jap. to poss. Philipp.; inshore–320 m+. Slender; narrow head, ant. nasal flaps do not reach mouth, lower lab. furr. only; D1 fin >> D2; 6–9 darker saddles, many irr. large dk. and lt. spots in larger spec.

Dwarf Catshark *Scyliorhinus torrei* **p254**
Bahamas and Virgin Is.; on/nr. bottom 229–550 m. Slender; ant. nasal flaps do not reach mouth, lower lab. furr. only; D1 fin >> D2; 7–8 darker brown saddles, ad. obscure, scattered large reg. white spots, no black.

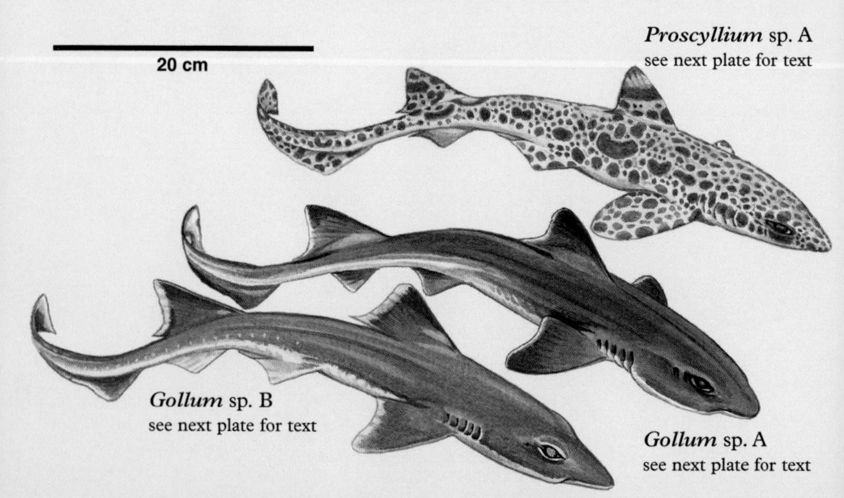

Proscyllium sp. A
see next plate for text

Gollum sp. B
see next plate for text

Gollum sp. A
see next plate for text

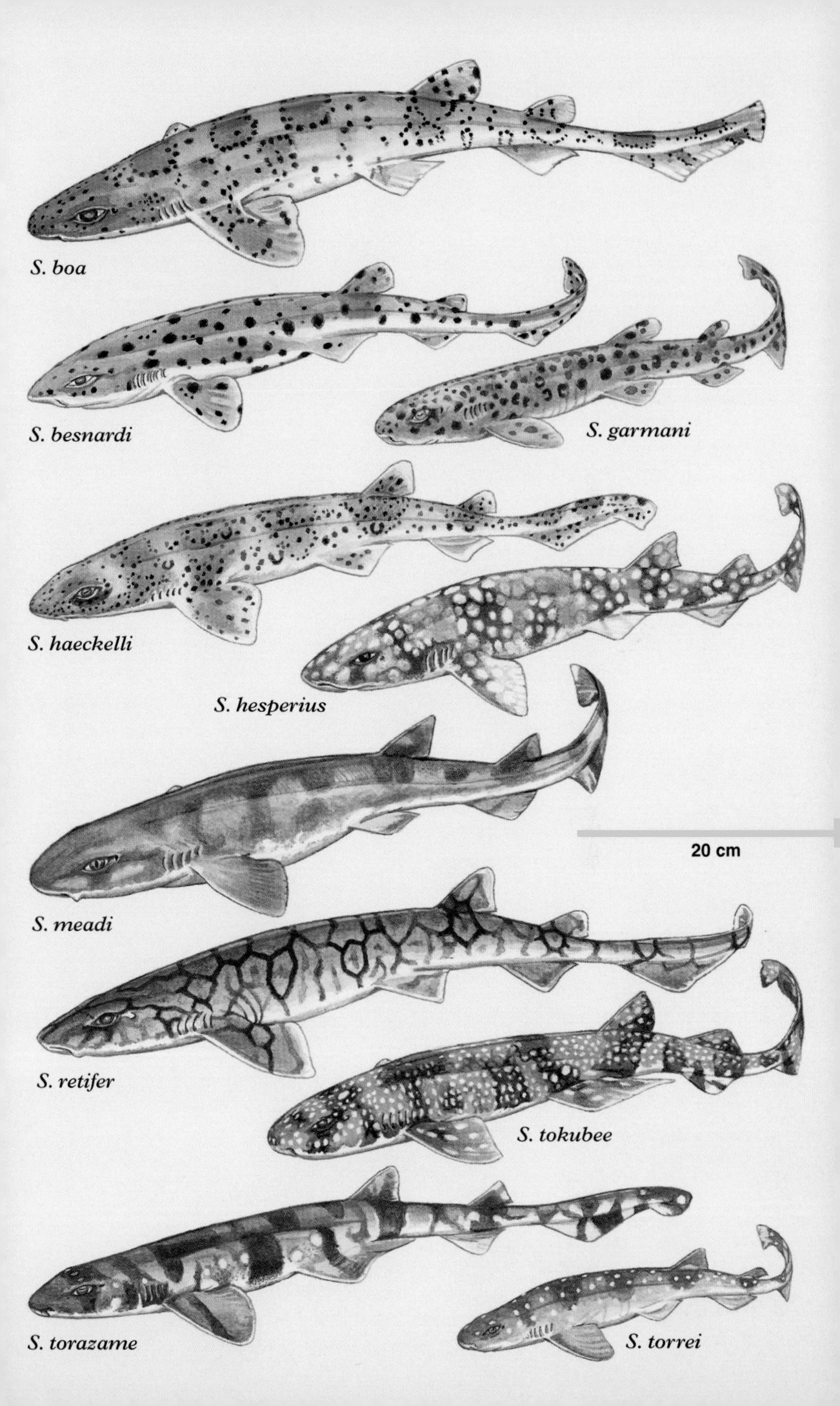
S. boa
S. besnardi
S. garmani
S. haeckelli
S. hesperius
20 cm
S. meadi
S. retifer
S. tokubee
S. torazame
S. torrei

Plate 43 Proscylliidae, Pseudotriakidae & Leptochariidae

Harlequin Catshark *Ctenacis fehlmanni* **p254**
Somalia; outer contin. shelf. Stoutish; nictating eyelids, ant. nasal flaps do not reach large triang. mouth, v. short lab. furr.; large red-brown saddle blotches, smaller round spots and vert. bars.

Cuban Ribbontail Catshark *Eridacnis barbouri* **p255**
Flor. Straits to N Cuba; bottom 430–613 m. Slender; nictating eyelids, ant. nasal flaps do not reach large triang. mouth, v. short lab. furr.; A fin 2/3 D fin ht.; faint dk. bands on ribbon-like C fin.

Pygmy Ribbontail Catshark *Eridacnis radcliffei* **p255**
Patchy IWP; mud bottom 71–766 m. Slender; nictating eyelids, ant. nasal flaps do not reach broad triang. mouth, no lab. furr.; A fin <1/2 D fin ht.; prom. dk. bands on ribbon-like C fin, black on D fins.

African Ribbontail Catshark *Eridacnis sinuans* **p256**
Mozam. to Tanz.; 180–480 m. Slender; nictating eyelids, ant. nasal flaps do not reach triang. mouth, short lab. furr.; A fin 1/2 D fin ht.; faint dk. bands on ribbon-like C fin, lt. margins on D fins.

Graceful Catshark *Proscyllium habereri* **p256**
W Pac.; contin. and insular shelves 50–100 m. Slender; large eyes, nictating eyelids, ant. nasal flaps nearly reach triang. mouth; indistinct dusky saddles, small to large dk. brown spots, occ. small white.

Clown or Magnificent Catshark *Proscyllium* sp. A **p257**
Andaman Sea; nr. edge outer contin. shelf >200 m. Similar to *P. habereri* but more variegated pattern of large and small spots, incl. 2 spots and bar forming 'clown face' pattern below D fins. See previous plate.

Slender Smoothhound or Gollumshark *Gollum attenuatus* **p258**
NZ; 129–724 m. Slender; v. long angular snout, nictating eyelids, ant. nasal flaps short, triang. mouth, short lab. furr., small spiracles; rel. short subtriang. D1 fin, D2 slightly > D1; grey above lt. below.

Sulu Gollumshark *Gollum* sp. A **p258**
Philipp.; prob. deepwater. Shorter snout than other *Gollum*; dk. grey-brown above lighter below, D P and V fins with paler post. margins, term. C lobe light with dk. to black edge. See previous plate.

Whitemarked Gollumshark *Gollum* sp. B **p259**
New Caledonia; prob. deepwater. Snout as ≥ *G. attenuatus*; grey-brown above pale brown below, charac. conspic. row of white spots on tail and bold white marks on head and fins. See previous plate.

False Catshark *Pseudotriakis microdon* **p259**
Patchy WW except S. Atl and E. Pac.; bottom 200–1890 m. Stocky soft body; short snout bell-shaped, nictating eyelids, ant. nasal flaps short, huge angular mouth, short lab. furr.; dk. brown to blackish.

Pygmy False Catshark (new genus and species) **p260**
Arab. Sea and Maldives; to 1120 m. Stocky soft body; short bell-shaped snout, nictating eyelids, ant. nasal flaps short, angular mouth, short lab. furr.; rel. short subtriang. D1 fin, D2 > D1; dk. grey-brown.

Barbeled Houndshark *Leptocharias smithii* **p260**
Angola to Maurit. poss Med.; nr. bottom 10–75 m. Slender; horizontal oval eyes, internal nictating eyelids, nostrils with slender barbels, long arched mouth, v. long lab. furr.; lt. grey-brown, lt. below.

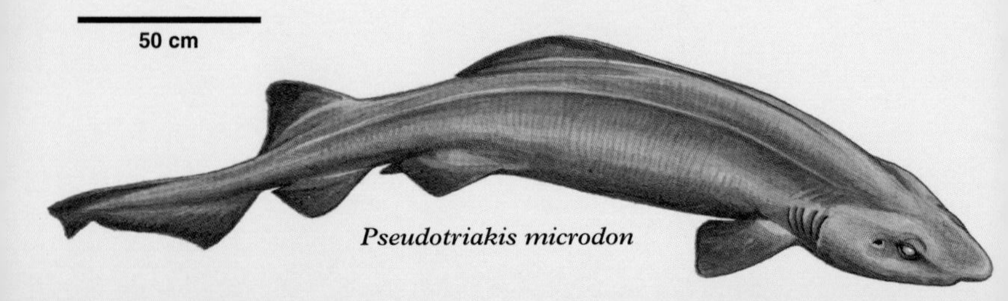

Pseudotriakis microdon

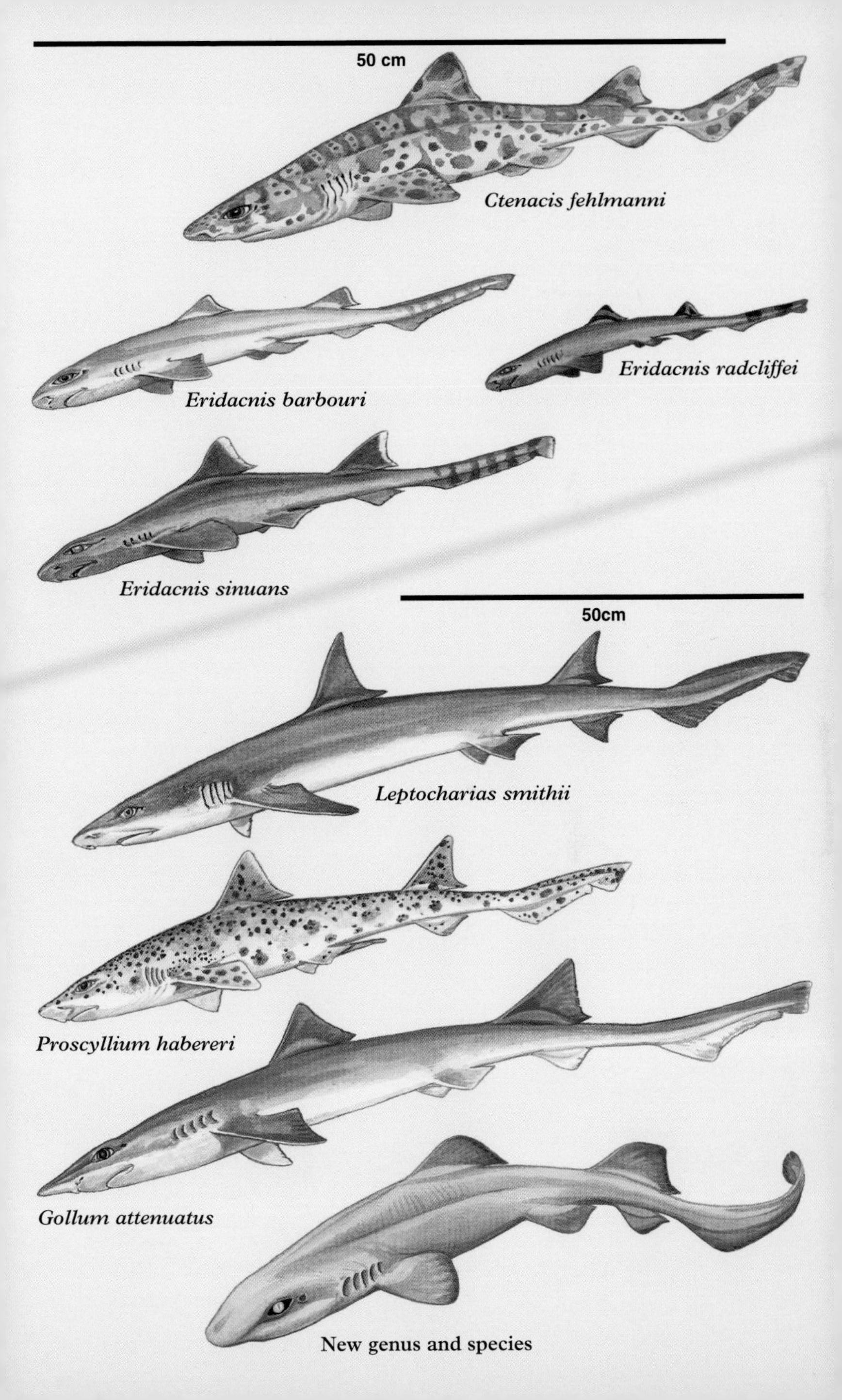
50 cm
Ctenacis fehlmanni
Eridacnis barbouri
Eridacnis radcliffei
Eridacnis sinuans
50cm
Leptocharias smithii
Proscyllium habereri
Gollum attenuatus
New genus and species

Plate 44 Triakidae I

Whiskery Shark *Furgaleus macki* **p261**
Austr.; shallow temp. contin. shelf on/nr. bottom. Stocky almost humpbacked; conspic. subocular ridges, dorsolateral eyes, only houndshark with ant. nasal flaps forming barbels, v. short arched mouth; variegated dk. blotches/saddles that fade with age.

Tope, Soupfin or School Shark *Galeorhinus galeus* **p262**
WW temp. waters; often on/nr. bottom 2–471 m. Slender; snout long and conical, no obvious subocular ridges, small ant. nasal flaps, large arched mouth; D1 fin >> D2, D2 = A fin, v. long terminal C lobe; grey above white below, black fin markings in young, sometimes few dusky spots.

Sailback Houndshark *Gogolia filewoodi* **p262**
N New Guinea; prob. nr. bottom 73 m. Preoral snout long; D1 fin v. large and triang. (length ≈ C fin); grey-brown above lighter below.

Darksnout or Deepwater Sicklefin Houndshark *Hemitriakis abdita* **p263**
Austr., poss. New Caled.; deepwater 225–400 m. Slender; snout long and parabolic, small ant. nasal flaps, arched mouth; falcate D, P and A fins, D1 fin = D2; grey-brown with dk. stripe under snout, distinct white D, P and A fin tips, juvs. prom. dk. bars and saddles.

Sicklefin Houndshark *Hemitriakis falcata* **p263**
NW Austr.; temp. to subtrop. 146–197 m. Slender; small ant. nasal flaps, arched mouth; strongly falcate D, P and A fins, D1 fin = D2, D1 fin origin over/behind P fin rear tip; grey-brown with dk. no stripe under snout, distinct white D fin tips, juvs. saddles and large spots.

Japanese Topeshark *Hemitriakis japanica* **p264**
China to Japan; temp. to subtrop. inshore–>100 m. Snout mod. long and parabolic, slit-like eyes with subocular ridge, short ant. nasal flaps, broad arched mouth; D1 fin ≈ D2, D1 fin origin over/behind P fin rear tip, D1 fin >> A fin; conspic. white-edged fins.

Whitefin Topeshark *Hemitriakis leucoperiptera* **p264**
Philipp.; coastal–48 m. Snout mod. long and parabolic, mod. long eyes with subocular ridge, small ant. nasal flaps, broadly arched mouth; strongly falcate fins, D1 fin > D2, D1 fin origin over/behind P fin rear tip, D1 fin >> A fin; no dk. stripe under snout, conspic. white-edged fins.

Ocelate Topeshark *Hemitriakis* sp. A **p265**
Philipp.; prob. inshore. Near term young distinctly patterned with dk. O-shaped spots., bars and saddles. Adult size and coloration unknown.

Blacktip Topeshark or Pencil Shark *Hypogaleus hyugaensis* **p265**
W IO, NW and SW Pac.; trop. and wm. temp. nr. bottom 40–230 m. Fairly slender; snout long and broadly pointed, large oval eyes with subocular ridges, small ant. nasal flaps, arched mouth; D1 fin > D2, D2 > A fin, rel. short terminal C lobe; bronzy to grey-brown above, lighter below, dusky D and upper C fin tips, esp. in young.

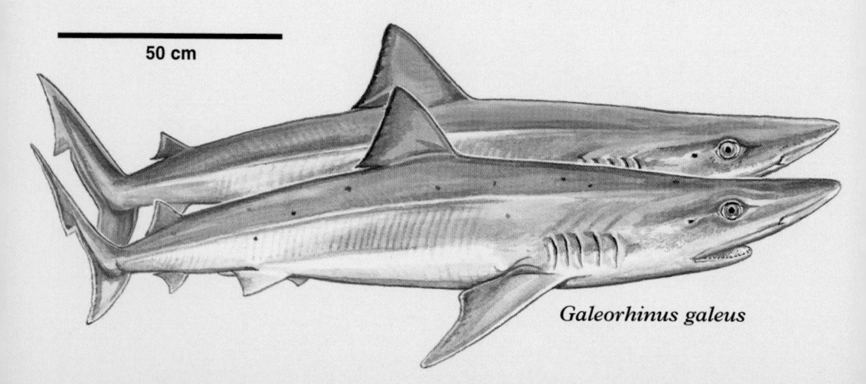

Galeorhinus galeus

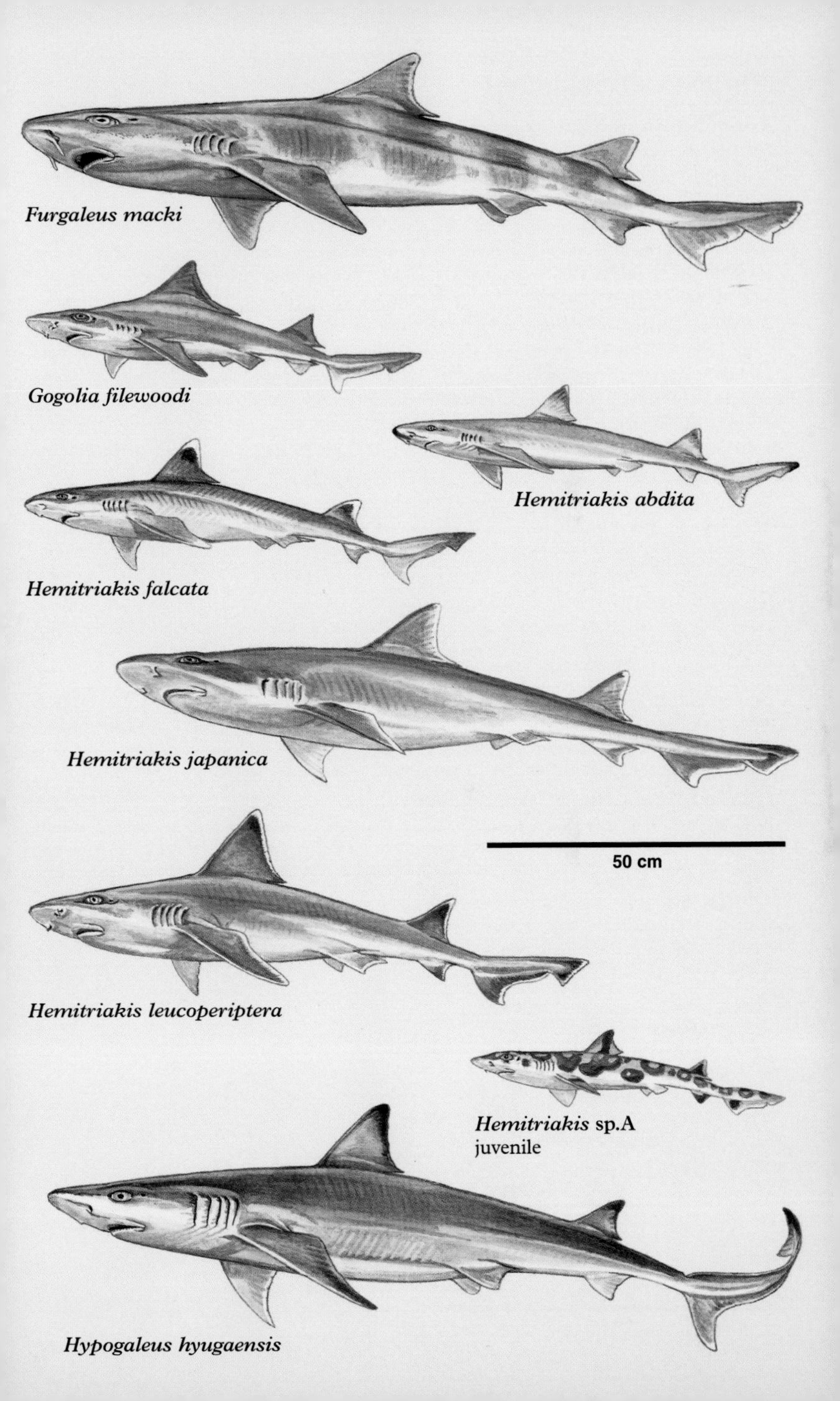
Furgaleus macki
Gogolia filewoodi
Hemitriakis abdita
Hemitriakis falcata
Hemitriakis japanica
50 cm
Hemitriakis leucoperiptera
Hemitriakis sp.A
juvenile
Hypogaleus hyugaensis

Plate 45 Triakidae II

Longnose Houndshark *Iago garricki* **p266**

SW Pac.; deepwater trop. 250–475 m. Slender; D1 fin v. slightly > D2, D1 fin origin over P fin inner margins; grey-brown, conspic. black D fin tips and upper post. margins, lower pale, pale tips and post. margins to P and C fins.

Bigeye Houndshark *Iago omanensis* **p266**

N IO, Red and Arabian Sea, on/nr. bottom <110–1000 m+. Slender; large eyes, large gills; D fins small, D1 fin origin over P fin bases, vent. C fin lobe small; grey-brown, lighter below, sometimes with darker D fin post. margins.

Lowfin Houndshark *Iago* sp. A **p267**

G of Aden to SW. India; outer contin. shelf and upper slope also semipelagic 0–330 m. V slender; short head, large eyes, large gills; D fins small and low, D1 fin origin over P fin bases, P fins small, vent. C fin lobe small; dk. brown, lighter below.

Flapnose Houndshark *Scylliogaleus quecketti* **p280**

S Afr.; surfline and close offshore. Short blunt snout, fused ant. nasal flaps large, covers mouth; D1 fin = D2, D2 fin >> A fin; grey above, cream below, newborns white post. margins to D, A and C fins.

Sharpfin Houndshark *Triakis acutipinna* **p280**

Ecuador; trop. contin. waters. Snout broadly rounded, widely separated ant. nasal flaps do not reach mouth, long upper lab. furr. reach lwr. symphysis; narrow fins, P fins falcate, D1 fin abrupt vert. post. margin; no spots or bands.

Spotted Houndshark *Triakis maculata* **p281**

Peru to N Chile and Galap.; inshore temp. contin. shelf. Stout; snout broadly rounded, widely separated lobate ant. nasal flaps do not reach mouth, long upper lab. furr. reach lwr. symph.; broad fins, P fins falcate with straight post margin, D1 fin backward sloping post. margin; many black spots.

Spotted Gully Shark *Triakis megalopterus* **p282**

S Angola to S Afr.; surfline to shallow inshore. Snout broad blunt, widely separated small lobate ant. nasal flaps do not reach mouth, upper lab. furr. does not reach lwr. symph.; broad large fins, P fins falcate with concave post. margin, D1 fin almost vert.; usually many small black spots.

Banded Houndshark *Triakis scyllium* **p282**

S Siberia to China; on/nr. bottom estuaries to close inshore. Short snout broadly rounded, widely separated lobate ant. nasal flaps do not reach mouth, long upper lab. furr. reach lwr. symph.; P fins almost triang., D1 almost vert.; sparsely scattered small black spots maybe absent, dusky saddles in young.

Leopard Shark *Triakis semifasciata* **p283**

Oregon to G of Calif.; on/nr. bottom 0–4 m records to 91 m. Snout broadly rounded, widely separated ant. nasal flaps do not reach mouth, upper lab. furr. reach lwr. symphysis; P fins falcate; unique colour pattern of saddles and spots.

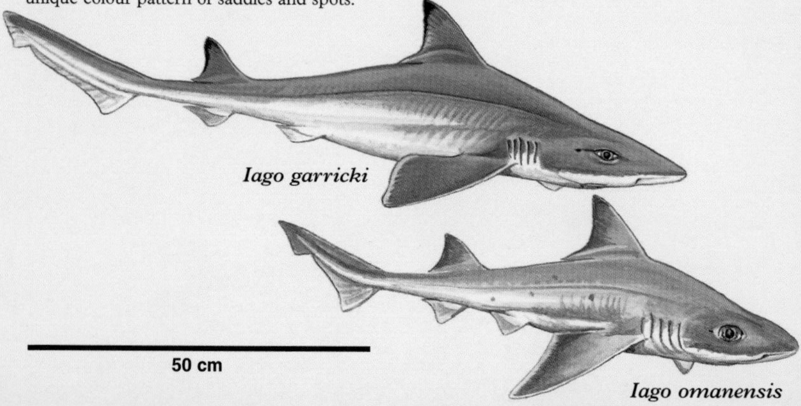

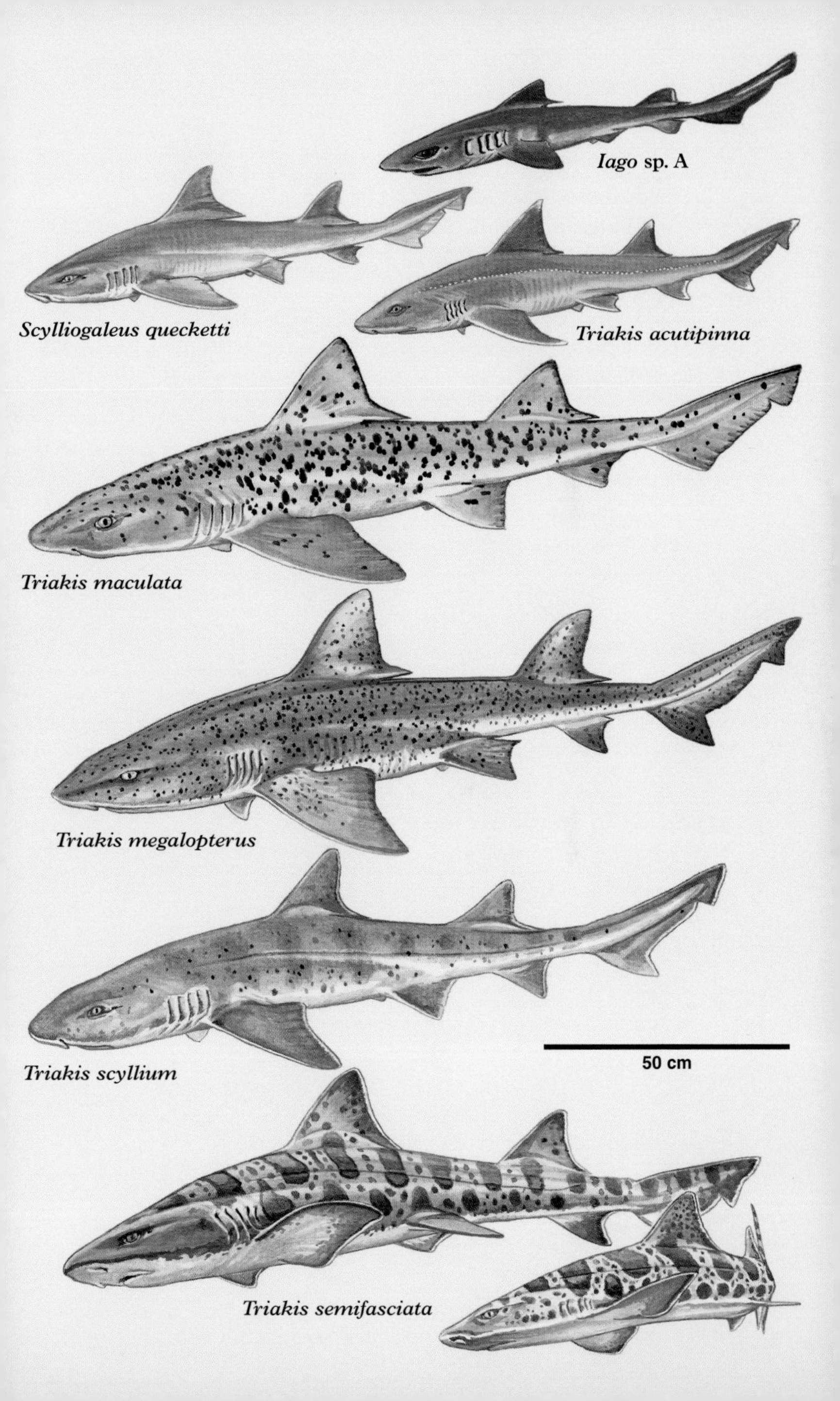

Iago sp. A
Scylliogaleus quecketti
Triakis acutipinna
Triakis maculata
Triakis megalopterus
Triakis scyllium
50 cm
Triakis semifasciata

Plate 46 Triakidae III

Starry Smoothhound *Mustelus asterias* **p268**
NE Atl. N Sea to Maurit., Can. Is. and Med.; on/nr. sand and gravel intertidal–100 m+. Nostrils closer together than sim. species in region; unfringed D fins, P and V fins fairly small; grey or grey-brown, only European smoothhound with numerous white spots.

Dusky Smoothhound *Mustelus canis* **p269**
NW Atl. Canada to Arg. in widely sep. discrete pop.; <18–200 m. Short head and snout, large close-set eyes, nostrils widely spaced, upper lab. furr. > lower; unfringed D fins, C fin deeply notched; grey, usually unspotted, newborn dusky tip D and C fins.
M. c. insularis has lower D fin and longer terminal C fin lobe, occurs 137–808 m.

Smoothhound *Mustelus mustelus* **p275**
Temp. E Atl.; intertidal–350 m. Short head and snout, large close-set eyes, nostrils widely spaced, upper lab. furr. slightly > lower; unfringed D fins, semifalcate C fin vent. lobe; grey to grey-brown, usually unspotted, occ. dk. spots.

Whitespot Smoothhound *Mustelus palumbes* **p276**
Namibia to S Mozamb.; on/nr. sand and gravel intertidal–360 m. Nostrils rel. widely spaced, upper lab. furr. > lower; unfringed D fins, rel. large P and V fins (> *M. asterias, M. manazo* and *M. mustelus*); grey to grey-brown, only S African smoothhound with white spots.

Blackspot Smoothhound *Mustelus punctulatus* **p277**
W Sahara and Med.; inshore contin. shelf on bottom. Short head and snout, narrow head, large eyes, nostrils close together, upper lab. furr. slightly > lower; prom. fringed D fins; grey above light below, usually black-spotted; often mistaken for *M. mustelus*.

Narrownose Smoothhound *Mustelus schmitti* **p277**
S Brazil to N Arg.; offshore contin. shelf 60–195 m. Short head, snout mod. long, nostrils close together, upper lab. furr. >> lower; D fins prom. fringed giving dk. frayed appearance; grey, white-spotted; easily distinguished from other white-spotted *Mustelus*.

Grey Gummy Shark *Mustelus* sp. A **p279**
N Austr.; trop. contin. shelf 100–300 m. Upper lab. furrows < than lower; deeply notched C fin; bronzy, unspotted.

Whitespotted Gummy Shark *Mustelus* sp. B **p279**
N Austr.; trop. contin. shelf 120–400 m. V. sim. to *M. antarcticus* but range does not overlap; grey-brown with numerous white spots, D fins sometimes have dk. tips, juvs. more strongly marked other fins have pale tips and post margins; slight var. between E. and W. forms.

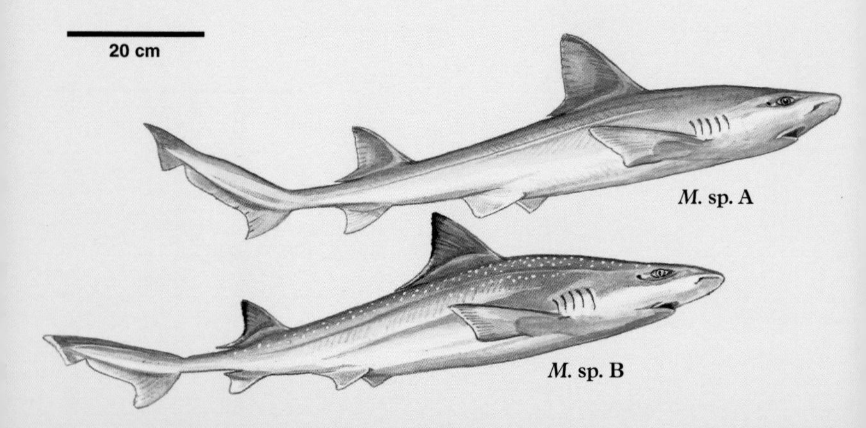

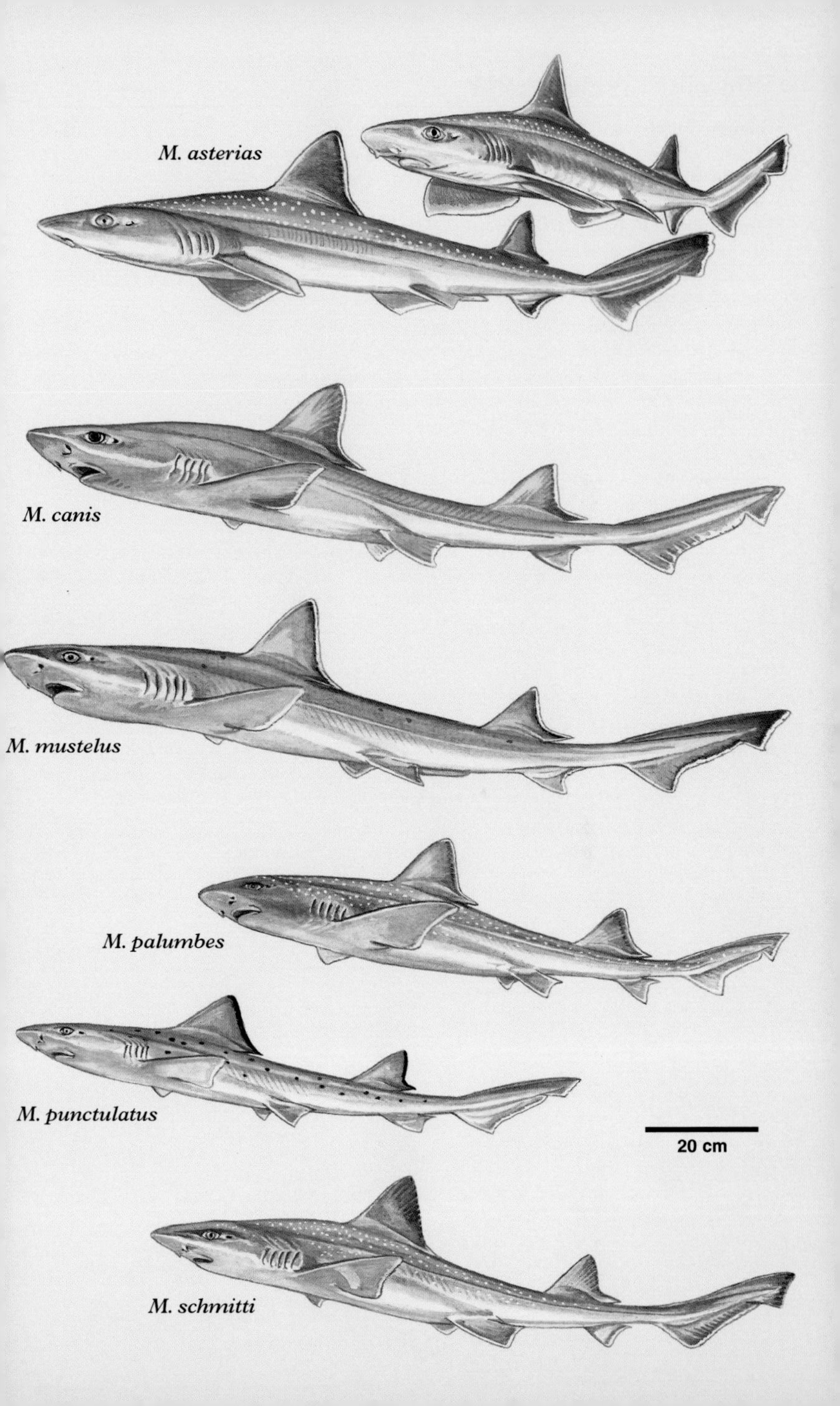

M. asterias
M. canis
M. mustelus
M. palumbes
M. punctulatus
20 cm
M. schmitti

Plate 47 Triakidae IV

Gummy Shark *Mustelus antarcticus* **p267**

Austr.; temp. on/nr. bottom shore–350 m. Only *Mustelus* in Austr. temp. waters; snout rel. long, large nostrils widely spaced, long upper lab. furr. slightly > lower; unfringed D fins, D1 fin slightly > D2, rel. small P and V fins; bronze to grey-brown, white spotted, sometimes black spots.

Grey Smoothhound *Mustelus californicus* **p268**

N Calif. to G of Calf.; bottom wm. temp. to trop. inshore and offshore. Short narrow head, fairly small eyes, nostrils widely spaced, short mouth, upper lab. furr. = lower; triangular D fins, vent. C fin lobe poorly developed; uniform grey, no spots, lt. below.

Sharpnose Smoothhound *Mustelus dorsalis* **p269**

Mex. to Ecuador; trop. inshore contin. shelf. Long acutely pointed snout, v. small wide-set eyes, large nostrils moderately spaced, upper lab. furr. > lower; broadly triangular D1 fin; uniform grey or grey-brown, no spots, lt. below.

Spotless Smoothhound *Mustelus griseus* **p271**

Japan to Vietnam; bottom inshore–51 m+. Short head and snout, small eyes, nostrils widely spaced, upper lab. furr. ≈ lower; unfringed D fins, vent. C fin lobe semifalcate; uniform grey or grey-brown, no spots, lt. below.

Brown Smoothhound *Mustelus henlei* **p271**

N Calif. to Mex., and Ecuador to Peru; contin. shelf intertidal–200m+. Short head, mod. long snout, large close-set eyes, nostrils widely spaced, upper lab. furr. > lower, caud. ped. long; dk. broadly fringed D fins; usually iridescent bronzy-brown occ. grey, no spots, white below.

Spotted Estuary Smoothhound or Rig *Mustelus lenticulatus* **p272**

NZ; cold temp. insular shelf close inshore–220 m. Short head, mod. long snout, fairly large wide-set eyes, nostrils widely spaced, long upper lab. furr. > lower; unfringed D fins, rel. large P and V fins; grey or grey-brown, white unspotted (only one in range), lt. below.

Starspotted Smoothhound *Mustelus manazo* **p273**

S Siberia to Vietnam, and Kenya; mud and sand inshore temp. to trop. contin shelf. Short head, mod. long snout, large close-set eyes, nostrils fairly close-set, upper lab. furr. >> lower; unfringed D fins, rel. small P and V fins; grey to grey-brown, white-spotted (only one in range), lt. below.

Speckled Smoothhound *Mustelus mento* **p274**

Galapagos to Peru and Argentina; inshore and offshore temp. contin. and insular shelves 16–50 m. Short blunt angular snout, fairly small mod. spaced eyes, nostrils widely spaced, upper lab. furr. slightly > lower, caud. ped. short; unfringed D fins; grey to grey-brown, white-spotted, lt. below.

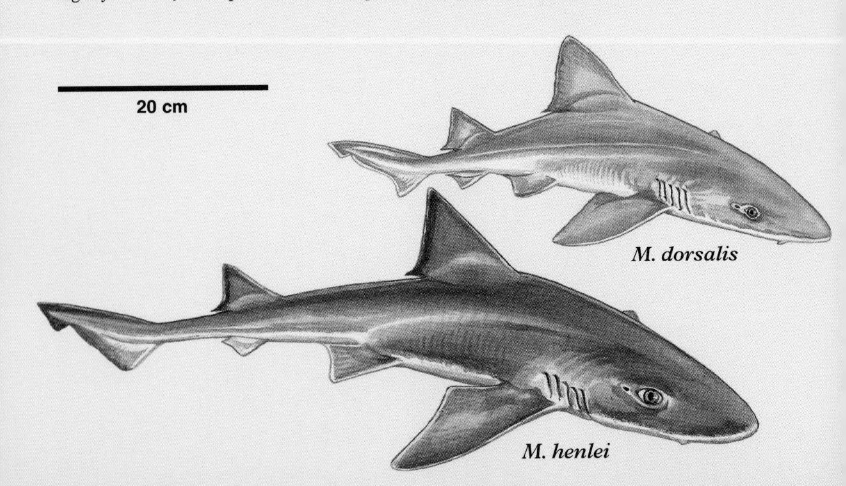

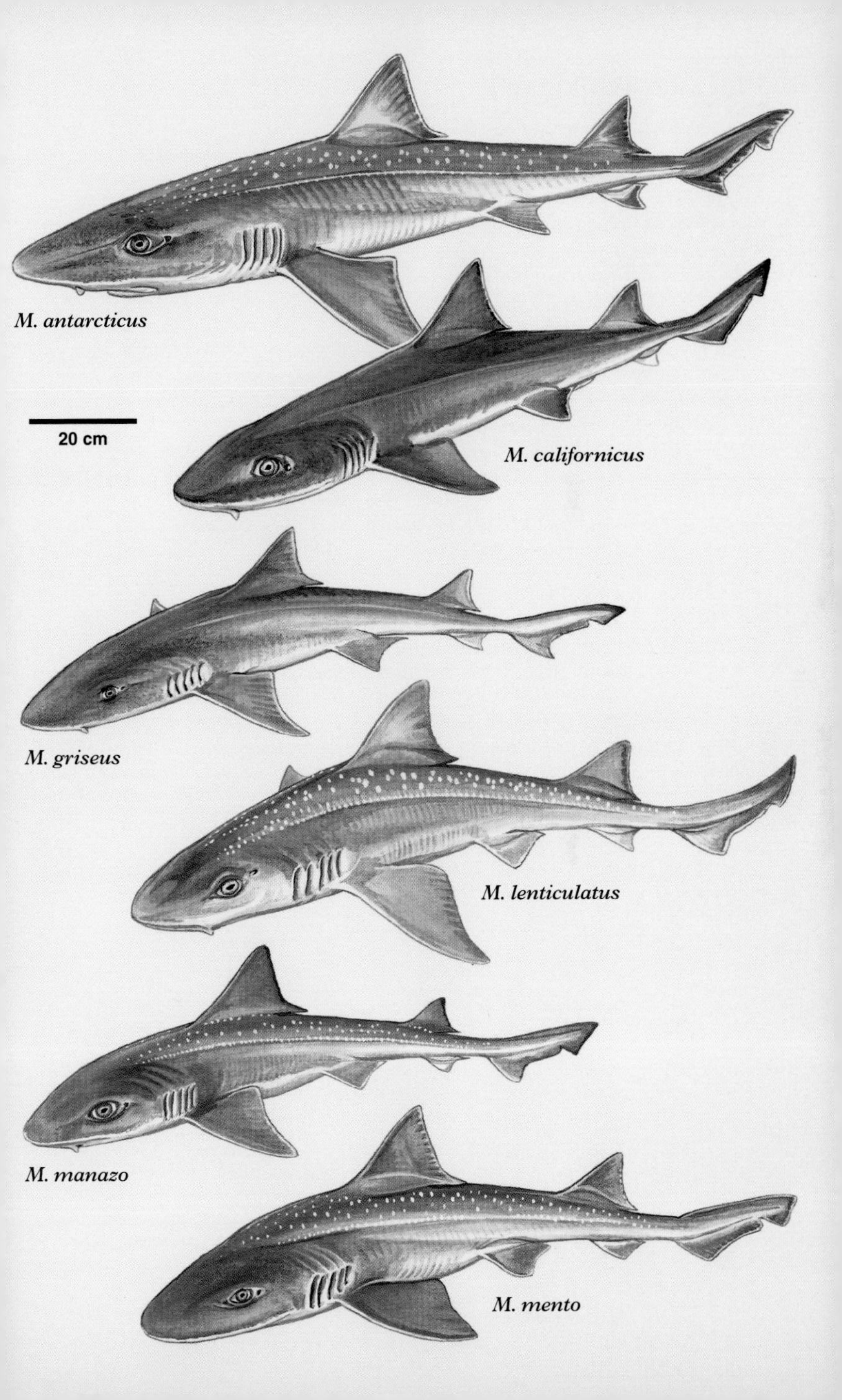

M. antarcticus
20 cm
M. californicus
M. griseus
M. lenticulatus
M. manazo
M. mento

Plate 48 Triakidae V

Striped Smoothhound *Mustelus fasciatus* **p270**
S Brazil to N Arg.; bottom temp. contin. shelf and uppermost slope poss. intertidal–500 m. Long head, long angular pointed snout, v. small eyes, nostrils widely spaced, upper lab. furr. >> lower, caud. ped. short; broadly triangular unfringed D fins; vert. dk. bars at least to juv., unspotted.

Smalleye Smoothhound *Mustelus higmani* **p272**
N G of M to Brazil; mud, sand and shell bottom close inshore to 1281 m. Long acutely pointed snout, v. small wide-set eyes, nostrils widely spaced, upper lab. furr. = lower; unfringed D fins, falcate D1 fin only slightly > D2; uniform grey or grey-brown, no spots, lt. below.

Sicklefin Smoothhound *Mustelus lunulatus* **p273**
S Calif. to Panama; wm. temp. to trop. contin. shelf close inshore to well offshore. Distinguished from sim. *M. californicus* by more pointed snout, more widely sep. eyes, shorter mouth, shorter upper lab. furr. and strongly falcate D fins; uniform grey to grey-brown, no spots, lt. below.

Venezuelan Dwarf Smoothhound *Mustelus minicanis* **p274**
Colomb. and Venez.; offshore outer contin. shelf 71–183 m. Short head and snout, v. large close-set eyes, nostrils widely spaced, upper lab. furr. slightly > lower; unfringed D fins, vent. C fin lobe poorly developed; uniform grey, no spots, lt. below, newborn D and C fin tips dusky.

Arabian, Hardnose or Moses Smoothhound *Mustelus mosis* **p275**
W IO, Red Sea and Gulf; bottom contin. shelf inshore and offshore, coral reefs. Short head and snout, large fairly close-set eyes, nostrils widely spaced, upper lab. furr. = lower; unfringed D fins, vent. C fin lobe semifalcate; uniform grey or grey-brown, no spots, lt. below, SA D1 fin tip white D2 and C fin black.

Narrowfin or Florida Smoothhound *Mustelus norrisi* **p276**
G of Mex.- USA, Columb., Venez. and S. Brazil; sand and mud bottom close inshore–84 m+. Short narrow head, rel. large eyes, nostrils close-set, rel. long mouth, upper lab. furr. < lower; strongly falcate fins; uniform grey, no spots, lt. below, newborn D and C fin tips dusky.

Gulf of Mexico Smoothhound *Mustelus sinusmexicanus* **p278**
G of Mex.- USA and Mex.; offshore contin. shelf and upper slope 36–229 m. Short head and snout, large close-set eyes, nostrils widely spaced, upper lab. furr. > lower; unfringed D fins, large D1 fin, vent. C fin lobe mod. expanded; uniform grey, no spots, lt. below, newborn D and C fin tips dusky.

Humpback Smoothhound *Mustelus whitneyi* **p279**
Peru to S Chile; contin. shelf 16–211 m, prefers island rocky bottom. Almost humpbacked; fairly long head and snout, fairly large eyes, nostrils widely spaced, upper lab. furr. >> lower; dk. fringed D fins, vent. C fin lobe v. slightly falcate; uniform grey, no spots, lt. below.

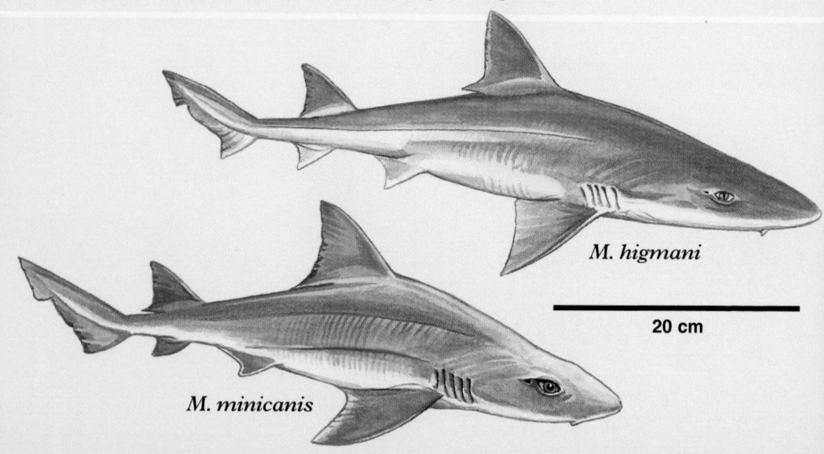

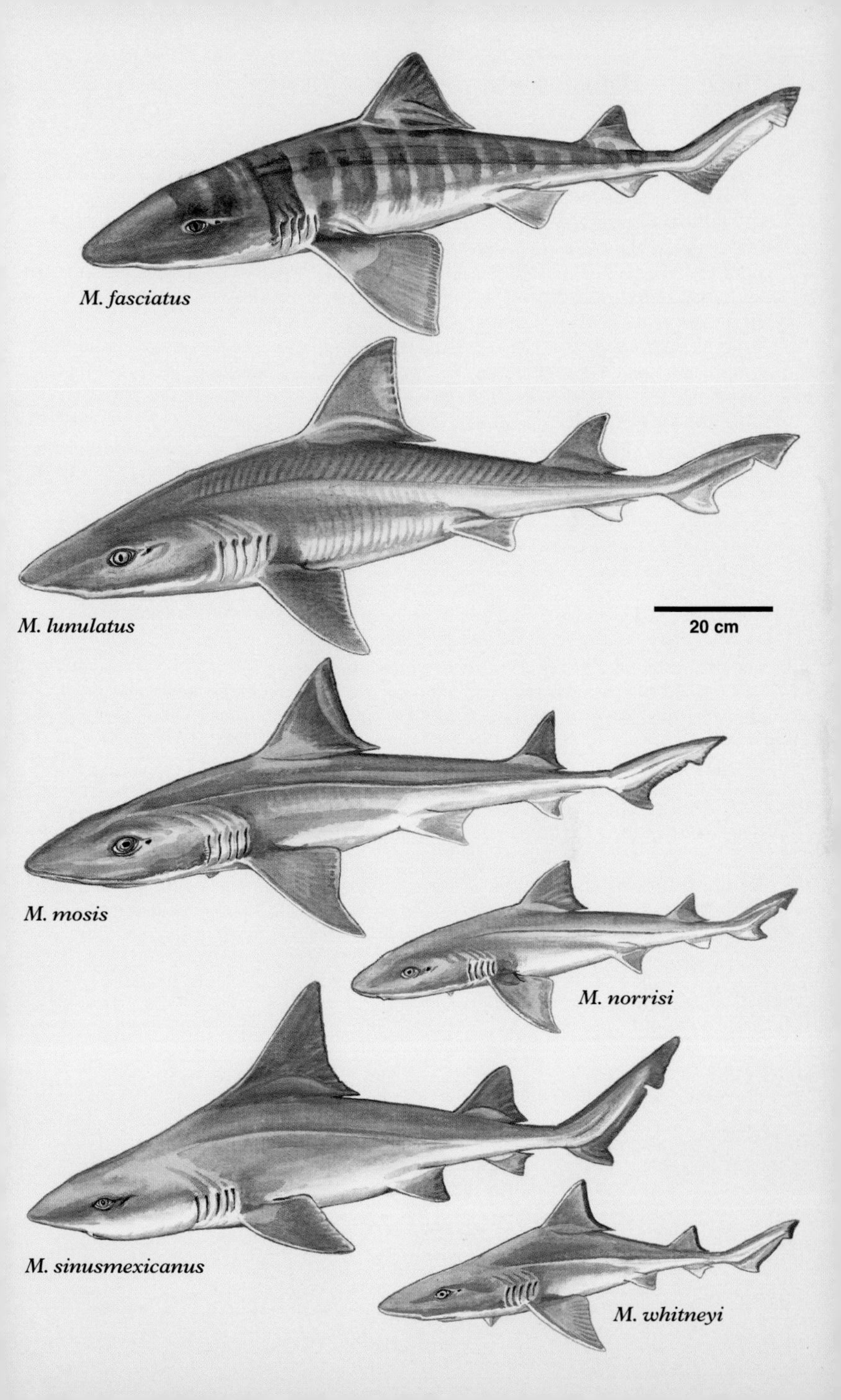
M. fasciatus
M. lunulatus
20 cm
M. mosis
M. norrisi
M. sinusmexicanus
M. whitneyi

Plate 49 Hemigaleidae

Hooktooth Shark *Chaenogaleus macrostoma* p284

The Gulf to Indonesia and China; contin. and insular shelves to 59 m. Fairly long angular snout, large lat. eyes with nictitating eyelid, small spiracles, v. long mouth, gills > twice eye length; D2 fin 2/3 D1 and opp./slightly ahead smaller A fin origin; lt. grey or bronzy, occ. black D2 tip and C fin post. tip.

Sicklefin Weasel Shark *Hemigaleus microstoma* p284

India to Philipp. and China; bottom contin. shelf. Fairly long rounded snout, eyes with nictitating eyelid, small spiracles, v. short arched mouth, gills short; D, P, V fins and vent. C fin lobe strongly falcate, D2 fin 3/4 D1 and opp. ≈ size A fin origin; lt. grey or bronzy, D fins with lt. tips and margins, occ. white spots on sides.

Australian Weasel Shark *Hemigaleus* sp. A p285

N Austr. and PNG; contin. shelf on/nr. bottom 12–167 m. V. similar to *H. microstoma* but has different coloration, unmarked D1 fin, dk. D2 tip and margin, dk. C fin post. tip, no white spots.

Snaggletooth Shark *Hemipristis elongatus* p285

S Afr. to N Austr., Philipp. and China; contin. and insular shelves to 132 m. Long broadly rounded snout, large lat. eyes with nictitating eyelid, small spiracles, teeth protrude from mouth, gills > 3 times eye length; strongly curved concave fins, D2 fin 2/3 D1 and ahead smaller A fin origin; lt. grey or bronzy, dusky D2 tip and C fin post. tip.

Whitetip Weasel Shark *Paragaleus leucolomatus* p286

E S Afr.; trop. coastal to 20 m. Long snout, large eyes with nictitating eyelid, small spiracles, fairly long mouth, gills ~ twice eye length; D, V and vent. C fin lobe not falcate, D2 fin ht. ≈ D1 and ahead smaller A fin origin; dk. grey, white below, prom. white tips and margins on most fins except black tipped D2, underside snout with broad dusky patches.

Atlantic Weasel Shark *Paragaleus pectoralis* p286

Cape Verdes, Maurit. to Angola; trop. shelf to 100 m. Mod. long snout, large oval eyes with nictitating eyelid, small spiracles, short small mouth, gills < 1.5 eye length; D2 fin 2/3 D1 and ahead much smaller A fin origin; striking long. yellow bands on lt. grey/bronze background, white below.

Slender Weasel Shark *Paragaleus randalli* p287

Arabian Sea; inshore contin. shelf shallow water. Snout with narrowly rounded tip, large lat. eyes with nictitating eyelid, small spiracles, long mouth, gills ≈ eye length; concave fins, D2 fin 2/3 D1 and slightly ahead smaller A fin origin; grey to grey-brown, lt. below, dk. fins with inconspic. lt. post. margins, underside snout with narrow black lines.

Straighttooth Weasel Shark *Paragaleus tengi* p287

G of Thailand to S China and Japan; inshore no depth reported. Mod. long rounded snout, large lat. eyes with nictitating eyelid, small spiracles, short arched mouth, gills ~1.2-1.3 eye length; D2 fin 2/3 D1 and slightly ahead smaller A fin origin; lt. grey, no prom. markings.

50 cm

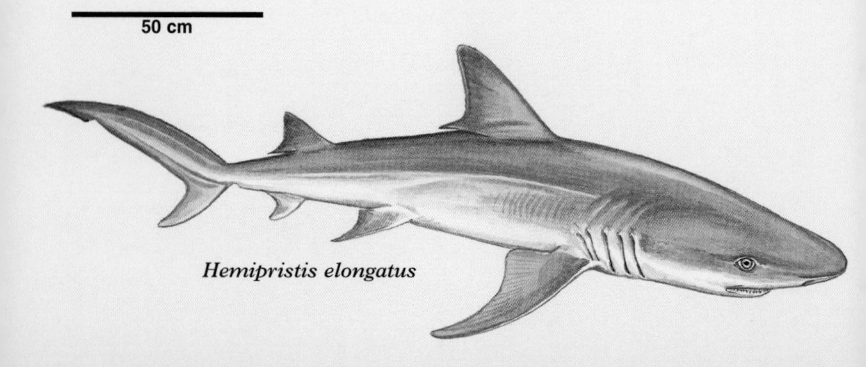

Hemipristis elongatus

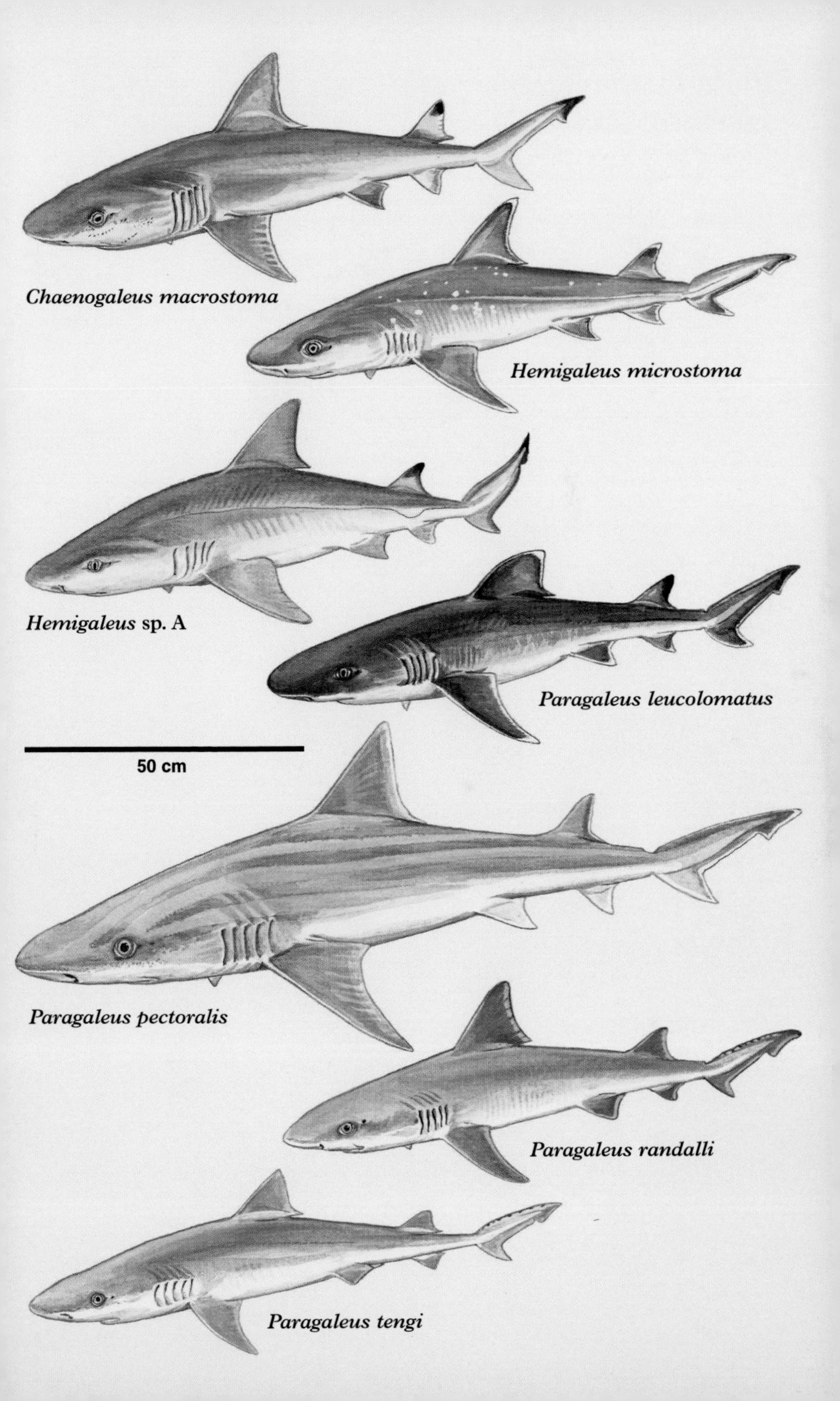
Chaenogaleus macrostoma
Hemigaleus microstoma
Hemigaleus sp. A
Paragaleus leucolomatus
50 cm
Paragaleus pectoralis
Paragaleus randalli
Paragaleus tengi

Plate 50 Carcharhinidae I

Tiger Shark *Galeocerdo cuvier* **p308**

WW temp and trop. seas; on/nr. contin. and insular shelves, surface and intertidal poss. to 140m. Huge; v. short broad bluntly rounded snout, eyes fairly large, big mouth with v. long lab. furr. upper >> lower, large spiracles; low caud. keels; interdorsal ridge v. prom, mod. broad semifalcate P fins, D1 fin > 2.5 times D2 ht., both D fins v. long rear tips; grey above with vert. black to dk. grey bars and blotches, bold in young fading in adults, white below.

Bull Shark *Carcharhinus leucas* **p299**

WW subtrop. and trop. seas; usually close inshore, hypersaline lagoons and river mouths, 1–30 m, can be found at 152 m+ and 100s km up rivers and lakes. Large thick-headed; v. short broad bluntly rounded snout, small eyes, upper lab. furrows v. short, no spiracles; weak caud. keels; no interdorsal ridge, large angular P fins, broad triang. D1 fin < 3.2 times D2 ht., both D fins short rear tips; grey to grey-brown, fin tips dusky but not conspic. marked except in young, white below.

Oceanic Whitetip Shark *Carcharhinus longimanus* **p300**

WW warm temp. and trop. seas; usually far offshore in open sea 18–28°C, surface–184 m+. Large; short bluntly rounded snout, small eyes, upper lab. furrows v. short, no spiracles; weak caud. keels; interdorsal ridge low, long paddle-like P fins, huge rounded D1 fin >> D2; grey or brown, D1 fin, P fins and sometimes C fin tips conspic. mottled white, juvs. black tips on some fins and black patches on caud. ped., white below.

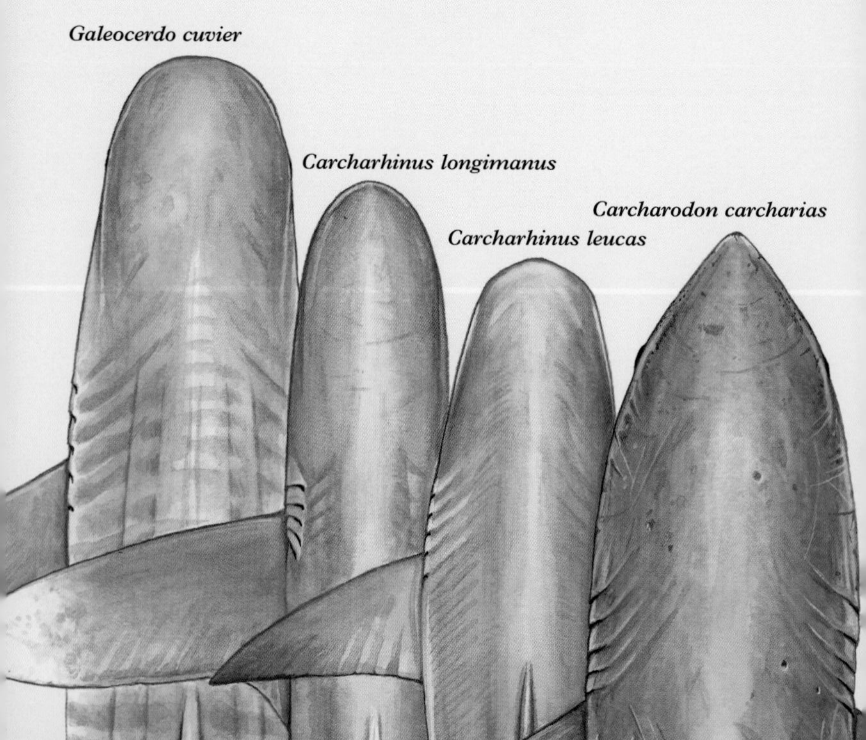

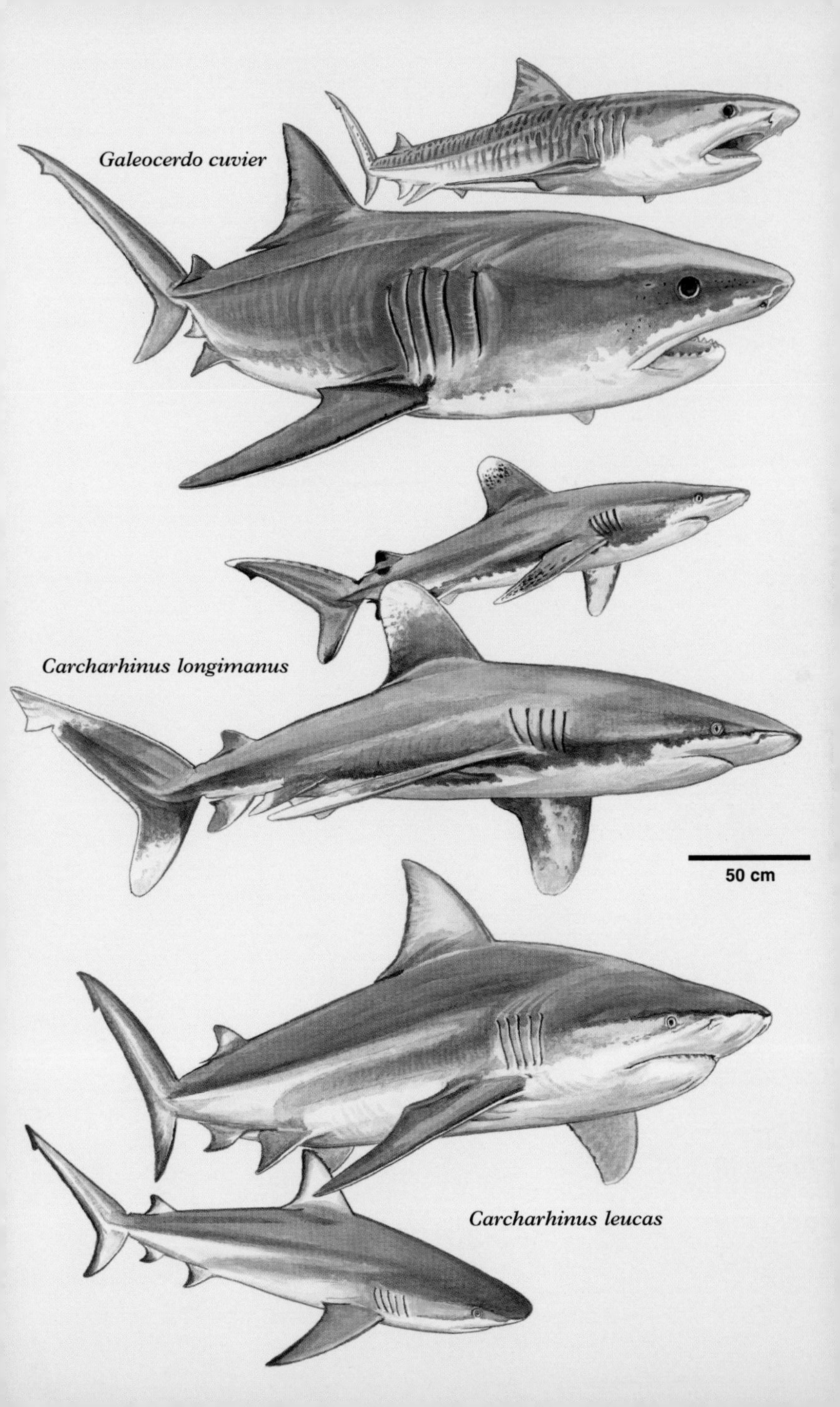

Galeocerdo cuvier
Carcharhinus longimanus
50 cm
Carcharhinus leucas

Plate 51 Carcharhinidae II

Silky Shark *Carcharhinus falciformis* **p295**

WW trop. seas; oceanic and epipelagic, surface–500 m+. Large, slim; fairly long flat rounded snout, large eyes, small mouth; no caudal keels; narrow interdorsal ridge present, long narrow P fins (comparatively shorter in young), D1 fin behind P fins, D2 fin low with greatly elongated inner margin and rear tip; dk. grey to grey-brown or nearly black, incospic. pale flank band, white below, fin markings inconspic. tips dusky except D1 fin.

Blue Shark *Prionace glauca* **p316**

WW temp. and trop. seas, poss. most wide-ranging shark; oceanic and epipelagic, surface–350 m. Large, graceful, slim; long conical snout, large eyes, small mouth; weak caudal keels; no interdorsal ridge, long narrow scythe-shaped P fins, D1 fin well behind P fins closer to V fin base, D2 fin < 1/3 D1; usually dk. blue back with bright blue flanks and sharply demarcated white below.

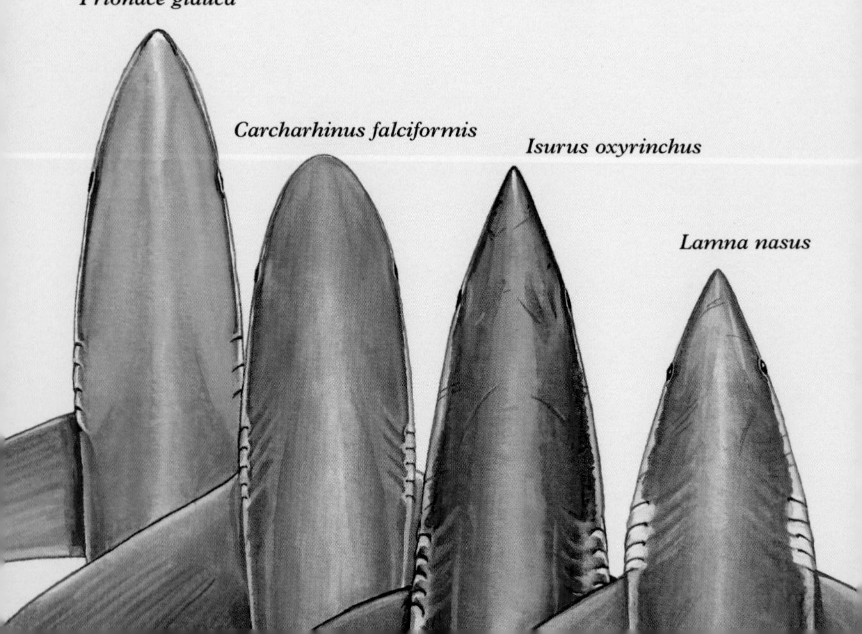

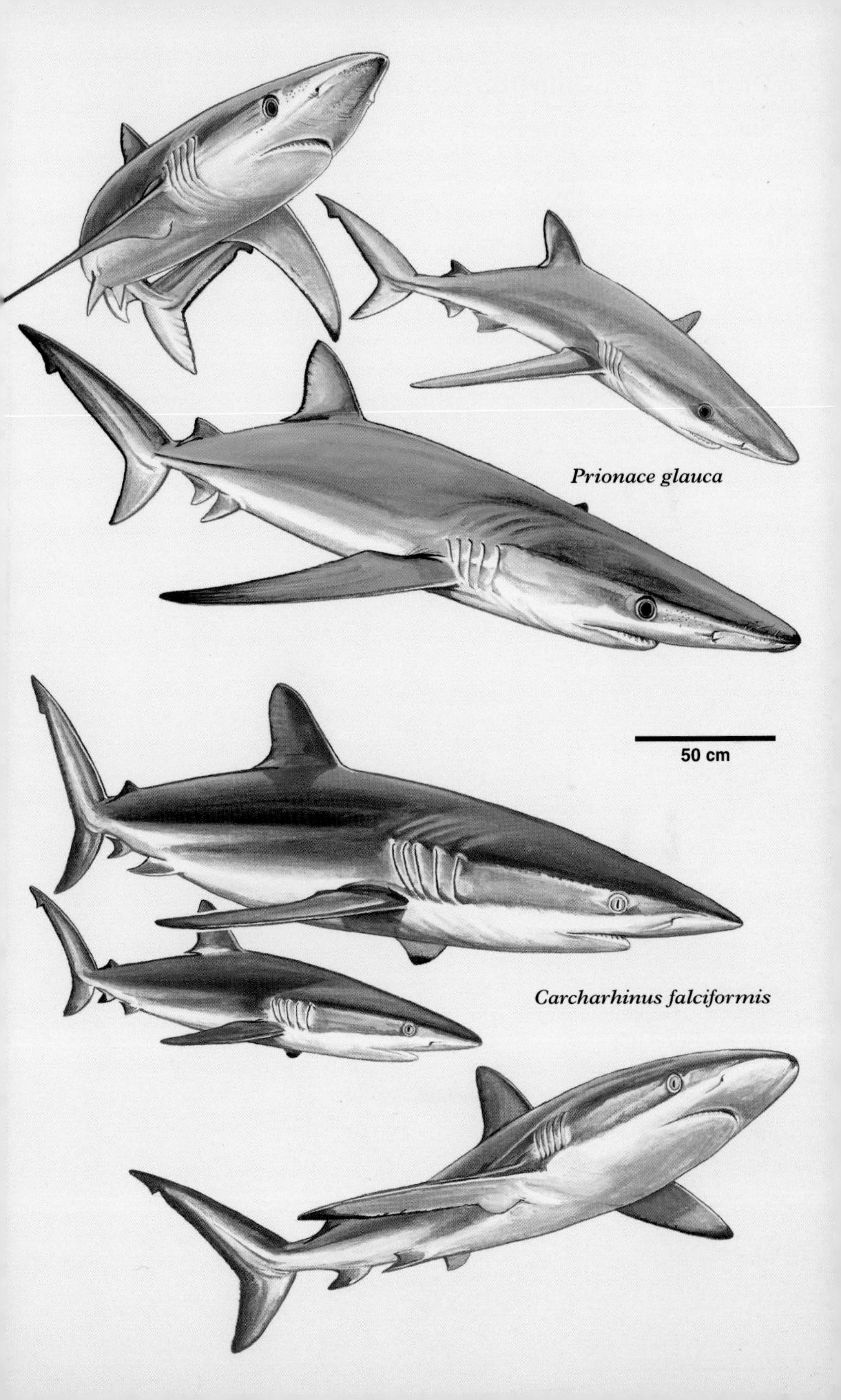

Prionace glauca
50 cm
Carcharhinus falciformis

Plate 52 Carcharhinidae III

Silvertip Shark *Carcharhinus albimarginatus* p289

Trop. IP, wide patchy distrib.; inshore and offshore, not oceanic, surface–800 m. Mod. long broadly rounded snout, eyes round; interdorsal ridge present, P fins with narrow tips, D1 fin apex narrowly rounded or pointed; dk. grey above occ. bronze-tinged, white below, striking white tips and post. margins on all fins.

Grey Reef Shark *Carcharhinus amblyrhynchos* p290

IP; coastal-pelagic and inshore–140 m, deeper than *C. melanopterus*. Mod. long broadly rounded snout, eyes usually round; no interdorsal ridge, P fins narrow and falcate, D1 fin apex narrowly rounded or pointed, D2 fin large with short rear tip; grey above, white below, D1 fin plain or irr. to prom. white edge, obv. broad black post. margin to entire C fin, blackish tips to other fins.

Blacktip Reef Shark *Carcharhinus melanopterus* p302

W Pac., IO and E Med.; v. shallow water around reefs, rarely close offshore. Short broadly rounded snout, horizontally oval eyes; no interdorsal ridge, P fins narrow and falcate, D1 fin apex rounded, D2 fin large with short rear tip; brown-grey above, white below, brilliant black fin tips highlighted by white.

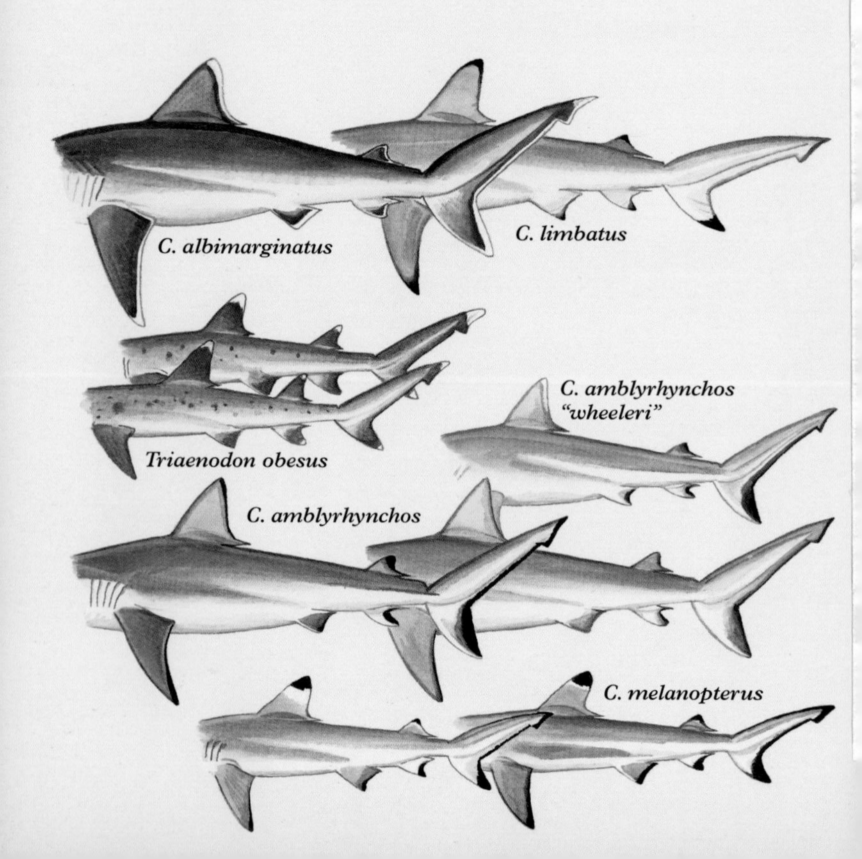

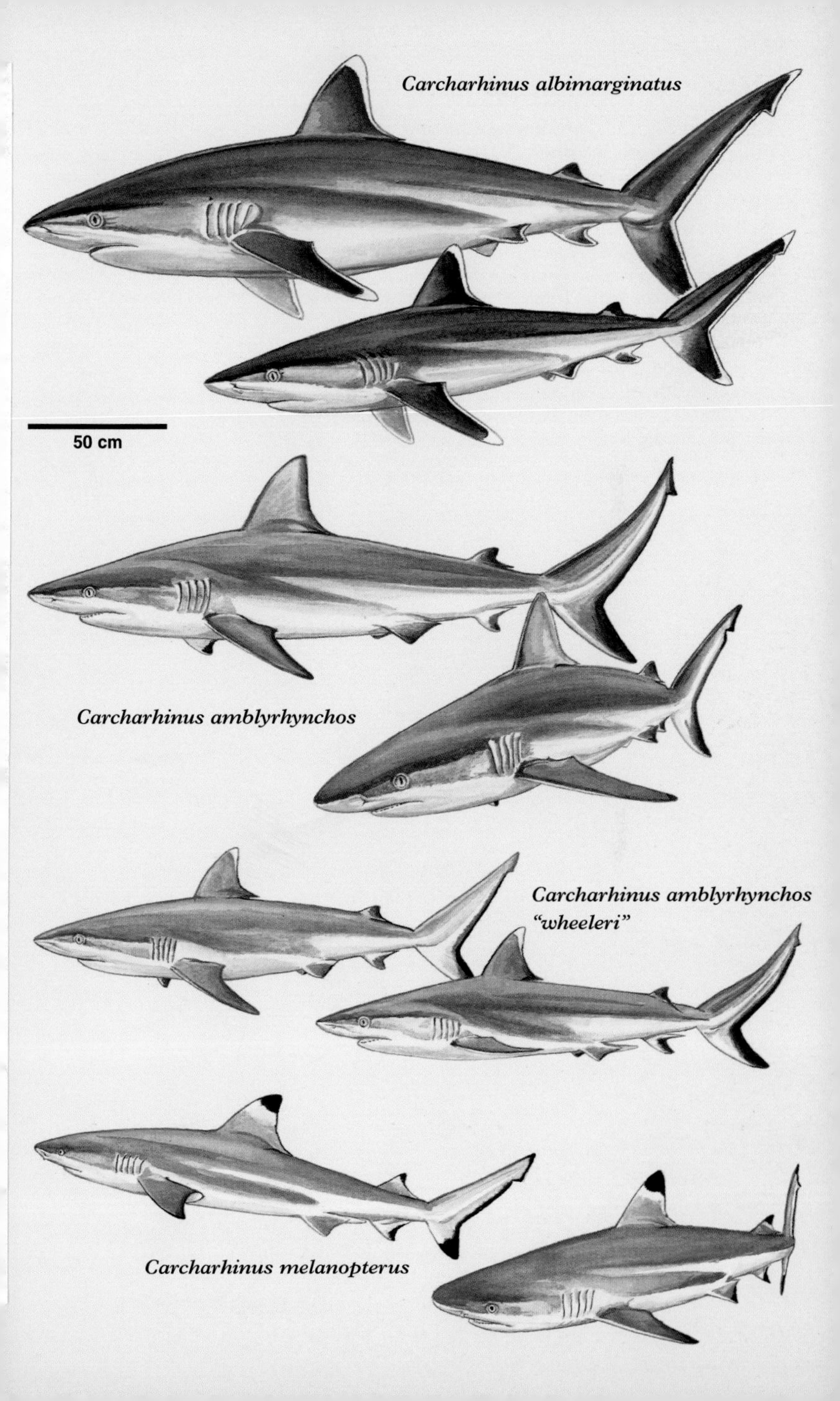
Carcharhinus albimarginatus
50 cm
Carcharhinus amblyrhynchos
Carcharhinus amblyrhynchos
"wheeleri"
Carcharhinus melanopterus

Plate 53 Carcharhinidae IV

Graceful Shark *Carcharhinus amblyrhynchoides* **p290**

G of Aden to Philipp. and NW. Austr.; contin. and insular shelves coastal-pelagic. Tubby; fairly short wedge-shaped pointed snout, fairly large eyes, large gills; no interdorsal ridge, mod, large P fins, large triang. D1 fin, both short rear tips; grey with conspic. white flank band, often black tipped fins.

Sharptooth Lemon Shark *Negaprion acutidens* **p315**

Trop. IW and central Pac.; inshore on/nr. bottom 0–30 m, prefers turbid still water bays, estuaries, outer reef shelves and lagoons. Stocky; broad blunt snout, fairly small eyes, mod. large gills; no interdorsal ridge, v. sim. to *N. brevirostris* usually D, P and A fins more falcate, D1 fin = D2; yellow-brown above, white below.

Whitetip Reef Shark *Triaenodon obesus* **p321**

IP; contin. shelves and island terraces, usually on/nr. bottom in crevices/caves in coral reefs and lagoons. Slender; v. short broad snout, oval eyes, mod. gills; no interdorsal ridge, fairly broad triang. P fins, D1 fin behind P fins, D1 fin ≈ D2; grey-brown, occ. scattered dk. spots on sides, brilliant v. conspic. white tips on D1 and term. C fin, lighter below.

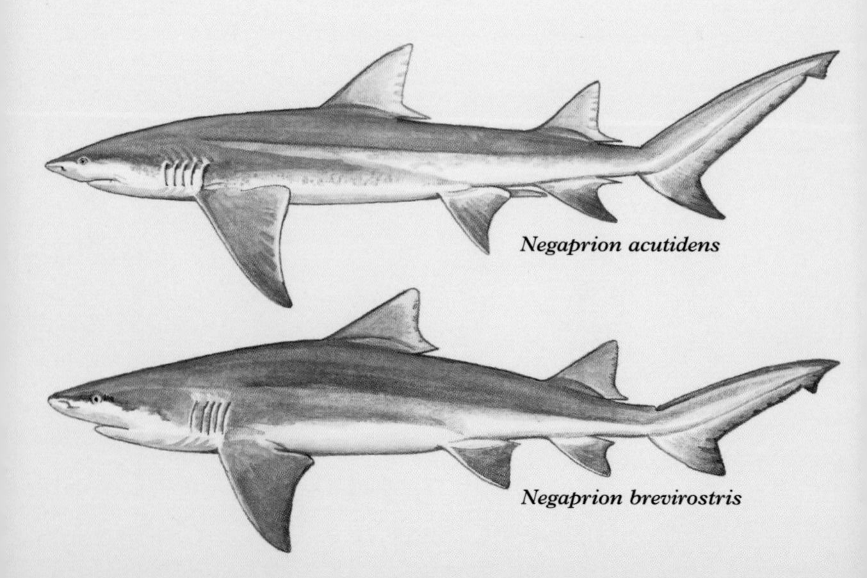

Negaprion acutidens

Negaprion brevirostris

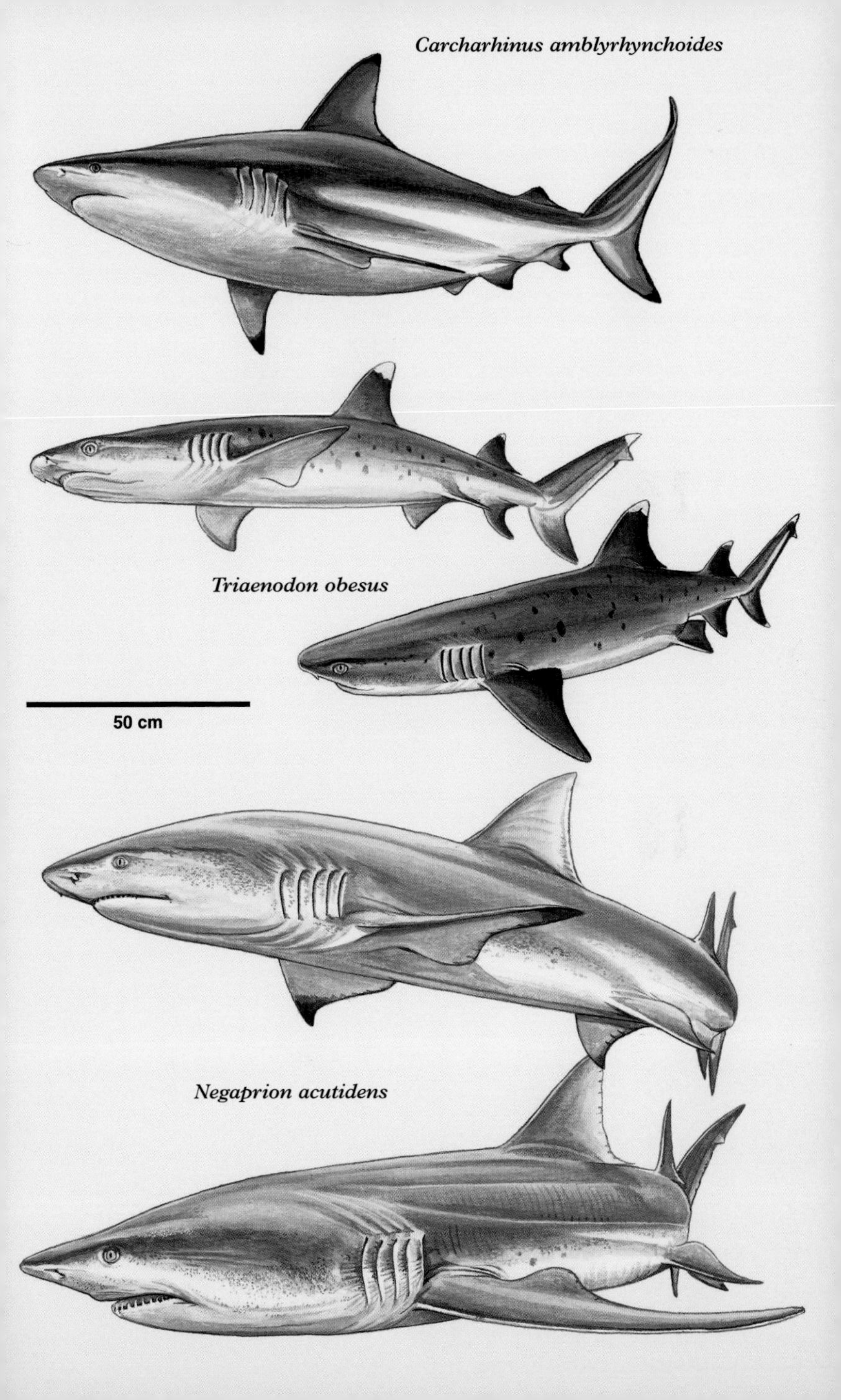

Carcharhinus amblyrhynchoides
Triaenodon obesus
50 cm
Negaprion acutidens

Plate 54 Carcharhinidae V

Pigeye/Java Shark *Carcharhinus amboinensis* p291

IWP: SA to Austr. and E Atl.: Nigeria; contin. and insular shelves 0–60 m. Large; massive thick head, v. short broad blunt snout, small eyes, short lab. furrows; no interdorsal ridge, large angular P fins, high erect triang. D1 fin, D2 fin < 1/3 D1, both short rear tips; grey above, white below, fin tips dusky not conspic. marked.

Smalltail Shark *Carcharhinus porosus* p304

W Atl. and E Pac.; shallow contin. shelf and estuaries nr. mud bottom 0–36 m. Small; long pointed snout, large circular eyes, short lab. furrows; no interdorsal ridge, small P fins, D1 fin large and falcate, D2 fin small origin over A fin midbase, deeply notched A fin; grey above, light below, fin tips P, D and C fins frequently dusky not conspic. marked, inconspic. white flank band.

Caribbean Sharpnose Shark *Rhizoprionodon porosus* p319

Caribb. and trop. S Am.; usually close inshore on contin. and insular shelves, also offshore–500m. Small; long snout, fairly large eyes, long lab. furrows; no interdorsal ridge, mod. P fins, D1 fin origin usually over/slightly behind P fin free rear tips, small D2 fin origin over A fin midbase; brown or grey-brown above, white below, sometimes with white spots on sides, white edged fins.

Atlantic Sharpnose Shark *Rhizoprionodon terraenovae* p320

NW Atl.; coastal from intertidal–280 m, usually > 10 m, often close to surf zone off sandy beaches. Sim. to *R. porosus*; small; long snout, fairly large eyes, long upper lab. furrows; no interdorsal ridge, mod. P fins, D1 fin origin usually over/slightly in front P fin free rear tips, D2 fin small origin over A fin midbase inserts.; grey to grey-brown above, white below, with small white spots on sides in larger specimens, P fins with white margins, D fin tips dusky.

Daggernose Shark *Isogomphodon oxyrhynchus* p313

Trop. N S Am.; turbid water in estuaries, mangroves, river mouths and shallow banks 4–40 m. Med.-sized; unmistakable extremely long flat pointed snout, tiny circular eyes, short prom. lab. furrows; no interdorsal ridge, large paddle-shaped P fins, D1 fin origin over P fins, D2 fin small ~ 1/2 D1, notched A fin; uniform grey or yellow-grey above, light below.

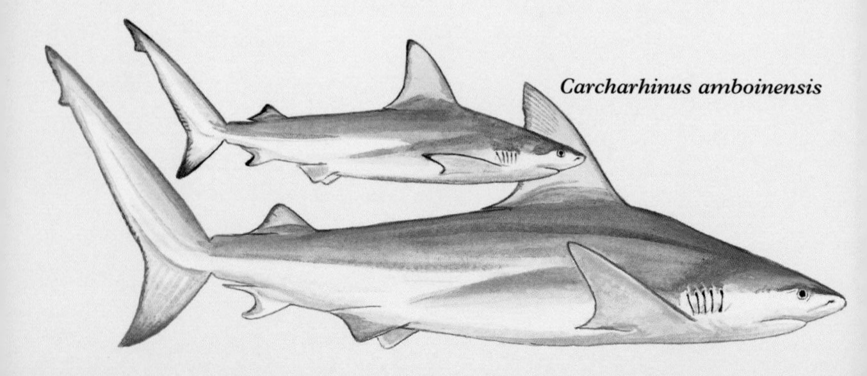

Carcharhinus amboinensis

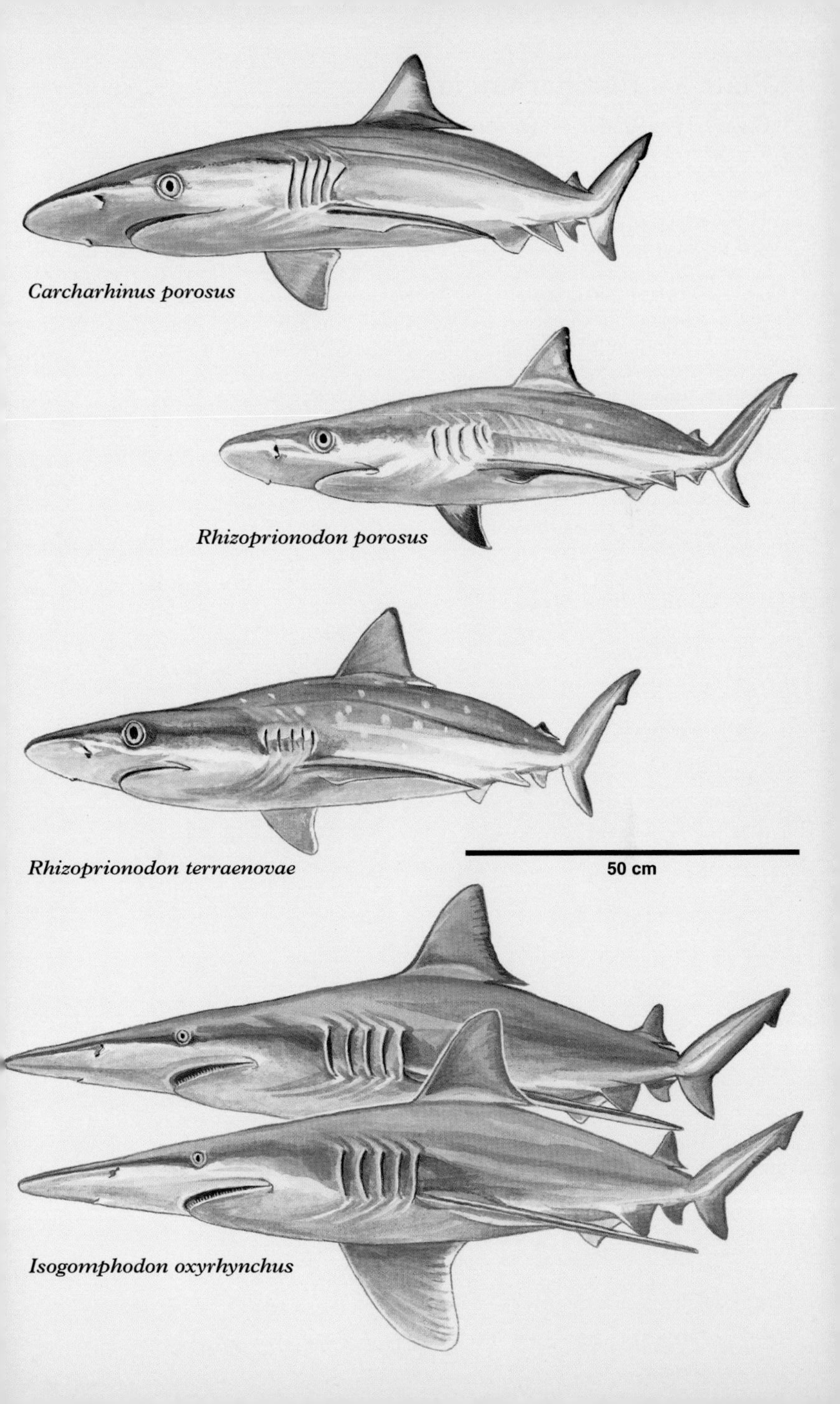

Carcharhinus porosus
Rhizoprionodon porosus
Rhizoprionodon terraenovae
50 cm
Isogomphodon oxyrhynchus

Plate 55 Carcharhinidae VI

Ganges Shark *Glyphis gangeticus* p309

Ganges-Hooghly river system; freshwater rivers. poss. estuaries and inshore. Large, stocky; short broadly rounded snout, tiny eyes; no interdorsal ridge, longi. upper precaud. pit, D1 fin origin over post. 1/3 P fin base, D2 ~1/2 ht. D1, deeply notched A fin; grey above with no conspic. markings, white below.

Speartooth Shark *Glyphis glyphis* p310

IWP: PNG and Bay of Bengal; inshore poss. rivers and estuaries. Large, stocky; short broadly rounded snout, tiny eyes; no interdorsal ridge, longi. upper precaud. pit, D1 fin origin over rear P fin base, D2 large ~3/5 height D1, A fin with deep notch; grey-brown above with no conspic. markings, white below.

Irrawaddy River Shark *Glyphis siamensis* p310

Irrawaddy River mouth; rivers and/or estuaries. Stocky; short broadly rounded snout, tiny eyes; no interdorsal ridge, longitudinal upper precaud. pit, D1 fin origin over rear P fin base, D2 ~1/2 height D1, A fin with deep notch; grey-brown above with no conspic. markings, white below.

Bizant River Shark *Glyphis* sp. A p311

IWP: PNG and Austr.; rivers and inshore. Sim. to *G. glyphis*, possibly not distinct (see text).

Borneo River Shark *Glyphis* sp. B p311

Borneo; turbid brackish to freshwater rivers. Large, stocky; short broadly rounded snout, tiny eyes; no interdorsal ridge, longitudinal upper precaud. pit, D1 fin origin over post. 1/3 P fin base, D2 large slightly >1/2 height D1, A fin with deep notch; grey-brown above with dusky to blackish fin margins and tips, conspic. dusky blotch on each flank and P fin base, white below.

New Guinea River Shark *Glyphis* sp. C p312

N Austr. and PNG; turbid brackish to freshwater rivers and adj. marine waters. Large, slender; flat headed, short broadly rounded snout, tiny eyes; no interdorsal ridge, longitudinal upper precaud. pit, D1 fin origin over post. 1/3 P fin base, D2 large ~2/3 height D1, A fin with deep notch; grey above with dusky fin margins, white below.

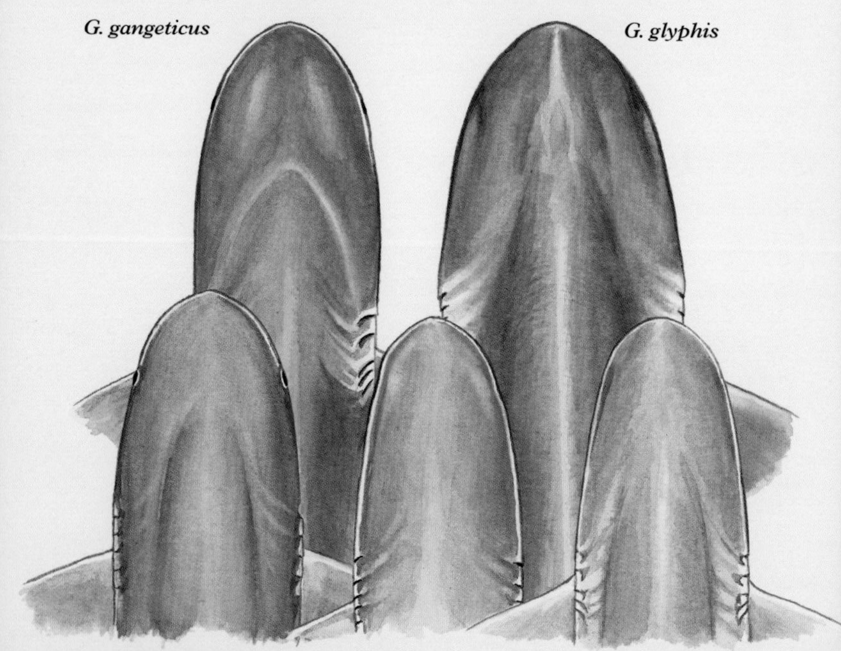

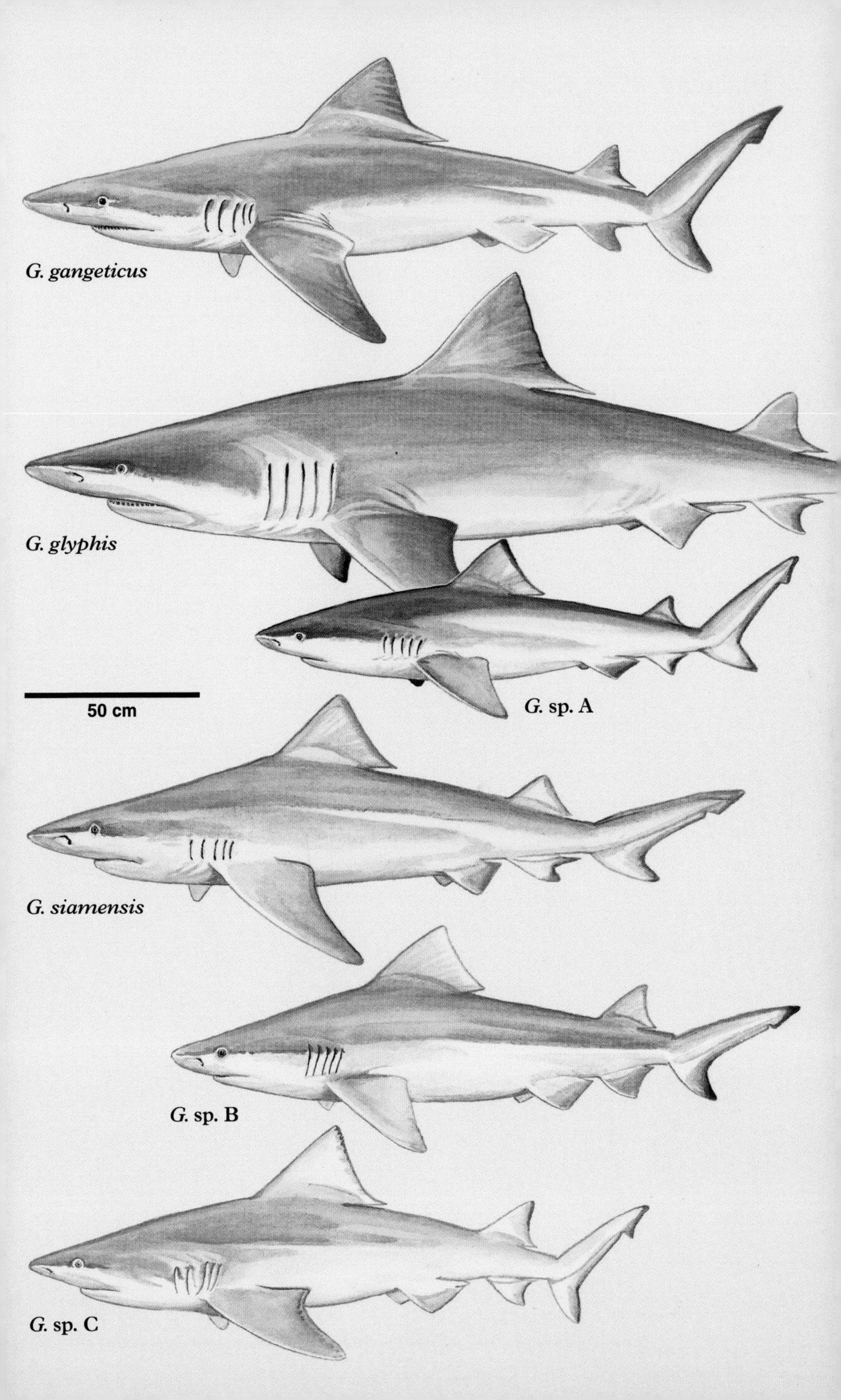

G. gangeticus
G. glyphis
G. sp. A
50 cm
G. siamensis
G. sp. B
G. sp. C

Plate 56 Carcharhinidae VII

Whitecheek Shark *Carcharhinus dussumieri* **p294**

IWP, Arabian Sea to Japan; trop. inshore contin. and insular shelves. Often confused with *C. sealei*, mod. long rounded snout, small wide-spaced nostrils, fairly large horiz.-oval eyes, upper lab. furrows short and inconspic.; interdorsal ridge present, small semifalcate P fins, small triang. D1 fin, short rear tips to D fins, no preanal ridge; grey to grey-brown with black or dusky spot on D2 fin, other fins with pale post. margins, inconspic. lt. flank stripe, white below.

Pondicherry Shark *Carcharhinus hemiodon* **p297**

IWP; trop. coastal on contin. and insular shelves. Mod. long pointed snout, small wide-spaced nostrils, fairly large eyes, upper lab. furrows short and inconspic.; interdorsal ridge present, small P fins, mod. large D1and D2 fins, short rear tips to both, no preanal ridge; grey with black tips to P, D2 and C fins, conspic. white flank stripe, white below.

Milk Shark *Rhizoprionodon acutus* **p317**

E Atl. and IWP; contin shelf midwater to nr. bottom 1–200 m. Long narrow snout, small wide-spaced nostrils, large eyes, only requiem shark in range with long upper and lower lab. furr.; interdorsal ridge present, small P fins, D1 fin origin well behind P fin origin, small low D2 behind larger A fin, long preanal ridge; bronze to grey, most fin tips slightly pale, juv. D and term. C fin tips dk. occ. adults, white below.

Pacific Sharpnose Shark *Rhizoprionodon longurio* **p318**

E Pac., Calif. to Peru; littoral to contin. and insular shelves inshore and offshore. Long snout, small wide-spaced nostrils, large eyes, long upper and lower lab. furr.; interdorsal ridge present, small P fins, D1 fin origin usually over or slightly in front P fin free tip, D2 origin well behind larger A fin origin, long preanal ridge; grey or grey-brown, lt. edged P fins, dusky tipped D fins, white below.

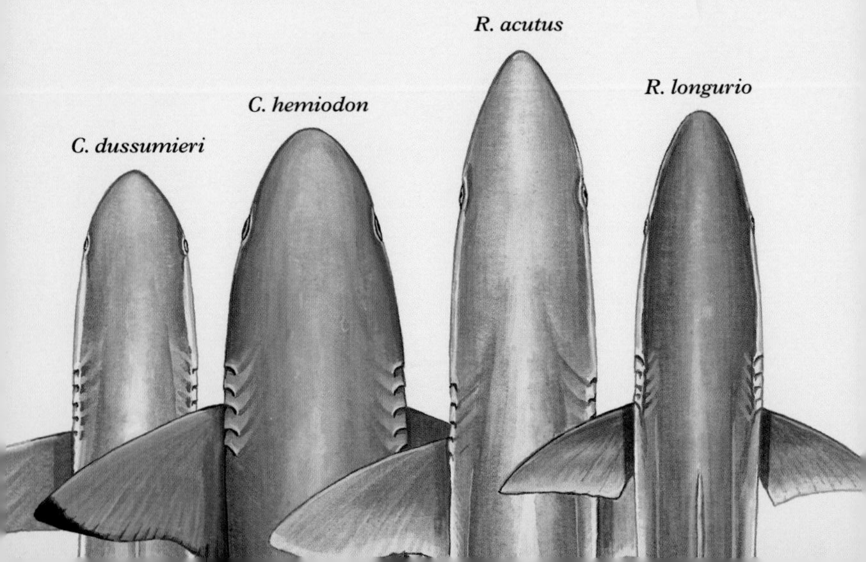

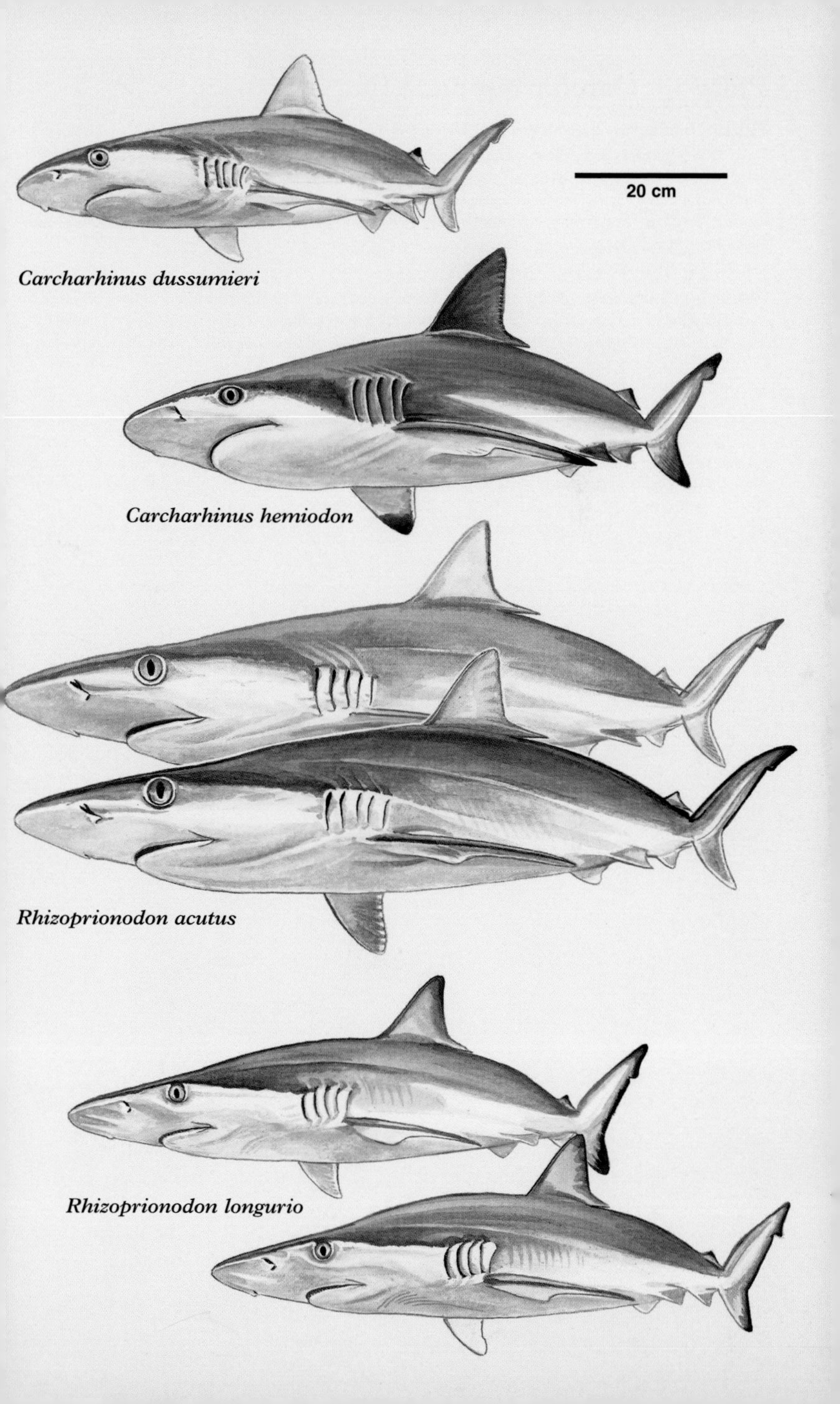
20 cm
Carcharhinus dussumieri
Carcharhinus hemiodon
Rhizoprionodon acutus
Rhizoprionodon longurio

Plate 57 Carcharhinidae VIII

Bignose Shark *Carcharhinus altimus* p289

Prob. WW in trop. and wm. seas; offshore deep contin. and insular shelf edges and uppermost slopes 90–430 m+. Heavy cylindrical body; large long broad snout, circular mod. large eyes, long ant. nasal flaps, upper lab. furrows short and inconspic., mod. long gills; prom. high interdorsal ridge, large straight P and D fins; grey occ. bronzy above with obscure dusky fin tips except V fins, inconspic. white flank stripe, white below.

Spinner Shark *Carcharhinus brevipinna* p293

Wm. temp. and trop. Atl., Med. and IWP; coastal-pelagic on contin. and insular shelves and close inshore, sirface–bottom, 75 m+. Slender; long narrow pointed snout, circular small eyes, short rel. inconspic. ant. nasal flaps, prom. lab. furrows, long gills; no interdorsal ridge, small P and D1 fins, D2 mod. large, both D fins with short rear tips; grey most fin tips of ad. and juvs. have conspic. black tips, inconspic. white flank stripe, white below.

Blacktip Shark *Carcharhinus limbatus* p300

Widespread trop. and subtrop. seas; contin. and insular shelves, usually close inshore rarely > 30 m, tolerates reduced salinity. Stout; long narrow pointed snout, circular small eyes, low triang. ant. nasal flaps, upper lab. furrows short and inconspic., long gills; no interdorsal ridge, mod. large falcate P fins, high D1 fin, D2 mod. large; grey to grey-brown, black fin tips to P, D2 and term. C fins, sometimes V and A fins, usually black edges on D1 apex and dors. C lobe, conspic. white flank stripe, white below.

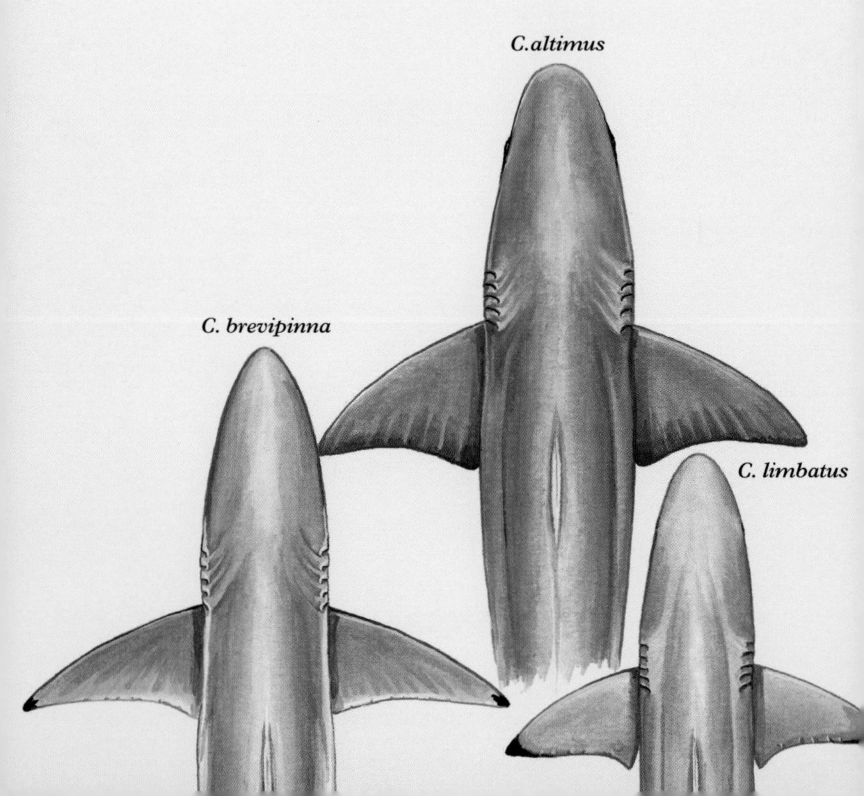

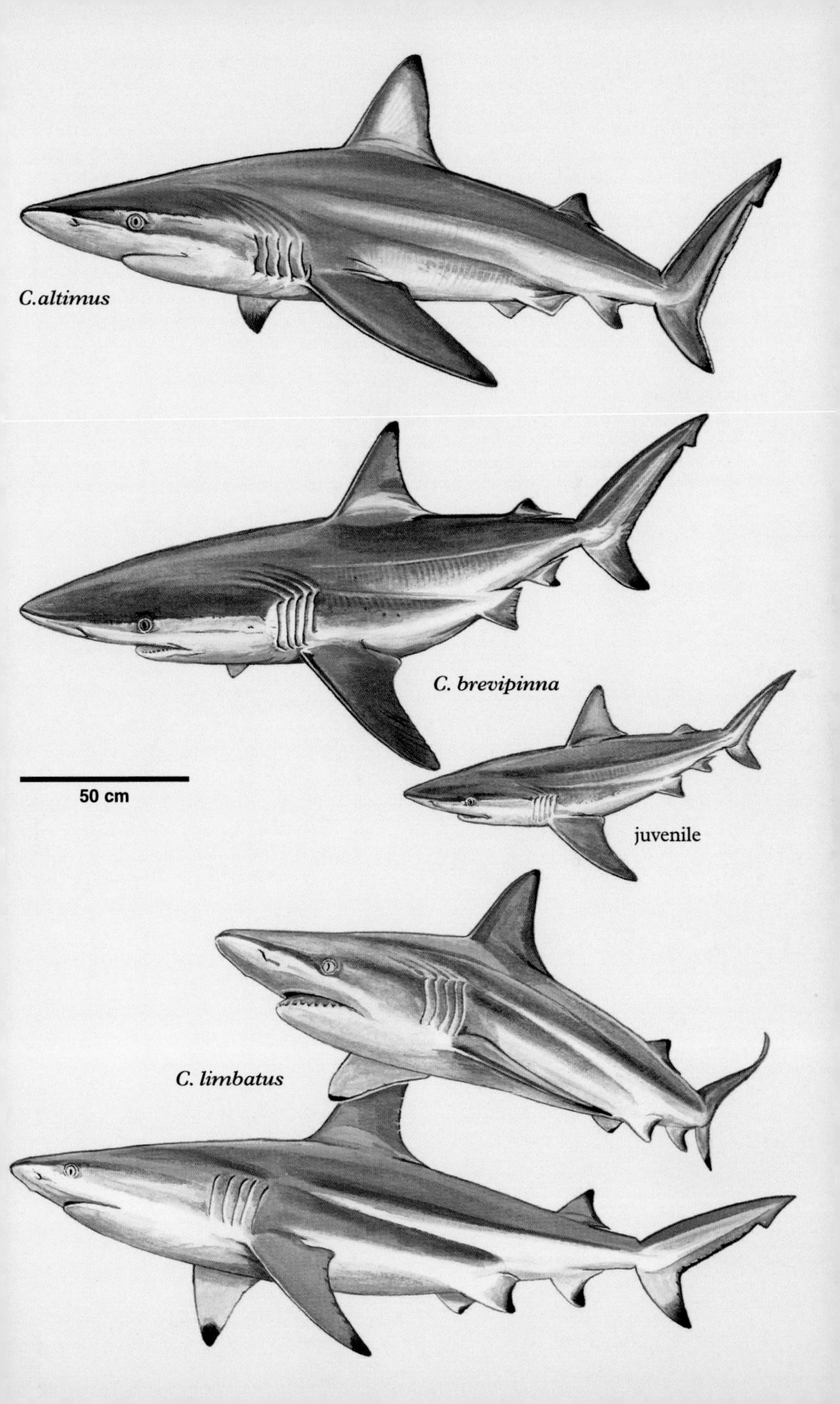
C.altimus
C. brevipinna
50 cm
juvenile
C. limbatus

Plate 58 Carcharhinidae IX

Galapagos Shark *Carcharhinus galapagensis* p296

WW patchy mainly around oceanic islands; coastal-pelagic, shallow inshore to well offshore 0–180 m. Fairly long broad snout, fairly large eyes, low ant. nasal flaps; low interdorsal ridge, large semifalcate P fins, mod. large D1 fin with short rear tip, D1 origin ~ opp. midlength P fins; grey-brown, most fins with inconspic. dusky tips, inconspic. white flank stripe, white below.

Dusky Shark *Carcharhinus obscurus* p302

Poss. WW in trop. and wm. temp. seas; contin. and insular shelves, shoreline to adj. oceanic waters 0–400 m. Broad rounded snout, fairly large eyes, low poorly developed ant. nasal flaps; low interdorsal ridge, mod. large curved P fins, mod. large falcate D1 fin, D1 origin over/ahead of P fin free rear tips; grey to bronzy, most fins with inconspic. dusky tips, inconspic. white flank stripe, white below.

Sandbar Shark *Carcharhinus plumbeus* p304

Poss. WW in trop. and wm. temp. seas; common in bays, harbours and river mouths, also offshore usually nr. bottom 1–280 m. Mod. long rounded snout, fairly large eyes, low short ant. nasal flaps; interdorsal ridge present, large P fins, v. large erect D1 fin, D1 origin over/slightly ahead P fin inserts.; grey-brown to bronzy, tips and post. margins of fins often inconspic. dusky, inconspic. white flank stripe, white below.

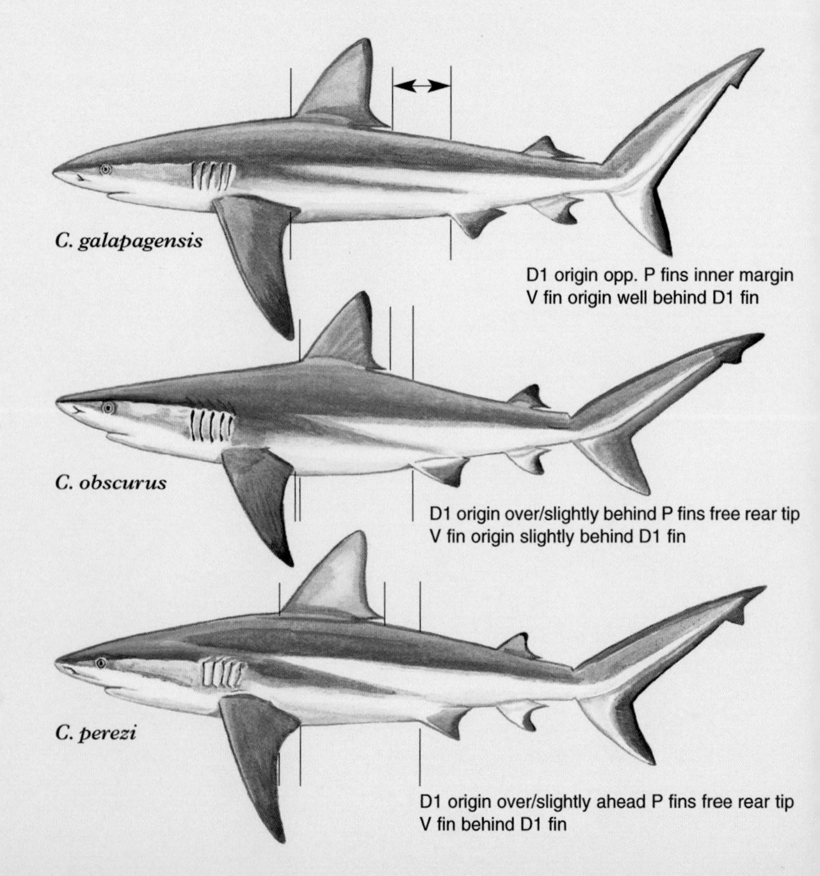

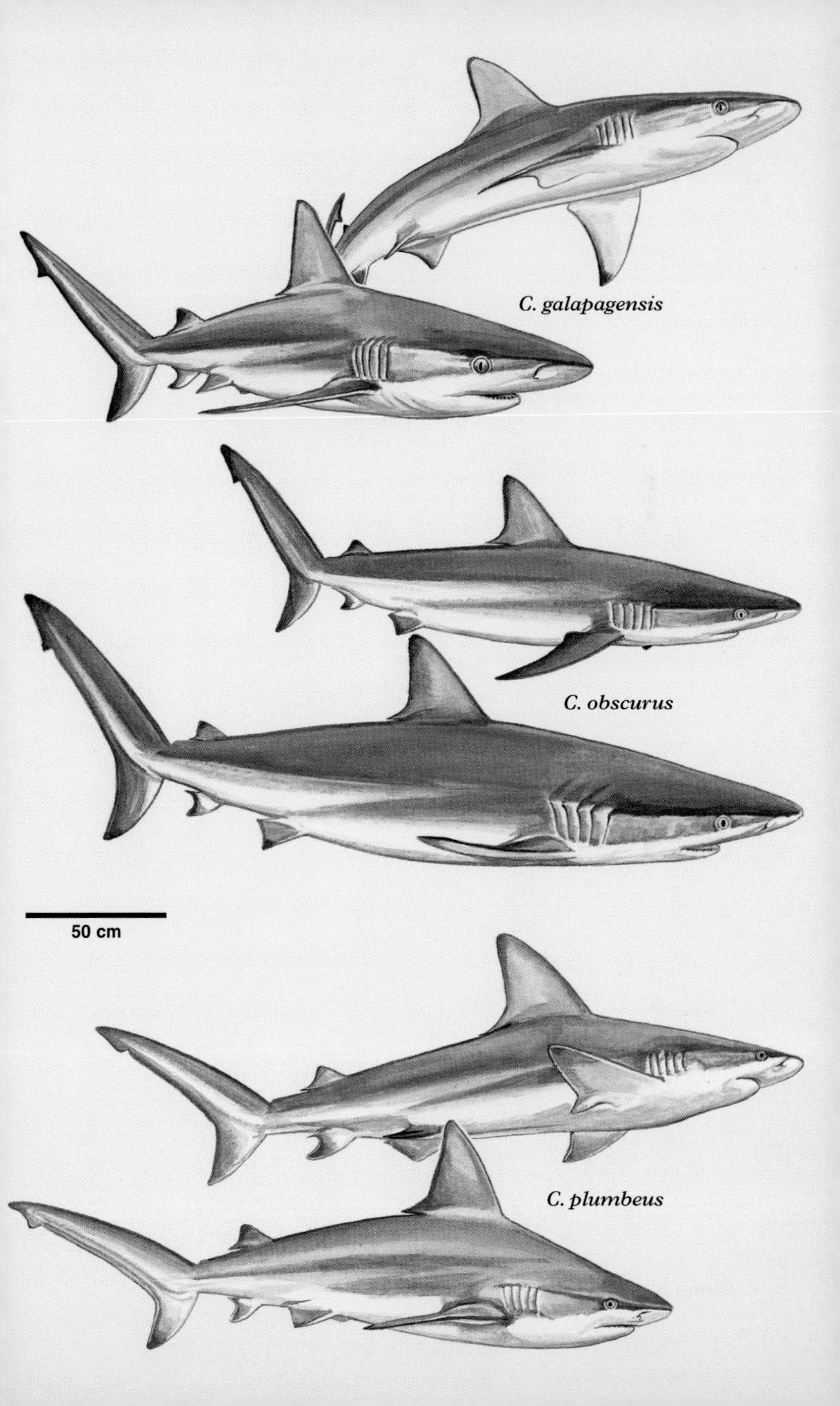
C. galapagensis
C. obscurus
50 cm
C. plumbeus

Plate 59 Carcharhinidae X

Nervous Shark *Carcharhinus cautus* **p294**

E IO and SW Pac.; shallow inshore water on contin. and insular shelves, coral reefs and estuaries, occ. deeper water. Short bluntly rounded snout, horiz. oval eyes, nipple-like ant. nasal flaps, short lab. furr., mod. large gills; no interdorsal ridge, mod. large D2 fin with short rear tip; grey to lt. brown with black D and C fin margins, C and P fins with black tips, conspic. white flank stripe, white below.

Creek Whaler Shark *Carcharhinus fitzroyensis* **p296**

N Austr.; mainly inshore, intertidal–40 m+. Long parabolic snout, mod. large round eyes, lobate ant. nasal flaps, short lab. furr., short gills; no interdorsal ridge, broad triang. fins, D1 fin origin over P fin rear tips, D2 ~ over A fin origin; grey-brown to bronze, lacks any conspic. markings, lt. below.

Spottail Shark *Carcharhinus sorrah* **p306**

Trop. IWP; shallow water contin. and insular shelves, coral reefs 0–140 m. Long rounded snout, large round eyes, long narrow nipple-like ant. nasal flaps, short inconspic. lab. furr., mod. short gills; interdorsal ridge present, v. low long D2 fin; med. grey with large conspic. black P, D2 and term. C fin tips, conspic. white flank stripe, white below.

Broadfin Shark *Lamiopsis temmincki* **p313**

Scattered IWP; inshore contin. shelf. Mod. long snout ≈ mouth width, small round eye, short broadly triang. ant. nasal flaps, short lab. furr., 5th gill ~ 1/2 gill 1; no interdorsal ridge, v. broad triang. P fins, A fin post. margin ~ straight; lt. grey or tan, lacks any conspic. markings, lt. below.

Whitenose Shark *Nasolamia velox* **p314**

E Pac., Baja Calif. to Peru; inshore and offshore contin. shelf 15–192 m. V. long conical snout, mod. large round eyes, close-set nostrils width nostril ≈ space between, v. short lab. furr.; no interdorsal ridge, mod. broad triang. P fins, D1 fin >> D2, A fin slightly > D2; grey-brown to lt. brown, prom. black spot outlined with white on upper snout, lt. below.

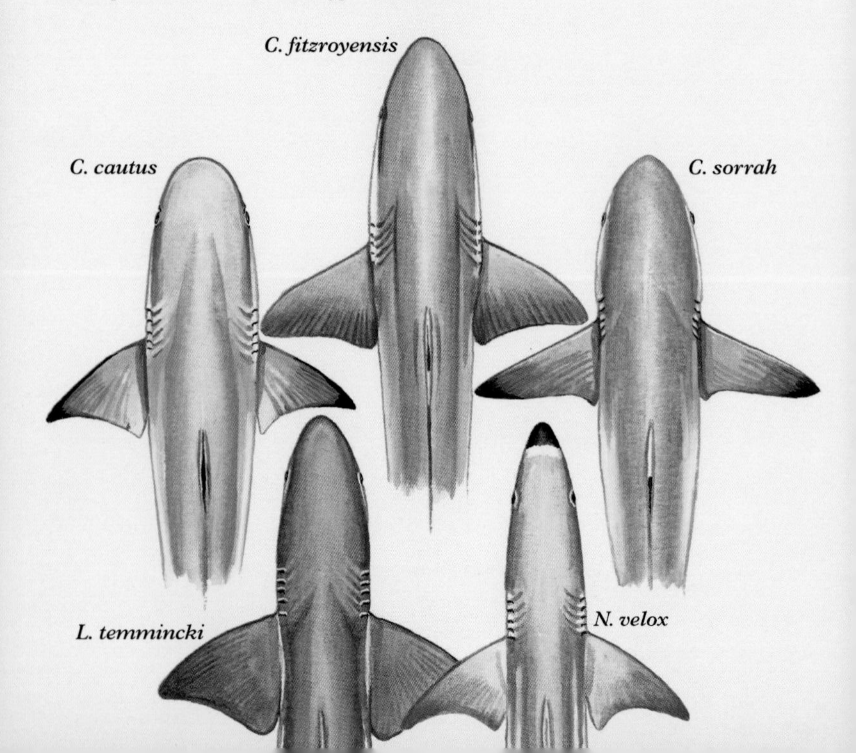

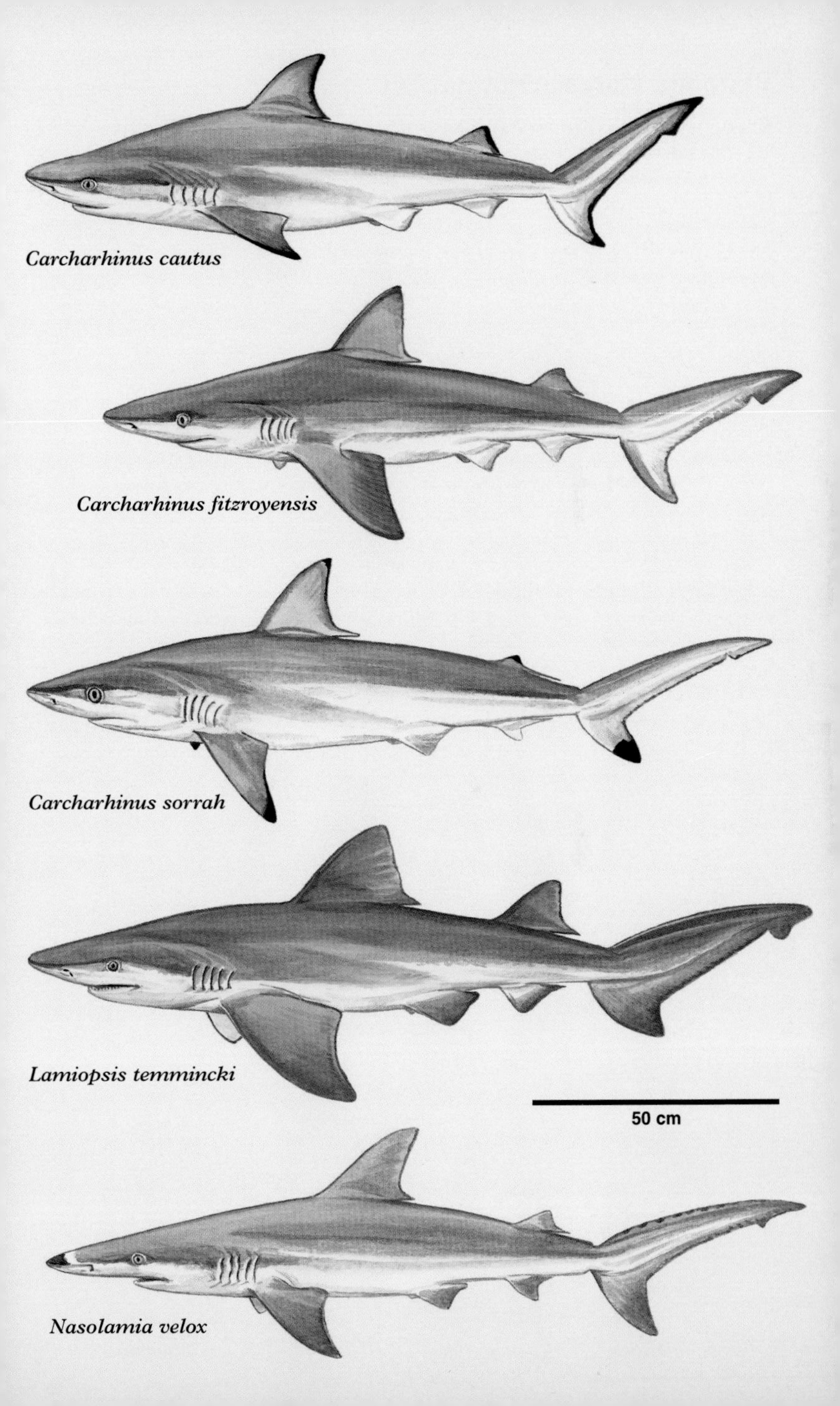
Carcharhinus cautus
Carcharhinus fitzroyensis
Carcharhinus sorrah
Lamiopsis temmincki
50 cm
Nasolamia velox

Plate 60 Carcharhinidae XI

Borneo Shark *Carcharhinus borneensis* **p292**
IWP; trop. inshore/coastal. Long pointed snout, large eyes; no interdorsal ridge, small P fins, small D fins with short tips; brown with dk. markings on D2 and term C fin tips, P, V, and A fins with lt. margins.

Hardnose Shark *Carcharhinus macloti* **p301**
IWP; close inshore–170 m. Long pointed hard snout, mod. large eyes; no interdorsal ridge, small D fins, D2 v. low with v. long rear tip; grey to grey-brown, fins with lt. margins, inconspic. lt. flank stripe.

Blackspot Shark *Carcharhinus sealei* **p305**
IWP; surfline–40 m. Long rounded snout, large oval eyes; no/small interdorsal ridge, small P and D1 fins; grey or tan with conspic. dusky/black tip on D2 fin, other fins with lt. margins, lt. flank stripe.

Sliteye Shark *Loxodon macrorhinus* **p314**
IWP; clear water 7–80 m. Long narrow snout, large eyes with rear notch; no/rudimentary interdorsal ridge, D1 fin >> D2, grey to brown, black margin on D1 and C fins, other fins with lt. margins.

Brazilian Sharpnose Shark *Rhizoprionodon lalandei* **p318**
W Atl.; contin. shelf 3–70 m. Long snout, large eyes, wide-spaced nostrils, long upper and lower lab. furr.; small D2 fin, long anal ridge; dk. grey to grey-brown, margins lt. to P fins and dusky to D fins.

Grey Sharpnose Shark *Rhizoprionodon oligolinx* **p319**
Trop. IWP; contin. and insular shelves, littoral to offshore. Long snout, large eyes, small wide-spaced nostrils, short lab. furr.; small D2 fin, long anal ridge; grey to bronzy, inconspic. dusky fin margins.

Australian Sharpnose Shark *Rhizoprionodon taylori* **p320**
Austr.; trop. inshore contin. shelf. V. sim. to *R. oligolinx*; D and C fins with dk. margins, term C fin dk, other fins with lt. margins.

Spadenose Shark *Scoliodon laticaudus* **p321**
IWP; close inshore. Unmistakable v. long flattened snout, small eyes; no interdorsal ridge, short broad triang. P fins, D1 fin >> D2; bronzy-grey with no conspic. markings.

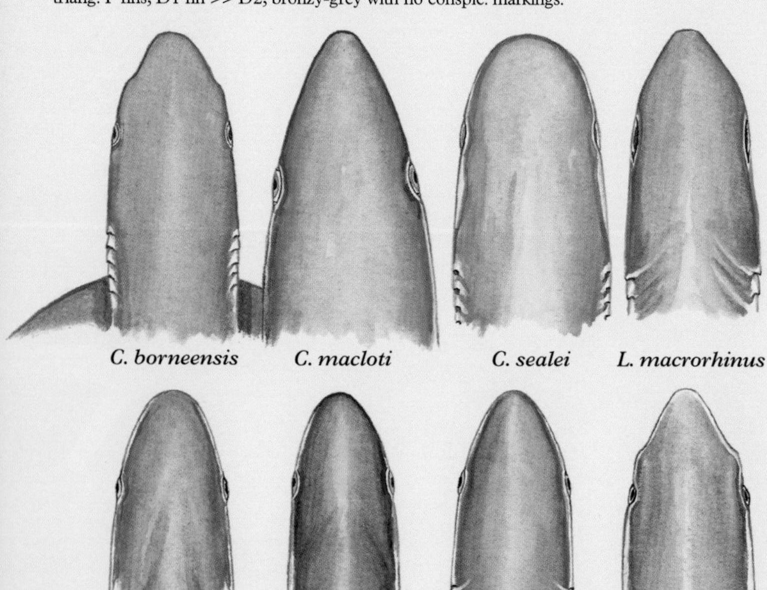

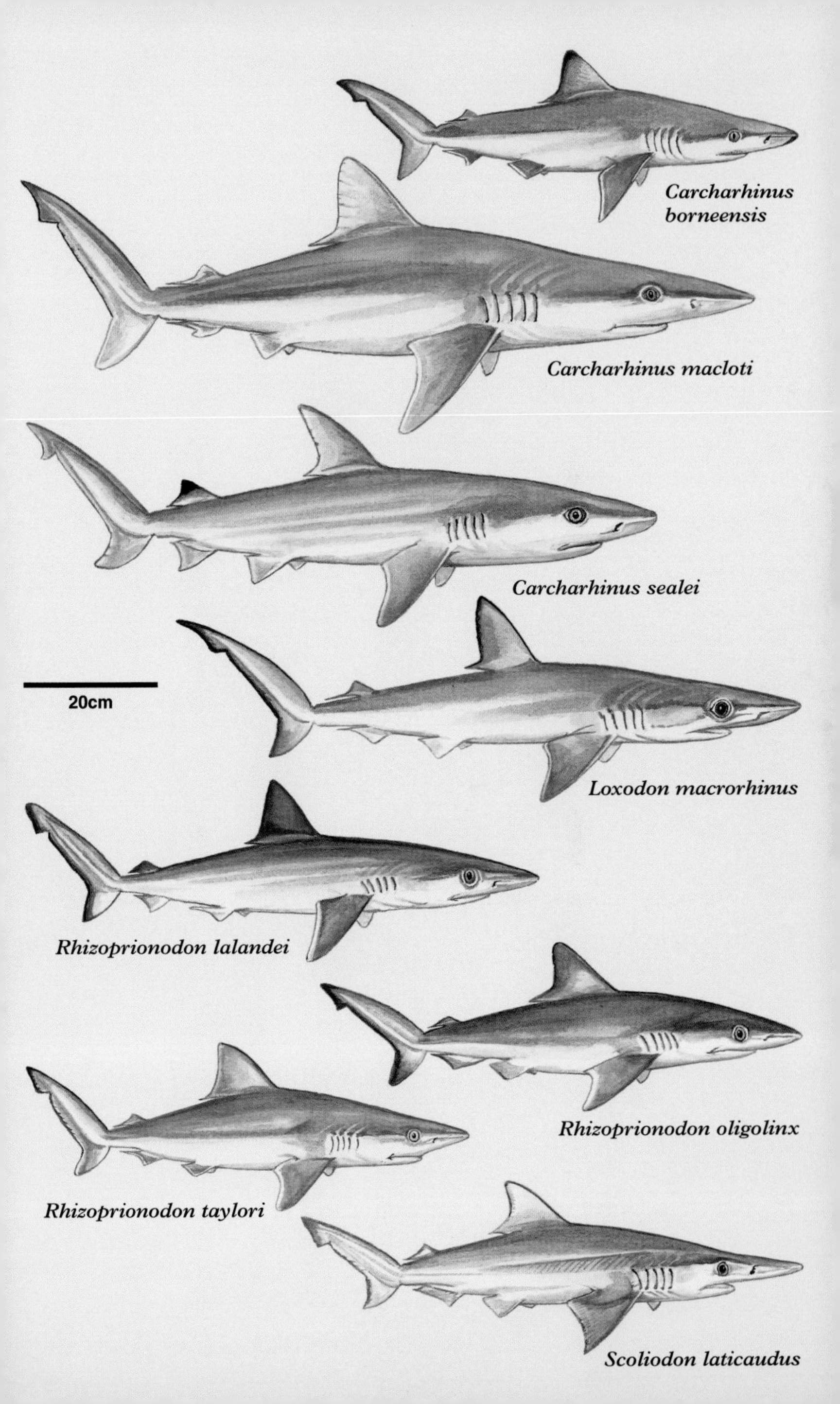

Carcharhinus borneensis
Carcharhinus macloti
Carcharhinus sealei
20cm
Loxodon macrorhinus
Rhizoprionodon lalandei
Rhizoprionodon oligolinx
Rhizoprionodon taylori
Scoliodon laticaudus

Plate 61 Carcharhinidae XII

Bronze Whaler *Carcharhinus brachyurus* **p292**

Most wm. temp. waters in IP, Atl. and Med.; close inshore–100 m+ offshore. Bluntly pointed broad snout; no interdorsal ridge, long P fins, small D fins with short rear tips; olive-grey to bronzy, most fins with inconspic. darker margins and dusky tips, fairly prom. white flank stripe, white below.

Caribbean Reef Shark *Carcharhinus perezi* **p303**

W Atl., S USA to Brazil inc. Caribb.; commonest Caribb. coral reef shark, nr. bottom to 30 m+, on hard bottom and mud nr. river deltas in Brazil. Bluntly rounded snout; interdorsal ridge present, large narrow P fins, small D1 fin and mod. large D2 with short rear tip; dk. grey or grey-brown, underside P, V, A and vent. C lobe dusky but not prom. marked, inconspic. white flank stripe, white below.

Lemon Shark *Negaprion brevirostris* **p315**

Trop. W Atl., W Afr. and E. Pac. (Mex. to Ecuador); inshore and coastal surface and intertidal–92 m+, adapted to low low oxygen shallow water environ. Short nosed; no interdorsal ridge, D, P and V fins weakly falcate, D1 fin ≈ D2; pale yellow-brown, no fin markings, no flank stripe, lt. below.

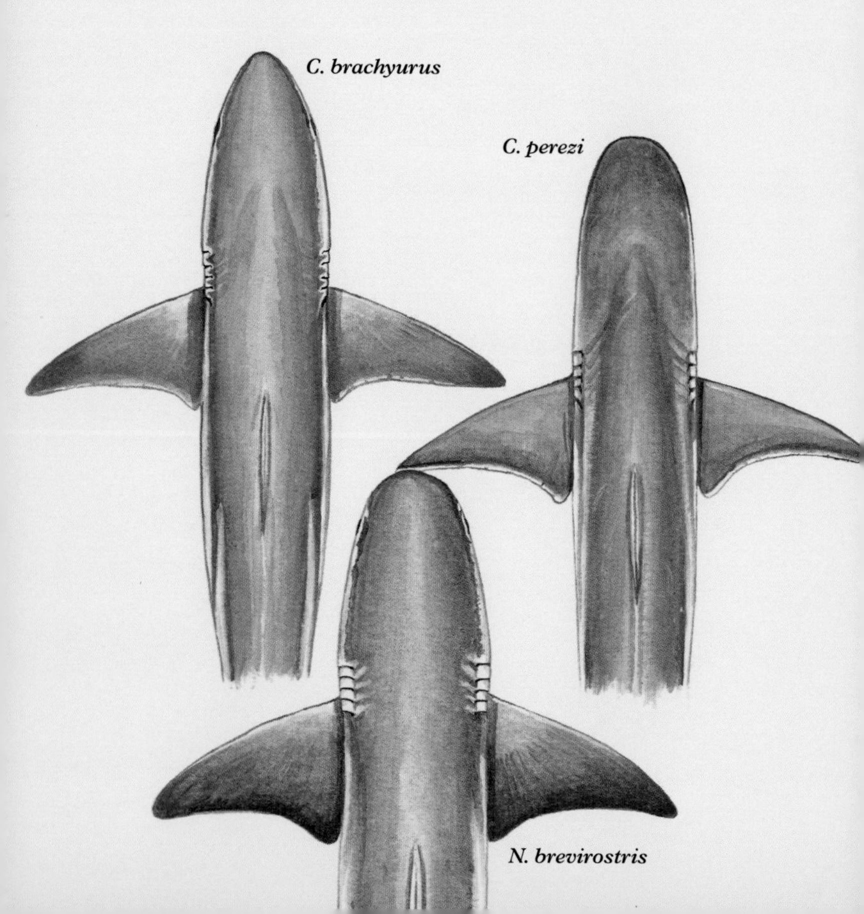

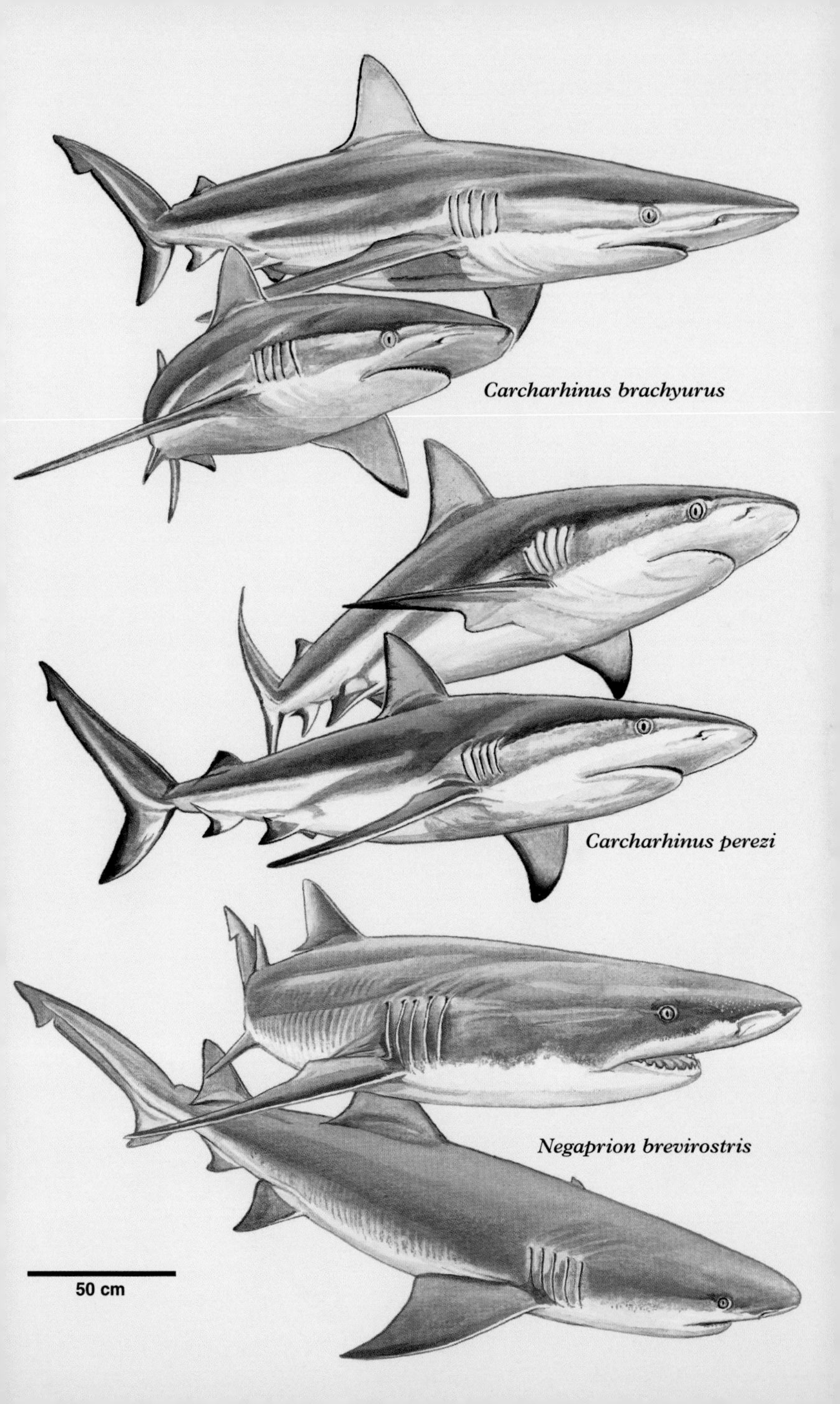
Carcharhinus brachyurus
Carcharhinus perezi
Negaprion brevirostris
50 cm

Plate 62 Carcharhinidae XIII

Blacknose Shark *Carcharhinus acronotus* **p288**

W Atl., S USA to S Brazil inc. Caribb.; coastal contin. and insular shelves, mainly over sand, shell and coral 18–64 m. Mod. long rounded snout; mod. large eyes, short inconspic. upper lab. furr., short gills; no interdorsal ridge, small P fins, D fins with short rear tips, D1 small, D2 mod. large, D2 fin origin ~ over A fin origin; grey, dk. tip to snout, D2 and term. C fin tips dk.

Finetooth Shark *Carcharhinus isodon* **p298**

W Atl., S USA to S Brazil inc. Caribb.; wm. temp. to trop. inner contin. shelf, intertidal–20 m. Long pointed snout; mod. large eyes, short inconspic. upper lab. furr., v. long gills; no interdorsal ridge, small P fins, D fins with short rear tips, D1 small, D2 mod. large, D2 fin origin over/post. A fin origin; dk. blue-grey, no prom. fin markings, inconspic. white flank stripe, white below.

Smoothtooth Blacktip *Carcharhinus leiodon* **p298**

G of Aden.; only one known specimen, prob. inshore. Sim. to *C. amblyrhynchoides*, short bluntly pointed snout; mod. large eyes, long gills; no interdorsal ridge, small P fins, D fins with short rear tips, mod. larg D1 and D2; conspic. black tips on all fins.

Night Shark *Carcharhinus signatus* **p306**

Trop. Atl.; deepwater coastal and semi-oceanic on/along outer contin. and insular shelves and off upper slopes 0–600 m. Long pointed snout; large eyes, short inconspic. upper lab. furr.; small P fins, interdorsal ridge present, D1 small with mod. long rear tip, D2 low with long rear tip, D1 fin origin over P fins; grey-brown with no conspic, fin markings, white below.

Australian Blacktip Shark *Carcharhinus tilsoni* **p307**

Trop. Austr.; contin. shelf, close inshore– ~150 m. Sim. to *C. limbatus*, long snout; fairly large eyes; no interdorsal ridge, mod. large P fins, D fins with short rear tips, D1 large, D2 mod. large, D1 fin origin ~ over P fin inserts.; grey to bronzy with black tipped fins, V and A fins occ. plain, pale flank stripe, pale below.

False Smalltail Shark *Carcharhinus* sp. A **p307**

S China Sea; only known from three specimens presumably inshore. Mod. long bluntly pointed snout; large eyes, short lab. furr.; no interdorsal ridge, mod. large P fins, D1 fin large and over rear 1/3 P fin bases, D2 small and origin over first 1/3 A fin base, A fin broadly notched; grey, P, D and C fin webs often inconspic. dusky, inconspic. white flank stripe, lt. below.

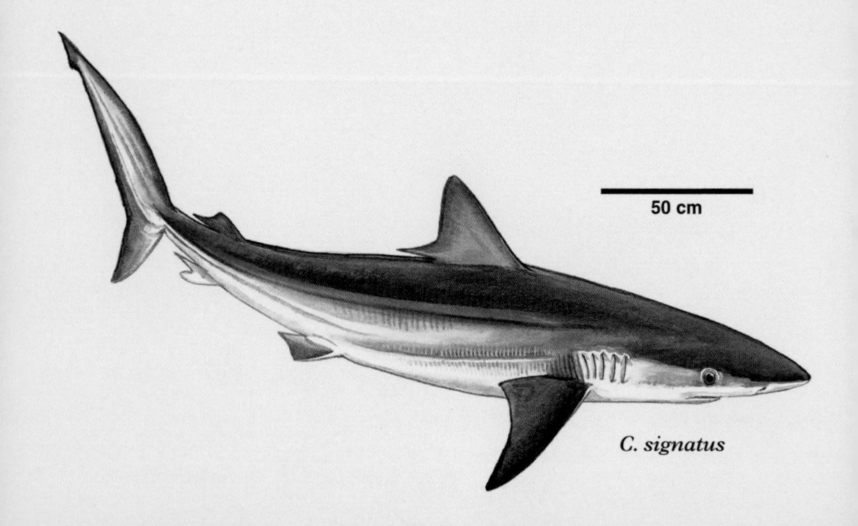

C. signatus

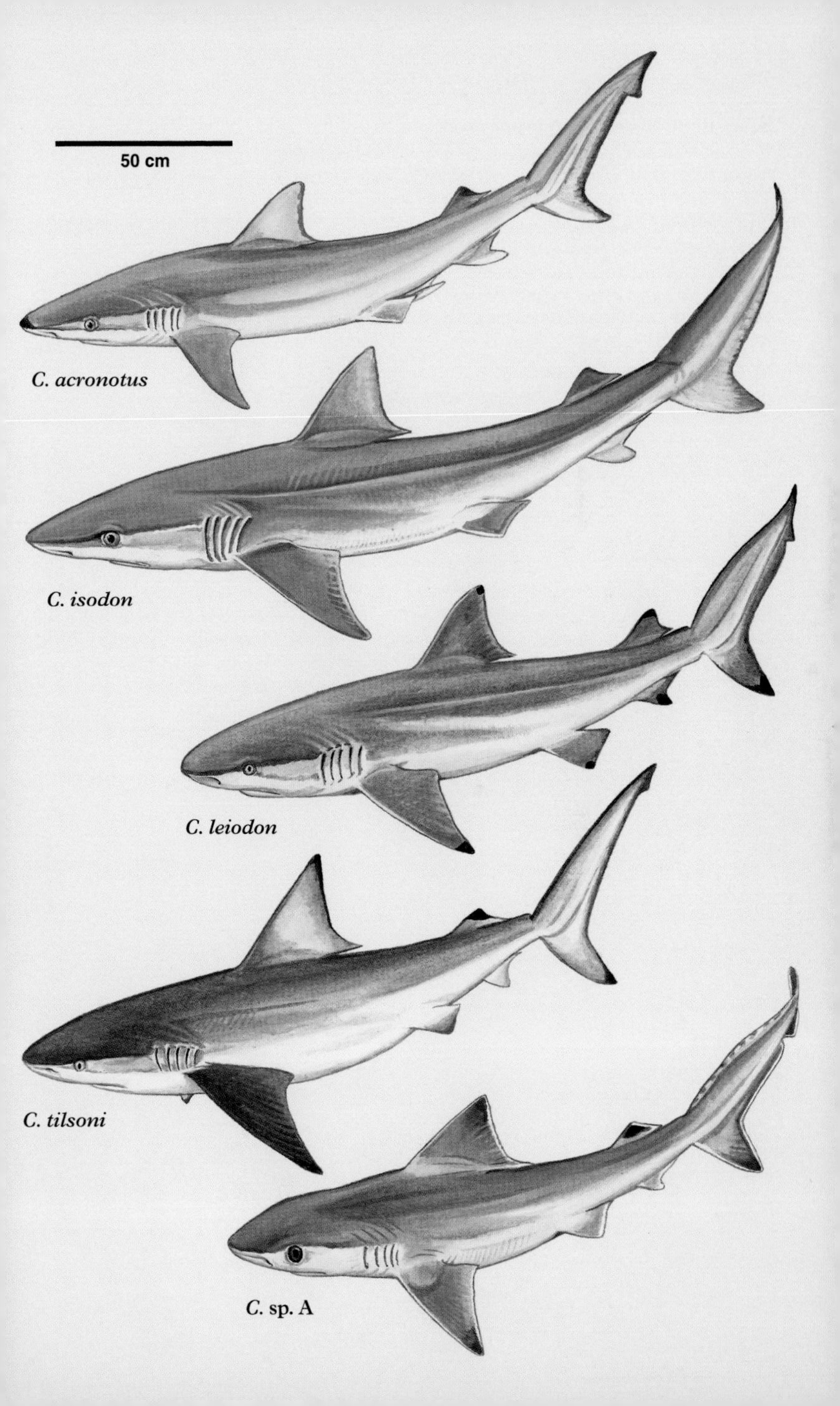
50 cm
C. acronotus
C. isodon
C. leiodon
C. tilsoni
C. sp. A

Plate 63 Sphyrnidae I

Winghead Shark *Eusphyra blochii* **p322**

N IO to Austr. and China; shallow water contin. and insular shelves. Immense wing-shaped head with narrow blades, width between eyes ~ 1/2 TL; D1 fin origin over P fin bases, futher forward than other hammerheads; upper precaudal pit longi. not crescentic; brown above, white below.

Mallethead Shark *Sphyrna corona* **p323**

E Pac., Mex. to Peru; presumably contin. shelf. Mod. broad ant. arched mallet-shaped head, medial and lateral indent. on ant. edge and trans. post. margin, no prenarial grooves, snout mod. long ~2/5 head width, small strongly arched mouth; D1 fin free rear tip over V fin insert., A fin post. margin ~ straight; upper precaudal pit trans. crescentic; grey, white below extending to back of head.

Scoophead *Sphyrna media* **p324**

W Atl., Pan. to S Brazil and E Pac., G of Calif. to N Peru; contin. shelf. Mod. broad ant. arched mallet-shaped head, weak medial and lateral indent. on ant. edge and trans. post. margin, no prenarial grooves, snout mod. short ~ 1/3 head width, mod. large broadly arched mouth; D1 fin free rear tip over V fin insert.; upper precaudal pit trans. crescentic; grey, lt. below.

Bonnethead *Sphyrna tiburo* **p325**

W Atl. and E Pac. trop. seas; contin. and insular shelves 0–80 m. Unique v. narrow smooth shovel-shaped head, no indent., no prenarial grooves, snout mod. long ~ 2/5 head width, broadly arched mouth; D1 fin free rear tip ant. to V fin origins, A fin post. margin shallowly concave; upper precaudal pit trans. crescentic; grey or grey-brown, often with spots, lt. below.

Smalleye Hammerhead *Sphyrna tudes* **p325**

W Atl., Venez. to Urug.; contin. shelf to 12 m+. Broad ant. arched mallet-shaped head, medial and lateral deep indent. on ant. edge none on post. margin, no prenarial grooves, snout short < 1/3 head width, mod, large broadly arched mouth; D1 fin free rear tip over V fin insert., A fin post. margin mod. concave; upper precaudal pit trans. crescentic; grey-brown to golden, lt. below.

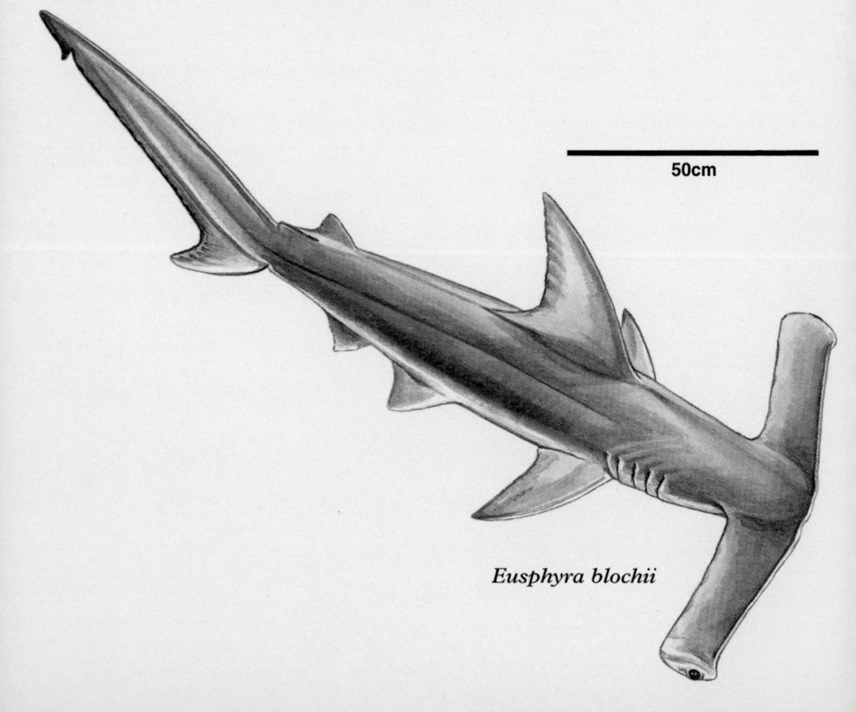

Eusphyra blochii

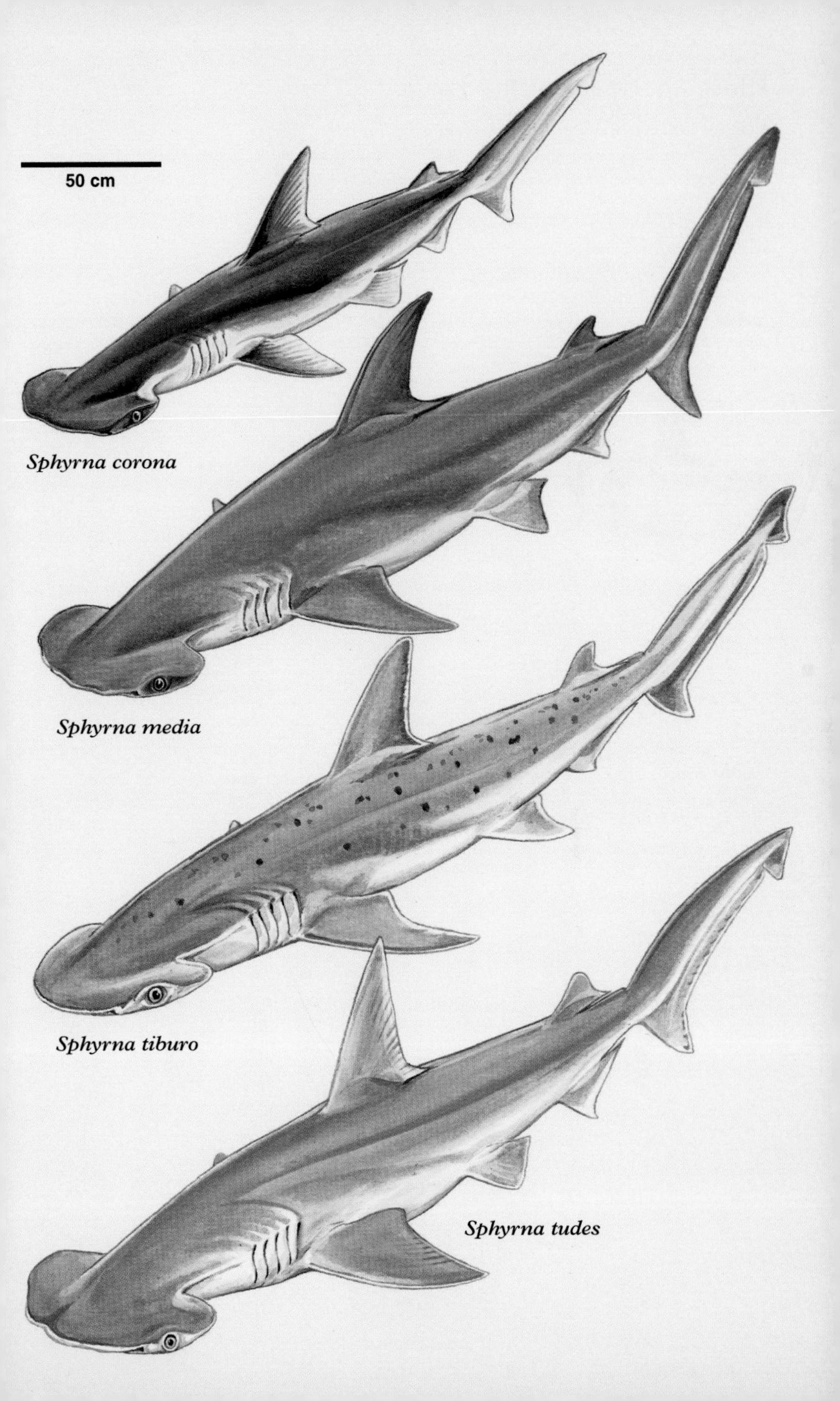
50 cm
Sphyrna corona
Sphyrna media
Sphyrna tiburo
Sphyrna tudes

Plate 64 Sphyrnidae II

Scalloped Hammerhead *Sphyrna lewini* p323

WW wm. temp. and trop. seas; over contin. and insular shelves and adj. water surface–275 m+. Broad narrow bladed head with arched ant. margin, prom. median indent. and lateral indent., well developed prenarial grooves, snout short 1/5–1/3 head width, broadly arched mouth; D1 fin mod. high, D2 and P fins low; upper precaudal pit trans. crescentic; lt. grey or bronzy with dusky P fin tips and dk. blotch on lower C fin lobe, white below.

Great Hammerhead *Sphyrna mokarran* p324

WW trop. seas; coastal-pelagic and semi-oceanic, close inshore to well offshore 1–80 m+. Broad narrow bladed head with nearly straight ant. margin, prom. median indent. and lateral indent., no/weakly developed prenarial grooves, snout short < 1/3 head width, broadly arched mouth; D1 fin v. high and falcate, D2 and P fins high; upper precaudal pit trans. crescentic; lt. grey or grey-brown, fins unmarked, white below.

Smooth Hammerhead *Sphyrna zygaena* p326

WW wm. temp. and trop. seas; contin. and insular shelves, close inshore to well offshore 0–20 m+. Broad narrow bladed head with broadly arched ant. margin, no median indent. but prom. lateral indent., well developed prenarial grooves, snout short 1/5–<1/3 head width, broadly arched mouth; D1 fin mod. high, D2 and P fins low; upper precaudal pit trans. crescentic; olive-grey or dk. grey-brown, undersides of P fin tips dusky, white below.

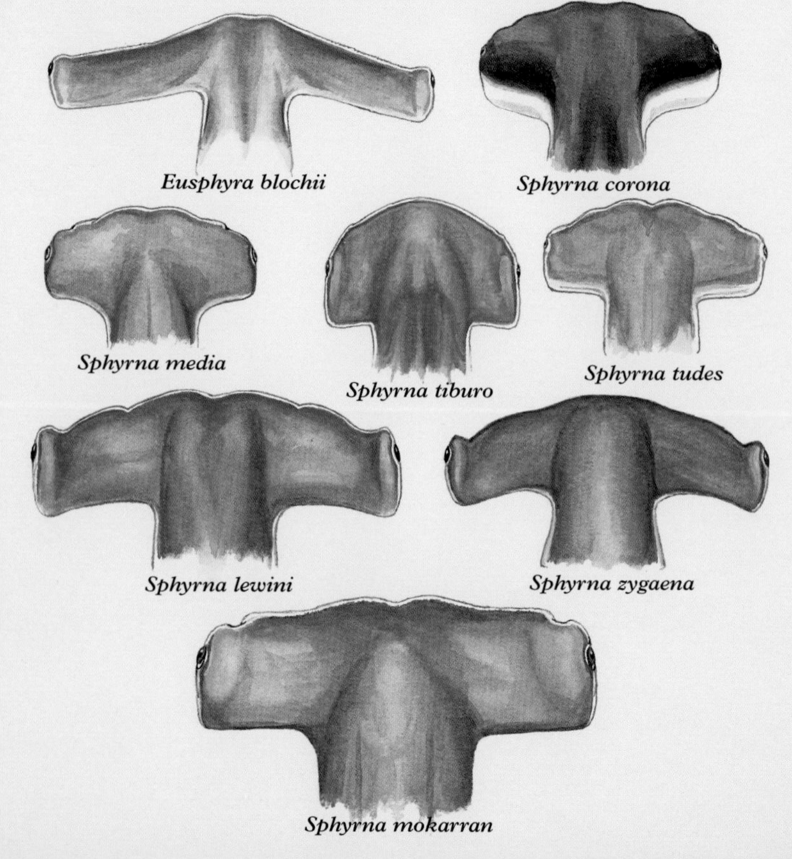

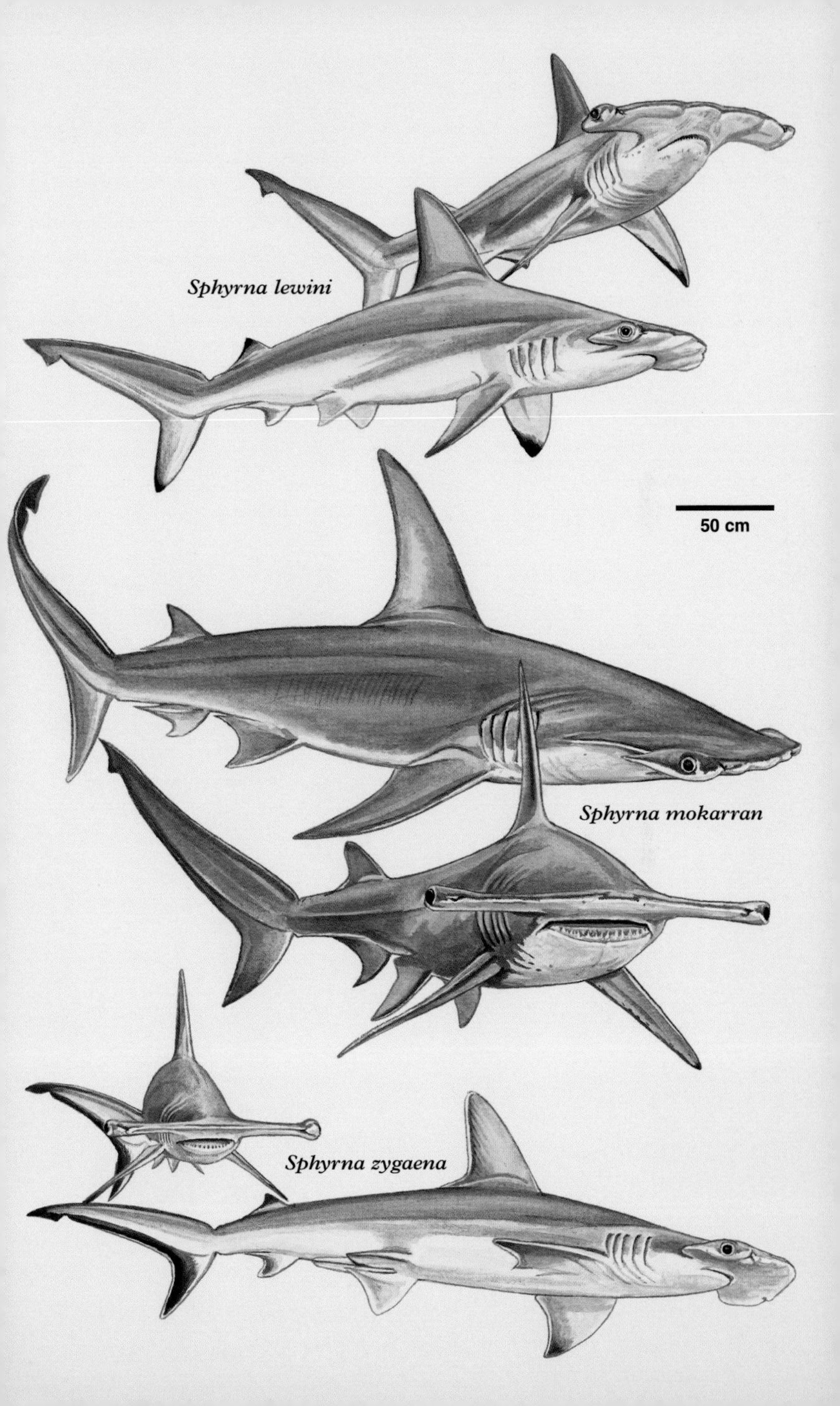
Sphyrna lewini
50 cm
Sphyrna mokarran
Sphyrna zygaena

Hexanchiformes: Cow and Frilled Sharks

This order includes two families of distinctive sharks: Chlamydoselachidae (Frilled Sharks) and Hexanchidae (Cow Sharks).

Identification. Six or seven pairs of gill slits in front of pectoral fins; one spineless dorsal fin over or behind pelvic fins; one anal fin. Vertebral column extends into the caudal fin's long dorsal lobe, ventral lobe short or absent. Large mouth. Eyes on the side of the head, spiracles very small, located well behind and above eyes.

Biology. Most species are widespread world-wide, from tropical to temperate and boreal waters. Mostly found in deep cold water in the tropics, but also inshore in temperate seas.

Status. Taken as bycatch and in targeted fisheries (trawls, nets, lines and by sports anglers).

Chlamydoselachidae: Frilled Sharks

All frilled sharks are very similar externally (with some local variability) and often considered to be one wide-ranging living species. Internally, specimens from the southern African coast (southeast Atlantic and southwest Indian Ocean) differ from sharks from Japan and Taiwan (northwest Pacific) and apparently are a distinct species. More anatomical information (e.g. chondrocranium of skull, vertebral and fin structure) is needed from specimens in the east and northwest Atlantic, and eastern Pacific.

Frilled Shark *Chlamydoselachus anguineus* Plate 1

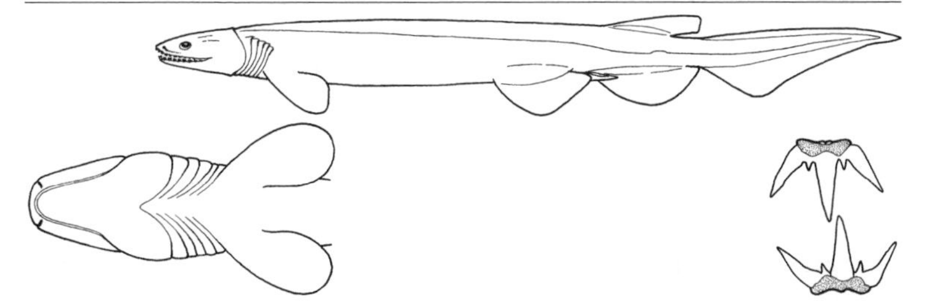

Measurements. Born: ~39 cm TL. Mature: ~92–163 ♂, 130–135 cm ♀. Max: 196 cm TL ♀.

Identification. Dark chocolate-brown, brownish-grey or brownish-black. Elongated eel-shape. Flattened snake-like head. Very short snout. Large terminal mouth with widely spaced, needle-sharp, slender three-cusped teeth. Six pairs of curved gill slits, lower ends of first pair connected under throat. Dorsal fin low and much smaller than anal fin, pectoral fins smaller than pelvic fins.

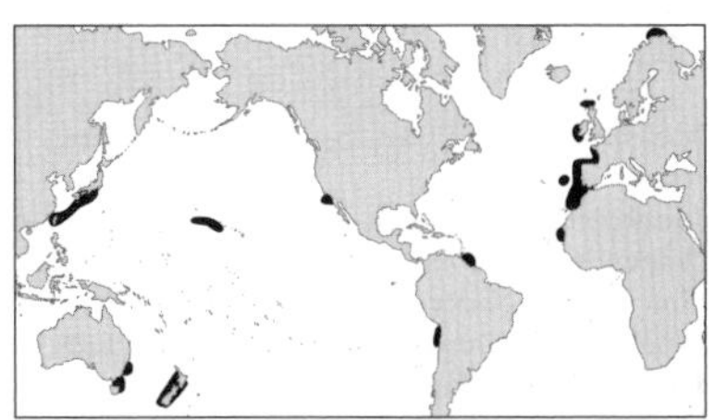

Distribution. Widely but patchily distributed world-wide, rare to uncommon, usually in deep water.

Habitat. Benthic, epibenthic and pelagic. Offshore shelves and upper continental and island slopes (50–1500 m). Occasionally at the surface.

Behaviour. Feeds on deepwater actively-swimming squid and fishes. Habit (in captivity) of swimming with mouth open may lure prey to conspicuous white teeth.

Biology. Ovoviviparous; 6–12 pups per litter feed on huge uterine eggs (11–12 cm long). Pregnant females have very large abdomens. May reproduce all year in deep water, but mates in spring off Japan. Gestation period probably very long (one to two years).

Status. IUCN Red List: Near Threatened. Bycatch of deep bottom trawls and gillnets. Utilised for fishmeal and meat. Occasionally kept in aquaria. Harmless to man.

Southern African Frilled Shark *Chlamydoselachus* sp. A Plate 1

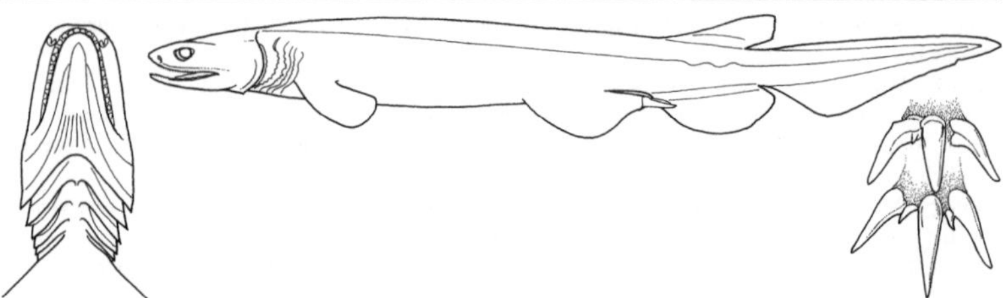

Identification. Difficult to distinguish externally from *Chlamydoselachus anguineus*. Different internal structure of chondrocranium and vertebrae, vertebral counts and calcification patterns, pectoral-fin skeletal morphology and radial counts, and intestinal valve counts.

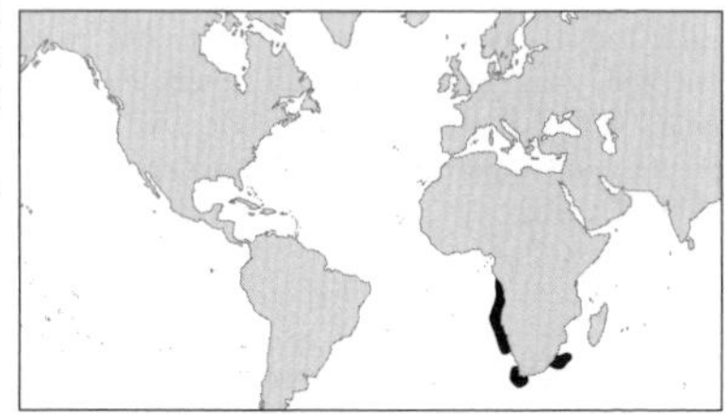

Distribution. Southern Africa (Angola, Namibia and South Africa).
Habitat. As for *C. anguineus*.
Behaviour. As for *C. anguineus*.
Biology. As for *C. anguineus*.
Status. IUCN Red List: Not Evaluated.

Hexanchidae: Cow Sharks

Moderately slender to stocky cylindrical sharks with six or seven pairs of gill slits (first pair not connected across the throat) in front of pectoral fins. Ventral mouth. Large compressed comb-like teeth in the lower jaw, smaller cuspidate teeth in the upper jaw. Single spineless dorsal fin, relatively high, angular and short. Pectoral fins angular, larger than pelvic fins. Single anal fin smaller than the dorsal fin. Marked sub-terminal notch in the caudal fin. Four species.

Sharpnose Sevengill Shark *Heptranchias perlo* Plate 1

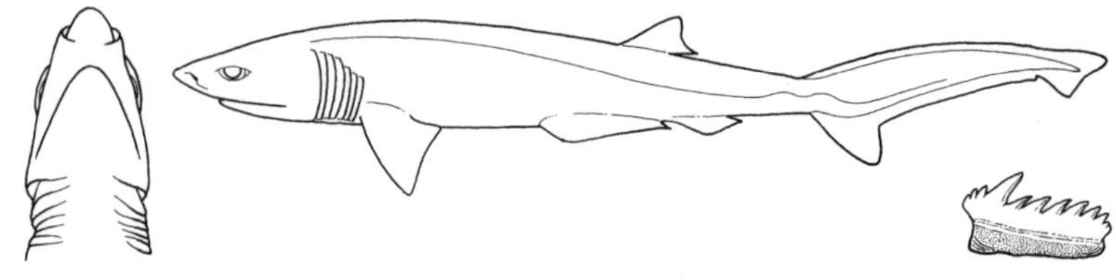

Measurements. Born: 26–27 cm TL. Mature: ~80 ♂, ~100 cm ♀. Max: 139 cm TL. (214 cm TL record an error?).
Identification. Acutely pointed head. Seven pairs of gill slits. Narrow mouth, five rows of comb-shaped teeth in lower jaw. Large eyes. Black blotch on tip of dorsal fin and upper caudal lobe prominent in young, faded or absent in adults.

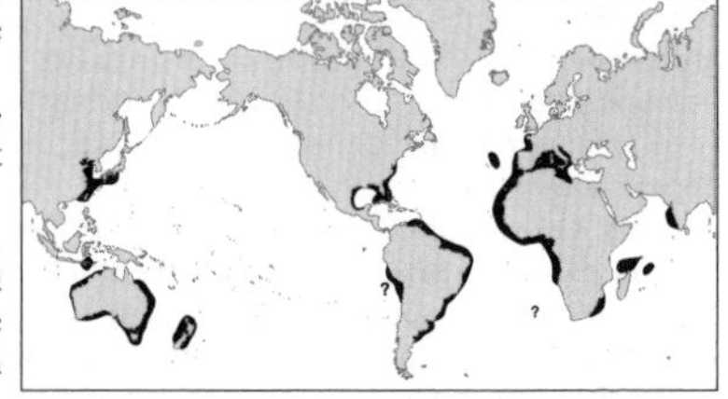

Distribution. Wide-ranging but patchily distributed. Tropical and temperate seas, not northeast Pacific.
Habitat. Mainly deepwater (27–720 m, max. 1000 m), continental and island shelves and upper slopes, occasionally shallower water close inshore. Benthic and epibenthic; may also swim well off the bottom.
Behaviour. Poorly known. Probably a strong active swimmer feeding on small to moderately large

demersal and pelagic fishes, small sharks, crustaceans, squid and cuttlefish. Bites when captured.
Biology. Ovoviviparous, 6–20 pups per litter. Apparently reproduces year round.
Status. IUCN Red List: Near Threatened. Relatively uncommon. Bycatch in bottom trawl and longline fisheries may have caused population declines. Occasionally kept in aquaria.

Bluntnose Sixgill Shark *Hexanchus griseus* Plate 1

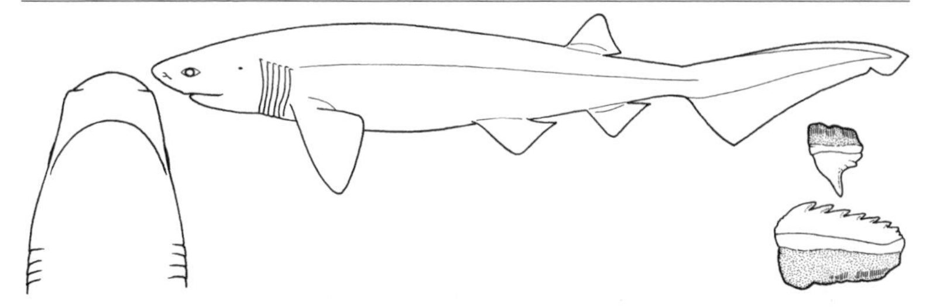

Measurements. Born: 65–74 cm TL. Mature: ~309–330 ♂, 350–420 cm ♀. Max: at least 482, probably 550 cm TL.
Identification. Skin grey or tan to blackish, sometimes darker spots on sides. Light coloured lateral line and trailing fin edges. Underside lighter than upper in newborns. Large heavy powerful body, soft supple fins. Broad head, ventral mouth (width >two x length), six rows of large comb-shaped lower teeth on each side. Small eyes, dark pupil ringed with white, fluorescent blue-green in life.

Distribution. Patchy distribution world-wide; possibly absent from Arctic and Antarctic.
Habitat. Shelves and slopes of continents, islands, seamounts and mid-ocean ridges, usually 500–1100 m, to at least 1875 m. Young may occur close inshore in cold water, and adults in shallow water close to submarine canyons.
Behaviour. Observed by divers, filmed from submersibles, and tracked for short distances. Occurs alone and in groups. A slow but strong swimmer, often sensitive to light. Adults passive, reported docile by divers, but small sharks aggressive when captured.
Biology. Feeds on squid, benthic and pelagic bony fishes, small sharks and rays. Large sharks (>2 m) take cetaceans and seals. Ovoviviparous, very large litters (22–108 pups). Young and adults may be segregated. Possibly long-lived.
Status. IUCN Red List: Near Threatened. Locally common in bycatch and target fisheries for food, fishmeal and oil, and in sports fisheries. Vulnerable to overfishing (fisheries collapsed in the Maldives, declined in Canada). Subject of dive ecotourism in British Columbia, Canada.

Bigeye Sixgill Shark *Hexanchus nakamurai* (=*H. vitulus*) Plate 1

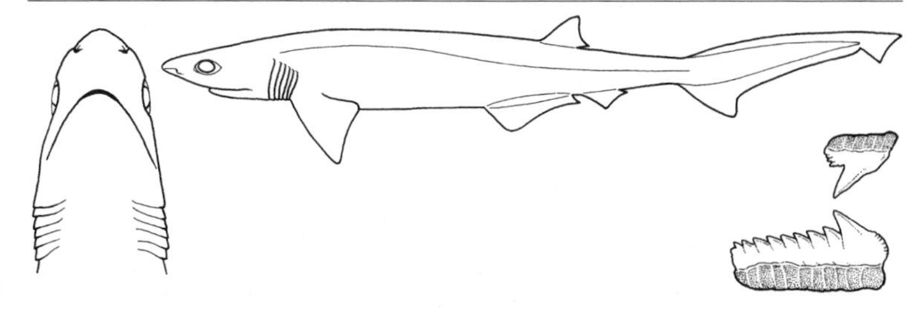

Measurements. Born: 40–43 cm TL. Mature: ~123 ♂, ~142 cm ♀. Max: >180 cm TL.
Identification. Colour sharply divided between dark dorsal and light ventral surface. Fins usually with white trailing edges and tips, sometimes dusky. Slender shark. Body and fins quite firm. Narrow head, narrow ventral mouth (width ~1.5 x length). Five rows of large comb-shaped teeth in lower jaw on each side. Large eyes. Short ventral caudal lobe (strong in adults, weak in young). Upper caudal lobe deeply notched.

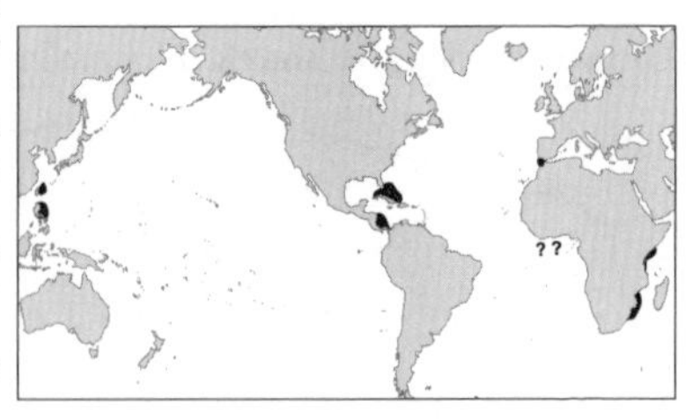

Distribution. Widely but patchily distributed in warm temperate and tropical seas, perhaps not in east Pacific.
Habitat. Continental and island shelves and slopes, on or near bottom, 90–621 m, occasionally near surface or inshore.
Behaviour. Little-known, primarily deepwater shark. Approaches submersibles cautiously.
Biology. Ovoviviparous, 13–26 pups/litter. Feeds on small to medium-sized bony fishes and crustaceans.

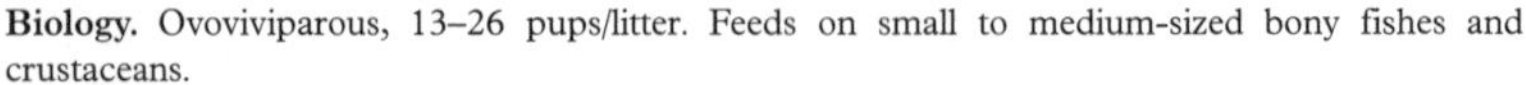

Status. IUCN Red List: Not Evaluted. Uncommon to rare, taken as bycatch in some fisheries, not commercially important. Formerly named *Hexanchus vitulus*.

Broadnose Sevengill Shark *Notorynchus cepedianus* Plate 1

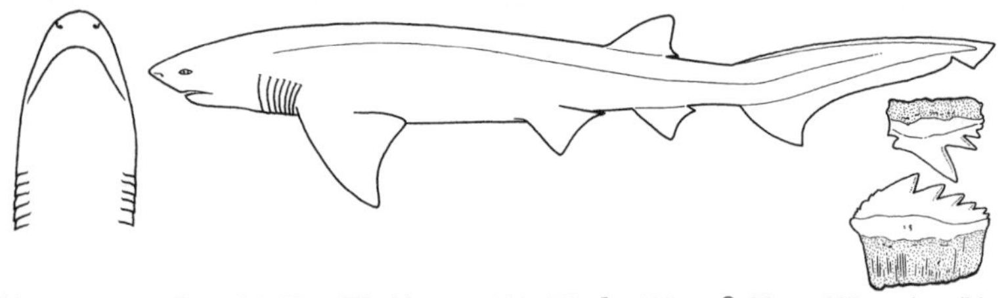

Measurements. Born: 34–45 cm TL. Mature: ~130–170 ♂, ~200 cm ♀. Max: ~290 cm (possibly 300–400 cm TL).
Identification. Usually many small black spots on grey/brown body, occasionally plain or with white spots. Newborns' black tips to dorsal fin and upper caudal lobe vanish with growth. Bluntly pointed broad head. Wide mouth, six rows of large comb-shaped lower teeth on each side. Small eyes.

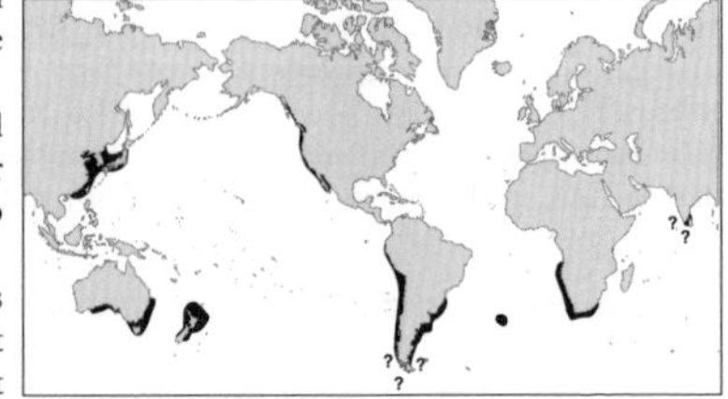

Distribution. Wide but disjunct, mostly in cool temperate inshore continental waters. Some populations may be isolated.
Habitat. Coastal, common in shallow bays and close to shore, to the surf line (<1–50 m). Larger sharks range into deeper channels and offshore to at least 136 m.
Behaviour. Active strong swimmer, often cruises steadily and slowly near the bottom, sometimes at the surface. Attacks prey at high speed. Most active at night, during overcast conditions, and in turbid water. Often moves in and offshore with rising and falling tide, and out of low salinity conditions. Apparently social; often found in aggregations. Probably migratory in temperate areas, but adjacent populations may be isolated and use different breeding grounds.
Biology. Powerful top predator on marine vertebrates (chondrichthyans, bony fishes and marine mammals such as seals and small cetaceans); also eats carrion. Some observed 'spy-hopping'. Ovoviviparous (aplacentally viviparous). Mating may occur in autumn/winter. Probably a one-year gestation followed by one year recovery. Half the adult female population gives birth each spring to 67–104 pups in shallow nursery bays. Males mature at 4–5 years and females 11–21 years, longevity 30–50 years. Newborn length doubles in 6 months; adults grow only 0–9 cm/year. Rarely harmful to

people (captive sharks have occasionally bitten divers and anglers), but few verified reports of unprovoked biting of people in the wild.

Status. IUCN Red List: Near Threatened globally; Vulnerable in the northeast Pacific (population depleted by fishing). Probably requires careful management. Important in target and bycatch commercial fisheries and for sports angling. Utilised for meat, hide, liver oil and display in aquaria. Formerly named *Notorynchus maculatus* and *N. pectorosus*.

Squaliformes: Dogfish Sharks

A large and varied order with 130 species in seven families: Echinorhinidae, bramble sharks (two species); Squalidae, dogfish sharks (at least 18 species); Centrophoridae, gulper sharks (15 species); Etmopteridae, lantern sharks (50 or more species, some undescribed); Somniosidae, sleeper sharks (15–17 species); Oxynotidae, roughsharks (five or six species); and Dalatiidae, kitefin sharks (nine species).
Identification. Two dorsal fins (with or without spines); no anal fin. Caudal fin with vertebral column elevated into a moderately long dorsal lobe; ventral lobe absent to strong. Five gill slits, all in front of pectoral fin origins. Nostrils not connected to the mouth by grooves. Spiracles behind and about opposite or above level of eyes. Eyes on side of head, no nictitating lower eyelids. Size ranges from dwarf to huge.
Biology. All species for which reproductive mode is known are ovoviviparous (aplacental viviparity), litter size from one or two to over 20 pups. Some dogfish sharks have the oldest known age at maturity, lowest fecundity (one foetus per litter in a few *Centrophorus*) and longest known gestation period (18–24 months for Spurdog, *Squalus acanthias*). Some species are solitary; others form huge nomadic schools that range long distances on annual migrations. Some hunt and feed cooperatively. A few (cookie-cutter sharks) are parasitic.
Status. Dogfish sharks occur in a wide range of marine and estuarine habitats and depths in all oceans world-wide and include the only sharks found in high latitudes close to the poles. Their greatest diversity occurs in deepwater (many species occur nowhere else). Many species are fished commercially for meat or liver oil. Those with slow growth and low reproductive capacity are highly vulnerable to depletion by overfishing.

Echinorhinidae: Bramble Sharks

Two species of large, sluggish, deepwater sharks.
Identification. Skin denticles very large and thorn-like. Stout cylindrical body, five gill openings in front of pectoral fin, fifth larger than others. Broad flat head and snout, very small spiracles far behind the eyes. Two similar-sized small dorsal fins, no spines, set close together and close to the caudal fin, origin of first dorsal slightly behind pelvic fin origin. No anal fin. Ventral caudal lobe poorly developed in adults, absent in young, subterminal caudal notch lacking or not obvious.
Biology. Widely distributed on soft bottom habitats on temperate and tropical continental and insular shelves and upper slopes, sometimes entering shallow cold water. Large mouth and pharynx may be used to suck in prey (bony fishes, small chondrichthyans and invertebrates) as it comes within range.
Status. Uncommon to rare and poorly known.

Bramble Shark *Echinorhinus brucus* Plate 5

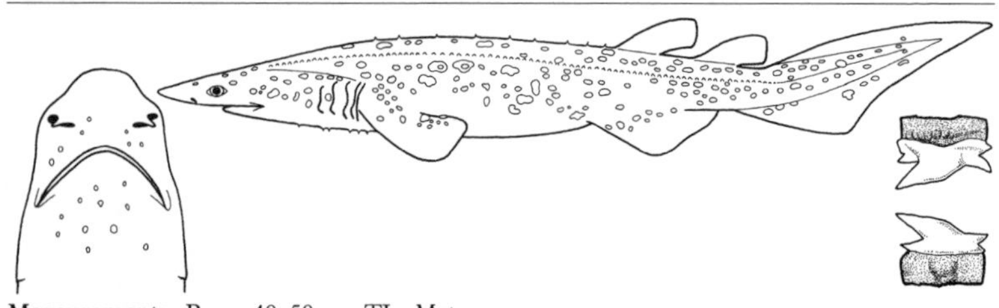

Measurements. Born: 40–50 cm TL. Mature: <150 ♂, 200–220 cm ♀. Max: 305–310 cm TL.
Identification. Grey, brownish or blackish, often lighter below. May have red or black spots or blotches on back and sides. Fin edges blackish.. Adults have large (>1 cm) sparsely and irregularly scattered whitish thorn-like denticles with smooth margins, some fused into multi-cusped plates. Young (<90 cm) have close-set small denticles below snout and around mouth,

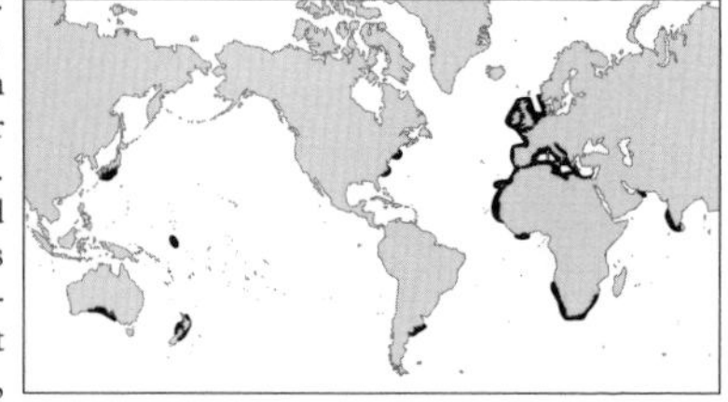

becoming large, scattered and conspicuous in larger sharks.
Distribution. East Atlantic, Mediterranean, Indo-west Pacific.
Habitat. Deepwater, continental and island shelves and slopes, on or near bottom (200–900 m). In shallow cold water upwelling areas.
Behaviour. Poorly known. Sluggish.
Biology. Ovoviviparous, 15–26 pups/litter. Eats bony fishes, small sharks, crustaceans.
Status. IUCN Red List: Data Deficient. Apparently rare, occasionally fished.

Prickly Shark *Echinorhinus cookei* Plate 5

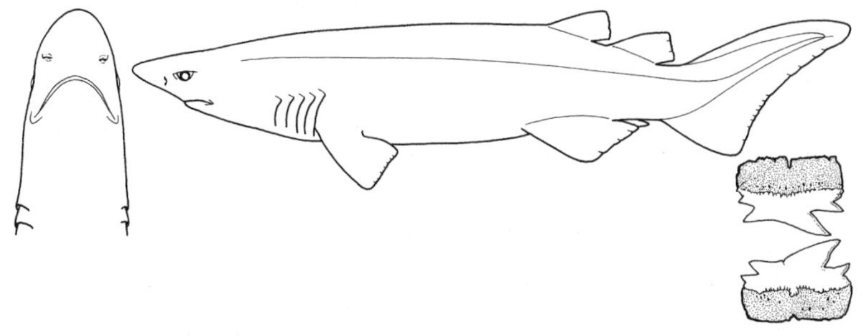

Measurements. Born: 45 cm TL. Mature: ~180–230 ♂, >254 cm ♀. Max: 400 cm TL.
Identification. Colour light to medium grey, grey-brown, or blackish, often lighter below. White around mouth and beneath snout, trailing fin edges blackish. Numerous regularly spaced (not fused) light coloured inconspicuous denticles with single cusps and scalloped bases (<5 mm). Very few below snout and around mouth.
Distribution. Pacific Ocean.
Habitat. Continental and insular shelves and upper slopes, seamounts. Close to bottom, prefer-ring soft sediments, 11–1100 m.
Behaviour. Sluggish and docile, seen stationary, swimming slowly alone and in groups (not school-ing). May investigate divers and submersibles.
Biology. Ovoviviparous, litter of 114 pups recorded. Eats bony and cartilaginous fishes, octopus and squid.
Status. IUCN Red List: Near Threatened. Rare, fragmented range, bycatch of deepwater fisheries.

Squalidae: Dogfish Sharks

Two genera, *Cirrhigaleus* (two species) and *Squalus*, with three main groups of species and/or subspecies: *S. acanthias* (one species, possible regional subspecies), *S. megalops* (two or three species) and *S. mitsukurii* (at least ten species, several recently discovered and not yet described). Dogfish sharks are recorded almost world-wide in boreal, temperate and tropical seas, except in the tropical eastern Pacific. (Tropical records may be incomplete because cooler offshore and deep water is poorly surveyed.)
Identification. Two dorsal fins with strong ungrooved spines. No anal fin. Snout short to moderately long, mouth short and transverse, low bladelike cutting teeth in both jaws. Spiracles large and close to eyes. Body cylindrical in cross section. First dorsal origin opposite or slightly behind the pectoral fins, second dorsal fin strongly falcate. Pelvic fins smaller than pectoral fins. Caudal peduncle with strong lateral keels, caudal fin without subterminal notch. Colour light grey to medium brown, not black, no luminous organs. *Cirrhigaleus* is stockier, with rough skin, enlarged nasal barbels, precaudal pits vestigial or absent, a short caudal peduncle, equal-sized first dorsal fins, and a weak ventral caudal lobe. *Squalus* is usually slenderer, smooth-skinned, has smaller nasal barbels, strong precaudal pits, an elongated caudal peduncle, the second dorsal fin much smaller

than the first, and a strong ventral caudal lobe. Many species of *Squalus* are very hard to distinguish without precise measurements of features and vertebral counts and the use of a more detailed key than can be provided here (see Compagno in prep. a).

Biology. Probably all ovoviviparous (aplacentally viviparous), 1–32 pups/litter. Several species are social. Some *Squalus* form huge highly nomadic schools that undertake local and long-distance annual migrations and feed communally, sometimes cleaning out or driving away local populations of prey species. Other species seem to be solitary or only occur in small groups. Prey includes mainly bony fishes, cephalopods and crustacea, also other cartilaginous fishes and invertebrates. All species have powerful jaws and can dismember prey larger than themselves. Some species are wide-ranging in both northern and southern hemispheres, but absent from intervening tropical seas (e.g. *S. acanthias*) or equatorial zone. Others are localized endemics. Most are benthic (although some have pelagic young). They occur from the intertidal (mainly in cool temperate waters) down to depths of 600 m, occasionally to 1446 m, but are usually replaced by other deepwater shark families below 700–1000 m).

Status. Despite the considerable (original) abundance and wide range of some species, they are highly vulnerable to overfishing because of their late maturity, longevity, low fecundity and long intervals between litters (*S. acanthias* has the longest known gestation period of any vertebrate: 18–24 months). Several stock declines have occurred and recovery of depleted populations is very slow. Rare species of no little or no commercial value that occur in heavily fished areas are taken as bycatch in fisheries for more common and abundant species. They may become threatened with extinction if restricted to small endemic or localised populations or isolated habitats, such as seamounts, targeted by fishing operations. Their abundance and ease of capture make some species of *Squalus*, particularly *S. acanthias*, among the most important species targeted by commercial shark fisheries. They are landed by up to 50 countries in directed fisheries or as utilised bycatch, mostly taken in bottom trawl fisheries, but also by hook and line, gill nets, seines, pelagic trawls and fish traps. They are of high value for their meat, liver oil, fins and occasionally leather. Sports anglers target some species, but they are not important game fish. A few species are regularly displayed in aquaria. Some species use their mildly toxic fin spines and teeth for defence; a hazard to human handlers. Some are considered 'trash fish' because they can cause damage to fishing gear and prey on or drive away more valuable fisheries species.

Roughskin Spurdog *Cirrhigaleus asper* Plate 2

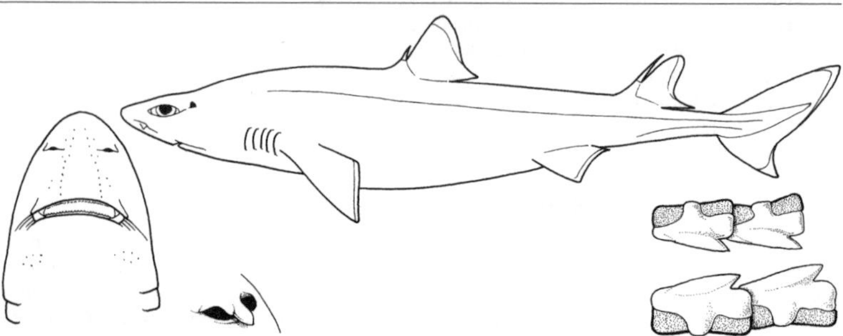

Measurements. Born: ~25–28 cm TL. Mature: 85–90 ♂, 89–118 cm ♀. Max: 118 cm TL.

Identification. Fins conspicuously white-edged without dark markings, no white spots on sides. Stocky rough-skinned body, broad flat head, short broadly rounded snout, anterior nasal flaps with short stubby medial barbel. Upper precaudal pit weak or absent. Dorsal fins equal-sized with very high stout spines, first dorsal origin behind pectoral bases.

Distribution. Warm temperate to tropical west Atlantic, Indian Ocean and central Pacific.
Habitat. On or near bottom, upper and outer continental and insular shelves and slopes (73–600 m), sometimes off bays and river mouths.
Behaviour. Unknown. Heavy body may indicate a moderately inactive benthic life.
Biology. Ovoviviparous, 18–22 pups/litter. Eats bony fish and squid.
Status. IUCN Red List: Not Evaluated. Not commercially fished, probably taken as bycatch.

Mandarin Dogfish *Cirrhigaleus barbifer* Plate 2

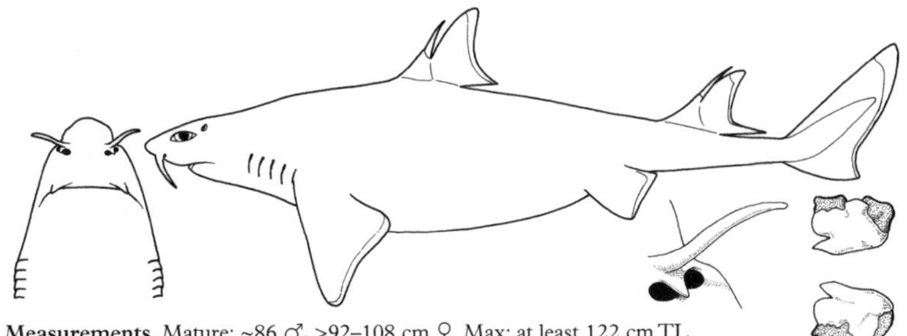

Measurements. Mature: ~86 ♂, >92–108 cm ♀. Max: at least 122 cm TL.
Identification. As for *Cirrhigaleus asper*, but barbels on anterior nasal flaps very elongated and moustache-like, reaching the mouth.
Distribution. Western Pacific. Patchy, southern Japan to New Zealand.
Habitat. Continental and insular upper slopes and outer shelves, 146–640 m, on or near bottom.
Behaviour. Unknown. Long barbels may contain chemosensory receptors for prey location.
Biology. Ovoviviparous. One reported litter of ten pups (five per uterus). Prey unknown.
Status. IUCN Red List: Near Threatened. Uncommon to rare, taken in bycatch.

Piked Dogfish *Squalus acanthias* Plate 3

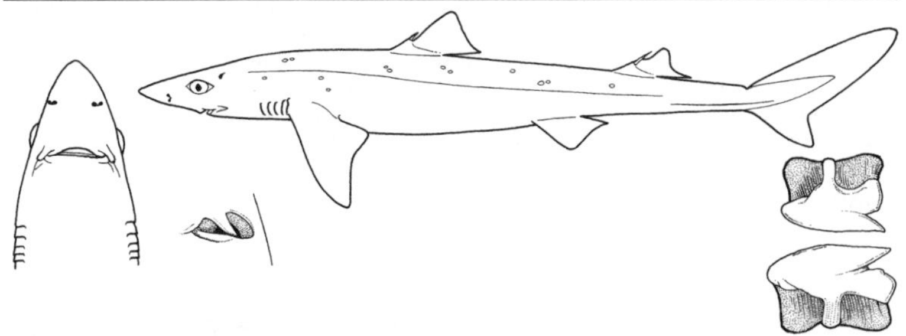

Measurements. Born: 18–33 cm TL. Mature: 52–100 ♂, 66–120 cm ♀. Max: 160–200 cm TL. Highly variable regionally.
Identification. See family. Slender dogfish with body grey to bluish-grey above, lighter to white below, often with white spots on sides; pectoral fins with light posterior margins in adults, dorsal fin tips and edges dusky or plain in adults, with black apices and white posterior margins and free rear tips in young; no conspicuous black blotches on fins. Narrow head, relatively long, pointed snout and no medial barbel on anterior nasal flaps. Pectoral fins with shallowly concave posterior margins and narrowly rounded rear tips. First dorsal fin low, origin usually behind or sometimes over pectoral free

rear tips, first dorsal spine slender and very short with origin well behind pectoral free rear tips.

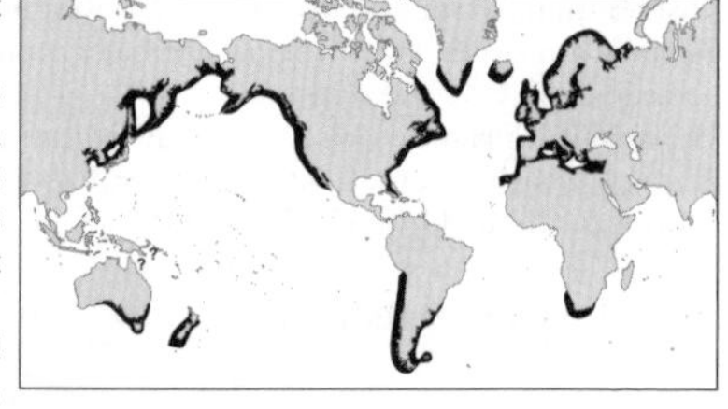

Distribution. Almost world-wide, except tropics and near poles. Little or no mixing between northern and southern hemisphere populations and limited genetic mixing between some stocks with overlapping range and feeding grounds but different migration patterns.

Habitat. Boreal to warm-temperate continental and insular shelves, occasionally slopes (0–600, possibly 1446 m), from surface to bottom. Epipelagic (0–200 m) in cold water. Usually near bottom on continental shelf, near surface in oceanic waters. Often on soft sediments in enclosed and open bays and estuaries, where most nursery grounds occur.

Behaviour. Mainly demersal, apparently also epipelagic, sometimes solitary or schooling with other small sharks, often forming immense dense feeding aggregations on rich feeding grounds. Segregates by size and sex into packs or schools of small juveniles (both sexes), mature males, larger immature females, or large mature females (often pregnant). Mixed adult schools occasionally reported. Pregnant females usually pup in shallow inshore nursery grounds; some populations pup in deepwater on outer shelves and upper slopes. Slow swimmer but undertakes long-distance seasonal migrations north–south or deep–shallow as water temperature changes (prefers 7–8 to 12–15°C). Other populations are resident year-round. Long-distance movements reported from tag returns include a 6500 km Pacific Ocean crossing, and at least 1600 km in the northwest Atlantic.

Biology. Extremely long-lived, slow-growing and late-maturing. Ovoviviparous. Litter size (1–32) varies regionally and larger females have more and larger pups. Gestation period varies regionally, from 18–24 months, to only 12 months in the Black Sea (where largest females, pups and litters occur). Mating may occur in winter, with pups born in winter, spring or summer. Age of younger fish can be measured from annual growth rings on fin spines (spines wear down in old fish) or seasonal calcium peaks in vertebrae. Two northeast Atlantic males tagged and measured when mature, and recaptured over 30 years later grew only 0.27–0.34 cm annually. Age at maturity, 10–25 years, and longevity, 70–>100 years, varies between populations. Feeds mainly on bony fishes and invertebrates, occasionally other chondrichthyans. Can capture and dismember prey larger than itself. Predators include large sharks, teleosts and marine mammals.

Status. IUCN Red List: Near Threatened globally (Endangered northeast Atlantic, Vulnerable northwest Atlantic). Extremely well-studied. Possibly once the most abundant shark and the most important commercial species, utilised for meat (high value in Europe), liver oil and fins and supporting large target trawl and line fisheries comparable to those for bony fishes. Its slow growth, late maturity, longevity and low reproductive capacity make it highly vulnerable to overfishing, particularly since aggregations of large pregnant females are usually targeted. Few fisheries are managed, some stocks are now very seriously depleted or collapsed and catches declining steeply. Also of commercial fisheries significance because large numbers may damage fishing gear and affect catches of other species. Targeted by sports anglers in some regions, displayed in public aquaria, and important for scientific research and teaching. Not dangerous to man except through lacerations from the mildly toxic dorsal spines.

Longnose Spurdog *Squalus blainvillei* Plate 3

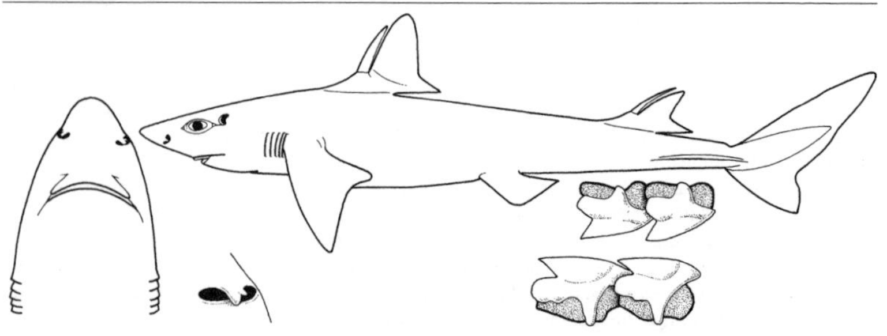

Measurements. Born: ~23 cm TL. Mature: ~50 ♂, 60 cm ♀. Max: 80–100 cm TL.
Identification. Greyish-brown heavy body, no white spots, no obvious dark dorsal or caudal fin markings, white-edged dorsal fins. Broad head, relatively short broad snout, large medial barbel on anterior nasal flaps. Posterior pectoral fin margins nearly straight to shallowly concave, narrowly rounded rear tips. High erect first dorsal fin, origin over or just behind pectoral bases, origin of very heavy long spine over pectoral inner margins.
Distribution. Temperate to tropical east Atlantic. (Possibly west Pacific?)

Habitat. On or near muddy bottom, continental shelves and upper slopes, 16–>440 m.
Behaviour. Forms large schools.
Biology. Ovoviviparous, three or four pups/litter. Eats bony fishes, crustaceans and octopi.
Status. IUCN Red List: Not Evaluated. Reportedly common, but possibly confused with *Squalus megalops* and *S. mitsukurii*. Fished with other *Squalus* species.

Cuban Dogfish *Squalus cubensis* Plate 3

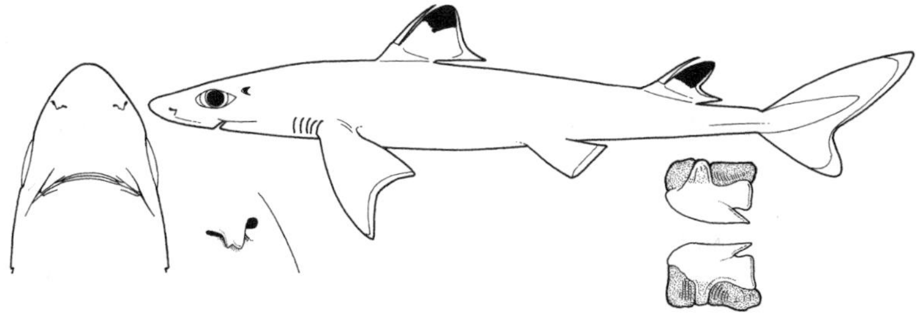

Measurements. Mature: ~50 cm TL. Max: 75–?110 cm TL.
Identification. Slender unspotted grey body, large black or dusky patches on dorsal fin tips, conspicuous white posterior margins on pectoral and caudal fins. Broad head, short rounded snout, small medial barbel on anterior nasal flaps. Deeply concave posterior pectoral fin margins and angular rear tips. First dorsal fin moderately high, spine slender and high, fin and spine origin over pectoral inner margins.
Distribution. Warm-temperate and tropical west Atlantic.

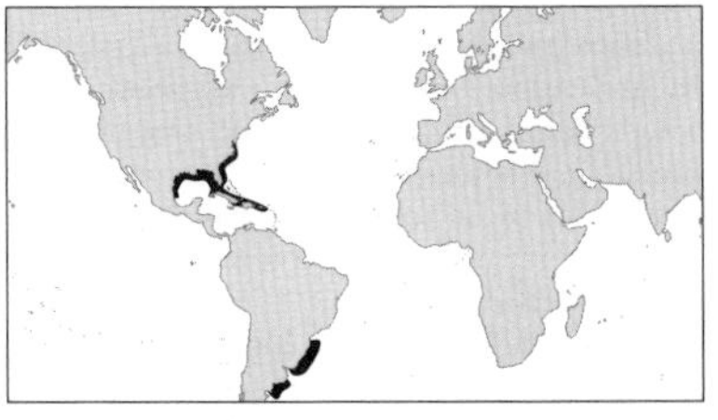

Habitat. Offshore, outer continental shelf and upper slopes, on or near bottom, 60–380 m. Young shallower than adults.
Behaviour. Occurs in large dense schools.
Biology. Ovoviparous, ~ten pups/litter.
Status. IUCN Red List: Not Evaluated. Apparently common. Fished for liver oil and vitamins.

Japanese Spurdog *Squalus japonicus* Plate 3

Measurements. Born: 19–30 cm TL. Mature: ~50–70 ♂, 56–80 cm ♀. Max: ~95 cm TL.
Identification. Fairly slender grey, reddish-grey or bluish-brown body without white spots. Dorsal fin apices, dorsal caudal margin, and subcaudal notch dusky but no conspicuous black patches; pectorals and caudal fin lobes with white posterior margins. Narrow head, very long pointed snout, small medial barbel on anterior nasal flaps. Shallowly concave posterior pectoral fin margins and narrowly rounded rear tips. First dorsal fin moderately high, long slender spine, origins over pectoral inner margins.

Distribution. Northwest Pacific.
Habitat. On or near bottom, temperate–tropical outer shelves and upper slopes, 120–340 m.
Behaviour. Unknown.
Biology. Ovoviparous, two to eight pups/litter (increasing with female size) each year.
Status. IUCN Red List: Not Evaluated. Common. Fished off Japan and Taiwan. Kept in aquaria.

Shortnose Spurdog *Squalus megalops* Plate 3

Measurements. Born: 23–25 cm TL. Mature: ~34–51 ♂, 37–62 cm ♀. Max: 77 cm TL.
Identification. Small slender grey-brown to dark brown dogfish without white spots. Dorsal fin apices, dorsal caudal margin, and area of posterior caudal notch dusky or blackish; pectoral and caudal fin lobes with white posterior margins; no conspicuous large black patches on fins. Broad head, short broad snout, and small medial barbel on anterior nasal flaps. Fairly concave posterior pectoral fin margins, angular rear tips. First dorsal fin moderately high, short slender spine, origins over pectoral inner margins.
Distribution. East Atlantic, Indo-west Pacific.
Habitat. Continental shelves and upper slopes, on or near seabed, 0–732 m. Nursery grounds on outer continental shelves.
Behaviour. Social, often in large dense schools, partly sexually segregated.
Biology. Ovoviviparous (aplacental viviparous), one to six pups/litter (generally two or three) after ~two-year gestation. Feeds on variety of bony fishes, invertebrates, sometimes batoids. 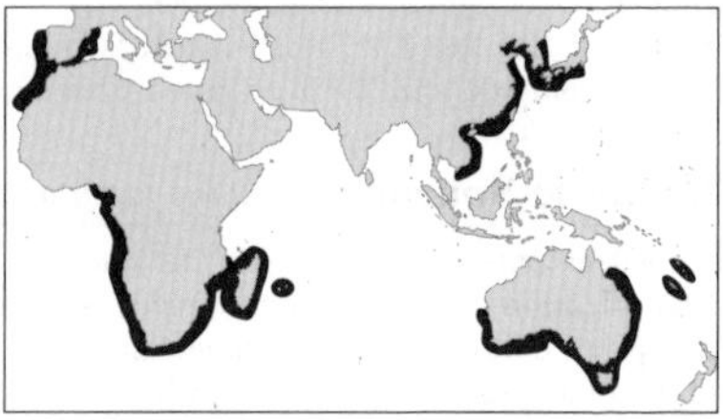
Status. IUCN Red List: Data Deficient. Common to abundant, important in some fisheries. Taxonomy uncertain, may be a species complex.

Blacktail Spurdog *Squalus melanurus* Plate 3

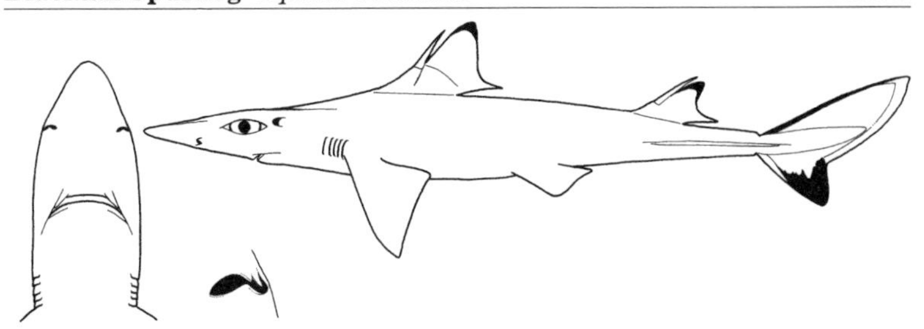

Measurements. Adults 67–80 cm TL.
Identification. Slender dark grey-brown body without white spots; black-tipped dorsal fin apices, partial black edge near end of dorsal caudal margin; prominent black patch on lower caudal lobe; pectorals, pelvics and upper caudal lobe with white posterior margins. Head broad, snout very long and broad, small medial barbel on anterior nasal flaps. Pectoral fins with straight posterior margins and narrowly rounded rear tips. First dorsal fin high, with origin over pectoral insertions, long slender first dorsal spine origin over pectoral inner margins.

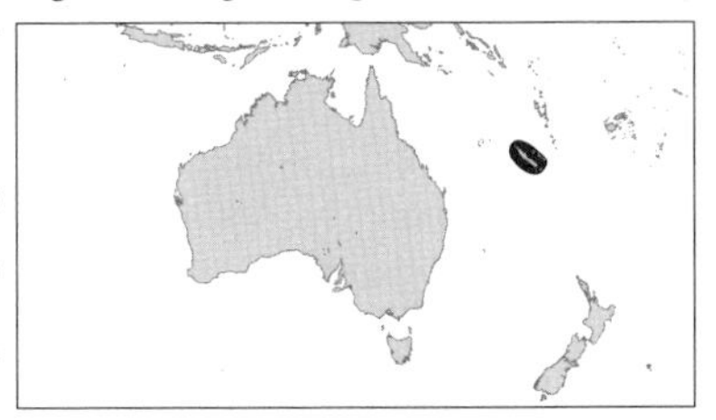

Distribution. Southwest Pacific: Noumea.
Habitat. Upper insular slopes, 34–480 m.
Behaviour. Vigorously whips body and long second dorsal spine when captured.
Biology. Probably ovoviviparous, three pups/litter. Feeds on bony fishes.
Status. IUCN Red List: Least Concern. Very little deepwater fishing within small range.

Shortspine Spurdog *Squalus mitsukurii* Plate 3

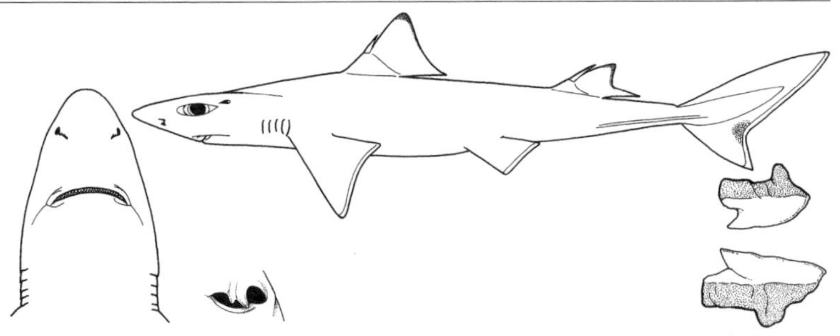

Measurements. Born: 21–30 cm TL. Mature: 47–85 ♂, 50–100 cm ♀. Max: 125 cm TL (varies in different localities).

Identification. Fairly slender dogfish with a grey or grey-brown body, pale below, no white spots. Dorsal fins with dusky tips and area of posterior caudal notch dusky; these areas blackish in young. Pectoral, pelvic, and caudal fins with white posterior margins. Head broad, snout relatively long and broad, large medial barbel on anterior nasal flaps. Pectoral fins with shallowly concave posterior margins and narrowly rounded

rear tips. First dorsal fin moderately high, origin over or just behind pectoral bases; first dorsal spine fairly stout and short (much less than fin base), origin over pectoral inner margins.
Distribution. Apparently extensive, patchy, but may be a species complex.
Habitat. Continental and insular shelves and upper slopes, submarine ridges and seamounts on or near bottom, 4–954 m, mostly 100–500 m.
Behaviour. Often in large aggregations or schools. Some sexual segregation by depth or latitude.
Biology. Ovoviviparous, 2–15 pups/litter (increasing with size of mother, varying regionally) after two year gestation. Age at maturity (4–11 years ♂ and 15–20 years ♀) varies between populations. Powerful predator on bony fishes and invertebrates.
Status. IUCN Red List: Data Deficient (because of taxonomic uncertainties). Extremely vulnerable to rapid depletion by fisheries, one population Endangered.

Cyrano Spurdog *Squalus rancureli* Plate 3

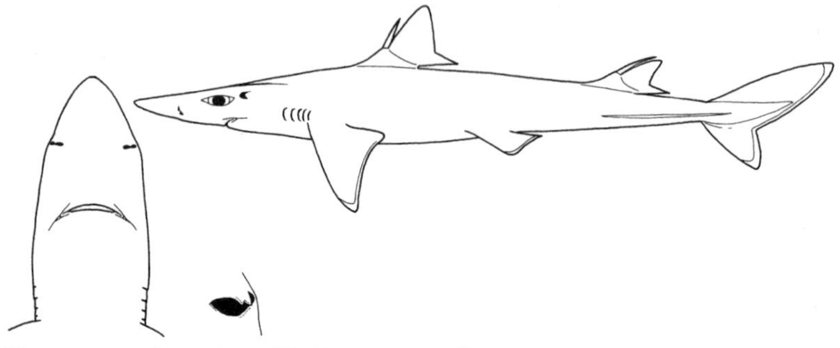

Measurements. Born: ~24 cm TL. Mature: ~65 cm ♀. Max: 77 cm TL.
Identification. Slender dogfish with body dark grey-brown above, abruptly light grey below, no white spots. Dorsal fin webs dusky and apices blackish, caudal fin web dusky. Pectorals, pelvics, and upper lobe of caudal fin with white posterior margins, lower caudal lobe with white tip. Head broad, snout extremely long and broad, small medial barbel on anterior nasal flaps. Pectoral fins with broadly concave posterior margins and very narrowly rounded rear tips. First dorsal fin high, first dorsal spine long, origins of both over pectoral inner margins.

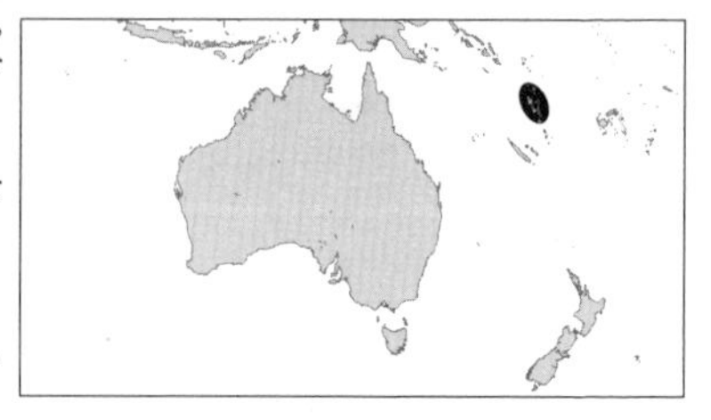

Distribution. Southwest Pacific: Vanuatu.
Habitat. Deep water (320–400 m) on insular slope.
Behaviour. Unknown.
Biology. Probably ovoviviparous, three pups.
Status. IUCN Red List: Near Threatened. Extremely small endemic range.

Bartail Spurdog *Squalus* sp. A Plate 2

Measurements. Largest is a 62 cm TL immature male.
Identification. Fairly slender dogfish, body greyish-brown above, white below, no white spots. Dorsal fins with dusky tips, dorsal caudal margin dark, posterior caudal fin margin white, obvious dark bar along caudal base, sometimes dark spot above tip of caudal base. Head broad, snout relatively short and broad, large medial barbel on anterior nasal flaps. Pectoral fins with deeply concave posterior margins and narrowly rounded rear tips. First dorsal fin high and short, fin and spine origin over pectoral inner margins; first dorsal spine fairly stout and high (~less than fin base).

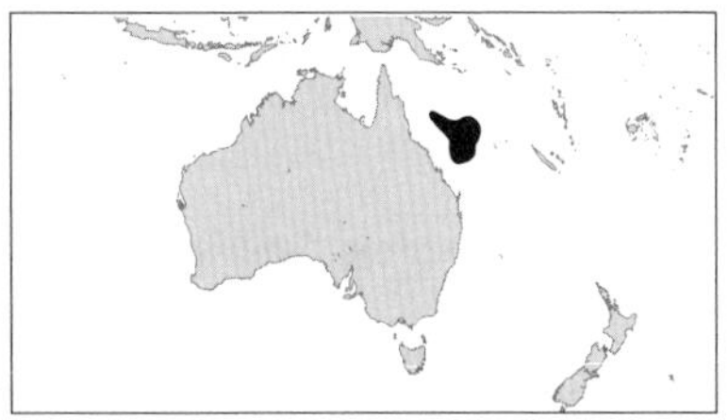

Distribution. Northeast Australia.
Habitat. Offshore, upper continental slope, 220–450 m.
Behaviour. Unknown.
Biology. Unknown. Presumably ovoviviparous.
Status. IUCN Red List: Data Deficient. Rare or uncommon endemic (known from very few specimens).

Eastern Highfin Spurdog *Squalus* sp. B Plate 2

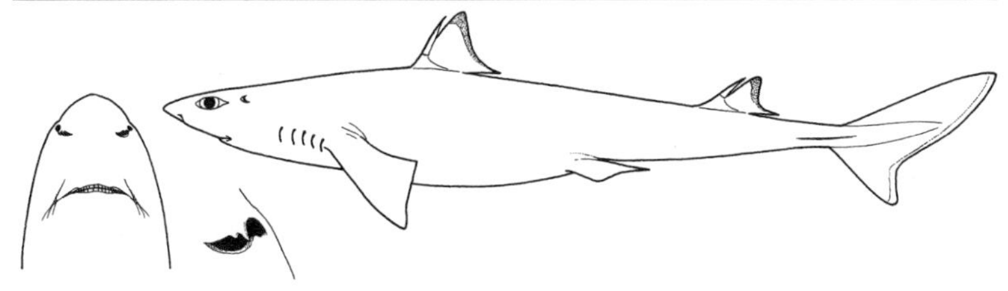

Measurements. Smallest mature ♂ 62 cm TL, largest specimen 65 cm TL.
Identification. Similar to *Squalus* sp. A. Distinguished by small nasal barbel, shorter first dorsal and caudal fins, thicker second dorsal fin spine, and no dark marks on caudal fin. Broad triangular slightly concave pectorals. Dorsal fins with dusky tips and edges; light rear caudal margin and ventral lobe.

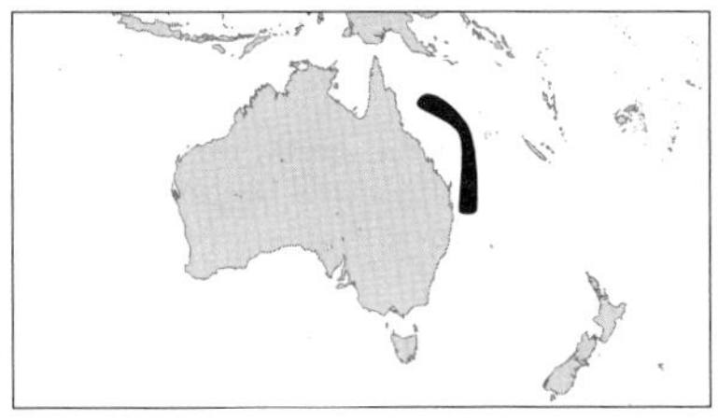

Distribution. Eastern Australia
Habitat. Upper continental slope, 240–450 m.
Behaviour. Unknown.
Biology. Unknown. Presumably ovoviviparous.
Status. IUCN Red List: Data Deficient. Rare to uncommon, endemic to small area.

Western Highfin Spurdog *Squalus* sp. C Plate 2

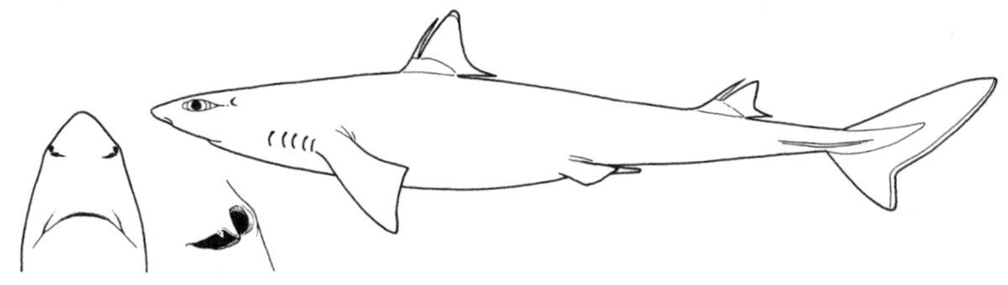

Measurements. Males mature at ~56 cm TL. Max: at least 78 cm TL.
Identification. Similar to *Squalus* sp. B apart from greyish dorsal fins with pale tips, slightly longer snout, slender second dorsal fin spine slightly lower than fin apex, and caudal fin with narrow pale

border (sometimes an inconspicuous dark bar on base).
Distribution. Western Australia.
Habitat. Upper continental slope, 220–510 m.
Behaviour. Virtually unknown. Aggregates by sex.
Biology. Virtually unknown. Presumably ovoviviparous.
Status. IUCN Red List: Data Deficient. Endemic, possibly discarded bycatch of deepwater trawl fisheries.

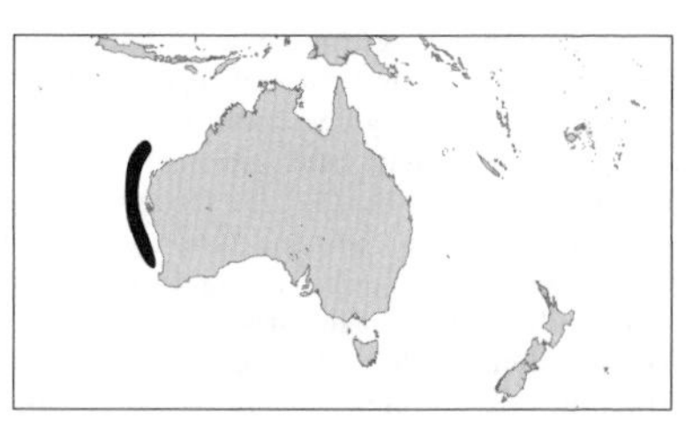

Fatspine Spurdog *Squalus* sp. D Plate 2

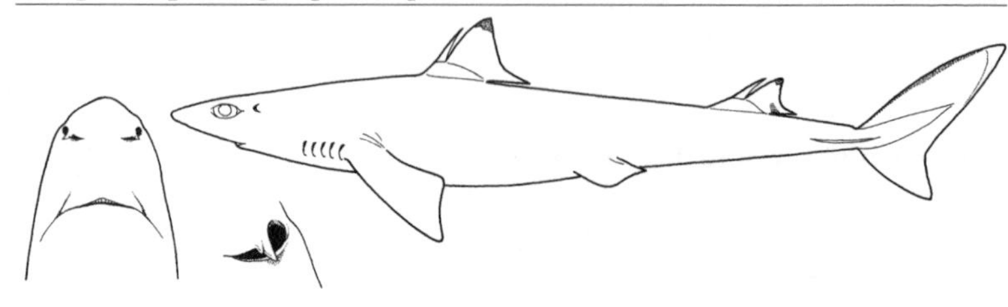

Measurements. Mature: ~44 cm ♂ TL. Max: ~56 cm TL.
Identification. Small slender dogfish with light grey body, paler below, no white spots. Pale fins with dusky dorsal fin tips and front edge of tail; ventral caudal lobe, upper caudal tip and trailing edge pale. Head broad, snout short and broad, small medial barbel on anterior nasal flaps. Pectoral fins with shallowly concave posterior margins and narrowly rounded rear tips. First dorsal fin moderately high, with origin and spine origin over pectoral inner margins; dorsal spines very stout (first shorter than fin base).
Distribution. Western Australia.
Habitat. Outer continental shelf and uppermost slope, 180–200 m.
Behaviour. Unknown.
Biology. Unknown. Presumably ovoviviparous.
Status. IUCN Red List: Data Deficient. Endemic to a lightly fished area.

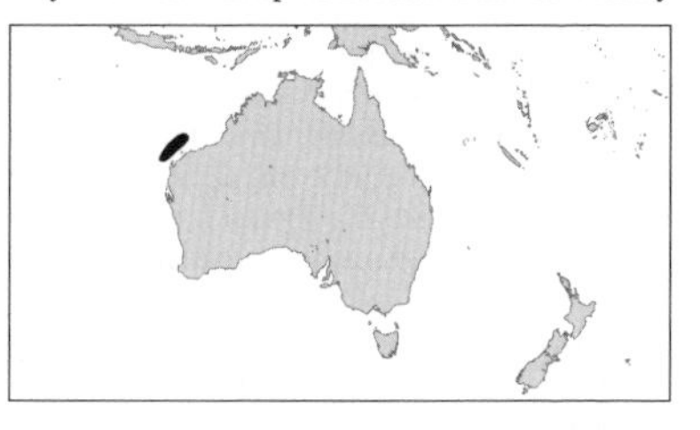

Western Longnose Spurdog *Squalus* sp. E Plate 2

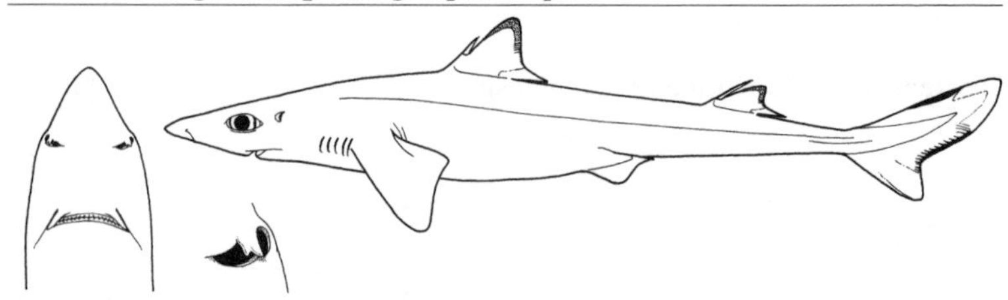

Measurements. Mature:~50 cm ♂ TL. Reach at least 55 cm TL.
Identification. Slender pale grey body, white below, no white spots. Dark tips and posterior margins to dorsal fins, dark blotch and margin on short caudal fin (less obvious in adults), pale posterior margins on pectorals. Head narrow, snout very long and narrow, small medial barbel on anterior nasal flaps. Pectoral fins with shallowly concave posterior margins and rounded rear tips. First dorsal fin moderately high, origin opposite pectoral inner margins and spine origin just behind them; first

dorsal spine slender and very short.
Distribution. West Australia. Similar (identical?) species recorded in South China Sea.
Habitat. Upper continental slope, 300–510 m.
Behaviour. Virtually unknown. Apparently solitary.
Biology. Unknown. Presumably ovoviviparous.
Status. IUCN Red List: Data Deficient. Possibly discarded bycatch of deepwater trawls.

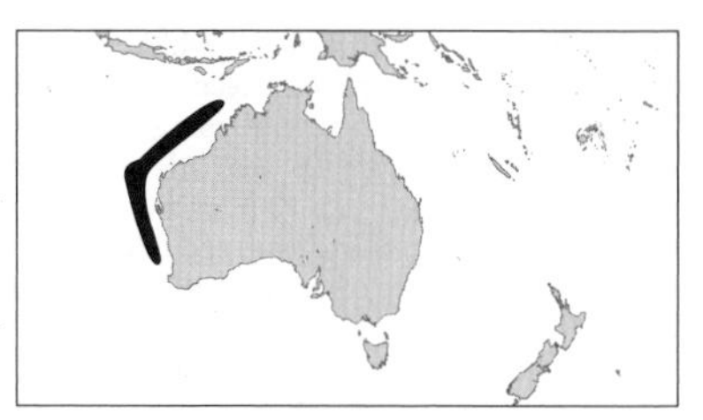

Eastern Longnose Spurdog *Squalus* sp. F Plate 2

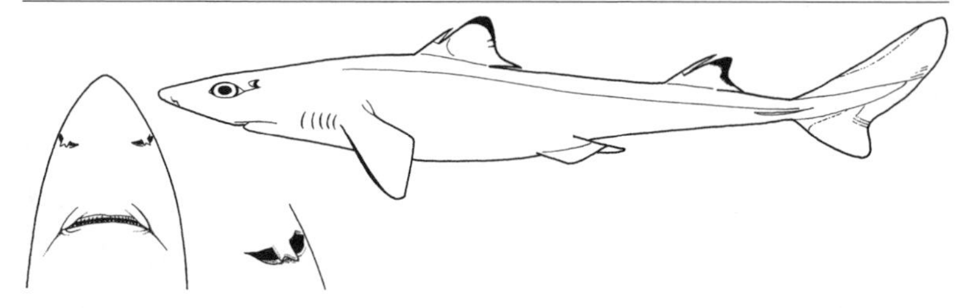

Measurements. Born: ~22 cm TL. Mature: 52 cm ♂ TL (♀ larger). Max: 64 cm TL.
Identification. Very similar to *Squalus* sp. E, but shorter snout, longer first dorsal spine (slightly shorter than second), lower first dorsal fin, and larger caudal fin. Conspicuous dark blotch opposite and above subcaudal notch on trailing edge of caudal fin, upper and lower caudal lobes white-tipped. Dark bar along dorsal and ventral caudal base.
Distribution. East Australia.
Habitat. Upper continental slope, 220–500 m.
Behaviour. Virtually unknown.
Biology. Virtually unknown.
Status. IUCN Red List: Data Deficient. Endemic, seriously depleted by a deepwater trawl fishery in part of range, but most of range unfished.

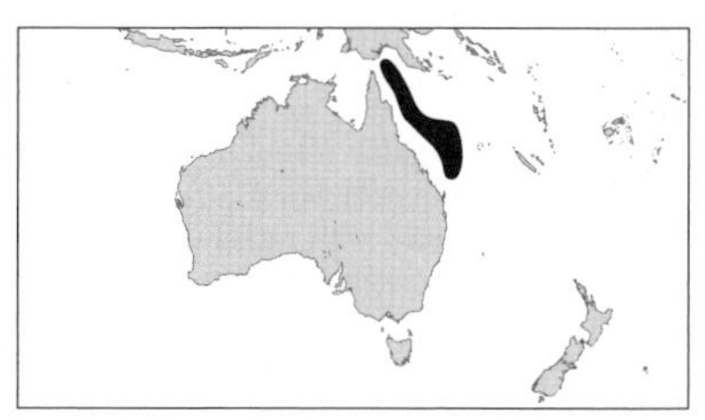

Centrophoridae: Gulper Sharks

About 16 mainly deepwater, bottom-dwelling species in genera *Centrophorus* and *Deania*, recorded almost world-wide from cold temperate to tropical seas, except in northeast Pacific and very high latitudes. Most diverse in warm waters and in the Indo-west Pacific, where some localised endemic species occur (although many species may be more widely distributed than currently recorded). Main depth range ~1000–1500 m, but a few records from as shallow as 50 m and one *Centrophorus* photographed below 4000 m. *C. squamosus* has been collected from the open ocean, between 0–1250 m in water about 4000 m deep.
Identification. Short- to long-nosed cylindrical sharks with huge green or yellowish eyes, blade-like teeth in both jaws, uppers without cusplets, lowers much larger. Two dorsal fins with grooved spines, first dorsal slightly smaller to much larger than second and originating in front of pelvic fin origins; second dorsal fin with a straight to weakly concave posterior margin. No anal fin. *Centrophorus* species have a moderate-sized snout, angular to greatly elongated pectoral free rear tips, and smoother skin with leaf-shaped or block-like denticles. *Deania* species have a very long snout and rough skin with tall slender dorsal dermal denticles topped by pitchfork-shaped crowns. *Centrophorus granulosus*, *C. harrissoni*, and probably other *Centrophorus* species have often been

identified as *Centrophorus uyato*, which was based on a species of spurdog (*Squalus*) from the Mediterranean Sea and is incorrectly applied to *Centrophorus*.
Biology. Little known. Several species are social, forming small schools or huge aggregations (aggregating species are or were among the commonest deep sea sharks). Reproduction is by aplacental ovoviviparity (1–12 pups/litter). They feed mainly on bony fishes and cephalopods, also crustacea, small sharks and rays, and tunicates.
Status. Poorly sampled and identified and imperfectly known, despite their importance in commercial target and bycatch fisheries. Caught by line gear, trawls and gill nets. Valuable meat and oily, squalene-rich livers. Conservation status poorly known, but of considerable concern because rapidly expanding and largely unmanaged and unmonitored deepwater fisheries are now taking large numbers of these species. Gulper sharks have a very limited reproductive capacity, with small litters, long gestation periods, slow growth and late maturity; fisheries can cause rapid population depletion and recovery is very slow if fishing pressure is lifted.

Needle Dogfish *Centrophorus acus* Plate 4

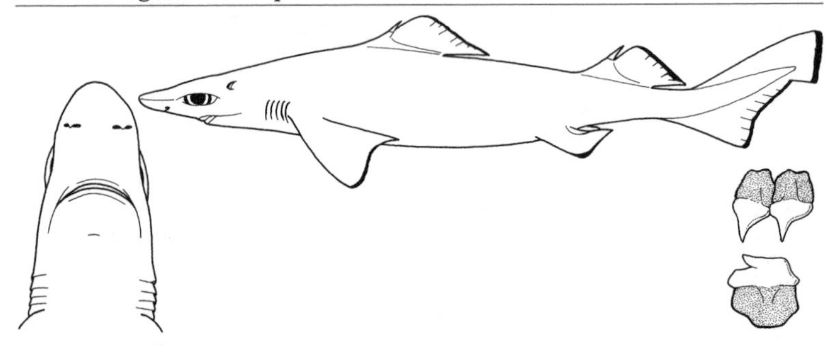

Measurements. Mature: ~100–105 cm TL ♂, 154–161 cm TL ♀.
Identification. Rough brown skin (covered with leaf-shaped, overlapping denticles), slightly lighter below. Fin webs dusky with darker edges, not prominently marked. Fairly long flat snout. Free rear tips of pectoral fins extend into long narrow angular lobes that reach past first dorsal spine origin. First dorsal fin low and long, second as high and almost as large.
Distribution. Northwest Pacific: Japan, Taiwan Island, Philippines. (West Atlantic records likely another species.)

Habitat. Outer continental shelves and upper slopes, 150–950 m, mostly >200 m, possibly to 1786 m.
Behaviour. May feed more actively at night.
Biology. Presumably ovoviviparous with small litters (five eggs in one female). Females mature >20 years old.
Status. IUCN Red List: Not Evaluated. Caught off Japan and Taiwan, utilised for liver oil.

Dwarf Gulper Shark *Centrophorus atromarginatus* Plate 5

Measurements. Born: ~30–33 cm TL. Mature: >40 ♂, >44 cm ♀. Max: at least 94 cm TL.
Identification. Grey or grey-brown above, lighter below, prominent blackish markings on all or most fins (not always on pelvic). Skin smooth (denticles block-shaped and widespaced, not overlapping). Fairly long thick snout. Rear tips of pectoral fins narrowly angular and greatly elongated. Two dorsal fins with large grooved spines, first dorsal fin short and higher than second, spine base of second over pelvic fin inner margins or rear tips.

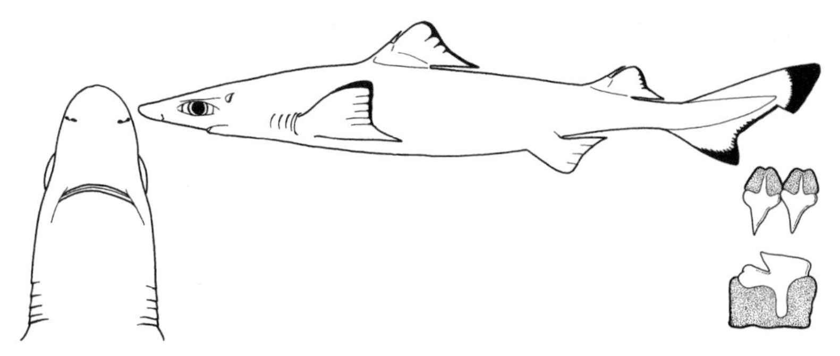

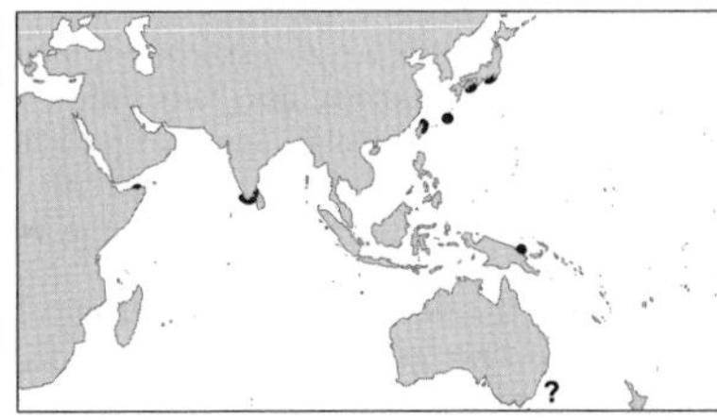

Distribution. Northwest Indian Ocean and West Pacific: Japan, Taiwan Island, northern Papua New Guinea. (Range is probably wider.)
Habitat. Outer continental and insular shelves, upper slopes, 150–450 m.
Behaviour. Unknown.
Biology. Unknown.
Status. IUCN Red List: Not Evaluated. Caught off Japan and Taiwan. Liver oil utilised

Gulper Shark *Centrophorus granulosus* Plate 4

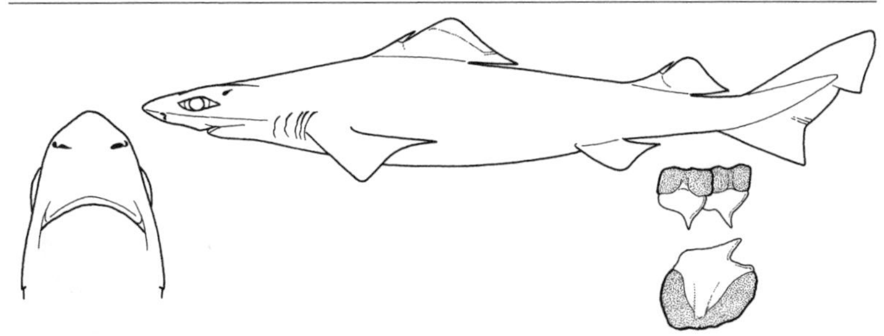

Measurements. Born: 30–42 cm TL. Mature: 60–80 cm ♂, >90 cm ♀. Max: 105–110 cm TL.
Identification. Dark grey or grey-brown above, lighter below, dusky fin webs, dark fin tips only in juveniles. Smooth skin (block-shaped denticles widespaced, not overlapping). Rather short, thick snout. Free rear tips of pectorals extend into acute lobe. First dorsal quite short and high, second almost as high with spine base over inner pelvic fin margins. Shallow notch in postventral caudal fin margin of adults, lower lobe moderately long.

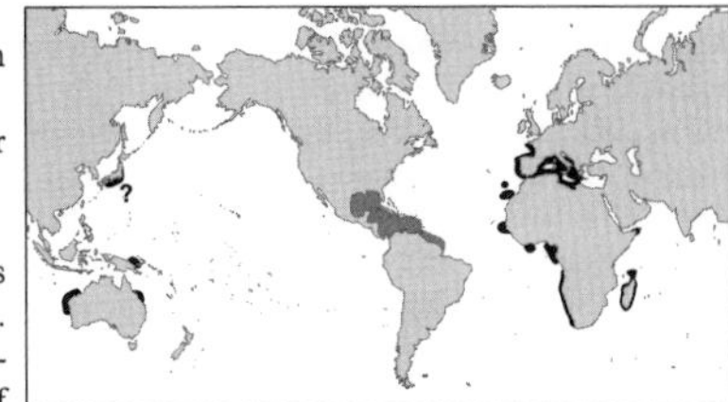

Distribution. Widespread: Atlantic, west Indian Ocean, west and possibly central Pacific.
Habitat. Continental shelves and slopes, on or near bottom, 50–1440 m, mostly 200–600 m.
Behaviour. Unknown.
Biology. Ovoviviparous, one or two pups/litter. Feeds mainly on bony fishes, also squid and crustaceans.
Status. IUCN Red List: Vulnerable. As *Centrophorus uyato*, Critically Endangered regionally off Australia, but species identification to be confirmed. Target and bycatch deepwater fisheries for liver oil and meat have caused population declines.

Longnose Gulper Shark *Centrophorus harrissoni* Plate 4

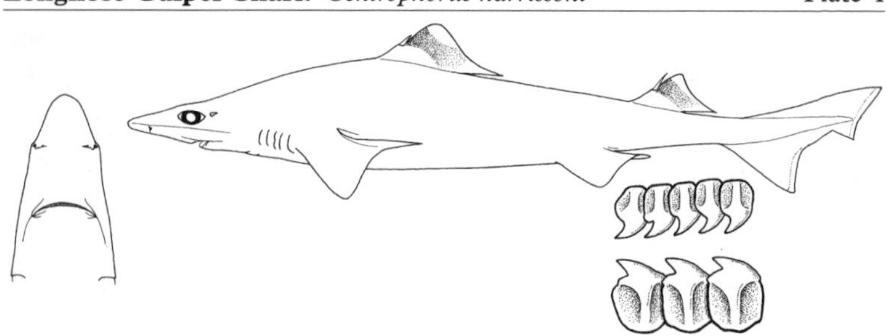

Measurements. Born: 32 cm TL. Mature: ~85 cm ♂. Max: 110 cm TL.
Identification. Light greyish, paler below, prominent dark oblique blotch or bar extending back from leading edge of dorsal fins, more diffuse mark on tail. Smooth skin (block-shaped denticles widespaced, not overlapping). Snout long, narrow and flat. Pectoral fin rear tips narrowly angular and greatly elongated. First dorsal fin short and high, second lower with spine base over pelvic fin inner margins or rear tips. Shallow notch in postventral caudal fin margin of adults, lower lobe moderately long.
Distribution. Eastern Australia. (The status of similar Indo-west Pacific (western Australia, South Africa to Taiwan), and northwest Atlantic species needs confirmation.)
Habitat. Continental slope, 250–790 m.
Behaviour. Unknown.
Biology. One or two pups/litter born every one to two years. Probably late-maturing and long-lived.
Status. IUCN Red List: Critically Endangered. Over 99% population decline since mid 1970s caused by demersal trawl fishery. Entire range fished.

Blackfin Gulper Shark *Centrophorus isodon* Plate 5

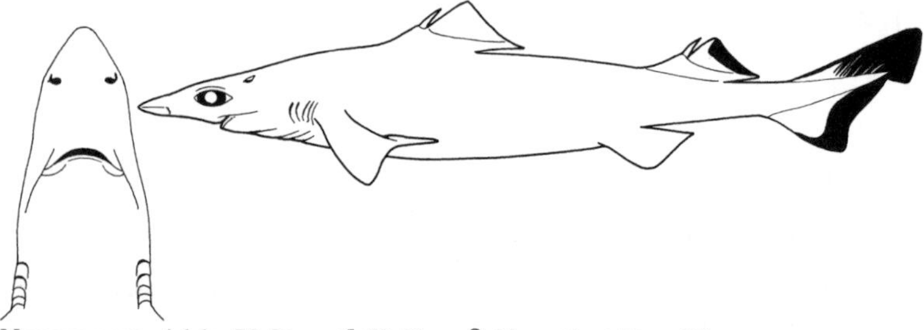

Measurements. Adults: 75–86 cm ♂, 78–93 cm ♀. Max: at least 93 cm TL.
Identification. Blackish-grey above, lighter below, blackish fin webs. Smooth skin (block-shaped denticles widespaced, not overlapping). Long, flat snout. Rear tips of pectoral fins narrowly angular and greatly elongated. First dorsal fin short and high, second lower with spine base over pelvic fin inner margins. Shallow notch in postventral caudal fin margin of adults, lower lobe moderately long.

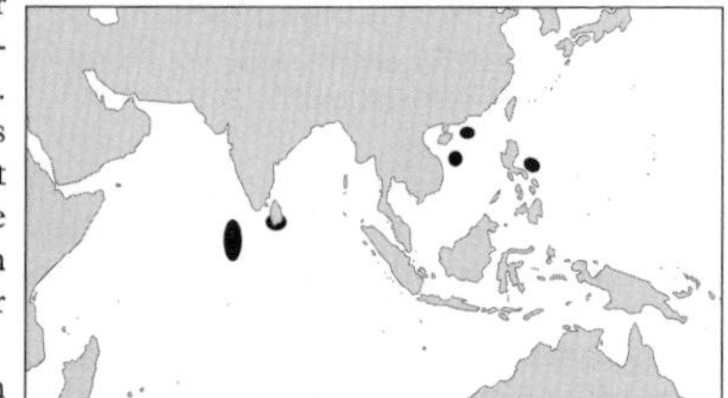

Distribution. Scattered records in Indian Ocean

and west Pacific. This rare species may be more wide-ranging than reported.
Habitat. Upper continental slopes, 760–770 m.
Behaviour. Unknown.
Biology. Unknown.
Status. IUCN Red List: Not Evaluated. Rare, and/or misidentified as other species of *Centrophorus.*

Lowfin Gulper Shark *Centrophorus lusitanicus* Plate 4

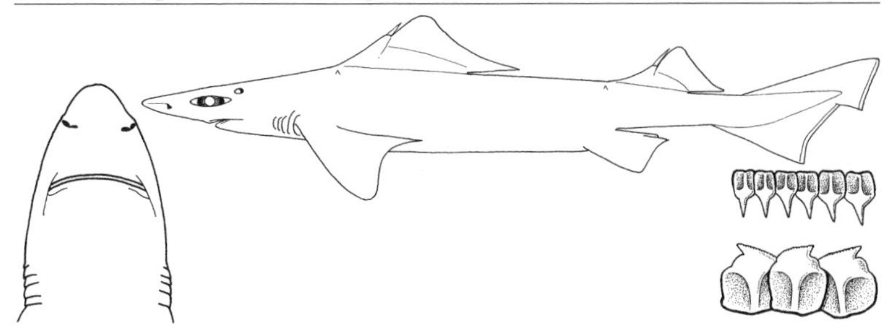

Measurements. Born: ~33–36 cm TL. Mature: ~75 cm ♂, 85 cm ♀. Max: 100 cm TL.
Identification. Grey-brown or grey above, slightly lighter below, dusky fin webs. Smooth skin (block-shaped denticles widespaced, not overlapping). Snout long and flat. Rear pectoral fins tips narrowly angular and greatly elongated. First dorsal fin very long and high, second one-third as long, but as high or higher. Shallow notch in postventral caudal fin margin of adults, lower lobe moderately long.
Distribution. East Atlantic: Portugal to Cameroon (not Mediterranean). West Indian Ocean: Mozambique. West Pacific: Taiwan Island.

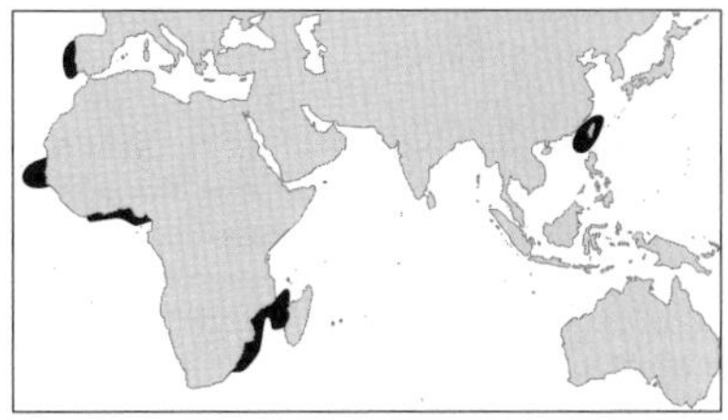

Habitat. Outer continental shelves and upper slopes, 300–1400 m, mostly 300–600 m.
Behaviour. Unknown.
Biology. Ovoviviparous, one pup/litter. Probably feeds on bony fishes, small sharks and cephalopods.
Status. IUCN Red List: Not Evaluated. Fished in east Atlantic (for meat) and off Taiwan.

Smallfin Gulper Shark *Centrophorus moluccensis* Plate 4

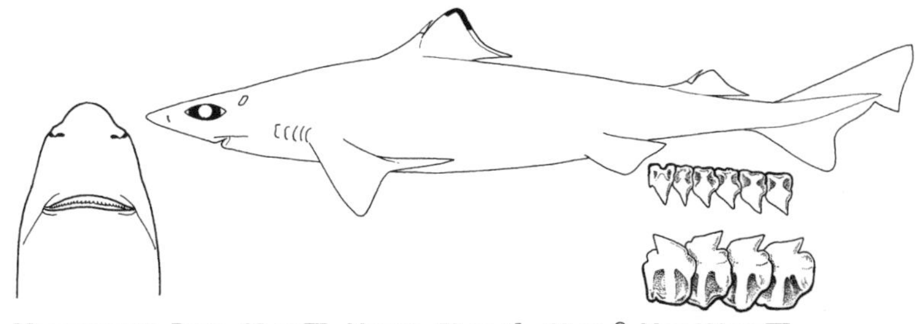

Measurements. Born: ~35 cm TL. Mature: ~70 cm ♂, ~90 cm ♀. Max: 100 cm TL.
Identification. Greyish-brown, much paler below, dusky fin webs, dark blotch below first dorsal fin tip in juveniles; narrow pale border to tail and sometimes to pectoral and pelvic fins. Smooth skin (block-shaped denticles widespaced, not overlapping). Short thick snout. Rear pectoral fin tips narrowly angular and greatly elongated. Short first dorsal fin, second less than half height of first, with spine base usually well behind pelvic fins. Deeply notched postventral caudal fin margin and deep lower lobe in adults.

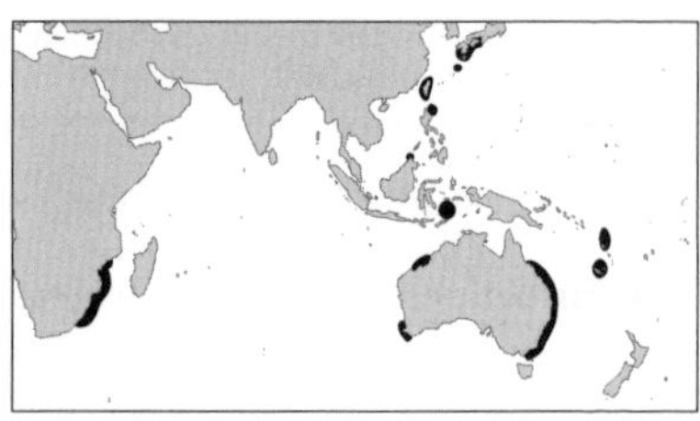

Distribution. Scattered, Indo-west Pacific. Distribution and taxonomy uncertain, may be a group of related species.
Habitat. Outer continental and insular shelves, upper slopes, 125–820 m.
Behaviour. Demersal.
Biology. Ovoviviparous, two pups/litter every two years. Feeds mainly on bony fishes and cephalopods, also elasmobranchs, crustaceans.
Status. IUCN Red List: Data Deficient. Endangered off east coast of Australia, where trawl fisheries depleted population >95% since 1970s, west coast not overfished.

Taiwan Gulper Shark *Centrophorus niaukang* Plate 4

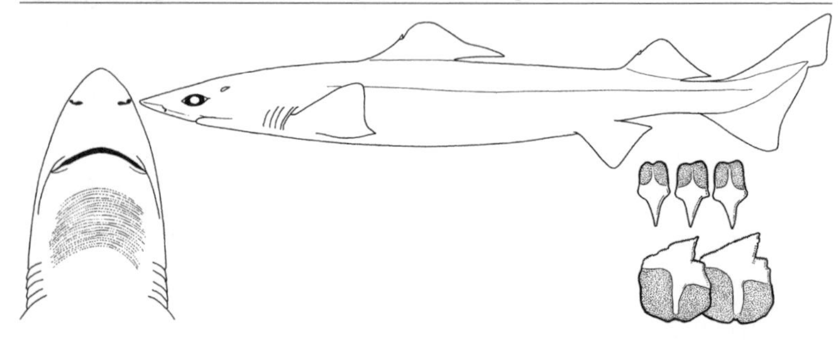

Measurements. Born: 35–45 cm TL. Mature: 90–110 cm ♂, 130–140 cm ♀. Max: 170 cm TL.
Identification. Dark grey or grey-brown, slightly lighter below, dusky fin webs, no prominent markings. Narrow tear-drop shaped lateral trunk denticles in adults, skin fairly rough. Short broad snout. Pectoral fin free rear tips short and angular. First dorsal fin very long and high, second almost as large. Posterior margin of caudal fin almost straight in adults, lower lobe weakly developed.

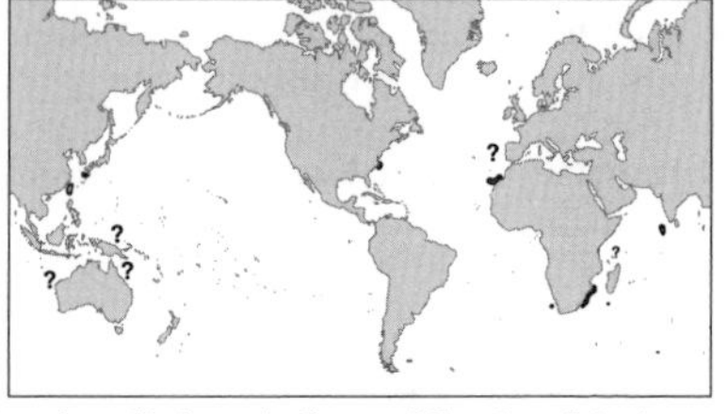

Distribution. Atlantic and Indo-Pacific: wide but patchy.
Habitat. Outer continental shelves and upper slopes, 98–~1000 m.
Behaviour. Unknown.
Biology. Ovoviviparous, one to six (mostly four to six) pups/ litter. Eats bony fishes, small chondrichthyans, squid and lobsters.
Status. IUCN Red List: Near Threatened. Not abundant, likely to decline rapidly when fished.

Leafscale Gulper Shark *Centrophorus squamosus* Plate 4

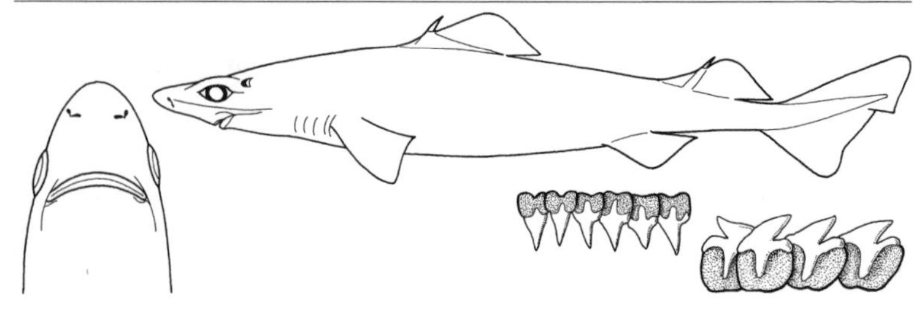

Measurements. Born: 35–40 cm TL. Mature: ~100 cm ♂, 110 cm ♀. Max: 160 cm TL.
Identification. Uniform grey, grey-brown or reddish-brown, dusky fins, no prominent markings. Rough skin (dermal denticle crowns leaf-shaped in adults, bristle-like in juveniles). Short thick or slightly flattened snout. Pectoral fin free edges short, not particularly angular and elongated. First dorsal very long and low, second shorter, higher, more triangular, with spine base usually opposite pelvic fin inner margins or free rear tips. Posterior margin of caudal fin almost straight or slightly concave in adults, lower lobe weakly developed.

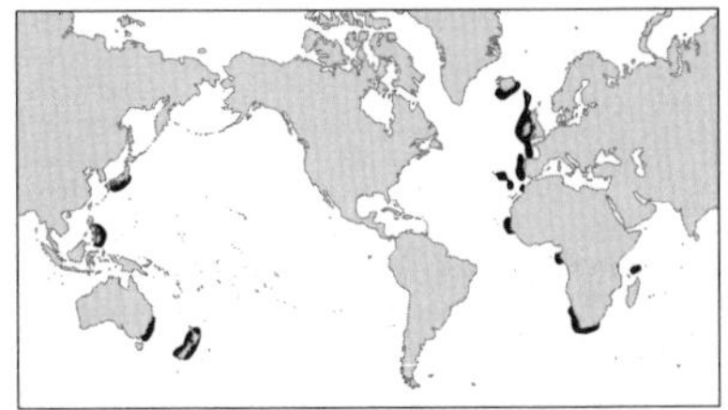

Distribution. Atlantic, west Indian, west Pacific Oceans.
Habitat. Continental slopes, 230–2400 m (rare above 1000 m in northeast Atlantic). Also 0–1250 m in water up to 4000 m deep.
Behaviour. Apparently demersal and pelagic.
Biology. Ovoviviparous, five to eight pups/litter.
Status. IUCN Red List: Vulnerable. Important in several deepwater fisheries.

Mosaic Gulper Shark *Centrophorus tesselatus* Plate 4

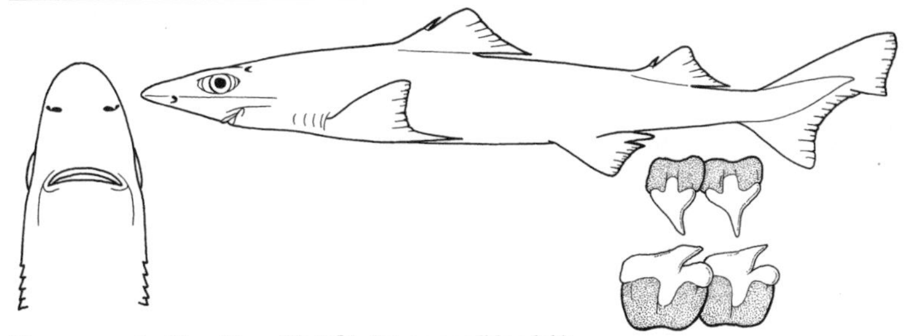

Measurements. Max: 89 cm TL (♂ holotype, possibly adult).
Identification. Light brownish above, white below, light margins to fins. Smooth skin (dermal denticles block-like and widespaced, not overlapping). Fairly long thick snout. Rear pectoral fin tips narrowly angular and greatly elongated. First dorsal fin fairly high and short, second almost as large with spine origin over free rear tips of pelvic fins. Caudal fin with shallowly notched rear margin.

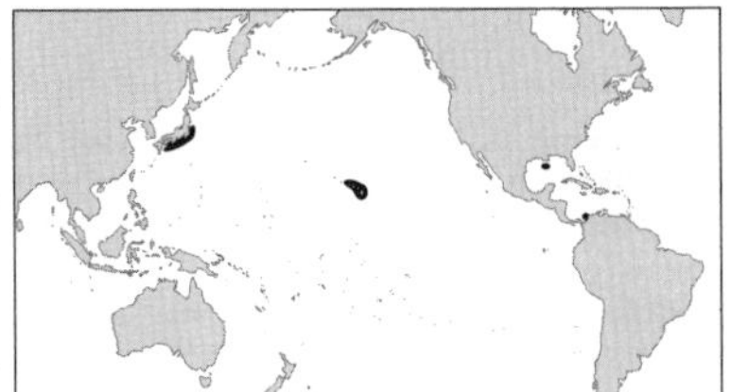

Distribution. West and central Pacific, northwest Atlantic, Indian Ocean (many records provisional).
Habitat. Insular slopes, on or near bottom, 260–730 m.
Behaviour. Unknown.
Biology. Unknown. Presumably ovoviviparous.
Status. IUCN Red List: Not Evaluated. Rare. Limited fisheries interest?

Birdbeak Dogfish *Deania calcea* Plate 5

Measurements. Born: >30 cm TL. Mature: ~80 cm ♂, 100 cm ♀. Max: 122 cm TL.
Identification. Grey to dark brown with darker fins; juveniles with dark posterior dorsal fin margins, dark patches above eyes and gill regions and on caudal fin lobes. Rough skin (lateral trunk denticles pitchfork-shaped, ~0.5 mm long). Extremely long flattened snout. No subcaudal keel beneath caudal peduncle. Extremely long, low first dorsal fin, much shorter, taller second dorsal with longer fin spine.

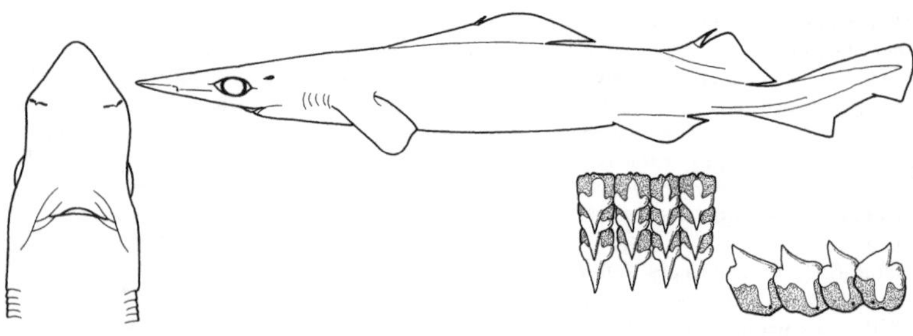

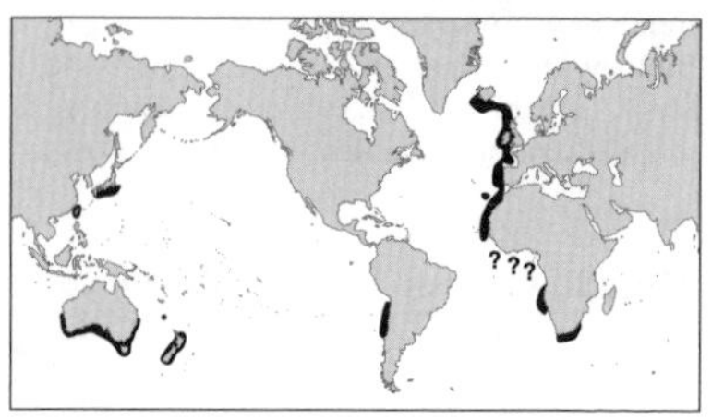

Distribution. East Atlantic: Iceland to South Africa. Pacific: Japan, Taiwan Island, Australia, New Zealand, Peru to Chile.
Habitat. Continental and insular shelves and slopes, 70–1470 m (usually 400–900 m).
Behaviour. Sometimes schooling.
Biology. Ovoviviparous, 6–12 pups/litter. Feeds on bony fishes and shrimps.
Status. IUCN Red List: Least Concern. Bycatch in deepwater fisheries, no declines reported.

Rough Longnose Dogfish *Deania hystricosa* Plate 5

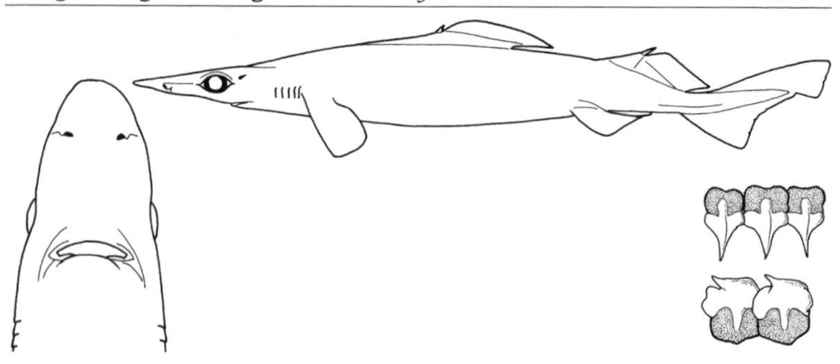

Measurements. Mature: ~84 cm ♂, 92–106 cm ♀. Max: 111 cm TL.
Identification. Blackish-brown to grey-brown. Skin very rough from large pitchfork-shaped lateral trunk denticles (~1 mm long). Very long flattened snout. Extremely long, low first dorsal fin, much shorter, taller second dorsal. No subcaudal keel beneath caudal peduncle.

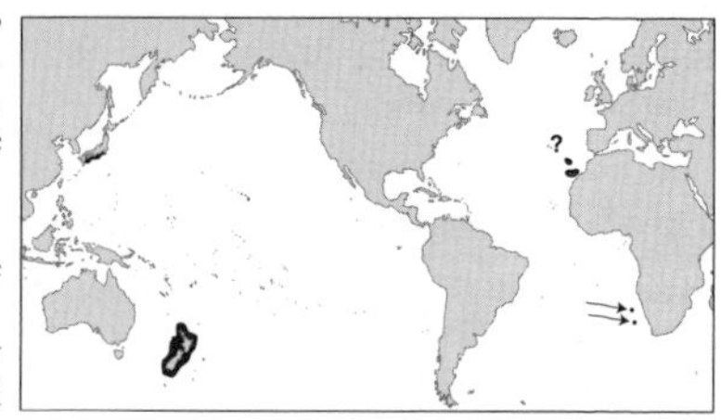

Distribution. East Atlantic: Madeira, Canaries, possibly Azores, Namibia and South Africa. Northwest Pacific: Japan, possibly New Zealand.
Habitat. Insular slopes, 470–1300 m, more abundant below 1000 m off South Africa.
Behaviour. Probably benthic and epibenthic.
Biology. Ovoviviparous, one female had 12 large eggs.
Status. IUCN Red List: Not Evaluated. Low fisheries interest, possible bycatch in deep trawl fisheries.

Arrowhead Dogfish *Deania profundorum* Plate 5

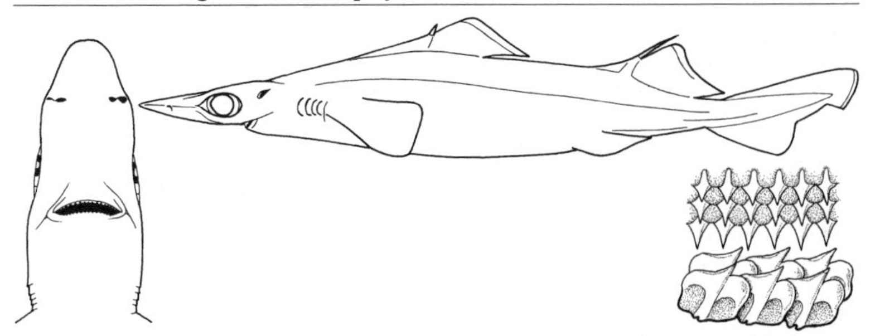

Measurements. Born: >31 cm TL. Mature: ~43 cm ♂, 62–80 cm ♀. Max: 97 cm TL (most smaller).
Identification. Dark grey or brown. Dermal denticles small (0.25 mm long). Smaller than other *Deania.* Extremely long flattened snout. Only *Deania* with a subcaudal keel beneath the caudal peduncle. First dorsal fin relatively short and high, second much taller with much higher fin spine.

Distribution. East Atlantic: western Sahara to South Africa. West Atlantic: Gulf of Mexico and Caribbean. Indo-west Pacific: South Africa, Gulf of Aden, Philippines.
Habitat. On or near bottom, upper continental and insular slopes, 275–1785 m.
Behaviour. Sometimes occurs in huge aggregations.
Biology. Ovoviviparous, five to seven pups/litter. Feeds on small bottom and midwater bony fishes, squid, crustaceans.
Status. IUCN Red List: Not Evaluated. Not commercially important, probably bycaught.

Longsnout Dogfish *Deania quadrispinosum* Plate 5

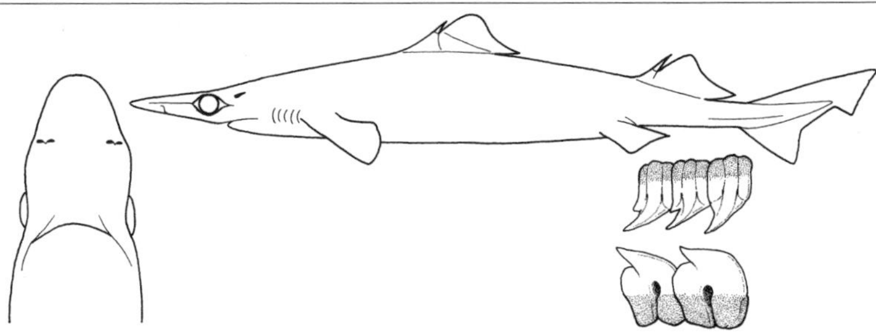

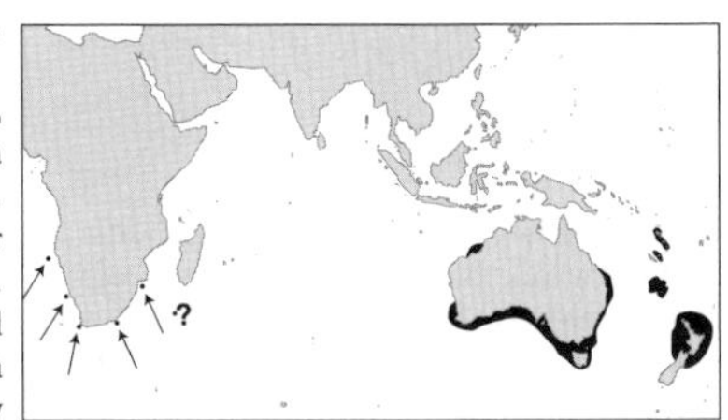

Measurements. Adult: 87 cm ♂, 110 cm ♀. Max: ~114 cm TL.
Identification. Grey or grey-brown to blackish, sometimes with white-edged fins; juveniles with dark blotch near leading edge of dorsal fins. Dermal denticles pitchfork-shaped (usually four spined) and quite large (0.75 mm long). Extremely long flattened snout. No subcaudal keel beneath caudal peduncle. First dorsal fin rather high, angular and short, second slightly taller and with much higher fin spine.

Distribution. Southern Africa, Australia, New Zealand, west Pacific islands.
Habitat. Outer continental shelf and slope, 150–1360 m (usually below 400 m).
Behaviour. Unknown.
Biology. Ovoviviparous. Eats bony fishes.
Status. IUCN Red List: Not Evaluated. Discarded bycatch from fisheries.

Etmopteridae: Lantern Sharks

The largest family of squaloid sharks, with >50 species in five genera (*Aculeola, Centroscyllium, Etmopterus, Miroscyllium, Trigonognathus*) occurring almost world-wide in deepwater, some wide-ranging, many endemic. New species are continually being discovered. The family includes what may be the smallest known species of sharks (*Etmopterus carteri* and *E. perryi*) that mature and give birth to litters of live young at 10–20 cm TL.
Identification. Dwarf to moderate-sized sharks (adults 10–107 cm long) with photophores, inconspicuous or forming distinct black marks on abdomen, flanks or tail (confined to or denser on ventral surface). Two dorsal fins with strong grooved spines (second fin and spine usually larger). No anal fin. No precaudal pits or lateral keels on caudal peduncle. *Centroscyllium* species (combtooth dogfishes) have short to moderately long snouts, comblike teeth with cusps and cusplets in both jaws, and strong grooved dorsal fin spines (second strikingly larger than first). *Aculeola* (hooktooth dogfish) is very similar except for small hook-like teeth in both jaws and very small equal-sized dorsal fin spines. Lantern sharks, *Etmopterus* species, often have dark markings (light organs – photophores) on underside. Their upper teeth have a cusp and one or more pairs of cusplets, very different in appearance to the blade-like, cutting lower teeth. Second dorsal fin and fin spine are much larger than the first. Some species have lines of denticles along their flanks and dorsal surfaces that give them an engraved appearance. Viper (*Trigonognathus*) and rasptooth (*Miroscyllium*) dogfish have very distinctive teeth. These deepwater sharks tend to be in poor condition when collected due to damage by fishing gear. It is also hard to confirm identifications without examining details of denticle and tooth structure, coloration and photophore patterns, and taking precise body measurements; it may not be possible to identify some lantern sharks to species using this guide alone. The *FAO Catalogue of Sharks of the World* should be consulted to confirm identifications.
Biology. Most species are bottom-dwelling in deep water, 200–1500 m, (range 50–4500 m); some semi-oceanic. Several etmopterid dogfishes are social, and form small to huge schools or aggregations. Reproduction, where known, is ovoviviparous (aplacental viviparous), with 3–20 pups/litter.
Status. Most species are common, but poorly known. Few are large enough to be of any commercial value and most are discarded if caught as bycatch.

Hooktooth Dogfish *Aculeola nigra* — Plate 6

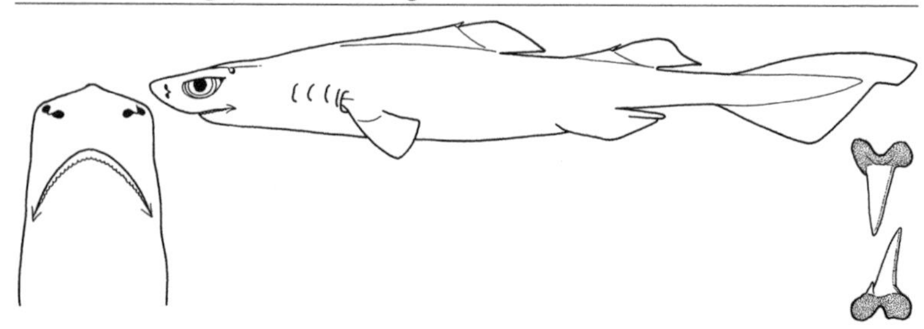

Measurements. Born: 13–14 cm TL. Mature: 42–46 ♂, 52–54 cm ♀ TL. Max: ~60 cm TL.
Identification. Blackish-brown stocky dogfish, broad blunt snout, thin lips, broad long arched mouth with small hook-like teeth in both jaws. Gill openings quite large. Very short grooved dorsal fin spines, much lower than low dorsal fins. First dorsal fin origin over pectoral inner margins; second slightly larger, origin opposite or slightly behind pelvic origins. Rounded free rear pectoral fin tips. Long upper caudal lobe, lower not differentiated.

Distribution. Eastern Pacific: Peru to Chile.
Habitat. Benthic and epibenthic, continental shelf and upper slope, 110–735 m.
Behaviour. Unknown.
Biology. Poorly known. Ovoviviparous, at least three pups/litter.
Status. IUCN Red List: Not Evaluated. Relatively common within limited range, possibly fisheries bycatch.

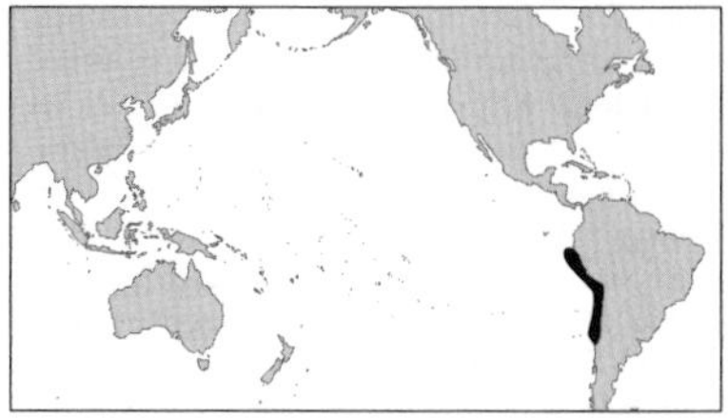

Highfin Dogfish *Centroscyllium excelsum* Plate 6

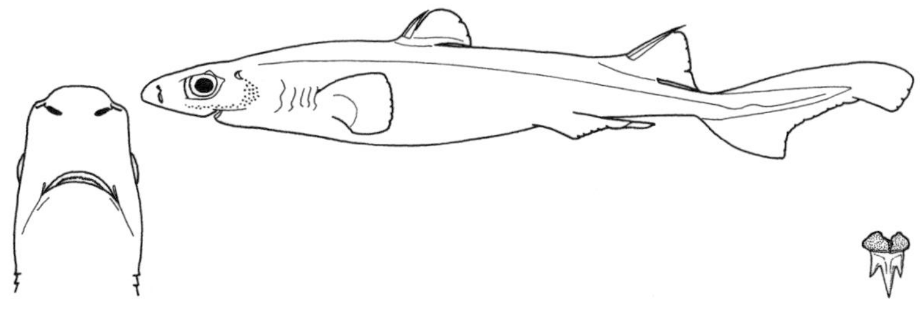

Measurements. Born: >8–9 cm TL. Adults: 52–64 cm TL.
Identification. Light brown above, darker below, lighter fin margins. Intense black markings around mouth and on lower surfaces of pectoral fins. Very high rounded first dorsal fin with short spine. Second dorsal much larger than first with very long spine reaching above and behind fin apex. Short caudal peduncle. Dorsal denticles sparse and irregular above, none below. Comb-like teeth in both jaws.
Distribution. Northwest Pacific.
Habitat. Deep seamounts, 800–1000 m.
Behaviour. Unknown.
Biology. Ovoviviparous, ten pups/litter. Eats bony fishes.
Status. IUCN Red List: Not Evaluated. Only 21 specimens known.

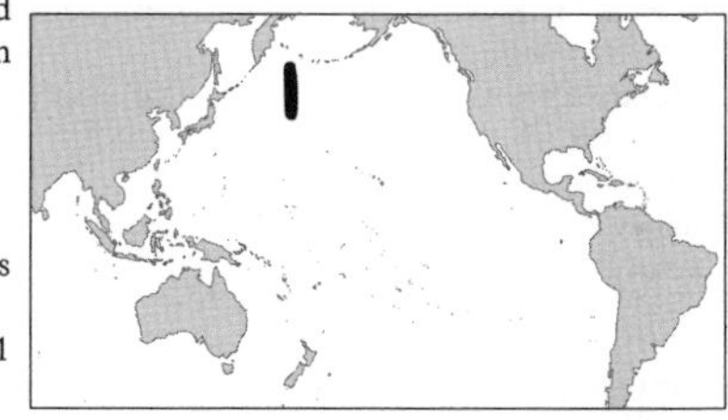

Black Dogfish *Centroscyllium fabricii* Plate 6

Measurements. Born: ~15 cm TL. Mature: 46–50 cm ♂, 50–66 cm ♀. Max: 84–107 cm TL.
Identification. Uniformly blackish-brown dogfish, no black or white marks. Numerous close-set denticles. First dorsal fin low, spine short. Second fin and spine larger. Fairly stout compressed long abdomen, short caudal peduncle. Arched mouth, comblike teeth in both jaws.
Distribution. Widespread in temperate Atlantic (tropical records uncertain).

Habitat. Outer continental shelves and slopes, 180–1600 m (mostly >275 m, possibly to 2250 m). Closer to surface at high latitudes and in winter.

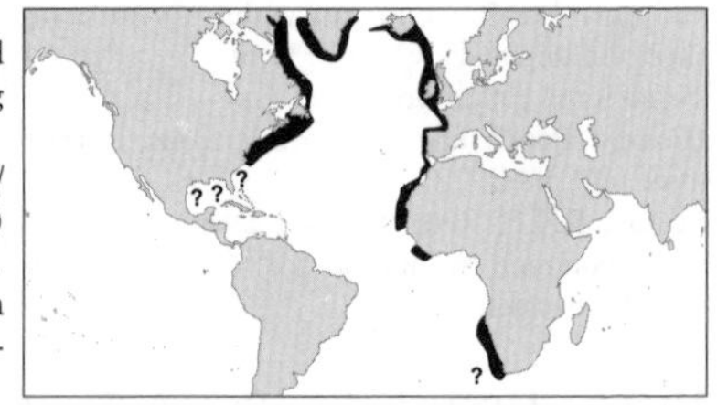

Behaviour. Schooling, segregating by sex and size. Larger schools in shallower water during winter and spring.

Biology Ovoviviparous, seven or eight pups/litter. Lumin-escent organs scattered (irregularly) in skin. Eats crustaceans, cephalopods, bony fishes.

Status. IUCN Red List: Not Evaluated. Often abundant. Commonly discarded fisheries by-catch.

Granular Dogfish *Centroscyllium granulatum* Plate 6

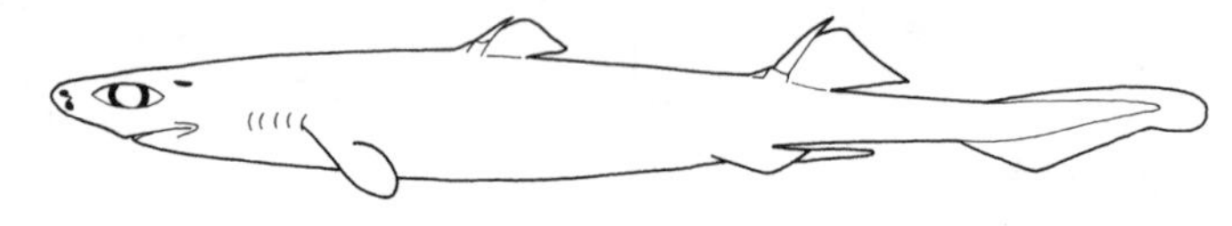

Measurements. Max: at least 28 cm TL.

Identification. Very small slender cylindrical dogfish, uniformly brownish-black body without conspicuous markings, closely covered by denticles with sharp, hooked cusps. Long abdomen, elongated caudal peduncle. Mouth narrowly arched, comb-like teeth in both jaws. First dorsal fin low, much smaller than second. First dorsal spine short, second very large, extending over and behind fin apex.

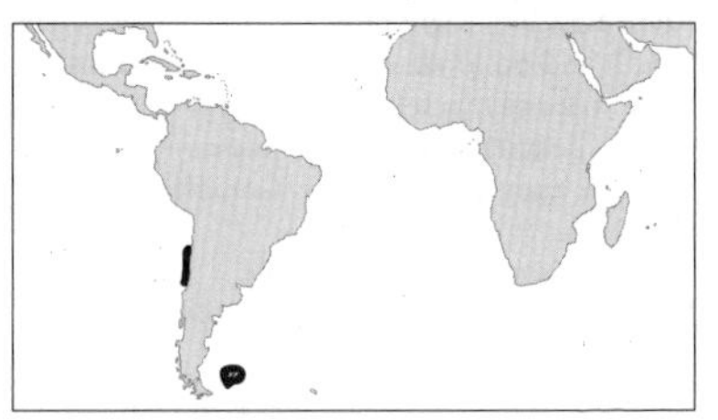

Distribution. South America: north-central Chile, Falkland/Malvinas Islands.

Habitat. Deepwater, upper continental slope, 400–448 m.

Behaviour. Unknown

Biology. Unknown. Presumably ovoviviparous.

Status. IUCN Red List: Not Evaluated. Unknown.

Bareskin Dogfish *Centroscyllium kamoharai* Plate 6

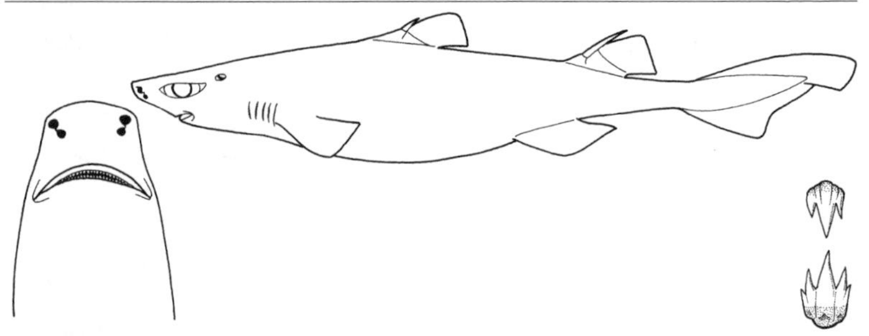

Measurements. Mature: 40–45 cm ♂, 42–44 cm TL ♀. Max: 60 cm TL.

Identification. Small stout uniform blackish dogfish (except for white patches where very delicate skin chafed during capture) fins with lighter rear margins. Skin smooth, almost naked, fragile and easily damaged (few, wide-spaced denticles). Mouth extremely short, very broadly arched with comblike teeth in both jaws. Caudal peduncle short. First dorsal fin very low and rounded with short spine. Second dorsal fin slightly larger than first, spine about as high as fin apex.

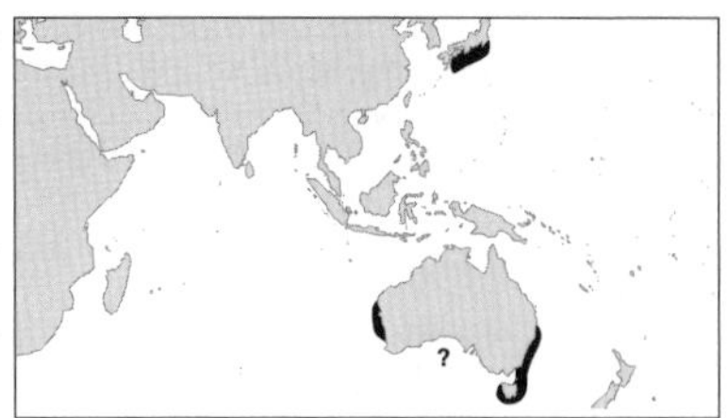

Distribution. West Pacific: Japan, Australia, possibly Philippines.
Habitat. Bottom, deep continental shelf, 500–1200 m, mostly >900 m.
Behaviour. Unknown.
Biology. Poorly known. Presumably ovoviviparous.
Status. IUCN Red List: Data Deficient. Unreported fisheries bycatch.

Combtooth Dogfish *Centroscyllium nigrum* Plate 6

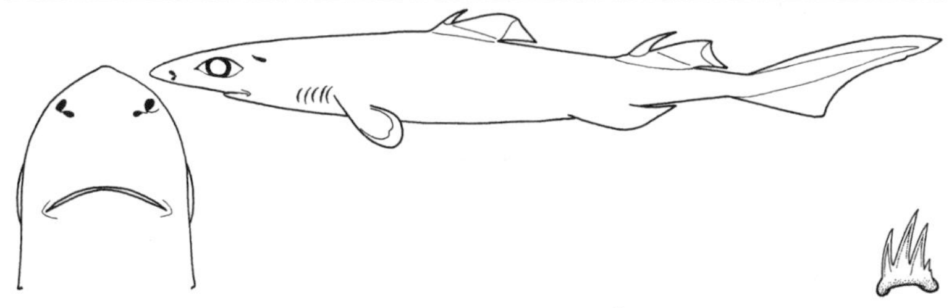

Measurements. Adult: 35–43 cm TL ♂, to at least 50 cm TL ♀.
Identification. Small dogfish with fairly stout uniformly blackish-brown body, fins with prominent white tips and margins. Mouth short, broadly arched, with comblike teeth in both jaws. Dorsal fins similar in size, first fin spine short, second longer (as high but in front of fin apex) with origin over or in front of pelvic fin insertions.

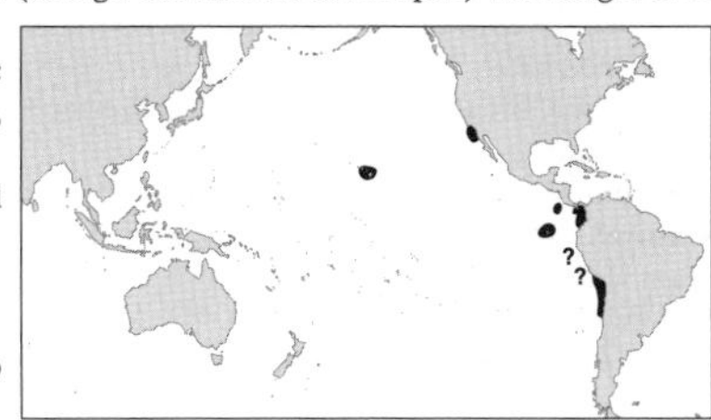

Distribution. Central and eastern Pacific (Hawaiian Islands, American continental shelf, Cocos and Galapagos Islands).
Habitat. On or near bottom on continental and insular slopes, 400–1143 m.
Behaviour. Unknown.
Biology. Presumably ovoviviparous.
Status. IUCN Red List: Not Evaluated. No fisheries interest, sometimes bycaught.

Ornate Dogfish *Centroscyllium ornatum* Plate 6

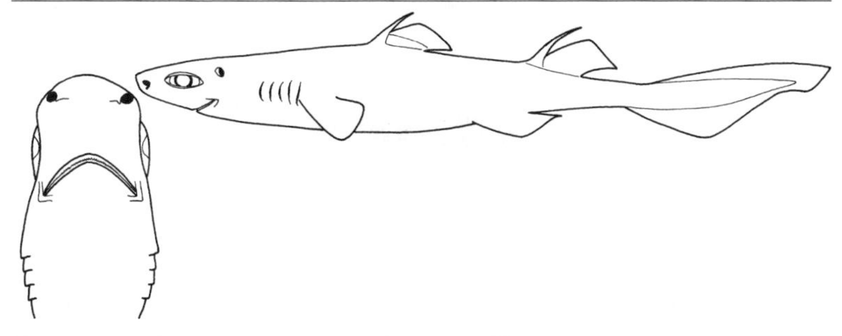

Measurements. Uncertain. Immatures to at least 30 cm TL.
Identification. Blackish body and fins without conspicuous markings. Denticles close-set and numerous. Low rounded first dorsal, very elongated spine, nearly as high as second dorsal spine,

reaching above fin apex. Second dorsal fin larger, very long spine extending above fin apex. Mouth narrowly arched with comb-like teeth in both jaws.
Distribution. Northern Indian Ocean (Arabian Sea, Bay of Bengal).
Habitat. Upper continental slope, near seabed, 521–1262 m.
Behaviour. Unknown.
Biology. Unknown. Presumably ovoviviparous.
Status. IUCN Red List: Not Evaluated. Possibly rare, no interest to fisheries.

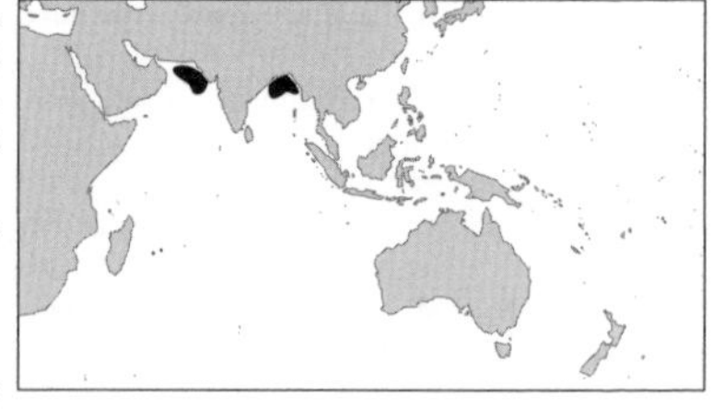

Whitefin Dogfish *Centroscyllium ritteri* Plate 6

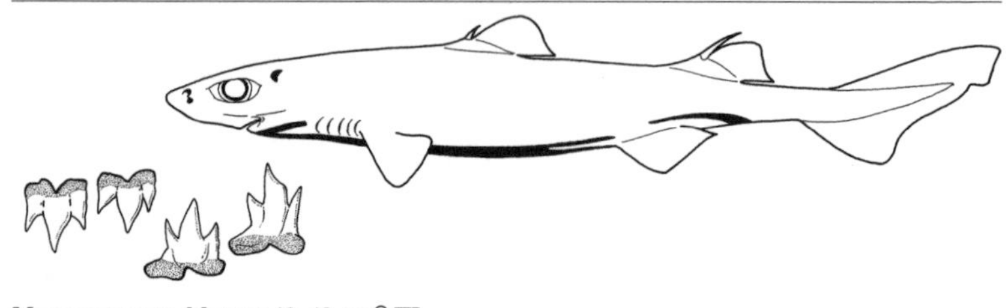

Measurements. Mature: 42–43 cm♀ TL.
Identification. Grey-brown above. Only *Centroscyllium* with striking black markings (concentrations of photophores) on undersides of head, abdomen and pectoral fins, black stripe beneath caudal peduncle extends over pelvic bases. Fins with white margins. Numerous close-set denticles. First dorsal fin low and rounded, spine very short. Second dorsal fin about same size, less rounded, spine moderately long, reaching above fin apex. Mouth very broadly arched with comblike teeth in both jaws.
Distribution. Northwest Pacific.
Habitat. Deepwater, continental slopes and seamounts, 320–1100 m.
Behaviour. Unknown.
Biology. Unknown. Presumably ovoviviparous.
Status. IUCN Red List: Not Evaluated. No interest to fisheries.

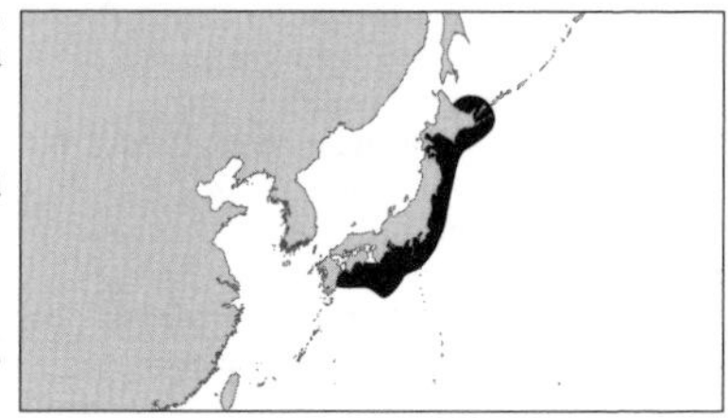

Giant Lanternshark *Etmopterus baxteri* Plate 8

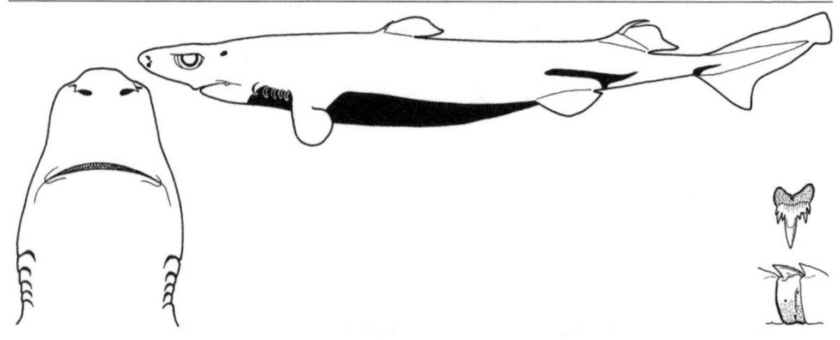

Measurements. Born: ~20 cm TL. Mature: 55–60 cm ♂, 64–69 cm TL ♀. Max: 88 cm TL.
Identification. Very large, stout lanternshark with uniformly dark brown to blackish rough-textured

body with short thick snout. Conspicuous lines of hooked denticles on rear of body and caudal fin. Vague dusky markings on underside of snout and abdomen, above and behind pelvic fins. Black mark at caudal fin base and along its axis. Inconspicuous triangular white blotch on top of head. Pointed multicusped upper teeth, knifelike single-cusped lower teeth. Two widely separated dorsal fins with spines, second fin about twice area first, second spine strongly curved to rear. Short broad tail.
Distribution. Southern Australia, New Zealand. Southern Africa. Possibly islands, seamounts and submarine ridges in south Atlantic and Indian Oceans.

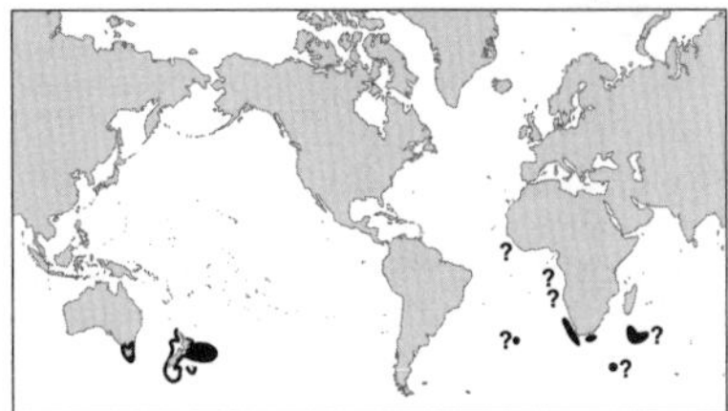

Habitat. Upper insular slopes, on or near bottom at 250–1500 m, possibly deeper.
Behaviour. Unknown.
Biology. Ovoviviparous, 6–16 pups/litter.
Status. IUCN Red List: Least Concern. Fairly common. Bycatch of commercial fisheries. Formerly misidentified as *Etmopterus granulosus*; possibly identical to a poorly known species, *E. tasmaniensis*, from Australia.

Blurred Smooth Lanternshark *Etmopterus bigelowi* Plate 8

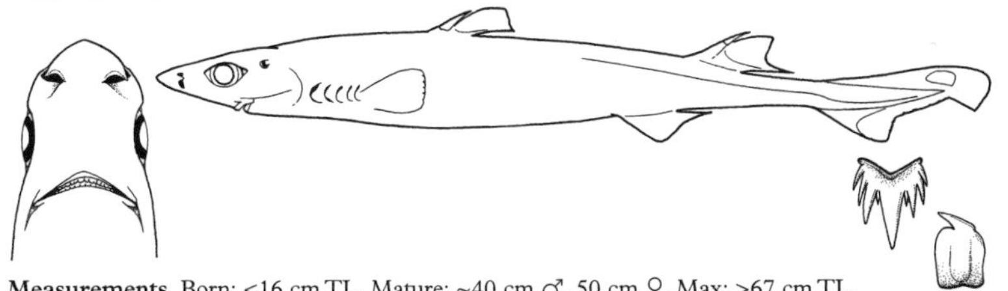

Measurements. Born: <16 cm TL. Mature: ~40 cm ♂, 50 cm ♀. Max: >67 cm TL.
Identification. Fairly large slender, dark brown or blackish (slightly darker below), broad-headed, long-tailed, lanternshark with long thick flat snout, small white blotch on top of head. This and *Etmopterus pusillus* are the only lanternsharks with flat, block-like dermal denticles and smooth skin texture. Multicuspid pointed upper teeth, knife-like lower teeth. Light edges to fins, photophores present but not as conspicuous dark bands on caudal fin and flanks. Two dorsal fins with fin spines, first smaller than second.
Distribution. Atlantic: Gulf of Mexico to Argentina, equatorial west Africa to South Africa. Indo-Pacific: open ocean off South Africa, Japan, seamounts and ocean ridges, Peru.

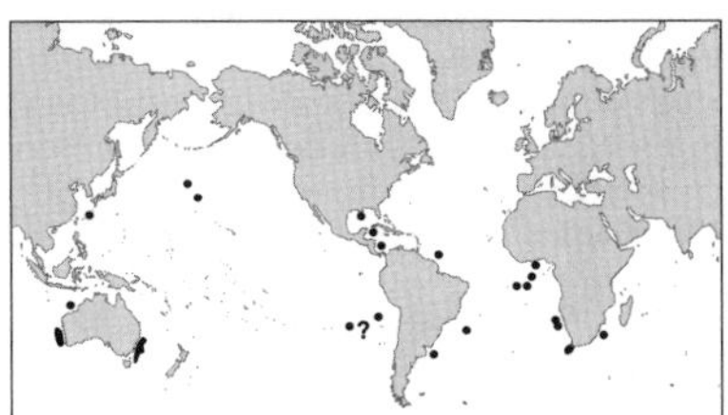

Habitat. Continental and insular shelves and slopes, submarine ridges, 163–1000+ m. Partly epipelagic (occurring near surface in open ocean 110–700 m).
Behaviour. Unknown.
Biology. Poorly known, apparently ovoviviparous.
Status. IUCN Red List: Not Evaluated. Unknown. No importance to fisheries, probably discarded bycatch. Formerly known as *Etmopterus* sp. A. in Australia.

Shorttail Lanternshark *Etmopterus brachyurus* Plate 8

Measurements. Mature: ~24–28 cm ♂, 42 cm ♀. Max: at least 42 cm TL.
Identification. Small, slender-bodied, broad-headed, short-tailed lanternshark with a short thick flat snout. Underside of body conspicuously darker than upper surface. Pointed upper teeth, knife-like lower teeth. Moderately rough skin with conspicuous lines of hooked denticles that run from snout tip to caudal fin. Two dorsal fins with spines, first smaller than second, long second spine curved weakly to rear in adults. Prominent ventral and very long tail markings.
Distribution. West Pacific: Philippines, Japan, Australia.

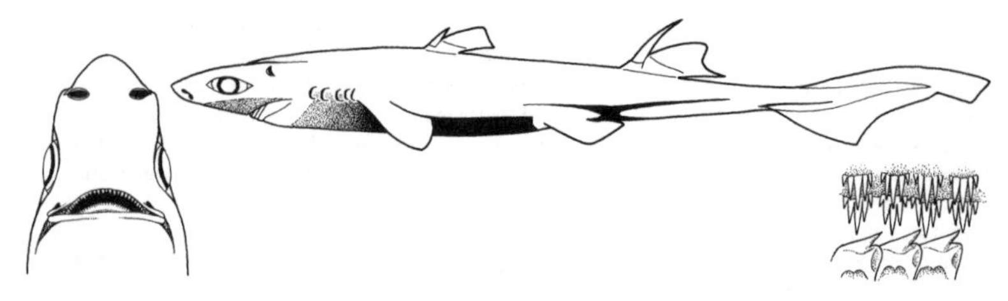

Habitat. Near bottom, 481 m.
Behaviour. Unknown.
Biology. Unknown. Presumably ovoviviparous.
Status. IUCN Red List: Not Evaluated. No interest to fisheries. The Sculpted Lanternshark, *Etmopterus* sp., of southern Africa has been referred to as *E. brachyurus* or *E.* cf. *brachyurus* but is darker, stouter, larger (adult ♂ to 48 cm, adult ♀ to 59 cm) and rougher-skinned. It is close to *E. bullisi* but may be distinct.

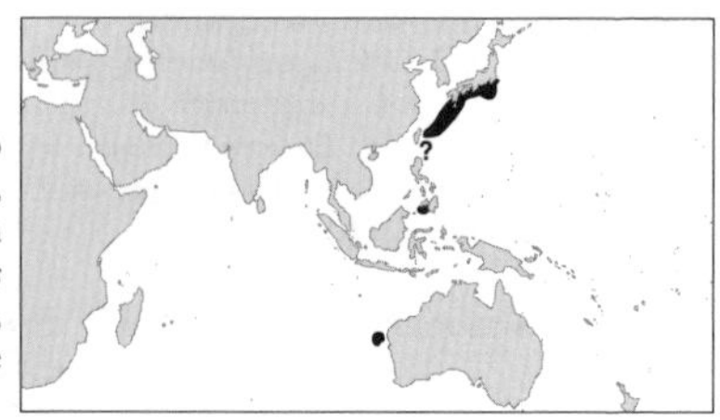

Lined Lanternshark *Etmopterus bullisi* Plate 7

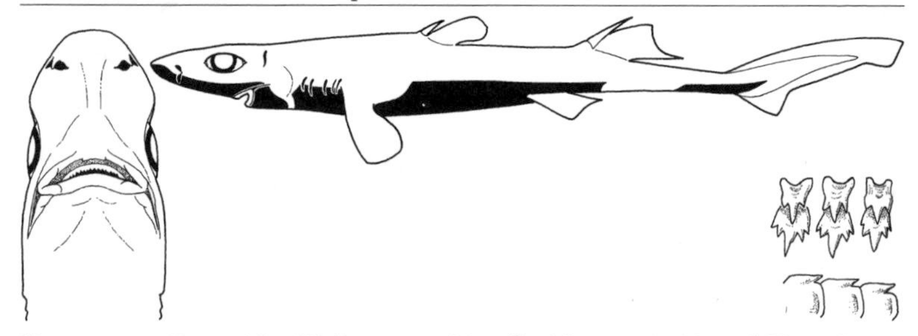

Measurements. Born: < 15 cm TL. Immature at 26 cm ♀, adolescent at 16-26 cm ♂. Max: unknown.
Identification. Dark sooty-grey slender body, underside black, lighter band on midline from eyes to first dorsal fin. Long tail. Conspicuous longitudinal rows of denticles on head, sides and back, extending to caudal fin base. Elongated narrow black mark running above and behind pelvic fins, others at caudal fin base and along axis. Gill openings very short. First dorsal origin over inner pectoral fin margins, base closer to pectoral bases than pelvics, very close to larger second dorsal.
Distribution. Northwest Atlantic: southern USA, Caribbean, Colombia.
Habitat. Continental slopes, on or near bottom, 275–824 m, mostly below 350 m.
Behaviour. Poorly known.
Biology. Poorly known. Presumably ovoviviparous.
Status. IUCN Red List: Not Evaluated. Probably discarded deepwater fisheries bycatch.

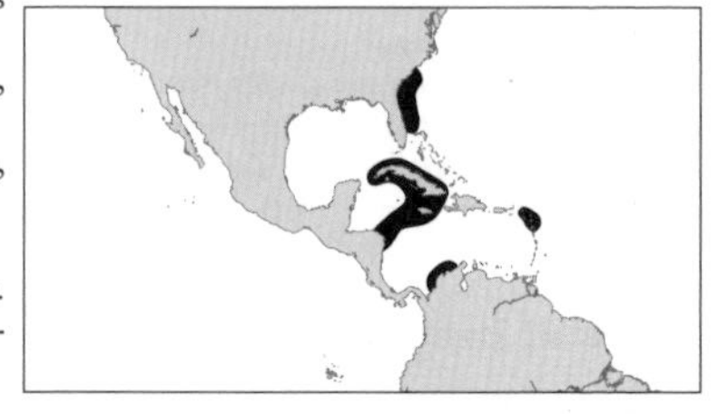

Cylindrical Lanternshark *Etmopterus carteri* Plate 8

Measurements. Mature: 18 cm (♂ & ♀). Max: 21 cm TL.
Identification. A tiny lanternshark with a uniformly dark body without concentrations of photophores, fins with pale webs. Head semicylindrical, nearly as deep as wide at eyes, snout very short and bluntly rounded. Gills broad.

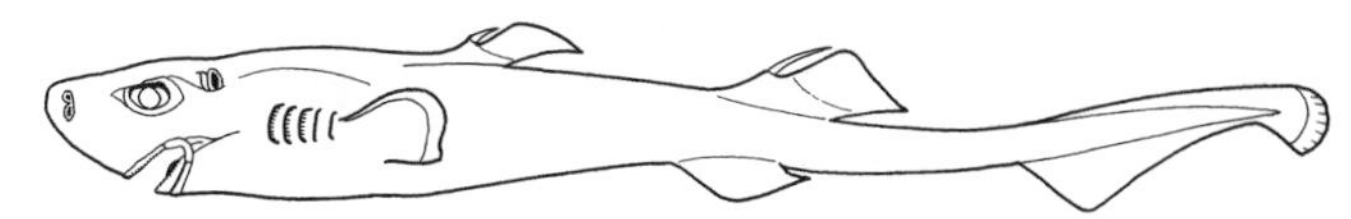

Distribution. Northwest Atlantic: Caribbean coast of Colombia.
Habitat. Upper continental slopes, 283–356 m, possibly also epipelagic.
Behaviour. Unknown.
Biology. Unknown. Presumably ovoviviparous.
Status. IUCN Red List: Not Evaluated. No known human impact.

Tailspot Lanternshark *Etmopterus caudistigmus* Plate 8

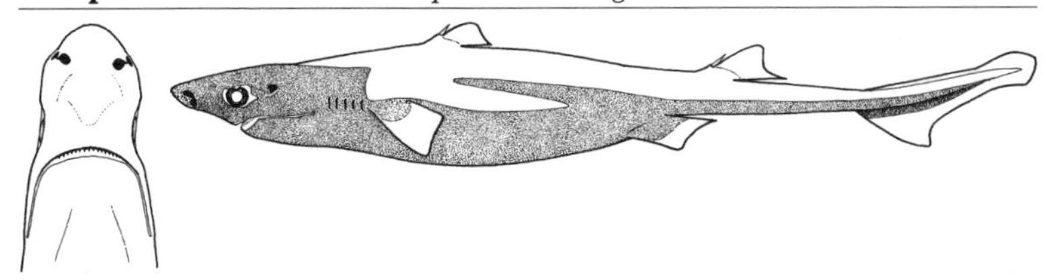

Measurements. Mature: 31 cm TL ♂. Max: 34 cm TL ♀.
Identification. A small dark lanternshark with black underside, slender body with narrow head and long thick narrow snout. Long tail with obvious photomarks. Small close-set lateral trunk denticles form regular longitudinal rows on body and tail, not head. Two dorsal fins with short fin spines; first smaller than second, second fin spine weakly curved, pointing posterodorsally in adults.
Distribution. Southwest Pacific: New Caledonia.
Habitat. Insular slope, 638–793 m.
Behaviour. Unknown.
Biology. Presumably ovoviviparous.
Status. IUCN Red List: Least Concern. Three specimens caught on longline in largely unfished area.

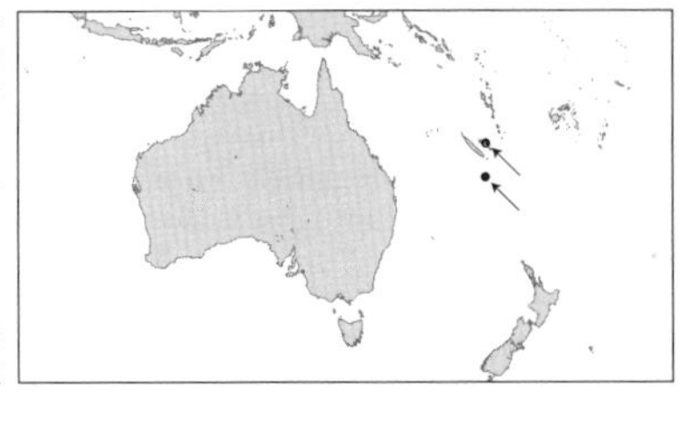

Combtooth Lanternshark *Etmopterus decacuspidatus* Plate 8

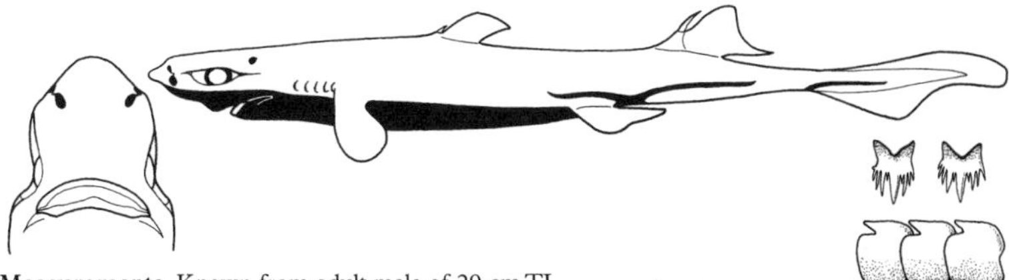

Measurements. Known from adult male of 29 cm TL.
Identification. Moderately slender body, brown above, black underside, elongated narrow black mark running above, in front, and behind pelvic fins, other elongated black marks at caudal fin base and along axis. Upper teeth with four or five pairs of cusplets on each side. Fairly long, broad tail. Gill openings very short. No regular rows of denticles on sides. First dorsal fin origin slightly behind

free pectoral rear tips, dorsal base much closer to pectoral bases than to pelvics. Second dorsal about twice area of first.
Distribution. Northwest Pacific: Hainan Island, China.
Habitat. Collected on or near bottom, 512–692 m.
Behaviour. Unknown.
Biology. Unknown. Presumably ovoviviparous.
Status. IUCN Red List: Not Evaluated. Known from only one specimen.

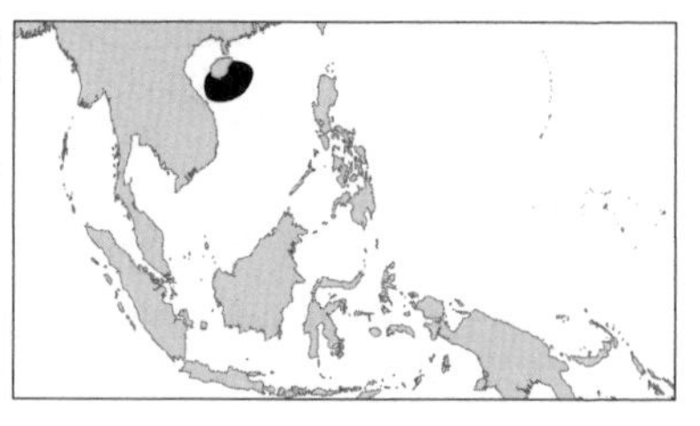

Pink Lanternshark *Etmopterus dianthus* Plate 9

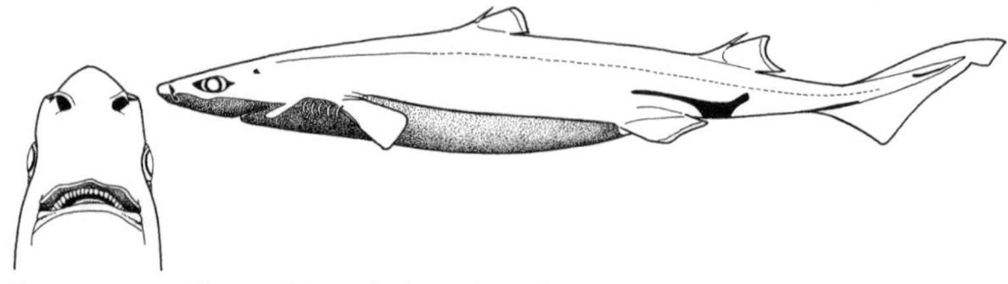

Measurements. Mature: >34 cm ♂. Max: at least 41 cm TL.
Identification. Stout lanternshark, pinkish above when fresh, brownish-grey after preservation, dusky to black below, distinctive black markings behind pelvic fin, on caudal peduncle, and upper caudal fin (upper fin tip dark). Fine, bristle-like denticles not arranged in rows. First dorsal fin small and low, short spine; second less than twice size of first, spine nearly as long as fin tip.
Distribution. Southwest Pacific: Australia and New Caledonia.
Habitat. Bottom of upper continental shelf, 700–880 m.
Behaviour. Unknown.
Biology. Unknown. Presumably ovoviviparous.
Status. IUCN Red List: Least Concern. Range unfished.

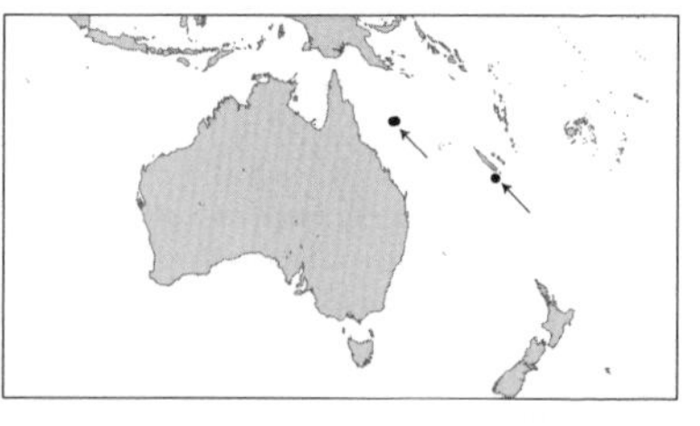

Lined Lanternshark *Etmopterus dislineatus* Plate 9

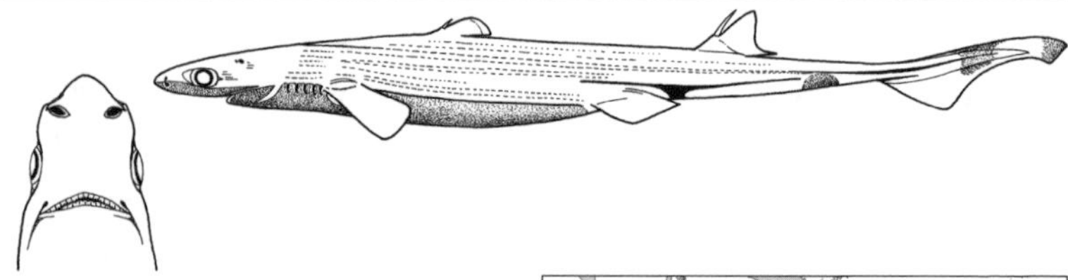

Measurements. Mature: ~34 cm ♂. Max: at least 45 cm TL.
Identification. Very elongated attractive lanternshark, light silvery-brown above, much darker below with a pattern of dark broken lines of dots and dashes along the upper flanks. Distinct black markings near pelvic fin and on the caudal peduncle, mid caudal fin and upper fin tip. Hooked, bristle-like denticles not arranged in

regular rows. First dorsal fin very low and small, about half size of second.
Distribution. Pacific, northeastern Australia.
Habitat. Upper continental slope, on or near bottom, 590–700 m.
Behaviour. Unknown.
Biology. Unknown. Presumably ovoviviparous.
Status. IUCN Red List: Least Concern. Small known range is unfished.

Blackmouth Lanternshark *Etmopterus evansi* Plate 9

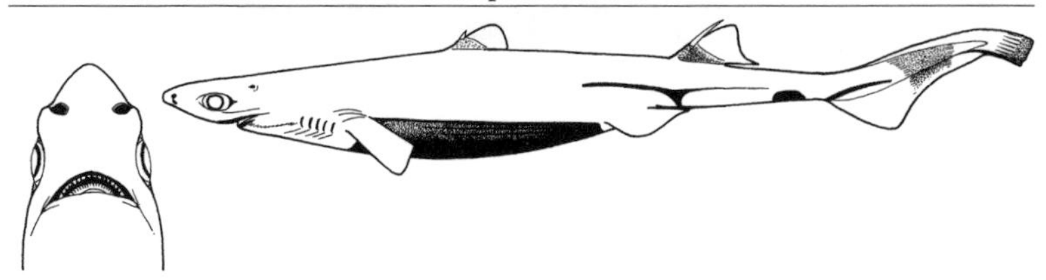

Measurements. Male of 26 cm TL was mature. Max: at least 30 cm TL.
Identification. Small light brown lanternshark, darker below, dark borders around mouth, above eyes, sometimes around gill slits. Distinct black markings behind the pelvic fin, on caudal peduncle (a dark ventral blotch) and upper caudal fin (dark bands through middle and tip of upper lobe). Hook-like denticles arranged in distinct but weakly defined rows along dorsal midline and on caudal peduncle, not on head.

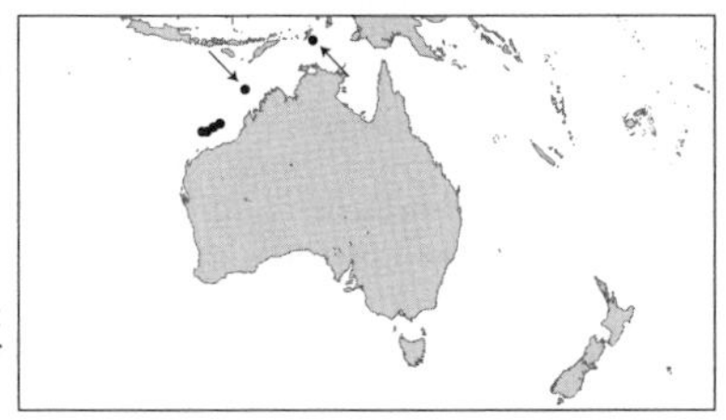

Distribution. Pacific: northwest Australia and Arafura Sea (Indonesia).
Habitat. Shoals and reefs on continental shelf, 430–550 m.
Behaviour. Unknown.
Biology. Unknown. Presumably ovoviviparous.
Status. IUCN Red List: Least Concern. Not currently fished, but could become bycatch of expanding deepwater fisheries.

Pygmy Lanternshark *Etmopterus fusus* Plate 9

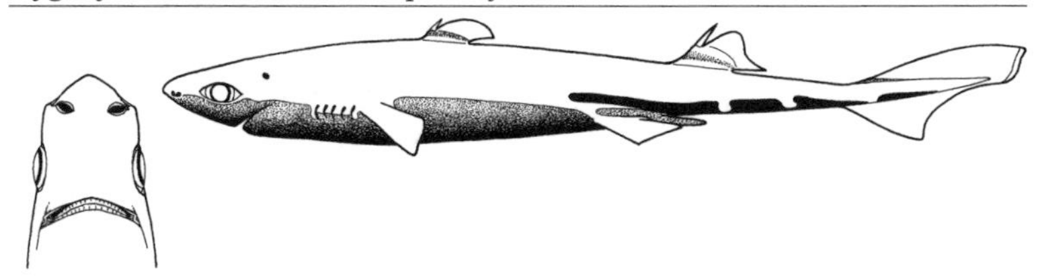

Measurements. Mature: 25–26 cm TL ♂. Max: at least 30 cm TL.
Identification. Dwarf lanternshark with firm cylindrical body; denticles in regular rows on flanks and caudal peduncle, but not on head. Dark greyish or black, faint dark markings on flank above and behind pelvic fins and on tail. Fins pale, with dark margins. Second dorsal fin more than twice height of first. Elongated caudal peduncle. Similar to *E. splendidus*.

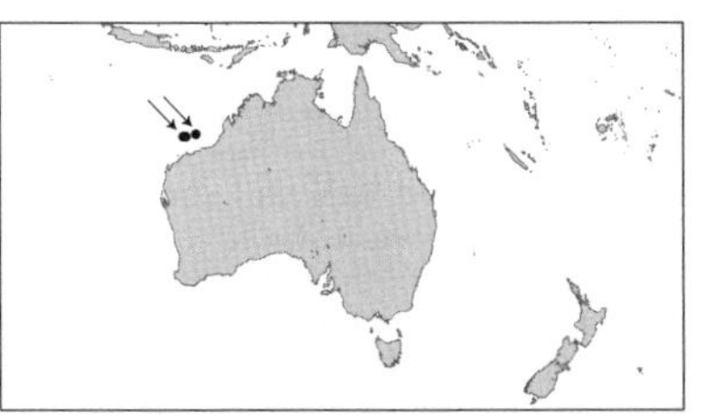

Distribution. East Indian Ocean: western Australia, possibly Java, Indonesia.
Habitat. Continental slope, 430–550 m (Australia), possibly 120–200 m (Indonesia).

Behaviour. Unknown.
Biology. Unknown. Presumably ovoviviparous.
Status. IUCN Red List: Least Concern. Only minor fishing activity in Australian range.

Broadband Lanternshark *Etmopterus gracilispinis* Plate 7

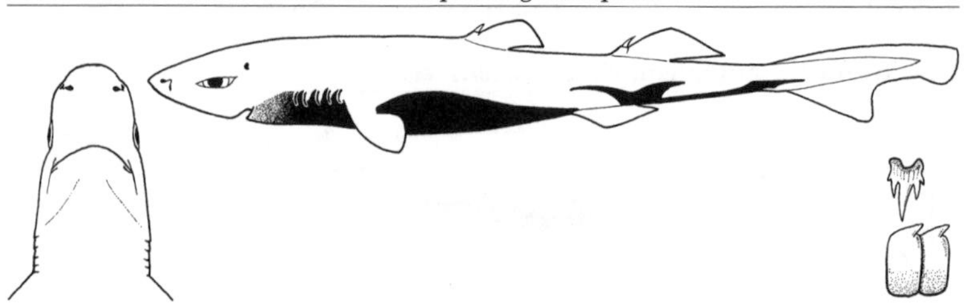

Measurements. Mature: ~26 cm ♂, 33 cm ♀. Max: ~33 cm TL.
Identification. Small stout lanternshark with blackish-brown body, grading to black below, short slender tail. Inconspicuous broad elongated black mark running above and behind pelvic fins, other elongated black marks at caudal fin base and along axis. Gill openings very short. No regular rows of lateral trunk denticles. First dorsal fin well behind pectoral rear tips; second dorsal about twice area of first.

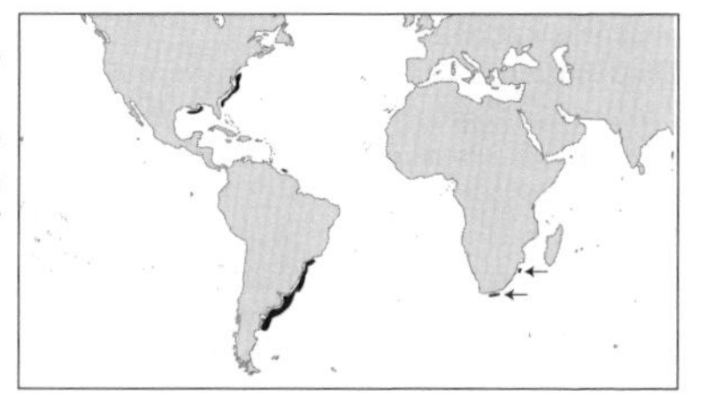

Distribution. West Atlantic: USA to Argentina. Western Indian Ocean: South Africa (possibly to southeast Atlantic).
Habitat. Outer continental shelves and upper-middle slopes, on or near bottom, 100–1000 m; epipelagic and mesopelagic at 70–480 m over water 2240 m deep off Argentina.
Behaviour. Unknown.
Biology. Unknown. Presumably ovoviviparous.
Status. IUCN Red List: Not Evaluated.

Southern Lanternshark *Etmopterus granulosus* Plate 7

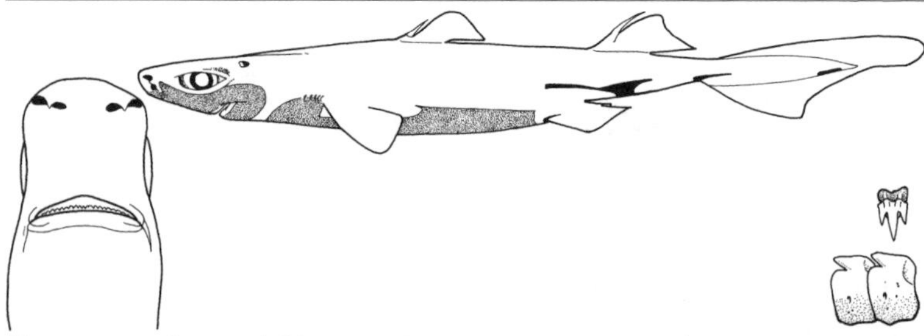

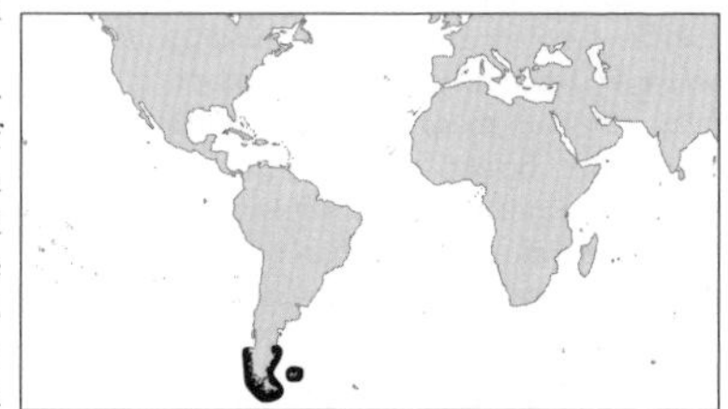

Measurements. Immature/adolescent: to 26 cm ♂. Max: size uncertain.
Identification. Grey-brown, heavy-bodied, big-headed lanternshark with conspicuous lines of large, rough conical denticles on body, not on head. Underside abruptly black. A short, broad black mark running above and slightly behind the pelvic inner margins, other elongated black marks at the caudal fin base. Gill openings very short. Denticles wide-spaced, randomly on head but in

conspicuous lines on sides; snout mostly bare, except for lateral patches of denticles. Second dorsal fin much larger than first. Short broad tail.
Distribution. South America (southern Argentina to southern Chile).
Habitat. Outermost continental shelves and upper slopes, 220–637 m.
Behaviour. Unknown.
Biology. Unknown. Presumably ovoviviparous.
Status. IUCN Red List: Not Evaluated. Confused with *Etmopterus baxteri* of the eastern hemisphere, but apparently confined to South America.

Caribbean Lanternshark *Etmopterus hillianus* Plate 7

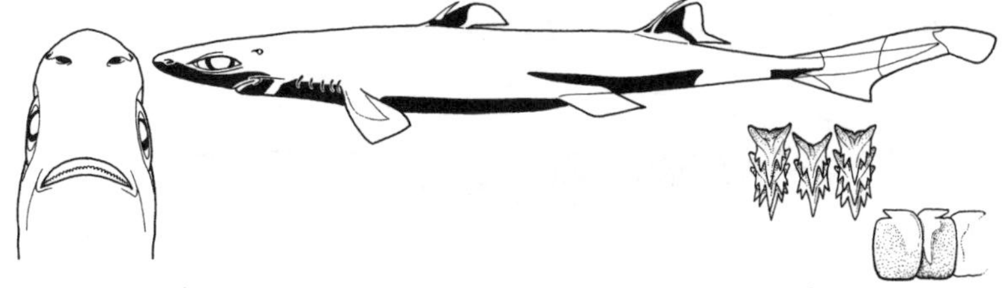

Measurements. Born: ~9 cm TL. Mature: ~20 cm ♂. Max: 26 ♂, 28 cm TL ♀.
Identification. A dwarf lanternshark with a stout body, moderately long tail. Grey or dark brown above, underside black. Elongated broad black mark above and behind pelvics, others at caudal fin base and along axis. Gill openings very short. Upper teeth generally with <three pairs of cusplets. Denticles largely cover snout; not arranged in rows on trunk. Interdorsal space short. Second dorsal fin much larger but less than twice area of first.

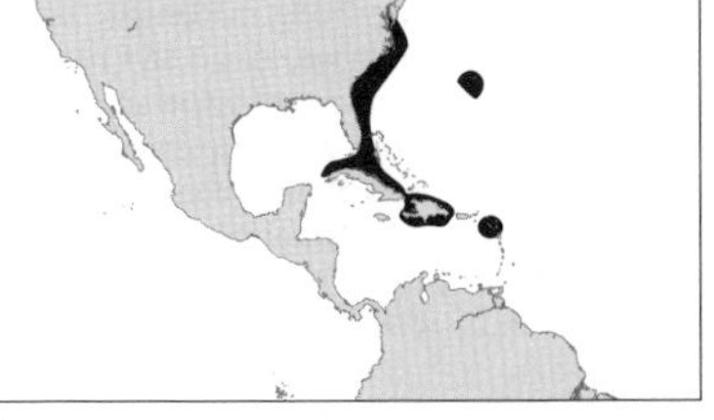

Distribution. Northwest Atlantic: Virginia to Florida (USA), Bahamas, Cuba, Bermuda, Hispanola, Lesser Antilles. Not known from western or southern Caribbean.
Habitat. Upper continental and insular slopes, on or near bottom, 311–695 m.
Behaviour. Unknown.
Biology. Ovoviviparous, four or five pups/litter.
Status. IUCN Red List: Not Evaluated. Low fisheries interest (caught by hook-and-line off Cuba).

Smalleye Lanternshark *Etmopterus litvinovi* Plate 9

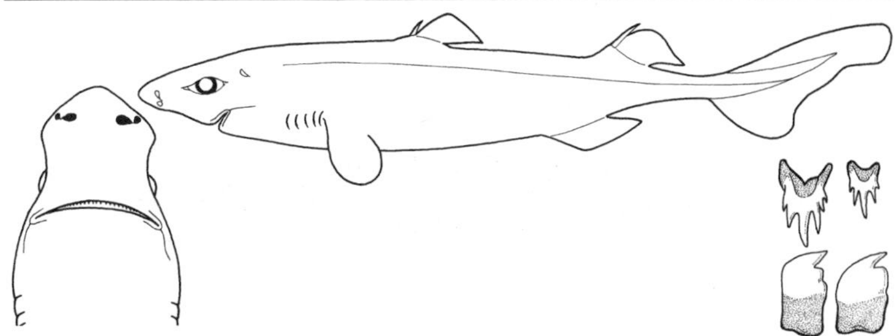

Measurements. Born: uncertain, less than 17 cm TL. Max: 61 cm TL.
Identification. A moderately large plain-coloured lanternshark with a large flat head, stout body, moderately long tail. No obvious markings. Gill openings long. Upper teeth generally with one to three pairs of cusplets. Denticles stout, hooked, conical and not arranged in rows on trunk. Interdorsal space short. First dorsal origin behind pectoral rear tips, second dorsal fin slightly larger than first.

Distribution. Southeastern Pacific: Nazca and Sala y Gomez Ridges off Peru and Chile.
Habitat. Upper slopes, on or near bottom, 630–1100 m.
Behaviour. Unknown.
Biology. Unknown.
Status. IUCN Red List: Not Evaluated. No fisheries interest. Poorly known.

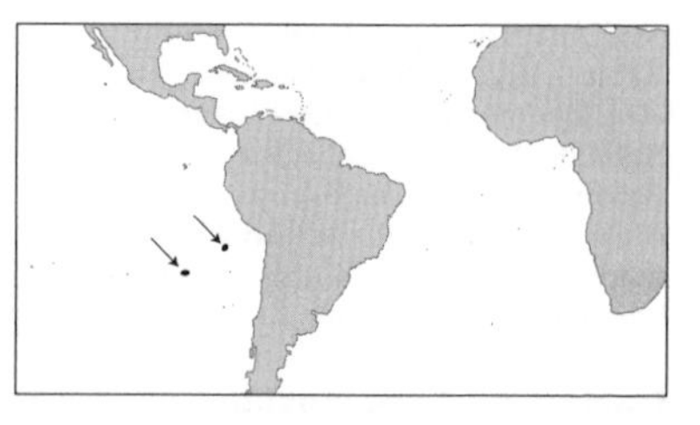

Blackbelly Lanternshark *Etmopterus lucifer* Plate 8

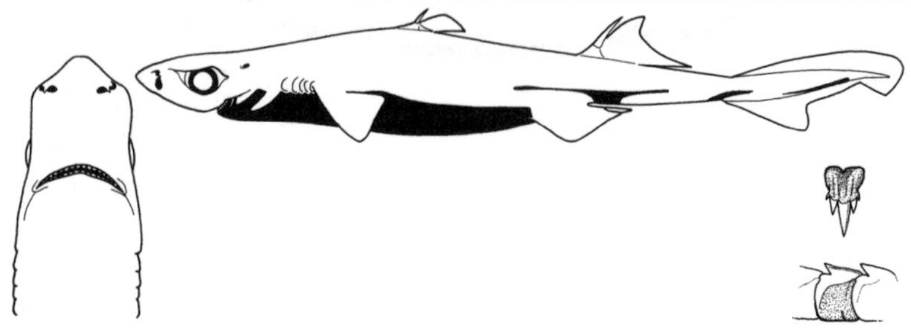

Measurements. Mature: 29 cm ♂, 34 cm ♀. Max: ~42 cm TL.
Identification. Stocky body, moderately long tail. Brown above, underside black, elongated narrow black mark running above, ahead and behind pelvics, others at tail base and along axis. Rows of denticles from snout tip to tail. Short interdorsal space, very large second dorsal fin. Teeth blade-like in lower jaw, with cusps and cusplets (generally <3 pairs) in upper. Gill openings moderately long.
Distribution. Western Pacific: confirmed from Japan, South China Sea, Australia, New Caledonia and New Zealand. Nominal records from southeast Pacific, south Atlantic, central Pacific, and some Indo-west Pacific localities require confirmation.
Habitat. Outer continental and insular shelves, upper slopes, on or near bottom, 158–1357 m.
Behaviour. Essentially unknown.
Biology. Presumably ovoviviparous. Eats mostly squid and small bony fishes, also shrimp.
Status. IUCN Red List: Not Evaluated.

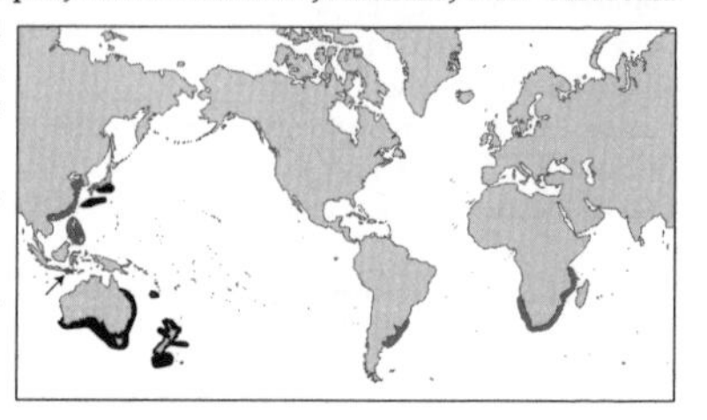

Slendertail Lanternshark *Etmopterus molleri* Plate 8

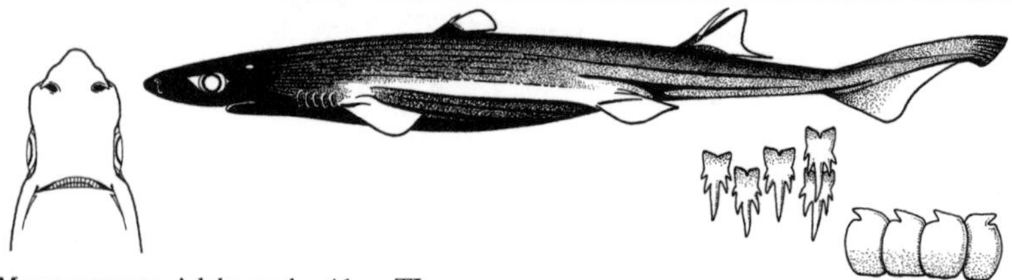

Measurements. Adults reach ~46 cm TL.
Identification. Slender with light brown back, dark brown flanks with black photomarks, underside abruptly black. Regular longitudinal rows of denticles on head, flanks, caudal peduncle, and caudal base, not above pectoral fins. No denticles on second dorsal fin, which is much larger than first.

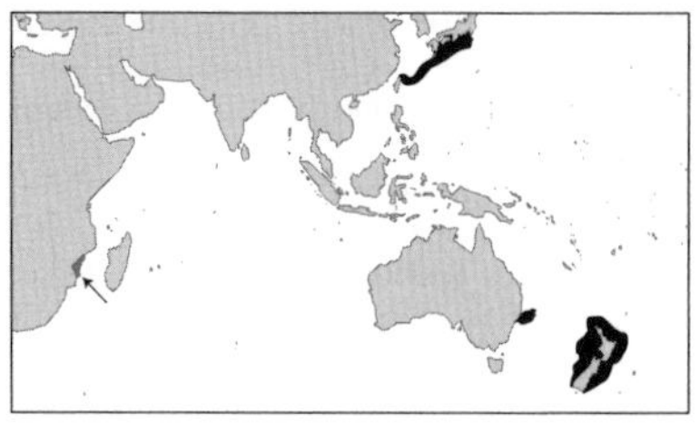

Similar to *Etmopterus brachyurus* and *E. lucifer*. Differs from former in having a largely naked second dorsal fin and the latter in having a taller second dorsal fin and a longer caudal peduncle.
Distribution. West Pacific: off Australia, New Zealand, Taiwan and Japan. Possibly western Indian Ocean off Mozambique.
Habitat. Outer continental and insular shelves, upper slopes, on or near bottom, 238–655 m.
Behaviour. Essentially unknown.
Status. IUCN Red List: Not Evaluated.

Dwarf Lanternshark *Etmopterus perryi* Plate 8

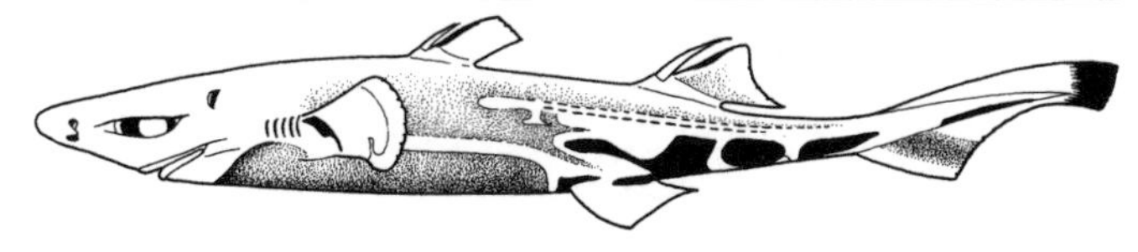

Measurements. Born: ~6 cm TL. Mature: ~16–17 cm ♂, 19 cm ♀. Max: ~21 cm TL.
Identification. One of the smallest living sharks. Long broad flat head. Brownish above, black below, without a band of white on the middle of the back. Conspicuous dark band at tip of terminal lobe of caudal fins and dark blotch on ventral caudal lobe. Second dorsal fin larger than first.

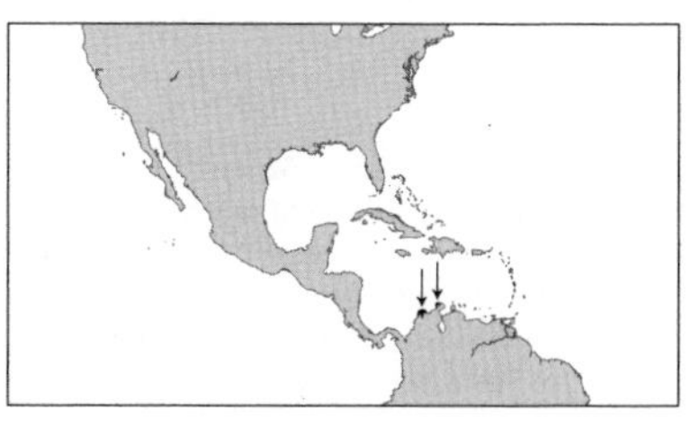

Distribution. Northwest Atlantic: Caribbean coast of Colombia.
Habitat. Upper continental slope, 283–375 m.
Behaviour. Unknown.
Biology. Ovoviviparous.
Status. IUCN Red List: Not Evaluated. No known human impact.

African Lanternshark *Etmopterus polli* Plate 7

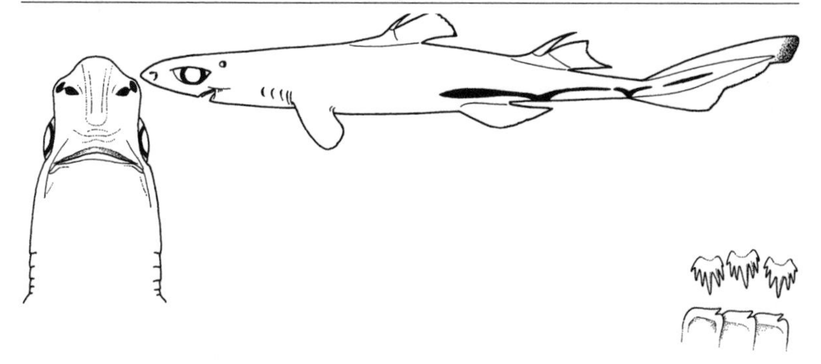

Measurements. Max: ~24 cm TL.
Identification. A dwarf lanternshark with a fairly stout dark grey body, fairly long tail. Gill openings short. Denticles wide-spaced, not in rows on sides, largely cover snout. Second dorsal fin as large or slightly larger than first. Underside of snout and abdomen blackish, with an elongated broad black mark running above, ahead and behind pelvic fins; other marks at base and along axis of tail.

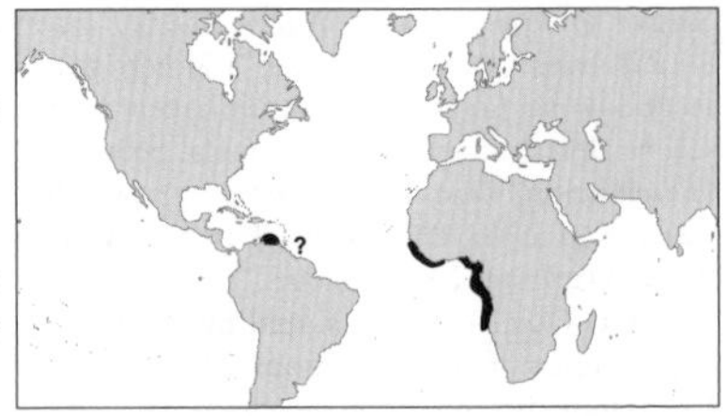

Distribution. East Atlantic: Guinea to Angola. ?West Atlantic: Venezuela.
Habitat. Upper continental slopes, on or near seabed, 300–1000 m.
Behaviour. Unknown.
Biology. Presumably oviviparous.
Status. IUCN Red List: Not Evaluated. Minor fisheries interest. Bycatch in offshore fisheries sometimes used for food or fishmeal.

Great Lanternshark *Etmopterus princeps* Plate 7

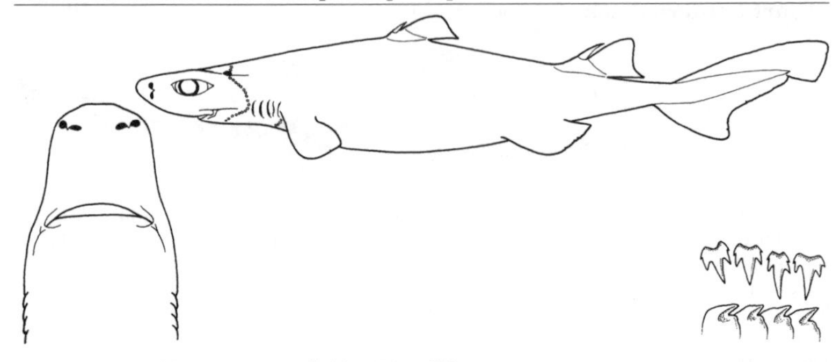

Measurements. Mature: ~65 cm ♂. Max: 75 cm TL.
Identification. A large stout lanternshark with a blackish body, no conspicuous dark markings. Moderately long, broad tail. Gill openings very long (half eye length). Denticles largely cover snout; widespaced (not in rows) on sides. Second dorsal fin much larger but less than twice area of first.
Distribution. Northwest Atlantic: Nova Scotia to New Jersey. Northeast Atlantic: between Greenland and Iceland to northwest Africa and Azores. Nominal range in south Atlantic and western Pacific uncertain.

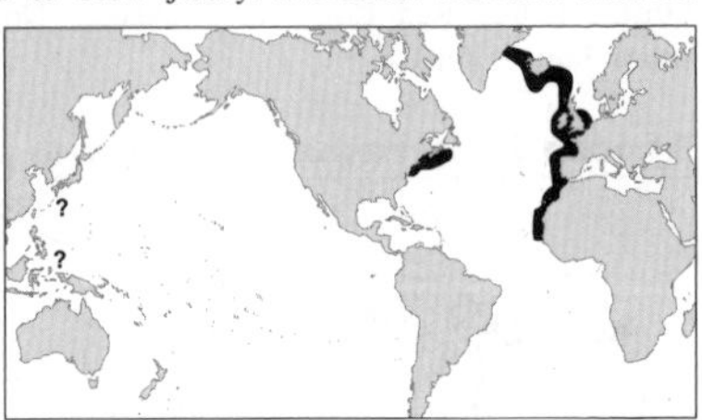

Habitat. Continental slopes, on or near bottom, 350–2213 m. 3750–4500 m on north Atlantic lower rise.
Behaviour. Unknown.
Biology. Poorly known. Presumably ovoviviparous.
Status. IUCN Red List: Not Evaluated. Probably a bycatch in east Atlantic slope fisheries.

False Lanternshark *Etmopterus pseudosqualiolus* Plate 9

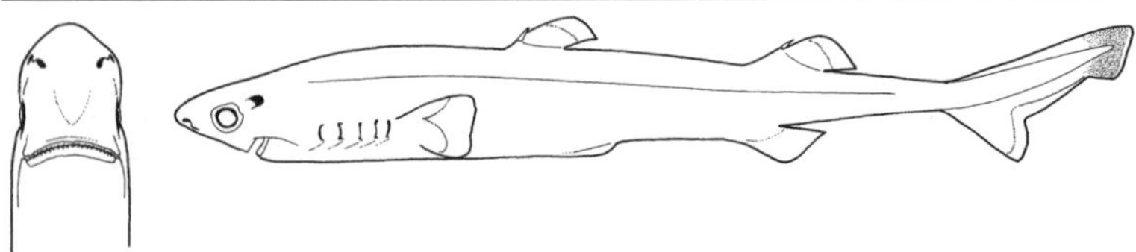

Measurements. Mature: 40-45 cm TL ♂.
Identification. Dark brown to black above, paler tail, underside dark but not abruptly black, posterior fin margins pale, conspicuous dark bands absent from middle of caudal fin, terminal caudal lobe dark. Greatly elongated but inconspicuous caudal base photomark on caudal fin.

Fusiform body, very short deep snout, small pectoral fins, short, rounded eyes. Teeth with three to five pairs of cusplets in males. Similar in body shape to *Etmopterus carteri.*
Distribution. West Pacific: Norfolk and Lord Howe ridges, near New Caledonia.
Habitat. Oceanic ridges, 1043–1102 m.
Behaviour. Possibly semi-oceanic.
Status. IUCN Red List: Least Concern. Deepwater fisheries very uncommon within limited range.

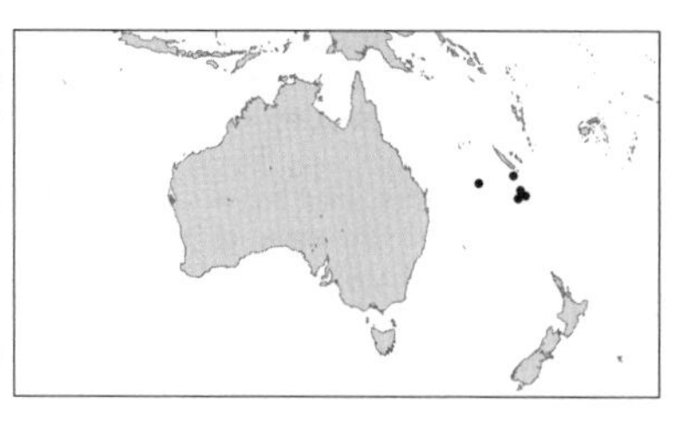

Smooth Lanternshark *Etmopterus pusillus* Plate 7

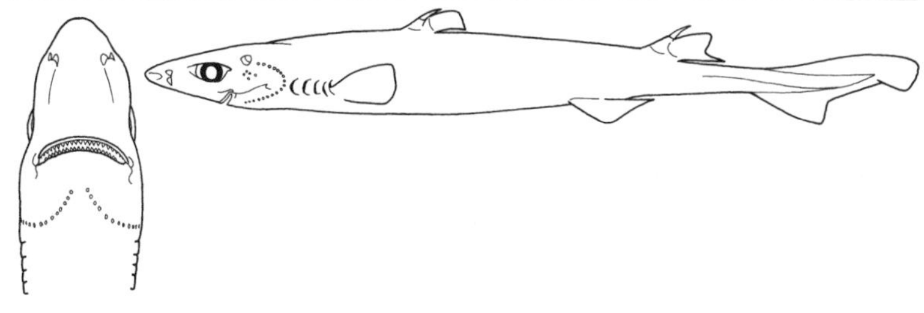

Measurements. Adults: 31–39 cm ♂, 38–47 cm ♀. Max: possibly 50–100 cm TL.
Identification. A small, smooth-skinned lanternshark with a fairly slender blackish-brown body, fairly short broad tail. Obscure broad black mark above, in front and behind pelvic fins. Gill openings rather long. Upper teeth generally with <three pairs of cusplets. Wide-spaced low-crowned cuspless denticles (as in *Etmopterus bigelowi*), not arranged in rows, cover snout. Second dorsal fin less than twice area of first.
Distribution. Widespread in Atlantic. Indo-west Pacific: South Africa and Japan.
Habitat. Continental slopes, on or near bottom, 274–1000 m (possibly 1998 m). Oceanic in south Atlantic, 0–708 m over deep water.
Behaviour. Unknown.
Biology. Ovoviviparous. Eats fish eggs, lanternfish, hake, squid, other small sharks.
Status. IUCN Red List: Not Evaluated. Utilised bycatch in east Atlantic bottom fisheries.

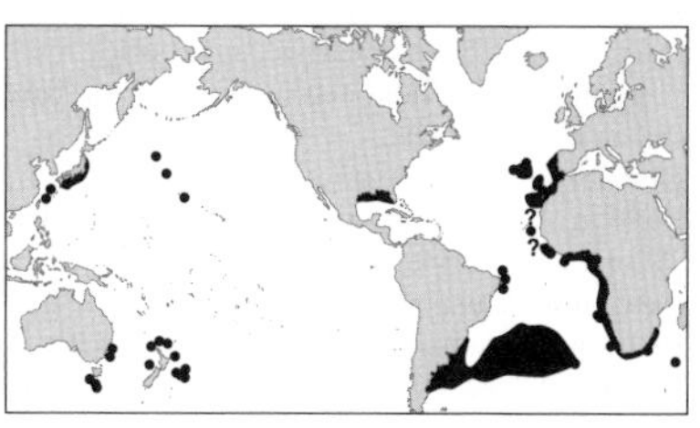

Densescale Lanternshark *Etmopterus pycnolepis* Plate 9

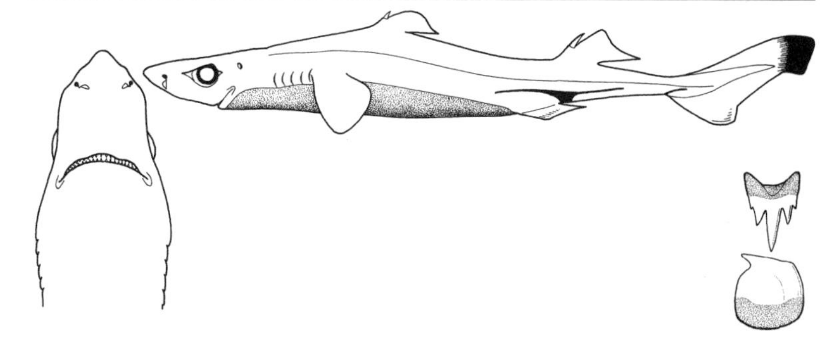

Measurements. Born: uncertain. Max: 44–45 cm TL.
Identification. A moderately small plain-coloured lanternshark with a narrow head, slender body and moderately long black-tipped tail. Conspicuous black photomarks on flanks, tail, and underside of body. Gill openings long. Upper teeth generally with one to three pairs of cusplets. Denticles hooked, conical, very small and arranged in dense rows on head, trunk and tail. Interdorsal space long. First dorsal origin ahead of pectoral rear tips, second dorsal fin larger than first. Similar to *Etmopterus brachyurus*, *E. molleri*, and *E. lucifer*, but with smaller, more close-set denticles and flank marks with front and rear extensions equal-sized.

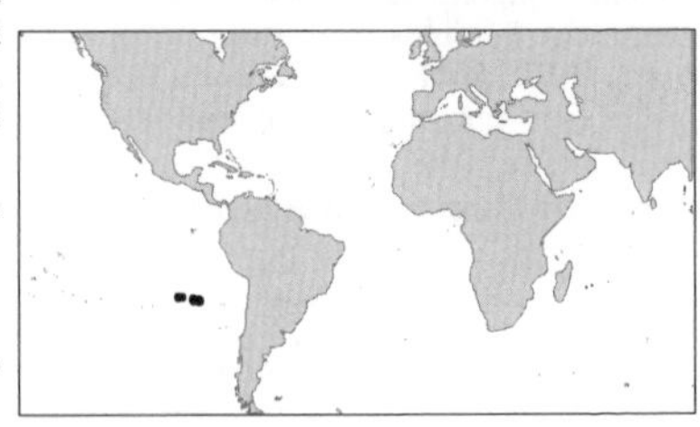

Distribution. Southeastern Pacific: Nazca and Sala y Gomez Ridges off Peru and Chile.
Habitat. Upper slopes, on or near bottom, 330–763 m.
Behaviour. Unknown.
Biology. Unknown.
Status. IUCN Red List: Not Evaluated. No fisheries interest.

West Indian Lanternshark *Etmopterus robinsi* Plate 8

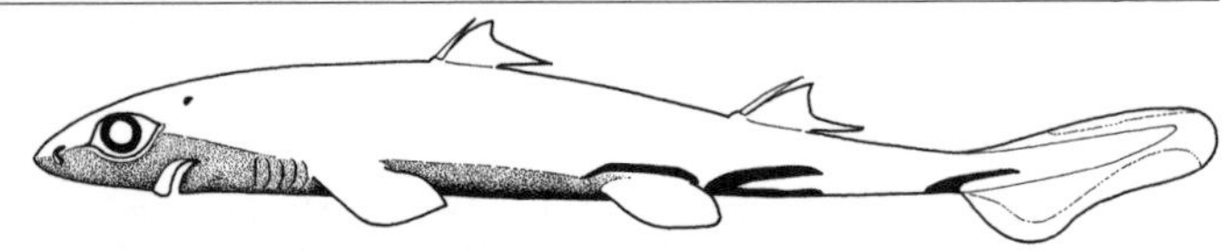

Measurements. Mature: 26 cm ♂. Max: 31 cm TL ♂, 34 cm TL ♀.
Identification. Moderately stout grey or dark brown body, moderately long tail. Gill openings about one third eye length. Upper teeth generally with less than three pairs of cusplets. Lateral denticles with slender, hooked conical crowns, wide-spaced, not in regular rows; denticles largely cover snout. Second dorsal fin much larger but less than twice area of first. Underside of snout and abdomen conspicuously dark, elongated broad black mark runs above and behind pelvic fins, others at base and along axis of tail.

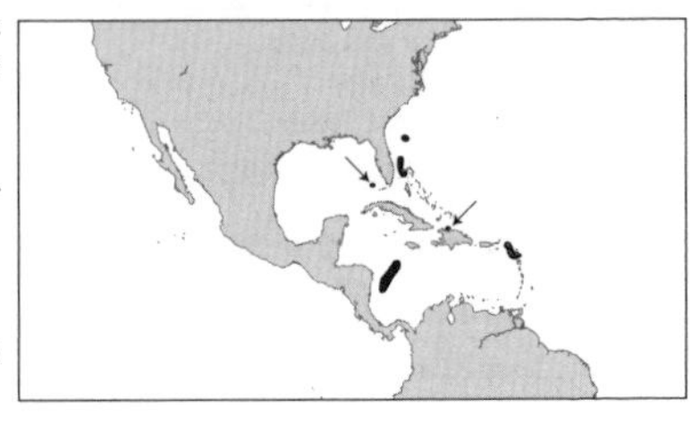

Distribution. West Atlantic: Caribbean.
Habitat. Continental and insular slopes, 412–787 m, mostly >549 m.
Behaviour. Unknown.
Biology. Unknown. Presumably ovoviviparous.
Status. IUCN Red List: Not Evaluated. No human impact.

Fringefin Lanternshark *Etmopterus schultzi* Plate 7

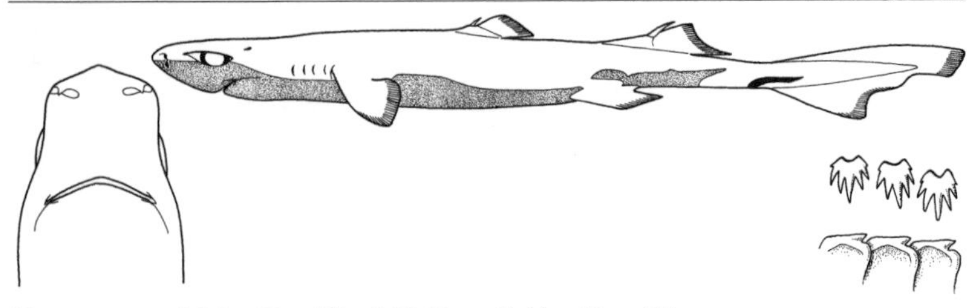

Measurements. Adults: 27 cm TL ♂, 28–30 cm ♀. Max: 30 cm TL.
Identification. Slender, moderately long tail (dorsal tail margin ~equal to head length). Light brown above, dusky-grey below; elongated narrow dusky mark above and behind pelvic fins; elongated

black marks at tail base and along axis. Gill openings very short. Very slender-crowned, hooked, lateral trunk denticles, wide-spaced, no rows of enlarged thorns on flanks; denticles largely cover snout. First dorsal far behind pectorals, ~equidistant between pectoral and pelvic bases, fin margins naked, prominently fringed margin of naked ceratotrichia. Second dorsal fin about twice area of first.

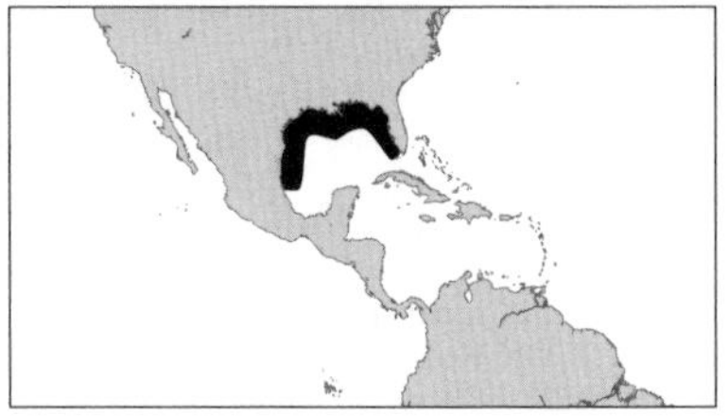

Distribution. West Atlantic.
Habitat. Upper continental slopes, on or near bottom, 220–915 m, mostly >350 m.
Behaviour. Unknown.
Biology. Presumably ovoviviparous.
Status. IUCN Red List: Not Evaluated. No interest to fisheries.

Thorny Lanternshark *Etmopterus sentosus* Plate 7

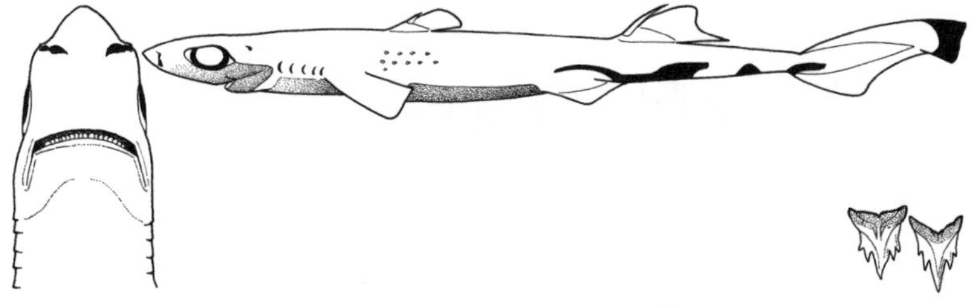

Measurements. Born: 6 cm TL. Mature: 22–26 cm ♂, 25–26 cm ♀. Max: ~27 cm TL.
Identification. A dwarf lanternshark with a slender greyish-black body, inconspicuously black on its underside. Two or three rows of unique enlarged hooklike denticles on flanks, absent from other lanternsharks. Elongated broad black mark above, in front and behind pelvic fins, others at tail base and along axis. Moderately long broad tail (dorsal caudal margin ~equal to head length). Gill openings quite long. Three or four pairs of cusplets on upper teeth. Distal margins of fins largely naked and more or less fringed with ceratotrichia. Second dorsal fin over twice area of first.

Distribution. West Indian Ocean.
Habitat. Near bottom, perhaps 200–500 m.
Behaviour. Unknown.
Biology. Unknown. Presumably ovoviviparous.
Status. IUCN Red List: Not Evaluated. Relatively common off Mozambique.

Velvet Belly *Etmopterus spinax* Plate 7

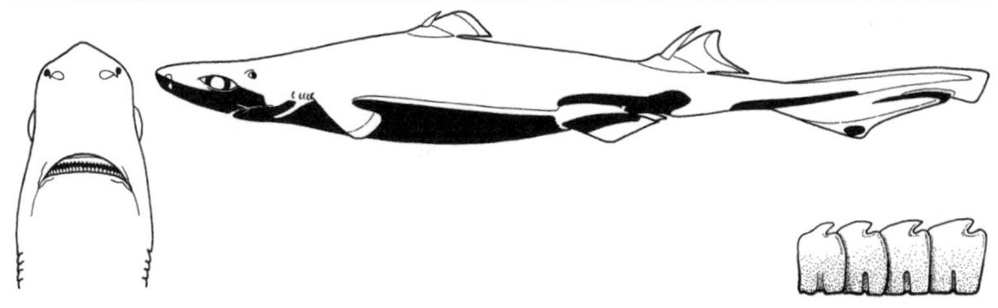

Measurements. Born: 12–14 cm TL. Mature: 33–36 cm. Max: 60 cm TL (rare >45 cm).
Identification. Long, fairly stout body with long tail, brown above, abruptly black below, elongated

narrow black mark above and behind pelvic fins, others at sides of tail. Gill openings very short. No lines of lateral trunk denticles, snout largely covered with denticles. Second dorsal fin about twice area of first.

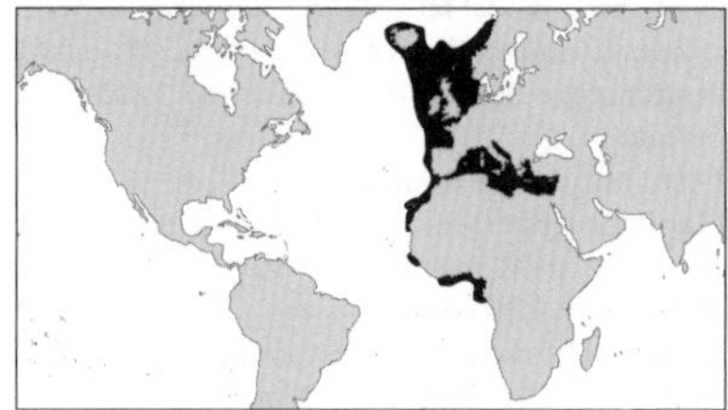

Distribution. East Atlantic: Iceland to Gabon. Western Mediterranean.
Habitat. Outer continental shelves and upper slopes, near or well above bottom 70–2000 m, mostly 200–500 m.
Behaviour. Unknown.
Biology. Ovoviviparous, 6–20 pups/litter. Feeds on small fishes, squid and crustaceans.
Status. IUCN Red List: Not Evaluated. Common. Caught offshore by bottom and pelagic trawls, utilized for fishmeal and food.

Splendid Lanternshark *Etmopterus splendidus* Plate 9

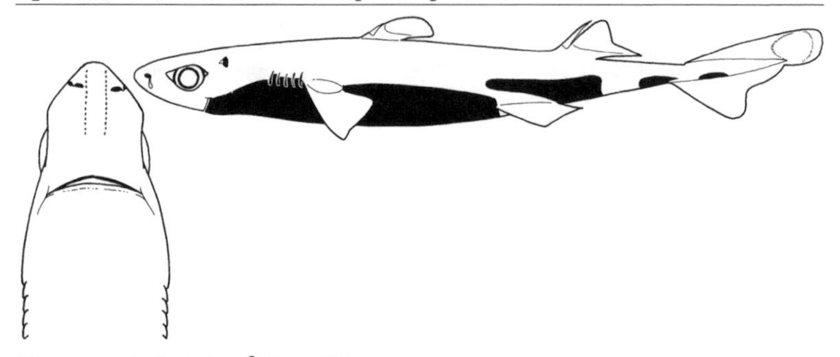

Measurements. Immature ♀: 25 cm TL.
Identification. A spindle-shaped dwarf lanternshark, similar to *Etmopterus fusus*, dorsal and ventral surfaces dark, back purplish-black, underside bluish-black in life (brownish-black in preservative), precaudal fins with pale reddish-brown webs; no conspicuous dark bands on caudal fin, but a lighter patch between dark base and terminal lobe of caudal.

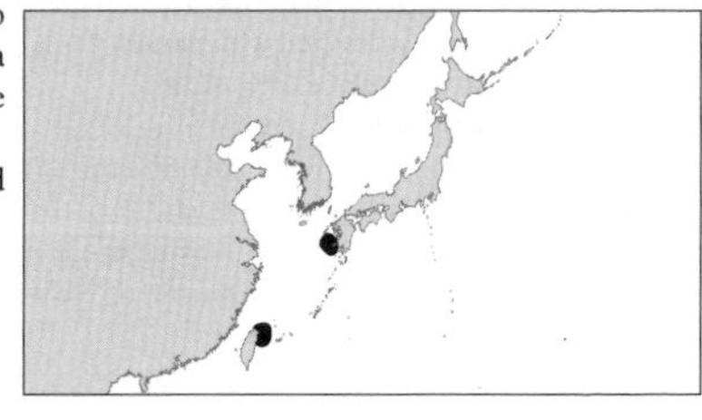

Distribution. Northwest Pacific: Japan and Taiwan. Possibly Java, Indonesia
Habitat. Continental slope, 210 m.
Behaviour. Unknown.
Biology. Unknown. Presumably ovoviviparous.
Status. IUCN Red List: Not Evaluated.

Brown Lanternshark *Etmopterus unicolor* Plate 9

Measurements. Born: ~17 cm TL. Mature: ~53 cm ♂. Max: 69 cm TL ♀.
Identification. Large robust grey-brown, dark brown or brownish-black lanternshark, dark beneath. Distinct broad, elongated caudal markings on tail, indistinct pelvic flank markings. Moderately long caudal fin (length of dorsal margin about equal to head length) and short caudal peduncle. Gill openings fairly large: half eye length. Upper teeth generally have <three pairs of cusplets. Bristle-like lateral trunk denticles with slender, hooked conical crowns, distributed irregularly, not in rows; denticles largely cover snout. Long low first dorsal fin with very short fin spine, second dorsal about twice height of first with strong spine, fin area about half as large again as that of first.
Distribution. Tentative range off southeast Atlantic and western Indian Ocean: South Africa, possibly southern Namibia. West Pacific: Japan, possibly southern Australia, New Zealand.
Habitat. Continental slope and seamounts, 402–1380 m.

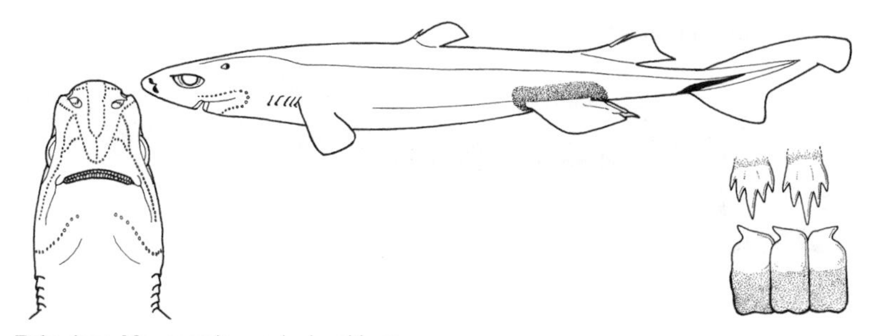

Behaviour. May sometimes swim in midwater.
Biology. Almost unknown, presumably ovoviviparous.
Status. IUCN Red List: Not Evaluated. Probably bycatch of trawl and other fisheries, utilization unknown. Similar large dark lanternsharks with bristly denticles from widespread areas tentatively included as a single species, formerly known as *Etmopterus compagnoi* off South Africa and Bristled Lanternshark, *E.* sp. B. off Australia.

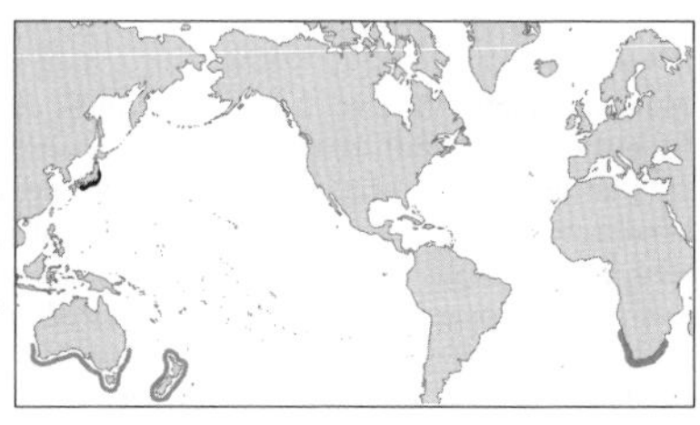

Hawaiian Lanternshark *Etmopterus villosus* Plate 8

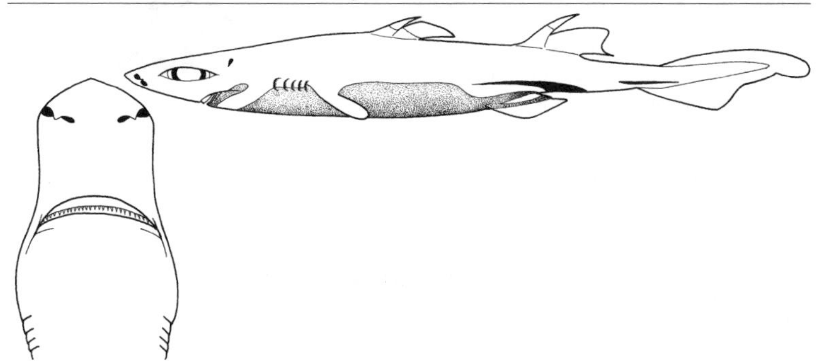

Measurements. Max: at least 46 cm TL.
Identification. Dark brown or blackish stout body, underside slightly darker, indistinct black mark above pelvic fins. Gill openings moderately long, about quarter eye length. Upper teeth with <3 pairs of cusplets. Lateral trunk denticles with slender, hooked conical crowns, wide-spaced, in regular rows on rear of trunk and tail, no rows of greatly enlarged denticles on flanks above pectoral fins; snout covered with denticles. Distal margins of fins largely covered with skin. Second dorsal fin much larger but less than twice area of first, with tall slightly curved spine.
Distribution. Pacific: Hawaiian Islands.
Habitat. Insular slopes, on or near bottom, 406–911 m.
Behaviour. Unknown.
Biology. Unknown, presumably ovoviviparous.
Status. IUCN Red List: Not Evaluated.

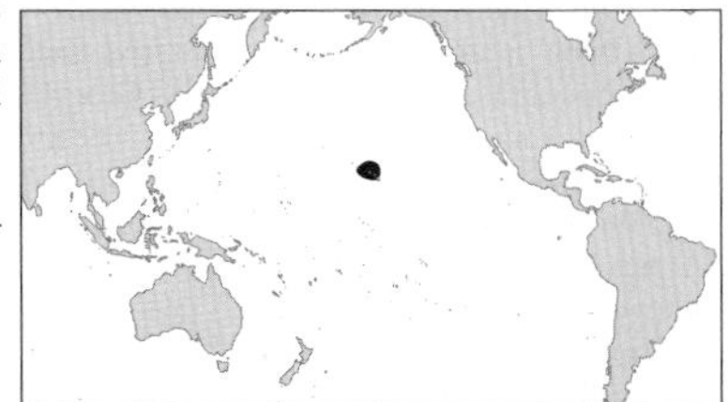

Green Lanternshark *Etmopterus virens* Plate 7

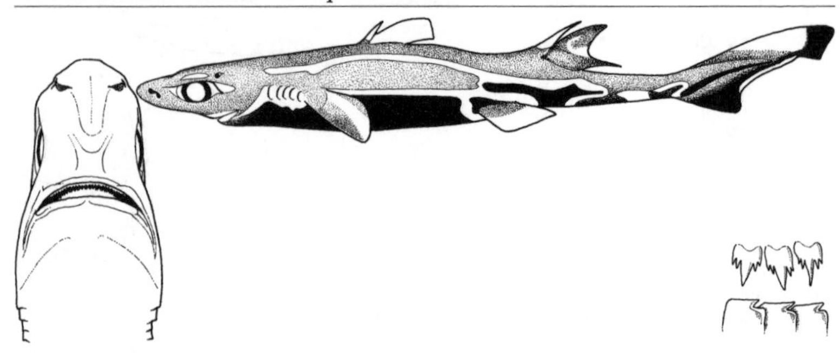

Measurements. Mature: 18 cm ♂, 22 cm ♀. Max: 26 cm TL.
Identification. Dwarf dark brown or grey-black moderately slender lanternshark with long narrow tail (dorsal margin = head length). Elongated broad black mark above and behind pelvic fins, others at tail base and along axis. Underside black. Gill openings very short, <one-third eye length. Upper teeth generally with <three pairs of cusplets. Lateral trunk denticles very short, stout, hooked, conical crowns; wide-spaced, not in rows; denticles largely cover snout. Second dorsal fin over twice area of first.
Distribution. Northwest Atlantic: northern Gulf of Mexico and Caribbean, possibly Brazil.

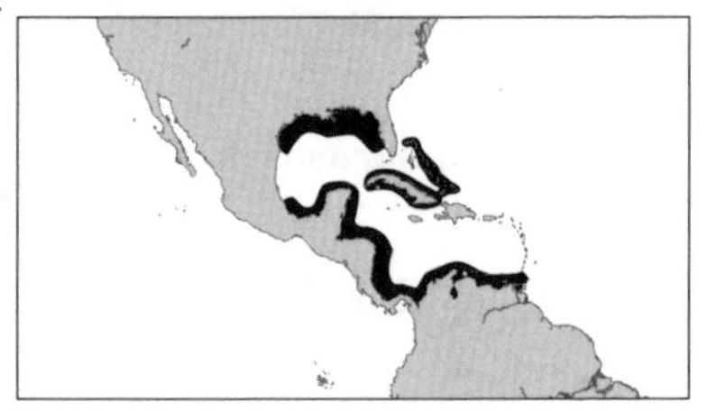

Habitat. Upper continental slopes, 196–915 m, most >350 m.
Behaviour. May school and communally attack very large prey (squid).
Biology. Presumably ovoviviparous.
Status. IUCN Red List: Not Evaluated. Relatively common, possibly discarded from deep-water bottom fisheries.

Rasptooth Dogfish *Miroscyllium sheikoi* Plate 9

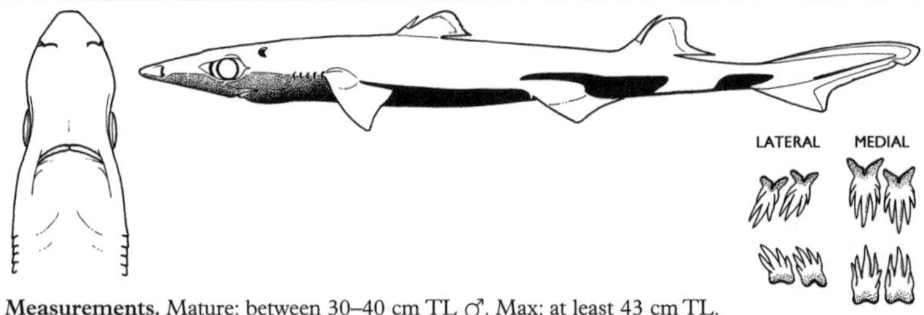

Measurements. Mature: between 30–40 cm TL ♂. Max: at least 43 cm TL.
Identification. Black below, dark brown above with black photomarks on caudal peduncle and caudal fin. Unmistakable long flat snout, short mouth with small comb-like compressed teeth with cusps and cusplets in both jaws. Grooved dorsal fin spines on both dorsal fins, second dorsal spine much larger than the first.

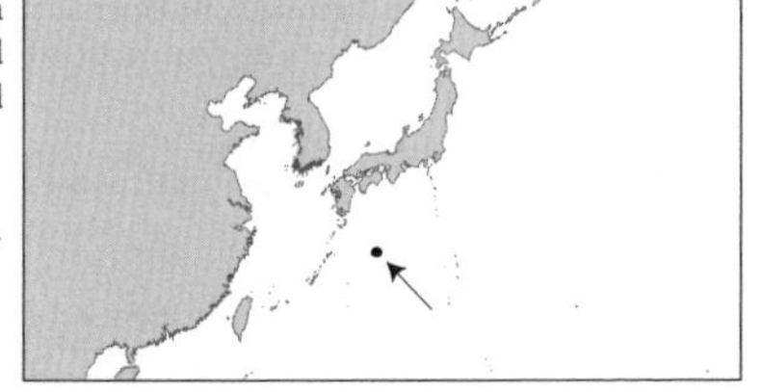

Distribution. Northwest Pacific, near Japan.
Habitat. Upper slope of submarine ridge, 340– 370 m.
Behaviour. Unknown.
Biology. Probably ovoviviparous.
Status. IUCN Red List: Not Evaluated.

Viper Dogfish *Trigonognathus kabeyai* Plate 9

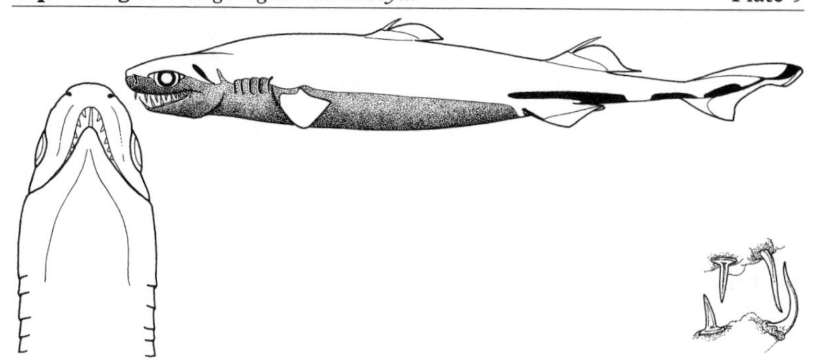

Measurements. Mature: 37-44 cm ♂, >44 cm TL ♀. Max: at least 44 cm TL.
Identification. Very long, narrow, terminal, snake-like mouth with huge curved fang-like teeth in front of both highly protrusible jaws. Deep pocket around front of upper jaws. Grooved spines on both dorsal fins. Body dark brown above, black below, with black photomarks on caudal peduncle and caudal fin. Very large diagonally elongated spiracle.
Distribution. North and central Pacific, Japan and Hawaiian Islands.

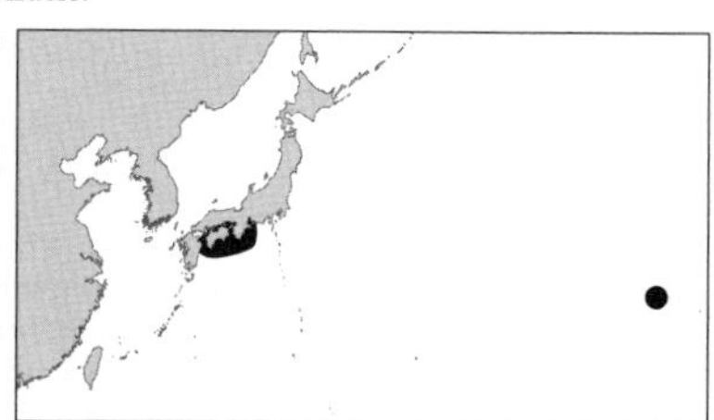

Habitat. On bottom, upper continental slopes, 330–360 m. Uppermost slope of seamount at 270 m. Possibly also oceanic?
Behaviour. Long narrow jaws can probably be protruded to impale relatively large prey (bony fishes and crustaceans) on fang-like teeth, then swallowed whole. Apparently luminescent.
Biology. Unknown. Possibly ovoviviparous.
Status. IUCN Red List: Not Evaluated. Apparently rare and localised.

Somniosidae: Sleeper Sharks

Eighteen deepwater benthic and oceanic species in seven genera (*Centroscymnus*, *Centroselachus*, *Proscymnodon*, *Scymnodalatias*, *Scymnodon*, *Somniosus*, and *Zameus*). Occur circumglobally in most seas, ranging from the tropics to Arctic and Antarctic oceans. Size ranges from small (40–69 cm) to gigantic (>6 m).
Identification. Fairly broad head and flat snout. Short thin-lipped almost transverse mouth. Small needle-like upper teeth, large compressed blade-like lower teeth. Spiracles large, close behind eyes. Lateral ridges on abdomen, not usually on caudal peduncle (except most *Somniosus*). Pectoral fins low, angular or rounded, not falcate, rear tips rounded and short. Pelvic fins subequal or larger than first dorsal and pectoral fins, and subequal to or smaller than second dorsal fin. Two small broad dorsal fins, origin of first in front of pelvic fin origins, space between greater than fin base length, second usually smaller or same size as first. Spines on both fins (may be covered by skin), or neither (*Scymnodalatias* and *Somniosus*). No anal fin. Caudal fin heterocercal with strong subterminal notch. No photophores.
Biology. Poorly known. Aplacentally ovoviviparous (four to at least 59 pups/litter). Mostly occur near the seabed on continental and insular slopes from 200 m to at least 3675 m; a few species are oceanic or semi-oceanic. In high northern latitudes, *Somniosus microcephalus* and *S. pacificus* occur on continental shelves, penetrate the intertidal, and occur at the surface. Sleeper sharks feed on bony fishes, other sharks, rays and chimaeras, cephalopods and other molluscs, crustaceans, seals, whale meat carrion, marine birds, echinoderms and jellyfish. At least one species (*Centroscymnus coelolepis*) is a 'cookie-cutter', biting chunks out of live whales, seals and bony fishes. Sleeper sharks do not seem to form large aggregations.

Status. Moderately common and an important component of commercial targeted and bycatch deepwater shark fisheries. Caught by line gear, demersal trawls, gillnets, traps, and even by spear or gaff. Their flesh is used for food or fishmeal. In Australia, many species have high mercury levels in their flesh and are discarded from deepwater fisheries. Their large very oily livers are processed for their high squalene content. Conservation status poorly known, but of considerable concern because expanding deepwater fisheries are now taking large numbers of deepwater sharks, including somniosids. Although the life history of these sharks is sketchily known, it is suspected that reproduction is limited and growth slow. If so, these species are highly vulnerable to overfishing. Several (e.g. *Scymnodalatias* species) are extremely poorly known, with only a few records and may be distinguished mainly by tooth counts and precise morphological measurements (see *FAO Catalogue of Sharks of the World* to confirm identifications and retain specimens for scientific study).

Portuguese Dogfish *Centroscymnus coelolepis* Plate 10

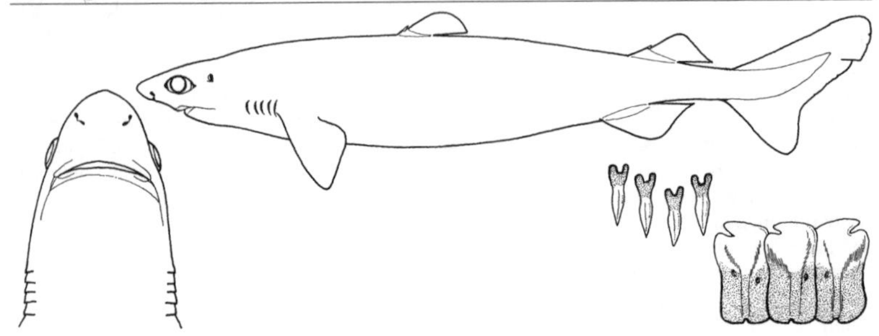

Measurements. Born: 30 cm TL. Mature: ~80 cm ♂, 100 cm ♀. Max: ~120 cm TL.
Identification. Uniformly blackish to golden-brown stocky short snouted dogfish. Large mouth, short labial furrows, slender upper teeth, short bent cusps on broader lower teeth. Large round flat overlapping denticles in adults. Very small spines (may be covered by skin) on small equal-sized dorsal fins, second close to asymmetrical caudal fin.

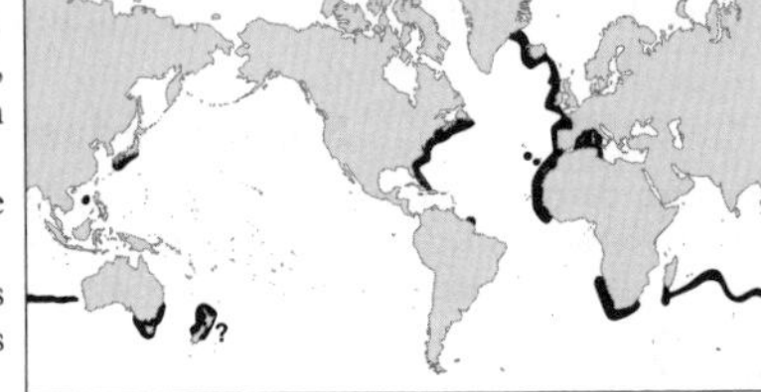

Distribution. Atlantic, Indian and Pacific Oceans.
Habitat. On or near bottom, continental slopes, upper and middle abyssal plain rises, 128–3675 m (mostly >400 m).
Behaviour. Can take bites out of live prey (like cookiecutter sharks, *Isistius*).
Biology. Ovoviviparous, 13–17 pups/litter. Eats bony fishes, other sharks, benthic invertebrates and cetacean and seal meat.
Status. IUCN Red List: Near Threatened. Long history of bycatch and target deepwater fisheries, used for fishmeal and squalene from liver oil.

Roughskin Dogfish *Centroscymnus owstoni* Plate 10

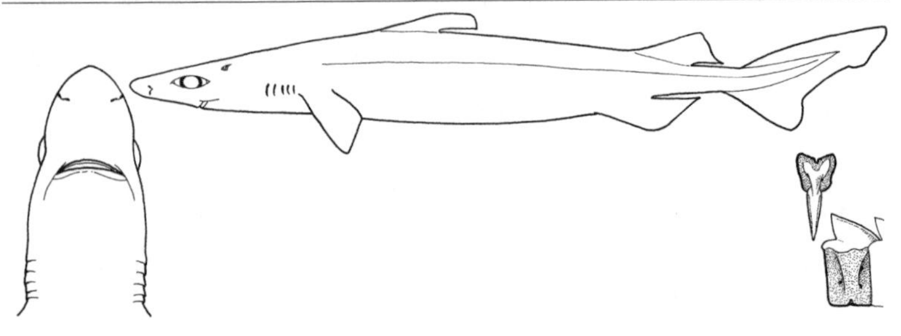

Measurements. Born: ~30 cm TL. Mature: ~72 cm ♂, 100 cm ♀. Max: 120 cm TL.
Identification. Similar to *Centroscymnus coelolepis* but longer snout, smaller denticles with small cusps in adults, longer, lower first dorsal fin, slightly taller, more triangular second dorsal fin. Dorsal fin spine tips often barely exposed.
Distribution. West Atlantic: Gulf of Mexico, Brazil, Uruguay. East Atlantic: Canary Islands and Madeira to South Africa. Pacific: Japan, Australasia and southeast.

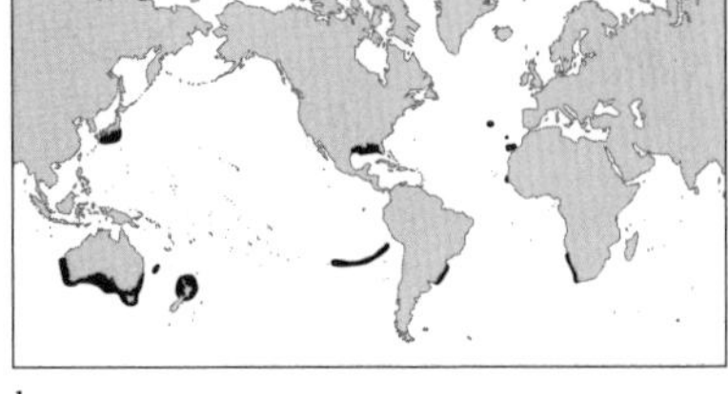

Habitat. On or near bottom, upper continental slopes and submarine ridges, 426–1459 m, mostly >600 m.
Behaviour. Sometimes in schools segregated by sex (females deeper than males off Japan).
Biology. Ovoviviparous, up to 34 eggs/female, litter size unknown. Eats bony fish and cephalopods.
Status. IUCN Red List: Least Concern. Moderately common. Localised fisheries interest (mainly bycatch), sometimes processed for fishmeal and squalene. Formerly recognised as *C. cryptacanthus* in east Atlantic.

Longnose Velvet Dogfish *Centroselachus crepidater* Plate 10

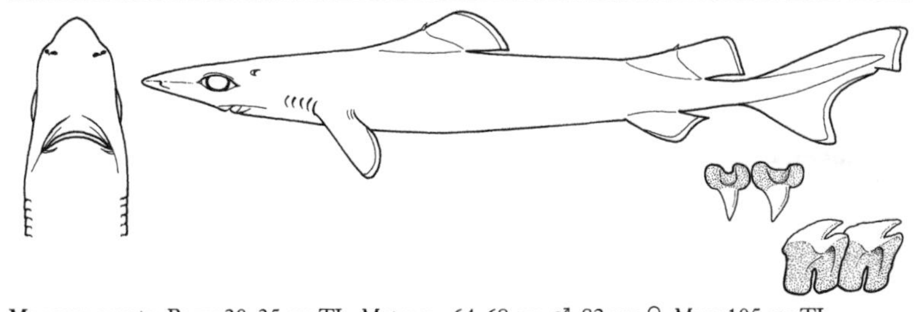

Measurements. Born: 30–35 cm TL. Mature: ~64–68 cm ♂, 82 cm ♀. Max: 105 cm TL.
Identification. Slender, black to dark brown with narrow light posterior fin margins. Very long snout, very long upper labial furrows almost encircle small mouth, round flat overlapping denticles with triple cusps in adults, lower teeth with fairly long bent cusps. Roughly equal dorsal fins, very small spine tips protruding. First dorsal base expanded forward as prominent ridge, origin over pectoral bases. Second dorsal free rear tip nearly reaches upper caudal origin.
Distribution. East Atlantic, Indo-Pacific, not northeast Pacific.

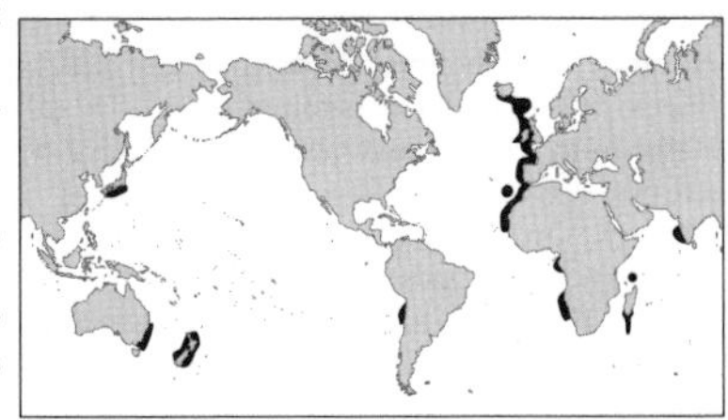

Habitat. On or near bottom, upper continental and insular slopes, 270–2080 m, mostly >500 m.
Behaviour. Unknown.
Biology. Ovoviviparous, four to eight pups/litter, breeds year-round. Eats fish and cephalopods.
Status. IUCN Red List: Least Concern. Common and wide-ranging. Limited fisheries interest, bycatch utilised for fishmeal and liver oil.

Largespine Velvet Dogfish *Proscymnodon macracanthus* Plate 10

Measurements. Holotype (♀): 68 cm TL.
Identification. Stocky dark brown or blackish body, fairly long snout, two dorsal fins with stout prominent spines, second dorsal fin higher than first. Thick fleshy lips, short upper labial furrows. Body tapering strongly behind large pectoral fins (apices almost or just reaching first dorsal spine when laid back). Trunk denticles small, not resembling bony fish scales, with triple cusps.
Distribution. Southwest Atlantic: Straits of Magellan. West Pacific: New Zealand (nominal).

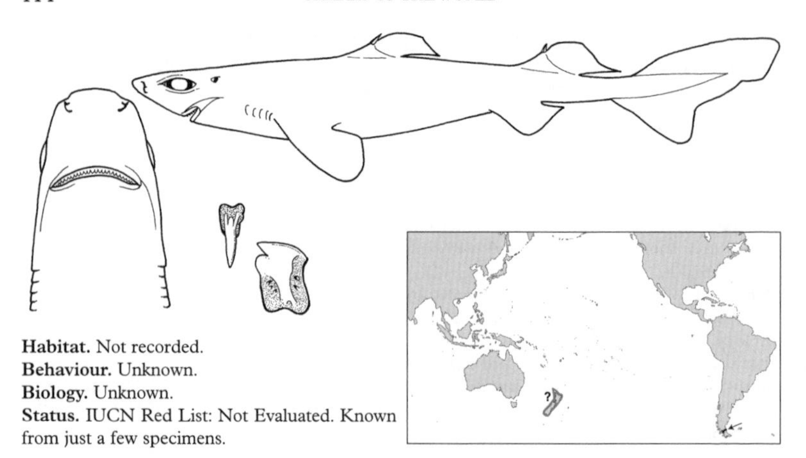

Habitat. Not recorded.
Behaviour. Unknown.
Biology. Unknown.
Status. IUCN Red List: Not Evaluated. Known from just a few specimens.

Plunket's Shark *Proscymnodon plunketi* Plate 10

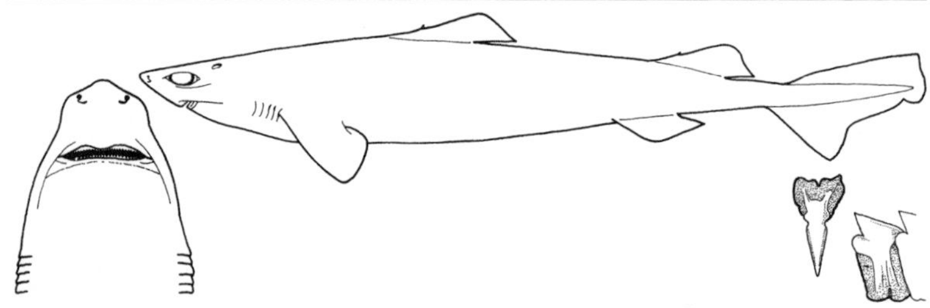

Measurements. Born: 32–36 cm TL. Mature: 100–131 cm ♂, 129 cm ♀. Max: 170 cm TL.
Identification. Dark grey-brown very short-snouted dogfish with a stocky body, tapering behind pectorals. Roughly equal-sized dorsal fins, tips of spines just protrude. First dorsal fin extends forwards in prominent ridge. Space between second dorsal and upper caudal fin about length of dorsal base, free rear tip well in front of upper caudal origin.
Distribution. Indo-Pacific: Australasia and southwestern Indian Ocean.
Habitat. Near bottom, continental and insular slopes, 219–1427 m (most common 550–732 m).
Behaviour. Forms large schools segregated by size and sex.
Biology. Ovoviviparous, up to 36 pups/litter. Feeds on cephalopods and bony fishes.
Status. IUCN Red List: Not Evaluated. Relatively uncommon. Deepwater fisheries bycatch, utilized in New Zealand for fish meal and squalene.

Whitetail Dogfish *Scymnodalatias albicauda* Plate 11

Measurements. Born: >20 cm TL. Adult ♀: 74–110 cm TL.
Identification. Dark brown or mottled greyish body, lighter below, whitish-grey fin margins, obvious white blotches and dark terminal lobe on caudal fin. Short, broad rounded snout. Horizontally elongated eyes. Long broadly arched mouth. Small upper teeth with very narrow acute erect cusps. Larger, blade-like, unserrated interlocked lower teeth, with high, erect cusps. Elongated pectoral fins. No dorsal fin spines, second dorsal slightly larger than first and very close to tail. Lower tail margin half as long as upper margin.

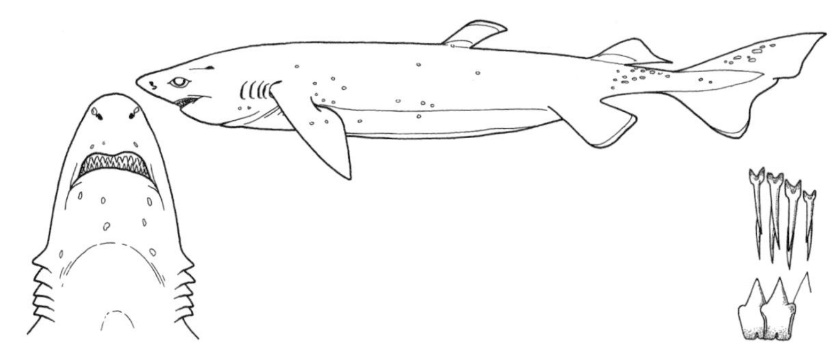

Distribution. Southern Ocean.
Habitat. Oceanic in epipelagic zone, 0–200+ m over 1400–4000 m depth; near bottom on submarine ridge, 512 m.
Behaviour. Possibly mesopelagic or bathypelagic, may rise to near surface at night.
Biology. Ovoviviparous, very large litters: at least 59 pups.
Status. IUCN Red List: Data Deficient. Very rarely caught by deepwater trawls and tuna longlines.

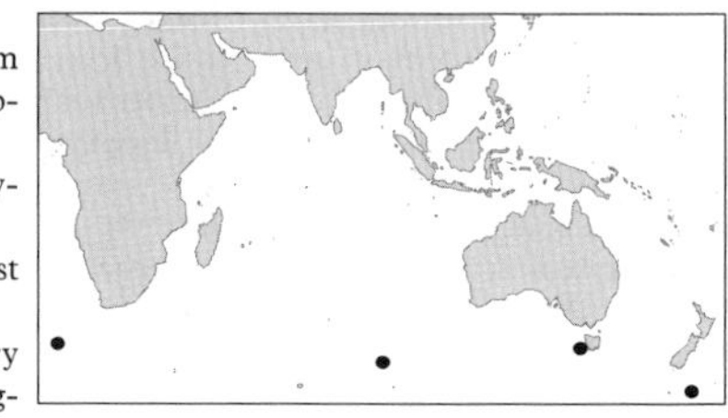

Azores Dogfish *Scymnodalatias garricki* Plate 11

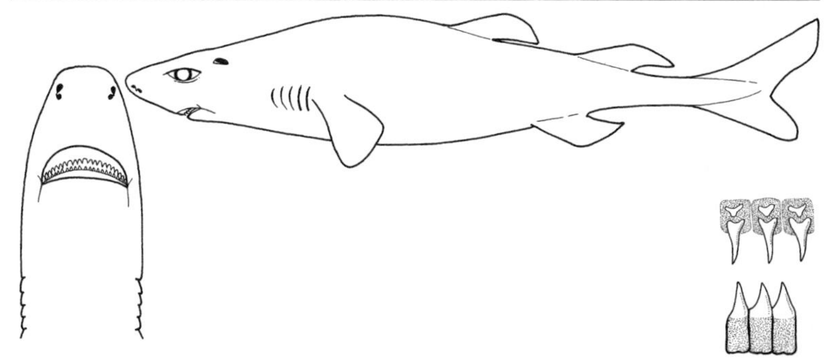

Measurements. Adolescent male (only known specimen) was 37.7 cm TL.
Identification. Very small uniform dark brown dogfish with long head. Snout long and broad, rounded. Eyes horizontally elongated. Mouth long and broadly arched. First dorsal fin located half way down body, free rear tip just in front of pelvic origins, no dorsal fin spines. Very small narrow upper teeth, lower teeth larger, blade-like and interlocked. Strong lower caudal lobe.
Distribution. North Atlantic Ridge.
Habitat. Apparently oceanic or deep-benthic. Caught at 300 m in open ocean over a seamount.
Behaviour. Unknown.
Biology. Unknown. Presumably ovoviviparous.
Status. IUCN Red List: Not Evaluated. Known from one specimen.

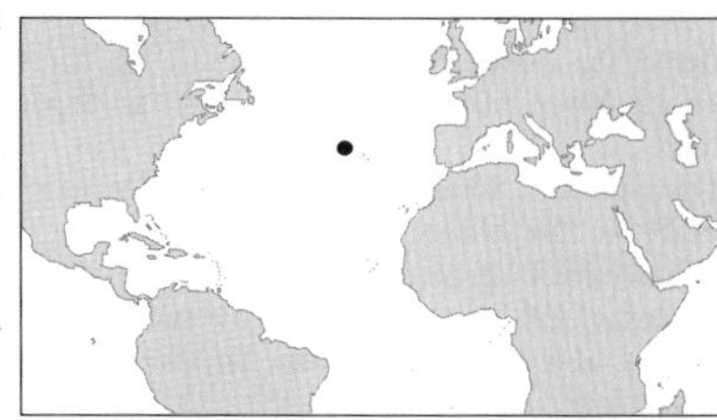

Sparsetooth Dogfish *Scymnodalatias oligodon* Plate 11

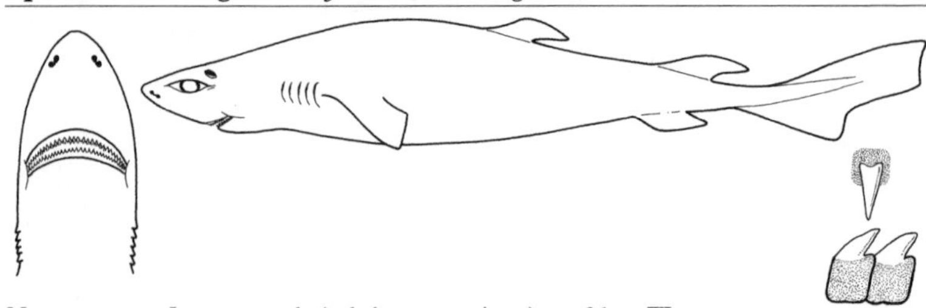

Measurements. Immature male (only known specimen) was 26 cm TL.
Identification. Uniformly dark brown. Long, broad, pointed snout. Eyes horizontally elongated. Mouth long and broadly arched. Larger lower teeth blade-like, unserrated and interlocked, with strongly oblique cusps. Fewer small upper teeth (tooth row counts 33/42). No dorsal fin spines. Ventral tail lobe weakly developed, lower margin about half as long as upper tail margin.

Distribution. Southeast Pacific: open ocean.
Habitat. Apparently oceanic. Caught near surface (0–200 m) in water 2000–4000 m deep.
Behaviour. Unknown.
Biology. Unknown, presumably ovoviviparous.
Status. IUCN Red List: Not Evaluated. Known from one specimen.

Sherwood Dogfish *Scymnodalatias sherwoodi* Plate 11

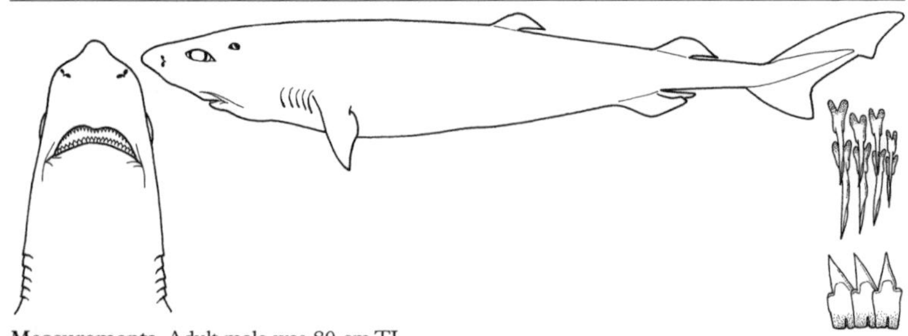

Measurements. Adult male was 80 cm TL.
Identification. Dark brown above, lighter below, light margins on gill slits and pectoral fins. Moderately long, flattened, pointed snout. Eyes horizontally elongated. Mouth long and broadly arched. Upper teeth small with very narrow acute erect cusps. Lower teeth larger, blade-like, unserrated and interlocked, with high, erect cusps. First dorsal in middle of back behind elongated, leaf-shaped pectoral fins, free rear tip in front of pelvic fins. Second dorsal origin above rear third of pelvic base. Pectoral fins relatively short. No dorsal fin spines. Asymmetric caudal fin, short strong lower lobe about half as long as upper lobe.

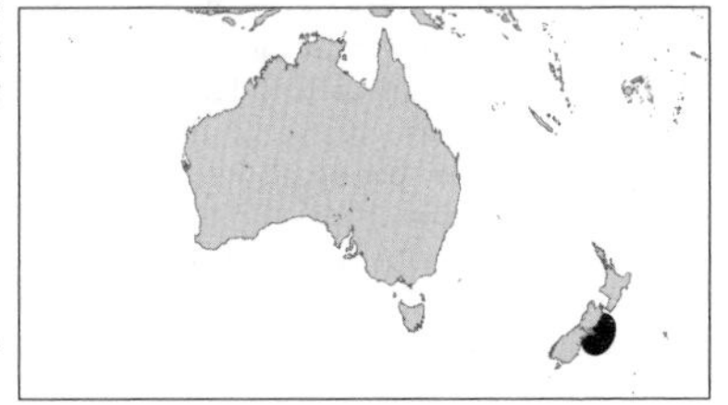

Distribution. Southwest Pacific: New Zealand.
Habitat. Deepwater, 400–500 m.
Behaviour. Unknown.
Biology. Unknown, presumably ovoviviparous.
Status. IUCN Red List: Data Deficient. Infrequent deepwater trawl bycatch.

Knifetooth Dogfish *Scymnodon ringens* Plate 11

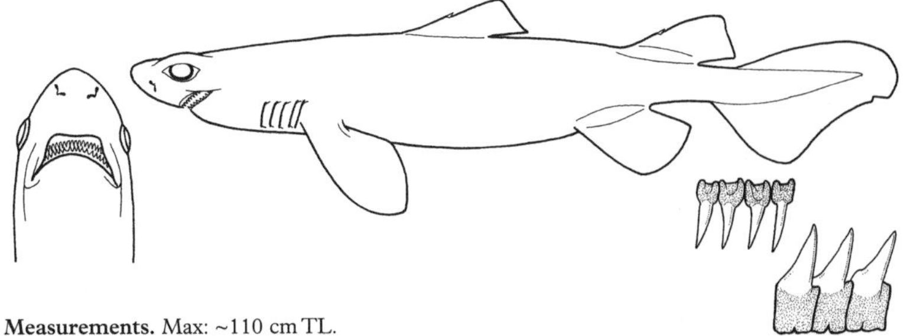

Measurements. Max: ~110 cm TL.
Identification. Uniformly black, no obvious fin markings. Rather thick, high head, broad short snout, very large, broadly arched mouth. Small lanceolate upper teeth, huge triangular cutting lower teeth. Rather long gill slits (>half eye length). Small dorsal fin spines, second dorsal fin slightly larger than first. Caudal fin asymmetric, not paddle-shaped, with weak subterminal notch and no lower lobe.
Distribution. East Atlantic: Scotland to Senegal. West Pacific: New Zealand?.

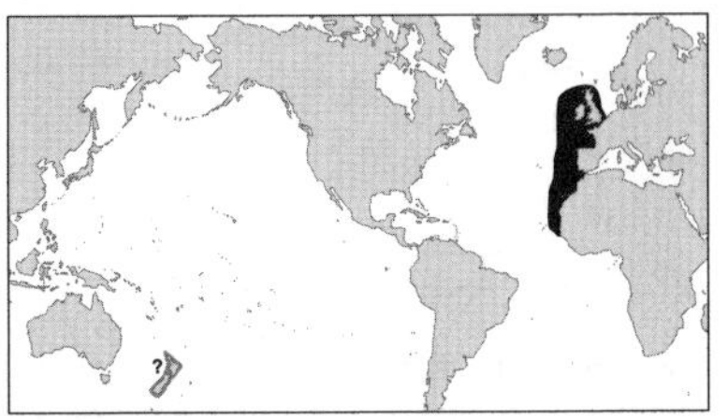

Habitat. Continental slope, on or near bottom, 200–1600 m.
Behaviour. Immense, triangular, razor-edged lower teeth suggest a formidable predator, capable of attacking and dismembering large prey.
Biology. Probably ovoviviparous.
Status. IUCN Red List: Not Evaluated. Relatively common in east Atlantic. Bycatch of bottom trawls, line gear and fixed bottom nets; dried salted for food and fishmeal.

Southern Sleeper Shark *Somniosus antarcticus* Plate 12

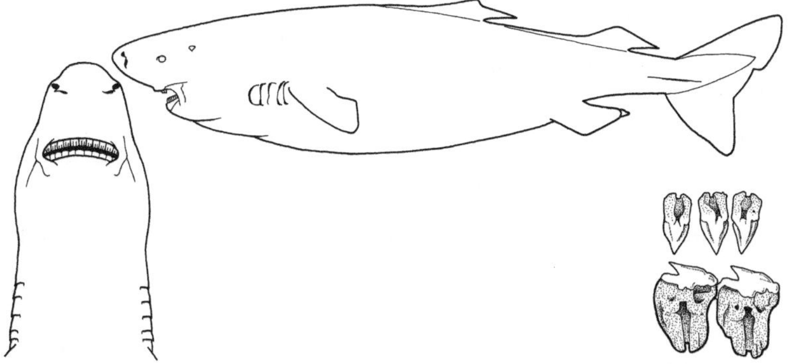

Measurements. Born: ~40 cm TL. Max: ~600 cm TL (one of the largest sharks, largest Antarctic fish).
Identification. Gigantic, uniform grey to pinkish body and fins. Short, rounded snout, heavy cylindrical body, very small precaudal fins. Spear-like upper teeth, slicing lower teeth with low bent cusps and high roots. Skin rough, denticles with strong hooklike erect cusps. Spineless, equal-sized, dorsal fins, first slightly closer to pelvics than to pectoral fins. Distance between dorsal fin bases (~80% of snout to first gill slits) and very low first dorsal fin distinguishes this species from *Somniosus microcephalus* and *S. pacificus*. Long lower caudal lobe, short upper lobe, short caudal peduncle, lateral keels on base of caudal fin.

Distribution. Southern oceans and Antarctic.
Habitat. Continental and insular shelves, upper slopes down to at least 1200 m. Water temperature 0.6–12°C.
Behaviour. Sluggish, slow-moving epibenthic shark, able to capture large and active prey (fishes and invertebrates). Also feeds on carrion.
Biology. Ovoviviparous; ten pups in uterus of 5 m female.
Status. IUCN Red List: Data Deficient. Discarded toothfish fishery bycatch.

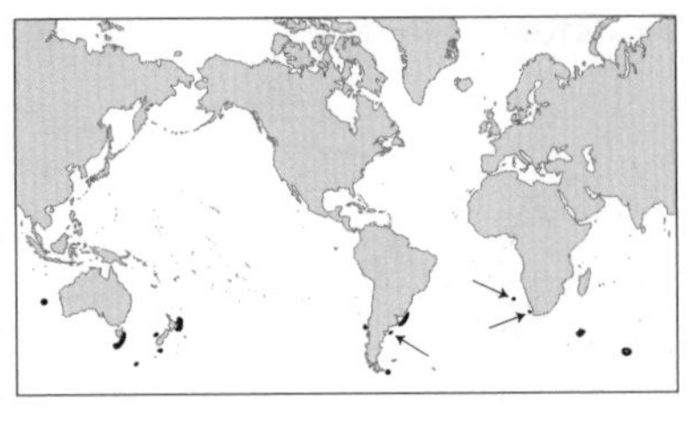

Frog Shark *Somniosus longus* Plate 12

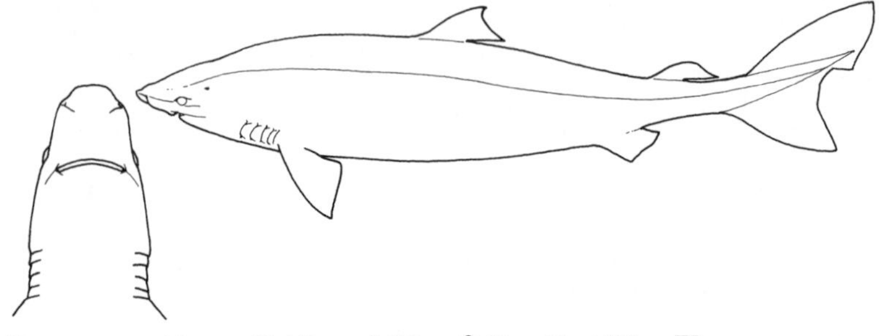

Measurements. Mature: ~98–110 cm ♂, 130 cm ♀. Max: at least 130 cm TL.
Identification. Similar to *S. rostratus*. Small, uniformly dark body and fins. Short, rounded snout, short head, slender cylindrical body. Spear-like upper teeth, slicing lower teeth with high semierect cusps and low roots. Skin smooth, denticles with flat cusps. Spineless dorsal fins, first about as long as second and closer to pectorals than pelvic fins. Short caudal peduncle, long lower caudal lobe, short upper lobe, lateral keels on base of caudal fin.
Distribution. West Pacific, southeast Pacific.
Habitat. Outer continental shelves and upper slopes, on or near bottom, 250–1160 m.
Behaviour. Unknown.
Biology. Ovoviviparous.
Status. IUCN Red List: Not Evaluated. Rare.

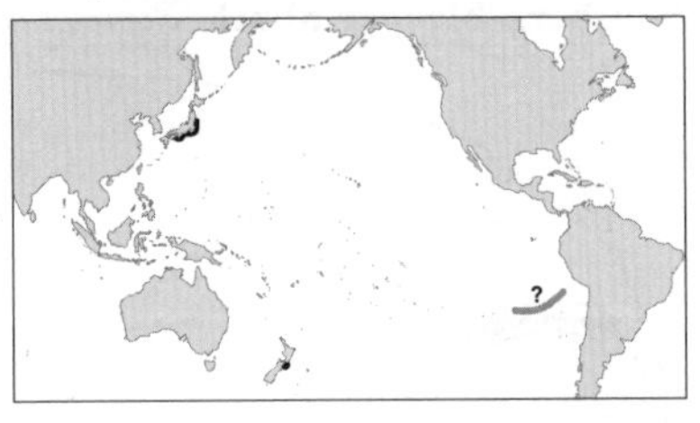

Greenland Shark *Somniosus microcephalus* Plate 12

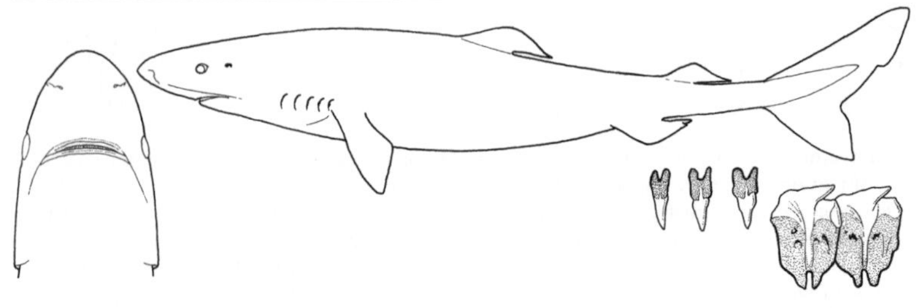

Measurements. Born: ~40 cm TL. Most adults: 244–427 cm. Max: >640, possibly 730 cm TL.
Identification. A gigantic medium grey or brown, heavy-bodied sleeper shark, sometimes with

transverse dark bands, small dark spots and blotches, and small light spots. Short, rounded snout, heavy cylindrical body, small precaudal fins. Spear-like upper teeth, slicing lower teeth with low bent cusps and high roots. Skin rough, denticles with strong hooklike erect cusps. Spineless, equal-sized, low dorsal fins, first slightly closer to pectoral fins than to pelvics. Distance between dorsal fin bases about equal to snout to first gill slits. Long lower caudal lobe, short caudal peduncle, lateral keels present on base of caudal fin.

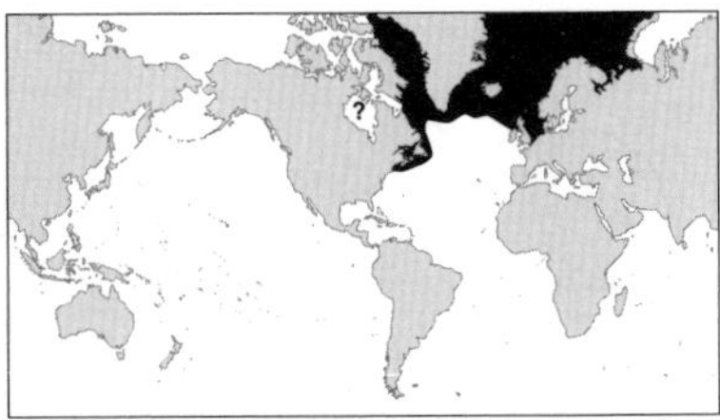

Distribution. North Atlantic and Arctic, occasionally to Portugal.
Habitat. Continental and insular shelves and upper slopes to at least 1200 m, water temperature 0.6–12°C. Inshore in Arctic winter (from intertidal and surface in shallow bays and river mouths), retreats to 180–550 m when temperature rises. May move into shallower water in spring and summer in north Atlantic.
Behaviour. Sluggish, offers little resistance to capture, easily fished through iceholes, but able to capture large active prey including fishes, invertebrates, seabirds and seals. Also feeds on dead cetaceans and drowned horses and reindeer. A highly conspicuous, possibly luminescent, copepod parasite often attached to cornea of eye is speculated to lure prey species to the shark under a mutualistic and beneficial relationship.
Biology. Ovoviviparous, ten pups/litter.
Status. IUCN Red List: Not Evaluated. Abundant. Traditionally fished by hook and line, longline or gaff for liver oil. Also taken in nets and cod traps. Meat toxic when fresh (unless washed); harmless dried or semi-putrid. Used for human and sled-dog food. Eskimos used skin for boots, and sharp lower dental bands as knives. Considered harmless, no confirmed incidents of these sharks biting people.

Pacific Sleeper Shark *Somniosus pacificus* Plate 12

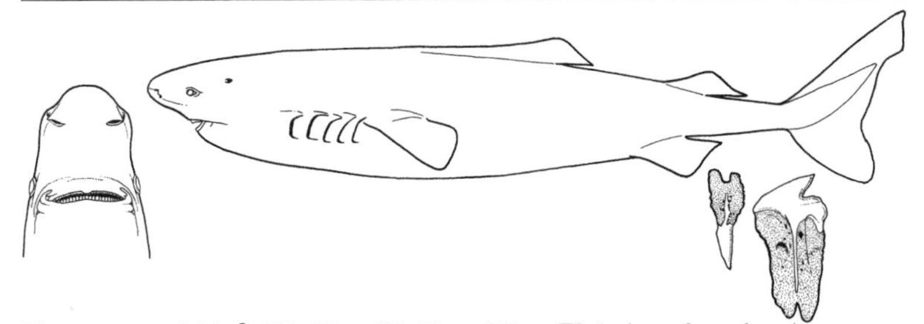

Measurements. Adult ♀: 370–430 cm TL. Max: >700 cm TL (estimate from photos).
Identification. Giant sleeper shark with uniform greyish body and fins. Short, rounded snout, heavy cylindrical body, small precaudal fins. Spear-like upper teeth, slicing lower teeth with low bent cusps and high roots. Skin rough and bristly, denticles with strong hooklike erect cusps. Spineless, equal-sized, low dorsal fins, first slightly closer to pelvics than to pectoral fins. Distance between dorsal fin bases about 70% of snout to first gill slits. Long lower caudal lobe, short upper lobe, short caudal peduncle, lateral keels variably present or absent on base of caudal fin.

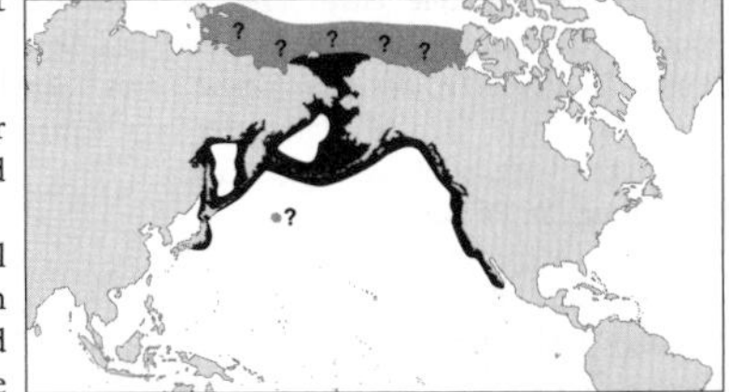

Distribution. North Pacific: Japan to Mexico.
Habitat. Continental shelves and slopes, to over 2000 m. Ranges into littoral in north (once found trapped in a tide pool). Very deep in south.
Behaviour. Lumbering sluggish sharks. Small mouth and large oral cavity suggest suction feeding. Takes a wide variety of surface and bottom animals. Seal remains in stomach may be

scavenged carrion or taken alive. Possible sexual segregation (pregnant females not recorded).
Biology. Probably ovoviviparous; up to 300 large eggs/female.
Status. IUCN Red List: Not Evaluated. Relatively common.

Little Sleeper Shark *Somniosus rostratus* Plate 12

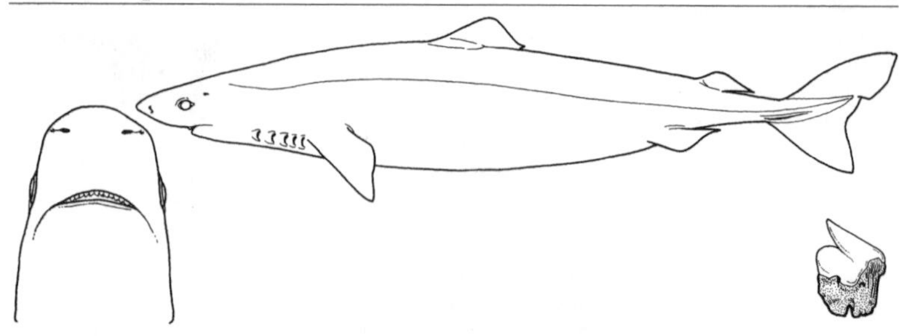

Measurements. Born: 21–28 cm TL. Mature: ~70 cm ♂, 80 cm ♀. Max: 140 cm TL.
Identification. Similar to *Somniosus longus* but smaller eyes and shorter second dorsal fin. Short, rounded snout, short head, blackish slender cylindrical body. Spear-like upper teeth, slicing lowers with high semierect cusps and low roots. Skin smooth: flat cusped denticles. Spineless dorsal fins, first higher, closer to pectorals than pelvic fins. Short caudal peduncle, long lower caudal lobe, short upper lobe, lateral keels on caudal fin base.
Distribution. Northeast Atlantic, west Mediterranean, northwest Atlantic (Cuba).
Habitat. Outer continental shelves, upper and lower slopes, on or near bottom, 200–2200 m.
Behaviour. Unknown.
Biology. Ovoviviparous.
Status. IUCN Red List: Not Evaluated. Rare to sporadically common. Minimal fisheries interest, caught on longlines and by bottom trawls, used for fishmeal and possibly food.

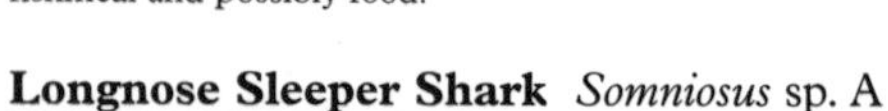

Longnose Sleeper Shark *Somniosus* sp. A Plate 12

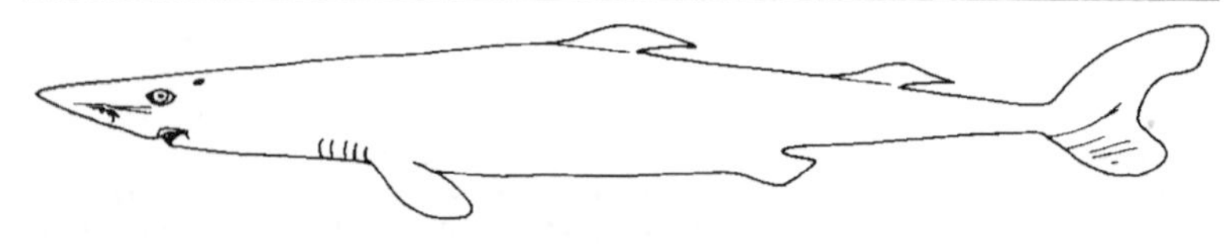

Measurements. Known from one specimen of 83 cm TL.
Identification. Sleeper shark with elongated, pointed, flattened snout, head moderately long. Spear-like upper teeth, cutting lower teeth with short, low, strongly oblique cusps and high, narrow roots. Two spineless, equal-sized dorsal fins, first dorsal fin insertion closer to pelvic bases than pectoral bases. Caudal peduncle very long, caudal fin short, but long ventral lobe.
Distribution. Portugal.
Habitat. Deep water off Portugal.
Behaviour. Unknown.
Biology. Unknown.
Status. IUCN Red List: Not Evaluated.

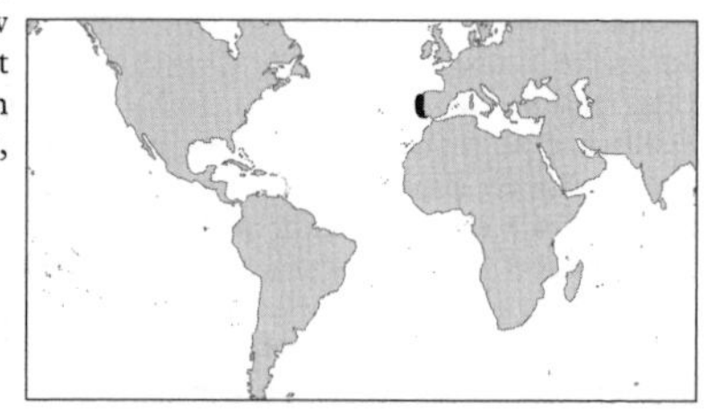

Japanese Velvet Dogfish *Zameus ichiharai* Plate 11

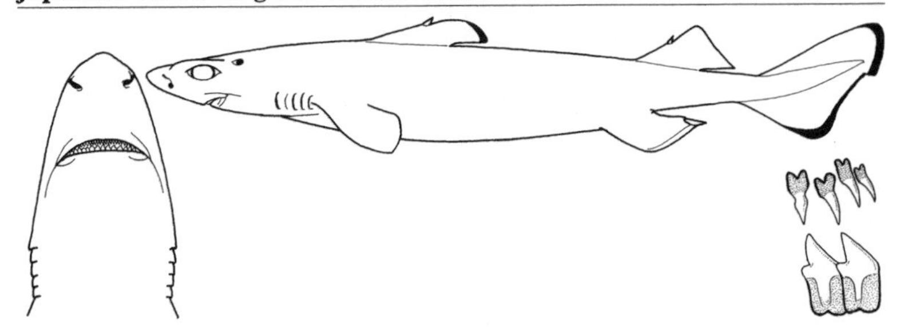

Measurements. Born: unknown. Adult ♂: 101 cm TL. Max: >101 cm TL.
Identification. Black dogfish, rather low flat head, moderately long snout. Mouth fairly narrow, short and nearly transverse. Small spear-like upper teeth, lowers large high, knife-cusped cutting teeth. Gill slits rather short (<half eye length). Two dorsal fins with small fin spines. Pectoral fins narrow and leaf-shaped; apices well in front of first dorsal spine. Pelvic fins small, about equal to second dorsal. Caudal peduncle long, caudal fin with strong subterminal notch and short lower lobe.

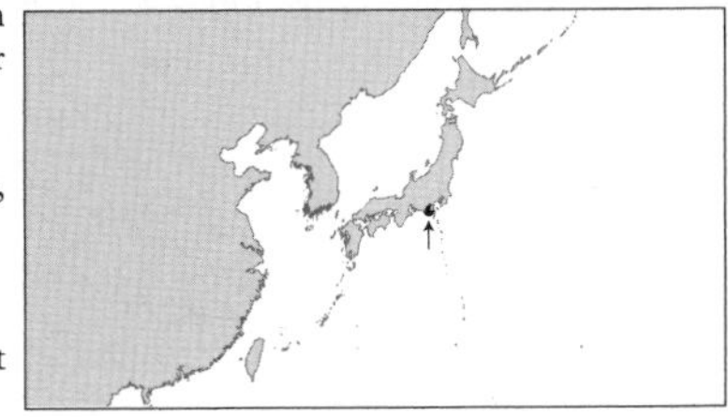

Distribution. Japan.
Habitat. Slope on or near bottom, Suruga Bay, Honshu, Japan, 450–830 m.
Behaviour. Unknown.
Biology. Unknown.
Status. IUCN Red List: Not Evaluated. Caught by deepset line fisheries.

Velvet Dogfish *Zameus squamulosus* Plate 11

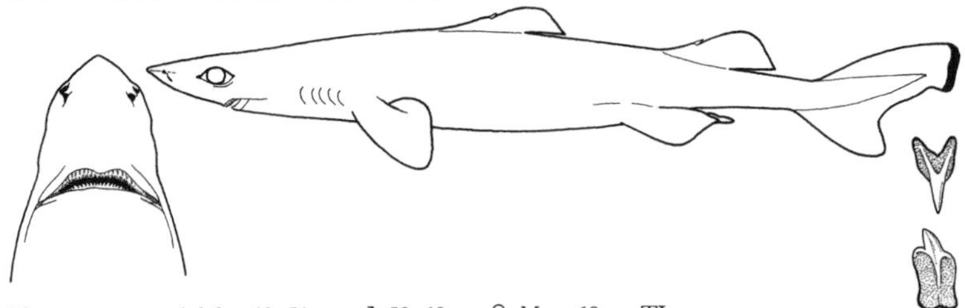

Measurements. Adults: 49–51 cm ♂, 59–69 cm ♀. Max: 69 cm TL.
Identification. Small slender black dogfish, low flat head, fairly long narrow snout. Short narrow mouth, post-oral grooves much longer than short upper labial furrows. Small spear-like upper teeth, high-cusped, knife-like lower teeth. Denticles tricuspidate with transverse ridges. Two dorsal fins with small fin spines, second larger, ~same size as small pelvic fins. Long tail, strong subterminal tail notch and short lower lobe.

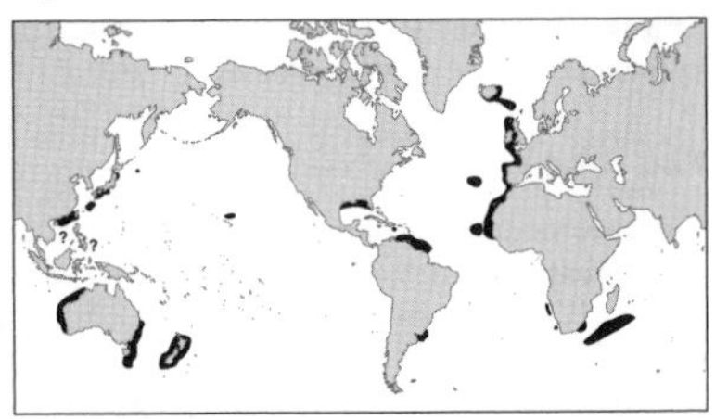

Distribution. Patchy world-wide, not east Pacific.
Habitat. Continental and insular slopes, on or near bottom, 550–1450 m. Epipelagic and oceanic off Brazil (0–580 m in water 2,000 m deep) and Hawaiian Islands (27–35 m in water 4000–6000 m deep).
Behaviour. Unknown.

Biology. Probably ovoviviparous.
Status. IUCN Red List: Not Evaluated. Minor bycatch in bottom fisheries. Used dried-salted for food and as fishmeal. Formerly recognised as *Scymnodon obscurus* in east Atlantic.

Oxynotidae: Roughsharks

Identification. Five unmistakable species of small sharks. Compressed body, triangular in cross-section with lateral ridges on abdomen and rough skin from large, prickly and close-set denticles. Two high sail-like spined dorsal fins. Rather broad flattened head with flat blunt snout, small thick-lipped mouth encircled by elongated labial furrows, close-set large nostrils, and large to enormous spiracles close behind eyes. Small, spearlike upper teeth form a triangular pad; lower teeth highly compressed to form a saw-like cutting edge, in only 9–13 rows. No anal fin.
Biology. Poorly known deepwater sharks. Scattered distribution, mainly on temperate to tropical continental and island shelves. May be weak swimmers, relying on large oily livers for buoyancy. Diet mainly small bottom-living invertebrates (worms, crustaceans and molluscs) and fishes. Females bear litters of 7–23 pups.
Status. Uncommon bycatch in deepwater bottom fisheries. May be processed for fish meal, liver oil or occasionally human food.

Prickly Dogfish *Oxynotus bruniensis* Plate 13

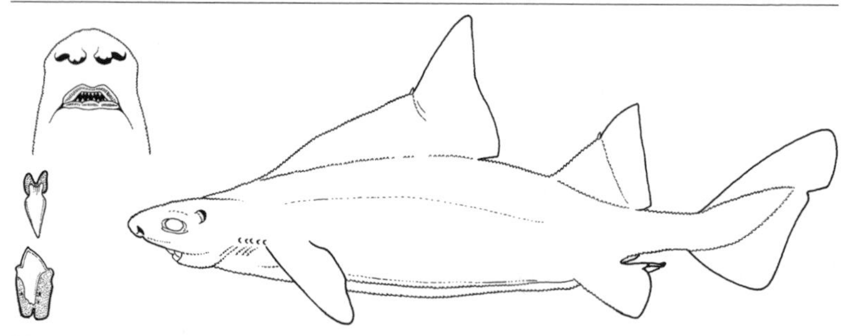

Measurements. Born: ~24 cm TL. Mature: ~60 cm ♂, 72 cm ♀. Max: 72–91 cm TL.
Identification. Uniform light grey-brown in colour, no prominent markings.Spiracles almost circular. Dorsal fins with triangular tips, trailing edges straight or slightly concave. First dorsal spine leans slightly forward.
Distribution. Temperate southwest Pacific. Locally common off New Zealand and adjacent submarine ridges and seamounts, occasionally caught off southern Australia.

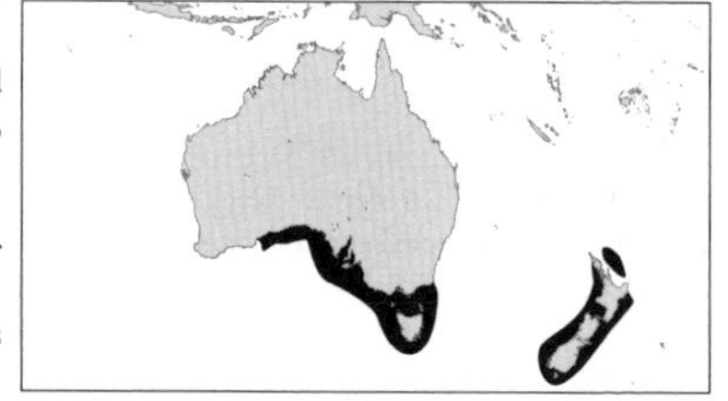

Habitat. Deepwater on outer continental and insular shelves and upper slopes, 46–1067 m, most common at 350–650 m.
Behaviour. Unknown.
Biology. Poorly known. Ovoviviparous, one litter of seven embryos reported.
Status. IUCN Red List: Data Deficient. Taken as bycatch in bottom trawls, probably discarded.

Caribbean Roughshark *Oxynotus caribbaeus* Plate 13

Measurements. Immature (♂ & ♀) at 20–21 cm. Max: >49 cm TL (adult ♂).
Identification. Grey or brownish, with pattern of dark bands on a light background, dark blotches and small spots on head, body, tail and fins, separated by prominent light areas over pectoral and pelvic fins. Spiracles relatively small and circular. Dorsal fins with narrowly triangular tips and concave trailing edges. First dorsal spine leans forward.

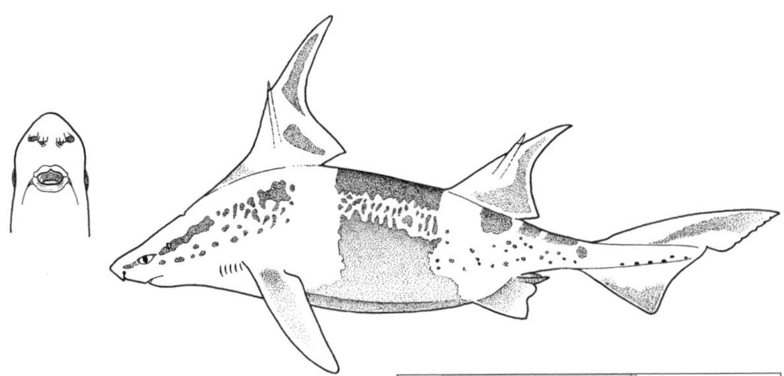

Distribution. West Atlantic, Gulf of Mexico, Caribbean coast of Venezuela.
Habitat. Bottom on upper continental slopes, depth 402–457 m, water temperature 9.4–11.1°C.
Behaviour. Unknown.
Biology. Unknown.
Status. IUCN Red List: Not Evaluated. Not important in fisheries.

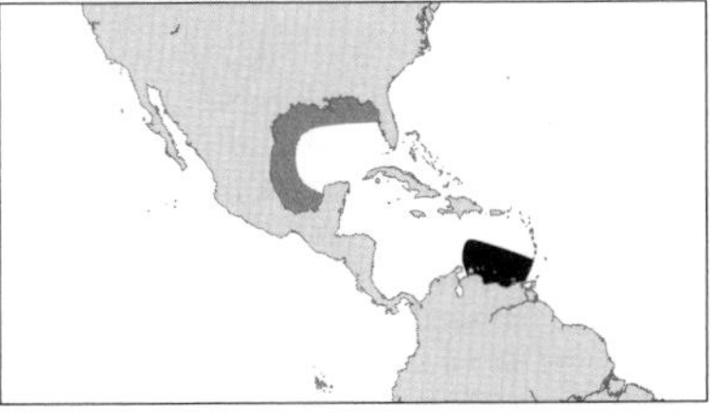

Angular Roughshark *Oxynotus centrina* Plate 13

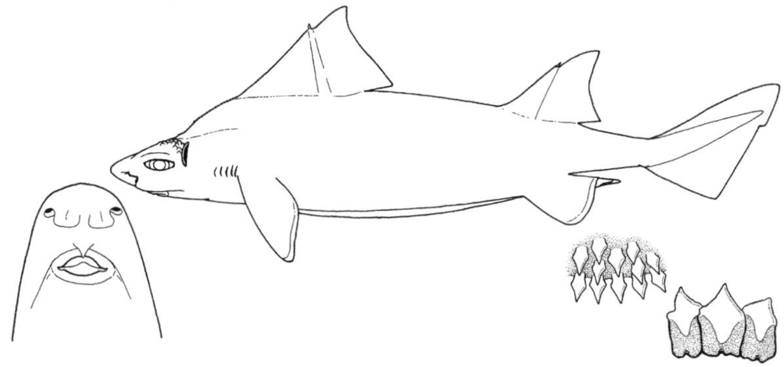

Measurements. Born: <25 cm TL. Mature: ~50 cm. Most records <1m. Max: ~150 cm TL.
Identification. Grey or grey-brown, darker blotches on head and sides (not always clear in adults); light horizontal line crosses cheek below eye. Ridges over eyes expanded into huge rounded knobs, covered with enlarged denticles (absent in other roughsharks). Spiracles very large and vertically elongated, almost as high as eye length. First dorsal spine leans slightly forward.
Distribution. East Atlantic and Mediterranean (not Black Sea); possibly off Mozambique, West Indian Ocean.
Habitat. Coralline algal and muddy bottom on continental shelves and upper slopes, 50–660 m, mostly below 100 m.

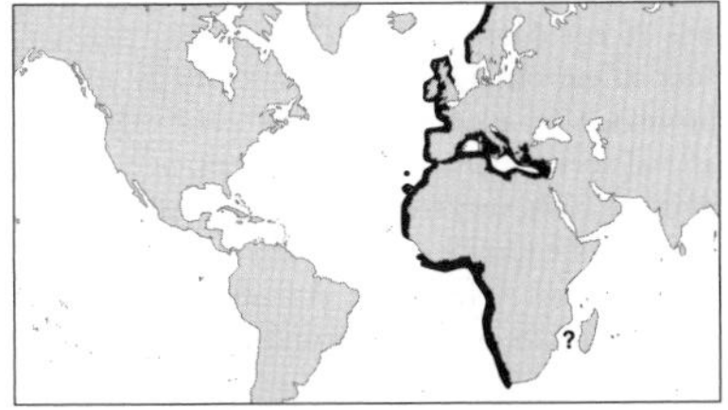

Behaviour. Unknown.
Biology. Litter size 7–8 (Angola) to 23 (Mediterranean). Eats worms, crustaceans and molluscs.
Status. IUCN Red List: Not Evaluated. Rare to uncommon. Minor offshore trawl bycatch.

Japanese Roughshark *Oxynotus japonicus* Plate 13

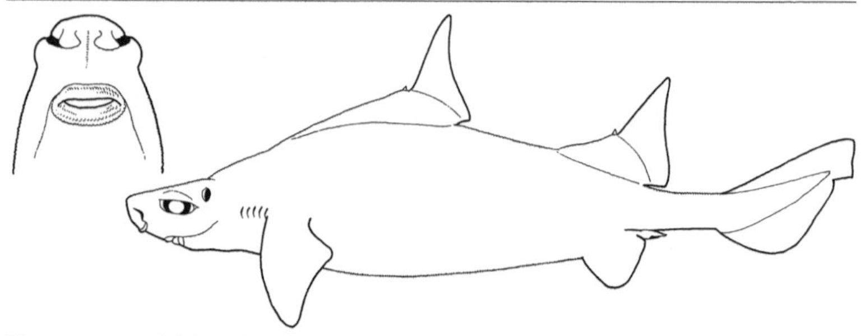

Measurements. Adult male was 54 cm TL.
Identification. Uniform dark brown body except for white lips, nasal flap margins, fin axils and inner clasper margins. Spiracles vertically oval, less than half eye length. Dorsal fins with narrow triangular tips and shallowly concave trailing edges. First dorsal spine leans slightly back.

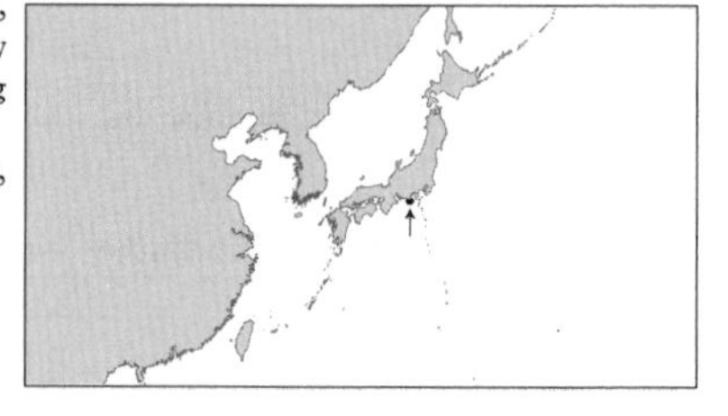

Distribution. West Pacific: Japan (Suruga Bay, Honshu).
Habitat. Upper continental slope, 225–270 m.
Behaviour. Unknown.
Biology. Unknown.
Status. IUCN Red List: Not Evaluated.

Sailfin Roughshark *Oxynotus paradoxus* Plate 13

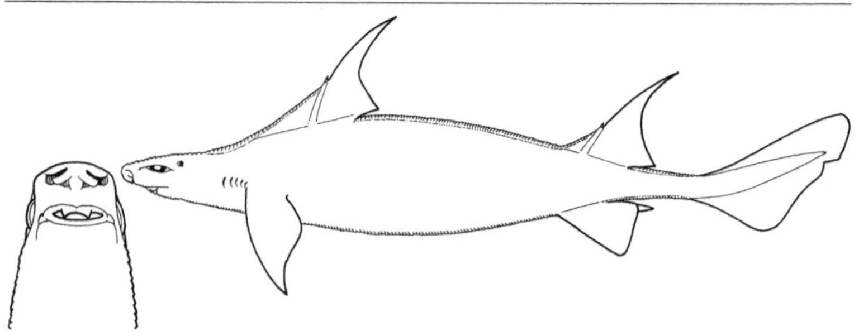

Measurements. Born: ~25 cm TL. Mature male was 75 cm. Max: ~118 cm TL.
Identification. Blackish or dark brown without prominent markings. Spiracles relatively small and almost circular. Dorsal fins tall, narrow and pointed with very concave trailing edges. First dorsal spine leans back.
Distribution. Northeast Atlantic: Scotland to Canaries, Azores, Senegal, and possibly Gulf of Guinea.

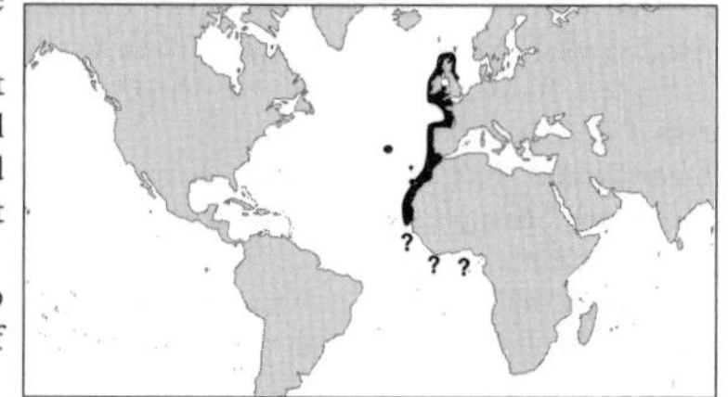

Habitat. Continental slope, 265–720 m.
Behaviour. Unknown.
Biology. Ovoviviparous.
Status. IUCN Red List: Not Evaluated. Uncommon bycatch of bottom trawls (possibly longliners) targeting deep-benthic squaloid dogfish, used for fish meal.

Dalatiidae: Kitefin Sharks

Ten species in seven genera of dwarf to medium-sized deepwater sharks (*Dalatias, Euprotomicroides, Euprotomicrus, Heteroscymnoides, Isistius, Squaliolus, Mollisquama*) distributed almost world-wide in open ocean or on bottom in mostly temperate to tropical seas. Some species wide-ranging, others restricted to single ocean basins or ridges, (but might prove to be more widespread with additional sampling).
Identification. Head narrow and conical, snout short. Strong jaws with small spear-like upper teeth, large blade-like interlocked lower teeth with smooth or serrated (*Dalatias* only) edges. Pectoral fins with short, broadly-rounded free rear tips. Two dorsal fins without fin spines or with first dorsal fin spine only (*Squaliolus* species). Second dorsal varies from slightly smaller to much larger than first. No anal fin. Caudal fin with long upper lobe, long to very short or absent lower lobe, and well-developed subterminal notch.
Biology. Poorly known. Ovoviviparous (aplacental viviparous) with up to 6–16 pups/litter in larger species. Some species are powerful predators, others ectoparasites; some solitary and some occurring in aggregations for at least part of their lifecycle.
Status. One species (*Dalatias licha*) is important in target and bycatch fisheries for meat (used for human consumption and/or fishmeal) and squalene oil from their large livers. Others are too small to be of value.

Kitefin Shark *Dalatias licha* Plate 14

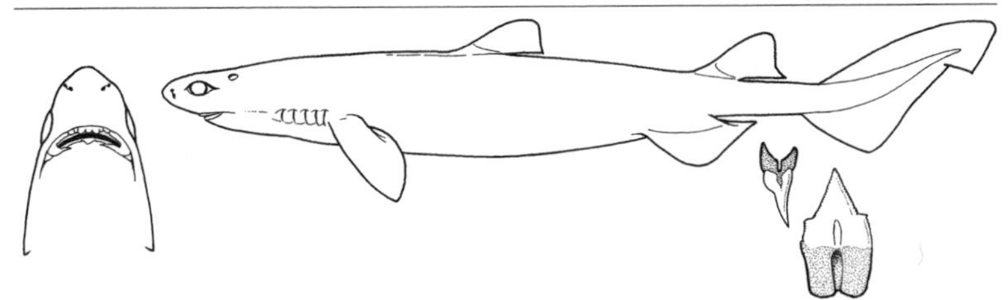

Measurements. Born: ~30 cm TL. Mature: 77–121 cm ♂, 117–159 cm ♀. Max: 159–182 cm TL.
Identification. A medium-sized cylindrical brown to blackish shark with a short blunt snout, thick fringed lips, and serrated lower teeth. Spineless dorsal fins, first originates behind pectoral fin rear tips with base closer to pectoral than pelvic fin bases, second larger. Weak ventral caudal fin lobe. Posterior margins of most fins are translucent.

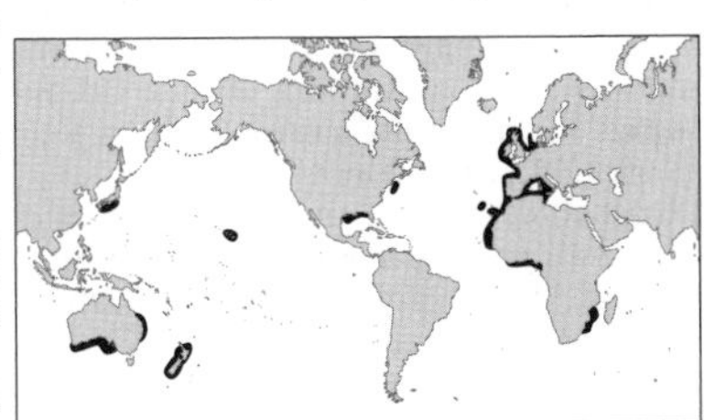

Distribution. Atlantic, Indian and Pacific Oceans.
Habitat. Deepwater (37–1800 m, mainly >200 m), warm-temperate and tropical outer continental and insular shelves and slopes, usually on or near bottom.
Behaviour. Hovers above bottom (large oil-filled liver provides neutral buoyancy) and swims well off the bottom. Solitary hunter of mainly deepwater fishes, may take bites out of large live prey.

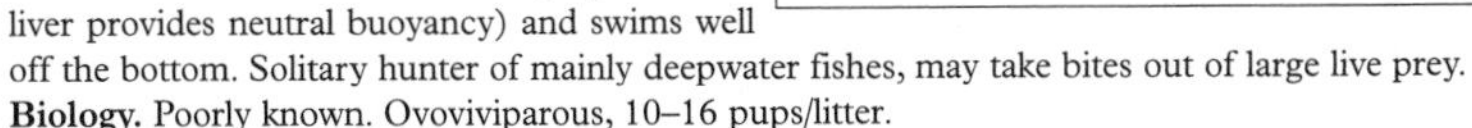

Biology. Poorly known. Ovoviviparous, 10–16 pups/litter.
Status. IUCN Red List: Data Deficient. Fisheries for meat and squalene may deplete populations rapidly.

Taillight Shark *Euprotomicroides zantedeschia* Plate 14

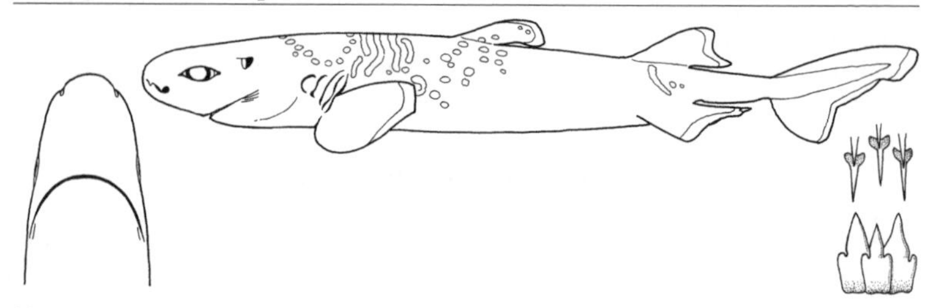

Measurements. Immature female: 17.6 cm TL. Adult male: 41.6 cm TL.
Identification. Compressed blackish-brown shark with a short snout, very wide fifth gill slit (>twice length of first), thick fringed lips, large lobate pectoral fins, a mid ventral keel on caudal peduncle, and the cloaca greatly expanded as a luminous gland containing yellow papillae. First dorsal fin insertion about equidistant between pectoral and pelvic fin bases, slightly smaller than second dorsal. Second dorsal origin in front of pelvic origins (behind it in other genera). Paddle-shaped caudal fin. All fins with prominent white margins.

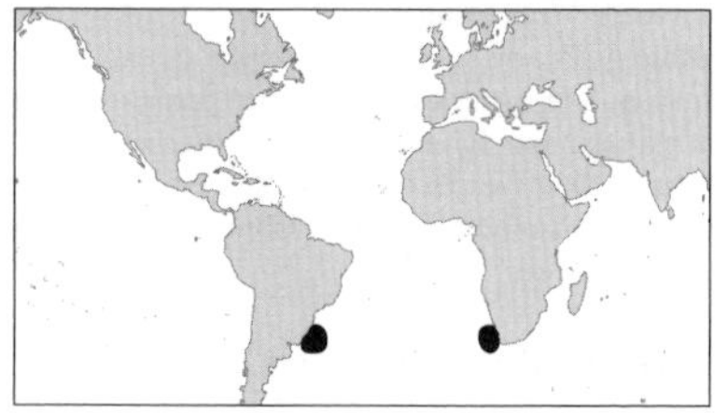

Distribution. South Atlantic.
Habitat. Epipelagic? One caught in bottom trawl on continental shelf, 458–641 m, one near surface (0–25 m) far offshore.
Behaviour. Unknown.
Biology. Unknown, probably ovoviviparous, few young.
Status. IUCN Red List: Not Evaluated. Two specimens recorded.

Pygmy Shark *Euprotomicrus bispinatus* Plate 14

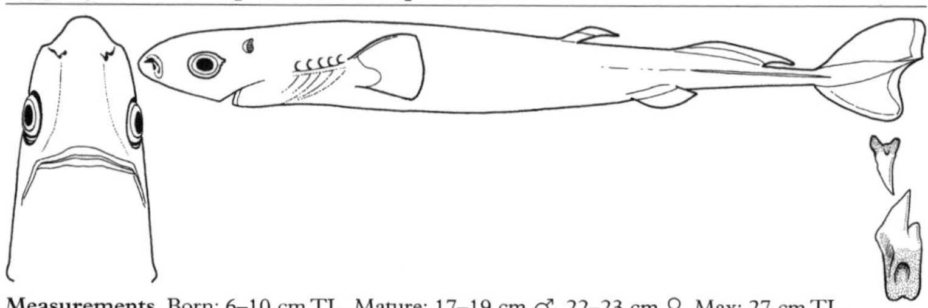

Measurements. Born: 6–10 cm TL. Mature: 17–19 cm ♂, 22–23 cm ♀. Max: 27 cm TL.
Identification. Tiny cylindrical black shark with a bulbous snout, large eyes, light-edged fins and luminous organs on underside of body. Gill slits tiny. Low lateral keels on caudal peduncle. Tiny flaglike first dorsal fin, quarter length of second, well behind pectoral fins. Caudal fin nearly symmetrical, paddle-shaped.

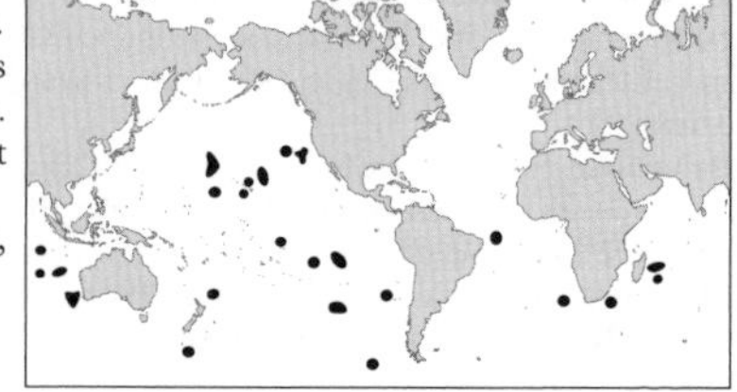

Distribution. Oceanic and amphitemperate. South Atlantic, south Indian and Pacific Oceans.
Habitat. Epipelagic, mesopelagic, perhaps bathypelagic in mid-ocean, 1829–9938 m. Migrates from surface at night to >1500 m (at least to midwater, perhaps bottom) by day.
Behaviour. Eats deepwater squid, bony fishes, some crustaceans, not large prey.
Biology. Ovoviviparous, eight pups/litter.
Status. IUCN Red List: Not Evaluated.

Longnose Pygmy Shark *Heteroscymnoides marleyi* Plate 14

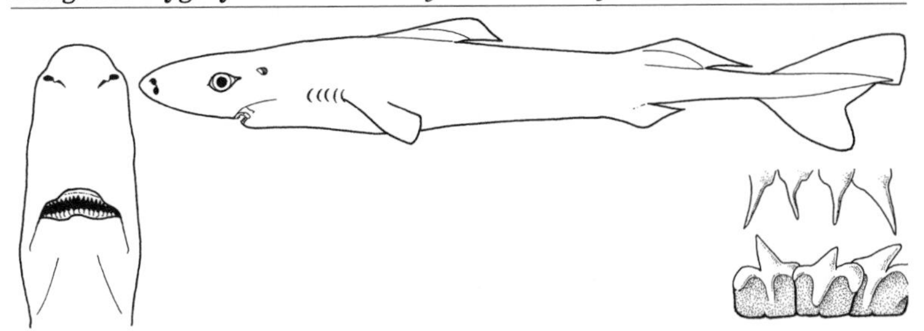

Measurements. Born: ~12 cm TL. Mature female was 33.3 cm TL. Max: ~37 cm TL ♂.
Identification. Dark brown dwarf shark with a very long bulbous snout, thin unpleated lips, and dark fins with light margins. Small gill slits. First dorsal fin far forward, originating over pectoral bases, second only slightly larger than first. Caudal fin almost symmetrical, paddle-shaped with a strong ventral lobe.

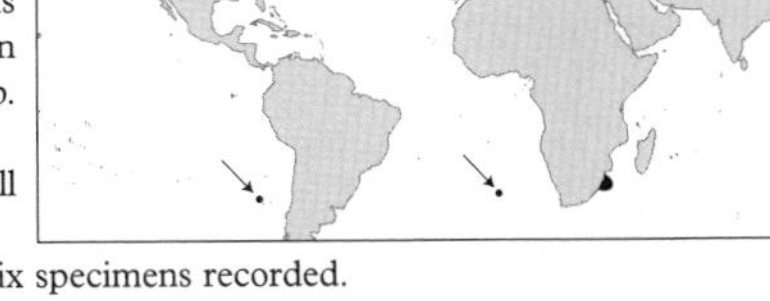

Distribution. Southern Ocean. Possibly circumglobal.
Habitat. Oceanic. Epipelagic in cold, subantarctic oceanic waters and cold current systems (Benguela and Humboldt). Recorded between surface and 502 m in areas 830 to >4000 m deep.
Behaviour. Unknown.
Biology. Presumably ovoviviparous, likely small litters. Food unknown.
Status. IUCN Red List: Not Evaluated. Only six specimens recorded.

Cigar or Cookiecutter Shark *Isistius brasiliensis* Plate 14

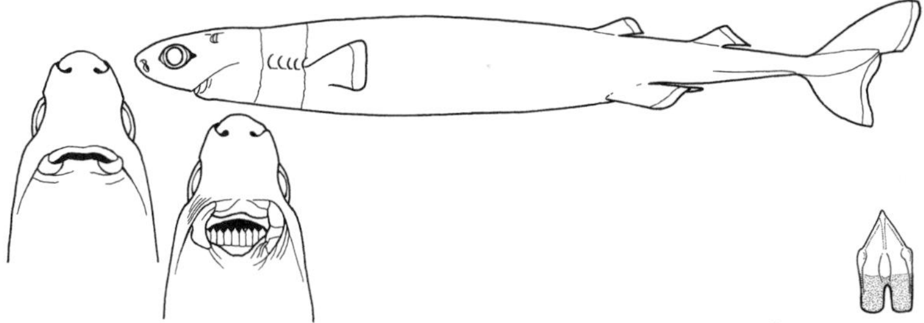

Measurements. Mature: ~31–37 cm ♂, 38–44 cm ♀. Max: >39 ♂, >50 cm TL ♀.
Identification. Small, medium grey or grey-brown cigar-shaped shark with light-edged fins and prominent dark collar mark around throat. Very short bulbous snout, large triangular lower teeth in 25–31 rows and suctorial lips. Dorsal fins set far back, first dorsal base above pelvic origins, pelvic fins larger than dorsal fins. Large, nearly symmetrical paddle-shaped caudal fin with long ventral lobe. Luminous organs covering lower surface except fins and collar can glow bright green.

Distribution. Atlantic, southern Indian Ocean and Pacific.
Habitat. Wide-ranging tropical oceanic shark,

epipelagic to bathypelagic. Only caught at night, sometimes at surface but usually deeper in water 85–3500 m, frequently near islands, (possibly pupping grounds or concentrations of prey).
Behaviour. Poor swimmer; probably migrating vertically from deep water (2000–3000 m) to midwater or surface at night. An ectoparasite on large fish and cetaceans, which are possibly lured to the shark by its bioluminescent light organs. Its thick lips and modified pharynx are used to attach itself to the prey, then razor-sharp lower teeth bite into the skin and twisting movements cut out a plug of flesh. Once cut out, the shark pulls free, holding the plug by its hook-like upper teeth; leaving behind a crater wound. Reported to have attacked rubber sonar domes on nuclear submarines. An anecdotal report of a mid-ocean swimmer nipped by a dwarf shark could be this species. Can be active and bite when caught.
Biology. Presumably ovoviviparous, ~six or seven pups/litter. Large oil-filled liver and body cavity and small fins suggest neutral buoyancy (or may compensate for highly calcified skeleton). Lower teeth swallowed as they are replaced in rows (perhaps to recycle calcium). Eats deepwater fish, squid and crustacea.
Status. IUCN Red List: Not Evaluated. Too small to be taken by most fisheries, but occasional bycatch in pelagic fish and squid fisheries.

South China Cookiecutter Shark *Isistius labialis* Plate 14

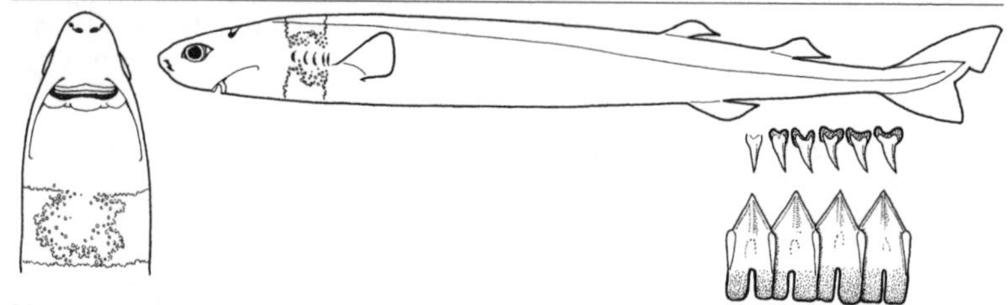

Measurements. Only specimen was 44 cm TL.
Identification. Very similar to *Isistius brasiliensis*, with dark collar marking well-developed over gill region, but upper teeth rows more numerous (43 versus 31–37 rows), eyes slightly further forward, and caudal fin less symmetrical and with a shorter ventral lobe.

Distribution. Western Pacific, South China Sea.
Habitat. Continental slope, 520 m, distance off bottom unknown.
Behaviour. Unknown.
Biology. Unknown.
Status. IUCN Red List: Not Evaluated. Known from one specimen.

Largetooth Cookiecutter Shark *Isistius plutodus* Plate 14

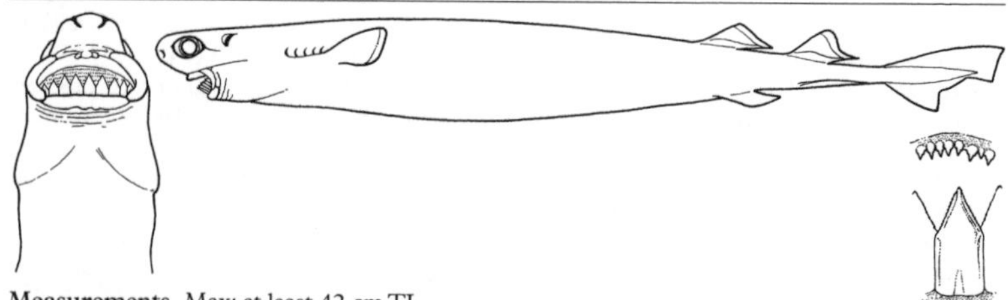

Measurements. Max: at least 42 cm TL.
Identification. Larger jaws and bigger mouth than *Isistius brasiliensis* and *I. labialis* and with only 17–19 rows of enormous lower teeth. Eyes set well forward on short-snouted head provide binocular vision. No collar marking or a weak collar marking on throat. Small asymmetric caudal fin with short ventral lobe.

Distribution. West Atlantic, northeast Atlantic, west Pacific.
Habitat. Epipelagic, possibly bathypelagic, 100–200 m. Known from a few scattered localities near land.

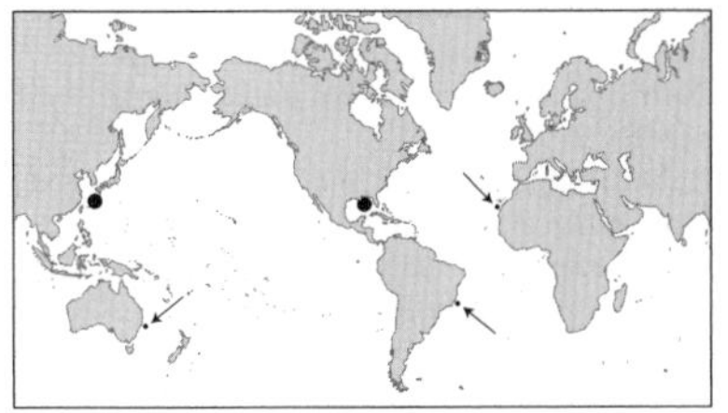

Behaviour. Smaller fins suggest this is a less active swimmer than *I. brasiliensis*. Feeds in a similar manner, can take even larger and more elongated (twice as long as diameter of mouth) bites out of bony fishes and may do likewise to other prey.
Biology. Unknown.
Status. IUCN Red List: Not Evaluated.

Pocket Shark *Mollisquama parini* — Plate 14

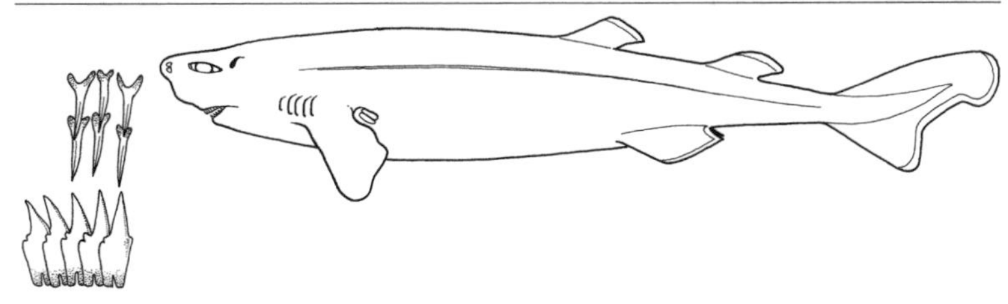

Measurements. Max: over 40 cm TL (adolescent female).
Identification. Small dark brown shark with a short blunt conical snout, thick fringed lips, a large pocket-like gland with a conspicuous slit-like opening just above each pectoral base (possibly secreting a pheromone or luminous fluid) and dark fins with light margins. Medium-sized gill slits, the fifth almost twice length of first. First dorsal fin well behind pectoral fins, base just in front of pelvic bases, second dorsal about as large as first. Caudal fin asymmetrical, not paddle-shaped, with a weak lower lobe.

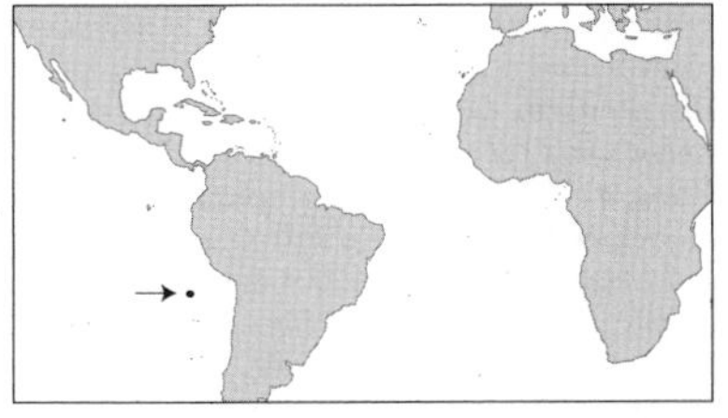

Distribution. Southeast Pacific.
Habitat. Only known specimen was caught on the Nasca submarine ridge (~1200 km east of Chile) at 330 m depth.
Behaviour. Unknown.
Biology. Unknown.
Status. IUCN Red List: Not Evaluated.

Smalleye Pygmy Shark *Squaliolus aliae* — Plate 14

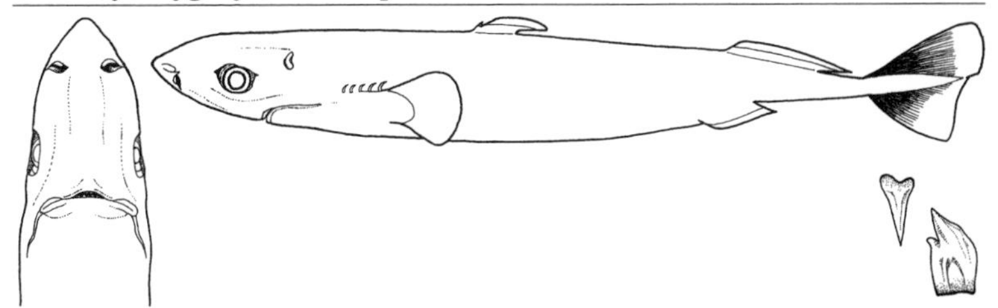

Measurements. One of the smallest living sharks. Mature: ~15 cm ♂. Max: ~22 cm TL.
Identification. Spindle-shaped, blackish dwarf shark with long, bulbous, conical snout and prominent light fin margins. Fin spine (sometimes covered by skin) only on first dorsal fin, origin opposite inner margins or rear tips of pectoral fins, second dorsal base over twice length of first.

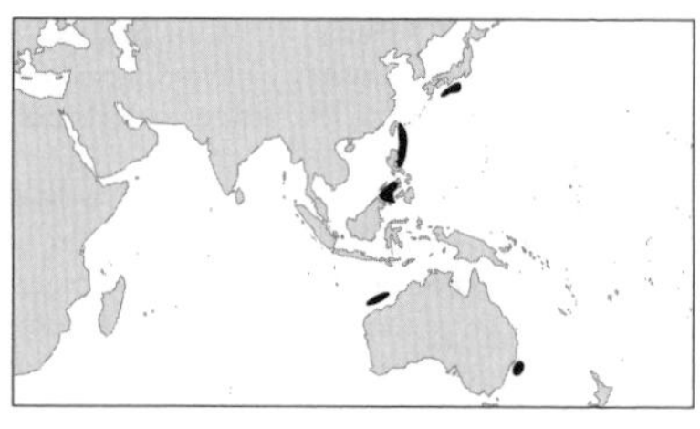

Paddle-shaped caudal fin. Eye smaller than in *Squaliolus laticaudus*, upper eyelid angular. Pair of lateral papillae on upper lip partially cover teeth. Ventral surface covered with photophores.

Distribution. Southeastern Indian Ocean and west Pacific.

Habitat. Epipelagic or mesopelagic near land, depth range 200–2,000 m.

Behaviour. May migrate daily from shallow depths at night to deeper water by day.

Biology. Virtually unknown. Ovoviviparous, litter size unknown.

Status. IUCN Red List: Least Concern. Wide-ranging, considered by some authorities too small to be threatened by fisheries, but a common bycatch of bottom trawl fisheries in the west Pacific.

Spined Pygmy Shark *Squaliolus laticaudus* Plate 14

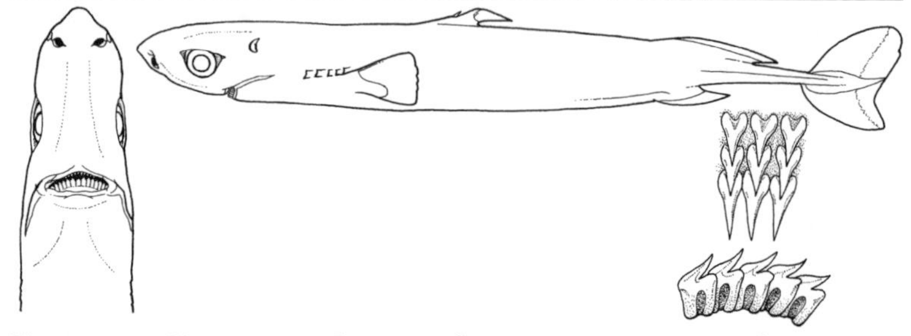

Measurements. Mature: ~15 cm ♂, 17–20 cm ♀. Max: ~22 cm ♂, ~28 cm TL ♀.

Identification. Very similar to *Squaliolus aliae* but has a larger eye with a broadly arched upper eyelid and lateral papillae absent or weakly developed on upper lip. Ventral surface covered with photophores.

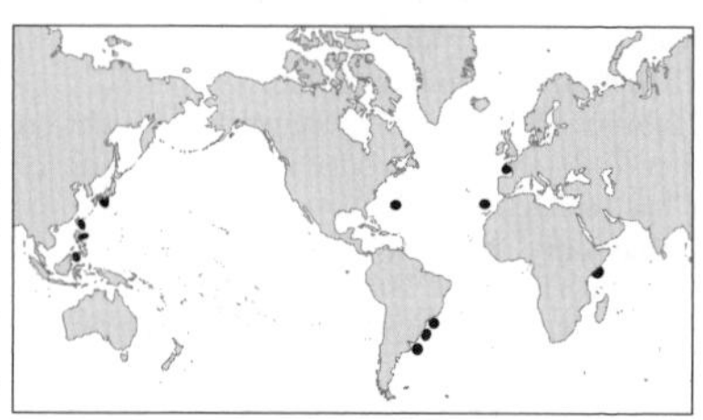

Distribution. Oceanic, nearly circumtropical, Atlantic and Indo-Pacific Oceans.

Habitat. Wide-ranging, tropical epipelagic species. Highly productive areas near continental and island land masses, sometimes over shelves, usually over slopes, not in mid-ocean.

Behaviour. Migrates vertically from 200 m (night) to 500 m (day); not caught at surface. Photophores may eliminate shadow caused by light from above.

Biology. Probably ovoviviparous. Feeds on vertically migrating deepwater squid and fishes.

Status. IUCN Red List: Not Evaluated.

Pristiophoriformes: Sawsharks

This Order contains one family of little-known small sharks, distributed world-wide in the fossil record, but now found only on the continental and insular shelves and upper slopes of the northwest and southeast Atlantic, west Indian and west Pacific Oceans, in shallow water in temperate regions, deeper in the tropics. Sometimes found in large schools or feeding aggregations. At least one species segregates by depth (adults in deeper water than young). Some have a very restricted distribution.

Identification. Small slender sharks (max. ~150 cm TL, most <70 cm) with cylindrical bodies, flattened heads and a long, flat saw-like snout with a pair of long string-like ventral barbels in front of the nostrils and close-set rows of lateral and ventral sawteeth. Eyes on the side of the head, large spiracles. Two spineless dorsal fins, no anal fin. Thick lateral ridges on the caudal peduncle, a long dorsal caudal lobe and no ventral lobe. May be confused with the sawfishes, which are batoids (rays) with flattened bodies, pectoral fins fused to the head, and gill slits underneath the head.

Biology. Ovoviviparous (aplacentally viviparous); foetuses gain all food from their yolk sac, which is reabsorbed just before the litters of 7–17 pups are born. The large lateral rostral teeth erupt before birth but lie flat against the rostrum until after birth. The tooth-studded rostrum has sensors to detect vibrations and electrical fields and is probably used to capture and kill prey. It may also be used for defence, or when competing or courting with other sawsharks (parallel cuts and scratches occasionally seen on adults are presumably from interactions with other sawsharks). The long rostral barbels may have taste, touch or other sensors and be trailed along the bottom to locate prey, including small fishes, crustaceans and squid.

Status. Harmless to man, despite very sharp (non-toxic) teeth; handle with care. Some species are or were common where they occur. Very vulnerable to bycatch if a restricted range coincides with fisheries; their saws may easily be entangled in fishing gear including nets. Taken as a bycatch in demersal gillnet and bottom trawl fisheries and marketed for food or discarded. Saws may be sold as curios.

Pristiophoridae: Sawsharks

Two genera. *Pristiophorus* have five lateral pairs of gill slits, *Pliotrema* have six.

Sixgill Sawshark *Pliotrema warreni* Plate 15

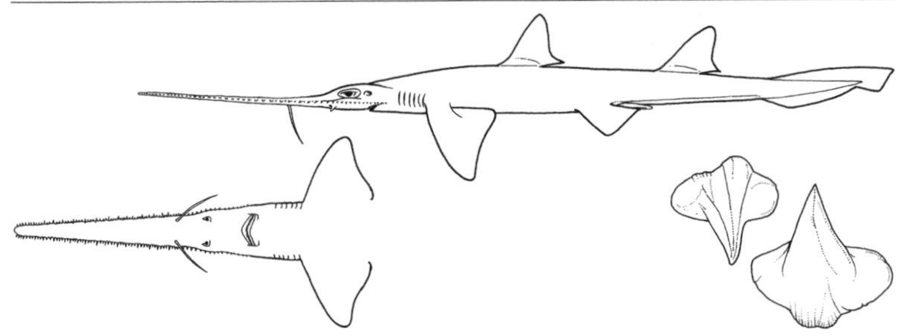

Measurements. Born: 35–37 cm TL. Mature: ~83 cm ♂, ~110 cm ♀. Max: 112 cm ♂, >136 cm TL ♀.

Identification. The only sawshark with six pairs of gill slits. Barbs on posterior edges of larger rostral sawteeth (absent in other sawsharks). Barbels closer to mouth than in other species.

Distribution. Southeast Atlantic (South Africa), southwest Indian Ocean (South Africa, south Mozambique, southeast Madagascar).

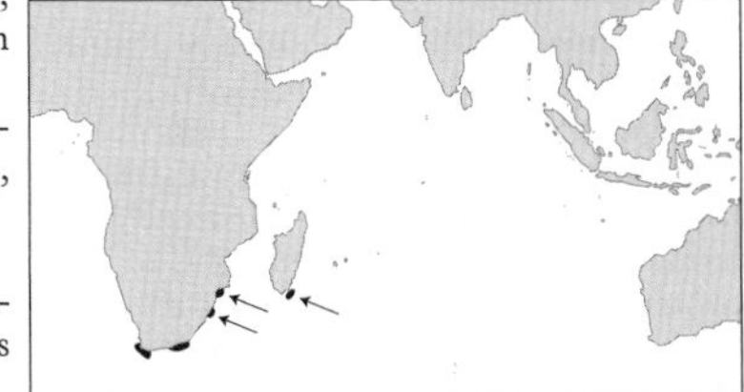

Habitat. On or near bottom, offshore continental shelves and upper slopes (37–500 m), adults deeper than pups.

Behaviour. May use inshore pupping grounds.

Biology. Five to seven pups/litter, 7–17 developing eggs/female. Feeds on small fish, crustaceans and squid. Predators include Tiger Sharks.

Status. IUCN Red List: Not Evaluated. Uncommon and localized in distribution. Discarded bycatch from bottom trawl fisheries.

Longnose Sawshark *Pristiophorus cirratus* Plate 15

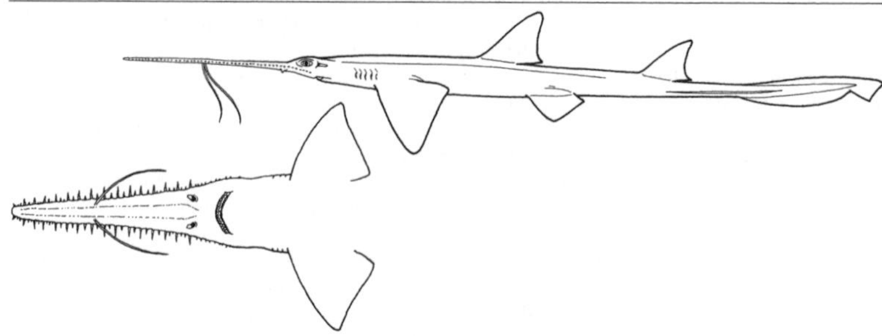

Measurements. Born: 31–34 cm TL. Mature: ~97 cm TL ♂. Max: 134–137 cm TL.

Identification. Pale yellow/greyish-brown above with varie-gated (sometimes faint) dark blotches, spots and bars; white below. Darker brown rostral midline and edges, tooth margins blackish; fin bases blotched. Large stocky sawshark. Long narrow rostrum (26–30% TL), barbels much closer to rostral tip than nostrils. 19–21 large lateral sawteeth, 9–11 in front of barbels, 9–10 behind. Juveniles have two or three smaller teeth between large sawteeth.

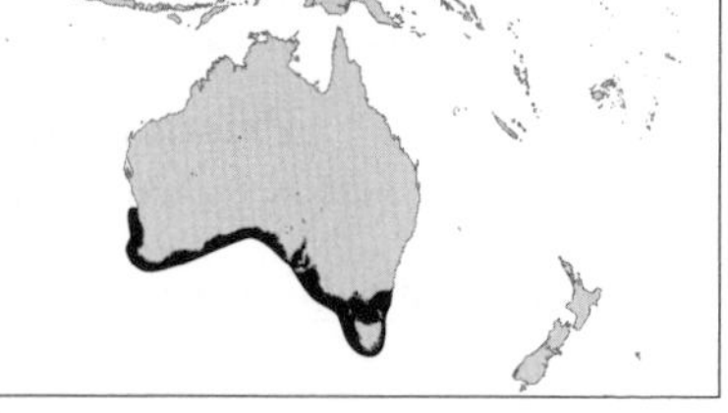

Distribution. Southern Australia.

Habitat. Continental shelf and upper slope, on or near sandy or gravel-sand bottom. Occasionally close inshore (bays and estuaries), usually offshore to at least 311 m.

Behaviour. Occurs in schools or feeding aggregations.

Biology. Litters of 6–19 pups every other winter. Feeds on small fishes and crustaceans.

Status. IUCN Red List: Least Concern. Moderately abundant. Fisheries bycatch levels stable, used for meat.

Japanese Sawshark *Pristiophorus japonicus* Plate 15

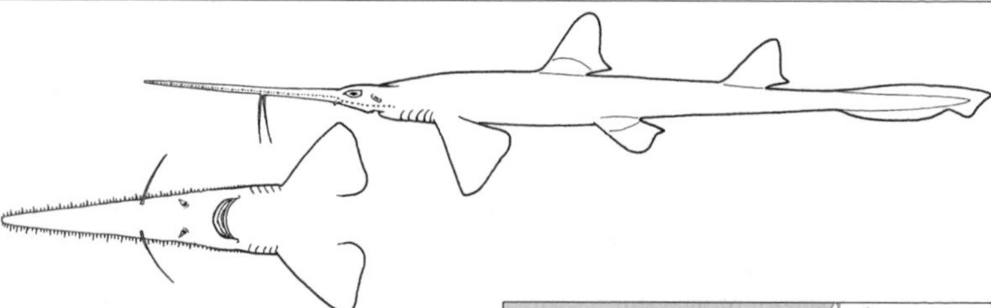

Measurements. Mature: ~80–100 cm ♂, ~100 cm ♀. Max: 136–153 cm TL.

Identification. Plain brown or reddish-brown above, white below. Darker brown rostral midline and edges. Large and stocky. Moderately long narrow rostrum (preoral length 26–30% TL), barbels closer to mouth than rostral tip. 23–43 large lateral rostral sawteeth, 15–26 before barbels, 8–17+ behind; 9–14 ventral teeth before

and 8–9 after.Juveniles with one or two smaller teeth between large lateral sawteeth and light trailing edges to pectoral and dorsal fins.
Distribution. Northwest Pacific: Japan, Korea, northern China, Taiwan Island.
Habitat. Temperate continental shelves and upper slopes, on or near sand or mud bottom.
Behaviour. Feeds on small bottom animals; pokes bottom with snout and barbels.
Biology. Usually 12 pups/litter.
Status. IUCN Red List: Not Evaluated. Common in coastal waters. Meat eaten in Japan. Kept in aquaria for short periods.

Shortnose Sawshark *Pristiophorus nudipinnis* Plate 15

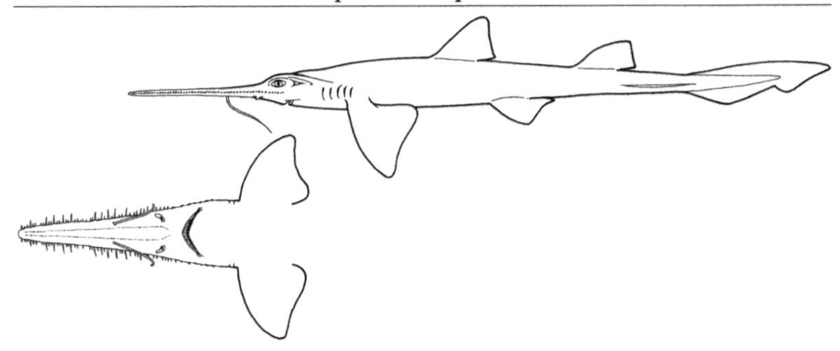

Measurements. Born: 25–32 cm TL. Mature: 107 cm ♂. Max: ~124 cm TL ♀.
Identification. Uniform slate-grey above, white below. Indistinct dusky stripes along rostrum midline and edges. Large stocky sawshark with short, broad, tapering rostrum (preoral length 22–24% TL). Barbels much closer to mouth than rostral tip. 17–19 large lateral rostral sawteeth, 12–14 before barbels, 6–8 behind. Juveniles usually with one smaller tooth between large sawteeth. 13–14 ventral teeth before barbels, four between barbels and nostrils. Nostrils diagonally oval and elongated.
Distribution. Southern Australia.
Habitat. Temperate-subtropical continental shelf, on or near bottom, close inshore to at least 70 m, possibly 165 m.
Behaviour. Hidden prey may be located with barbels and uprooted by snout.
Biology. Ovoviviparous, 7–14 pups produced every other year.
Status. IUCN Red List: Least Concern. Moderately common within range. Bottom trawl and gillnet bycatch, used for food.

Bahamas Sawshark *Pristiophorus schroederi* Plate 16

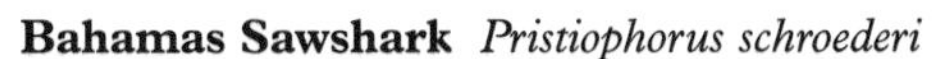

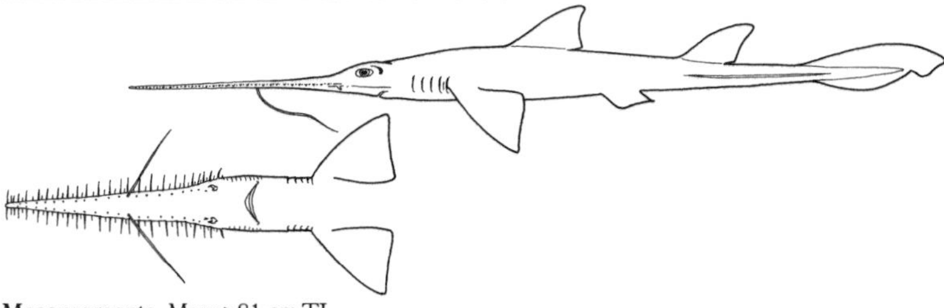

Measurements. Max: >81 cm TL.
Identification. Uniform unpatterned light grey above, whitish below. Darker brownish stripes along

rostrum midline and edges. Light edged pectoral fins, dark anterior edge to dorsals in juveniles. Slender, with a very long, narrow, tapering rostrum (preoral length 31–32% TL) with barbels halfway between mouth and rostral tip and a concave pre-barbel edge. 23 large lateral sawteeth, 13 before and 10 behind barbels. Juveniles usually with one smaller tooth between large lateral teeth.

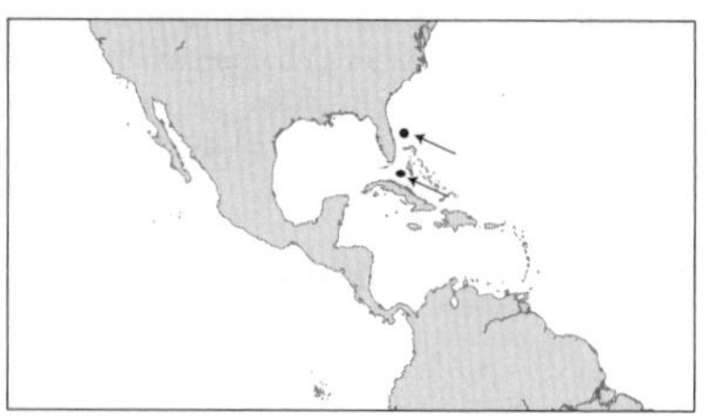

Distribution. Northwest Atlantic: between Cuba, Florida and Bahamas.
Habitat. On or near bottom, continental and insular slopes, 438–952 m.
Behaviour. Unknown.
Biology. Unknown.
Status. IUCN Red List: Not Evaluated. Possibly bycatch of deepwater fisheries.

Eastern Sawshark *Pristiophorus* sp. A — Plate 16

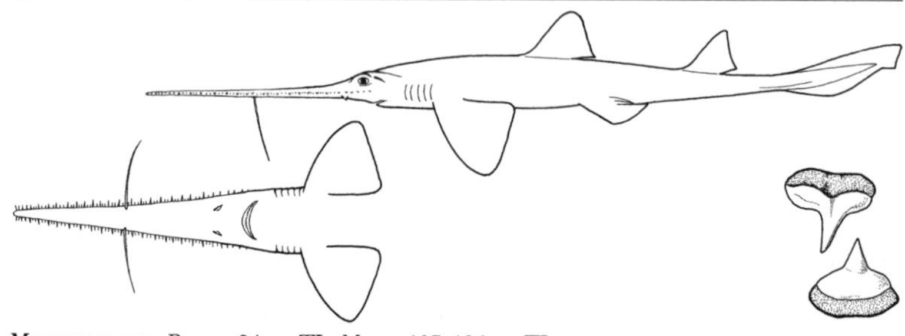

Measurements. Born: >34 cm TL. Max: >107–136 cm TL.
Identification. Plain grey-brown colour above (no spots or bars), white below. Rostrum with darker brownish stripes along midline and on edges; pectoral and dorsal fins with dark anterior margins in juveniles (absent in adults). Large stocky sawshark with moderately long, narrow rostral saw (preoral length 25–30% TL). Barbels closer to rostral tip than mouth, juveniles usually with one smaller tooth between large lateral rostral teeth. 20–25 large lateral sawteeth, 11–15 before barbels, 9–10 behind.

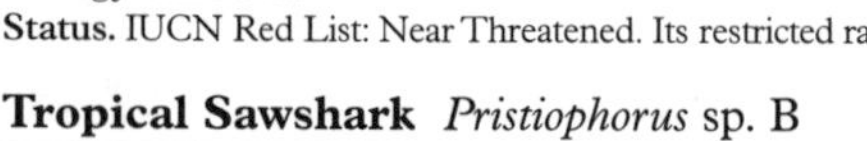

Distribution. Southeast Australia.
Habitat. Offshore continental shelf and upper slope, 100–630 m.
Behaviour. Unknown.
Biology. Unknown.
Status. IUCN Red List: Near Threatened. Its restricted range is heavily fished and bycatch is a threat.

Tropical Sawshark *Pristiophorus* sp. B — Plate 16

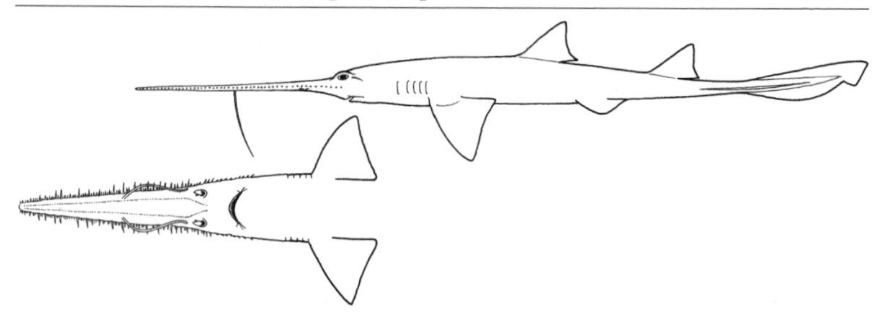

Measurements. Mature: ~62 cm ♂. Max: >84 cm TL

Identification. Uniform pale yellow-brown above, no spots or bars, white below. Small, slender, with long narrow tapering straight-sided saw (preoral length 29–31% TL, length and width at nostrils 5.2–6.0 times preorbital length). Rostral barbels slightly closer to rostral tip than mouth or equidistant between. Juveniles usually have two or three smaller teeth between large lateral rostral teeth.

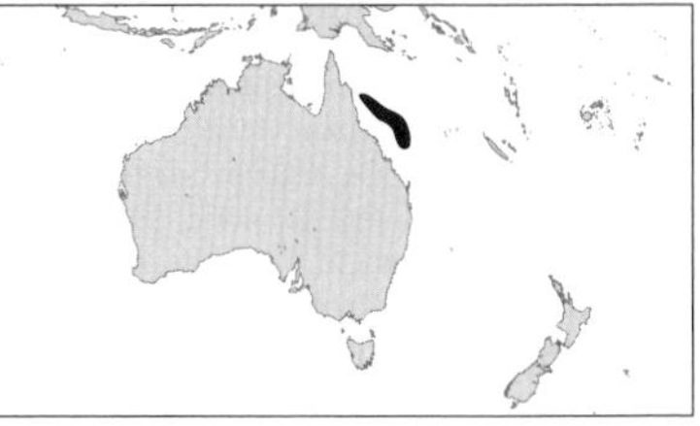

Distribution. East Australia (Queensland).

Habitat. Continental slope, 300–400 m.

Behaviour. Unknown.

Biology. Presumably ovoviviparous, otherwise unknown.

Status. IUCN Red List: Least Concern. Its range is largely unfished.

Philippine Sawshark *Pristiophorus* sp. C — Plate 16

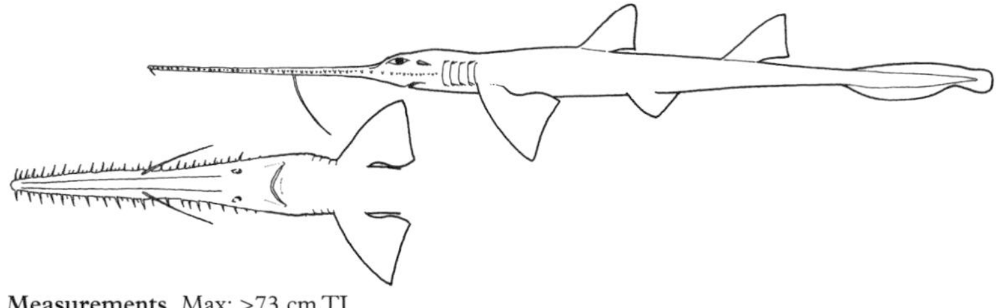

Measurements. Max: >73 cm TL.

Identification. Small, slender, with long and very narrow rostrum (preoral length 29–31% TL, width at nostrils 5.2–6.0 times preorbital length), slightly concave between barbels and nostrils. Barbels slightly closer to mouth than to rostral tip, juveniles with two or three smaller teeth between ~21 large lateral teeth (13–14 in front of barbels, 7–8 between barbels and nostril). Uniform dark brown above, white below, no spots or bars.

Distribution. Philippine Islands: off Apo Island and southern Luzon.

Habitat. Upper tropical continental slopes on bottom, 229–593 m.

Behaviour. Unknown.

Biology. Unknown.

Status. IUCN Red List: Not Evaluated.

Dwarf Sawshark *Pristiophorus* sp. D — Plate 16

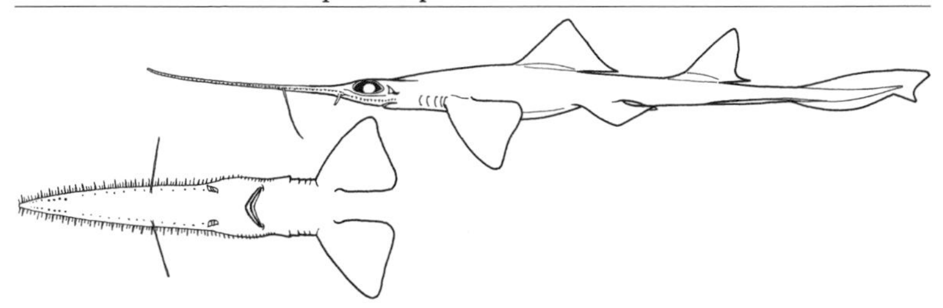

Measurements. Mature: ~45–50 cm ♂, 57 cm ♀. Max: >62 cm TL.

Identification. Very small. Broad, triangular first dorsal fin, rear tip extending behind pelvic fin midbases. Two rows of four or five enlarged pits on underside (pre-barbel) of rostrum. Barbels much closer to mouth than rostral tip. Prominent ridges on base of large lateral rostral teeth. Uniform brown above, white below, pale rostrum with dark brown stripes on middle and edges. Pectoral and dorsal fins with dark anterior margins (obvious in juveniles) and prominent light trailing edges.

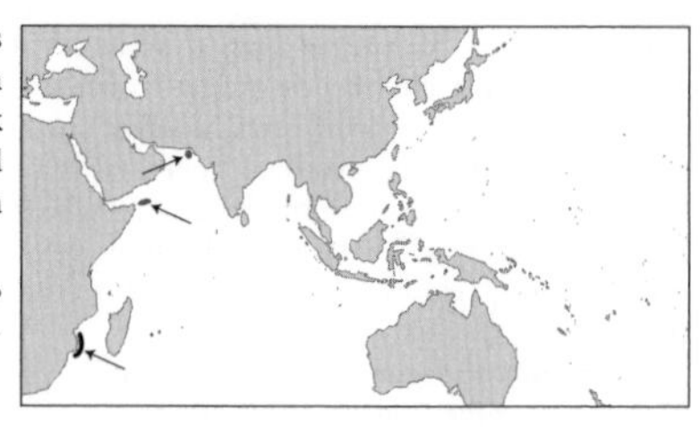

Distribution. West Indian Ocean: Mozambique, possibly Somalia, to the Arabian Sea off Pakistan.
Habitat. Upper continental slope, 286–500 m
Behaviour. Unknown.
Biology. Virtually unknown.
Status. IUCN Red List: Not Evaluated. Possibly a bycatch of deepwater trawl fisheries.

Squatiniformes: Angelsharks

Once referred to as Monkfish, Guillame Rondelet in 1555, for their similarity to a monk, they are found mainly on mud and sand on cool temperate continental shelves, from intertidal to continental slopes, deeper in tropical water. Absent from most of the Indian Ocean and central Pacific.

Identification. Medium-sized (mostly <1.6 m) bizarrely shaped sharks. May be heavily patterned on dorsal surface, most are uniformly pale below. Similar to rays, with a broad flattened body, short snout and large fins, but with gill openings on the sides of the head, not beneath, and very large pectoral fins not attached to the head opposite the gills (hindmost gill opening is in front of pectoral fin origins, but covered by triangular anterior fin lobes). Eyes on top of head, close to and level with large spiracles. Large mouth and nostrils (barbels on anterior nasal flaps) at front of snout are separate, mouth extends at sides to opposite or slightly behind the eyes. Very large labial furrows. Two spineless dorsal fins set back on precaudal tail, first over or behind the free rear tips of the very large pelvic fins. No anal fin. Short very thick keels at the base of the caudal peduncle, caudal fin with long dorsal lobe and even longer expanded ventral lobe. Some species are very hard to distinguish.

Biology. Poorly known, outside Europe and northeast Pacific. Reproduction ovoviviparous (aplacental viviparous); litters of 1–25 pups obtain all nourishment from a yolk sack before birth. Often lie buried by day in mud and sand (except for eyes and spiracles). Ambush feeders, using their unusually flexible 'necks' to raise their heads and protruding trap-like jaws to snap up prey at high speed. Food includes small bony fishes, crustaceans, squids, gastropods and clams. At least some species are nocturnally active and move off the seabed. Do not swim far; populations may easily be isolated by deepwater or areas of unsuitable habitat.

Status. Harmless, unless disturbed or provoked. Many species are intensively fished for food (the have valuable flesh), also oil, fishmeal and leather. Very vulnerable target and bycatch species in bottom trawl, line gear and fixed bottom nets. Significant population reductions reported for many heavily fished species. Recolonisation of depleted populations from adjacent areas may be very slow. Some tropical species are very poorly known. Divers have photographed an apparently undescribed species (not included in this book) on a coral reef in the Andaman Sea.

Squatinidae: Angelsharks

Single genus containing 20 benthic species, some undescribed.

Sawback Angelshark *Squatina aculeata* Plate 17

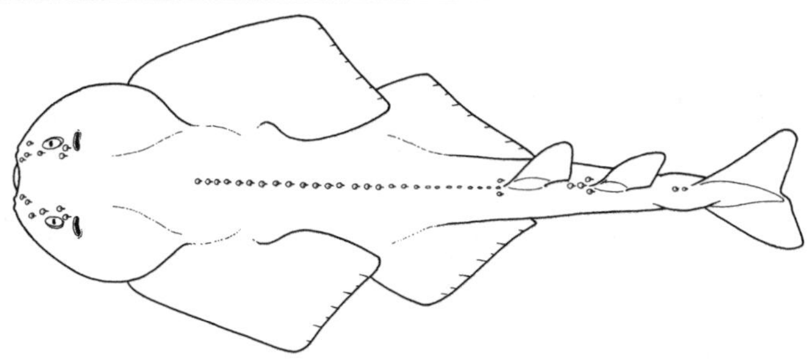

Measurements. Mature ~124 cm. Max: 188 cm TL.

Identification. Dull grey or light brown back sparsely scattered with small irregular white spots and regular small dark brownish spots. No ocelli. Large dark blotches on head, back, fin bases and tail. Large thorns on head and in a row along back. Concave between eyes, eye–spiracle distance <1.5 x eye length. Heavily fringed nasal barbels and anterior nasal flaps.

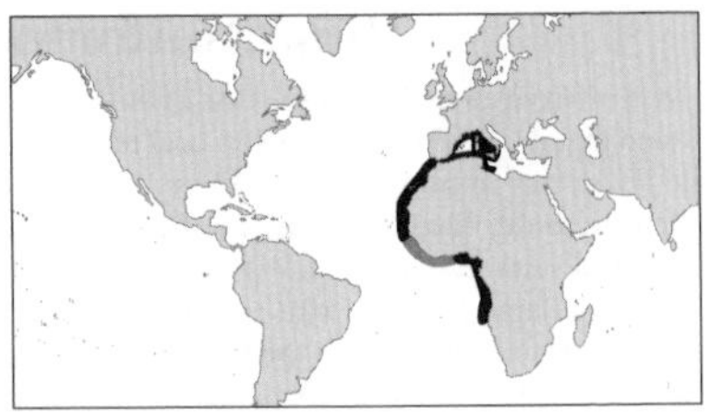

Distribution. East Atlantic: western Mediterranean to Nigeria, Gabon to Namibia.
Habitat. Offshore, outer continental shelf and upper slope, usually on mud, 30–500 m.
Behaviour. Unknown.
Biology. Poorly known. Feeds on small sharks, bony fishes, cuttlefish and crustaceans.
Status. IUCN Red List: Not Evaluated. Uncommon. Caught in target fisheries and as bycatch. Disappearing from Mediterranean.

African Angelshark *Squatina africana* Plate 17

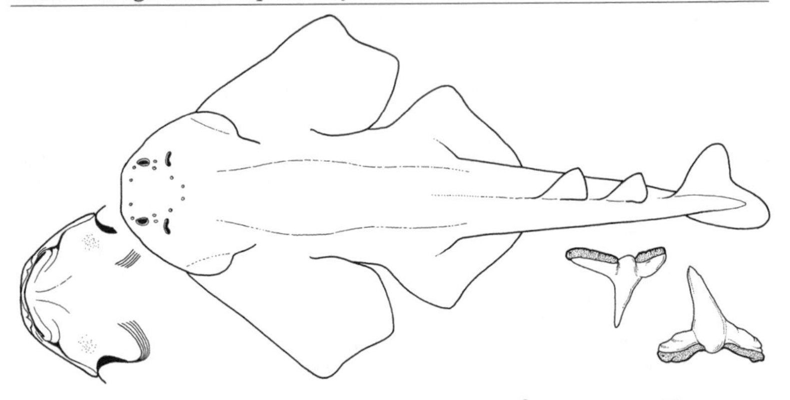

Measurements. Born: 28–30 cm TL. Mature: ~80cm ♂, 90 cm ♀. Max: ~122 cm TL.
Identification. Greyish or reddish-brown, many light and dark spots, often large granular-centred ocelli in young. Larger symmetrical dark bands or saddles, blotches on broad, angular, high pectoral fins. Dark tail base, white margins. Simple flat nasal barbels, tips tapering or spatulate. Anterior nasal flaps smooth or slightly fringed. No angular lobes on lateral dermal flaps. Enlarged thorns on head, not back. Concave between eyes.

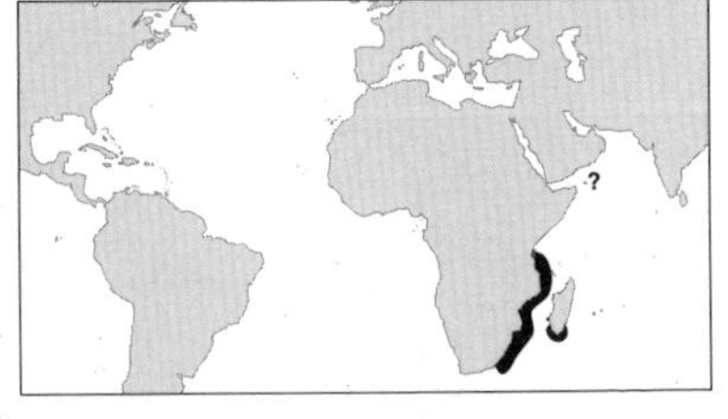

Distribution. East and southern Africa: South Africa to Mozambique, Tanzania and Madagascar, possibly Somalia. Nominal west African records possibly based on other species.
Habitat. Continental shelf and upper slope, sand and mud, surf line to 494 m (mainly 60–300 m).
Behaviour. Lies buried to ambush prey.
Biology. Seven to eleven pups/litter. Eats small bony fishes, cephalopods and shrimp.
Status. IUCN Red List: Not Evaluated. Common only on east coast of South Africa (kwaZulu-Natal). Trawl fishery bycatch.

Argentine Angelshark *Squatina argentina* Plate 17

Measurements. Born: unknown. Mature: ~120 cm. Max: 138 (?170) cm TL.
Identification. Purplish-brown with many scattered dark brown spots (no white), mostly in circular groups around a central darker spot, no ocelli. Paler dorsal fins. Simple, spatulate nasal barbels, slightly fringed or smooth anterior nasal flaps, no triangular lobes on lateral head folds. Concave between eyes. Enlarged thorns on snout, not back. Large, broad, obtusely angular pectoral fins, convex leading edge forming a distinct 'shoulder'.
Distribution. Southwest Atlantic: Brazil to Patagonia.

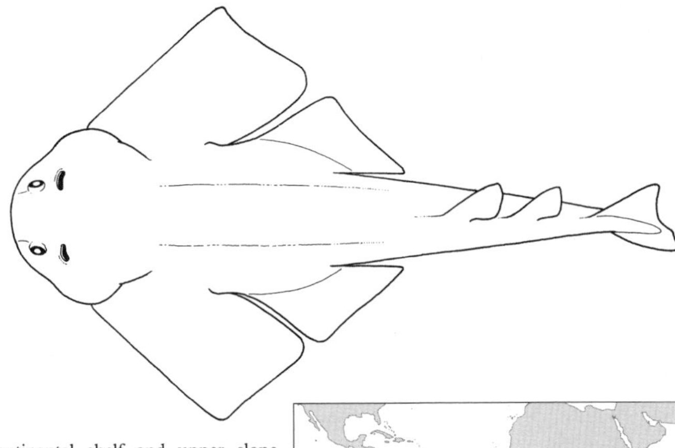

Habitat. Continental shelf and upper slope, 51–320 m (mostly 120–320 m).
Behaviour. Unknown.
Biology. Seven to eleven pups/litter. Feeds on demersal fishes, shrimp and squid.
Status. IUCN Red List: Data Deficient. Common. Target and bycatch species, with *Squatina guggenheim* and *S. punctata*.

Chilean Angelshark *Squatina armata* Plate 18

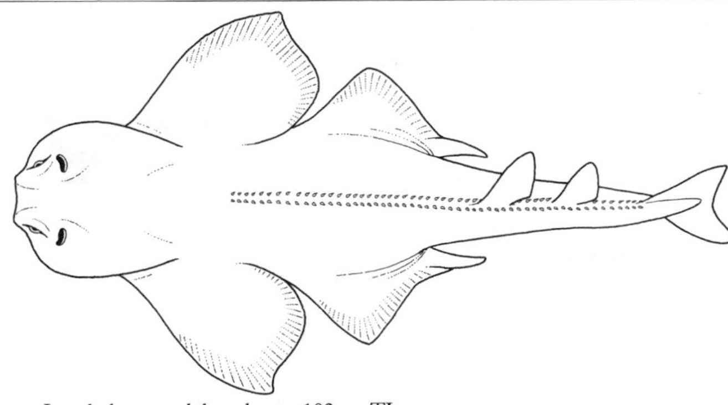

Measurements. Lost holotype, adult male was 103 cm TL.
Identification. Reddish-brown to grey above, paler below. Narrow head. The holotype had very heavy thorns on snout and between eyes and large spiracles, a double row of large hooked thorns on midline of back and between and behind dorsal fins, and enlarged thorns on the leading edge of the pectoral fins. First dorsal origin probably behind free rear tips of pelvic fins.

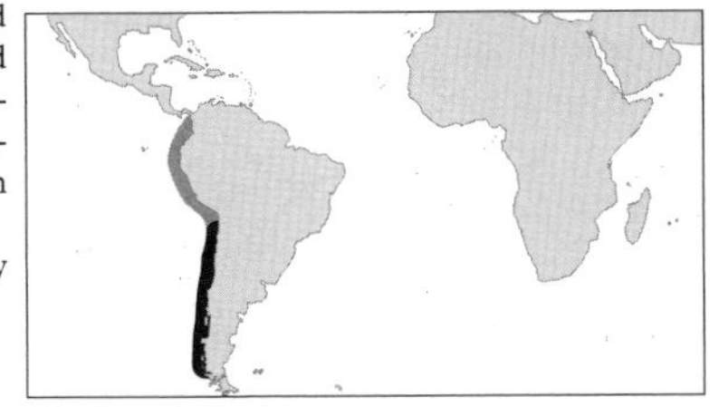

Distribution. Southeast Pacific: Chile, probably also Colombia, Ecuador and Peru.
Habitat. Continental shelf.
Behaviour. Unknown.

Biology. Unknown.
Status. IUCN Red List: Not Evaluated. Apparently distinct from *Squatina californica.* The Peruvian angelshark fishery may take more than one species.

Australian Angelshark *Squatina australis* Plate 18

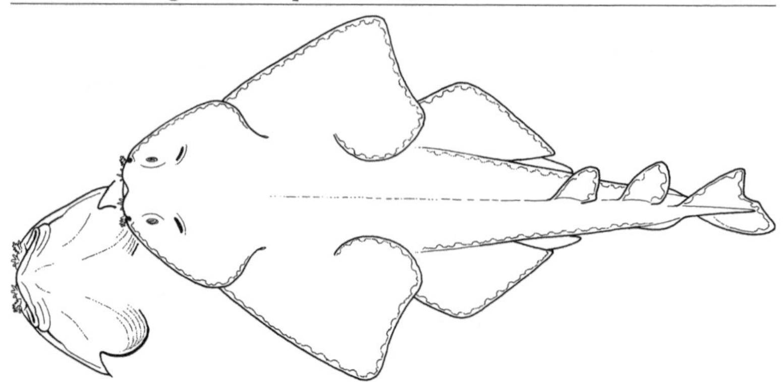

Measurements. Mature: ~90 cm ♂, 97 cm ♀. Max: >152 cm TL.
Identification. Dull greyish-brown with dense white spots and small darker brown spots. No large ocelli. White-edged fins, spots on leading edge of pale dorsal fins and lower tail lobe. Head flat or convex between eyes, small spiracles. No large thorns in adults, enlarged denticles on snout, head and multiple pre-dorsal rows in young. Heavily fringed nasal barbels and anterior nasal flaps, no triangular lobes on lateral head folds.

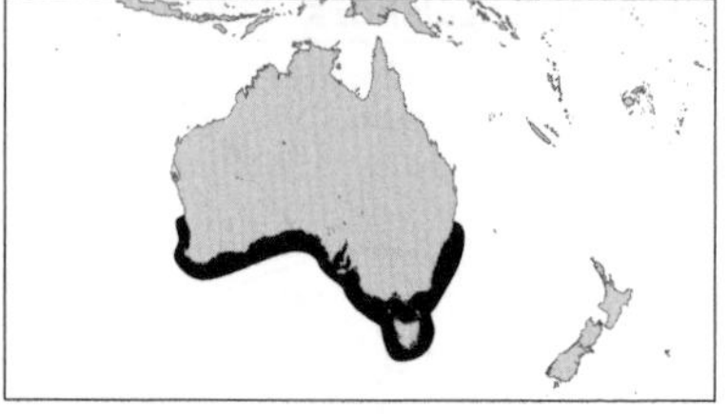

Distribution. Southern Australia.
Habitat. Sand and mud, often in seagrass or near rocky reefs, 0–130 m.
Behaviour. Lies buried by day, active at night.
Biology. Up to 20 pups/litter in autumn. Feeds on fishes and crustacea.
Status. IUCN Red List: Least Concern. Commercially valuable, trawled, but unfished and common in much of its range.

Pacific Angelshark *Squatina californica* Plate 18

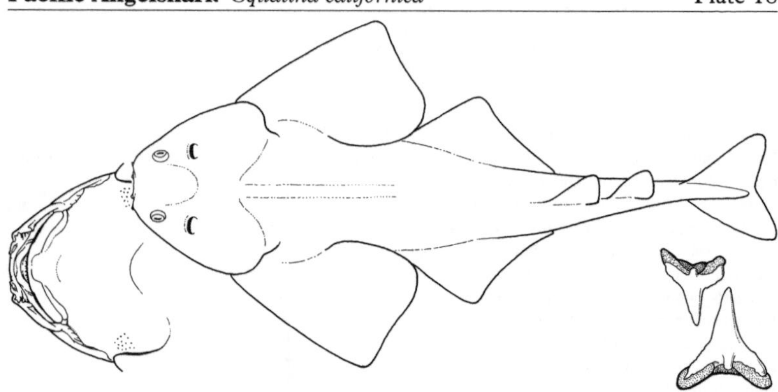

Measurements. Born: 25–26 cm TL. Mature: ~100 cm (California, smaller in Mexico). Max: 120–152 cm TL.

Identification. Reddish-brown to dark brown or blackish, scattered light spots (set around dark blotches in adults). Large paired dark blotches on back and tail form large ocelli in young. White edged pectoral and pelvic fins. Pale dorsal fins with dark blotches at base. Dark spot at base of pale, spotted tail. Simple, conical nasal barbels with spatulate tips, weakly fringed anterior nasal flaps, no triangular lobes on lateral head folds. Concave between large eyes, eye–spiracle space <1.5 x eye length. Thorns prominent in young, small or absent in adults. Fairly broad, long, high pectoral fins.

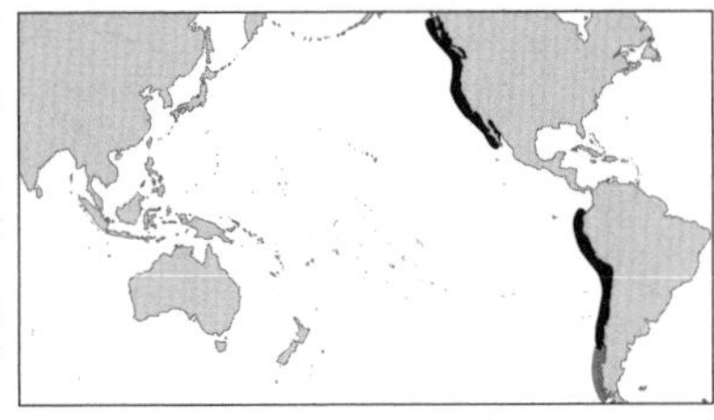

Distribution. Northeast Pacific, possibly southeast Pacific.

Habitat. Continental shelf, 1–200 m. Often around rocks, sometimes near kelp.

Behaviour. Lies buried in flat sand or mud by day to ambush prey. Active at night. Does not swim long distances.

Biology. Six to ten pups/litter after nine to ten month gestation. About 20% survive to maturity at 10–13 years old.

Status. IUCN Red List: Near Threatened. Abundant in California until fishery for meat caused population collapse in early 1990s. Gill net ban ended fishery. Taken as bycatch elsewhere. Subject of dive tourism in California.

Sand Devil *Squatina dumeril* Plate 17

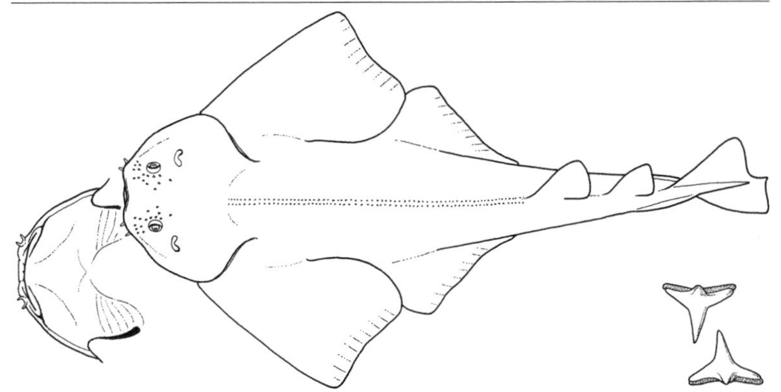

Measurements. Mature: 92–107 cm ♂. Max: 152 cm TL.

Identification. Almost plain bluish to ashy-grey, no elaborate markings or ocelli, dusky or blackish spots irregularly present or absent. Small white spots often present in young. Underside white with red spots and reddish fin margins. Dorsal and caudal fins darker, light bases. Dorsal rear tips light. Simple, tapering nasal barbels, weakly fringed or smooth anterior nasal flaps, lateral head folds low, no triangular lobes. Strongly concave between large eyes, eye–spiracle space <1.5 x eye length. Fairly broad, posteriorly angular pectoral fins. Few discrete small prominent thorns on snout, between eyes and spiracles in young, more numerous and forming patches in adults. Thorns along back of young are reduced and inconspicuous in adults.

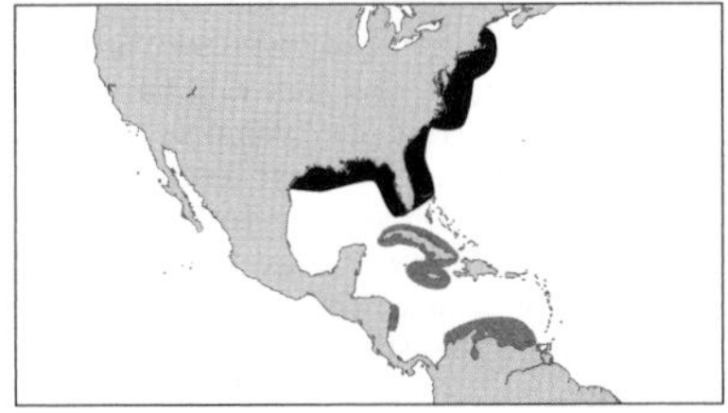

Distribution. Northwest Atlantic: New England to Gulf of Mexico. Unconfirmed: Cuba, Nicaragua, Jamaica and Venezuela.

Habitat. Continental shelf and slope, on or near bottom, close inshore to 1290 m, mostly 40–250 m.
Behaviour. Appears in inshore shallow water in spring and summer off USA, disappears (into deeper water?) in winter.
Biology. Up to 25 pups/litter born in summer. Eats small bottom fishes, crustaceans and bivalves.
Status. IUCN Red List: Not Evaluated. Not targeted by fisheries, some bycatch. Aggressive when captured.

Taiwan Angelshark *Squatina formosa* Plate 18

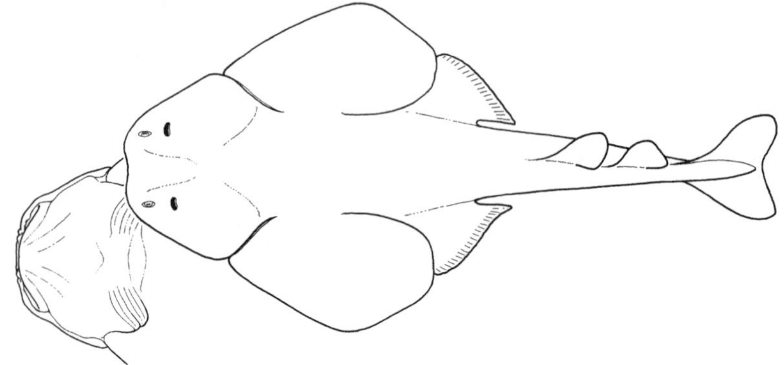

Measurements. Born: ~33 cm TL. Immature at 46 cm TL ♀. Adults presumably >100 cm TL.
Identification. Yellow-grey or brown with small paired dark ocelli. Small light spots between head and first dorsal fin, many small dark brown spots, larger irregular blotches, and a saddle or band alongside dorsal fins. Simple, very flat nasal barbels, tapering round tips. Weakly fringed or smooth anterior nasal flaps, no triangular lobes on lateral head folds. Head concave between large eyes, eye–spiracle space < eye length. Very broad, low, rounded pectoral fins. Patches of enlarged denticles on snout, between eyes, and in rows on back (in young).

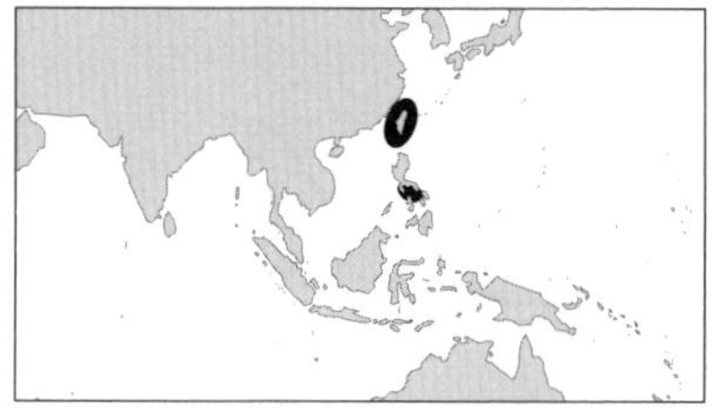

Distribution. Northwest Pacific: Taiwan Island, Philippines.
Habitat. Outer continental shelf and upper slope, 183–385 m.
Behaviour. Unknown.
Biology. Poorly known. Described from newborn and immatures.
Status. IUCN Red List: Not Evaluated.

Hidden Angelshark *Squatina guggenheim* Plate 19

Measurements. Born: ~33 cm TL. Mature: ~110 cm. Max: 130 cm TL.
Identification. Uniform dark tan with numerous small yellowish spots (no dark edges) and larger blackish marks. A few ocelli on pectoral fins. Nasal barbels with cylindrical bases, expanded slightly spatulate unfringed tips. Anterior nasal flaps barely fringed. Head concave between eyes, eye–spiracle space < 1.5 x eye length. Pectoral fins, quite large, high and angular; leading edge nearly straight, posterior edge concave. Short stout thorns in symmetrical groups on snout and between eyes, a pair between spiracles, no medial dorsal row.

Distribution. Southwest Atlantic: Brazil to Uruguay.

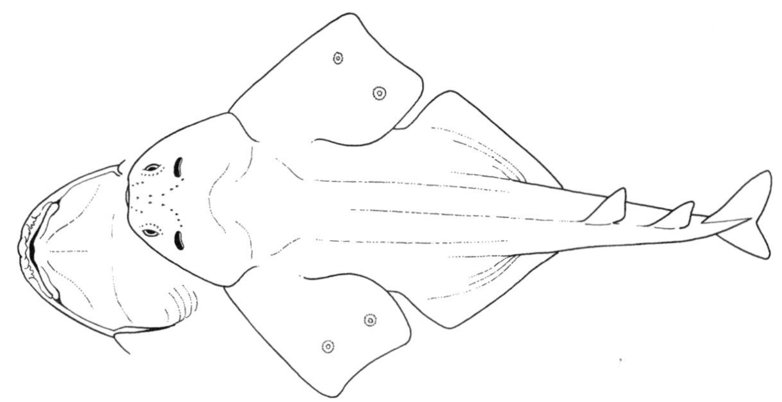

Habitat. Continental shelf, 35–115 m (mostly 35–93 m).
Behaviour. Unknown.
Biology. Six to eight pups/litter in spring after eleven month gestation. Eats fishes and shrimps.
Status. IUCN Red List: Endangered. Seriously depleted by fisheries. Formerly termed *Squatina occulta*, but correct name is *S. guggenheim*.

Japanese Angelshark *Squatina japonica* Plate 18

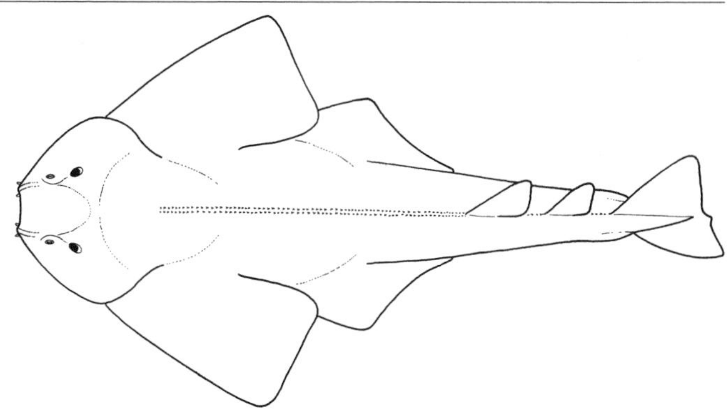

Measurements. Reaches up to 200 cm TL.
Identification. Rusty or blackish-brown, dense dark and small irregular white spots on dorsal surface. Large paired dark red-brown spots from base of head to pelvic fins. No ocelli on back. White below, with darker margins on fins and tail. Cylindrical nasal barbels with slightly expanded tips and weakly fringed or smooth nasal flaps, lateral head folds without triangular lobes. Concave between large eyes. Small thorns on snout, between eyes and spiracles and in a row along back. Denticles make surface rough. Fairly broad, high, rounded pectoral fins.
Distribution. Northwest Pacific.
Habitat. Poorly known.
Behaviour. Poorly known.
Biology. Poorly known.
Status. IUCN Red List: Not Evaluated. Taken in trawl fisheries.

Clouded Angelshark *Squatina nebulosa* Plate 19

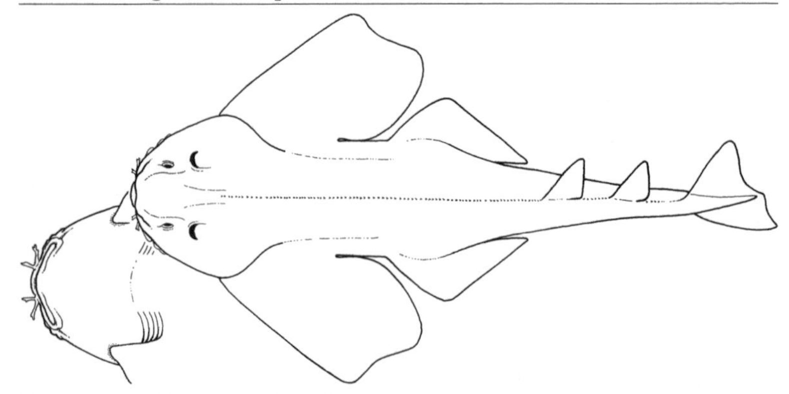

Measurements. Max: at least 163 cm TL.
Identification. Brownish to bluish-brown with scattered light spots and many small black dots. Large rounded dark spot at base of pectoral fin, dark blotches below dorsal fins. Ocelli absent or small, scattered and obscure (light rings, dark centres). Pale underside with darker pectoral fin margins. Light dorsal fin margins. Simple, tapering nasal barbels, weakly fringed or smooth anterior nasal flaps, lateral head folds with two triangular lobes each side. Large eyes, eye–spiracle space <1.5 x eye length. No enlarged thorns. Very broad, low, obtuse, posteriorly rounded pectoral fins.

Distribution. Northwest Pacific.

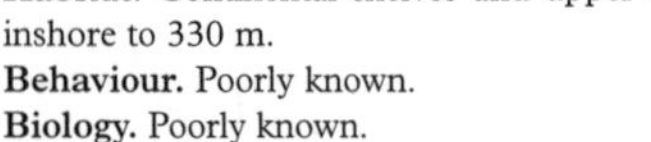

Habitat. Continental shelves and upper slopes, inshore to 330 m.
Behaviour. Poorly known.
Biology. Poorly known.
Status. IUCN Red List: Not Evaluated. Taken in demersal fisheries.

Smoothback Angelshark *Squatina oculata* Plate 19

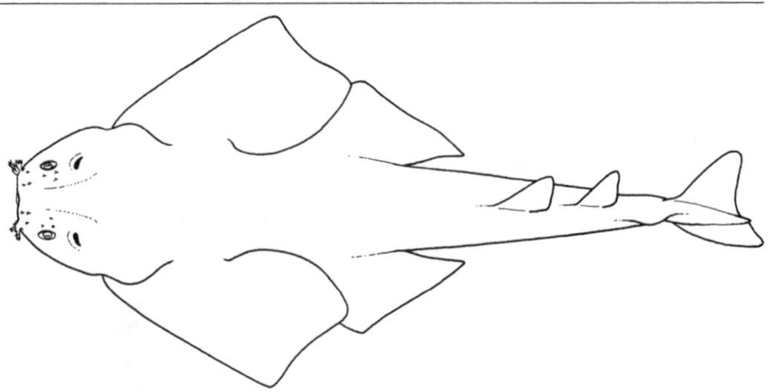

Measurements. Born: 24–27 cm TL. Max: 160 cm TL.
Identification. Grey-brown with small round white and blackish spots. White nuchal spot. Symmetrical large dark blotches or spots on base and rear tip of pectoral fins, tail base and under dorsal fins. Sometimes symmetrical white-edged dark ocelli. Dorsal and caudal fin margins white,

pectoral and pelvic fin margins dusky. Weakly bifurcated or lobed nasal barbels and weakly fringed anterior nasal flaps. Strongly concave between eyes, eye–spiracle space <1.5 x eye length. Large thorns on snout and above eyes.

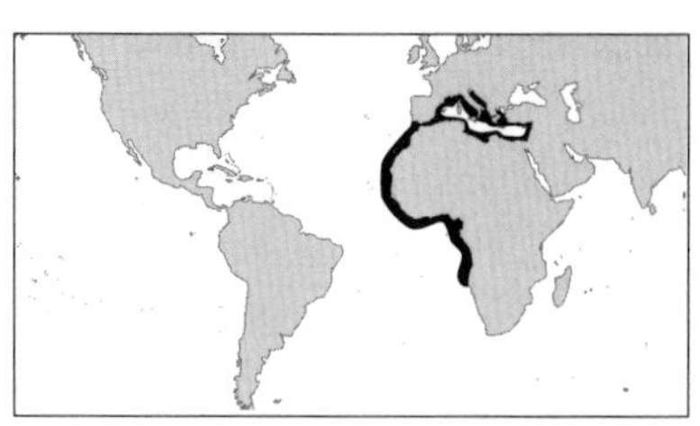

Distribution. East Atlantic and Mediterranean.
Habitat. Continental shelves and upper slopes, 20–500 m (mainly 50–100 m, deeper in tropics).
Behaviour. Unknown.
Biology. Eats small fishes, squid, octopus, shrimp and crabs.
Status. IUCN Red List: Not Evaluated. Fished off the African coast, bycatch elsewhere. Depleted in Mediterranean.

Angular Angelshark *Squatina punctata* Plate 18

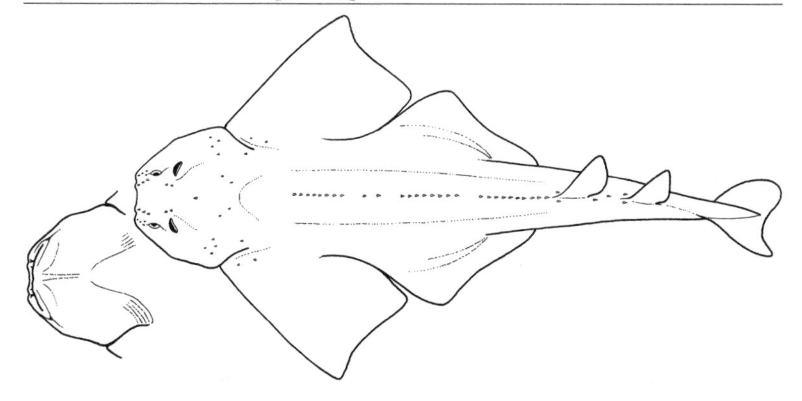

Measurements. Born: 25 cm TL. Mature: ~70–80 cm. Max: 91 cm TL.
Identification. Uniform dark tan above, pale below. Dorsal surface with small irregular dark spots present or absent in fresh specimens (turning white with fixation), regular pattern of several small to largish blackish spots, no ocelli. Nasal barbels with expanded, slightly spatulate, unfringed tips. Anterior nasal flaps weakly fringed. No triangular lobes on lateral head folds. Head broadly concave between eyes, eye–spiracle space < 1.5 x eye length. Pectoral fins relatively small, high, and angular, nearly straight anterior margin. Short stout thorns in symmetrical groups on snout, interorbital space, and a pair between spiracles; a median dorsal row of spines present.
Distribution. Southwest Atlantic: southern Brazil to Uruguay and Argentina.

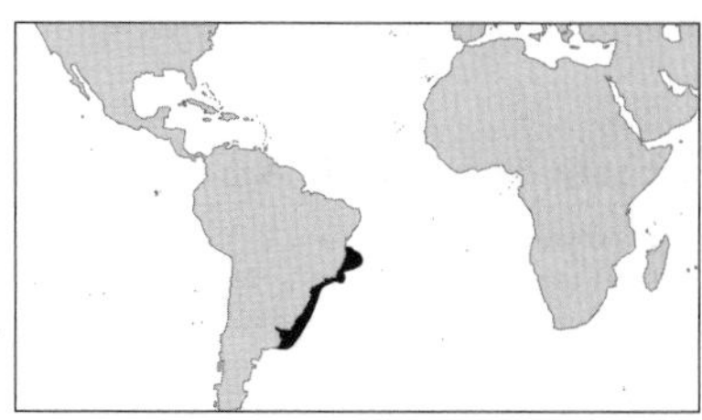

Habitat. Continental shelf, 10–80 m.
Behaviour. Females migrate in spring to shallow coastal waters to give birth, small juveniles remain inshore all year.
Biology. Three to eight (usually five or six) pups/litter develop in uteri for four months, then move into the greatly enlarged cloaca for remainder of eleven month gestation. Eats demersal fishes and shrimps.
Status. IUCN Red List: Vulnerable. Taken in target fisheries and bycatch, with *Squatina argentina* and *S. guggenheim*. Serious decline in southern Brazil. Formerly confused with *S. guggenheim*, but a separate species.

Angelshark *Squatina squatina* Plate 17

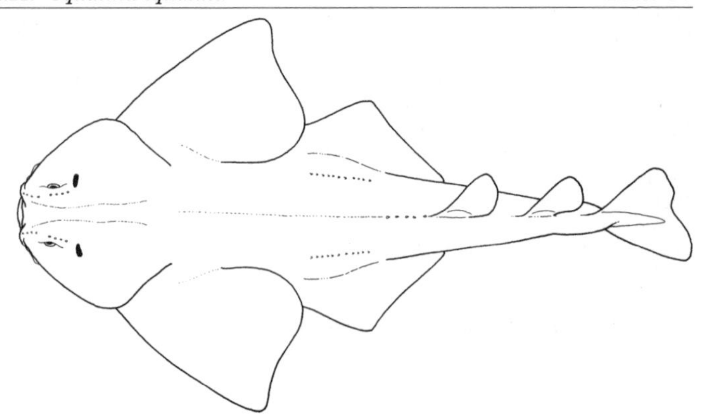

Measurements. Born: 24–30 cm TL. Mature: 126–167 cm ♀. Max: 183 cm ♂, possibly to 244 cm TL ♀.
Identification. Very large and stocky. Grey to reddish or greenish-brown back with scattered small white spots and blackish dots and spots. White nuchal spot may be present, no ocelli. Young often have white reticulations and large dark blotches, adults plainer. Dorsal fins with dark leading edge, pale trailing edge. Nasal barbels simple, with straight or spatulate tip, smooth or weakly fringed anterior nasal flaps, lateral head folds with a single triangular lobe each side. Small thorns on middle of back in young disappear with growth, but skin of back very rough. Patches of small thorns on snout and between eyes. Very high broad pectoral fins.
Distribution. Northeast Atlantic: historically from Norway to Mauritania, Canary Islands, Mediterranean and Black Sea, now vanished from some areas.

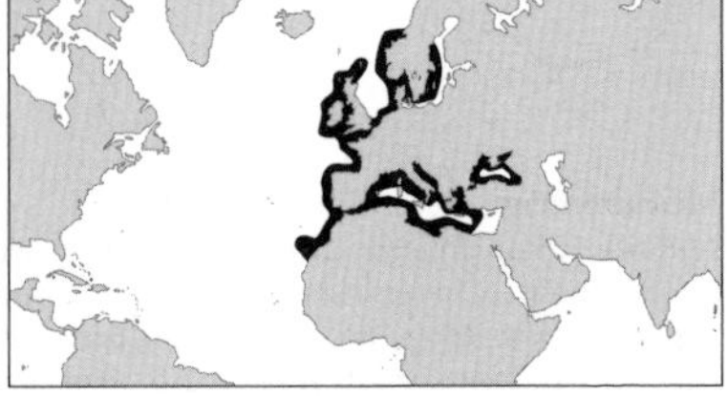

Habitat. Mud or sand, inshore (5 m) on coasts and estuaries to >150 m on continental shelf.
Behaviour. Torpid by day, lying buried with eyes protruding. Swims strongly off the bottom at night. Seasonally migratory in colder water, moving northwards in summer.
Biology. 7–25 pups/litter, increasing with female size. Gestation eight to ten months, born December–February in Mediterranean, July in England. Feeds mainly on flatfishes, skates, crustaceans and molluscs. One record of a cormorant swallowed.
Status. IUCN Red List: Vulnerable. Range and abundance declining severely throughout its range. Proposed for legal protection in Britain.

Ornate Angelshark *Squatina tergocellata* Plate 19

Measurements. Born: 33–42 cm TL. Mature: ~81–91 cm ♂, 115–125 cm ♀. Max: 140 cm TL.
Identification. Pale yellow-brown with grey-blue or white spots, no dorsal thorns. Three pairs of large ocelli, dark rings around centres with a mitotic pattern. Fins pale with spots or blotches. Strongly fringed nasal barbels and anterior nasal flaps. Strongly concave between large eyes, eye–spiracle space <1.5 x eye length.
Distribution. Southwestern Australia.

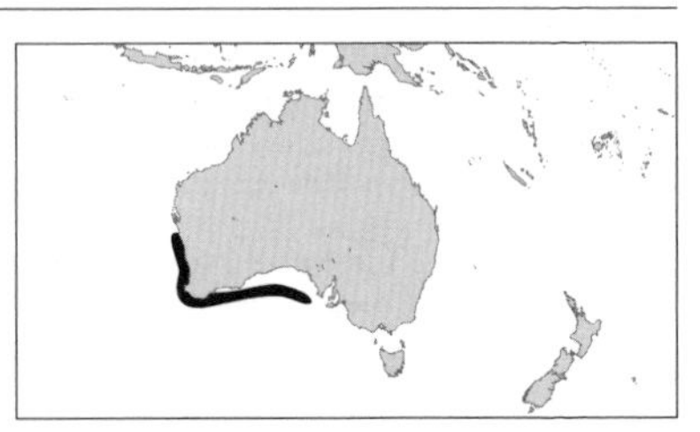

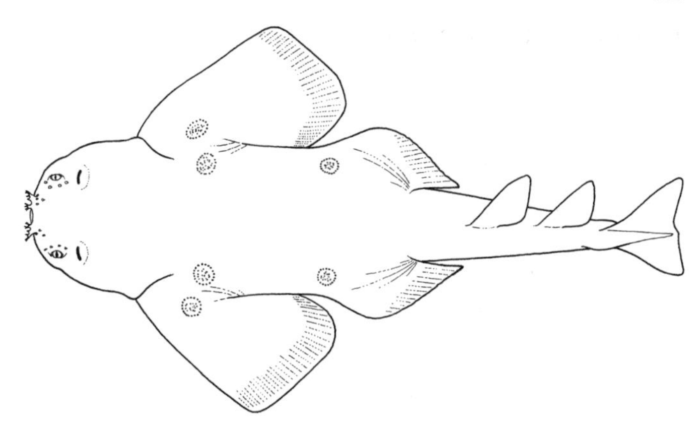

Habitat. Continental shelf and upper slope, on or near bottom, 130–400 m (most ~300 m, young shallower).
Behaviour. Partial sexual segregation. May swallow mud to buffer prey toxins.
Biology. Two to nine pups/litter, probably every two years, 6–12 month gestation. Feeds on fish and squid.
Status. IUCN Red List: Least Concern. Stable bycatch in small fishery, much of range untrawled.

Ocellated Angelshark *Squatina tergocellatoides* Plate 19

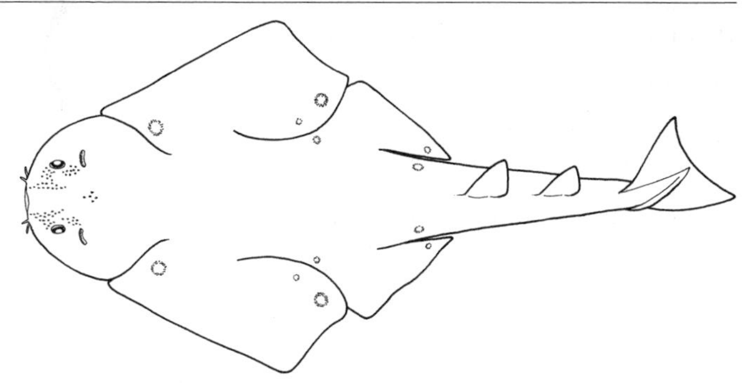

Measurements. One female described, 63 cm TL.
Identification. Strongly but finely fringed nasal barbels. Strongly fringed anterior nasal flaps. Lateral head folds have two low rounded lobes each side. Concave space between eyes. No enlarged predorsal thorns on back. Dorsal colour light yellowish-brown with a dense scattering of small round white spots and six pairs of large ocelli (dark rings around light centres) on pectoral and pelvic fins and on tail base. Dorsal fins with black base and leading edge.
Distribution. Northwest Pacific: Taiwan Straits.
Habitat. Unknown.
Behaviour. Unknown.
Biology. Unknown.
Status. IUCN Red List: Not Evaluated.

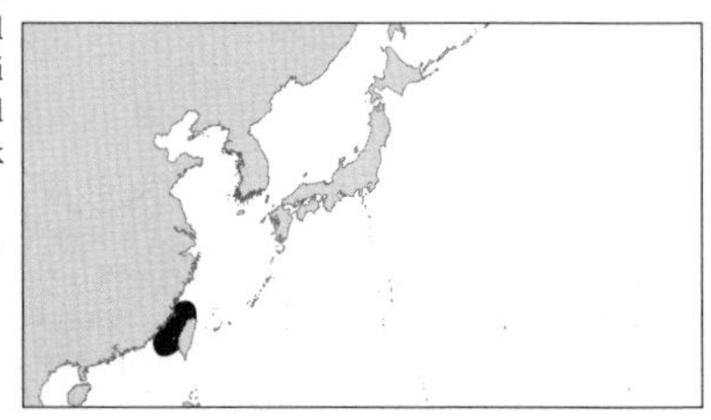

Eastern Angelshark *Squatina* sp. A Plate 19

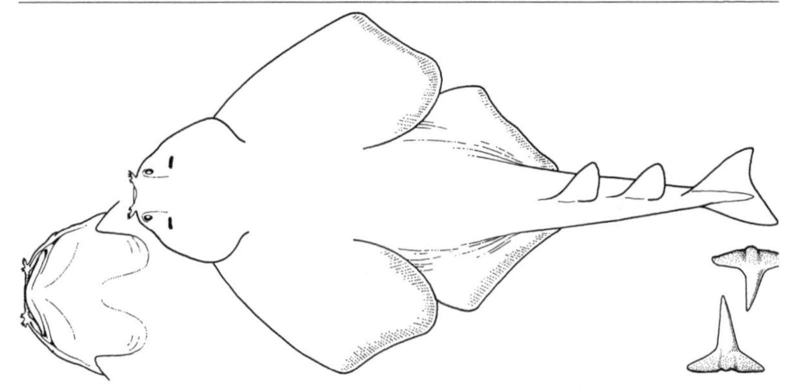

Measurements. Born: 30 cm TL. Mature: ~91 cm ♂, 107 cm ♀. Max: 110 cm ♂, 130 cm TL ♀.
Identification. Yellow-brown to chocolate-brown, dense pattern of symmetrical small white, dark-edged spots, many large brownish blotches, white nuchal spot (no ocelli). Light unspotted unpaired fins. Very short snout, concave interorbital space and heavy orbital thorns distinguish this from *Squatina australis*. Nasal barbels with expanded tips and lobate fringes. Low lateral head folds. Spiracles close to eyes, wider than eye-length. Strong orbital thorns, no medial row of predorsal thorns.

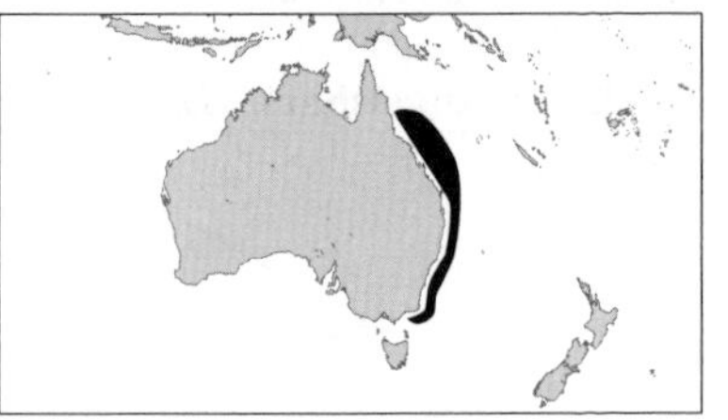

Distribution. Eastern Australia. Extends further north than *S. australis*.
Habitat. Outer continental shelf and upper slope (130–315 m), occasionally up to 60 m.
Behaviour. Unknown.
Biology. Up to 20 pups/litter.
Status. IUCN Red List: Vulnerable. Depleted in heavily-fished south, unfished in north.

Western Angelshark *Squatina* sp. B Plate 19

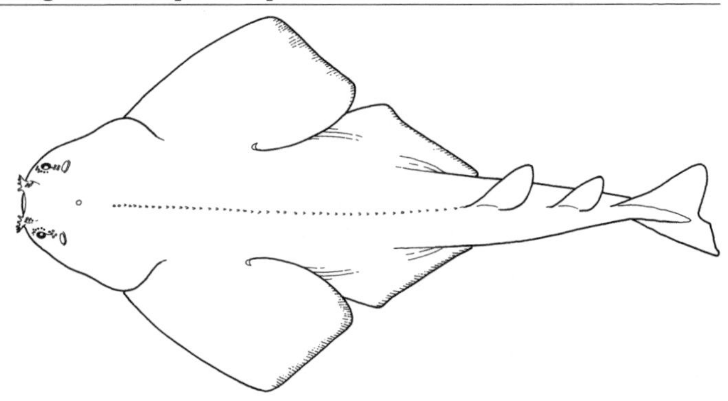

Measurements. Reaches at least 64 cm TL.
Identification. Differs from *Squatina tergocellata* in colour pattern and dorsal thorns. Concave interorbital space, very short snout, nasal barbels with expanded tips and lobate fringes, strong orbital thorns, a medial row of predorsal thorns. Dorsal surface medium to pale brownish or greyish

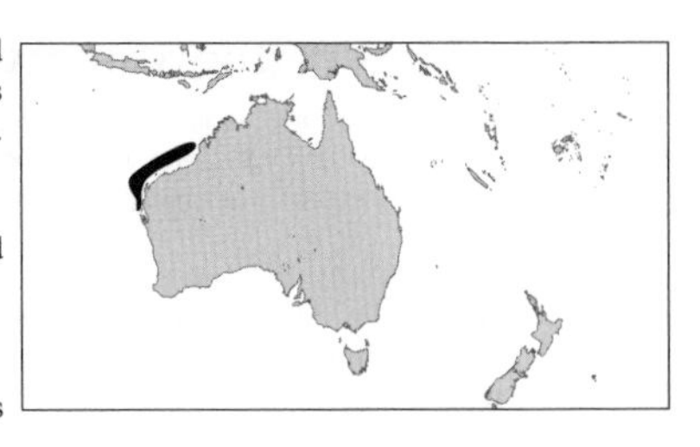

with a pattern of widely spaced blue spots and brown blotches, no symmetrical small white spots or ocelli, but a single small white nuchal spot. Light unpaired fins without dark spots.
Distribution. Western Australia.
Habitat. Tropical outer continental shelf and uppermost slope, 130–310 m.
Behaviour. Unknown.
Biology. Unknown.
Status. IUCN Red List: Data Deficient. Taken as bycatch, but very low fishing effort in range.

Heterodontiformes: Bullhead Sharks

An ancient order with a long fossil record (almost to the beginning of the Mezozoic Era), now represented by one living family, but many fossil species. The taxonomic name *Heterodontus*, 'different-teeth', refers to the front small pointed holding teeth and rear large blunt teeth for crushing invertebrates.

Identification. Small (165 cm TL max., mostly <100 cm), stout-bodied, with two dorsal fins (both spined) and an anal fin. All have blunt, 'pig-like' snouts, a small mouth, enlarged first gill slits, prominent eye ridges, rough skin and paddle-like paired fins.

Biology. Sluggish, night-active, benthic sharks. Swim slowly or crawl over rocky, kelp-covered and sandy bottom. Some rest by day in rocky crevices and caves. Oviparous (egg laying), with unique large screw-shaped eggcases that often become lodged firmly in crevices on the bottom. Large young (>14 cm) hatch over five months later. At least two species lay eggs in particular 'nesting' sites. At least one species is migratory when adult, returning each year after long migrations to its breeding sites; most have a restricted distribution. Mainly feed on benthic invertebrates (sea urchins, crabs, shrimp, marine gastropods, oysters and worms), rarely on small fish.

Status. Rare to uncommon, not important in commercial fisheries but often taken as bycatch. Also caught by sports fishers and divers. Some do well in aquaria and can be bred successfully under good conditions.

Heterodontidae: Bullhead or Horn Sharks

One genus with nine species occurring in Indian and Pacific Oceans, mainly in shallow waters on warm-temperate to tropical continental shelves, with a generally restricted distribution.

Horn Shark *Heterodontus francisci* Plate 20

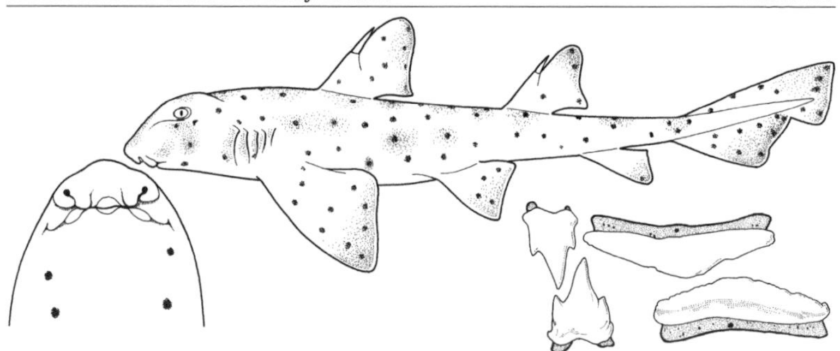

Measurements. Born: 15–16 cm TL. Eggcase: 10–12 x 3–4 cm. Mature: 58–59 cm TL ♂, ♀ larger. Max: 122 cm TL.

Identification. Dark to light grey or brown, usually with small (<one-third eye diameter) dark spots. Small dark spots on dusky patch below eye. No light bar between eye ridges. No harness pattern. First dorsal fin origin over pectoral fin bases, no white fin tips. Young more brightly coloured, with obvious dark saddles.

Distribution. East Pacific: USA (California, mainly south), Mexico (Baja California, Gulf of California), probably Ecuador and Peru.

Habitat. Intertidal to at least 150 m (mainly 2–11 m). Rock (often in deep crevices and caves), kelp, sandy gullies and flats. Juveniles often shelter on sand.

Behaviour. Nocturnal. Seldom moves from preferred resting place by day. Poor, sluggish swimmer, small home range, possibly a limited winter migration into deep water in the north. Uses mobile muscular paired fins to crawl over seabed. Mainly solitary (small aggregations reported).

Biology. Mates December–January. Eggs deposited February–April under rocks or in crevices, hatch in seven to nine months. In captivity, two eggs laid every 11–14 days for four months. Young start feeding one month after hatching. Feed on benthic invertebrates and, rarely, small fishes.

Status. IUCN Red List: Least Concern. Minimal fisheries interest. Caught by divers for sport (fin spines used for jewellery), may bite when provoked. Hardy and easy to maintain and breed in captivity.

Crested Bullhead Shark *Heterodontus galeatus* Plate 20

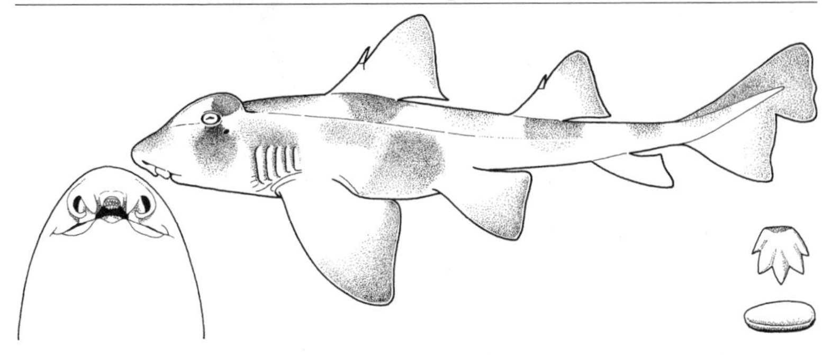

Measurements. Born: 17–22 cm TL. Mature: 55–60 cm ♂, 70 cm ♀. Max: 152 cm TL.

Identification. Extremely high short eye-ridges, depth between ridges about equal to eye length. Light brown or yellowish-brown with dark broad bands or saddles, no light or dark spots. Dark bar between eyes, broad dark blotch under eye.

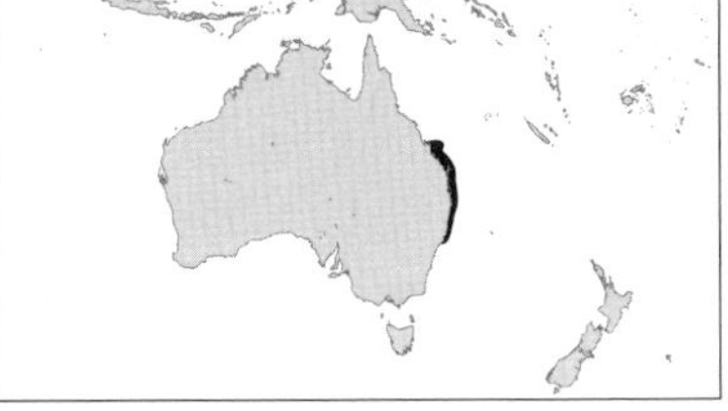

Distribution. Eastern Australia (southern Queensland, New South Wales).

Habitat. Intertidal to 93 m (more common deeper), rocky reefs, seagrass beds, and in seaweed.

Behaviour. Forces its way between rocks to find prey.

Biology. Lays 10–16 eggs/year in seaweeds or sponges during winter, hatch five to eight months later. Prey mainly sea urchins, also crustaceans, molluscs and small fishes.

Status. IUCN Red List: Least Concern. Relatively uncommon. Occurs only rarely in bycatch, usually released alive. Bred in captivity.

Japanese Bullhead Shark *Heterodontus japonicus* Plate 20

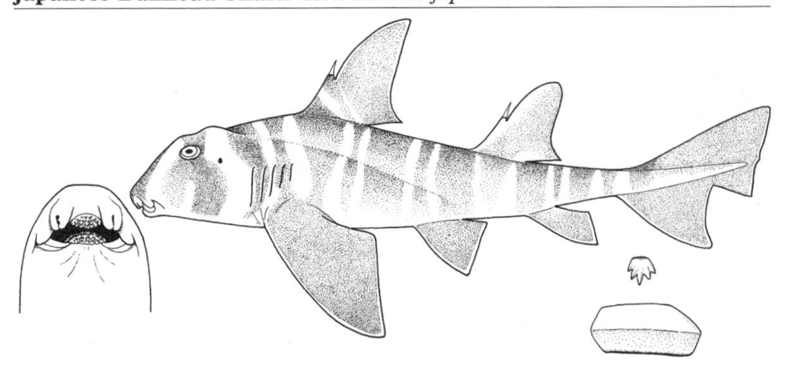

Measurements. Born: 18 cm TL. Mature: ~69 cm ♂. Max: 120 cm TL.

Identification. Tan to brown, about 12 irregular dark saddles and bands. No spots. Light bar between eyes, dark blotch under eye (indistinct in large adults). Hatchlings brighter coloured. First dorsal fin origin over pectoral fin bases.

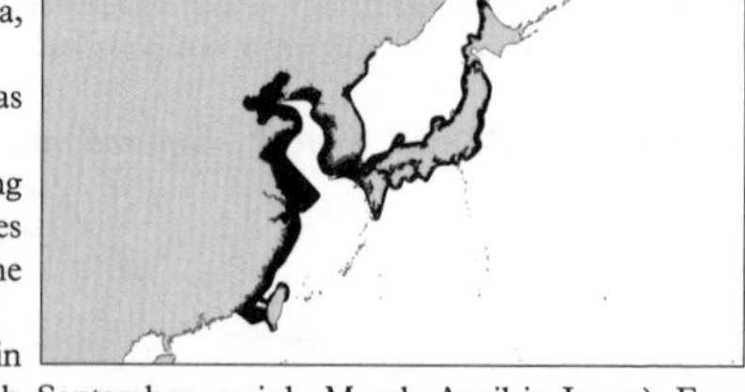

Distribution. Northwest Pacific: Japan, Korea, North China, Taiwan Island.

Habitat. Bottom, 6–37 m. Prefers rocky areas and kelp.

Behaviour. Moves by slow sluggish swimming and 'walking' with mobile paired fins. Protrudes jaws to capture prey. Several females share one 'nest', but do not guard eggs.

Biology. Lays pairs of eggs among rocks or in kelp, at 8–9 m, during 6–12 spawnings. (March–September, mainly March–April in Japan). Eggs hatch in about a year. Feeds on invertebrates and small fishes.

Status. IUCN Red List: Not Evaluated. Low fisheries importance. Kept in aquaria.

Mexican Hornshark *Heterodontus mexicanus* Plate 21

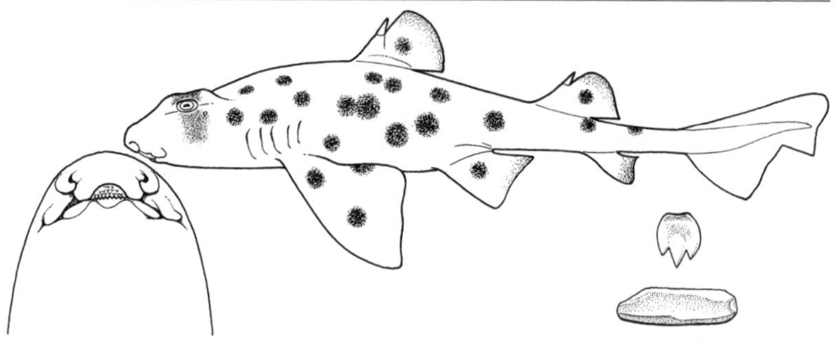

Measurements. Born: 14 cm TL. Eggcase: 8–9 cm long. Mature: 40–50 cm ♂. Max: 55 cm ♂, 70 cm TL ♀.

Identification. Light grey-brown to dark grey, large black spots (>half eye diameter). Light bar between eye ridges, one or two indistinct blotches under eye. First dorsal fin origin over pectoral fin bases. Eggs with long tendrils and rigid 'T'-shaped spiral flanges.

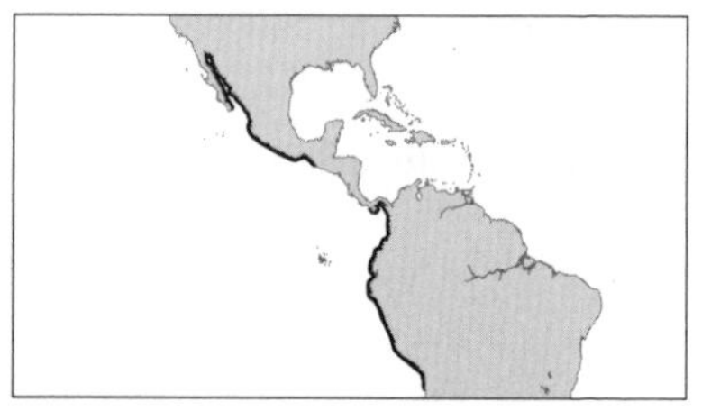

Distribution. East Pacific: Mexico to Colombia, probably Ecuador, and Peru.

Habitat. Rock, coral reefs, seamounts and sand, from close inshore to 20–50 m.

Behaviour. Tendrils may anchor eggcases.

Biology. Feeds on crabs and demersal fishes.

Status. IUCN Red List: Not Evaluated. Minimal interest to fisheries.

Port Jackson Shark *Heterodontus portusjacksoni* Plate 20

Measurements. Born: 23–24 cm TL. Eggcase: 13–17 long x 5–7 cm (broad end). Mature: 70–80 ♂, 80–95 cm ♀. Max: 165 (rare >237) cm TL.

Identification. Grey to light brown or whitish with unique distinctive black striped 'harness' marking. No spots. Dark band between and under eyes.

Distribution. Southern Australia. One (vagrant?) New Zealand record.

Habitat. Temperate waters, intertidal to at least 275 m.

Behaviour. Rest by day in groups on sand in a few favoured caves and gullies (these can be

relocated by sharks moved several km away). Traditional collective egg-laying sites used. Hatchlings move to nursery grounds nearby until adolescent, when they move well offshore and segregate by sex, joining the adult population a few years later. Adults segregate by sex, undertaking complex seasonal breeding migrations. Some males and all females move to inshore reefs in July to breed, returning offshore after mating (males) and egg laying (females). Some adults remain offshore in summer, others migrate south up to 850 km from breeding areas.

Biology. Mating and egg laying occurs in winter. 10–16 eggs laid in pairs every 8–17 days in rock crevices (pointed edge wedged down) on sheltered rocky reefs in 1–5 m (occasionally to 20–30 m) mainly in August–September. Eggs hatch 12 months later. Feed mainly on sea urchins, other benthic invertebrates, also small fish.

Status. IUCN Red List: Least Concern. Abundant. Taken as bycatch but mostly returned alive. Kept and bred in captivity.

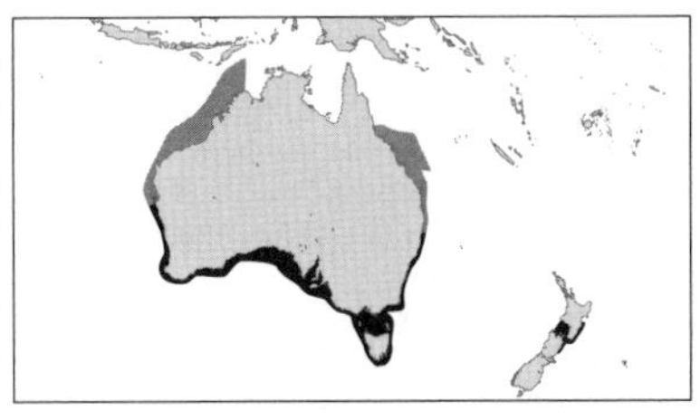

Galapagos Bullhead Shark *Heterodontus quoyi* Plate 21

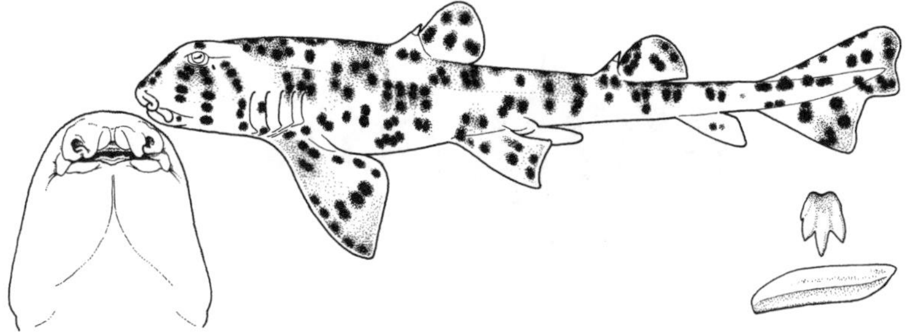

Measurements. Born: 17 cm TL. Eggcase: 11 cm. Mature: 48 cm ♂. Max: 61 cm TL.

Identification. Light grey or brown with large black spots (>half eye diameter), smaller and less distinct in young. Mottled dark spots or blotches under eyes; no light bar between eyes. First dorsal fin origin over pectoral fin inner margins.

Distribution. East Pacific: Peru (coast and islands), Galapagos Islands.

Habitat. Rocky and coral reefs, rests on ledges of vertical rock surfaces 16–30 m.

Behaviour. Nocturnal, very poorly known.

Biology. Feeds on crabs.

Status. IUCN Red List: Not Evaluated.

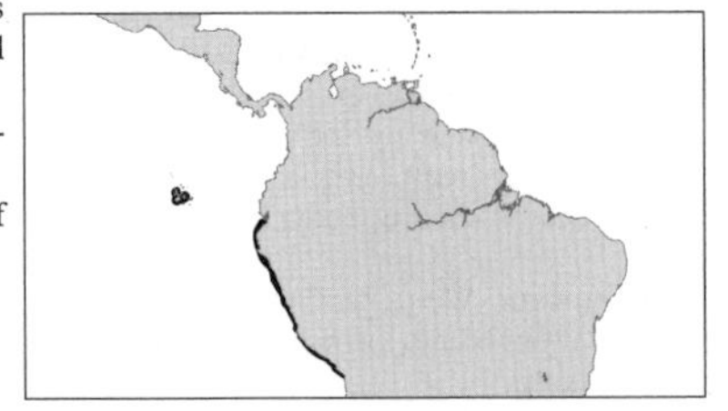

Whitespotted Bullhead Shark *Heterodontus ramalheira* Plate 21

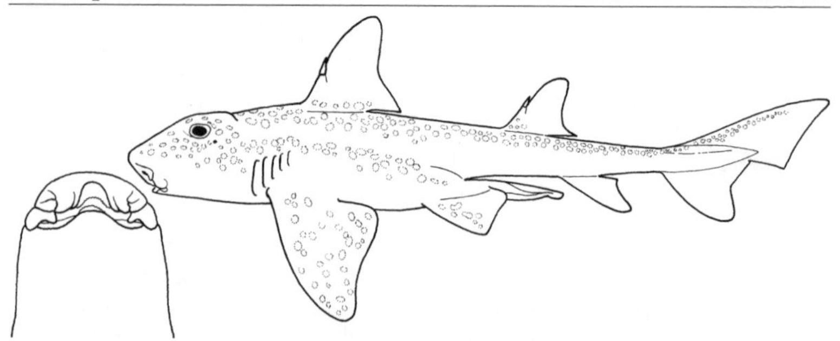

Measurements. Born: 18 cm TL. Mature: ~60 cm ♂, <75 cm ♀. Max: >83 cm TL.
Identification. Dark reddish-brown with white spots and darker indistinct saddles in adults. Hatchlings with striking unique whorled pattern of thin curved parallel dark lines, lost with growth. Parallel dark lines between and under eyes in hatchlings change to a dusky patch in larger juveniles, lost in adults.

Distribution. North and West Indian Ocean: South Africa (KwaZulu-Natal), Mozambique, Somalia, eastern Arabian Peninsula, southern Oman.
Habitat. Outer continental shelf and uppermost slope in deepish water (40–275 m, mostly >100 m).
Behaviour. Unknown.
Biology. Feeds on crabs.
Status. IUCN Red List: Not Evaluated. Apparently uncommon. Taken as bycatch.

Zebra Bullhead Shark *Heterodontus zebra* Plate 21

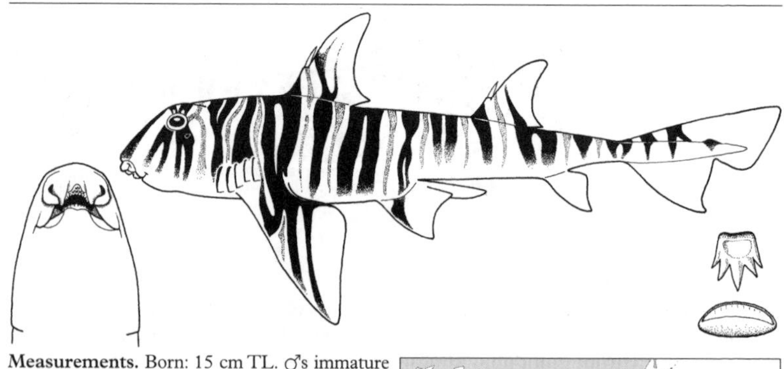

Measurements. Born: 15 cm TL. ♂s immature at 44 cm, mature at 64 cm. Max: 122 cm TL.
Identification. White or cream with striking zebra-striped colour pattern of numerous black or dark brown (red-brown in luveniles) vertical saddles and bands over body, tail and head. No spots.
Distribution. West Pacific: Japan, Koreas, China, Taiwan Island, Vietnam, Indonesia, northwest Australia.

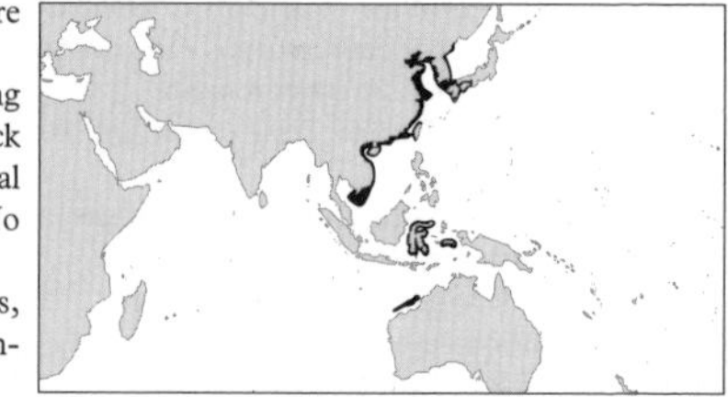

Habitat. Continental and insular shelves of the western Pacific from inshore down to at least 50 m in the South China Sea, deeper (150–200 m) off western Australia.
Behaviour. Unknown.
Biology. Poorly known.
Status. IUCN Red List: Least Concern. Apparently common. Taken in bycatch.

Oman Bullhead Shark *Heterodontus* sp. A — Plate 21

Measurements. Mature: 52 cm TL ♂, 61 cm TL ♀.
Identification. Tan to brown with four or five broad dark brown saddles, dark bar between eyes and blotch under eyes. No spots on body. Dark-tipped fins with a white spot on dorsal fin tips. Hatchling colour pattern unknown. First dorsal fin origin over pectoral fin inner margins. Distinguishable from *Heterodontus japonicus* by colour pattern and very low dorsal fins (possibly smaller size).

Distribution. Northern Indian Ocean: Oman.
Habitat. 80 m, presumably on soft bottom because obtained from trawl bycatch.
Behaviour. Unknown.
Biology. Unknown.
Status. IUCN Red List: Not Evaluated.

Orectolobiformes: Carpetsharks

This order contains about 33 species in seven families: *Parascyllidae* (collared carpetsharks), *Brachaeluridae* (blind sharks), *Orectolobidae* (wobbegongs), *Hemiscyllidae* (longtailed carpetsharks), *Ginglymostomatidae* (nurse sharks), *Stegostomatidae* (Zebra Shark) and *Rhincodontidae* (Whale Shark).

Identification. Two spineless dorsal fins and an anal fin. Nostrils have barbels (rudimentary in whale shark) and are connected via nasoral grooves to short mouths that end in front of the eyes (no nictitating eyelids).

Distribution. World-wide, warm-temperate and tropical seas, from intertidal to deep water, with the greatest diversity and endemism in the tropical Indo-west Pacific. Smallest are mainly sluggish bottom-living species. Larger species tend to be more active and wide-ranging (e.g. the circumglobal pelagic Whale Shark).

Biology. Varied reproductive strategies, including oviparity (egg-laying), ovoviviparity or aplacental vivipary (foetuses retained inside the female and nourished by their yolk sac until birth), and oophagy (egg-eating).

Parascylliidae: Collared Carpetsharks

Two genera in west Pacific, from inshore to deepish continental shelf. All four *Parascyllium* species are Australian endemics. Genus *Cirrhoscyllium* (three species) occurs from Vietnam to Taiwan Island and Japan.

Identification. Small slender sharks (<1 m). First dorsal fin originates behind pelvic fin bases, second well behind anal fin origin. Mouth entirely in front of eyes; tiny spiracles. *Cirrhoscyllium* have unique cartilage-cored paired barbels on throat, dark saddles and no spots or collar markings. *Parascyllium* have no barbels on throat and a pattern of saddles and spots.

Biology. Little-known benthic species can apparently change colour to match the seabed. Some (possibly all) are oviparous, laying elongated flattened eggcases.

Status. Some are hardy in captivity. May be taken in bycatch and may be threatened in heavily-fished areas.

Barbelthroat Carpetshark *Cirrhoscyllium expolitum* Plate 22

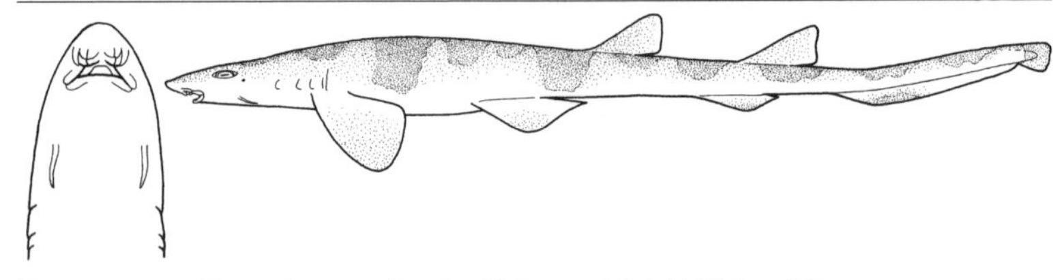

Measurements. Known from two females, 30.6 cm and (adult) 33.5 cm TL.

Identification. Six to ten diffuse saddle marks on back, not C-shaped. An elongated rounded saddle on each side of back and tail between pectoral and pelvic fins bases extends over pelvic fin bases. Cartilage-cored paired barbels on throat, nasoral grooves, mouth in front of eyes. Head length three times first dorsal-fin base.

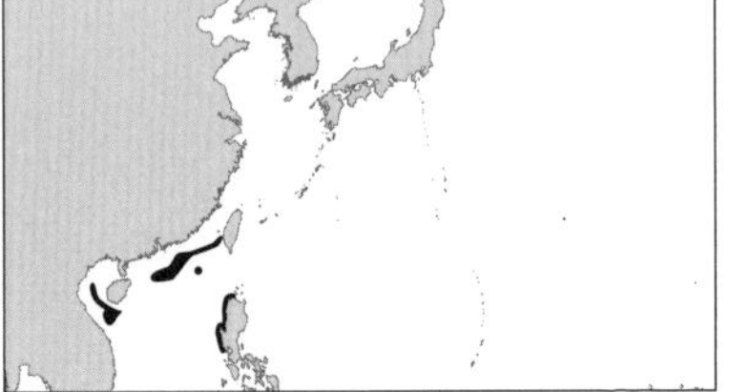

Distribution. South China Sea, northwest Pacific. China to Luzon, Philippines, Vietnam (Gulf of Tonkin).

Habitat. Bottom, outer continental shelf, 183–190 m.

Behaviour. Unknown.

Biology. Presumed oviparous.

Status. IUCN Red List: Not Evaluated. Presumably rare or uncommon, taken as bycatch.

Taiwan Saddled Carpetshark *Cirrhoscyllium formosanum* Plate 22

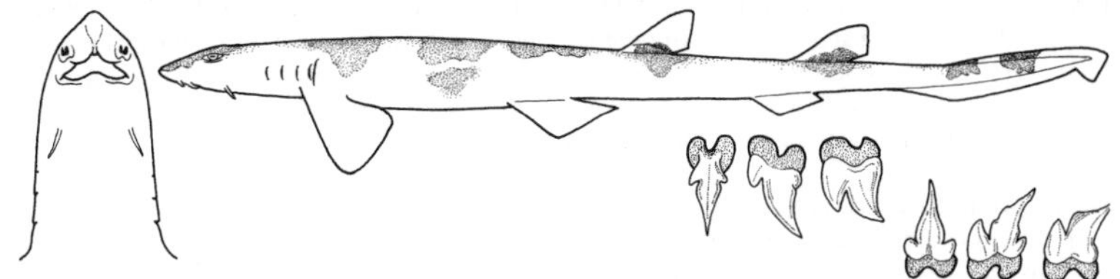

Measurements. Mature: 35 cm TL ♂ TL. Max: known 39 cm TL.
Identification. Six diffuse saddle marks on back, not C-shaped. An elongated rounded saddle on each side of back between pectoral and pelvic fins bases extends over pelvic fin bases. Cartilage-cored paired barbels on throat, nasoral grooves, mouth in front of eyes.Head length 2.3–2.6 times first dorsal fin base.

Distribution. Northwest Pacific: Taiwan Island.
Habitat. Outer shelf at about 110 m.
Behaviour. Unknown (only 12 specimens collected on bottom longlines).
Biology. Unknown.
Status. IUCN Red List: Not Evaluated. Possibly trawl bycatch.

Saddled Carpetshark *Cirrhoscyllium japonicum* Plate 22

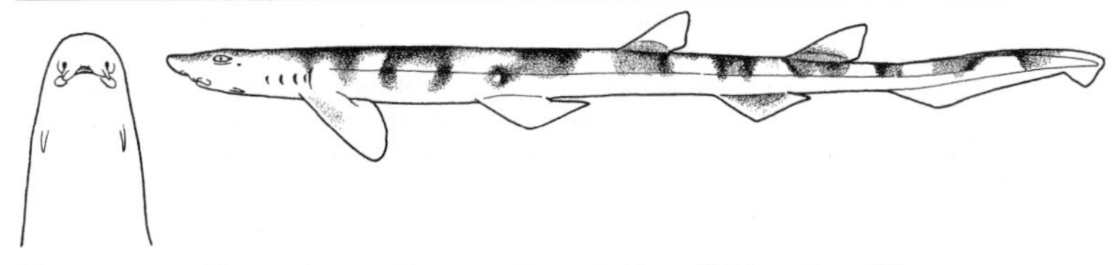

Measurements. Born: unknown. Mature: ~37 cm ♂, 44 cm ♀. Max: 49 cm TL.
Identification. Nine boldly-marked saddle marks on sides of body, one of them C-shaped between pectoral and pelvic-fin bases. Cartilage-cored paired barbels on throat, nasoral grooves, mouth in front of eyes, long snout.
Distribution. Northwest Pacific: southwest Japan (Shikoku and Kyushu to Yakushima Island, possibly Ryukyu Islands).
Habitat. Uppermost slope at 250–290 m.
Behaviour. Unknown.
Biology. Poorly known. Apparently oviparous (cased eggs discovered in a 45 cm female)
Status. IUCN Red List: Not Evaluated.

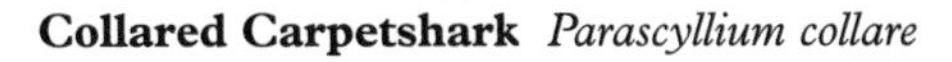

Collared Carpetshark *Parascyllium collare* Plate 22

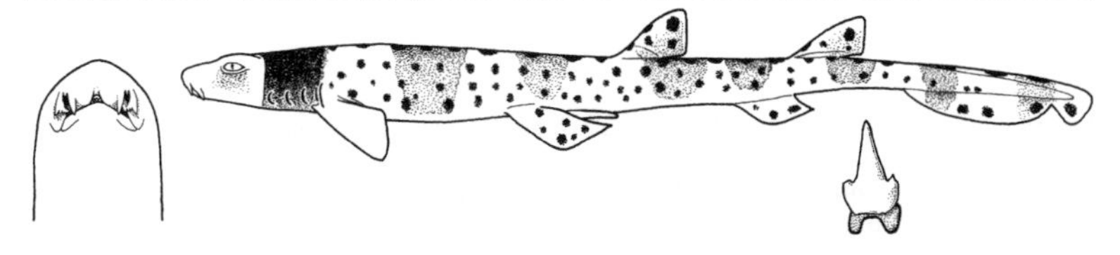

Measurements. Born: unknown. Mature: 80–85 cm ♂, 85–87 cm TL ♀.
Identification. Light yellowish to reddish-brown. Dark, unspotted, sharp-edged collar over gills, five dusky saddles on back. Mouth in front of eyes, nasal barbels, nasoral and circumnarial grooves. First dorsal fin origin behind pelvic fin bases, anal fin origin well in front of second dorsal origin. Large dark spots on body, tail and fins (except pectorals), <six spots on sides of tail between dorsal fins.

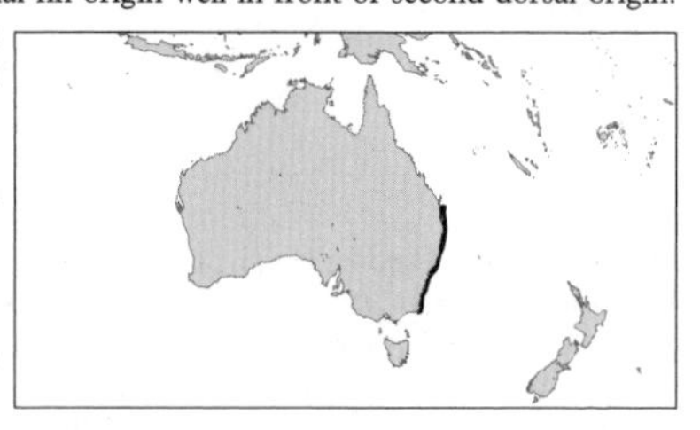

Distribution. East Australia.
Habitat. Rocky reefs and hard-bottomed trawl grounds on continental shelf, 20–160 m.
Behaviour. Unknown.
Biology. Lays flattened elongate eggcases.
Status. IUCN Red List: Least Concern. Discarded bycatch of trawlers, high survival rates.

Rusty Carpetshark *Parascyllium ferrugineum* Plate 22

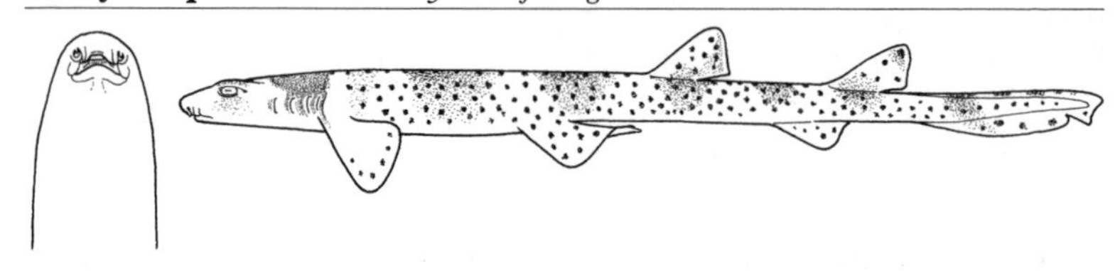

Measurements. Born: ~17 cm TL. Mature: ~60 cm ♂, 75 cm ♀. Max: 80 cm TL.
Identification. Grey-brown, indistinct dark collar around gills. Six or seven dusky saddles, dark spots on body, tail and fins (>six on sides of tail between dorsal fins; Tasmanian specimens more heavily spotted). Mouth in front of eyes, nasal barbels, nasoral and circumnarial grooves. First dorsal fin origin behind pelvic fin bases, anal fin origin in front of second dorsal origin.
Distribution. Southern Australia.

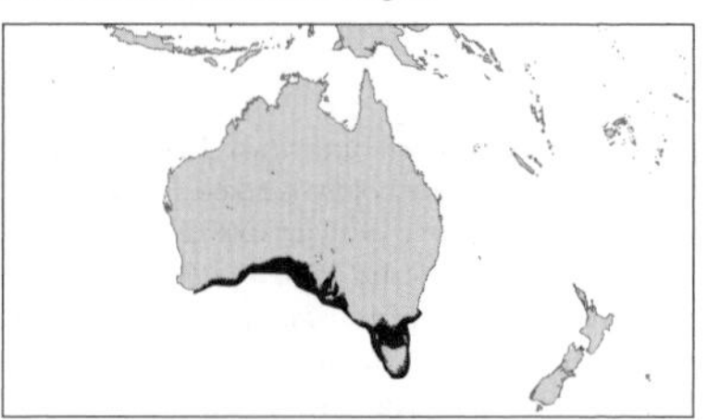

Habitat. On or near bottom near rocks, river mouths, in algae on reefs or in seagrass, 5–150 m.
Behaviour. Nocturnal. Hides in rocky caves and ledges by day.
Biology. Yellow eggcases with long tendrils laid in summer. Feeds on bottom-dwelling crustaceans and molluscs.
Status. IUCN Red List: Least Concern. Not targeted by fisheries, rare in bycatch.

Necklace Carpetshark *Parascyllium variolatum* Plate 22

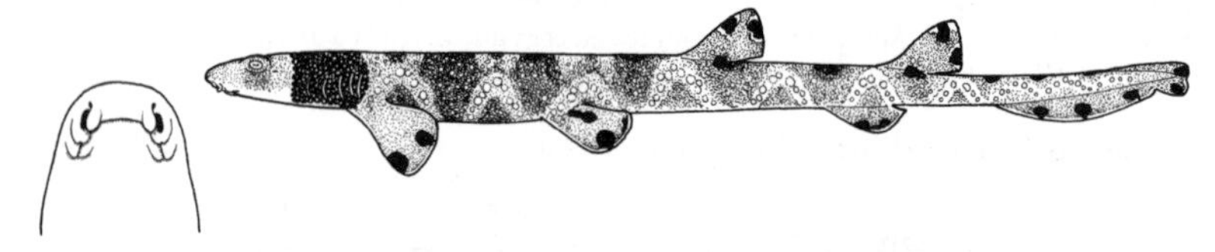

Measurements. Max: ~90 cm TL.
Identification. Dark greyish to chocolate-brown small shark with unmistakable, highly variable beautiful pattern: broad, dark, white-spotted collar over gills, obvious black spots on all fins, dark blotches and dense white spots on body. (Another white-spotted form from western Australia may be an undescribed species.)

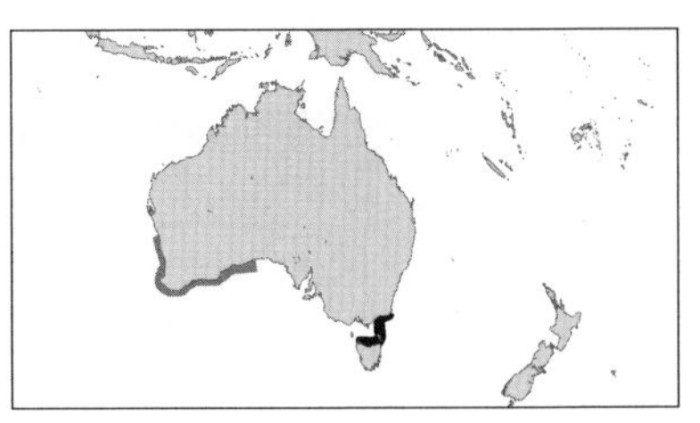

Distribution. Southern Australia. (Eastern and western forms may be more than one species.)
Habitat. A variety of habitats on continental shelf to 180 m, including sand, rocky reefs, kelp and seagrass beds.
Behaviour. Nocturnal. Juveniles hide under rocks and bottom debris in shallow water.
Biology. Virtually unknown, probably oviparous.
Status. IUCN Red List: Least Concern. Not targeted by fisheries, rare in bycatch.

Ginger Carpetshark *Parascyllium sparsimaculatum* Plate 22

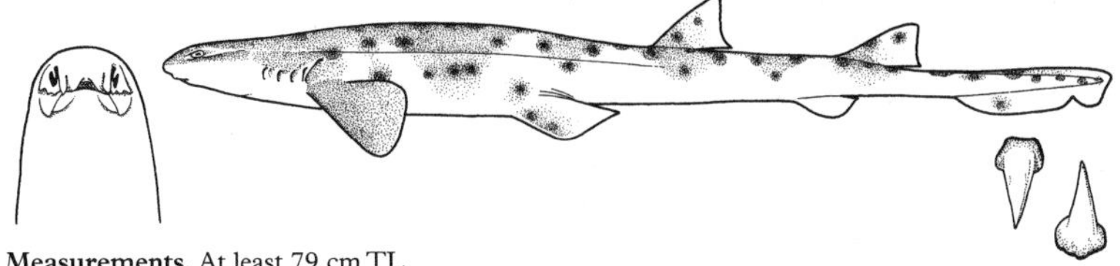

Measurements. At least 79 cm TL.
Identification. Pale brownish or greyish above, lighter below. Inconspicuous unspotted dusky half-collar around gills. Five indistinct dark saddles on back and tail. Sparse large dark spots and blotches on body and fins (<six on the sides of the tail between dorsal fins). First dorsal fin origin behind pelvic fins, anal fin origin well in front of second dorsal origin.

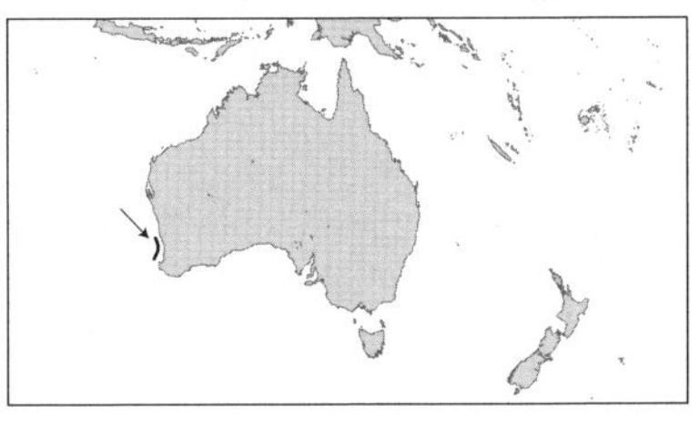

Distribution. Western Australia. Known from only a very small area.
Habitat. Deepwater, upper continental slope, 245–435 m
Behaviour. Unknown.
Biology. Unknown.
Status. IUCN Red List: Data Deficient (only three specimens known).

Brachaeluridae: Blind Sharks

Named because they close eyelids out of water. Endemic to eastern Australian coast, from a few cm deep to 137 m, on rocky or coral reefs or seaweed. Two mono-specific genera, *Brachaelurus* and *Heteroscyllium.*
Identification. Small stout sharks (most <80 cm) with two spineless dorsal fins set far back. Anal fin origin well behind origin of second dorsal fin, gap between anal and lower caudal fins shorter than length of anal fin. Large spiracles below and behind eyes, nasoral and circumnarial grooves, long barbels, small transverse mouth in front of eye, no lateral skin flaps on head.
Biology. Ovoviviparous (litters of six to eight finish absorbing their large yolk-sacks just before birth). Feed on small fishes, crustaceans, squid and sea anemones. At least one species can survive long periods out of water.
Status. Very restricted geographic range, one species very rare. Survive well in captivity.

Blind Shark *Brachaelurus waddi* Plate 22

Measurements. Born: 17 cm TL. Mature: <60 cm ♂, <66 cm ♀. Max: 120 cm TL.
Identification. Usually brown above with white spots, dark saddles in young, obsolete in adults, light yellowish underneath. Small, stout shark. Similar-sized dorsal fins set far back, origin of first over pelvic fin bases, second well in front of anal fin origin, anal and lower caudal fins almost touch. Large spiracles, nasoral and circumnarial grooves, long tapering barbels, small mouth in front of eye.
Distribution. East Australia.

Habitat. Rocky shores (tidepools), reefs, sea-grass beds, 0–140 m. Juveniles in high-energy surge zones.

Behaviour. Nocturnal; hides in caves and under ledges by day, feeds at night.
Biology. Seven or eight pups/litter in November. Feed on small fishes, crustaceans, squid and sea anemones.
Status. IUCN Red List: Least Concern. Relatively common. Collected for aquaria. Rarely taken in fisheries, extremely hardy and survives release well.

Bluegrey Carpetshark *Heteroscyllium colcloughi* Plate 22

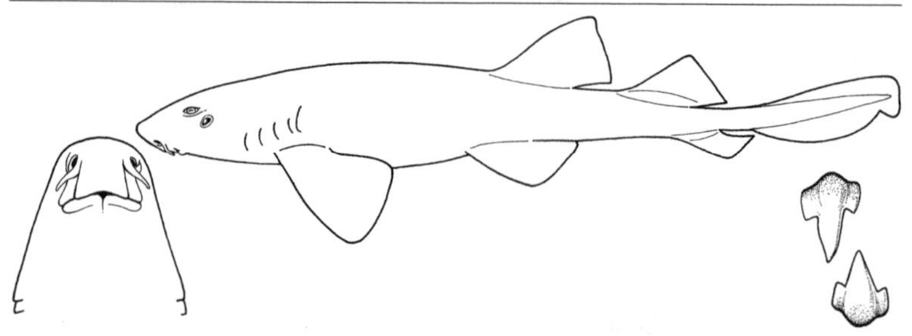

Measurements. Born: 17–18 cm TL. Mature ~50 cm ♂, 65 cm ♀. Max: >75 cm TL.
Identification. Greyish above, white below, no light spots. Conspicuous black markings on back, dorsal fins and caudal fin of young fade in adults. Small stout shark. Pair of long barbels with posterior hooked flap. Short mouth ahead of eyes, large spiracles well behind eyes. Two spineless dorsal fins, first larger and originating over pelvic fin bases. Short precaudal tail and caudal fin.
Distribution. East Australia.

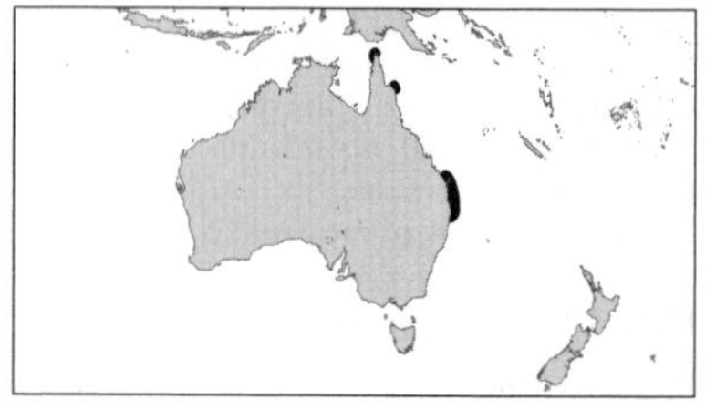

Habitat. Bottom in very shallow (<6 m) inshore areas.
Behaviour. Unknown.
Biology. Aplacentally viviparous, six to eight pups per litter.
Status. IUCN Red List: Vulnerable. Known from about 20 specimens in a small, well-surveyed area heavily used by fisheries and for recreation. Presumed rare.

Orectolobidae: Wobbegongs

Three genera, seven species, but additional undescribed species probably present within west Pacific range of the family, Australia to Japan. Live on the bottom in warm-temperate to tropical continental waters, intertidal to >110 m, often on rocky and coral reefs or on sandy bottom.

Identification. Very distinctive, flattened, highly patterned, well-camouflaged sharks. Dermal flaps along sides of broad, flat head, long barbels, short mouths in front of eyes and almost at the very front of the short snout. Heavy jaws, two rows of enlarged, sharp, fang-like teeth in upper jaw, three in lower. Nasoral grooves, circumnarial grooves and flaps, symphysial grooves. Spiracles larger than upward-facing eyes. Two spineless dorsal fins and an anal fin. First dorsal fin origin over pelvic fin bases.

Biology. Ovoviviparous, large litters of 20 or more young. Powerful seabed predators on bottom animals (fishes, crabs, lobsters and octopi), lurking camouflaged by colour patterns and dermal lobes around head, sucking in and impaling prey on large teeth. Use paired fins to clamber around the bottom, even out of the water.

Status. Potentially dangerous bite if provoked. Some species kept and bred in aquaria. Some are important in fisheries.

Tasselled Wobbegong *Eucrossorhinus dasypogon* Plate 23

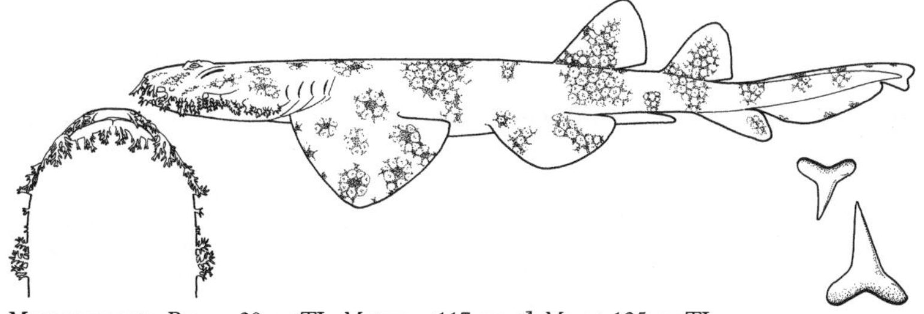

Measurements. Born: ~20 cm TL. Mature: <117 cm ♂. Max: >125 cm TL.

Identification. Reticulated pattern of narrow dark lines on a light background, scattered symmetrical enlarged dark dots at line junctions, indistinct saddles. Many highly branched dermal lobes on head and 'beard' on the chin Very broad paired fins.

Distribution. Southwest Pacific: Indonesia (Waigeo, Aru), New Guinea, Australia. (Malaysia?)

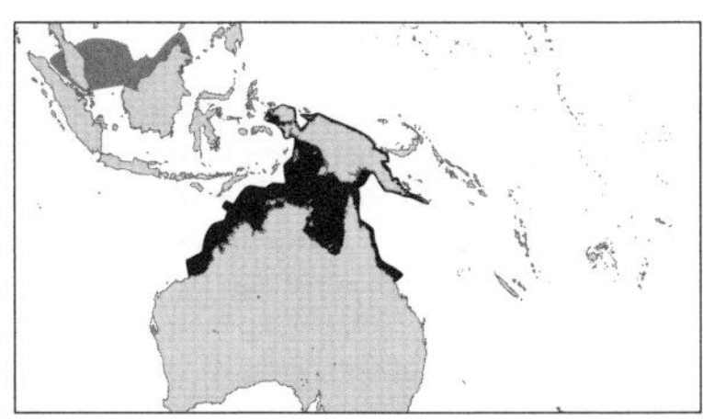

Habitat. Inshore coral reefs (coral heads, channels, reef faces).

Behaviour. Nocturnal, possibly solitary. Small home range. Rests by day with curled tail on bottom in caves and under ledges.

Biology. Largely unknown. Presumably ovoviviparous. Feeds on bottom fishes, possibly invertebrates. Catches nocturnal fishes sharing its caves.

Status. IUCN Red List: Near Threatened. Reef destruction and fisheries a threat in much of range (Least Concern in Australia). Reported to bite divers.

Japanese Wobbegong *Orectolobus japonicus* Plate 23

Measurements. Born: 21–23 cm TL. Mature: <100 cm ♀. Max: >107 cm TL.

Identification. Very obvious colour pattern of broad dark dorsal saddles with spots and blotches and dark (but not black) corrugated edges separated by lighter areas with dark broad reticular lines. Five dermal lobes below and in front of eyes on each side of head. Long, branched nasal barbels.

Distribution. Northwest Pacific: Japan, Korea, China, Taiwan Island, Vietnam. (Philippines record questionable.)

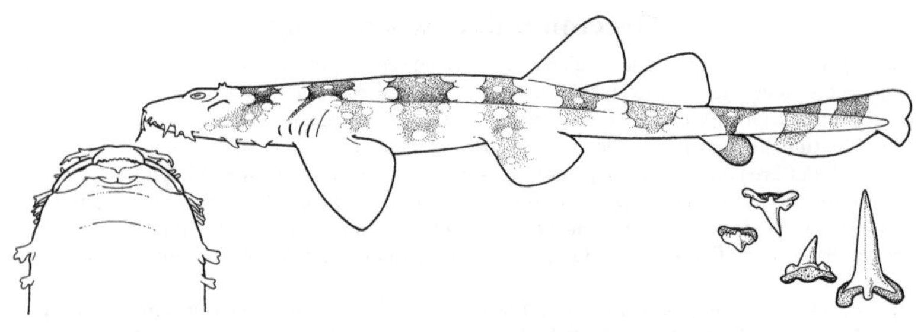

Habitat. Tropical inshore rocky and coral reefs.
Behaviour. Nocturnal, rarely seen by divers.
Biology. Ovoviviparous, up to 20–23 pups/litter born in spring (March–May) in captivity in Japan. One year gestation. Mainly eats benthic fish, also skates, shark eggcases, cephalopods and shrimp.
Status. IUCN Red List: Not Evaluated.

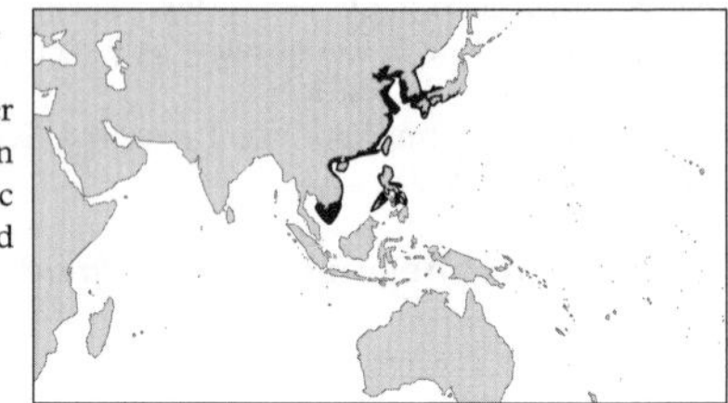

Spotted Wobbegong *Orectolobus maculatus* Plate 24

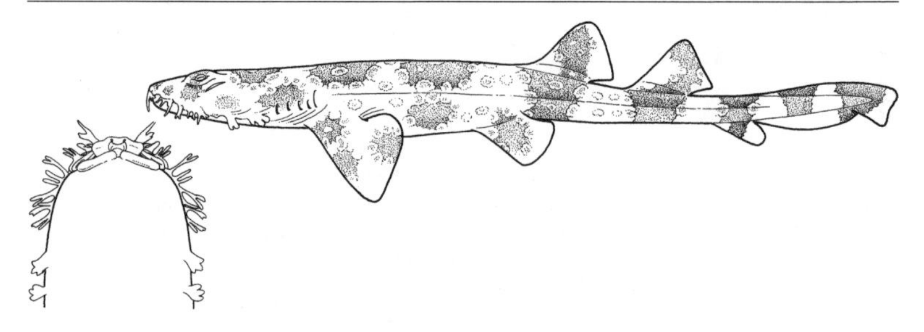

Measurements. Born: 21 cm TL. Mature: ~60 cm ♂. Max: ~320 cm TL.
Identification. Dark back, broad darker dorsal saddles with white 'O-shaped' spots and blotches and corrugated edges, separated by lighter areas with dark broad reticular lines. Six to ten dermal lobes below and in front of eyes, long branched nasal barbels.
Distribution. Southern Australia.
Habitat. Coral and rocky reefs, bays, estuaries, seagrass, tidepools, under piers, on sand. Intertidal to >110 m.
Behaviour. Possibly nocturnal. Sluggish and inactive by day in caves, under overhangs and in channels, singly or in aggregations. May return to resting sites. Can make short trips well above the seabed and climb (with back above water) between tidepools. Males fight in captivity during breeding season (July in New South Wales).

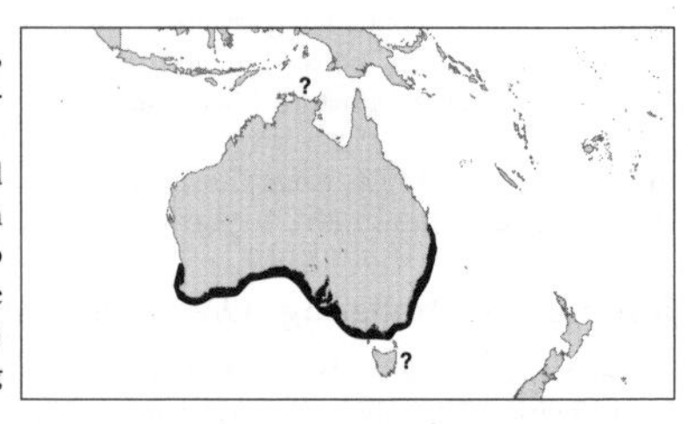

Biology. Ovoviviparous, large litters (up to 37 pups). Eats bottom invertebrates, bony fishes, sharks and rays.
Status. IUCN Red List: Near Threatened. Caught in many fisheries, declined ('Vulnerable') in New South Wales. Displayed in aquaria. Dangerous bite if provoked. Probably a species complex.

Ornate Wobbegong *Orectolobus ornatus* Plate 24

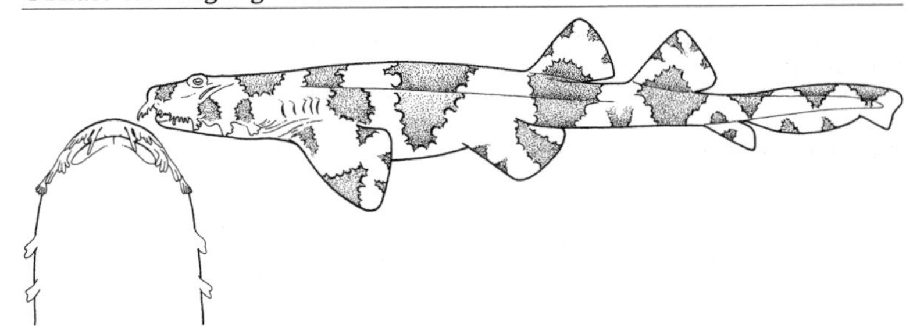

Measurements. Born: 20 cm TL. Mature: ~63–175 cm (two species?). Max: 288 cm TL.
Identification. Strongly variegated pattern of obvious broad dark, dorsal saddles with light spots and conspicuous black, corrugated borders, interspaced with lighter areas and conspicuous dark, light-centred spots. Nasal barbels with a few branches. Five dermal lobes below and in front of each eye, those behind spiracles unbranched or only weakly branched and broad.
Distribution. Australia. Records from Philippines, New Guinea, Indonesia and Japan may be different and undescribed species.
Habitat. Bays, seaweed-covered rock and coral reefs on coast and around offshore islands. Lagoons, reef flats and faces, and reef channels, intertidal to >100 m. Prefers clearer water than *Orectolobus maculatus*.
Behaviour. Nocturnal, resting singly and piled in aggregations by day in caves, under ledges and in trenches. Prowls at night.
Biology. Ovoviviparous, at least 12 pups/litter. Feeds on bony fishes, sharks, rays, cephalopods and crustaceans.
Status. IUCN Red List: Near Threatened. Caught in several fisheries. Declined ('Vulnerable') in New South Wales.

Northern Wobbegong *Orectolobus wardi* Plate 23

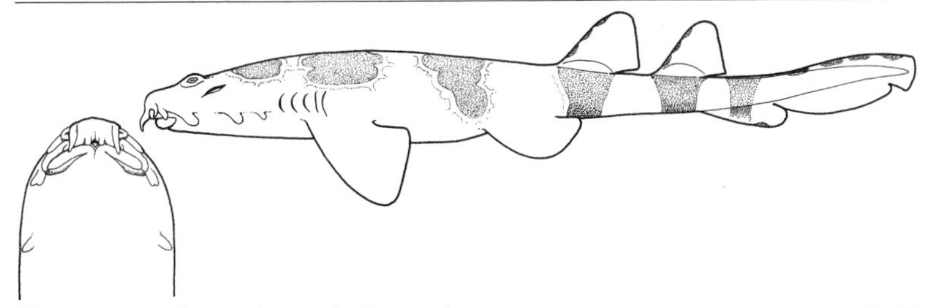

Measurements. Born: unknown. A 45 cm male was mature. Max: 63, possibly 100 cm TL.

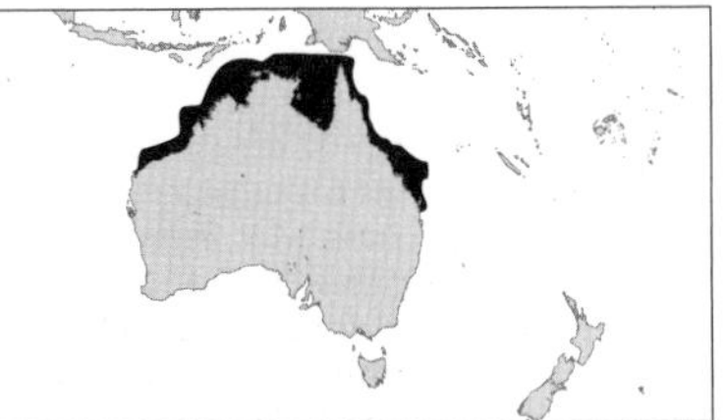

Identification. Simple sombre colour pattern: a few dark spots, dusky mottling, and three large dark, light-edged (ocellate) rounded saddles separated by broad dusky areas without spots or reticular lines in front of first dorsal fin. Two dermal lobes below and in front of each eye, dermal lobes behind spiracles unbranched and broad, unbranched nasal barbels.

Distribution. Australia.
Habitat. Shallow reefs <3 m deep, often in turbid water.
Behaviour. Nocturnal. Inactive (sometimes with head under a ledge) by day.
Biology. Unknown.
Status. IUCN Red List: Least Concern. Not fished, possibly common.

Western Wobbegong *Orectolobus* sp. A Plate 24

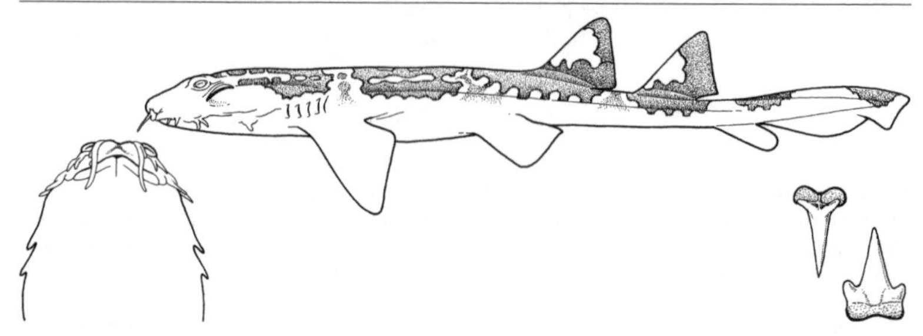

Measurements. Born: ~22 cm TL. An 85 cm male was mature. Max: 200 cm TL.
Identification. Strongly contrasting conspicuous broad dark, rectangular dorsal saddles with light spots and deeply corrugated edges (not black edged), separated by lighter areas with numerous broad dark blotches (without numerous light O-shaped rings). Four dermal lobes below and in front of each eye, dermal lobes behind spiracles unbranched or weakly branched and slender. Long nasal barbels with one small branch.

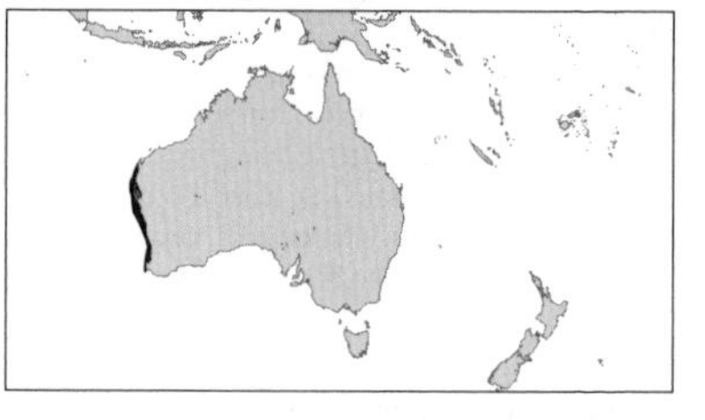

Distribution. Western Australia.
Habitat. Reefs and seagrass beds, intertidal to <100 m.
Behaviour. Unknown.
Biology. Poorly known.
Status. IUCN Red List: Least Concern. Discarded alive from lobster pots and gill nets.

Cobbler Wobbegong *Sutorectus tentaculatus* Plate 23

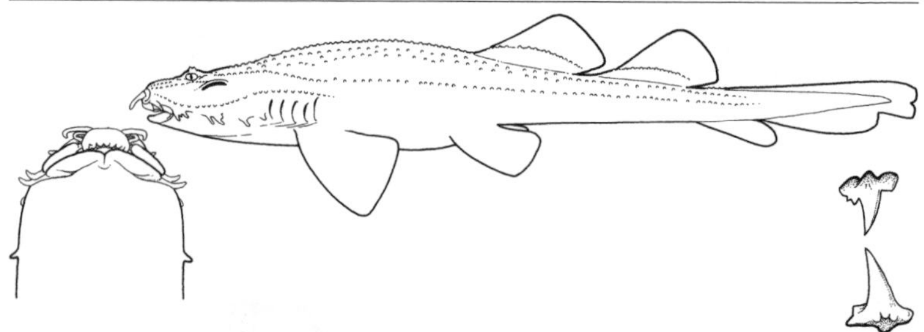

Measurements. Born: ~22 cm TL. Mature: ~65 cm ♂. Max: recorded 92 cm TL.
Identification. Striking colour pattern of broad dark, dorsal saddles with jagged, corrugated edges, separated by light areas with irregular dark spots. Rather slender, less flattened than most wobbegongs, rather narrow head. Chin smooth. A few slender, short, unbranched dermal lobes on sides of head form isolated groups broadly separated from one another, in four to six pairs. Simple unbranched nasal barbels. Rows of large warty dermal tubercles in rows on back and bases of very low, long dorsal fins (height half base length).

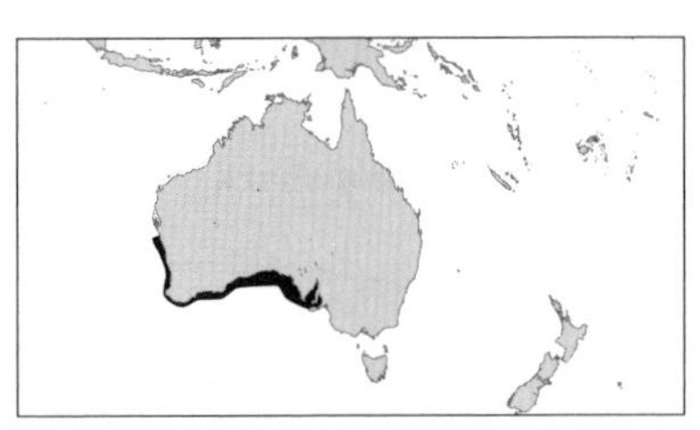

Distribution. South and western Australia.
Habitat. Rocky and coral reefs, in seaweeds. Depths unrecorded.
Behaviour. Unknown.
Biology. Unknown.
Status. IUCN Red List: Least Concern. Common, bycatch usually returned alive.

Hemiscylliidae: Longtailed Carpetsharks

Two Indo-west Pacific genera, *Chiloscyllium* (seven wide-ranging species) and *Hemiscyllium* (five, possibly six species, mainly in the western Pacific, also Seychelles). Occur in intertidal pools, very shallow water, rocky and coral reefs close inshore and on sediments, inshore and offshore in bays.
Identification. Small (mostly <1 m) and slender with very long tails, two equal-sized unspined dorsal fins, origin of second well ahead of origin of long, low rounded anal fin, which is separated by a notch from lower caudal fin. Small transverse mouth well in front of dorsolateral eyes, large spiracles below eyes, nasoral and circumnarial grooves, short barbels. Colour patterns of young often different and bolder than adults. *Chiloscyllium* species without a black hood on head or large dark spots on sides of body, mouth closer to eyes than snout tip. *Hemiscyllium* species with spots or hood, nostrils at end of snout, and obvious ridges above eyes.
Biology. Poorly known. Some (presumably all) are oviparous, laying oval eggcases. Distinct colour patterns of young suggest different habitat preferences from adults. Strong, muscular, leg-like paired fins used to clamber on reefs and in crevices. Large epaulette spots on *Hemiscyllium* species may be eyespots to intimidate predators. Food includes small bottom fishes, cephalopods, shelled molluscs and crustaceans.
Status. Taken in multispecies fisheries and as bycatch, sometimes in large numbers. Hardy, attractive and bred in captivity. Often common to abundant, but some species are rare, with limited distribution in threatened habitats.

Arabian Carpetshark *Chiloscyllium arabicum* Plate 25

Measurements. Hatch: <10 cm. Mature: ~45–54 cm. Max: 70 cm TL.
Identification. Unpatterned but for light spots on juveniles' fins. Dorsal and anal fins set far back on very long thick tail. Almost straight dorsal fin trailing edges, origin of first opposite or just behind pelvic fin insertions, second usually with longer base than first. Prominent ridges on back.

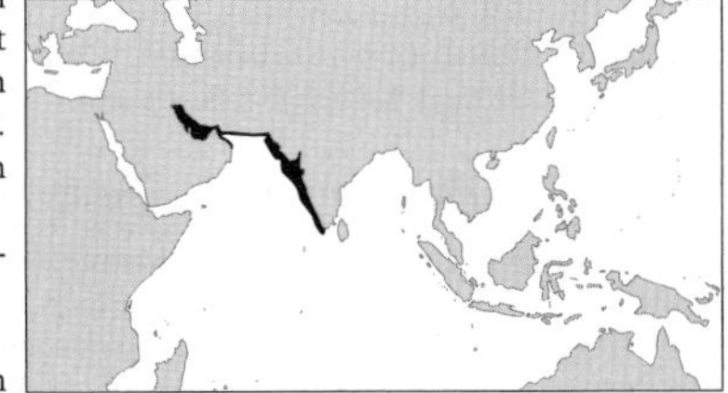

Distribution. Northwest Indian Ocean: Persian Gulf to India.
Habitat. Coral reefs, lagoons, rocky shores, mangrove estuaries; 3–100 m on bottom.
Behaviour. Poorly known.
Biology. Up to four eggs laid on coral reefs hatch

in 70–80 days. Feeds on squid, shelled molluscs, crustaceans and snake eels.
Status. IUCN Red List: Not Evaluated. Common in summer in Gulf. Has bred in captivity.

Burmese Bambooshark *Chiloscyllium burmensis* Plate 25

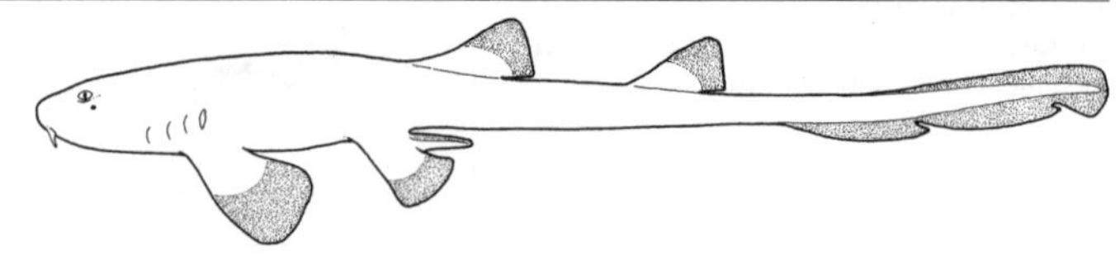

Measurements. Only known specimen is a 57 cm adult male.
Identification. Adults with dark fin webs, juvenile coloration unknown. Very small eyes. Dorsal fins (with straight or convex rear margins) and long, low anal fin set far back on very long thick precaudal tail, origin of first dorsal roughly opposite pelvic fin insertions. No lateral ridges on body.

Distribution. Northeast Indian Ocean: Burma (Myanmar).
Habitat. Unknown. (Inshore off Irrawaddy River delta?) Collected off Rangoon/Yangon.
Behaviour. Unknown.
Biology. Virtually unknown. Eats small bony fishes.
Status. IUCN Red List: Not Evaluated.

Grey Bambooshark *Chiloscyllium griseum* Plate 25

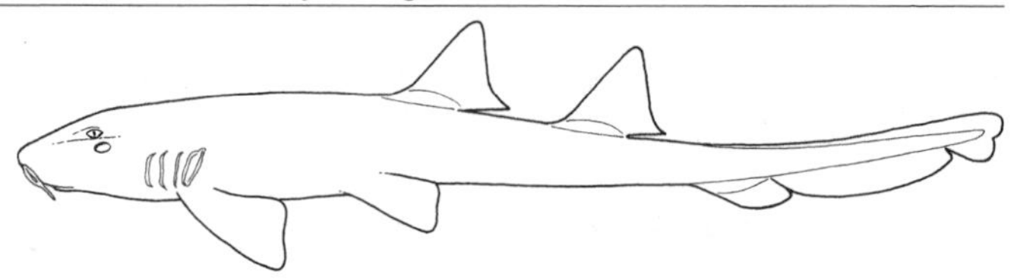

Measurements. Hatch: <12 cm TL. Mature: 45–55 cm TL ♂. Max: 77 cm TL.
Identification. Adults often unpatterned; young with obvious dark saddle-marks and transverse bands, no black edging. Dorsal fins (with straight or convex rear margins) and long, low anal fin set far back on very long thick tail, origin of first dorsal over rear of pelvic fin bases. No body ridges.

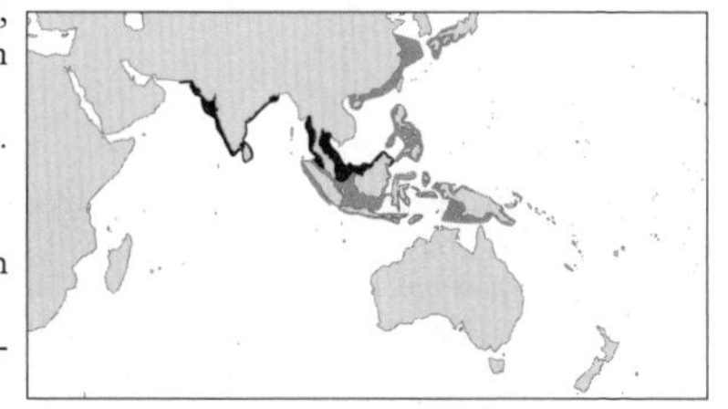

Distribution. Indo-west Pacific: Pakistan, India, Sri Lanka, Malaysia, Thailand. (Records from further east may be *Chiloscyllium hasselti*.)
Habitat. Inshore, on rocks and in lagoons. 5–80 m.
Behaviour. Unknown.
Biology. Oviparous, lays small oval eggcases on bottom. Probably feeds mainly on invertebrates.
Status. IUCN Red List: Not Evaluated. Common. Fished for food. Kept in public aquaria.

Indonesian Bambooshark *Chiloscyllium hasselti* Plate 25

Measurements. Hatch: 9–12 cm TL. Mature: 44–54 cm ♂. Max: 61 cm TL.
Identification. Adults often unpatterned except for dusky fins. Young have prominent saddle marks (broad dusky patches with conspicuous black edging separated by light areas and blackish spots) and dark blotches on fins. Dorsal fins (straight or convex rear margins) and long, low anal fin set far back

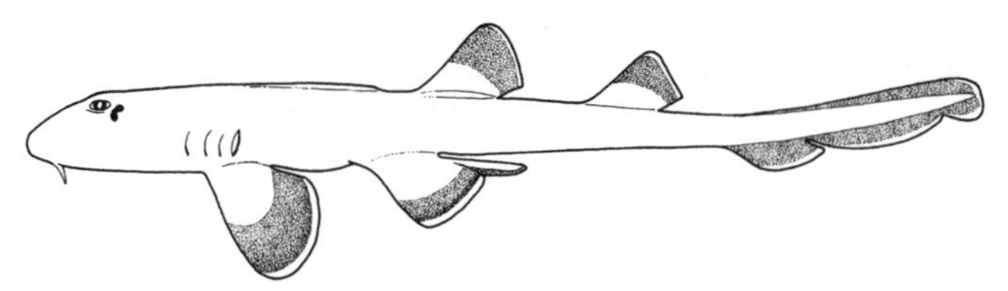

on very long thick tail. Origin of first dorsal fin over rear of pelvic fin bases. No body ridges.

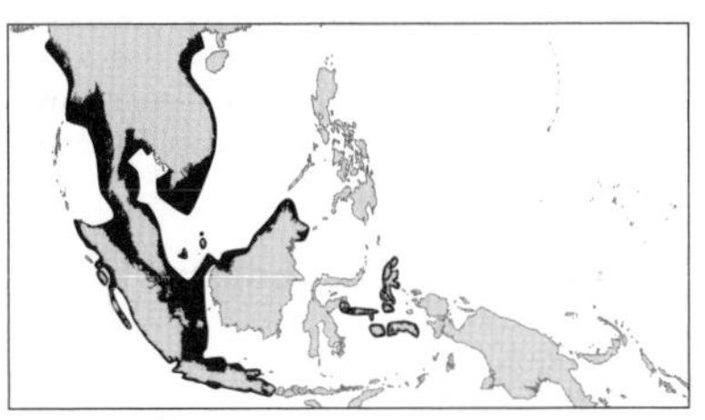

Distribution. Indo-west Pacific: Burma to Vietnam, Indonesia (Sumatra to New Guinea).
Habitat. Probably close inshore, to 12 m.
Behaviour. Eggs are attached to benthic marine plants.
Biology. Oviparous, eggs hatch in about December.
Status. IUCN Red List: Not Evaluated. Fished, status unknown.

Slender Bambooshark *Chiloscyllium indicum* Plate 25

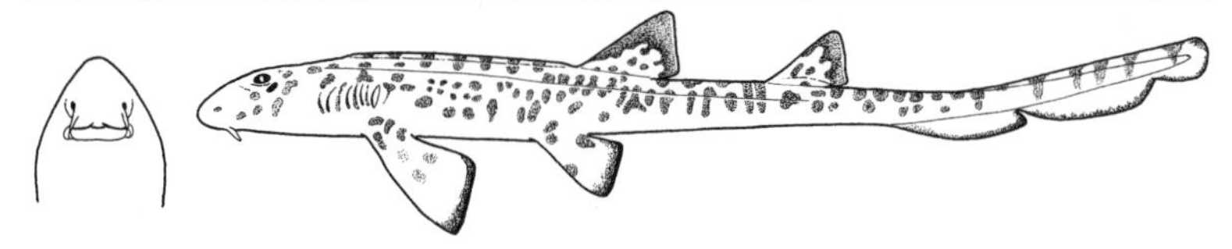

Measurements. Hatch: <13 cm TL. Mature: 39–42 cm ♂, 43 cm ♀. Max: 65 cm TL.
Identification. Numerous small dark spots, bars/saddles and dashes on light brown background, no prominent black edges to saddles in juveniles. Very slender body with lateral ridges. Dorsal fins (straight or convex rear margins) and anal fin set far back on very long slender tail. Origin of first dorsal opposite or just behind pelvic fin insertions.
Distribution. Indo-west Pacific: Sri Lanka to Indonesia, China and Taiwan Island. Possibly Arabian Sea to Solomon Islands, Korea and Japan.
Habitat. Inshore on bottom. Possibly in fresh water (lower Perak River, Malaysia).
Behaviour. Little-known.
Biology. Little-known. Oviparous.
Status. IUCN Red List: Near Threatened. Important in inshore fisheries, used for food. May be overfished, inshore habitat under threat.

Whitespotted Bambooshark *Chiloscyllium plagiosum* Plate 25

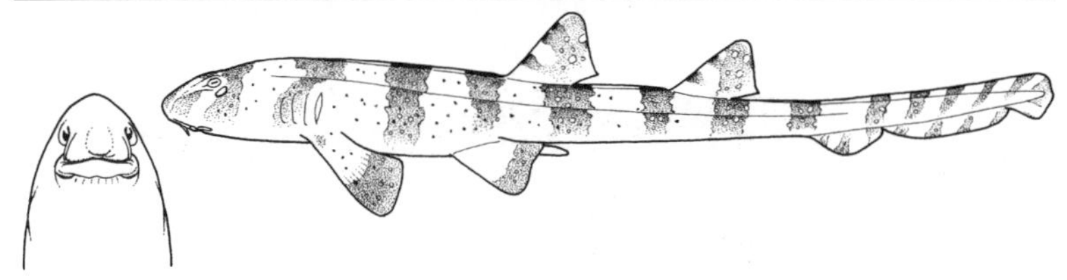

Measurements. Hatch: ~9–12 cm TL. Mature: ~50–60 cm ♂. Max: 95 cm TL.
Identification. Body dark, with numerous light and dark spots, dark bands and saddles not

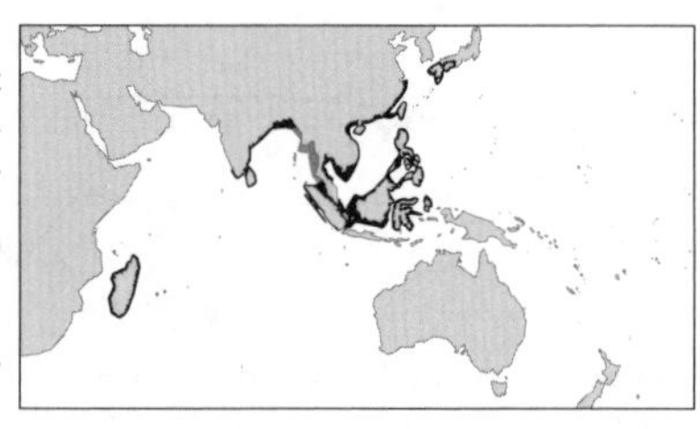

conspicuously edged with black. Dorsal fins (straight or convex rear margins) and anal fin set far back on very long thick tail, first dorsal fin origin opposite or just behind pelvic fin insertions. Lateral ridges on trunk.
Distribution. Indo-west Pacific: Madagascar to Indonesia, Philippines and Japan.
Habitat. Inshore on bottom, reefs in tropics.
Behaviour. Nocturnal. Rests in reef crevices by day, feeds at night.
Biology. Oviparous. Eats bony fishes and crustaceans.
Status. IUCN Red List: Not Evaluated. Common and important in inshore fisheries. Used for food. Popular aquarium species.

Brownbanded Bambooshark *Chiloscyllium punctatum* Plate 25

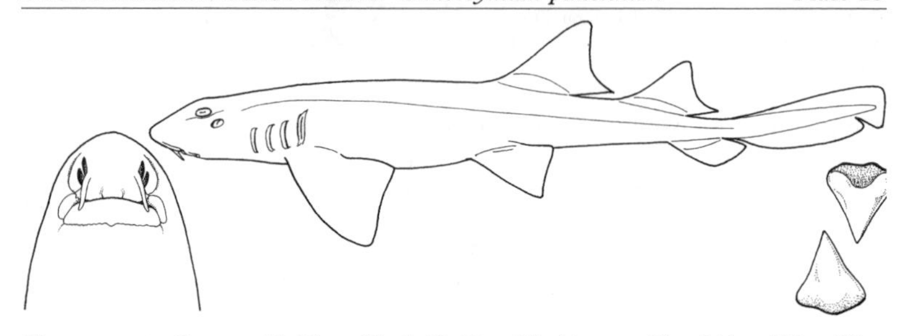

Measurements. Eggcases: 11x15 cm. Hatch: 13–17 cm TL. Mature: ~60 cm? Max: 105 cm TL.
Identification. Young with obvious dark bands (not black edged) and scattered small dark spots, which fade in light brown adults. Very light gill slit margins. Dorsal fins (distinctly concave rear margins, elongated free rear tips) and anal fin set far back on long thick tail. First dorsal origin over front half of pelvic fin bases.
Distribution. Indo-west Pacific: east India to north Australia and Japan.
Habitat. Coral reefs (intertidal pools, tidal flats, reef faces). Possibly also on soft bottom offshore to >85 m.
Behaviour. Hides in crevices and under corals. Can survive out of water for up to half a day.
Biology. Oviparous, lays rounded eggcases. Feeds on bottom invertebrates, possibly fish.
Status. IUCN Red List: Near Threatened. Common, regularly taken in inshore food fisheries. Displayed and bred in aquaria. May nip if provoked.

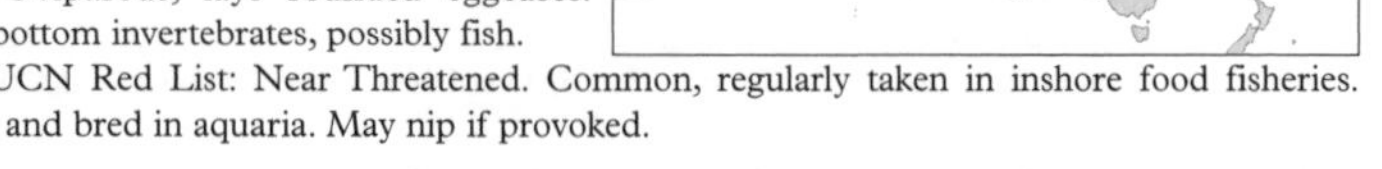

Indonesian Speckled Carpetshark *Hemiscyllium freycineti* Plate 25

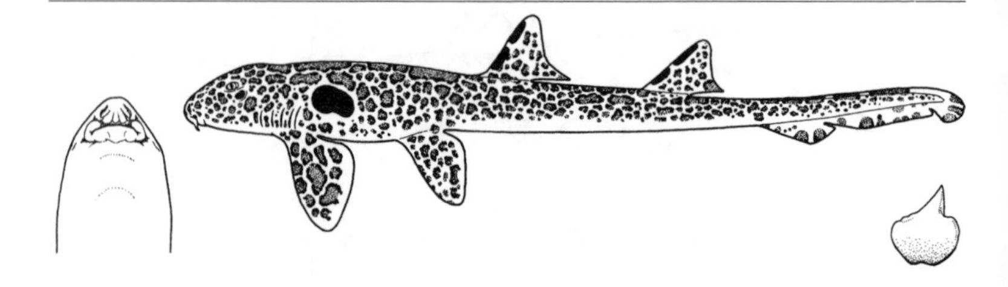

Measurements. Born: <19 cm TL. Mature: 37–62cm ♂. Max: 72 cm TL ♀.
Identification. Small dark spots on snout. Large, sparse dark spots on body, none white, no reticulate pattern. Moderately large black epaulette spot above pectorals without white ring or curved black marks around rear half. Dark paired fins with light edges in young, changing to scattered small and large dark spots in adults. Broad dark bands under head and completely encircling tail in young, lost in light undersides of adults. Dark blotches on anterior edges of dorsal fins. Dorsal and anal fins set very far back on extremely long thick tail.

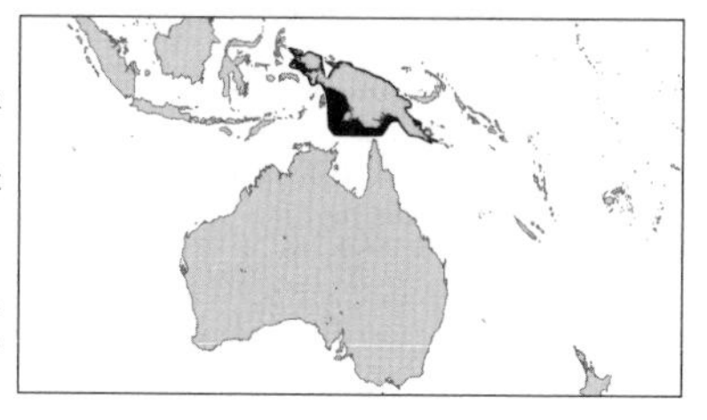

Distribution. New Guinea.
Habitat. Coral reefs, on sand and in seagrass in shallow water.
Behaviour. Hides in reef crevices by day, feeds at night.
Biology. Almost unknown
Status. IUCN Red List: Near Threatened. Restricted habitat affected by expanding fisheries and pollution.

Papuan Epaulette Carpetshark *Hemiscyllium hallstromi* Plate 25

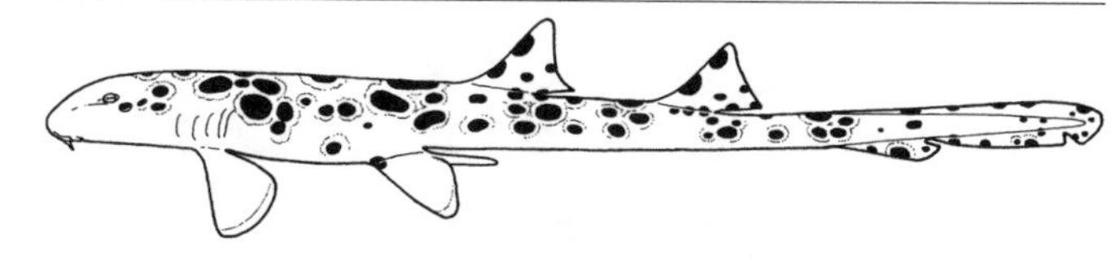

Measurements. Hatch: <19 cm TL. Mature: 47–64 cm ♂. Max: 77 cm TL.
Identification. Dorsal and anal fins set far back on extremely long thick tail. No dark spots on snout. Dark wide-spaced spots on body, some almost as large or larger than the conspicuous white-ringed black epaulette spot above pectoral fins. Epaulette spot partly ringed (above and behind) by smaller black spots. No white spots or reticular pattern. Paired fins unspotted; black webs and light edges in young fade to dusky in adults. Dark rings around tail in young are lost from uniformly pale underside of adults.
Distribution. Southern Papua New Guinea.
Habitat. Inshore seabed, possibly coral reefs.
Behaviour. Unknown.
Biology. Unknown.
Status. IUCN Red List: Vulnerable. Very limited range affected by fisheries and pollution.

Epaulette Carpetshark *Hemiscyllium ocellatum* Plate 25

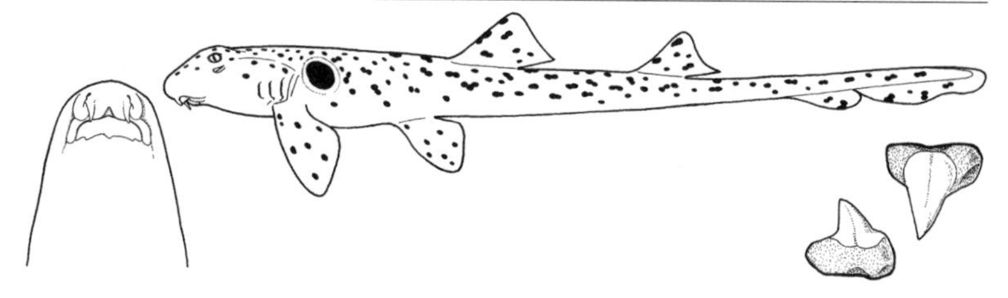

Measurements. Hatch: ~15 cm TL. Mature: 59–62 cm ♂, ~64 cm ♀. Max: 107 cm TL.
Identification. Dorsal and anal fins set far back on extremely long thick tail. No spots on snout. Dark spots on body and unpaired fins much smaller than conspicuous large black epaulette spot

(ringed with white, inconspicuous small dark spots behind and below). No white spots or reticular network. Pale-margined dark paired fins in young fade in adults. Sometimes a few small dark spots on adult paired fins. Dark bands around tail in young, adults with uniform light ventral tail surface.
Distribution. Southwest Pacific: New Guinea and Australia (possibly to Solomon Islands and Malaysia).

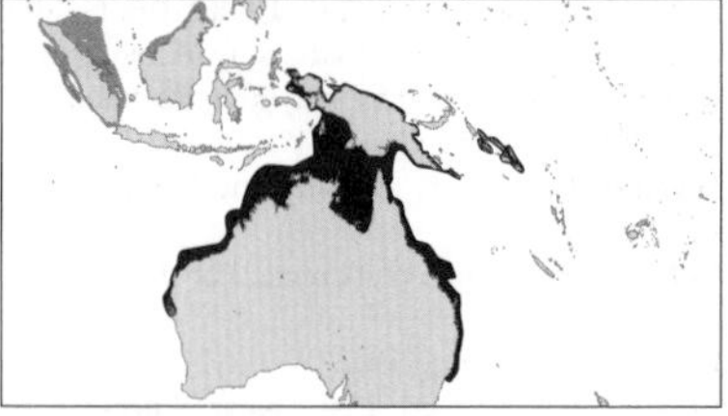

Habitat. Coral (particularly staghorn) in shallow water and tidepools, sometimes barely submerged.
Behaviour. More active at dusk and by night. Often feeds at low tide. Crawls, clambers and swims about. Thrashes tail when snout digs in sand. Unafraid of man, may nip when captured.
Biology. Oviparous, eggs hatch in ~120 days. Eats worms, crustaceans and small fishes.
Status. IUCN Red List: Least Concern. Abundant on the Great Barrier Reef; may be threatened in New Guinea.

Hooded Carpetshark *Hemiscyllium strahani* Plate 25

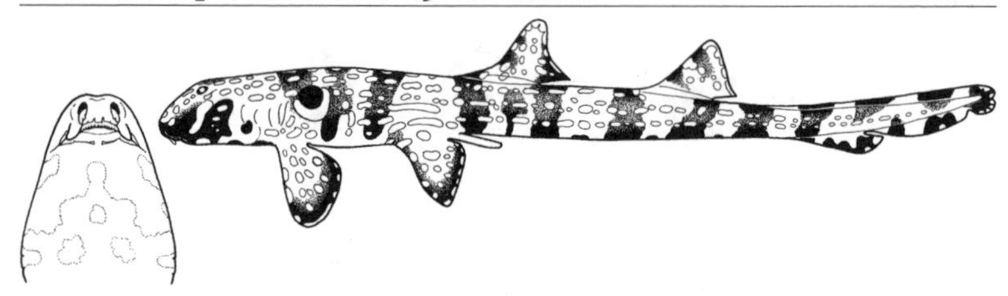

Measurements. Hatch: unknown. Mature: before 59 cm ♂, 73 cm ♀. Max: 80 cm TL.
Identification. Dorsal and anal fins set far back on extremely long thick tail. Black mask on adult snout and head. Black spots and bands beneath head, not on snout. Black epaulette spot partially merged with shoulder saddle, not completely surrounded by white ring. Dark saddles and blotches on body, many white spots on body and fins. No reticular pattern. Margins of paired fins white-spotted on black. Dark rings around tail.

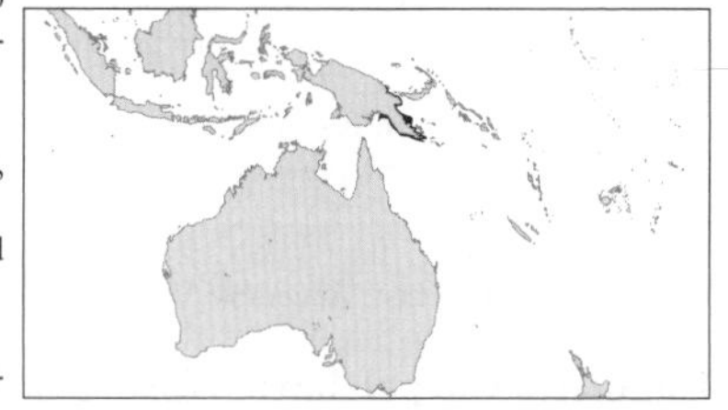

Distribution. Eastern Papua New Guinea.
Habitat. Inshore on coral reef faces and flats, mainly 3–13 m.
Behaviour. Nocturnal, hides in crevices and under table corals by day.
Biology. Unknown.
Status. IUCN Red List: Vulnerable. Small, fragmented range polluted and dynamite fished.

Speckled Carpetshark *Hemiscyllium trispeculare* Plate 25

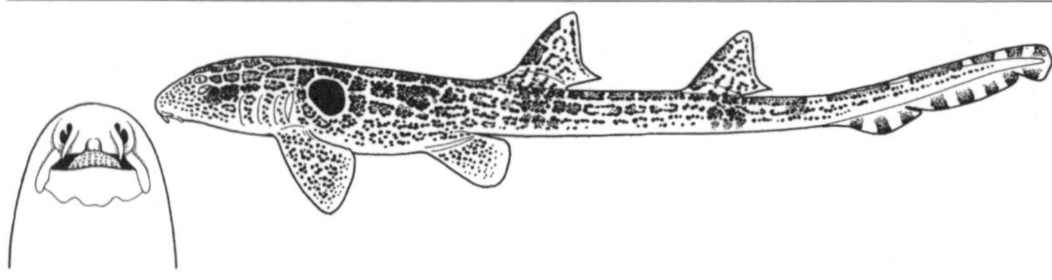

Measurements. Hatch: unknown. Mature: <56 cm. Max: 79 cm TL.

Identification. Small dark spots on snout, uniformly light under head. Large black epaulette spot with conspicuous white ring and two curved black marks around the posterior half partly surrounded by smaller black spots. Body and fins covered with numerous small and large dark spots separated by a reticular light network. No white spots. Dark saddles on back and tail extend around ventral surface.

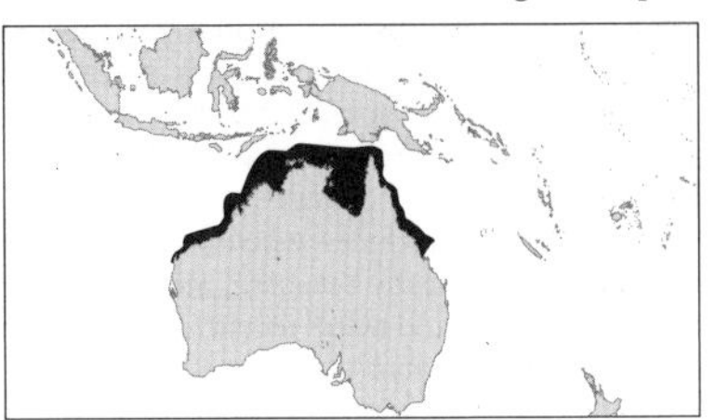

Distribution. Northern Australia, possibly Indonesia.
Habitat. Coral reefs in shallow water (often under table corals) and in tide pools.
Behaviour. Poorly known.
Biology. Poorly known.
Status. IUCN Red List: Least Concern.

Ginglymostomatidae: Nurse Sharks

Three mono-specific genera of subtropical and tropical continental and insular waters, including coral and rocky reefs, sandy areas, reef lagoons and mangrove keys. Intertidal and surf zone (sometimes barely covered) to at least 70 m.
Identification. Head broad and flattened, no lateral skin flaps. Snout rounded or truncated. Transverse, subterminal mouth in front of eyes, long nasoral grooves. Nostrils with barbels. Small spiracles behind eyes. Small gill slits, 5th almost overlaps 4th. Two spineless dorsal fins, second level with and about same size as anal fin; latter close to lower caudal fin. Precaudal tail much shorter than head and body. Caudal fin elongated with a strong terminal lobe and subterminal notch but no or very short ventral lobe. Unpatterned or a few dark spots in young.
Biology. Ovoviviparous (or possibly so in *Pseudoginglymostoma*). Nocturnal. Social, rest on the bottom in small groups. Cruise and clamber on bottom, mouths and barbels close to bottom, searching for food. Short small mouths with large cavities suck in a variety of bottom invertebrates and fishes, including active reef fish.
Status. Larger species are or were formerly common and often caught in local inshore fisheries for food, liver oil and tough leather. Some local extirpations reported. *Ginglymostoma* and *Nebrius* are hardy and can survive well in large aquaria. Not usually aggressive, but can bite hard and hang on if provoked. Popular for dive tourism.

Nurse Shark *Ginglymostoma cirratum* — Plate 26

Measurements. Born: 27–30 cm TL. Mature: ~210 cm ♂, 230–240 cm ♀. Max: ~300 (?430) cm TL.
Identification. Mouth (long barbels, nasoral grooves) in front of dorsolateral eyes. Tiny spiracles. Dorsal fins broadly rounded, first much larger than second dorsal and anal fins. Precaudal tail shorter than head and body, caudal fin >25% of total length. Adults uniform yellow- to grey-brown, young with small dark, light-ringed ocellar spots and obscure saddle markings.
Distribution. East Pacific: Mexico to Peru. West Atlantic: USA to Gulf, Caribbean and Brazil. East Atlantic: Cape Verde Islands, Senegal, Cameroon to Gabon (rarely north to France).
Habitat. Rocky and coral reefs, channels between mangrove keys and sand flats on tropical and

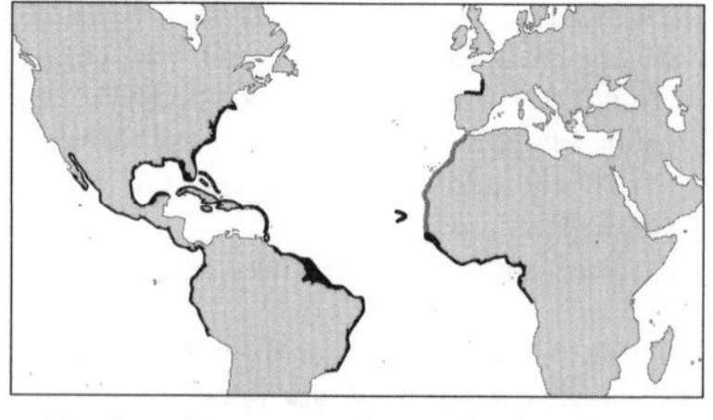

subtropical continental and insular shelves, from <1–12 m (40–130 m off Brazil).

Behaviour. Nocturnal, social shark. Rest in groups (even piles) by day in preferred shallow water locations on sand or in caves. Strong-swimming and active at night. Small home range. Can use muscular pectoral fins to clamber on bottom and snout to root out prey. Courtship and mating behaviour includes synchronized parallel swimming, sides nearly touching, male(s) beside or slightly behind and below female. Male bites one of the female's pectoral fins and both roll upside-down on seabed to mate.

Biology. Ovoviviparous: 20–30 pups/litter with large yolk-sacs. Born late spring/summer after five to six month gestation. Females reproduce every other year. Juvenile nursery areas in shallow turtle-grass beds and coral reefs. Males mature at 10–15, females 15–20 years old. Feed on bottom invertebrates, bony fishes and stingrays, sucked in rapidly with small mouth and large pharynx. Can extract conch snails from intact shell.

Status. IUCN Red List: Not Evaluated. Historically common in many areas, but small home ranges and aggregating habits make it highly vulnerable to local extirpation. Docile, popular with divers and hardy in aquaria, but will bite if provoked.

Tawny Nurse Shark *Nebrius ferrugineus* Plate 26

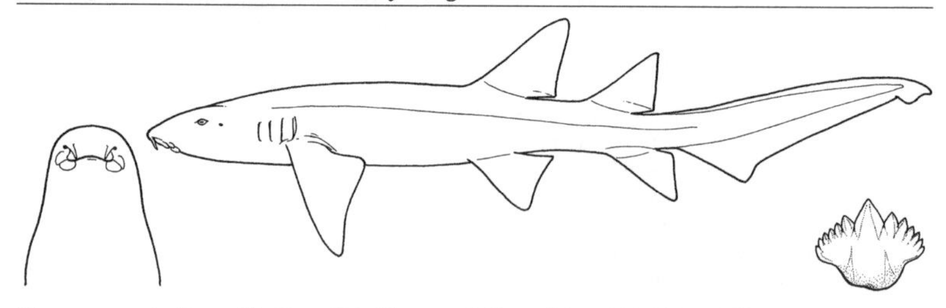

Measurements. Born: 40–60 cm TL. Mature: ~250 cm. Max: 314–320 cm TL.

Identification. Mouth (fairly long barbels) in front of lateral eyes. Tiny spiracle. Large first dorsal fin base over pelvic fin bases. Angular fins. Caudal fin fairly long (>25% TL). Colour may slowly change between shades of brown, depending on habitat.

Distribution. Wide-ranging, tropical Indo-Pacific: South Africa to Red Sea and Gulf, East Asia north to Japan, Australia to Marshall Islands and Tahiti.

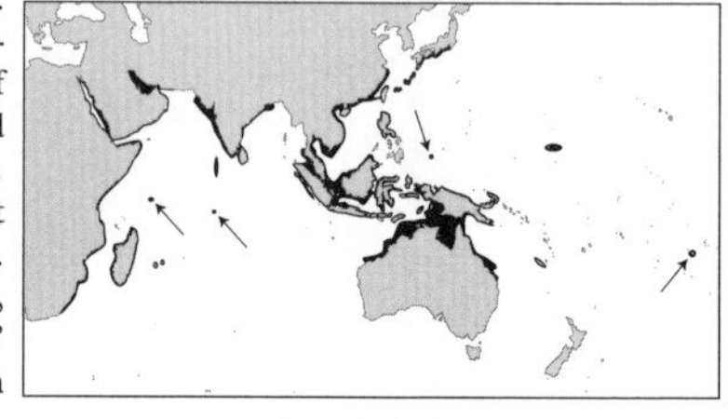

Habitat. On or near bottom in sheltered areas: lagoons (particularly juveniles), channels, crevices and caves in outer coral and rocky reef edges, seagrass and sand on and near reefs and off beaches. Intertidal to >70 m, mainly 5–30 m.

Behaviour. Mainly nocturnal, prowl reefs at night searching for prey to suck out of crevices. Aggregate in shelter by day. Limited home range, often return to same resting place. May 'spit' water when caught and spin on the line when hooked.

Biology. Ovoviviparous (aplacental viviparous). Young feed inside the uterus on large infertile yolky eggs (oophagy). Litter size uncertain (one to four, depending upon competition in the uterus). Feeds on corals, crustaceans, cephalopods, sea urchins and reef fish, occasionally sea snakes.

Status. IUCN Red List: Vulnerable. Fished through much of its range. Local extirpations reported. Small litters and limited dispersion will prevent rapid recovery from overfishing. Docile and popular with divers, may bite if harassed. Hardy in aquaria.

Shorttail Nurse Shark *Pseudoginglyostoma brevicaudatum* Plate 26

Measurements. Mature: ~55–60 cm ♀. Max: 75 cm TL. Adult ♀, 33 years in captivity, is 70 cm long.
Identification. Colour dark brown, no spots or markings. Colour of young unknown. Young with striking pattern of dark saddles on light background, long caudal fin and broad pectorals. Distinguished from other nurse sharks by short nasal barbels, short precaudal tail and caudal fin (<20% TL), and equal size of two rounded dorsal fins and anal fin.
Distribution. West Indian Ocean: Tanzania, Kenya and Madagascar, possibly Mauritius and the Seychelles.

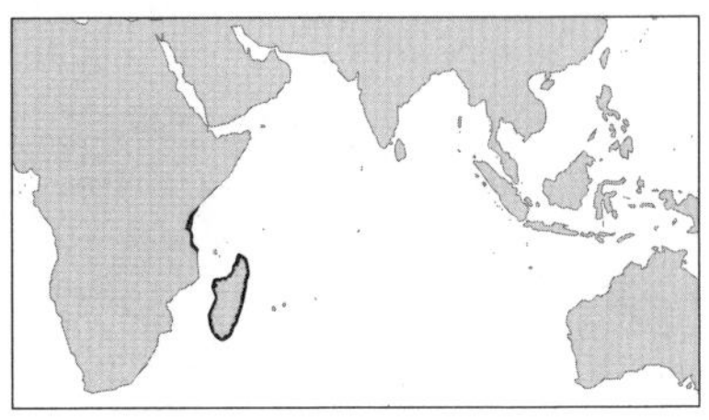

Habitat. Coral reefs, depth unknown.
Behaviour. Nocturnal in captivity, otherwise poorly known. Reported to survive for several hours out of water.
Biology. Poorly known. Presumed oviparous (egg-laying observed in captivity).
Status. IUCN Red List: Not Evaluated.

Stegostomatidae: Zebra Shark

One species.

Zebra Shark *Stegostoma fasciatum* Plate 26

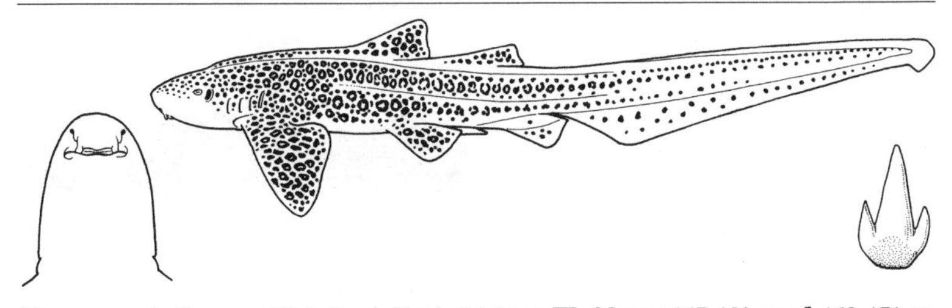

Measurements. Eggcase: (17x8x5 cm). Hatch: 20–36 cm TL. Mature: 147–183 cm ♂, 169–171 cm ♀. Max: possibly 354 cm TL, mostly <250 cm.

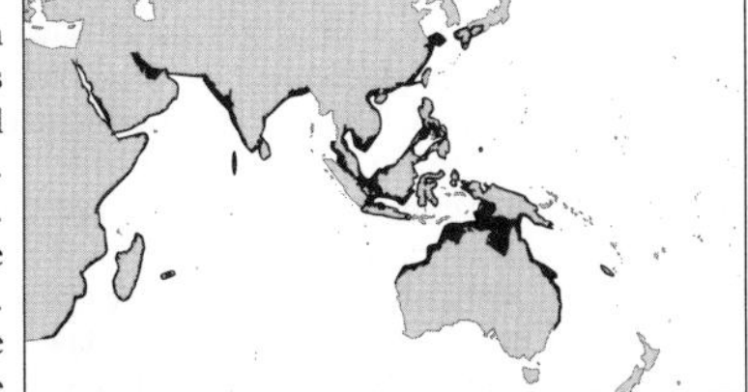

Identification. Young dark brown above, yellowish below, with vertical yellow stripes and spots separating dark saddles, which break up into small brown spots on yellow in sharks 50–90 cm long. Spots more uniformly distributed on large sharks. Large, slender flexible ridged body with a unique banded (juvenile) or variable spotted (adult) pattern. Broad tail fin as long as body. Small transverse mouth in front of lateral eyes. Small barbels, large spiracles. First dorsal fin set forwards on back, much larger than second. Anal fin close to tail.

Distribution. Indo-west Pacific: tropical continental and insular shelves, eastern Africa to Japan, New Caledonia and Palau.
Habitat. Coral reefs and offshore sediments, intertidal to 62 m. Adults and large spotted juveniles rest in coral reef lagoons, channels and faces. Striped young rarely seen, may be deeper (>50 m).
Behaviour. Poorly known. May rest propped up on pectorals, mouth open, facing current. Usually solitary, aggregations rare. Sluggish by day, more active at night or when food present. Can swim strongly and squirm into crevices to search for food.
Biology. Oviparous, lays large dark brown or purplish-black eggcases, anchored to bottom with fine tufts of fibres. Feeds on molluscs, crustaceans, small bony fishes, possibly sea snakes.
Status. IUCN Red List: Vulnerable. Relatively common, but taken in many fisheries, and its coral reef habitat is threatened. Not aggressive. Kept in captivity.

Rhincodontidae: Whale Shark

One species.

Whale Shark *Rhincodon typus* Plate 28

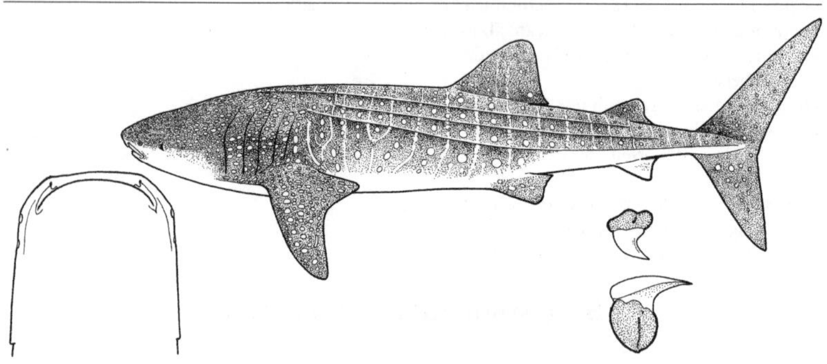

Measurements. Born: 55–64 cm TL. Mature: >600 cm ♂, >800 cm ♀. Max: possibly 1700–2100 cm TL.

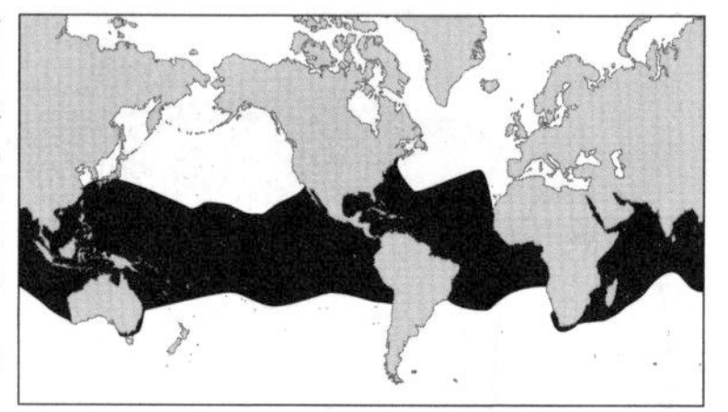

Identification. Unmistakable huge filter-feeder with checkerboard pattern of yellow or white spots on grey, bluish or greenish-brown back (white or yellowish underside). Broad, flat head, short snout, huge transverse almost terminal mouth in front of eyes. Prominent ridges on body, lowest ending in a keel on the caudal peduncle. Lunate, unnotched tail fin.
Distribution. Circumglobal, all tropical and warm temperate seas except Mediterranean.
Habitat. Pelagic, open ocean to close inshore off beaches, coral reefs and islands, surface to >700 m. Location of pupping and nursery grounds unknown.
Behaviour. Highly migratory. Tagging and photo-identification indicate regular visits to favoured feeding sites to feed at annual, seasonal or lunar fish and invertebrate spawning events. The high density of plankton produced on these occasions is consumed by suction feeding and gulping, often while hanging vertically. Long-distance, long-term migrations are undertaken, longest so far 13,000 km (in one direction only) in over 37 months.
Biology. Ovoviviparous. A female from Taiwan had ~300 pups in her uterus. Feeds on planktonic crustaceans and fish eggs.
Status. IUCN Red List: Vulnerable. Has been fished for meat, apparently unsustainably, in many areas. Steep declines in yields reported from the Philippines, Taiwan, Maldives and India. Legally protected in many states. Listed on Appendix II of CITES (to ensure that international trade is sustainable) and Appendix II of the Convention on Migratory Species (to encourage international management).

Lamniformes: Mackerel Sharks

Seven families and 15 species of mainly large active pelagic sharks.

Identification. Cylindrical body, two spineless dorsal fins (first originates over abdomen, well in front of pelvic fin origins) and anal fin. Vertebral axis extends into long upper tail lobe. Conical head, fairly short snout, five broad gill openings (hind two in front or above pectoral fin origin), large mouth extending behind eyes, nostrils free from mouth, no barbels or grooves. Very small spiracles well behind eyes.

Biology. World-wide in geographic distribution, mostly in warm water (some prefer cold) in a wide range of marine habitats, from the intertidal to at least 1600 m and in the open ocean; none occur in fresh water. Behaviour is very varied, from slow coastal sharks to fast oceanic swimmers, top predators to carrion and filter feeders. Some are highly migratory. Many are social and some will hunt cooperatively. Reproduction is ovoviviparous (aplacental viviparous); young eat eggs in the uterus (at least one species also cannibalises other embryos). Very varied diets, from marine mammals, birds and reptiles to other sharks, rays, bony fishes and invertebrates.

Status. Several species are very important in coastal commercial and sport fisheries and highly valued for sport, flesh and fins. Others are rare and not often recorded. Some of the larger species occasionally bite people, but are also important for dive ecotourism and film-makers, who view them with or without shark cages. Only one species is commonly kept in aquaria. Most are under threat from overfishing; some are completely unmonitored.

Odontaspididae: Sandtiger or Raggedtooth Sharks

Large heavy-bodied sharks with pointed snouts, upper precaudal pits present but no lower pits, caudal fins asymmetrical and without keels. Three species.

Sandtiger Shark, Spotted Raggedtooth or Grey Nurse Shark

Carcharias taurus Plate 27

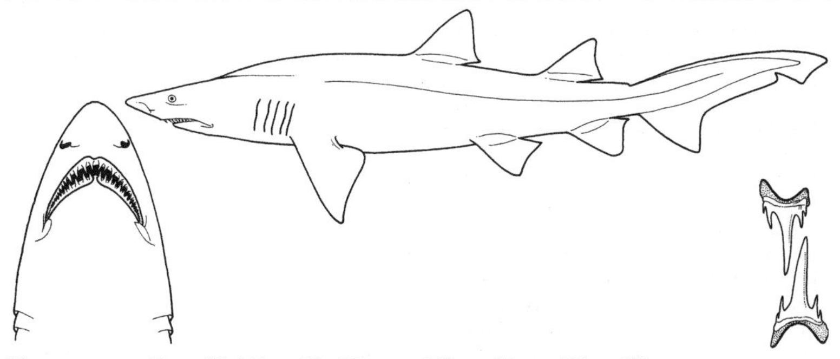

Measurements. Born: 95–105 cm TL. Mature: ~220 cm. Max: >430 cm TL.

Identification. Large, heavy, light brown body, often with scattered darker spots. Flattened conical snout, long mouth extends behind eyes, large slender pointed teeth. Long gill openings in front of pectoral fins. Large dorsal and anal fins similar in size, first dorsal closer to pelvic than to pectoral fins. Asymmetrical tail, short lower lobe.

Distribution. Warm-temperate and tropical Atlantic, Mediterranean and Indo-west Pacific. (Not central and eastern Pacific.)

Habitat. Coastal waters, from the surf zone (<1 m) to offshore reefs at least 191 m; mostly 15–25 m. Associated with underwater caves, gullies, and reefs. Usually on or near bottom, occasionally midwater or surface.

Behaviour. A slow but strong swimmer, more active at night. Air swallowed at the surface and held in stomach provides neutral buoyancy, enabling the shark to hover in the water. Complex social, courtship and mating behaviour studied in captivity and the wild. May aggregate in schools of 20–80 for feeding (observed to herd prey fishes), courtship, mating and birth. Some are highly migratory, moving to cooler water in summer.
Biology. Two young born every other year, one from each uterus. Each surviving embryo kills and eats smaller embryos and feeds on unfertilised eggs during 9–12 month pregnancy. One unborn pup bit an investigating scientist. Feeds on a wide range of fishes and invertebrates.
Status. IUCN Red List: Vulnerable. Many populations seriously depleted. Critically Endangered in NSW, Australia, after large numbers killed in sports and commercial fisheries and by divers. Legally protected in many countries. Kept in many public aquaria, docile and has bred in captivity. Important for dive ecotourism in South Africa and Australia, may bite if approached too closely.

Smalltooth Sandtiger or Bumpytail Raggedtooth *Odontaspis ferox*

Plate 27

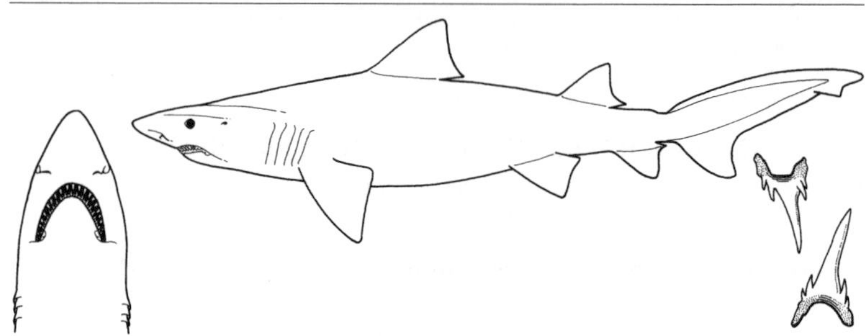

Measurements. Born: >105 cm TL. Mature: 275 cm ♂, 364 cm ♀. Max: >410 cm TL.
Identification. Differs from *Carcharias taurus* in long conical snout, fairly large eyes, first dorsal fin closer to pectoral fin bases than pelvics and larger than second dorsal and anal fins. Grey or grey-brown above, lighter below, often with darker spots.

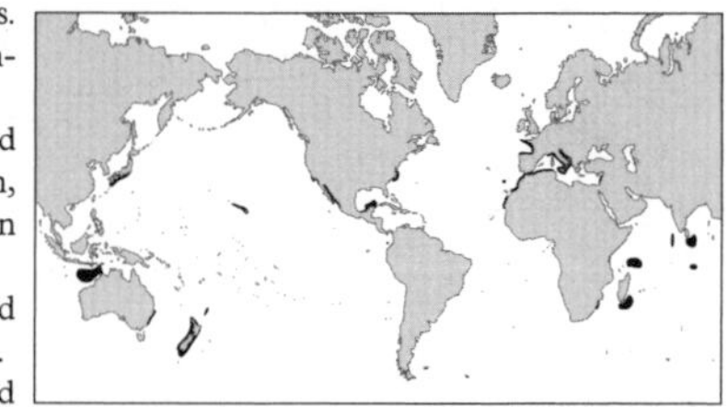

Distribution. May be world-wide in warm-temperate and tropical deep water.
Habitat. On or near bottom, continental and insular shelves and upper slopes, 13–420 m, possibly epipelagic, 140–180 m. Sometimes seen near coral reef dropoffs, rocky reefs and gullies.
Behaviour. Active offshore swimmer, reported alone and in small groups near reefs and gullies.
Biology. Reproduction poorly known. Presumed viviparous, pups nourished by oophagy. Litter size unknown. Prey: small bony fishes, squid and shrimp.
Status. IUCN Red List: Vulnerable. Rare and declining.

Bigeye Sandtiger *Odontaspis noronhai*

Plate 27

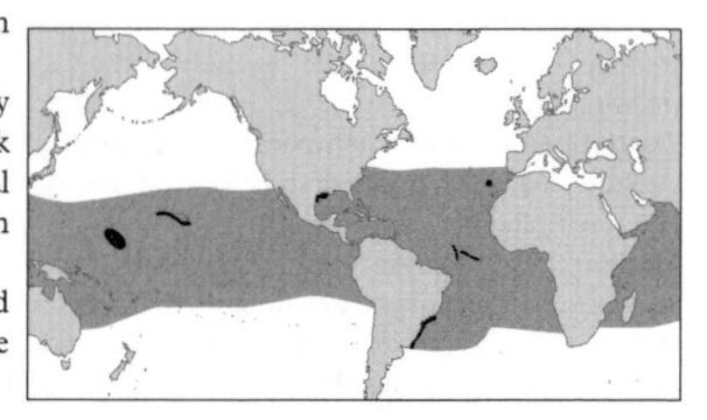

Measurements. Born: unknown. Mature: >220 cm ♂, ~325 cm ♀. Max: at least 360 cm TL.
Identification. Differs from other sandtigers by uniform unspotted dark reddish-brown to black colour above and below. Large eyes. First dorsal fin larger than second dorsal and anal fins, often has white blotch on tip.
Distribution. Known from a very few confirmed records in Atlantic and Central Pacific. May be world-wide in deep warm seas.

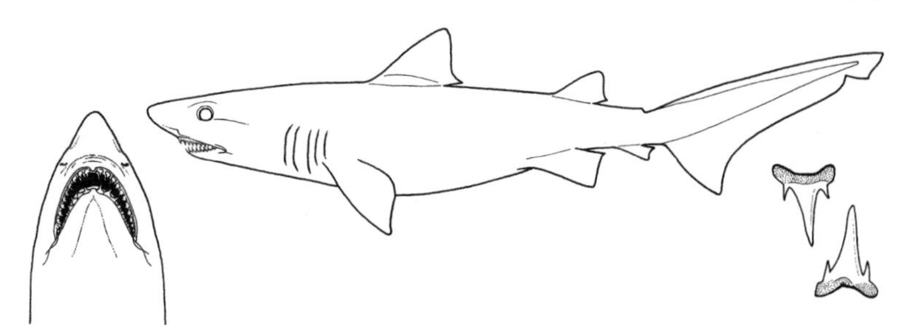

Habitat. Midwater in the open ocean, near bottom on continental and island slopes, 600 to >1000 m. Uniform dark colour suggests this is an oceanic midwater species.
Behaviour. Poorly known. May migrate vertically in mid-ocean (near surface by night, deep water by day).
Biology. Unknown.
Status. IUCN Red List: Data Deficient.

Pseudocarchariidae: Crocodile Shark

One species.

Crocodile Shark *Pseudocarcharias kamoharai* Plate 27

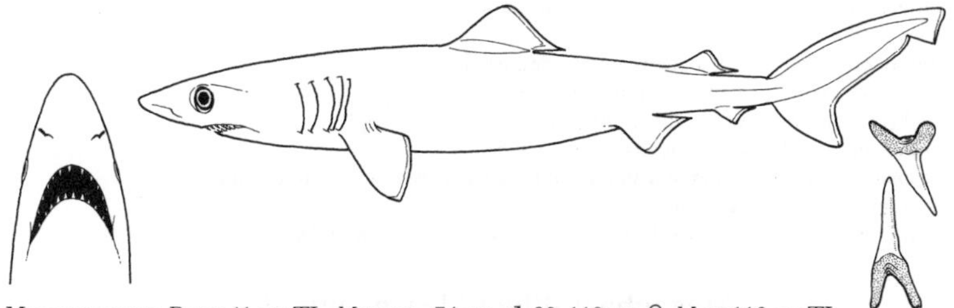

Measurements. Born: 41 cm TL. Mature: ~74 cm ♂, 90–110 cm ♀. Max: 110 cm TL.
Identification. Grey or grey-brown back, lighter below, light-edged fins. Very distinctive small, slender-bodied oceanic shark with small fins, huge eyes, long gill slits, and prominent long slender teeth on highly protrusable jaws.
Distribution. World-wide, oceanic tropical waters.

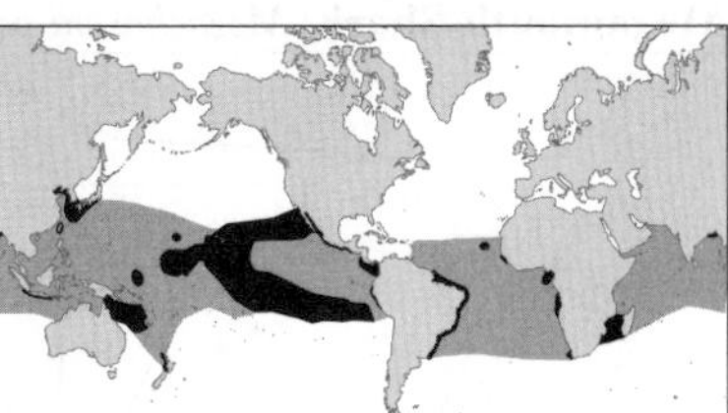

Habitat. Usually well offshore, far from land, from surface to at least 590 m.
Behaviour. Probably a strong active swimmer. May migrate vertically to surface at night, deeper water by day.
Biology. Ovoviviparous, four pups/litter feed on unfertilised eggs and possibly cannibalise other young before birth.
Status. IUCN Red List: Near Threatened. Population depletion very likely as a result of bycatch in pelagic longline fisheries.

Mitsukurinidae: Goblin Shark

One species.

Goblin Shark *Mitsukurina owstoni* Plate 27

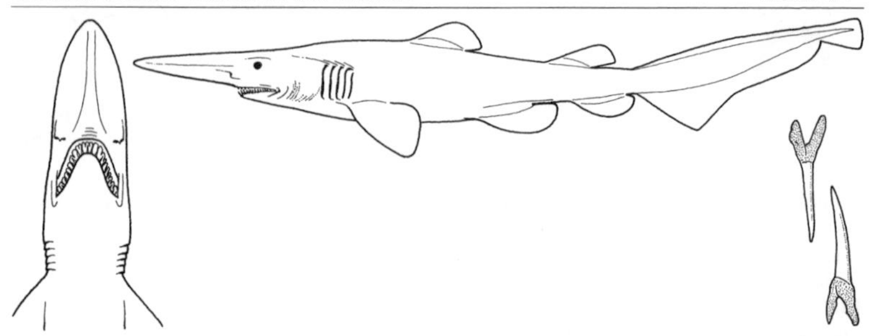

Measurements. Born: unknown, <107 cm. Mature: <264 cm ♂, <335 cm ♀. Max: over 500 cm.
Identification. Body soft, flabby and pinkish-white. Unmistakable flat elongated snout, protrusable jaws, long-cusped slender teeth. Long tail fin, no ventral lobe.

Distribution. Atlantic, western Indian Ocean and Pacific. Patchy distribution.
Habitat. Deepwater, outer continental shelves, upper slopes and off seamounts. Very rarely at surface or 95–137 m, mainly 270–960 m, to at least 1300 m.
Behaviour. Body form suggests poor swimmer. Blade-like snout may be used to detect prey. Highly specialised jaws can shoot forward rapidly to snap up prey. Slender front teeth suggest a diet of small, soft-bodied fishes and squid, but back teeth are modified to crush food.
Biology. Very poorly known.
Status. IUCN Red List: Not Evaluated. Occasionally bycatch in deepwater fisheries.

Megachasmidae: Megamouth Shark

One species.

Megamouth Shark *Megachasma pelagios* Plate 28

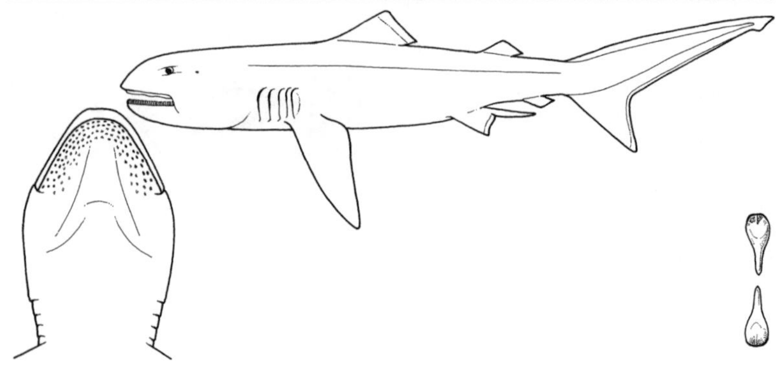

Measurements. Born: unknown. Mature: ~400cm ♂, ~500 cm ♀. Max: >550 cm TL.
Identification. Body grey above (light margins to blackish pectoral and pelvic fins), white below, dark spotting on lower jaw. Unmistakable large long head with short rounded snout. Huge terminal mouth extending behind eyes with numerous small hooked teeth.
Distribution. Probably world-wide in the tropics (not many records).

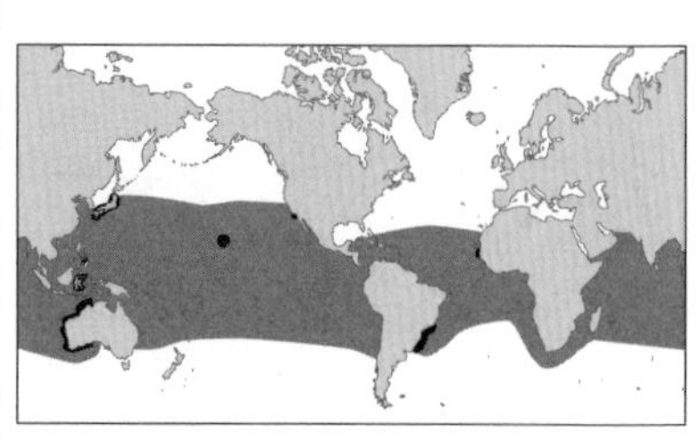

Habitat. Oceanic, coastal and offshore, 5–40 m on continental shelf, 8–166 m offshore over very deep water.
Behaviour. Probably migrates vertically with plankton, close to surface at night, deeper by day. May have luminescent tissue inside mouth to attract prey.
Biology. Feeds on plankton, particularly shrimp, possibly by suction. Reproduction unknown, presumed viviparous with oophagy.
Status. IUCN Red List: Data Deficient. Very rarely recorded.

Alopiidae: Thresher Sharks

Three species of large-eyed sharks with small mouths, large pectoral, pelvic, and first dorsal fins, tiny second dorsal and anal fins, and elongated, curved whip-like caudal fins as long as their bodies.

Pelagic Thresher *Alopias pelagicus* Plate 29

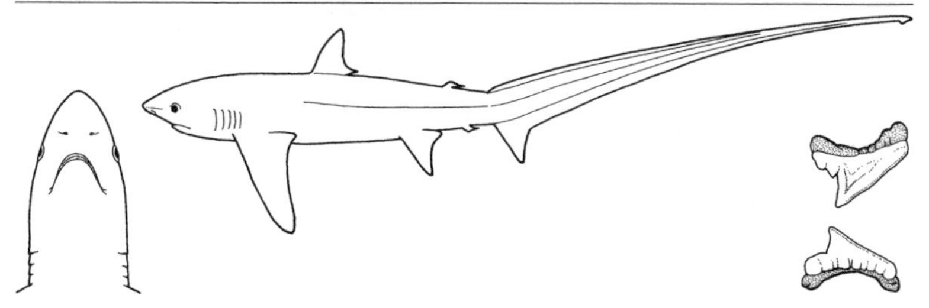

Measurements. Born: 130–160 cm TL. Mature: ~250–300 cm. Max: 365 cm TL.
Identification. Body deep blue above, white below (no white above pectoral fins). Long curving tail, upper lobe nearly as long as rest of shark. Fairly large eyes, very narrow head with straight forehead and arched profile, no labial furrows. Straight broad tipped pectoral fins.
Distribution. Indo-Pacific: South Africa to Australia, Tahiti, China, Japan, USA, Mexico and Galapagos.

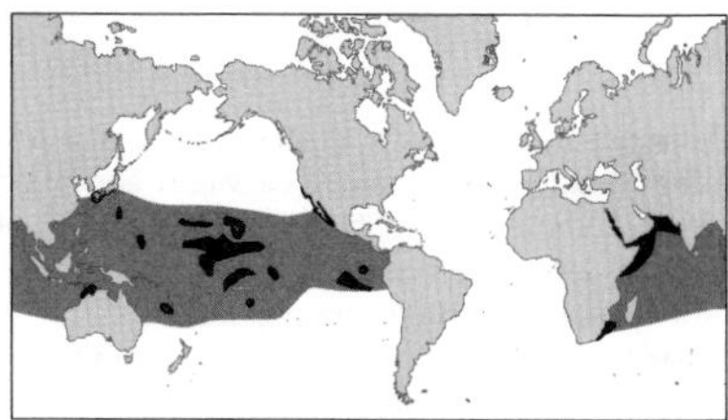

Habitat. Oceanic, wide-ranging, usually offshore, sometimes nearshore on narrow continental shelf, 0–152 m. Sometimes near coral reefs, drop-offs and seamounts.
Behaviour. Poorly-known, active, strong swimmer. Probably migratory. May repeatedly leap out of the water.
Biology. Ovoviviparous, two pups/litter (one from each uterus) feed on unfertilised eggs. Prey unknown, presumably small fishes and squid.
Status. IUCN Red List: Not Evaluated. Highly vulnerable to oceanic fisheries and likely depleted.

Bigeye Thresher *Alopias superciliosus* Plate 29

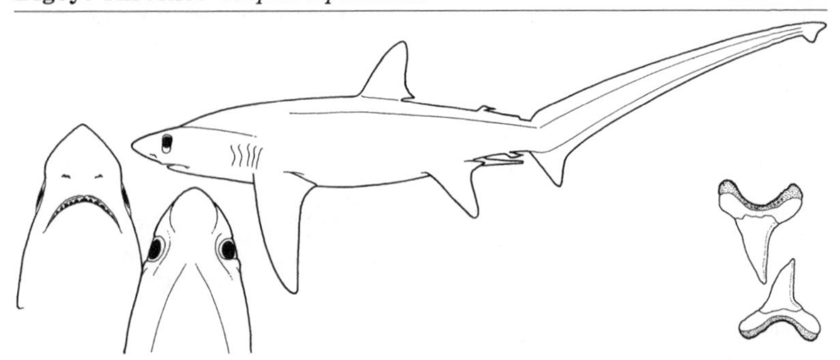

Measurements. Born: 100–140 cm TL. Mature: ~300 cm ♂, 300–350 cm ♀. Max: >460 cm TL.
Identification. Body purplish-grey or grey-brown above, light grey to white below, not extending above pectoral fins bases. Long curving, upper tail lobe nearly as long as rest of shark. Huge eyes extend onto almost flat-topped head. Deep horizontal groove above gills. Very long narrow pectoral fins, large pectoral fins.

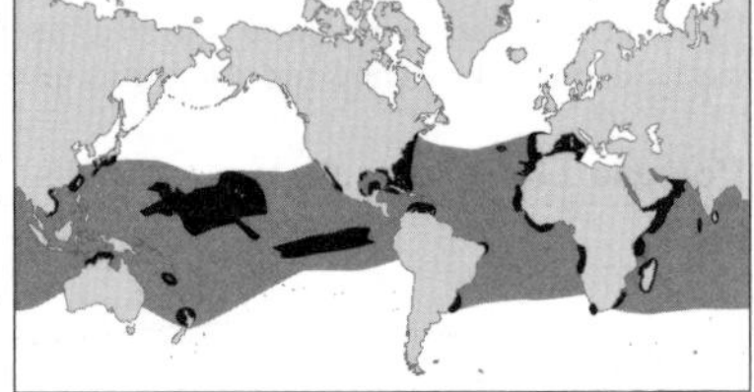

Distribution. World-wide, oceanic and coastal.
Habitat. Tropical and temperate seas, close inshore to open ocean, surface to >500 m, mostly >100 m.
Behaviour. Uses its tail to stun the pelagic fishes on which it feeds.
Biology. Ovoviviparous, two to four pups per litter.
Status. IUCN Red List: Not Evaluated. Highly vulnerable to oceanic fisheries and likely depleted.

Thresher Shark *Alopias vulpinus* Plate 29

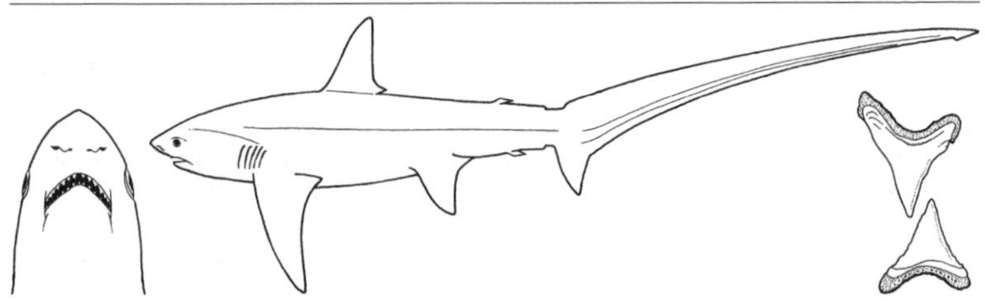

Measurements. Born: 114–160 cm TL. Mature: ~300 cm ♂, 370–400 cm ♀. Max: ?610 cm TL.
Identification. Blue-grey to dark grey above, sides silvery or coppery, underside white, extending in patch above pectoral fins. White dot on tips of narrow, pointed pectoral fins. Upper lobe of tail about as long as rest of shark. Fairly large eyes on side of head. Labial furrows present.

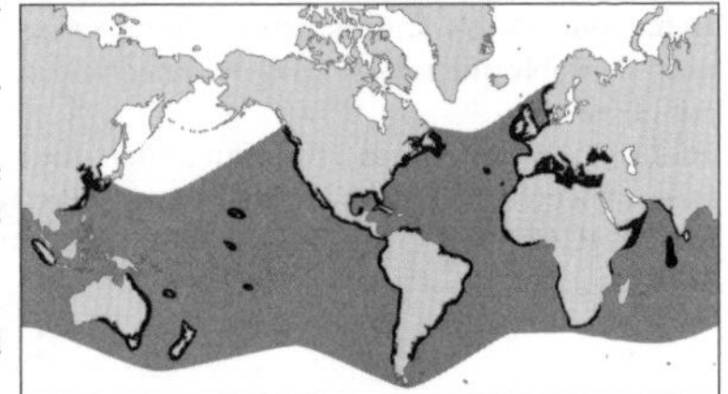

Distribution. Oceanic and coastal, almost world-wide in tropical to cold-temperate seas.
Habitat. From nearshore to far offshore, surface to at least 366 m. Most abundant near land (pups use inshore nurseries) and in temperate water.
Behaviour. Migrates seasonally along the coast. May leap out of the water. Herds and stuns small fishes with its tail, sometimes cooperatively.

Biology. Ovoviviparous, two to six (usually four) pups/ litter feed on infertile eggs.
Status. IUCN Red List: Data Deficient. Near threatened in Californian waters. Highly vulnerable to fisheries, likely depleted.

Cetorhinidae: Basking Shark

One species.

Basking Shark *Cetorhinus maximus* Plate 28

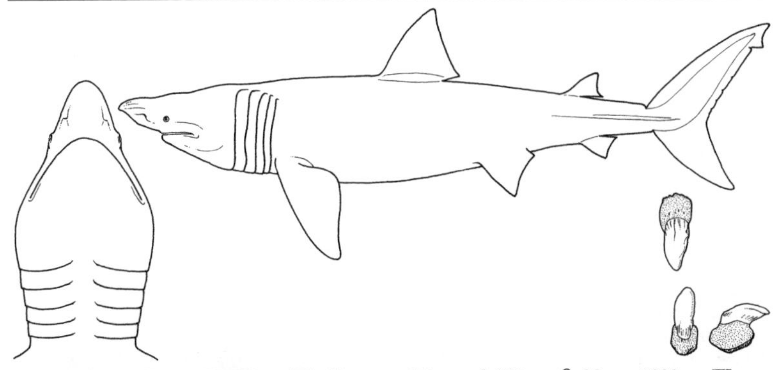

Measurements. Born: 150–170 cm TL. Mature: ~570 cm ♂, 800 cm ♀. Max: >1000 cm TL.
Identification. Colour variable, darker above than below, often with mottled pattern on back and sides, white blotches under head. Unmistakeable very large shark. Pointed snout, huge mouth with tiny teeth, huge gill slits almost encircle the head. Strong lateral keels on caudal peduncle, lunate tail.

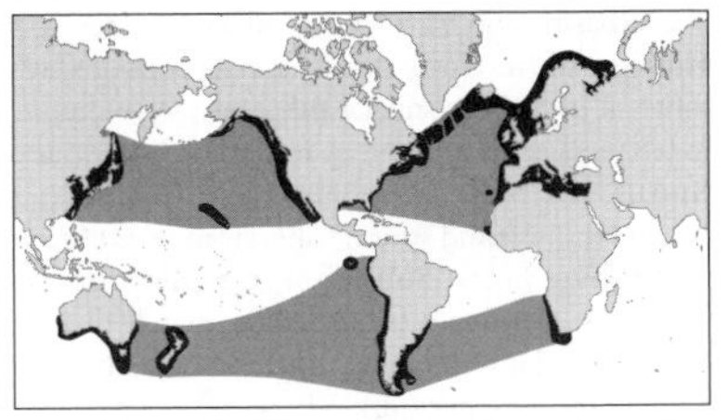

Distribution. World-wide, cold to warm-temperate water.
Habitat. Coast to continental shelf edge and slope. Associated with coastal and oceanic fronts.
Behaviour. Often seen feeding on surface aggregations of plankton, moving slowly forward with open mouth, sometimes in large groups (>100). Feeds on deepwater plankton concentrations in winter. Highly migratory. May shed gill rakers, but no evidence for hibernation in winter. Com-plex courtship behaviour reported. Can leap out of the water.
Biology. Poorly known plankton-feeder. Huge liver provides buoyancy. One litter of six pups reported, presumably oophagous.
Status. IUCN Red List: Vulnerable. Endangered regionally, where seriously depleted by target fisheries for oil, meat and fins. Listed on CITES Appendix II, protected by several countries and in the Mediterranean.

Lamnidae: Mackerel Sharks

Five species of large spindle-shaped sharks with large teeth, conical heads, long gill openings, and crescent-shaped caudal fins with strong caudal keels.

White Shark *Carcharodon carcharias* Plate 30

Measurements. Born: 110–160 cm TL. Mature: 350–400 cm ♂, 450–500 cm ♀. Max: ~600 cm TL.
Identification. Heavy, long-snouted spindle-shaped body, long gill slits, large first dorsal fin with dark free rear tip, tiny second dorsal and anal fins, strong keels on caudal peduncle and crescent-

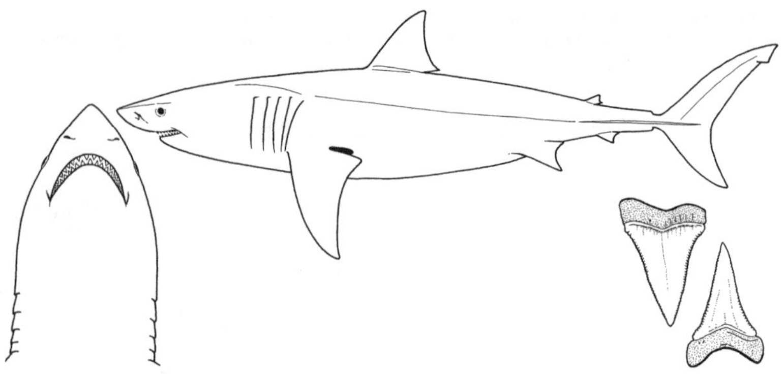

shaped tail. Huge, flat, triangular, serrated teeth. Very black eye. Very sharp colour change on flanks from greyish back to white underside. Black tip underneath pectoral fins, usually a black spot where rear edge joins body.

Distribution. Very wide-ranging through most oceans; among the greatest habitat and geographic range of any fish.

Habitat. Very shallow water inshore, to open ocean and oceanic islands, 0–1300 m, usually seen around rocky reefs near colonies of prey.

Behaviour. Intelligent and inquisitive shark with highly complex social behaviour. Very effective predator, may breach out of the water when attacking prey. Satellite and genetic studies indicate that these sharks are highly migratory, crossing ocean basins and travelling annually between Mexico and Hawaii.

Biology. Warm-blooded, maintaining constant high body temperature even in cold water. Feeds on wide range of prey, from smaller fishes (when young) to large marine mammals when mature. Litters of two to ten pups are nourished by unfertilised eggs during ~12 month gestation at two to three year intervals.

Status. IUCN Red List: Vulnerable. Rare, with evidence of depletion through commercial bycatch, beach meshing and sports fisheries in several parts of the world. Occasionally bites people and jumps into fishing boats. Subject of cage dive tourism and intensive film-making. Protected in several countries. Nominally protected through listing on the Convention for the Conservation of Migratory Species, and international trade regulated through Appendix III of CITES. Not kept successfully in aquaria for long periods.

Shortfin Mako *Isurus oxyrinchus* Plate 30

Measurements. Born: 60–70 cm TL. Mature: ~200–215 cm ♂, 275–290 cm ♀. Max: >400 cm TL.

Identification. Back brilliant blue or purple, underside usually white. Front half of pelvic fins dark, rear and undersides white. Long pointed snout, 'U'-shaped mouth, large blade-like teeth. Underside of snout and mouth white in adults, but dusky in Azores 'marrajo criollo'. Pectoral fins shorter than head. Strong keels on caudal peduncle, crescent-shaped tail fin.

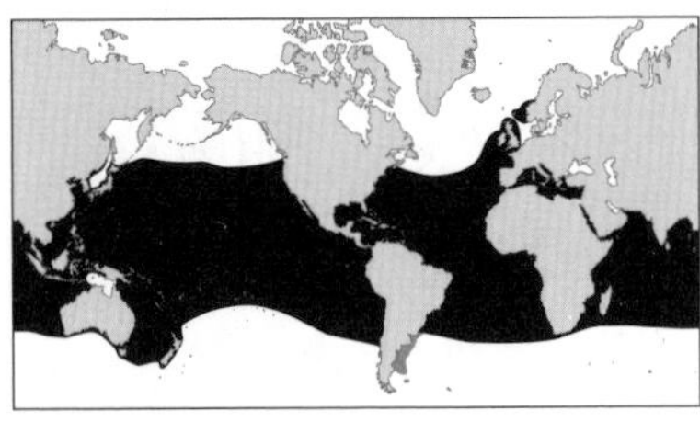

Distribution. World-wide in all temperate and tropical seas.
Habitat. Coastal and oceanic, 0–500 m in water >16°C.
Behaviour. Possibly the fastest shark in the world. Highly migratory. Very active, jumps out of the water.
Biology. Ovoviviparous, 4–25 (mostly 10–18, possibly up to 30) pups/litter feed off unfertilized eggs. Larger females have larger litters. Feeds mainly on fishes and squid, but very large sharks may take small cetaceans.
Status. IUCN Red List: Near Threatened. Taken in bycatch and target fisheries for valuable meat with moderate population declines reported. Important 'big game' fish. May attack if provoked. Subject of dive ecotourism.

Longfin Mako *Isurus paucus* Plate 30

Measurements. Born: 97–120 cm TL. Mature: <245 cm. Max: 417 cm TL.
Identification. Distinguished from *Isurus oxyrinchus* by less pointed snout, pectoral fins as long as head and relatively broad-tipped, and dusky underside to snout and mouth in adults.

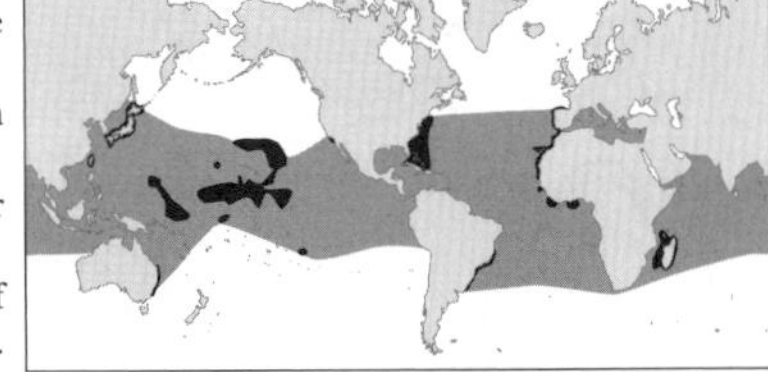

Distribution. Oceanic and tropical, probably world-wide (poorly recorded), common in western Atlantic and possibly Central Pacific, rare elsewhere.
Habitat. Poorly known. Possibly epipelagic in deepwater in the open ocean.
Behaviour. Poorly known. May be a slower swimmer than *I. oxyrinchus*.
Biology. Poorly known. Warm-blooded. Litters of two to eight pups (less fecund than *I. oxyrinchus*). Feeds on fishes and squid
Status. IUCN Red List: Not Evaluated. Presumed vulnerable to fisheries and likely depleted.

Salmon Shark *Lamna ditropis* Plate 30

Measurements. Born: 40–50 cm TL. Mature: ~180 cm ♂, 220 cm ♀. Max: >300 cm TL.
Identification. Dark grey or blackish, underside white with dusky blotches and dark underside of snout in adults. White patch over pectoral fin bases. Heavy body, short conical snout, long gill slits.

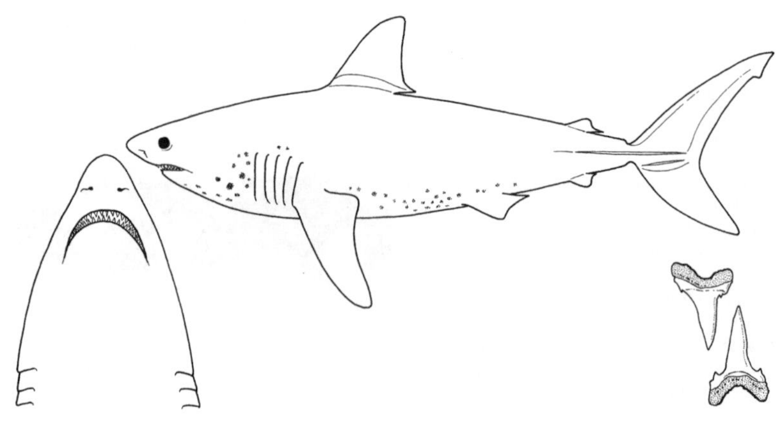

First dorsal fin with dark free rear tip. Strong keels on caudal peduncle, short secondary keels on base, crescent tail.

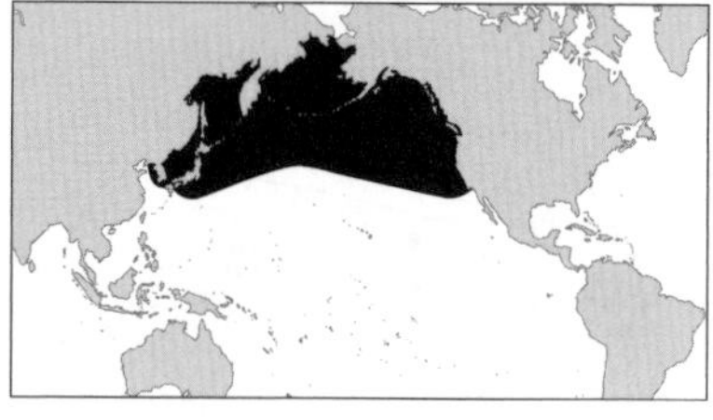

Distribution. North Pacific (males common in west, females in east).

Habitat. Cool coastal and oceanic water, 0 to at least 225m.

Behaviour. Seasonally migratory (following prey). Segregate by age and sex (adults move further north than young).

Biology. High body temperature enables these sharks actively to hunt prey in very cold water. Litters of two to five pups born in spring in nursery grounds. Feed on schooling fishes (salmon, herring, sardines).

Status. IUCN Red List: Data Deficient. Bycatch in pelagic fisheries largely unrecorded.

Porbeagle Shark *Lamna nasus* Plate 30

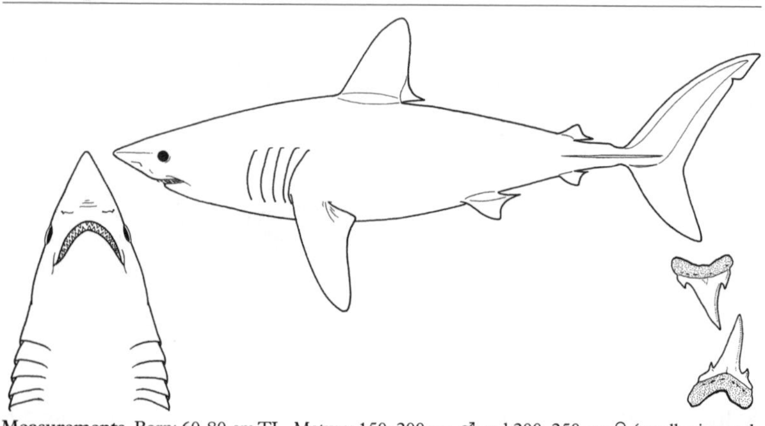

Measurements. Born: 60-80 cm TL. Mature: 150–200 cm ♂ and 200–250 cm ♀ (smaller in south Pacific). Max: >300 cm TL.

Identification. Very similar to Salmon Shark, except for coloration. No white patch above pectoral fin, but a very distinctive white free rear tip to first dorsal fin.

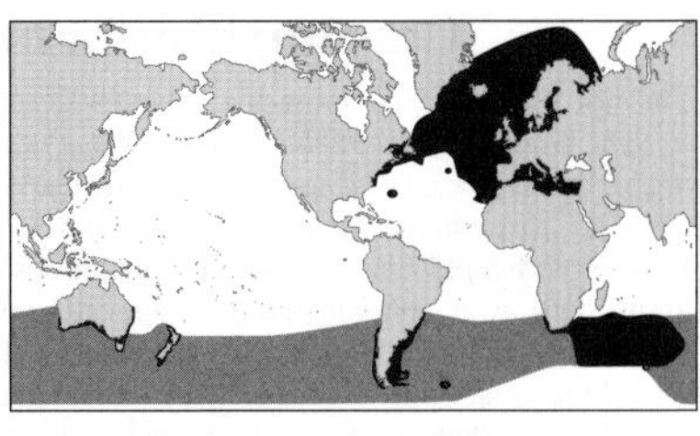

Distribution. North Atlantic and cool water (1–18°C) in southern hemisphere, not in equatorial seas.
Habitat. Inshore to continental offshore fishing banks, occasionally open ocean, <1–700 m.
Behaviour. Migratory, moves inshore and to surface in summer, winters offshore in deeper water. Populations segregated by age (size) and sex. Inquisitive, may approach boats and divers but not dangerous.
Biology. Mostly four pups/litter, feed on unfertilised eggs. Feeds on small fishes, dogfish, tope and squid.
Status. IUCN Red List: Near Threatened. North Atlantic populations seriously depleted (~90%) by commercial fisheries for high value meat. Assessed as Endangered by the Committee on the Status of Endangered Wildlife of Canada, where quotas have been reduced and breeding grounds closed to fisheries. Probably more seriously threatened in northeast Atlantic where fisheries are unmanaged. Southern hemisphere status unknown. Also a game fish.

Carcharhiniformes: Ground Sharks

This order is the largest, most diverse and widespread group of sharks. It contains about 225 species in eight families: *Scyliorhinidae* (catsharks), *Proscylliidae* (finback catsharks), *Pseudotriakidae* (false catsharks), *Leptochariidae* (barbeled houndsharks), *Triakidae* (houndsharks), *Hemigaleidae* (weasel sharks), *Carcharhinidae* (requiem sharks), and *Sphyrnidae* (hammerhead sharks). Most are small and harmless to people, but this order also includes some of the largest predatory sharks.

Identification. A very wide range of appearances, from strange bottom-living deepwater sharks, to typical large sharks. All have two spineless dorsal fins and an anal fin. A long mouth extends to or behind the eyes, which are protected by nictitating lower eyelids. Nasoral grooves are usually absent (broad and shallow when present in a few catsharks). If barbels are present, these are developed from anterior nasal flaps of nostrils. Largest teeth are well lateral on dental band, not on either side of symphysis, with no gap or intermediate teeth separating the large anterior teeth from still larger teeth in upper jaw. Intestine usually has a spiral or scroll valve.

Distribution. World-wide, from cold to tropical seas, intertidal to deep ocean, to pelagic in the open ocean. Some are poor swimmers and restricted to small areas of seabed, others are strong long-distance swimmers and highly migratory.

Biology. Very varied reproductive strategies, also very poorly known in some groups. Many species are oviparous (egg-laying), some depositing eggs on the seabed with embryos developing for up to a year before hatching, others retaining the eggs until close to hatching. More evolutionarily advanced families retain foetuses inside the female, nourished by the yolk sac or by a placenta, until live young are born.

Scyliorhinidae: Catsharks

By far the largest shark family; at least 160 species in 16 genera (~133 species covered here), with considerable taxonomic research still needed. More species are continually being discovered and described as commercial fisheries and research efforts move into deeper water – we unfortunately cannot include all recently discovered species here. Catsharks are found world-wide, from tropical to arctic waters, usually on or near the seabed, from the intertidal to the deep sea (over 2000 m), but often restricted to relatively small ranges. Many are rarely seen deepwater species, known from very few specimens (over a third of *Apristurus* spp. are known from only one scientifically described specimen and in several cases the original and only specimen has been lost).

Identification. Usually small (<80 cm long; some may mature at ~30 cm, a few reach ~160 cm). Elongated body, two small spineless dorsal fins (first dorsal base over or behind pelvic bases) and an anal fin. Long arched mouth reaches past the front end of the cat-like eyes. Members of some genera (e.g. *Apristurus* demon catsharks) are very difficult to tell apart. The plates group the most similar species to make comparisons easier, but in cases of difficulty readers should refer to Volume III of the *FAO Catalogue of Sharks of the World* (Compagno in prep.) and, if possible, preserve any specimens obtained.

Biology. Mostly poorly known (particularly deepwater species). Many are oviparous (egg laying). The most primitive species lay many pairs of large eggs (one from each oviduct), protected by tough eggcases with corner tendrils, onto the seabed. Hatching may take nearly a year. More advanced species retain their eggs until the embryos' development is almost complete, laying larger numbers of eggs about a month before hatching. A few retain the eggs until the embryos are fully developed and give birth to live young (ovoviviparity). There are poor swimmers that do not undertake long distance migrations. Some inshore species are nocturnal. They may sleep in groups in crevices by day, moving out to feed at night. They eat benthic invertebrates and small fishes.

Status. None are dangerous to people. A few are important in fisheries; many are taken as a bycatch. Some are regularly kept and breed in aquariums.

White-bodied Catshark *Apristurus albisoma* Plate 31

Measurements. Mature: ~40–50 cm. Max: 57 cm ♀, 60 cm TL ♂.

Identification. Whitish colour. Broad flattened head, elongated snout, moderately large nostrils, long arched mouth reaching front of very small eyes, very long labial furrows (uppers reaching upper symphysis, lowers about as long as uppers). First dorsal fin base over pelvic fin bases and

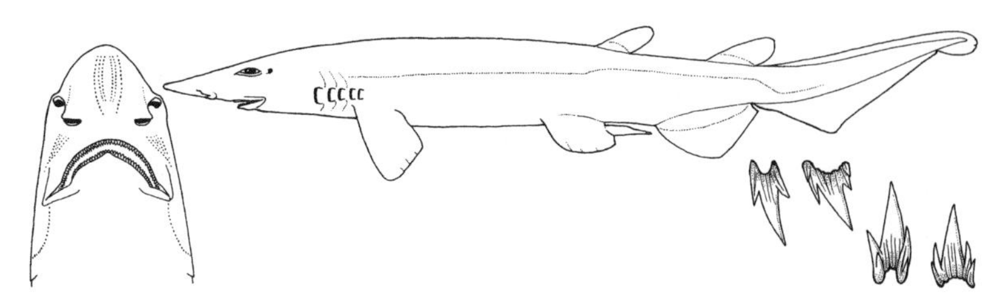

slightly smaller than second. Large short high anal fin separated from elongated tail fin by small notch.

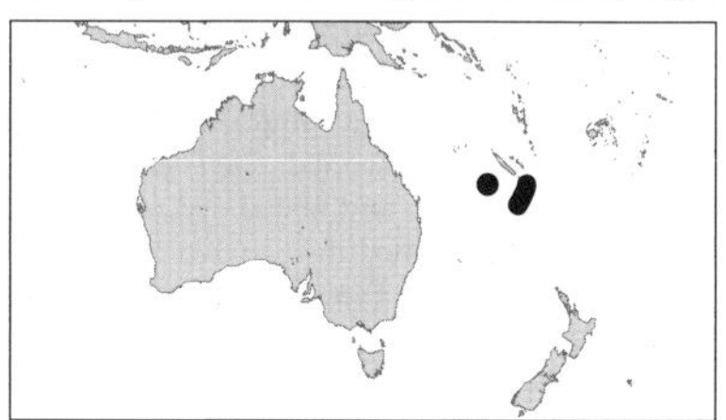

Distribution. Southwest Pacific: off New Caledonia (Norfolk and Lord Howe Ridges). A similar species occurs on ridges south of Madagascar.
Habitat. Deep slope, 935–1564 m, possibly on soft bottom.
Behaviour. Unknown.
Biology. Poorly known. Eats shrimp and cephalopods.
Status. IUCN Red List: Near Threatened. Caught with bottom trawls.

White Ghost Catshark *Apristurus aphyodes* Plate 31

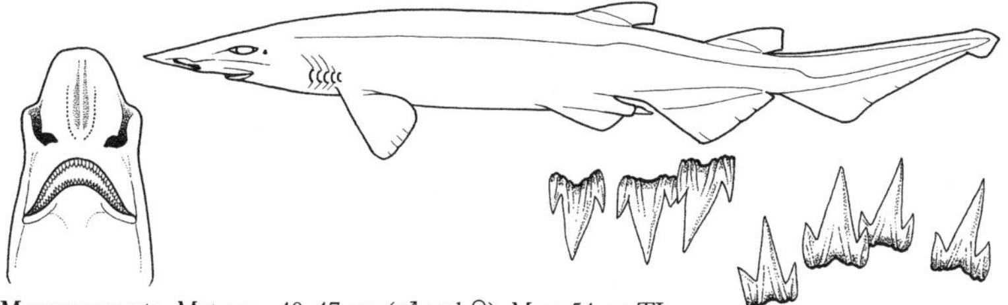

Measurements. Mature: ~40–47 cm (♂ and ♀). Max: 54 cm TL.
Identification. Pale grey, slightly darker grey edges to some fins. Broad flattened head, elongated snout, large nostrils, long arched mouth extending well in front of eyes, very long labial furrows (uppers reaching upper symphysis, lowers about as long as uppers). First dorsal fin base over or behind pelvic fin bases, about as large as second. Large moderately high and short anal fin separated from elongated tail fin by a small notch.

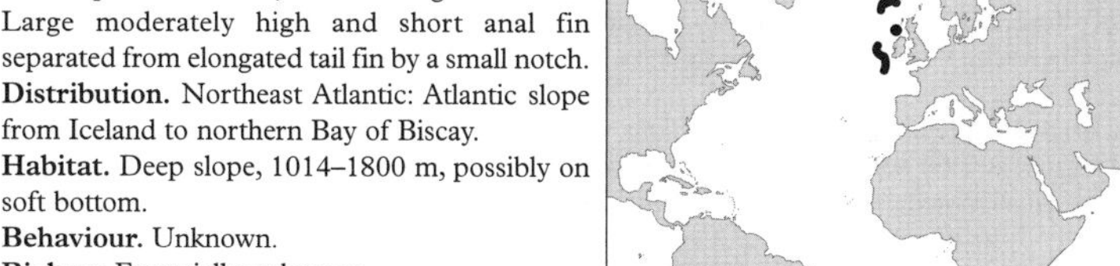

Distribution. Northeast Atlantic: Atlantic slope from Iceland to northern Bay of Biscay.
Habitat. Deep slope, 1014–1800 m, possibly on soft bottom.
Behaviour. Unknown.
Biology. Essentially unknown.
Status. IUCN Red List: Not Evaluated. Caught with bottom trawls.

Brown Catshark *Apristurus brunneus* Plate 31

Measurements. Eggs: ~5 x 2.5 cm. Hatch: 7–9 cm TL. Mature: 45–50 cm ♂, 42–48 cm ♀. Max: 69 cm TL.
Identification. Dark brown body, obvious light posterior margins on fins and upper tail edge. Broad flattened head, long snout, large nostrils, long arched mouth extending about opposite front of eyes,

very long labial furrows (lower shorter than upper). Gill slits < adult eye length. Dorsal fins equal sized, first origin over pelvic midbases. Small notch between very large elongated anal fin and elongated tail fin.

Distribution. East Pacific. Southeast Alaska to north Mexico. Possibly Panama, Ecuador and Peru.

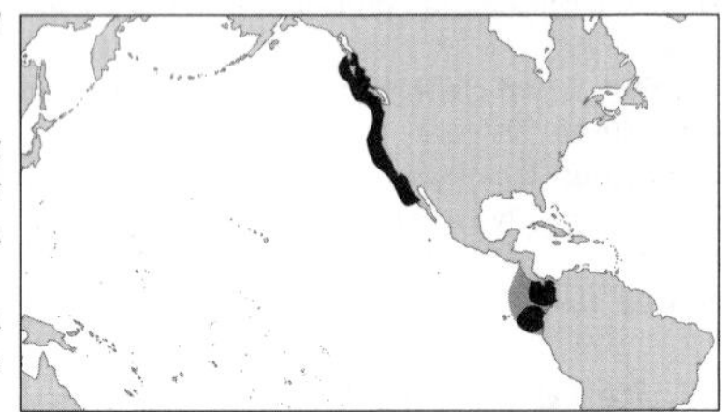

Habitat. Outer continental shelf and upper slope, 33–1298 m, on and well above bottom.

Behaviour. Unknown.

Biology. Oviparous. Pairs of eggs with long tendrils laid in spring and summer (Canada) may take one year to hatch. Feed well off bottom on small shrimp, squid and small fishes.

Status. IUCN Red List: Not Evaluated. Bycatch in deepwater fisheries. Probably two species recorded as *Apristurus brunneus.*

Hoary Catshark *Apristurus canutus* Plate 31

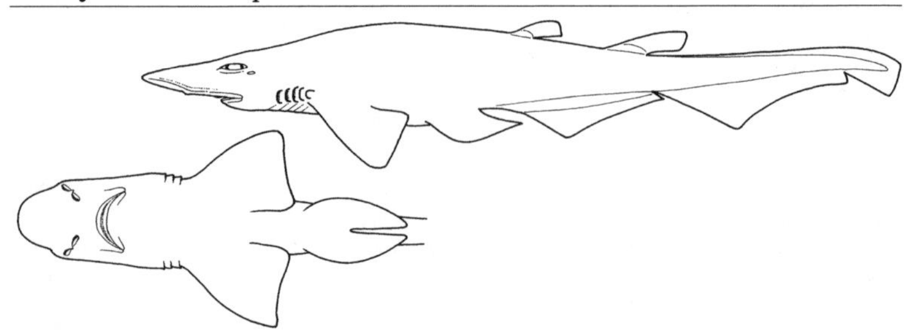

Measurements. Mostly unknown. Adult: 40–45 cm TL. Max: ~46 cm TL.

Identification. Dark grey with blackish fin margins. Broad flattened head, rounded snout, large nostrils, long arched mouth mostly under eyes, very long labial furrows (uppers reaching upper symphysis, lowers shorter than uppers). Eyes > widest gill slit. First dorsal much smaller than second, originating behind pelvic fin insertions. Pectoral and pelvic fins fairly close together. Very large elongated anal fin separated from elongated tail fin by small notch, anal fin base 2.5–3 x height.

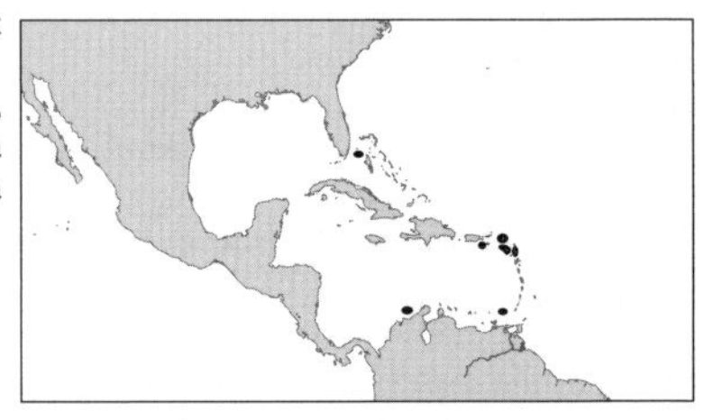

Distribution. Caribbean: Straits of Florida, Leeward Islands, Netherlands Antilles, Colombia and Venezuela. Nominal records from China uncertain.

Habitat. Insular slopes, 521–915 m.

Behaviour. Unknown.

Biology. Unknown.

Status. IUCN Red List: Not Evaluated.

Flaccid Catshark *Apristurus exsanguis* Plate 31

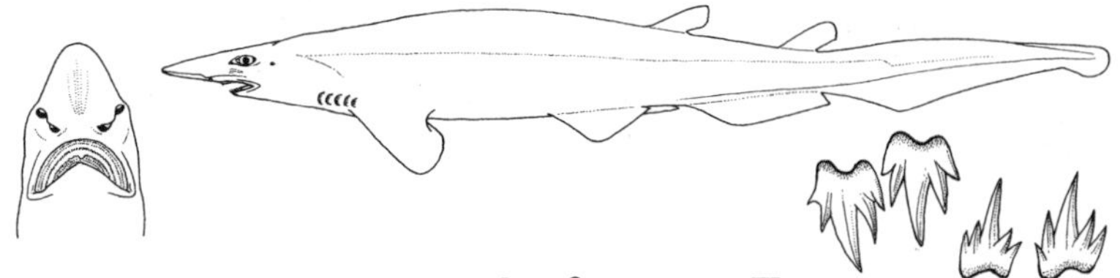

Measurements. Mature: ~65–70 cm TL (♂ and ♀). Max: 91 cm TL.
Identification. Pale grey to pale brown flaccid (seemingly bloodless) body. Broad flattened head, moderately elongated snout, large nostrils, long arched mouth extending slightly in front of eyes, very long labial furrows (uppers reaching upper symphysis, lowers shorter than uppers). Dorsal fins similar sized, base of first over rear of pelvic fin bases. Very long low elongated anal fin separated from elongated tail fin by a small notch.

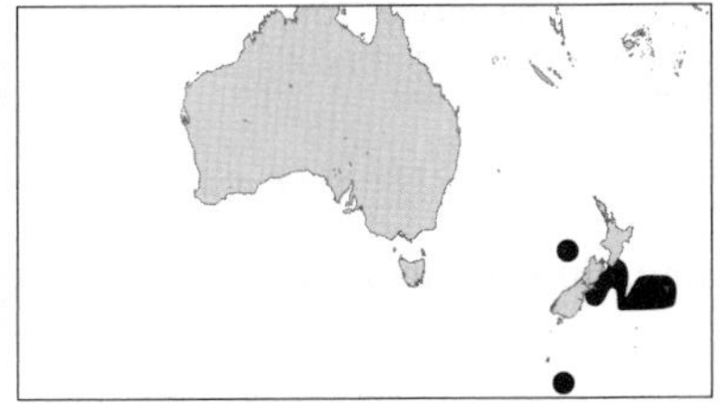

Distribution. Widespread around New Zealand and surrounding ridges and islands, including Chatham, Stewart and Campbell Islands.
Habitat. Insular slopes, 573–1200 m.
Behaviour. Unknown.
Biology. Little-known. Development oviparous, lays eggs in large grooved capsules with long tendrils.
Status. IUCN Red List: Least Concern.

Stout Catshark *Apristurus fedorovi* Plate 33

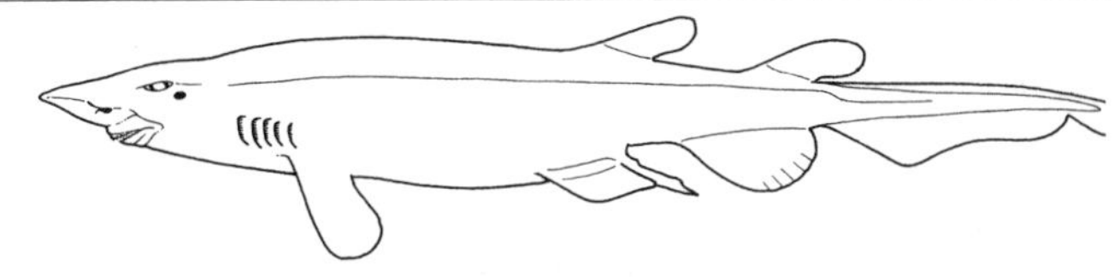

Measurements. Mostly unknown. Adult ♂: ~55 cm TL. Max: at least 68 cm TL.
Identification. Dark brown body. Broad flattened head, elongated snout, very small eyes, large nostrils, long arched mouth extending well in front of eyes, very long labial furrows (uppers reaching upper symphysis, lowers possibly as long as uppers). First dorsal slightly smaller or subequal to second, base over pelvic fin bases. Very large high rounded anal fin separated from elongated tail fin by a small notch.

Distribution. Northern Japan, including Honshu Island, 810–1430 m.
Habitat. Unknown.
Behaviour. Unknown.
Biology. Unknown.
Status. IUCN Red List: Not Evaluated. Close to *Apristurus microps* from the Atlantic.

Humpback Catshark *Apristurus gibbosus* Plate 33

Measurements. Mostly unknown. Type specimens 39–41 cm TL. Max: probably larger.
Identification. Dark (dusky) body, no light fin margins. Broad flattened head, small eyes, elongated and very broad spatulate snout, prominent humped back behind very flat head, large nostrils, long arched mouth reaching front of eyes, very long labial furrows (uppers reaching upper symphysis, lowers shorter than uppers). First dorsal fin base over pelvic fin bases, slightly smaller than second

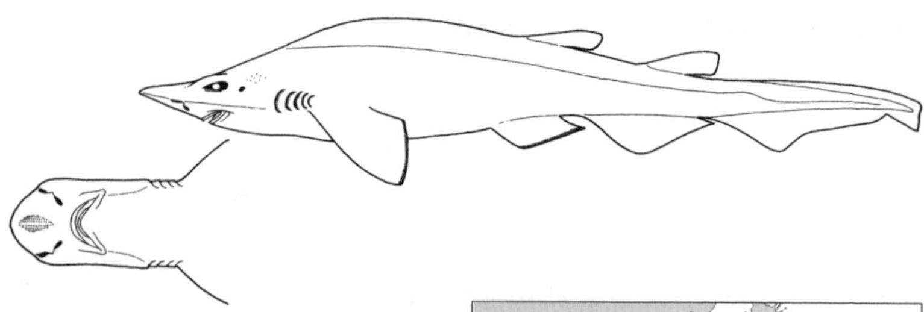

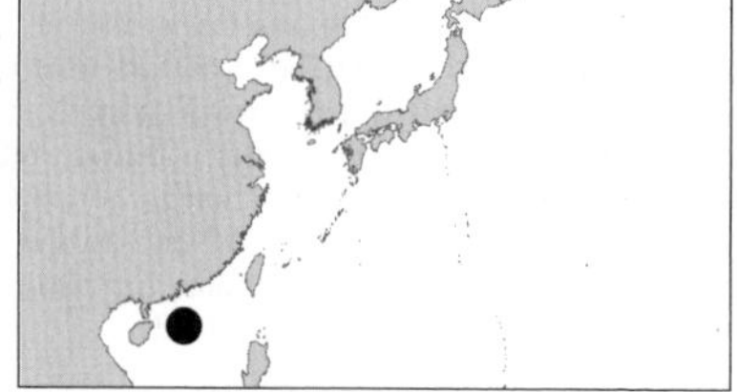

dorsal. Very large moderately elongated, high and angular anal fin, separated from elongated tail fin by a small notch.
Distribution. Types collected from the South China Sea off southern China.
Habitat. Continental slope, 913 m.
Behaviour. Unknown.
Biology. Unknown.
Status. IUCN Red List: Not Evaluated.

Longfin Catshark *Apristurus herklotsi* Plate 32

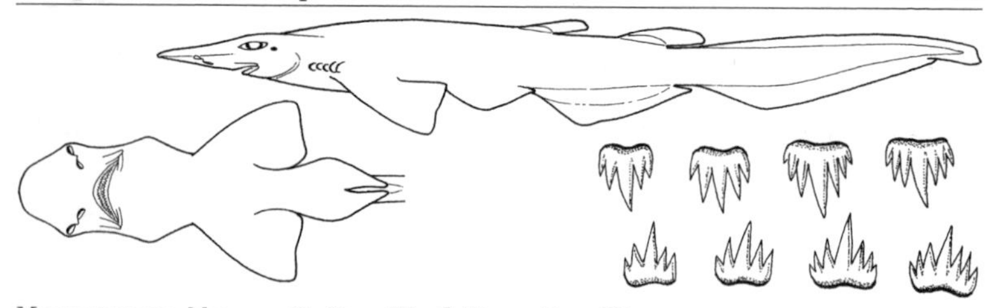

Measurements. Mature: ~40–43 cm TL ♂. Max: ~49 cm TL.
Identification. Small plain brownish to blackish-brown body. Broad flattened head, greatly elongated rounded snout, large nostrils, short mouth reaching front of eyes, very long labial furrows (uppers reaching upper symphysis, lowers shorter), small closely set teeth, short gill slits and incised gill septa. Unusually short abdomen (pectoral and pelvic fins very close together). First dorsal fin ~one-third size of second, originating in front of or about opposite pelvic insertions. Long narrow caudal fin and very long low, angular anal fin separated by small notch.

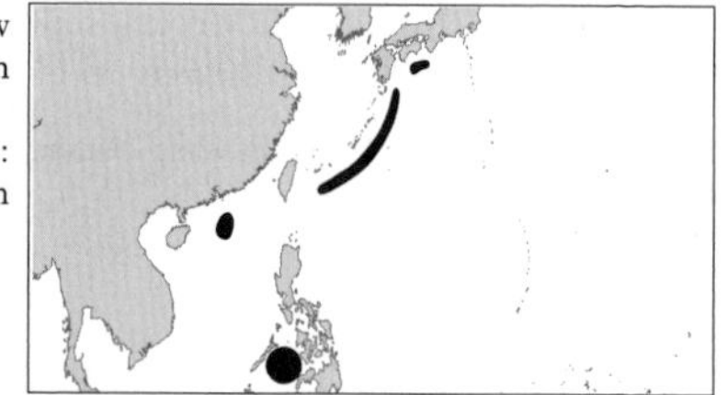

Distribution. South and East China Seas: China, Philippines, Okinawa Trough, southern Japan (Shikoku).
Habitat. Bottom in deepwater, 520–910 m.
Behaviour. Unknown.
Biology. Unknown.
Status. IUCN Red List: Not Evaluated.

Smallbelly Catshark *Apristurus indicus* Plate 31

Measurements. Mostly unknown. Immature: up to 34 cm TL.
Identification. Brownish or blackish body with paired fins, anal fin and caudal fin very close together. Broad flattened head, elongated snout, large nostrils, very short mouth reaching front of eyes, very long labial furrows (uppers reaching upper symphysis, lowers shorter than uppers), gill slits less than adult eye length. Two small spineless dorsal fins, first lower than second and extending

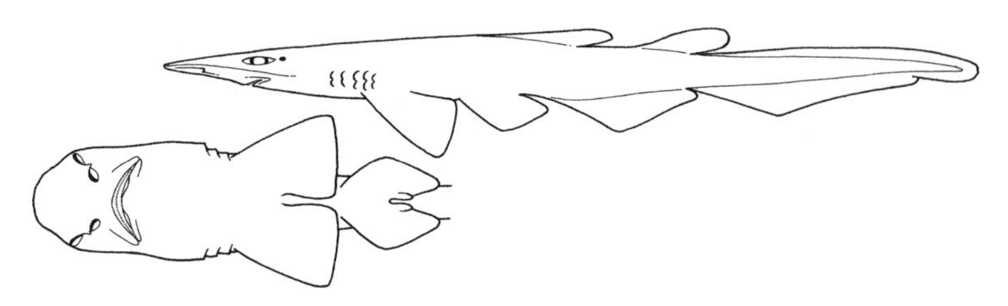

forward as long, low ridge on back.

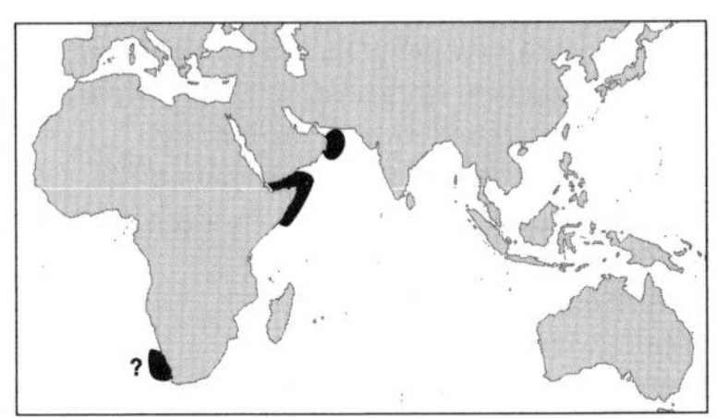

Distribution. West Indian Ocean: Somalia, Gulf of Aden, Oman. Records from European seas and west coast of southern Africa possibly erroneous.
Habitat. Continental slopes, deepwater (1289–1840 m) on bottom.
Behaviour. Unknown.
Biology. Unknown.
Status. IUCN Red List: Not Evaluated. Not fished.

Shortnose Demon Catshark *Apristurus internatus* Plate 32

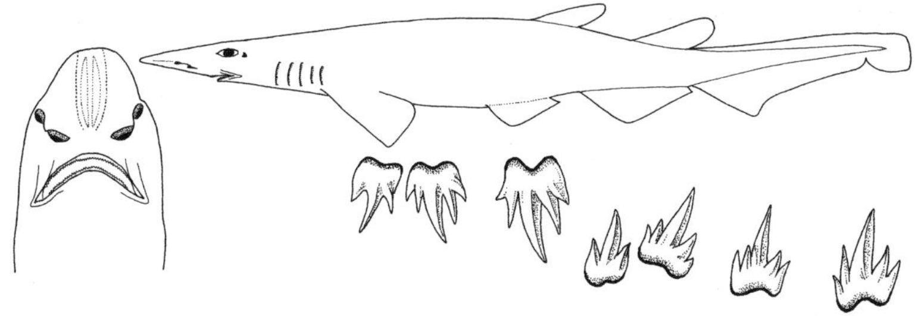

Measurements. Mostly unknown. Type specimens 40–42 cm TL (♀?), probably not mature.
Identification. Dark body, no light fin edges. Broad flattened head, relatively short and very broad snout, large nostrils, long arched mouth reaching front of eyes or slightly in front of them, with very long labial furrows (uppers reaching upper symphysis, lowers shorter). First dorsal base over pelvic fin bases, slightly smaller than second dorsal. Very large moderately elongated, high and angular anal fin, separated from elongated tail fin by a small notch.

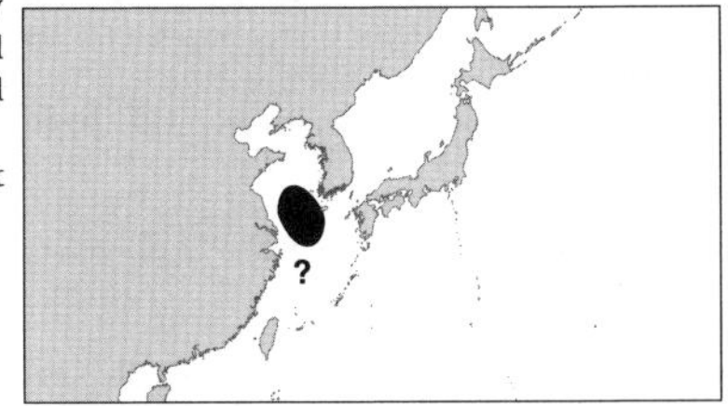

Distribution. East China Sea off China (exact locality of type specimens uncertain).
Habitat. Continental slope, 670 m.
Behaviour. Unknown.
Biology. Unknown.
Status. IUCN Red List: Not Evaluated.

Broadnose Catshark *Apristurus investigatoris* Plate 33

Measurements. Mostly unknown. Female type specimen: 24 cm TL.
Identification. Plain medium brown, no prominent fin markings, fairly well-developed caudal crest of denticles. Very broad flattened head, elongated snout, large nostrils, very short small mouth that extends slightly in front of eyes, very long labial furrows (uppers reaching upper symphysis, lowers about as long as uppers). Gill slits < adult eye length, abdomen short. First dorsal fin ~two-thirds

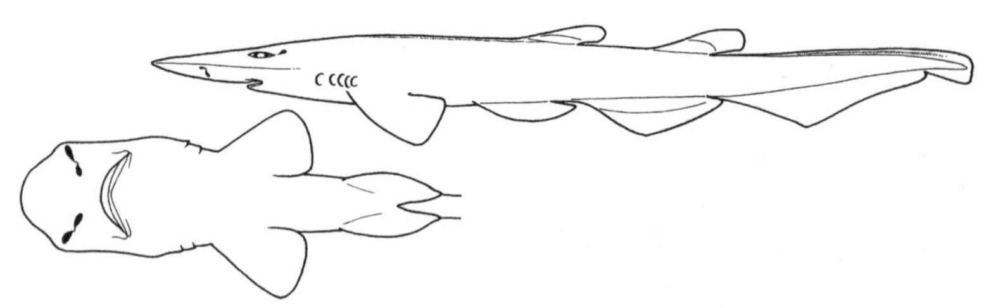

area of second, extending forward as a long low ridge to nearly over pelvic fin origins. Very large low rounded-angular elongated anal fin separated from elongated tail fin by a small notch.

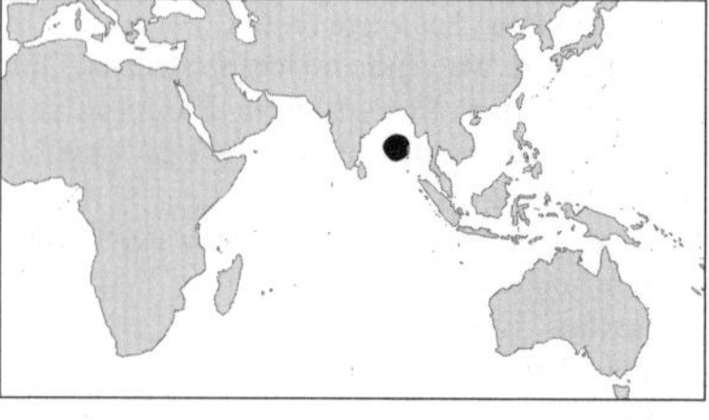

Distribution. Indian Ocean: Andaman Sea.
Habitat. Continental slope, deepwater (1040 m), on bottom.
Behaviour. Unknown.
Biology. Unknown.
Status. IUCN Red List: Not Evaluated. A broad-nosed short-bodied species similar to this and *Apristurus indicus* occurs south of Madagascar on submarine ridges.

Japanese Catshark *Apristurus japonicus* Plate 32

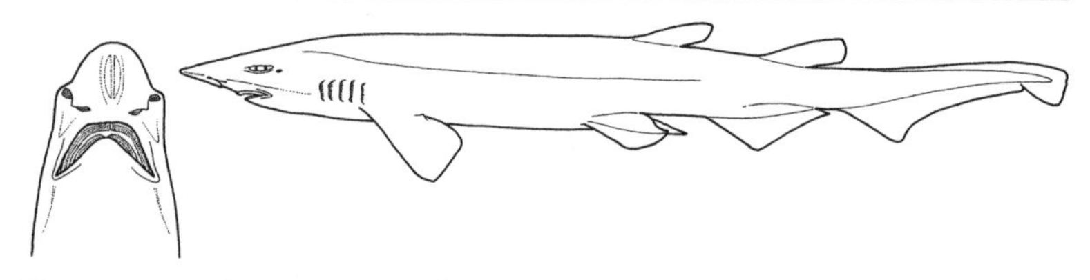

Measurements. Adults at least 65 cm ♀, 65–71 cm TL ♂. Max: 71 cm TL.
Identification. Blackish-brown. Extremely long abdomen (pectoral and pelvic fins widely separated) and short snout (for *Apristurus*). Broad flattened head with large nostrils, mouth large and extending slightly in front of eyes, very long labial furrows (uppers reaching upper symphysis, lowers shorter). Gill slits narrower than adult eye length. Dorsal fins equal-sized, origin of first over pelvic midbases. Very large moderately elongated angular anal fin separated from elongated tail fin by a small notch.

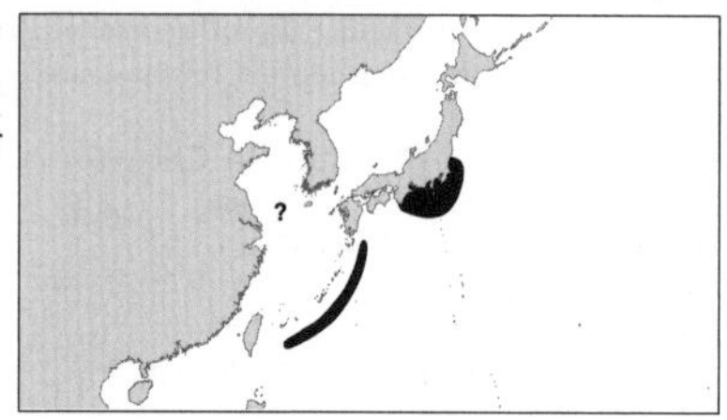

Distribution. Northwest Pacific: Off Japan (Honshu and Okinawa Trough), possibly off China (East China Sea).
Habitat. Slope near bottom, 820–915 m.
Behaviour. Unknown.
Biology. Oviparous.
Status. IUCN Red List: Not Evaluated. Reported to be abundant in limited range.

Longnose Catshark *Apristurus kampae* Plate 32

Measurements. Hatch: ~14 cm TL. Mature: 50 cm ♂, 48–52 cm TL ♀. Max: at least 57 cm TL.
Identification. Blackish or dark brown to grey body and fins, precaudal fins with light edges or nearly uniform coloration. Broad flattened head, elongated snout, large narrow nostrils, long arched mouth extending well anterior to eyes, very long labial furrows (uppers reaching upper symphysis,

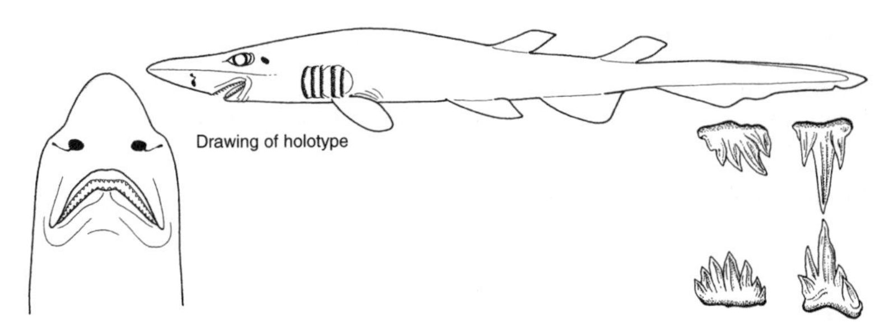

lowers about as long), very wide gill slits (>adult eye length). Similar-sized dorsal fins, first origin behind pelvic fin insertions. Long pectoral and pelvic fin bases. Very high rounded anal fin (length ~2 x height) separated from elongated tail fin by small notch.

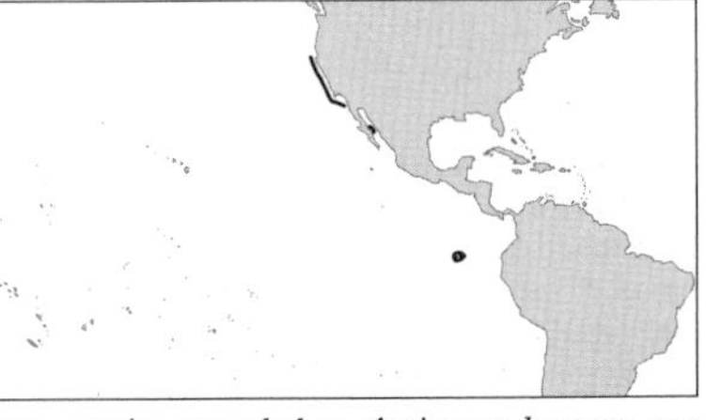

Distribution. Northeast Pacific: USA and Mexico (Oregon to Gulf of California), nominal from Galapagos Islands.
Habitat. Outer continental shelf and upper slope, deepwater (180–1888 m).
Behaviour. Unknown.
Biology. Oviparous, lays pairs of eggs (one per oviduct). Feeds on deepwater shrimp, cephalopods, and small oceanic bony fishes.
Status. IUCN Red List: Not Evaluated. Bycatch in deepwater trawls and sablefish traps. Possibly two species recorded as *Apristurus kampae*, one undescribed with prominent white fin edges.

Iceland Catshark *Apristurus laurussoni* Plate 31

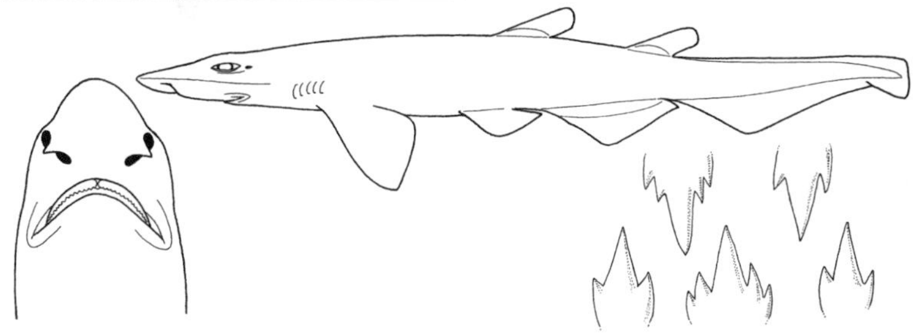

Measurements. Mostly unknown. Adults: ~68 cm ♂, 67 cm TL ♀.
Identification. Dark brown. Broad flattened head, fairly short snout, rather broad nostrils, short arched mouth extending to anterior ends of eyes, very long labial furrows (uppers reaching upper symphysis, lowers shorter). Narrow gill slits (<adult eye length) without prominent medial projections on gill septa. Small eyes. First dorsal slightly larger than second; space between > first dorsal base. Very large moderately elon-gated angular anal fin separated from elongated tail fin by small notch.

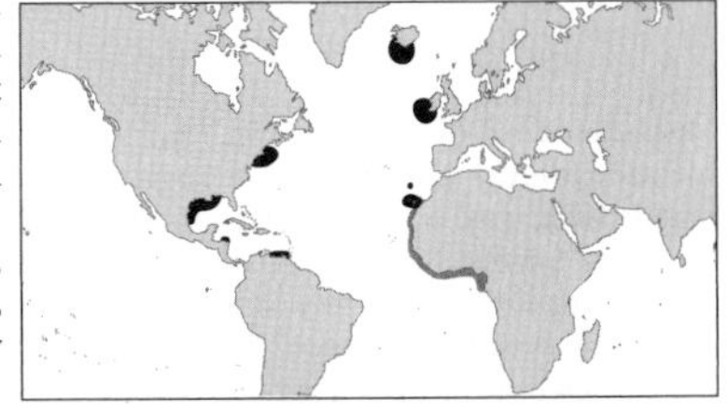

Distribution. West Atlantic: USA (Massachusetts, Delaware), northern Gulf of Mexico, Honduras, Venezuela. East Atlantic: Iceland, Ireland, Canary Islands and Madeira south to equatorial Africa.

Habitat. Continental slopes, deepwater (560–2060 m), on or near bottom.
Behaviour. Unknown.
Biology. Unknown. Presumably oviparous.
Status. IUCN Red List: Not Evaluated. Apparently fairly common. Probably discarded from deep-water trawls.

Longhead Catshark *Apristurus longicephalus* Plate 32

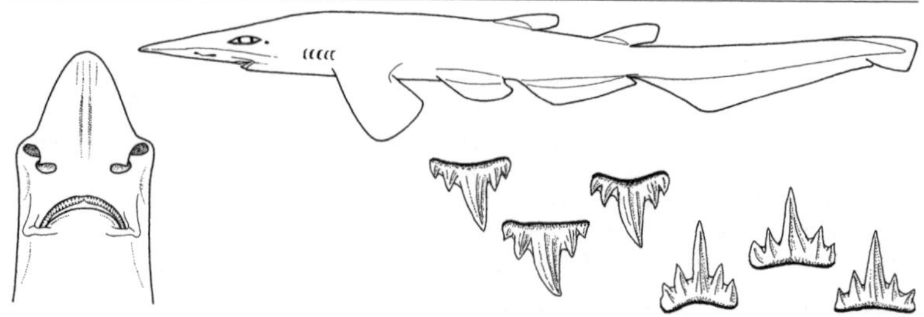

Measurements. Mature: >41 cm TL ♂. Max: at least 50 cm TL.
Identification. Grey-black or dark brown, no conspicuous markings. Broad flattened head, very long snout (preoral length ~12% total length). Small gill slits (<adult eye length). Nostrils rather broad. Short arched mouth extending to anterior ends of eyes, very small wide-spaced teeth, very long labial furrows (uppers reaching upper symphysis, lowers shorter). First dorsal fin above pelvic fin base, close to and slightly smaller than second. Pelvic, pectoral and anal fins close together. Large elongated angular anal fin separated from long, narrow caudal fin by a small notch.

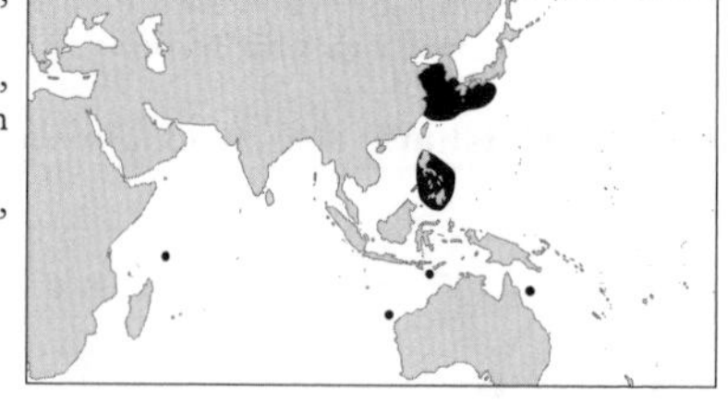

Distribution. West Pacific: Japan, China, Taiwan, Philippines, northern Australia. West Indian Ocean: Seychelles.
Habitat. Deepwater, probably near bottom, 500–1140 m.
Behaviour. Unknown.
Biology. Unknown.
Status. IUCN Red List: Not Evaluated.

Flathead Catshark *Apristurus macrorhynchus* Plate 32

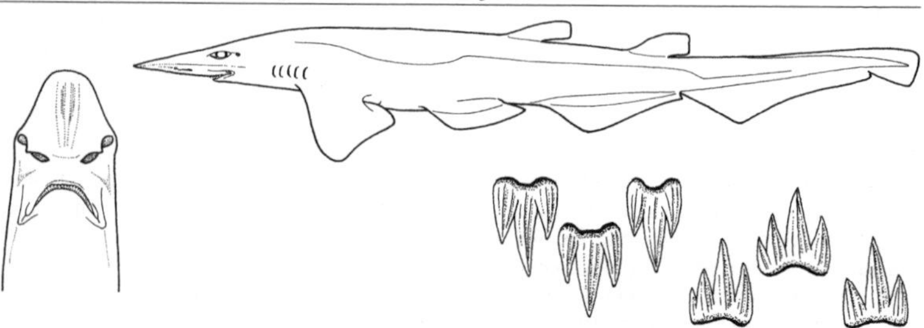

Measurements. Mostly unknown. Max: ~66 cm TL (adult ♀).
Identification. Light grey-brown above, whitish below and on fins. Very broad flattened head, rounded elongated snout, large nostrils, short arched mouth extending to anterior ends of eyes, very long labial furrows (uppers reaching upper symphysis, lowers shorter). Gill slits fairly small (< adult eye length). First dorsal originating over last quarter of long pelvic fin bases, two-thirds size of second, separated by a space > first dorsal base. Very large elongated angular anal fin separated by small notch from elongated tail fin.

Distribution. Northwest Pacific: off Japan (southeastern Honshu to Okinawa Trough), China and Taiwan Island.
Habitat. Island and continental slopes, on the seabed in deepwater (220–1140 m).
Behaviour. Unknown.
Biology. Oviparous, lays pairs of eggs.
Status. IUCN Red List: Not Evaluated. Probably discarded from deepwater trawl bycatch.

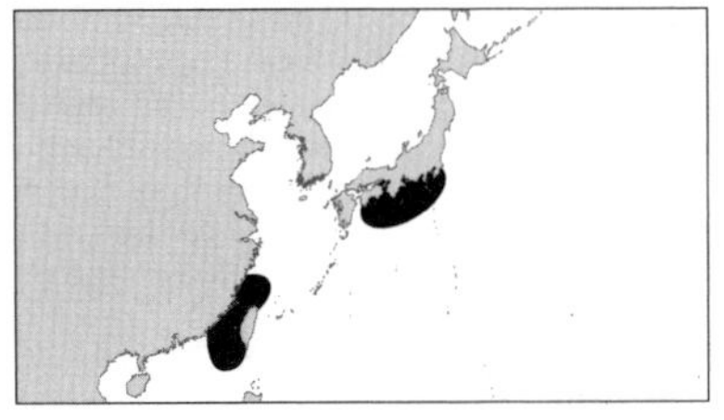

Broadmouth Catshark *Apristurus macrostomus* Plate 33

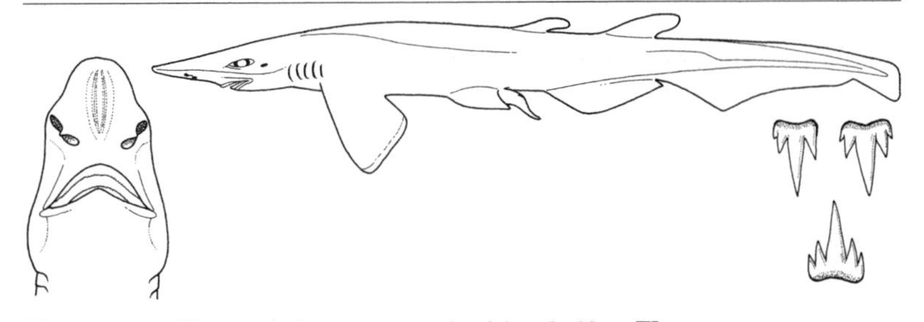

Measurements. Known only from an apparently adult male, 38 cm TL.
Identification. Possibly dark brown or grey-brown above and below, rear fin margins blackish. Broad flattened head, rounded elongated snout, large nostrils, short very large arched mouth extending slightly in front of anterior ends of eyes, very long labial furrows (uppers reaching upper symphysis, lowers shorter than uppers). Gill slits fairly small (< adult eye length). Pectoral fins very large. First dorsal fin originating well behind pelvic fin bases, < half size of second, separated from second by a space greater than first dorsal base. Very large elongated angular anal fin separated by small notch from elongated tail fin.
Distribution. South China Sea.
Habitat. Holotype captured from 913 m.
Behaviour. Unknown.
Biology. Unknown.
Status. IUCN Red List: Not Evaluated.

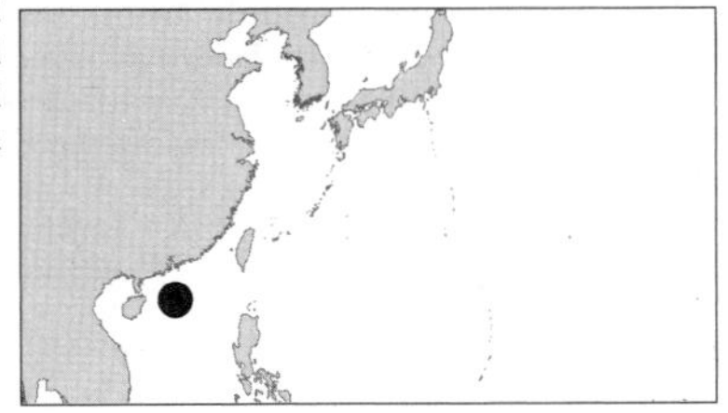

Ghost Catshark *Apristurus manis* Plate 31

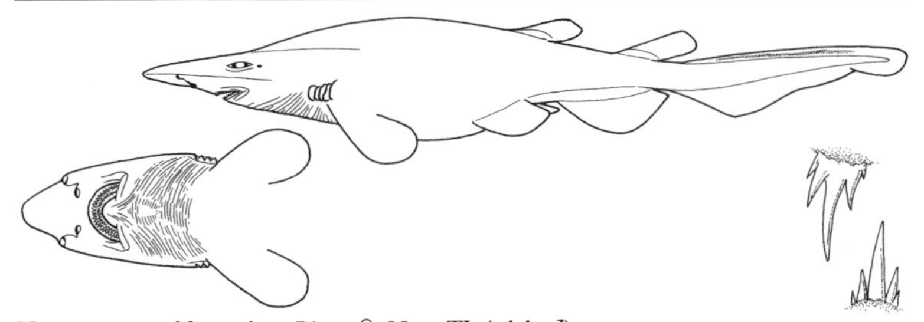

Measurements. Max: at least 76 cm ♀, 85 cm TL (adult ♂).
Identification. Dark grey or blackish, fin tips sometimes whitish (particularly in juveniles), lateral trunk denticles very sparse. Distinctively stout body strongly tapering in a wedge towards broad flattened head and elongated snout. Broad nostrils with circular apertures. Very large long mouth

expanded in front of small eyes, enlarged dental bands, very long labial furrows (uppers reaching upper symphysis, lowers ~ same length). Gill slits < adult eye length. Equal-sized dorsal fins, first originating over pelvic midbases, interdorsal space > first dorsal base. Large broadly rounded anal fin separated by small notch from narrow caudal fin with a crest of enlarged denticles.

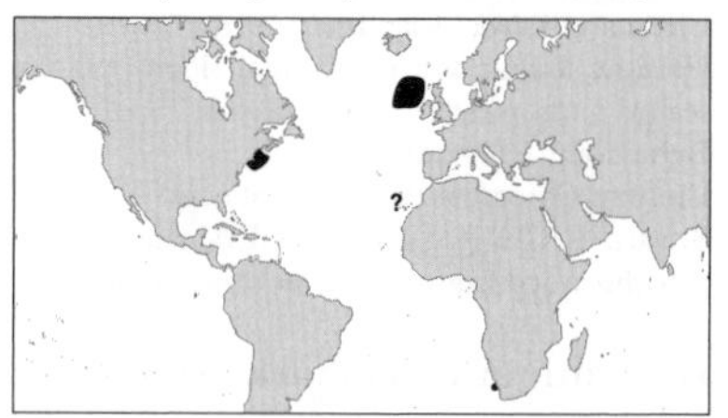

Distribution. North and southeast Atlantic Ocean.
Habitat. Continental slopes, 658–1740 m.
Behaviour. Unknown.
Biology. Unknown.
Status. IUCN Red List: Not Evaluated.

Smalleye Catshark *Apristurus microps* Plate 31

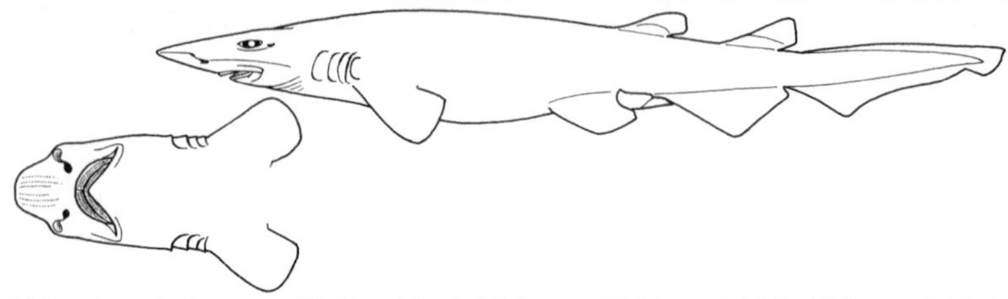

Measurements. Immature: 35–50 cm TL ♂. Adolescent: 47–51 cm ♂. Adult: 47–61 cm ♂, 49–57 cm TL ♀. Max: 61 cm TL.
Identification. Stout dark brown or grey-brown to purplish-black body, no obvious fin markings. Thick long broad snout, mouth projecting forward well in front of extremely small eyes, long labial furrows (uppers reaching upper symphysis, lowers ~same length). Moderately large gill slits (~adult eye length). Very short pectoral fins. Fairly equal-sized dorsal fins, origin of first over rear of pelvic bases, interdorsal space = first dorsal base. Small notch between very large elongated rounded-angular anal fin and elongated tail fin. Crest of denticles above tail.

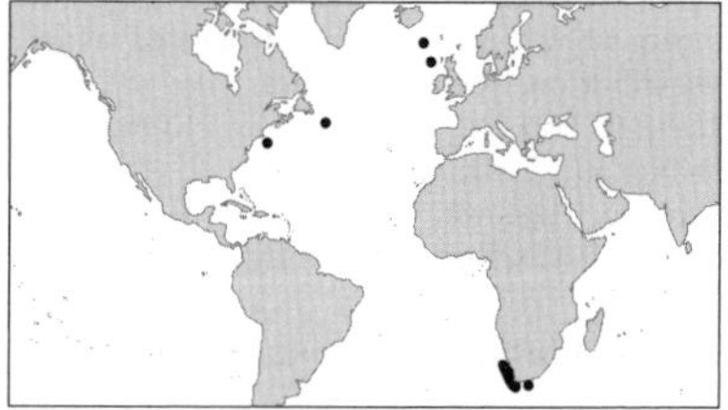

Distribution. North Atlantic (USA, Canada, Irish slope of Europe). South Atlantic. Southwest Indian Ocean (South Africa).
Habitat. Continental slopes, 700–2200 m.
Behaviour. Unknown.
Biology. Eats small bony fishes, shrimp, squid and other small sharks.
Status. IUCN Red List: Not Evaluated. No fisheries interest, probably discarded bycatch from deep trawls.

Smalldorsal Catshark *Apristurus micropterygeus* Plate 33

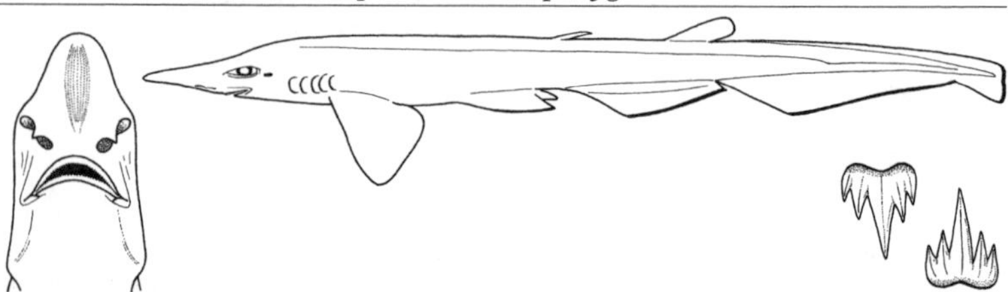

Measurements. Known from one adolescent male, 37 cm TL.
Identification. Possibly grey-brown body, fins not conspicuously marked. Broad flattened head,

rounded rather elongated snout, large nostrils, short arched mouth extending to anterior ends of eyes, very long labial furrows (uppers reaching upper symphysis, lowers shorter). Gill slits fairly small, slightly less than eye length. First dorsal tiny (~one-ninth size of second), originating just behind long pelvic fin bases. Interdorsal space much greater than first dorsal base. Very large elongated angular anal fin separated by small notch from elongated tail fin.

Distribution. South China Sea off China.
Habitat. Taken from 913 m.
Behaviour. Unknown.
Biology. Unknown.
Status. IUCN Red List: Not Evaluated.

Largenose Catshark *Apristurus nasutus* Plate 32

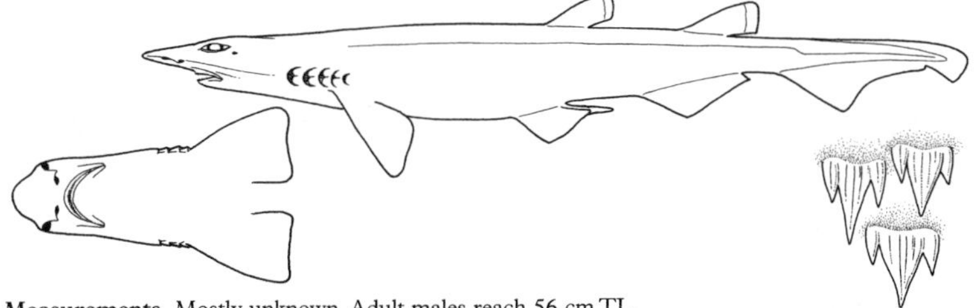

Measurements. Mostly unknown. Adult males reach 56 cm TL.
Identification. Medium brown, grey or grey-blackish body, posterior fin margins pale. Broad flattened head, elongated snout. Mouth extending a short distance in front of small eyes, long labial furrows (uppers reaching upper symphysis, lowers shorter). Gill slits < adult eye length. First dorsal fin slightly smaller than second and originating over pelvic fin midbases. Large angular anal fin separated from elongated tail fin by a small notch.

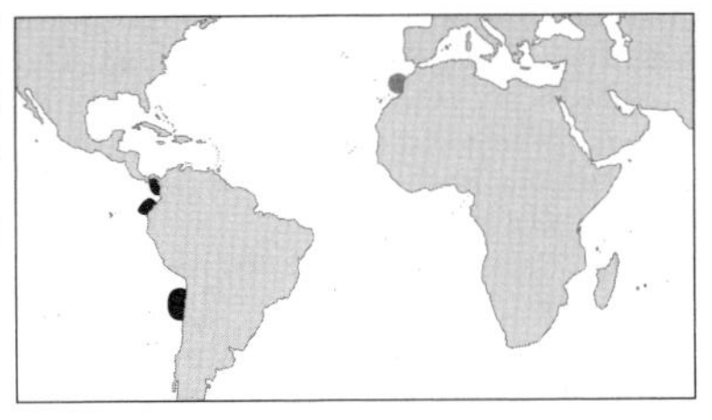

Distribution. East Pacific. Panama, Ecuador and Chile. (Record off Morocco may be an error, based on another species.)
Habitat. Upper continental slopes, on or near bottom, 400–925 m.
Behaviour. Unknown.
Biology. Unknown.
Status. IUCN Red List: Not Evaluated.

Smallfin Catshark *Apristurus parvipinnis* Plate 31

Measurements. Adults to 48 cm ♂, 52 cm TL ♀.
Identification. Grey-brown to blackish. Broad flattened head and broad rounded snout. Mouth

mostly under eyes, long labial furrows (uppers reaching upper symphysis, lowers shorter). Eye length much greater than widest gill slit. First dorsal fin extremely small, originating behind pelvic insertions. Large elongated anal fin separated from tail fin by a small notch. Dorsal caudal margin with fairly prominent crest of enlarged denticles.

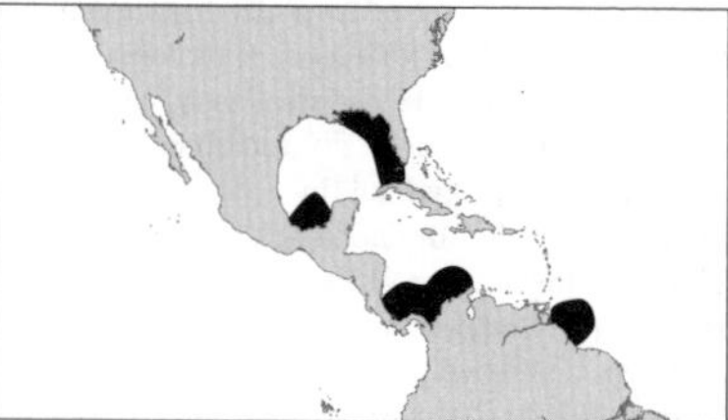

Distribution. West Atlantic: Gulf of Mexico and mainland Caribbean (USA, Mexico, Honduras, Panama, Colombia, Surinam, French Guiana). Nominal Indian Ocean records possibly not this species.
Habitat. Continental slope, on or near bottom, 635–1135 m.
Behaviour. Unknown.
Biology. Oviparous.
Status. IUCN Red List: Not Evaluated. Relatively common in deep trawl catches from Gulf of Mexico.

Fat Catshark *Apristurus pinguis* Plate 33

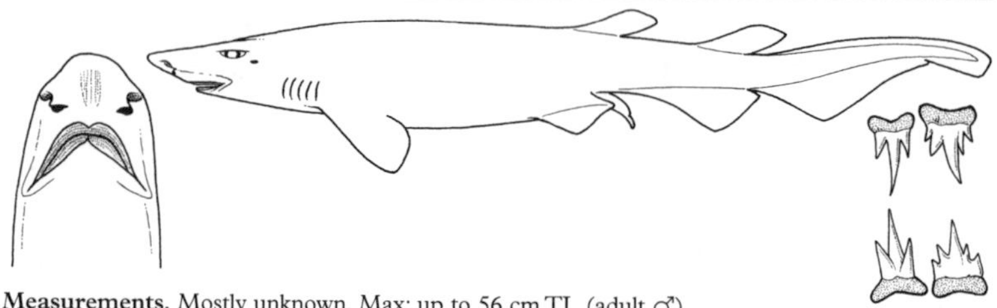

Measurements. Mostly unknown. Max: up to 56 cm TL (adult ♂).
Identification. Stout greyish-brown body, no obvious fin markings. Thick long broad snout, mouth projecting forward well in front of small eyes, long labial furrows (uppers reaching upper symphysis, lowers about same length). Moderately wide gill slits equals eye length in adults). Very short pectoral fins. Dorsal fins fairly equal-sized, interdorsal space roughly equal to first dorsal base, first dorsal origin over rear of pelvic bases. Large elongated, high, rounded-angular anal fin separated from elongated tail fin by a small notch. Possibly a crest of denticles on dorsal caudal margin (uncertain).

Distribution. East China Sea: between China and Ryu-Kyu Islands, Japan.
Habitat. Captured from 200–1040 m.
Behaviour. Unknown.
Biology. Unknown.
Status. IUCN Red List: Not Evaluated.

Spatulasnout Catshark *Apristurus platyrhynchus* Plate 32

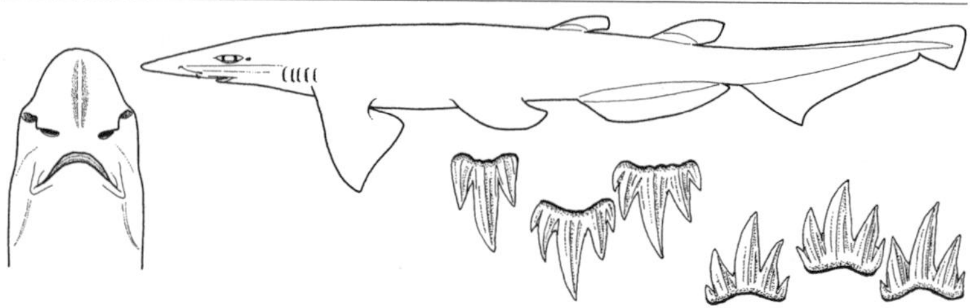

Measurements. Mature: ~60 cm TL (♂ and ♀). Max: 80 cm TL.
Identification. Light brown or grey to dark brown with blackish fin margins. Very broad flattened

rounded snout. Mouth mostly under large eyes (much longer than widest gill slit), long labial furrows (uppers reaching upper symphysis, lowers slightly shorter). First dorsal much smaller than second, originating behind pelvic insertions. Pectoral fins large. High rounded pelvic fins. Small notch separates large elongated angular anal fin from elongated tail fin. Deep notch near top of caudal fin.

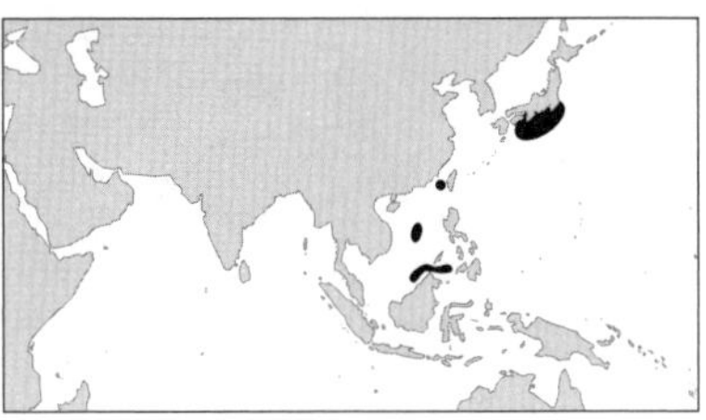

Distribution. West Pacific: southern Japan, China, near Taiwan, Philippines, North Borneo, East and South China Seas, Sulu Sea.
Habitat. Continental and insular slopes, deepwater (594–985 m).
Behaviour. Unknown.
Biology. Oviparous.
Status. IUCN Red List: Not Evaluated. Poorly known, apparently not taken in fisheries. *Apristurus acanutus* (China, South China Sea) and *A. verweyi* (Sabah, Borneo) are junior synonyms.

Deepwater Catshark *Apristurus profundorum* Plate 31

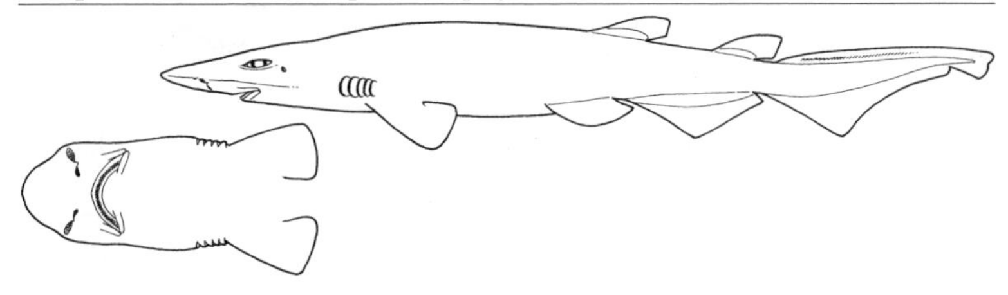

Measurements. Mostly unknown. Adolescent: 51 cm TL.
Identification. Slender brownish body. Erect denticles give skin a fuzzy texture. Thick flattened snout. Elongated nostrils with narrow apertures. Long large mouth expanded in front of small eyes with enlarged dental bands and long labial furrows (uppers reaching upper symphysis, lowers about same length). Gill slits rather large (slightly less than adult eye length). High rounded fins. Two equal-sized dorsal fins, first originating over pelvic midbases, interdorsal space slightly greater than first dorsal base. Large elongated rounded-angular anal fin separated from elongated tail fin by a small notch. Caudal fin with a crest of enlarged denticles.

Distribution. North Atlantic (west, possibly east). Nominal Indian Ocean record uncertain.
Habitat. Continental slope, ~1500 m.
Behaviour. Unknown.
Biology. Unknown.
Status. IUCN Red List: Not Evaluated. Close to *Apristurus microps*.

Broadgill Catshark *Apristurus riveri* Plate 31

Measurements. Adult: 43–46 cm ♂, 40–41 cm TL ♀.
Identification. Small, slender dark body with unmarked fins. Broad flattened head, elongated snout. Small nostrils (space between 1.5 x width). Very long mouth expanded in front of eyes, very long labial furrows (uppers reaching upper symphysis, lowers about same length). Males have much larger conical teeth and much longer and wider mouths and jaws. Enlarged gill slits, widest nearly adult eye length. Very small first dorsal fin originating in front of pelvic fin insertions. Fairly high angular anal fin, separated by small notch from narrow caudal fin.

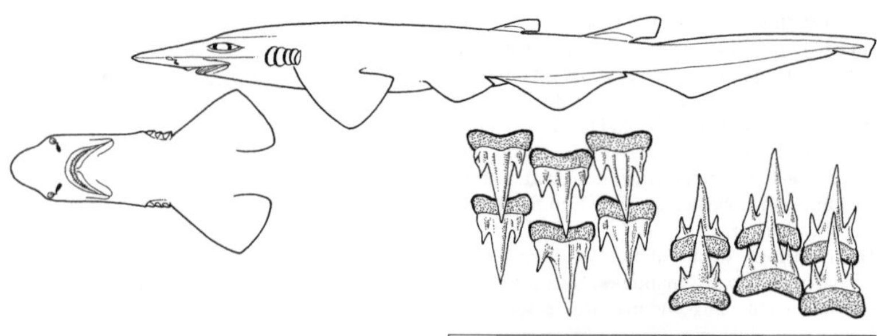

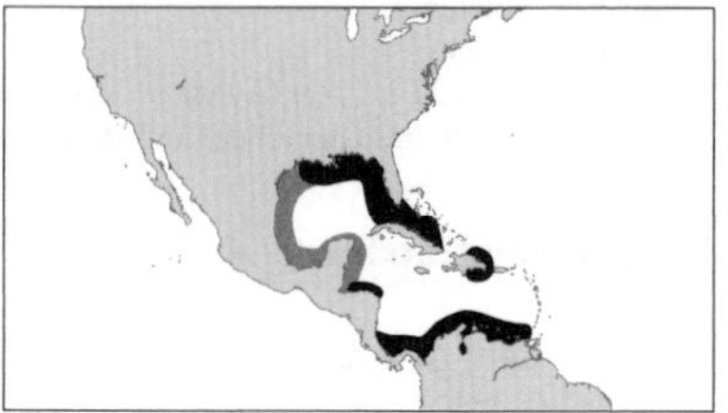

Distribution. Gulf of Mexico and Caribbean: Cuba, USA (Florida to Mississippi), Mexico, Honduras, Panama, Colombia, Venezuela, Dominican Republic.
Habitat. Continental slopes, on or near bottom, 732–1461 m.
Behaviour. Unknown.
Biology. Oviparous. Sexually dimorphic.
Status. IUCN Red List: Not Evaluated.

Saldanha Catshark *Apristurus saldanha* Plate 31

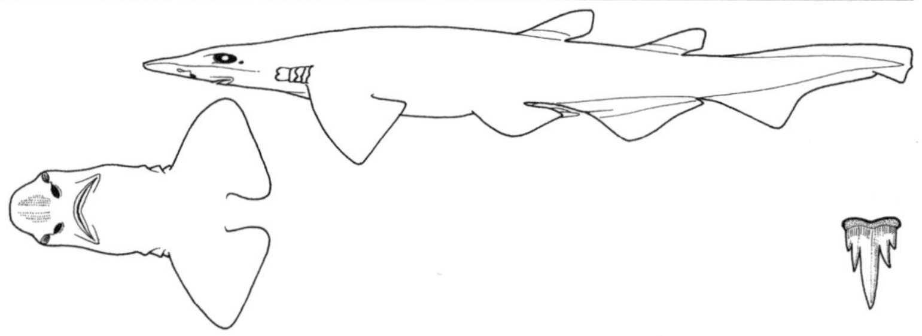

Measurements. Adolescent: 68 cm ♂. Adult: 74–88 cm ♂, 72–77 cm ♀. Max: reported 88 cm TL.
Identification. Plain dark slate-grey or grey-brown. Stout catshark with long thick broad snout, rather broad nostrils and fairly large eyes. Mouth not projecting in front of eyes, long labial furrows (uppers reaching upper symphysis, lowers shorter). Gill slits < adult eye length. Large pectoral fins. First dorsal slightly smaller or about equal to widely separated second and originating over pelvic fin midbases or slightly behind them. Elongated angular anal fin separated from tail fin by a small notch.

Distribution. Namibia and South Africa (eastern, western and northern Cape coasts).
Habitat. Continental slope, 344–1009 m.
Behaviour. Unknown.
Biology. Oviparous. Eats small bony fishes and cephalopods.
Status. IUCN Red List: Not Evaluated.

Pale Catshark *Apristurus sibogae* Plate 32

Measurements. Mostly unknown. One juvenile was 21 cm TL.
Identification. White or reddish-white with broad flattened head and narrow pointed snout. Mouth

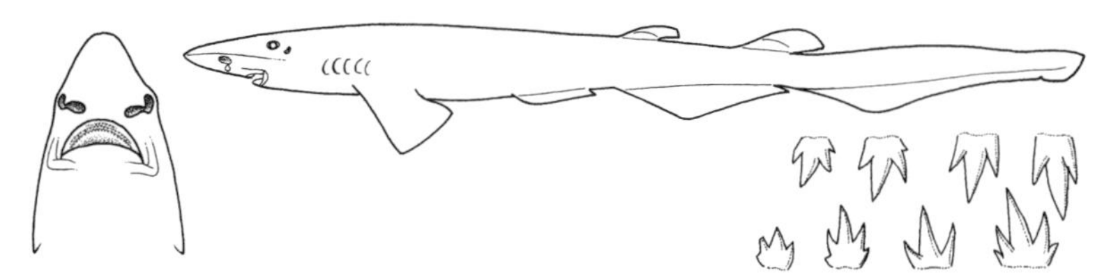

extending well in front of very small eyes (about equal to longest gill slit), long labial furrows (uppers reaching upper symphysis, lowers about same length). Very large pectoral fins, small low pelvic fins. First dorsal fin much smaller than second, originating behind pelvic fin insertions. Elongated angular anal fin separated from very long and narrow tail fin by a small notch.

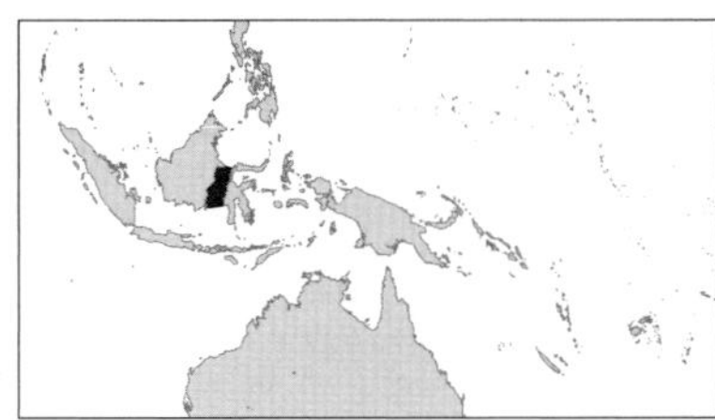

Distribution. Indonesia (Makassar Straits between Borneo and Sulawesi).
Habitat. Straits slope, 655 m.
Behaviour. Unknown.
Biology. Unknown.
Status. IUCN Red List: Not Evaluated. Known from one specimen.

South China Catshark *Apristurus sinensis* Plate 32

Measurements. Immature male was 42–45 cm TL. Max: presumably at least 50 cm TL.
Identification. Colour dark with no obvious markings. Broad flattened head, pointed angular snout, short mouth not expanded in front of eyes with very long labial furrows (uppers reaching upper symphysis, lowers shorter). Short gill slits (widest much less than eye length), short medial projections on gill septa. First dorsal fin about half the size of the second, originating over last quarter of pelvic fin bases. Pectoral and pelvic fins well separated. Large angular anal fin separated from tail fin by a small notch.
Distribution. South China Sea, off China.
Habitat. Taken from continental slope, 537–1000 m.
Behaviour. Unknown.
Biology. Unknown.
Status. IUCN Red List: Not Evaluated. Known from three specimens.

Spongehead Catshark *Apristurus spongiceps* Plate 32

Measurements. Mostly unknown. Gravid adult female was 50 cm TL.
Identification. Unmistakable dark brown stout catshark with area around gills and throat covered with grooves and pleats. Broad flattened head, rather broad snout and nostrils. Mouth extending well

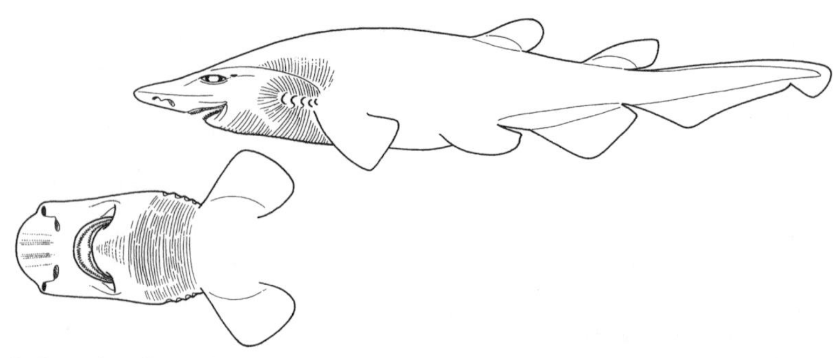

in front of small eyes, long labial furrows (uppers reaching upper symphysis, lowers about same length). Gill slits small, < adult eye length. Fins high and rounded. Two small (roughly equal-sized) spineless dorsal fins. Very large anal fin separated from elongated tail fin by a small notch.

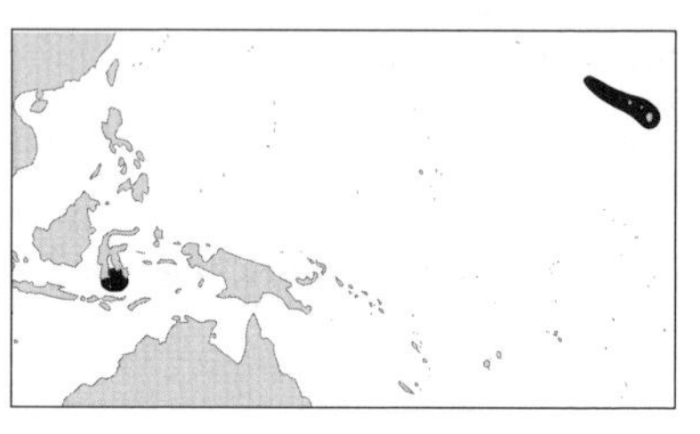

Distribution. Pacific. Hawaiian Islands and Banda Sea (off southern Sulawesi).
Habitat. Island slopes, on or near bottom, 572–1482 m.
Behaviour. Unknown.
Biology. Probably oviparous.
Status. IUCN Red List: Not Evaluated. One adult and one juvenile collected, one specimen photographed underwater.

Panama Ghost Catshark *Apristurus stenseni* Plate 32

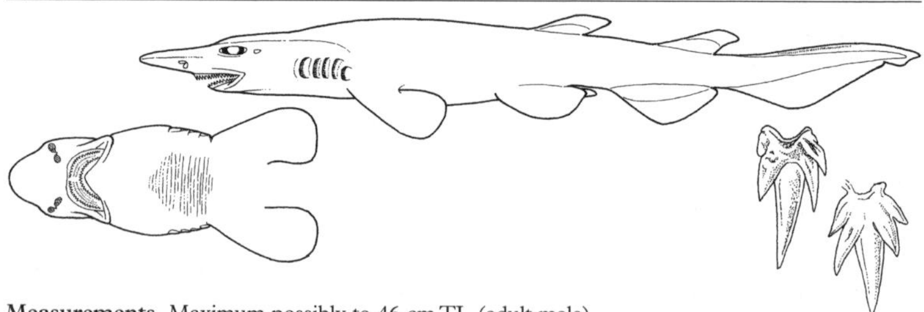

Measurements. Maximum possibly to 46 cm TL (adult male).
Identification. Slender blackish body, no obvious markings. Broad flattened head, elongated snout, narrow widely-set nostrils, extremely large mouth extending far in front of eyes, with very long labial furrows (uppers reaching upper symphysis, lowers about same length). Very small eyes and very wide gill slits (>adult eye length). Dorsal fins roughly equal size, first originating over pelvic midbases. Distance between pectoral and pelvic fin bases much less than preorbital snout length. Moderately high rounded-angular anal fin (length >three to four times height) separated by a small notch from the long, low caudal fin with a prominent dorsal crest of denticles.

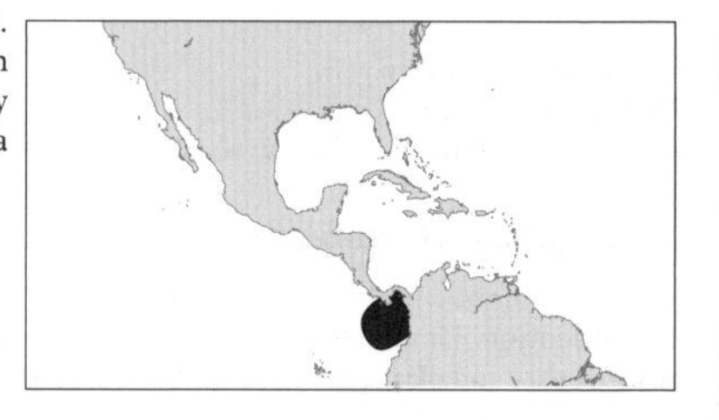

Distribution. East Pacific. Panama.
Habitat. Continental slope, 915–975 m.
Behaviour. Unknown.
Biology. Unknown.
Status. IUCN Red List: Not Evaluated.

Freckled Catshark *Apristurus* sp. A Plate 33

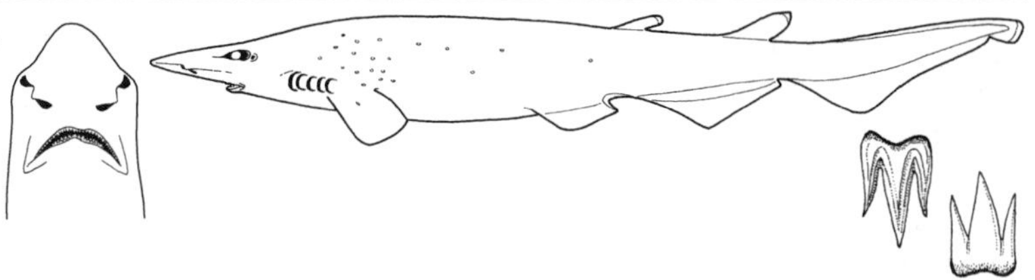

Measurements. Mature: 51–64 cm TL ♂. Max: at least 74 cm TL.
Identification. Body brownish in adults, often with a scattering of pale flecks, slightly greyish in juveniles. Fin margins translucent or black. Broad flattened head, fairly elongate, flattened snout, very small greyish teeth. Mouth extending to opposite front ends of small eyes, long labial furrows (uppers reaching upper symphysis, lowers shorter). Gill openings < eye diameter. First dorsal fin much smaller than second. Widely separated pectoral and pelvic fins. Long-based triangular low anal fin separated from elongated tail fin by a small notch.

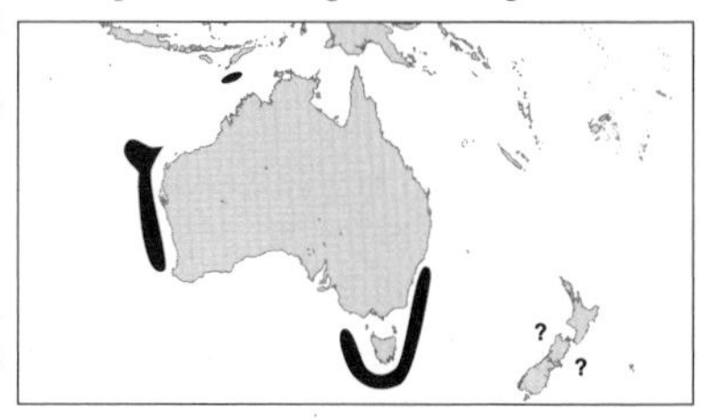

Distribution. Pacific: off southeast and west Australia, possibly also off New Zealand.
Habitat. Deepwater, 940–1290 m.
Behaviour. Unknown.
Biology. Unknown.
Status. IUCN Red List: Data Deficient. Deepwater demersal trawling is expanding within its range.

Bigfin Catshark *Apristurus* sp. B Plate 33

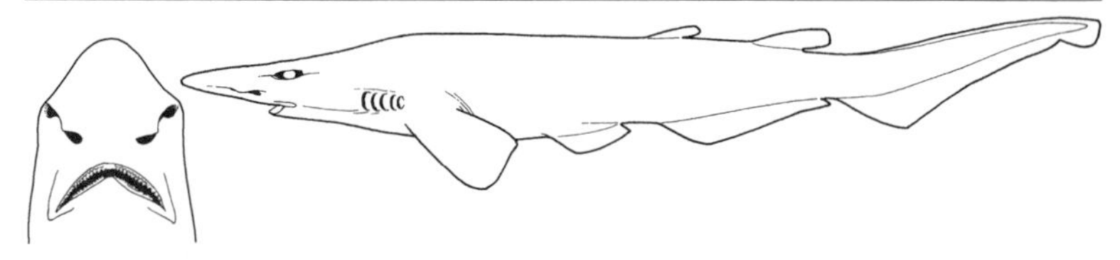

Measurements. Mature: at 53–57 cm TL ♂. Max: at least 67 cm TL.
Identification. Uniformly greyish to dark brownish, with black gill membranes and naked areas on fins. Slender, compressed body, fairly elongated flattened broad snout, large nostrils, gill openings much smaller than eye diameter. Mouth extending to opposite front ends of small eyes, long labial furrows (uppers reaching upper symphysis, lowers shorter). First dorsal fin originating behind pelvic fin origins, second much larger. Large pectoral fins almost extend to pelvic fins. Very long low angular anal fin separated from elongated tail fin by a small notch.

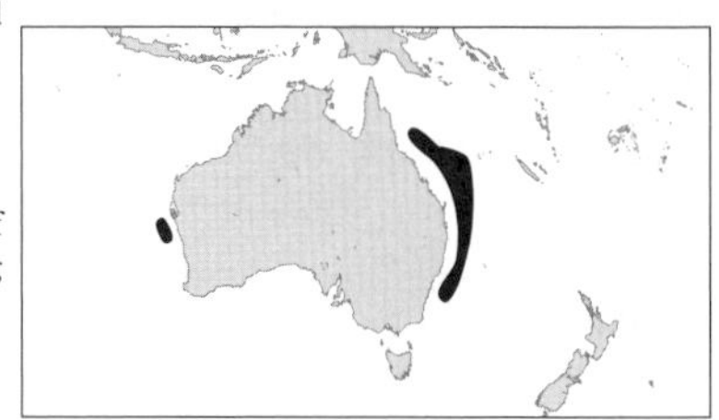

Distribution. Australia (warm temperate and tropical east and west coasts).
Habitat. Continental slopes from 730–1000 m.
Behaviour. Unknown.
Biology. Unknown.
Status. IUCN Red List: Data Deficient. Most of range unfished, but deepwater demersal trawling could expand in future.

Fleshynose Catshark *Apristurus* sp. C Plate 33

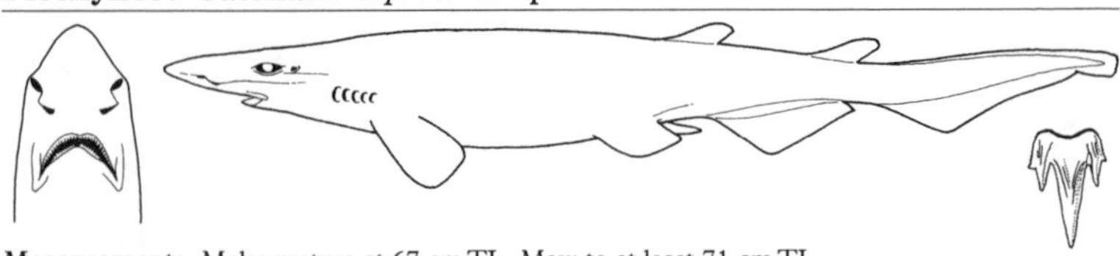

Measurements. Males mature at 67 cm TL. Max: to at least 71 cm TL.
Identification. Dark brown body, black naked fin tips, irregular scattering of pale flecks on most individuals. Slender body with fairly long, slender flattened fleshy snout. Mouth extending to just in front of small eyes, long labial furrows (uppers reaching upper symphysis, lowers shorter). Gill openings slightly smaller than eye diameter. Widely separated pectoral and pelvic fins. Dorsal fins equal size. Deep long-based angular anal fin separated from tail fin by a small notch.

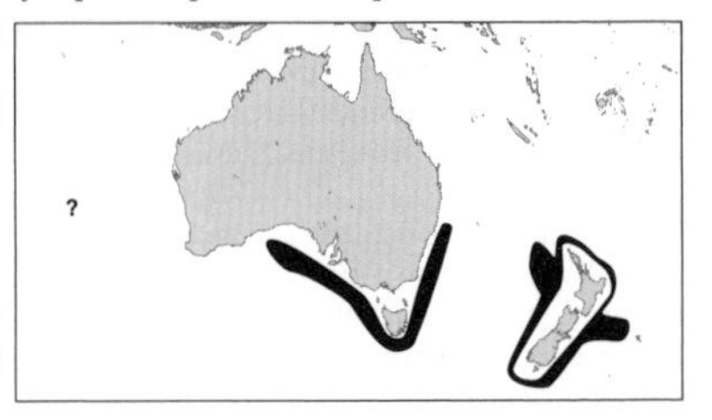

Distribution. Pacific: southern Australia and New Zealand.
Habitat. Continental shelf, 900–1150 m.
Behaviour. Unknown.
Biology. Unknown.
Status. IUCN Red List: Data Deficient. Distribution includes some intensive and expanding fisheries.

Roughskin Catshark *Apristurus* sp. D Plate 33

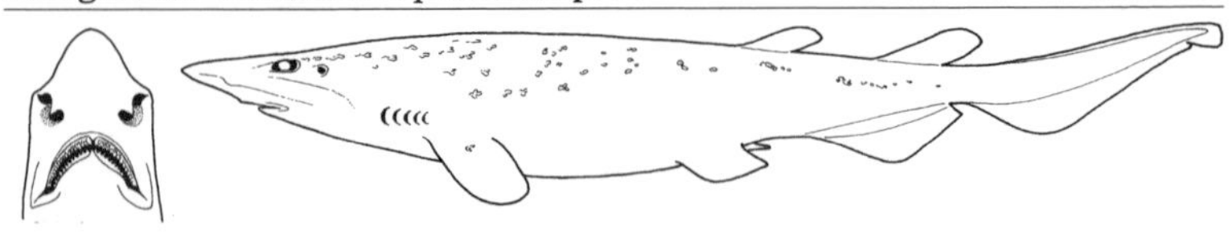

Measurements. Mature: ~67 cm TL ♂. Max: at least 86 cm TL.
Identification. Large dark brownish to black catshark with scattered paler spots and squiggles and widely-spaced denticles. Pale tip to outer caudal fin. Bulky flattened head, elongated broad snout, large teeth. Large mouth extending well in front of small eyes, long labial furrows (uppers reaching upper symphysis, lowers about same length). First dorsal fin slightly smaller than second. Pectoral and pelvic fins well separated. Relatively short, rounded anal fin separated from elongated tail fin by a small notch.

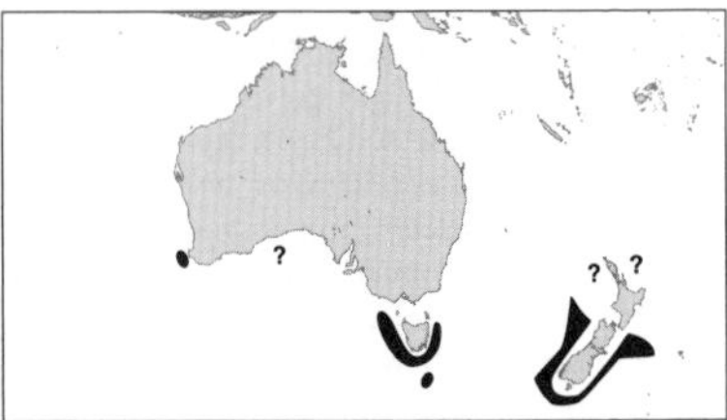

Distribution. Pacific: New Zealand and southern Australia.
Habitat. Continental shelf, 840–1380 m.
Behaviour. Unknown.
Biology. Unknown.
Status. IUCN Red List: Data Deficient. Possibly quite rare. Distribution includes intensive and expanding fisheries. Somewhat similar to *Apristurus manis* in Atlantic.

Bulldog Catshark *Apristurus* sp. E Plate 33

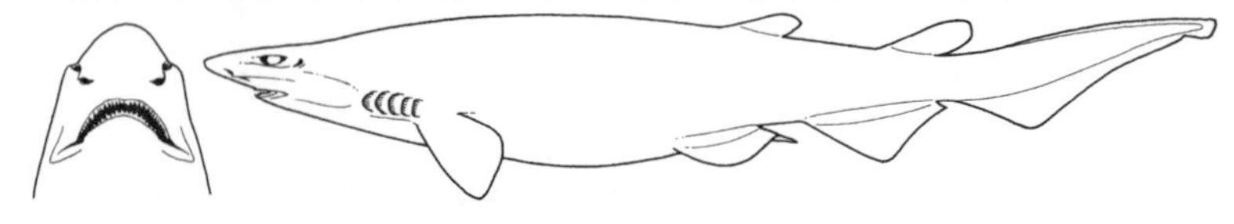

Measurements. Males mature at ~50 cm TL. Max: at least 63 cm TL.
Identification. Stout brownish body with stocky flattened head and short broad snout. Upper surface slightly lighter than underside, may have indistinct speckles, head may have indistinct pale blotches. Large mouth extending well in front of small eyes, long labial furrows (uppers reaching upper symphysis, lowers about same length), fairly large gill slits, small eyes and nostrils. Second dorsal fin slightly larger than first. Pectoral and pelvic fins broad and well separated. Anal fin short and very deep, separated from elongated tail fin by a small notch.

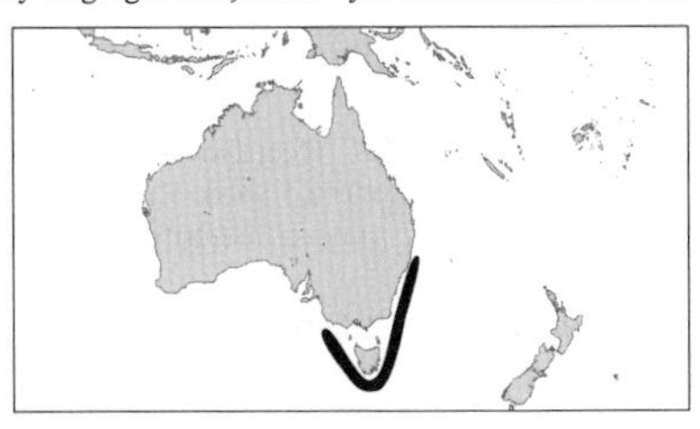

Distribution. Southeastern Australia.
Habitat. Continental slope, 1020–1500 m.
Behaviour. Unknown.
Biology. Unknown.
Status. IUCN Red List: Data Deficient. Distribution includes expanding deepwater fisheries.

Bighead Catshark *Apristurus* sp. F — Plate 33

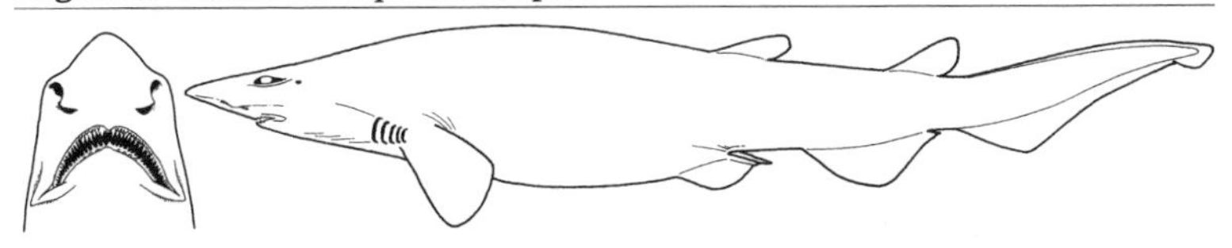

Measurements. Mostly unknown. Reaches at least 73 cm TL.
Identification. Very stout uniform greyish-brown catshark with black rear margins to anal and caudal fins. Extremely broad flattened head and very short snout. Pleats on throat. Mouth extending well in front of small eyes, long labial furrows (uppers reaching upper symphysis, lowers about same length). Fins rather broad and rounded. First dorsal fin lower than second. Pectoral and pelvic fins widely separated. Short, very deep anal fin separated from elongated tail fin by a small notch.

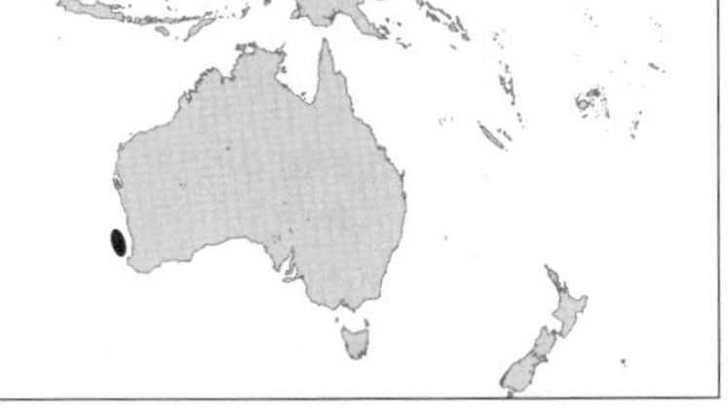

Distribution. West Australia (off Perth).
Habitat. Continental slope. Collected from 1030–1050 m.
Behaviour. Unknown.
Biology. Unknown.
Status. IUCN Red List: Data Deficient. Only three specimens recorded from one location.

Pinocchio Catshark *Apristurus* sp. G — Plate 33

Measurements. Males mature at ~51 cm TL. Reaches at least 61 cm TL.
Identification. Slender pale greyish-brown to brown body, long narrow flattened head and very long narrow pointed snout. Mouth extending opposite small eyes, long labial furrows (uppers reaching upper symphysis, lowers slightly shorter). First dorsal fin smaller than second. Large pectoral fins, small pelvic fins, very low angular long-based anal fin separated from elongated tail fin

by a small notch. Throat darker than body, tips and borders of fins white or with broader pale margins in the northeast population.

Distribution. Australia.

Habitat. Continental slope and seamounts, 590–1000 m.

Behaviour. Unknown.

Biology. Unknown.

Status. IUCN Red List: Data Deficient. Range includes fished and unfished areas. Relationship with *Apristurus herklotsi* needs assessment.

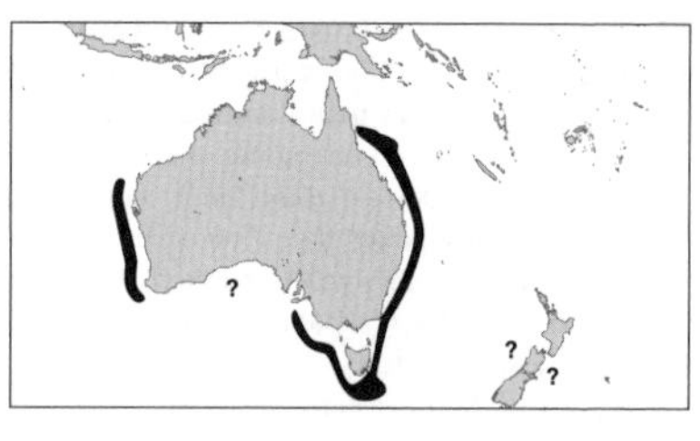

Grey Spotted Catshark *Asymbolus analis* Plate 34

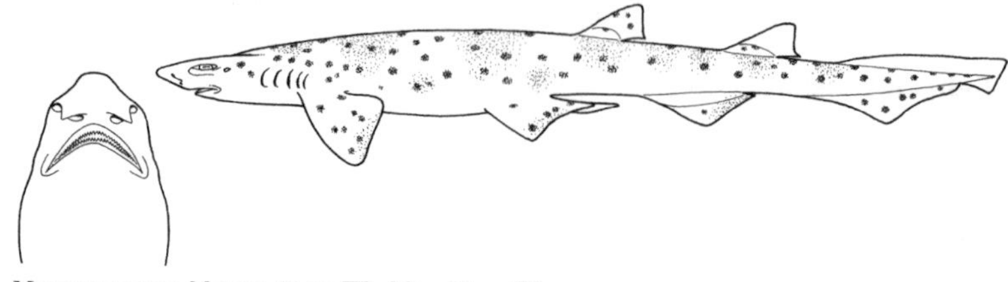

Measurements. Mature: 46 cm TL. Max: 61 cm TL.

Identification. Greyish with evenly-sized widely-spaced dark brown spots and obscure dark saddle-like blotches on back and sides. Some whitish specks. Short, slightly flattened pointed-rounded head and short thick snout. Narrow ridges below eyes. Short labial furrows along jaws, upper teeth exposed. Two small dorsal fins behind pelvic bases. Pelvic inner fin margins fused to form apron over claspers in adult males. Anal fin short and angular, short broad caudal fin.

Distribution. Australia, southern Queensland to Victoria.

Habitat. Bottom, continental shelf, close inshore (40 m) to offshore (175 m).

Behaviour. Unknown.

Biology. Unknown.

Status. IUCN Red List: Data Deficient. Uncommon, taken as trawl bycatch.

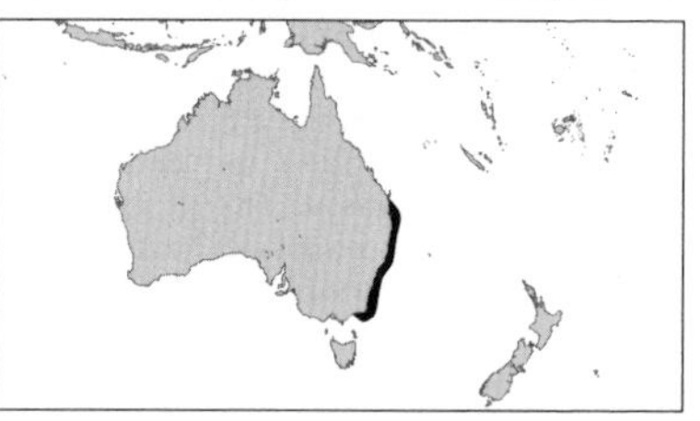

Blotched Catshark *Asymbolus funebris* Plate 34

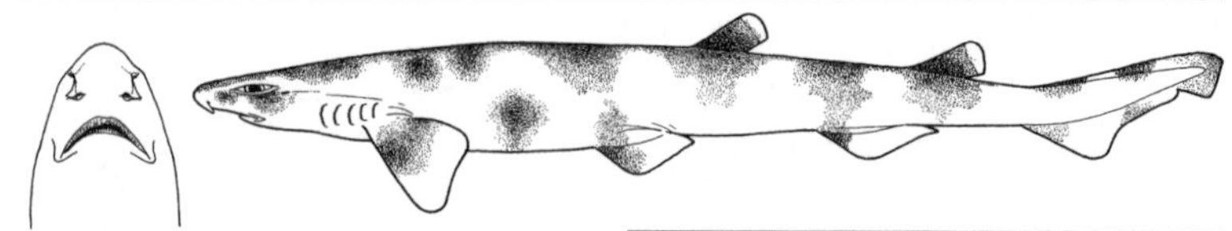

Measurements. Known from one 44 cm female.

Identification. Small, brown catshark with large dark brown blotches and saddles (no small spots). Three pre-dorsal saddles, bars beneath each dorsal fin, one between the fins. Ventral surface only slightly paler. Short, slightly flattened pointed-rounded head and short thick snout. Short labial furrows along jaws, upper teeth exposed. Narrow ridges below eyes. Dorsal

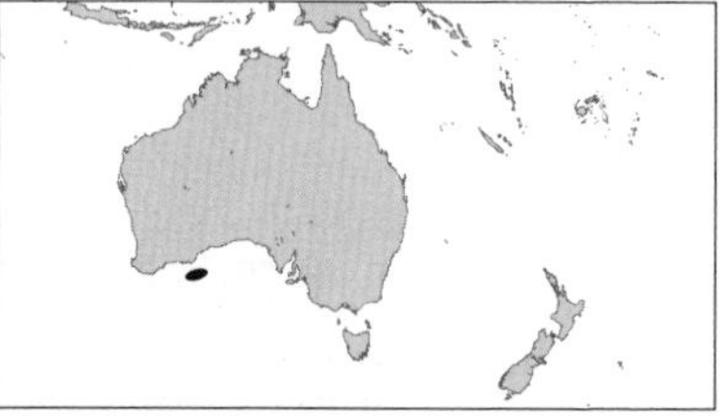

fins set back behind pelvic fins, inner margins of pelvic fins presumably fused into apron over adult male claspers. Anal fin short and angular, short broad caudal fin.
Distribution. Western Australia, off the Recherche Archipelago.
Habitat. Outer continental shelf, 195 m depth.
Behaviour. Unknown.
Biology. Unknown.
Status. IUCN Red List: Data Deficient. Known from only one specimen.

Western Spotted Catshark *Asymbolus occiduus* Plate 34

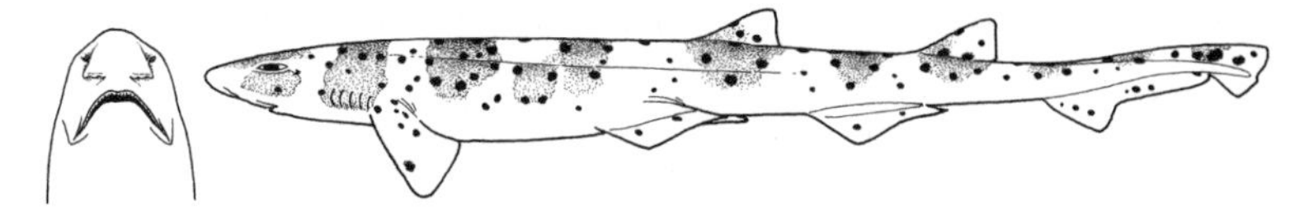

Measurements. Mature: 58 cm TL ♂. Max: at least 60 cm TL.
Identification. Bright yellowish-green, with similar-sized brownish-black spots and eight or nine distinct saddles (most obvious in juveniles, which have fewer spots). Short, slightly flattened head and short thick snout. Short labial furrows along jaws, upper teeth exposed. Narrow ridges and often a dark spot below eyes. One dark spot in front of each dorsal fin, dorsals behind pelvic fin bases. Inner pelvic fin margins fused into apron over adult male claspers. Anal fin short and angular, short broad caudal fin. No white spots, no spots on underside.

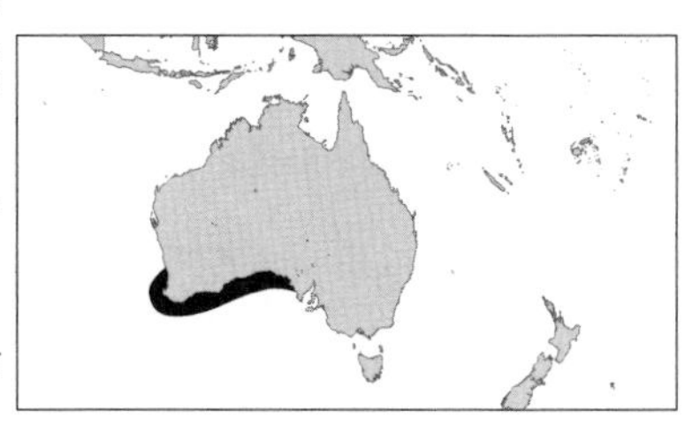

Distribution. South and western Australia.
Habitat. Bottom, outer continental shelf, 98–250 m.
Behaviour. Unknown.
Biology. Unknown.
Status. IUCN Red List: Least Concern. Most of range unfished.

Pale Spotted Catshark *Asymbolus pallidus* Plate 34

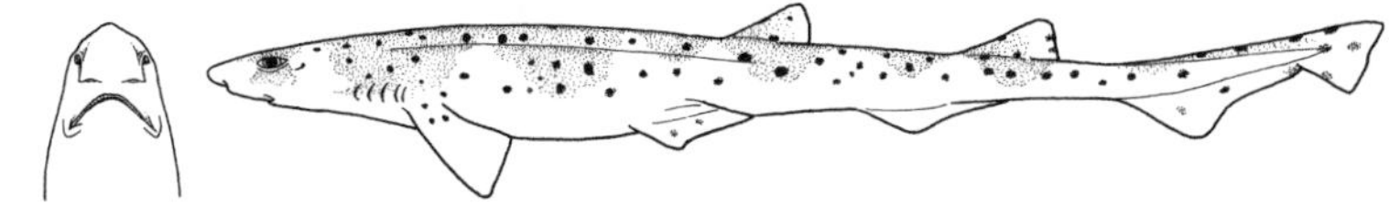

Measurements. Hatch: 19 cm TL. Mature: 32 cm ♂. Max: 46 cm TL.
Identification. Very small, pale yellowish with obvious even-sized dark brown spots (none on underside or beneath eye). No distinct saddles, bands, or white spots. Short, slightly flattened head, short thick snout. Short labial furrows along jaws, upper teeth exposed. Narrow ridge below eye. Usually a pair of spots in front of each dorsal fin (set far back behind pelvics), one spot at the centre of each dorsal fin base. Inner pelvic fin margins fused into apron over adult male claspers. Anal fin short and angular, short broad caudal fin.

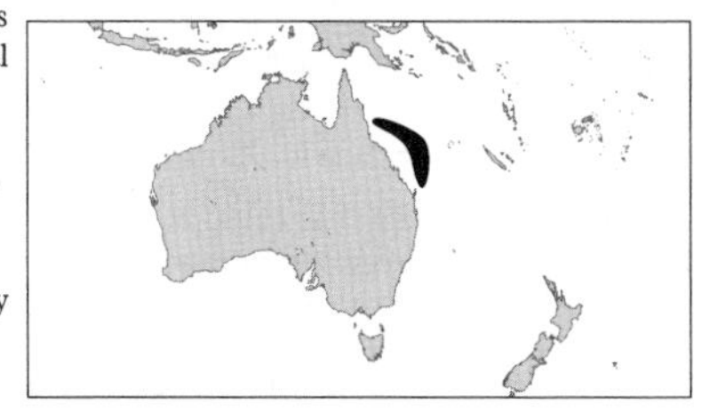

Distribution. Northeast Australia.
Habitat. Bottom, continental shelf, 270–400 m.
Behaviour. Unknown.
Biology. Unknown.
Status. IUCN Red List: Least Concern. Unlikely to be caught in commercial fisheries.

Dwarf Catshark *Asymbolus parvus* Plate 34

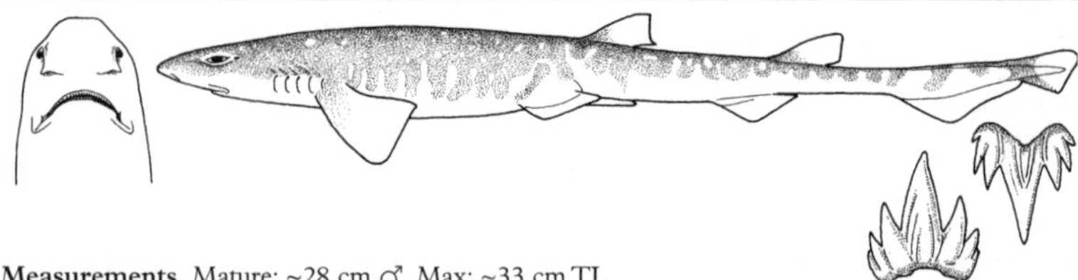

Measurements. Mature: ~28 cm ♂. Max: ~33 cm TL.

Identification. Very small, pale brown catshark with many white spots and lines and faint dark saddles or bands (most obvious on tail). Short, slightly flattened pointed-rounded head and short thick snout. Short labial furrows along jaws, tiny teeth. Narrow ridges below eyes. No dark spots or blotches, no spots on underside. Two small dorsal fins set behind pelvic fins and an anal fin. Inner margins of pelvic fins fused over adult male claspers, forming an apron. Anal fin short and angular, short broad caudal fin.

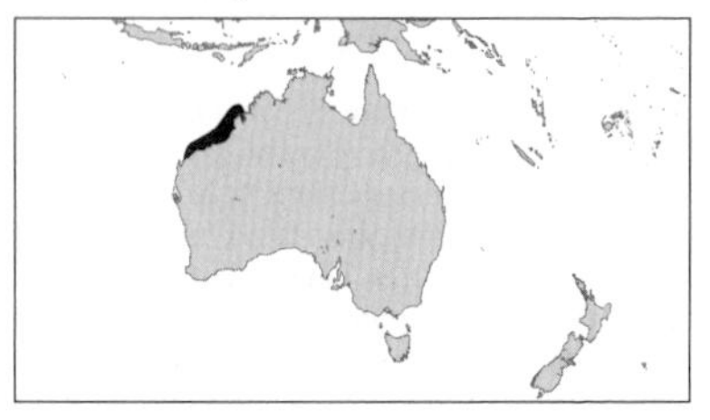

Distribution. Tropical northwestern Australia.

Habitat. Bottom, outer continental shelf and upper slope, 59–252 m.

Behaviour. Unknown.

Biology. Unknown.

Status. IUCN Red List: Least Concern. Infrequent bycatch. Thought to survive well when discarded.

Orange Spotted Catshark *Asymbolus rubiginosus* Plate 34

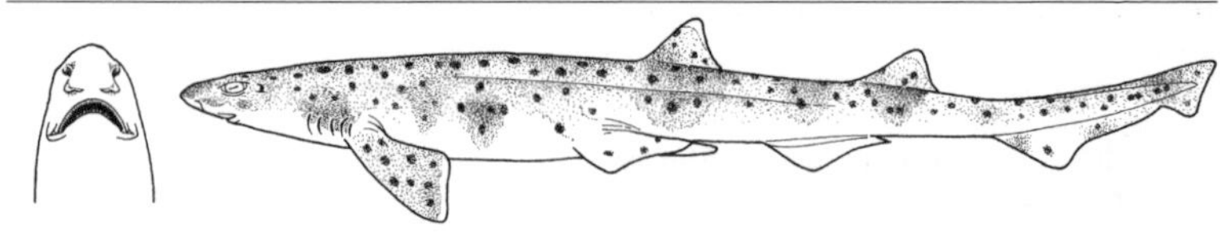

Measurements. Adult males 37 cm. Max: >53 cm TL.

Identification. Pale brown, many dark brown spots with orange-brown borders on back and sides, no spots on pale underside. Obscure dark saddles separated by clusters of spots along spine. Indistinct brownish blotch and ridge below eye. Usually a dark mark on leading and trailing edges of dorsal fin. Flank spots larger but less clear. Dorsal fins behind pelvics, inner margins of pelvic fins fused into apron over adult male claspers. Anal fin. Short broad caudal fin.

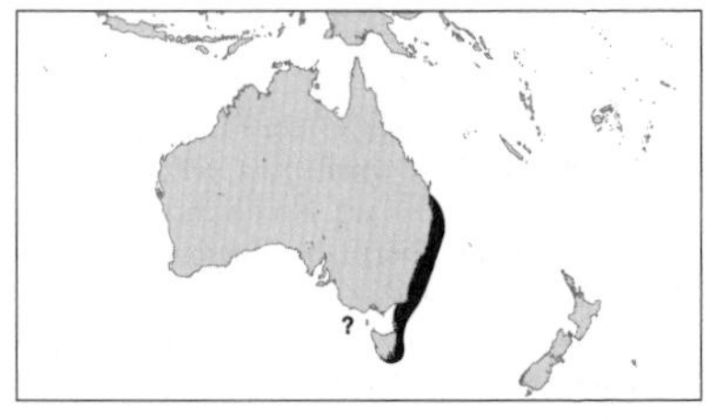

Distribution. Eastern Australia, southern Queensland to Tasmania.

Habitat. Bottom, continental shelf and upper slope, 25–540 m.

Behaviour. Unknown.

Biology. Thought to lay eggs year-round.

Status. IUCN Red List: Least Concern. Discarded bycatch in some trawl fisheries.

Variegated Catshark *Asymbolus submaculatus* Plate 34

Measurements. Reaches 43 cm TL.

Identification. Small, greyish-brown catshark with many small black spots. Darker irregular rusty-brown saddle-like blotches on back, bluish-grey blotches on sides. Small black and grey spots on lower surfaces of head, abdomen and tail, undersides otherwise pale grey. Short, slightly flattened rounded head, short thick snout. Long arched mouth, short labial furrows along jaws, large upper

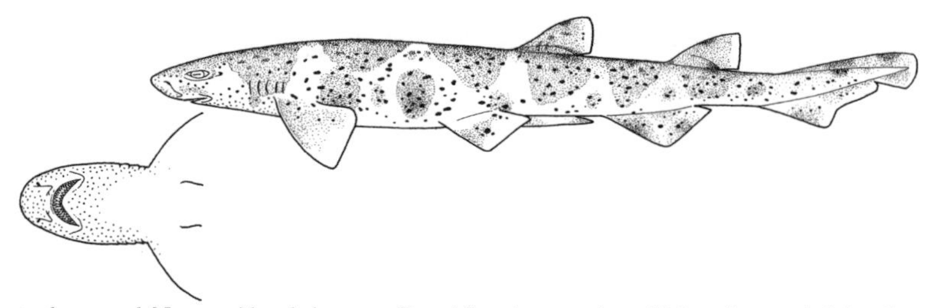

teeth exposed. Narrow ridges below eyes. Dorsal fins close together with broadly rounded tips, first dorsal base close to pelvic bases. Inner pelvic fin margins fused into an apron over adult male claspers. Anal fin short and angular, short broad caudal fin.

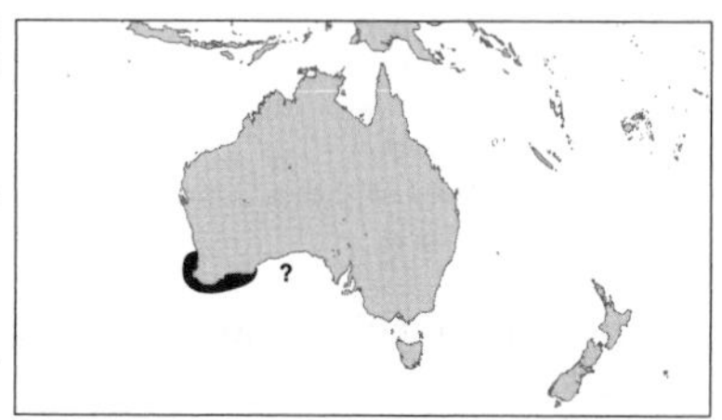

Distribution. Western Australia.
Habitat. Continental shelf down to 150 m, in caves and on ledges.
Behaviour. Nocturnal.
Biology. Unknown.
Status. IUCN Red List: Least Concern. Rarely seen, unlikely to be caught in fisheries.

Gulf Catshark *Asymbolus vincenti* Plate 34

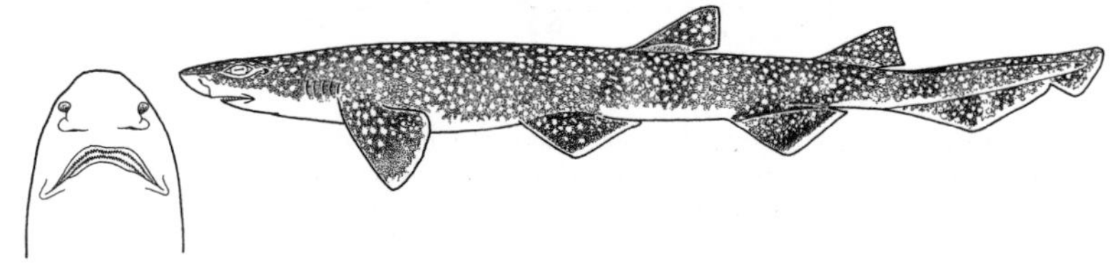

Measurements. Eggcase: 5 x 2 cm. Mature: 38 cm ♂. Max: >56 cm TL.
Identification. Mottled greyish-brown or chocolate, seven or eight dark saddles, many small faint white spots. Pale unspotted underside. Short, slightly flattened head with short thick snout. Short labial furrows along jaws, upper teeth exposed. Narrow ridges below eyes. Two small dorsal fins behind pelvic fins, inner pelvic fin margins fused into apron over adult male claspers. Anal fin short and angular. Short broad caudal fin.

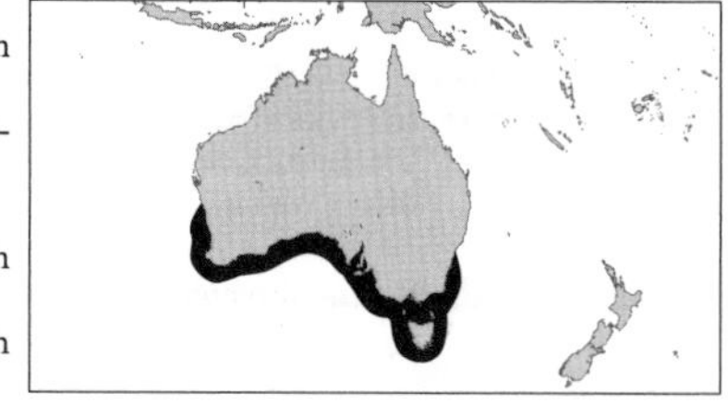

Distribution. Southern Australia (most common in Great Australian Bight).
Habitat. Bottom <100 m in east, often in seagrass beds. 130–220 m in Bight.
Behaviour. Unknown.
Biology. Oviparous. Lays pairs of eggcases with long filaments.
Status. IUCN Red List: Least Concern. Bycatch in trawl fisheries in only part of its range.

Banded Sand Catshark *Atelomycterus fasciatus* Plate 34

Measurements. Eggcase: 67 mm long. Mature: ~33 cm ♂, 35 cm ♀. Max: 45 cm TL.
Identification. Slender, brown saddles on a light background, a few scattered small black and sometimes small white spots. Narrow head, greatly expanded anterior nasal flaps extending to long mouth, very long labial furrows, nasoral grooves. Dorsal fins broadly triangular, much larger than anal fin, origin of first above rear third of pelvic bases.

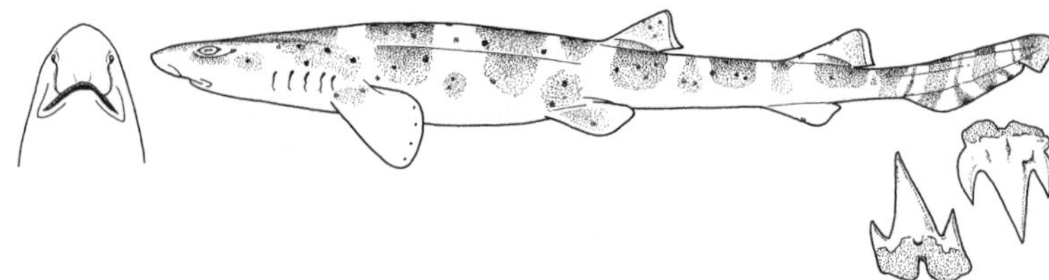

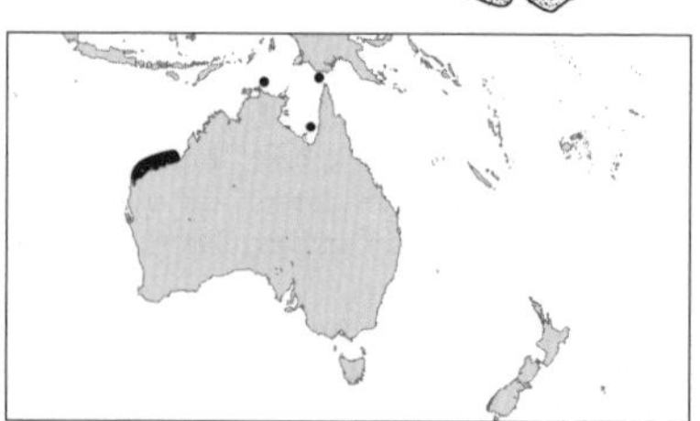

Distribution. Western Australia (isolated records from Northern Territory and Queensland).
Habitat. Bottom on sand and shelly sand, continental shelf, 27–122 m.
Behaviour. Unknown.
Biology. Oviparous, laying pairs of eggs (one per oviduct). Presumably feeds on small fishes and crustaceans.
Status. IUCN Red List: Least Concern. Occurs in largely unfished area.

Australian Marbled Catshark *Atelomycterus macleayi* Plate 34

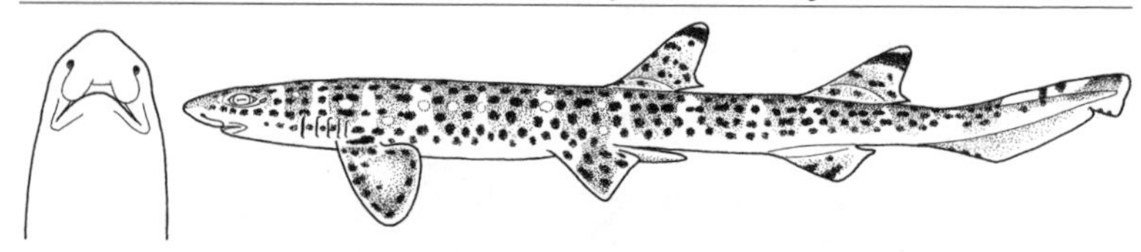

Measurements. Hatch: ~10 cm TL. Eggcase: 70 mm. Mature: ~48 ♂, 51 cm ♀. Max: 60 cm TL.
Identification. Slender light grey to grey-brown, narrow-headed catshark with darker grey or brown saddles, outlined and partly covered in adults by many small black spots, also scattered on flanks, but no white spots; hatchling young without spots. Greatly expanded anterior nasal flaps extend to mouth, nasoral grooves, and very long labial furrows. Dorsal fins much larger than anal fin, first dorsal origin opposite pelvic fin insertions.

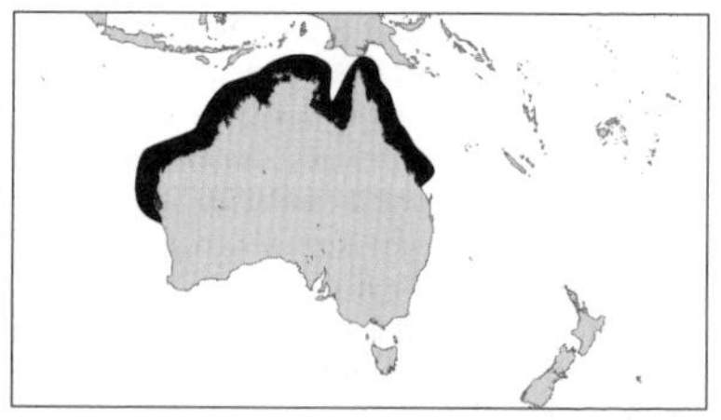

Distribution. Australia (Western, Northern Territory, possibly Queensland).
Habitat. Sand and rock, very shallow water (0.5–3.5 m).
Behaviour. Unknown.
Biology. Oviparous, lays pairs of eggcases.
Status. IUCN Red List: Least Concern. Apparently common, no fisheries within its range.

Coral Catshark *Atelomycterus marmoratus* Plate 34

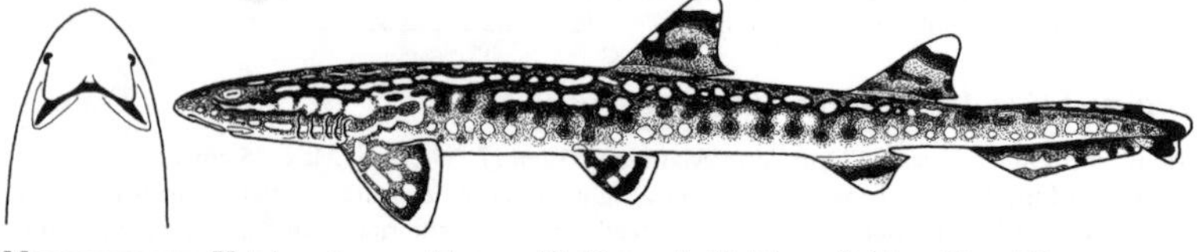

Measurements. Hatch: unknown. Mature: 47–62 cm ♂, 49–57 cm ♀. Max: 70 cm TL.
Identification. Dark narrow-headed slender shark. No clear saddle markings. Enlarged black spots

often merge to form dash and bar marks bridging saddle areas; scattered large white spots on sides, back and fin margins. Greatly expanded anterior nasal flaps extending to long mouth, nasoral grooves, very long labial furrows. Dorsal fins much larger than anal fin, origin of first opposite or slightly in front of pelvic fin insertions.

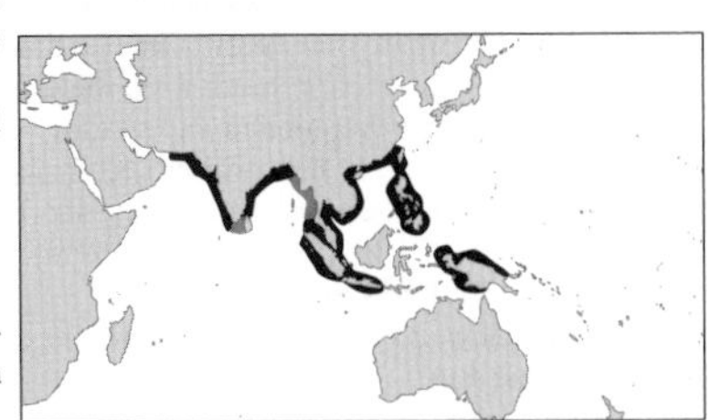

Distribution. Indo-west Pacific: Pakistan and India to New Guinea and southern China.
Habitat. Crevices and holes on coral reefs.
Behaviour. Unknown.
Biology. Oviparous, lays pairs of eggcases.
Status. IUCN Red List: Near Threatened. Common in artisanal fisheries. Habitat threatened in much of range.

New Caledonia Catshark *Aulohalaelurus kanakorum* Plate 35

Measurements. Adult male was 79 cm TL.
Identification. Fairly slender, thick-skinned elongated dark grey cylindrical body with a variegated colour pattern of large, dark, close-set blotches on body and fins surrounding numerous large white blotches. White-bordered fins. Moderately short broad caudal fin and short, slightly flattened, narrowly rounded head. Long arched mouth reaches past the front end of the cat-like eyes. Roughly equal-sized dorsal fins, origin of first over pelvic fin origins.
Distribution. New Caledonia, western Pacific.
Habitat. Coral reef, 49 m.
Behaviour. Unknown.
Biology. Unknown.
Status. IUCN Red List: Vulnerable. Presumed endemic. Known from one specimen and two photographs.

Blackspotted Catshark *Aulohalaelurus labiosus* Plate 35

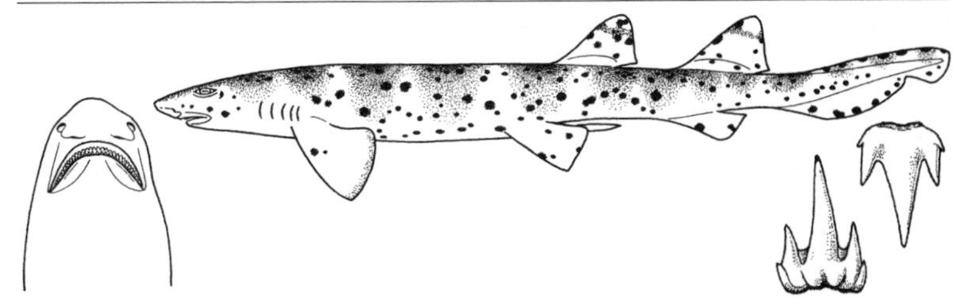

Measurements. Mature: 54–62 cm ♀. Max: 67 cm TL.
Identification. Fairly slender, light greyish to yellowish-brown, thick-skinned elongated cylindrical body with variegated pattern of small to large black spots or blotches and dark saddles on sides, back and fins, very few small white spots. White fin tips highlighted by dark blotches on dorsal, caudal and anal fins. Relatively narrow head with slightly flattened short narrowly rounded snout. Equal-sized dorsal fins, origin of first over or slightly in front of pectoral fin insertions. Moderately short broad caudal fin.
Distribution. Southwestern Australia.
Habitat. Shallow coastal waters and offshore reefs to at least 4 m.
Behaviour. Unknown.
Biology. Virtually unknown. Probably oviparous.
Status. IUCN Red List: Least Concern. Common and unfished within limited range.

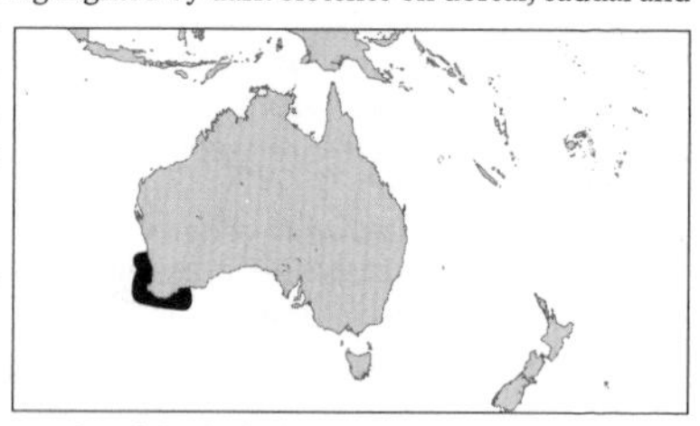

Arabian Catshark *Bythaelurus? alcocki* Not illustrated

Measurements. Unknown, presumably less than 30 cm.
Identification. Similar to *Bythaelurus hispidus* but with a longer snout, smaller eyes, second dorsal fin slightly larger than first, smaller anal fin, and black in colour with a hoary grey surface and white tips to some fins.
Distribution. Arabian Sea.
Habitat. Deep continental slope, on or near the bottom in water 1134–1262 m deep.
Behaviour. Unknown.
Biology. Unknown.
Status. IUCN Red List: Not Evaluated. Only one specimen recorded, and apparently lost. Placement in this genus tentative.

Dusky Catshark *Bythaelurus canescens* Plate 35

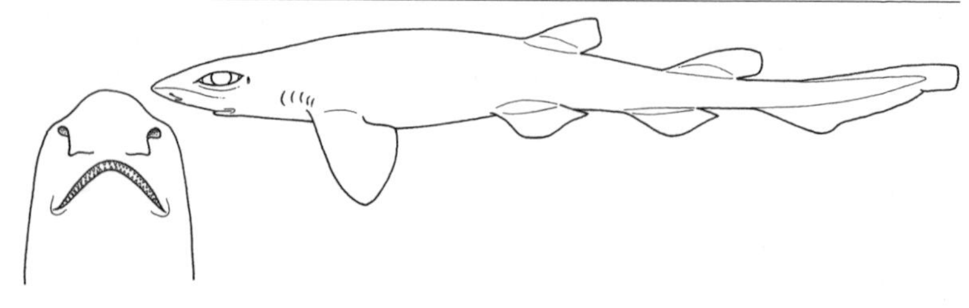

Measurements. Adults mature at 59 cm. Max: 70 cm TL.
Identification. Fairly large plain dark brown catshark, no markings in adults, young have white fin tips. Short rounded snout and fairly short tail. Long arched mouth reaches to slightly in front of the large cat-like eyes. Two small dorsal fins, first dorsal base over pelvic fin bases, anal fin almost as large as second dorsal.
Distribution. Southeast Pacific: Peru, Chile, Straits of Magellan.
Habitat. Deepwater on mud and rock of upper continental slope, 250–700 m.
Behaviour. Unknown.
Biology. Oviparous, apparently laying pairs of eggs. Feeds on bottom invertebrates.
Status. IUCN Red List: Not Evaluated. Common in deep water, and abundant bycatch in deepwater trawls.

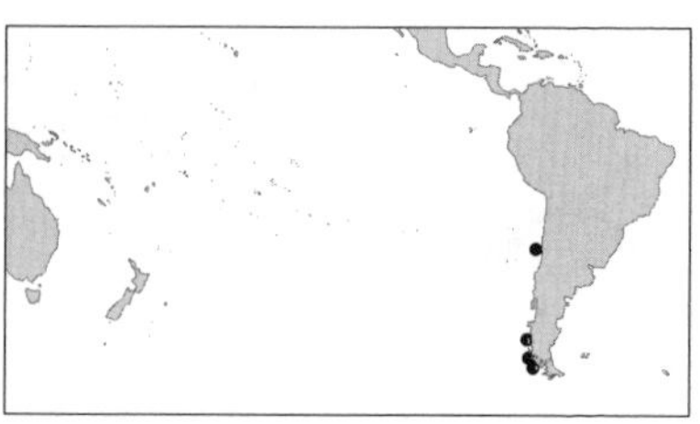

Broadhead Catshark *Bythaelurus clevai* Plate 35

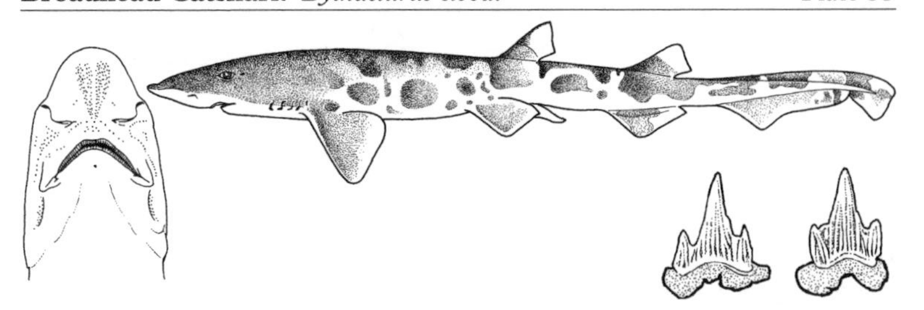

Measurements. Born: ~14 cm. Mature: ~36–39 cm ♂, 35–39 cm ♀. Max: ~39 cm TL.
Identification. Small grey catshark, white below, with few large conspicuous dark brown blotches and saddles and small spots on sides and upper surface of trunk and tail, but head nearly plain. Pectoral, pelvic, dorsal and anal fins with dark bases and light margins. Longish snout, narrow and pointed in side view, broad and bell-shaped from above, mouth long and arched, reaching past front ends of the small cat-like eyes. Two small dorsal fins, first dorsal base mostly over pelvic fin bases, anal fin larger than second dorsal, high and triangular, caudal fin short.

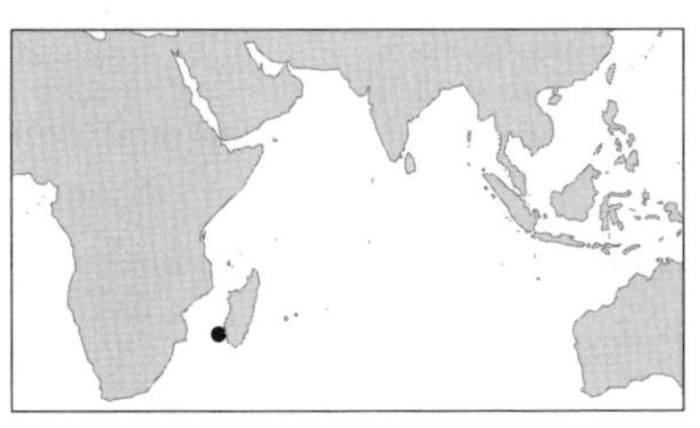

Distribution. West Indian Ocean, southwest Madagascar, common off Tulear.
Habitat. Upper insular slopes, 400–500 m.
Behaviour. Unknown.
Biology. Ovoviviparous, two pups per litter. Eats shrimp.
Status. IUCN Red List: Not Evaluated.

New Zealand Catshark *Bythaelurus dawsoni* Plate 35

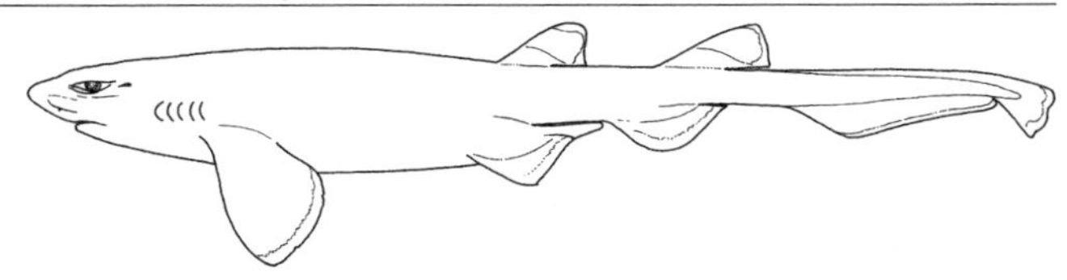

Measurements. Born: ~11 cm TL. Mature: ~35 cm ♂, 37 cm ♀. Max: ~42 cm TL.
Identification. Fairly short light brown or grey body, paler below. A line of white spots on sides of small animals, white fin tips. Dark bands on caudal fin. Broad flattened head. Elongated lobate anterior nasal flaps. Long arched mouth reaches past the front end of the large cat-like eyes. Two dorsal fins, first dorsal base over pelvic bases, second larger, and anal fin short and angular.

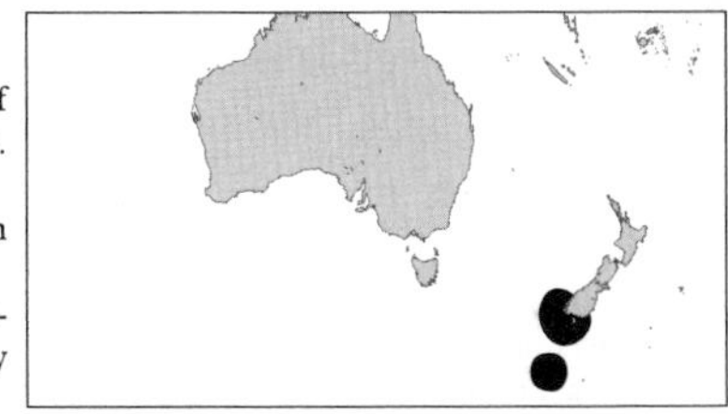

Distribution. New Zealand.
Habitat. On or near bottom, upper slopes of New Zealand and Auckland Islands, 371–420 m.
Behaviour. Unknown.
Biology. Ovoviviparous. Feeds on bottom crustaceans.
Status. IUCN Red List: Data Deficient. Apparently fairly common in deepwater but rarely recorded.

Bristly Catshark *Bythaelurus hispidus* Plate 35

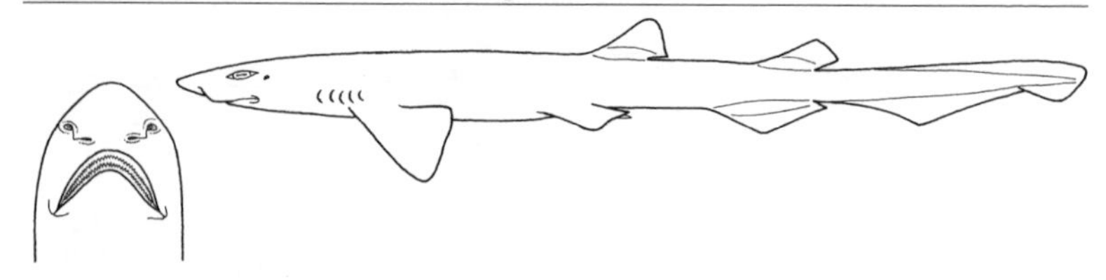

Measurements. Mature: ~22–24 cm. Max: 29 cm TL.
Identification. Very small, elongated pale brown or whitish catshark, sometimes with faint grey crossbands, white or dusky spots. Bristly skin. Short rounded snout, long arched mouth. Two dorsal fins (origin of first over rear of pelvic fin bases) and an anal fin.

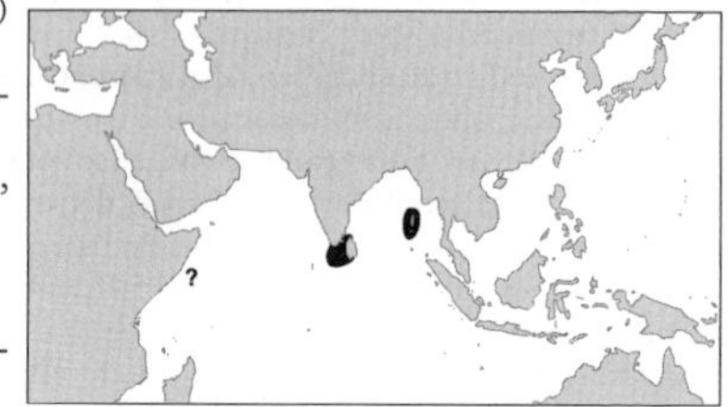

Distribution. Northern Indian Ocean, Southeastern India, Andaman Islands.
Habitat. Bottom on upper continental slopes, 293–766 m.
Behaviour. Unknown.
Biology. Eats small fishes, squid and crustacea
Status. IUCN Red List: Not Evaluated. Apparently common in deep water.

Spotless Catshark *Bythaelurus immaculatus* Plate 35

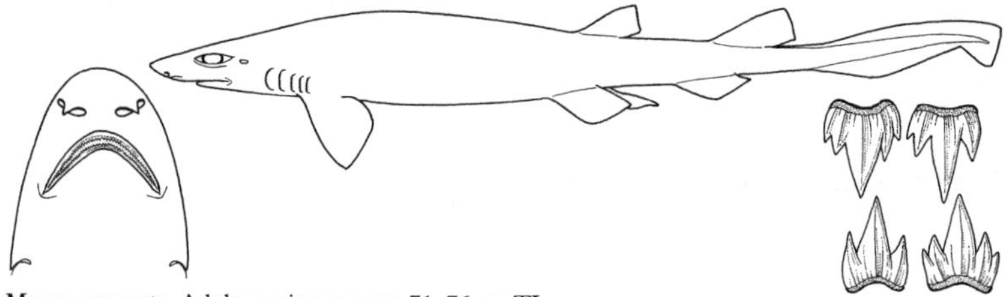

Measurements. Adult specimens were 71–76 cm TL.
Identification. A large drab yellowish-brown catshark with no markings, similar to *Bythaelurus dawsoni.* Rounded snout, long abdomen, two dorsal fins, first smaller, over pelvic fin bases, second much larger and over anal fin base.

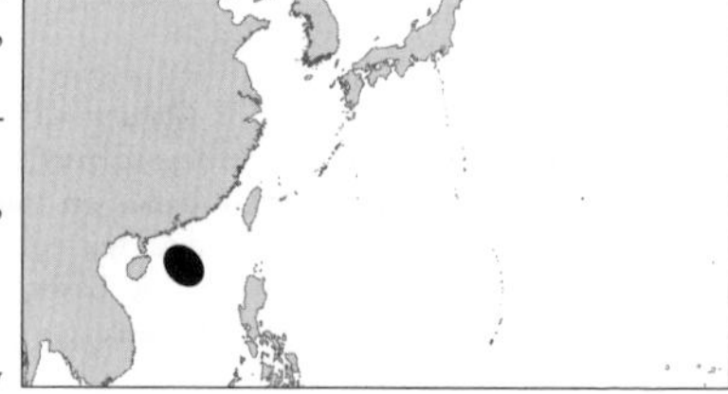

Distribution. South China Sea, about 380–400 km east of Hainan Island.
Habitat. Bottom, on the continental slope, 534–1020 m.
Behaviour. Unknown.
Biology. Unknown.
Status. IUCN Red List: Not Evaluated. Recently described.

Mud Catshark *Bythaelurus lutarius* Plate 35

Measurements. Born: ~14 cm TL. Mature: ~31 cm, gravid ♀ 33–37 cm. Max: ~39 cm TL.
Identification. Small plain grey-brown catshark, sometimes with dusky saddle bands, light underside. Short rounded snout. Long arched mouth reaches past the front end of the cat-like eyes. Two small dorsal fins, base of first roughly over pelvic fin insertions, second (slightly larger) origin over mid base of anal fin.

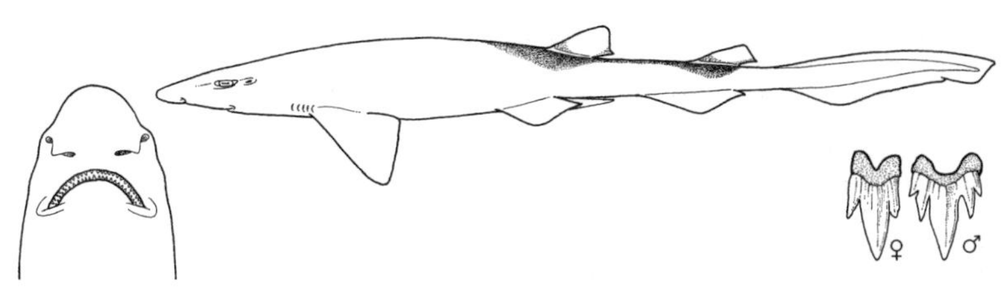

Distribution. Mozambique and Somalia.
Habitat. Deepwater on tropical continental slope, on or just above muddy bottom, at 338–766 m.
Behaviour. Unknown.
Biology. Ovoviviparous, two pups per litter. Eats cephalopods, small bony fishes, and crustaceans.
Status. IUCN Red List: Not Evaluated. Presumably not taken in fisheries.

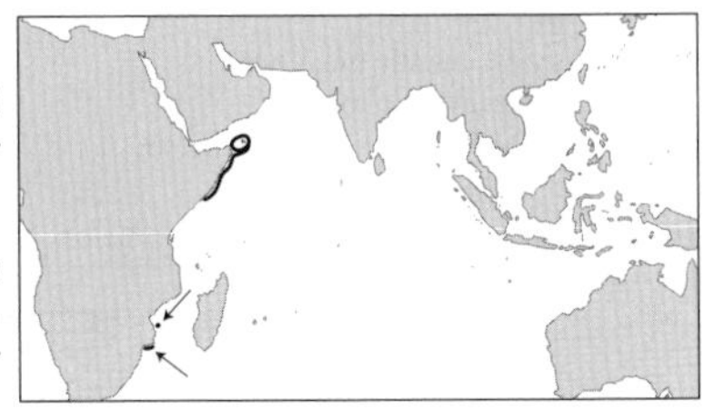

Dusky Catshark *Bythaelurus* sp. A. Plate 35

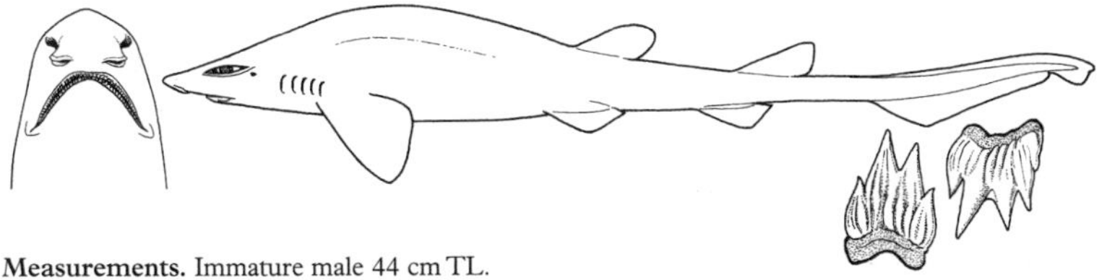

Measurements. Immature male 44 cm TL.
Identification. Uniform dark greyish-brown bristly skin with a few pale blotches beneath. Very broad flattened head, short rounded snout and long soft body. Two well separated similar-sized tall rounded dorsal fins, base of first over pelvic bases, second over anal fin.
Distribution. Northwestern Australia.
Habitat. Continental slope, 900 m.
Behaviour. Unknown.
Biology. Unknown.
Status. IUCN Red List: Data Deficient. Known from only one specimen. Occurs too deep to be caught in fisheries.

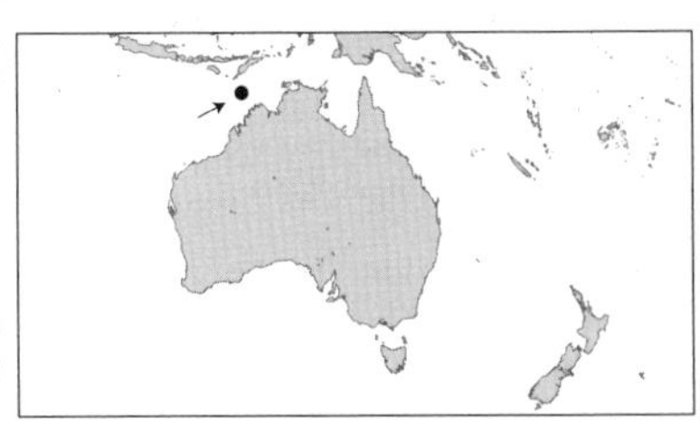

Galapagos catshark *Bythaelurus* sp. B Plate 35

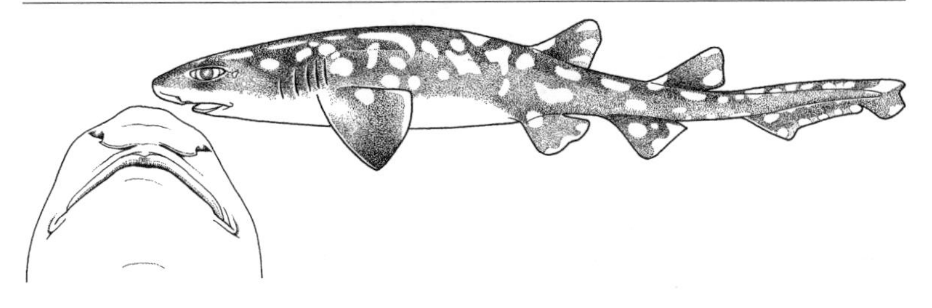

Measurements. Mature: ~45 cm ♂. Max: > 45 cm TL.
Identification. Small variegated catshark with a striking pattern of large white spots and blotches on grey background. Short, flattened rounded snout and short tail. Long arched mouth reaches past the front end of the cat-like eyes. Two medium-sized, rounded dorsal fins and an anal fin. First dorsal base over pelvic bases, anal fin about as large as second dorsal.

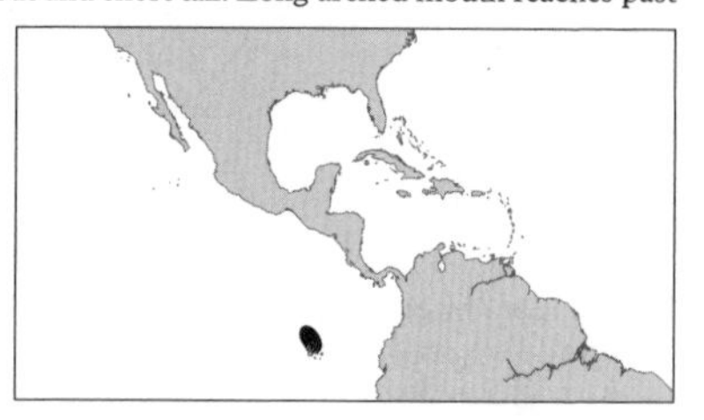

Distribution. Eastern Pacific, Galapagos Islands.
Habitat. Insular slopes on bottom between 400–600 m.
Behaviour. Essentially unknown.
Biology. Essentially unknown.
Status. IUCN Red List: Not Evaluated.

Reticulated Swellshark *Cephaloscyllium fasciatum* Plate 36

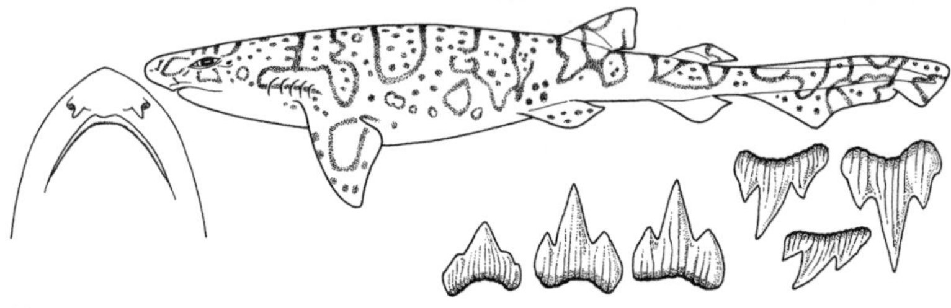

Measurements. Hatch: ~12 cm TL. Mature: ~36 cm ♂, 42 cm ♀. Max: >42 cm TL.
Identification. Small catshark with inflatable stomach. Adults have a striking pattern of dark lines forming open-centred saddles, loops, reticulations and spots on light greyish back and sides (spots absent in young). Underside spotted. Ridges over eyes, second dorsal fin much smaller than first anal fin.

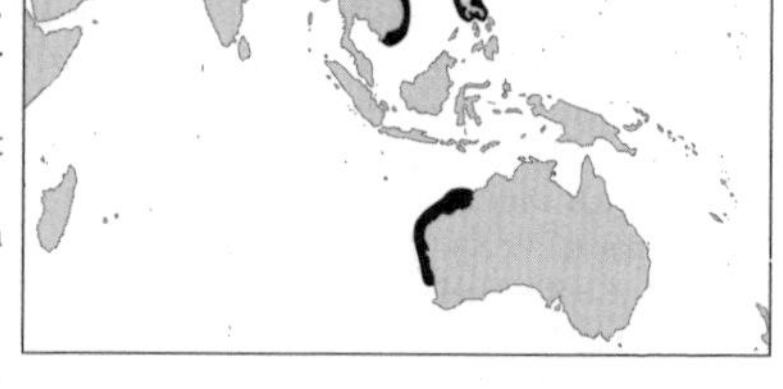

Distribution. Western Pacific: Vietnam, China (Hainan Island), Philippines (Luzon), NW Australia.
Habitat. On or near muddy bottom, uppermost slope, 219–450 m.
Behaviour. Can expand itself with air or water in an attempt to frighten predators.
Biology. Oviparous.
Status. IUCN Red List: Data Deficient globally. Taken as trawl bycatch. Least Concern in Australia where fishing effort is limited.

Draughtsboard Shark *Cephaloscyllium isabellum* Plate 36

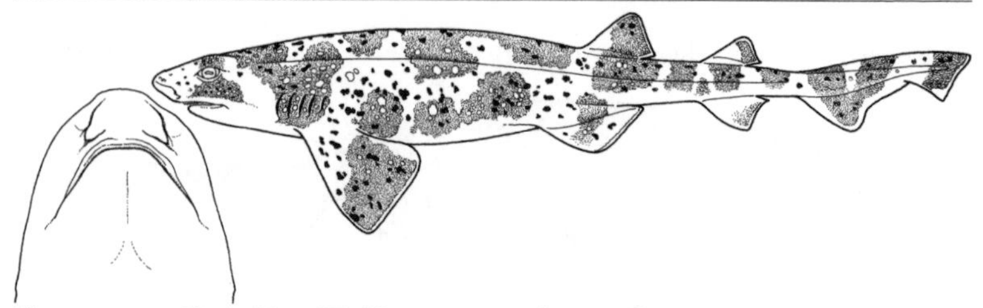

Measurements. Hatch: 16 cm TL. Mature: ~60 cm ♂, 80 cm ♀. Max: 150 cm TL.
Identification. A large stocky strongly patterned catshark with inflatable stomach and up to eleven dark brown irregular saddles and alternating blotches on its sides in a checkerboard pattern. Ridges over

eyes, second dorsal fin much smaller than the first. Anal fin. Eggcases smooth with tendrils.
Distribution. New Zealand, Japan and Taiwan specimens may be *C. umbratile* or undescribed species.
Habitat. Rocky and sandy bottom, shore to 673 m, most <400 m.
Behaviour. Can expand body with air or water.
Biology. Eggs laid in pairs. Eats crabs, worms, other invertebrates, and probably bony fishes.
Status. IUCN Red List: Least Concern. Bycatch in deepwater trawl fisheries, survives discard well.

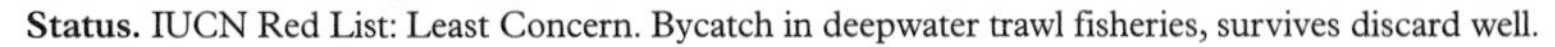

Australian Swellshark *Cephaloscyllium laticeps* Plate 36

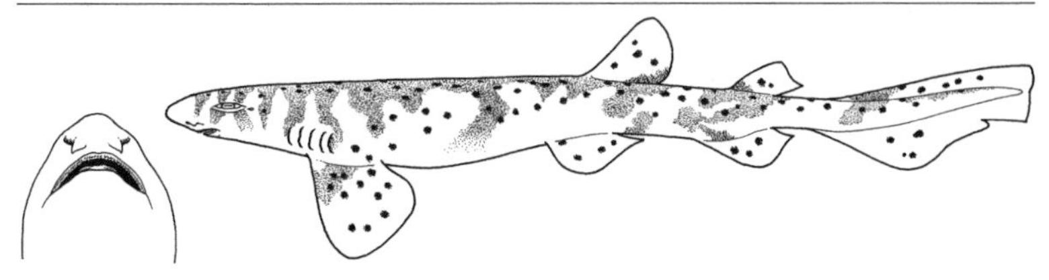

Measurements. Hatch: ~14 cm TL. Eggcase: ~13 x 5 cm. Mature: ~82 cm ♂. Max: >100 cm TL.
Identification. Strongly variegated pattern of dark brown or greyish close-set dark saddles and blotches, many dark spots and occasional light spots on lighter grey or chestnut background. Broad dark stripe between eye and pectoral origin, dark below eye. Underside cream, usually with dark stripe down belly in adults. No conspicuous light fin margins. Inflatable stomach. Ridges over eyes. Larger first dorsal fin over pelvic fin bases, second over anal fin.
Distribution. Southern Australia.

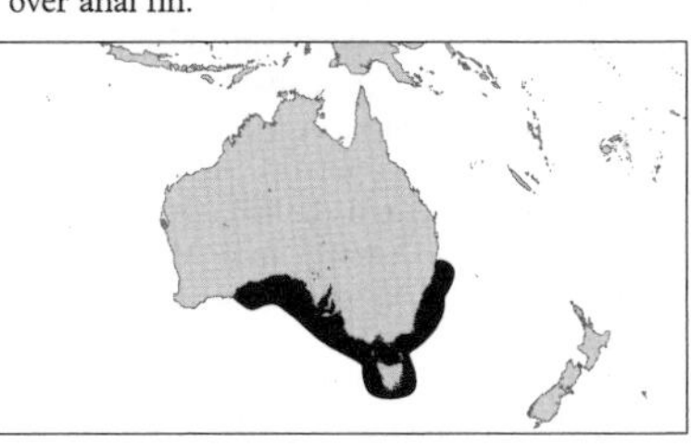

Habitat. Inshore on continental shelf to at least 60 m.
Behaviour. Lays ridged cream-coloured eggcases attached to seaweed and benthic invertebrates.
Biology. Feeds on small reef fish, crustaceans and squid.
Status. IUCN Red List: Least Concern. Discarded bycatch from shark gill net fishery; survives well. *Cephaloscyllium nascione* from New South Wales is synonymized with this species.

Indian Swellshark *Cephaloscyllium silasi* Plate 36

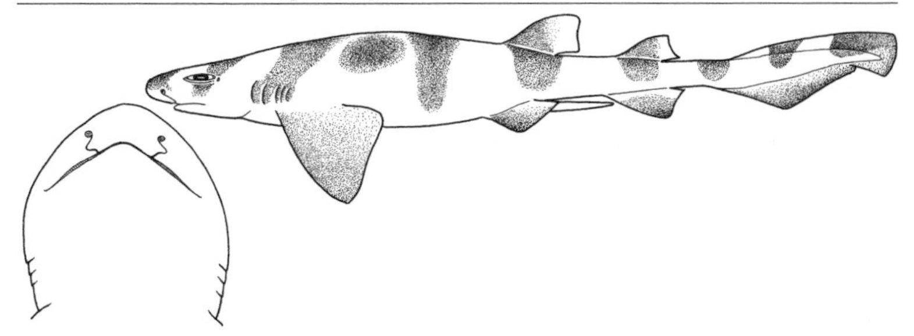

Measurements. Largest specimen is a 36 cm adult male.
Identification. Dwarf species. Pattern of seven moderately broad dark brown saddles on a light

brown background, an obscure darker blotch over the pectoral inner margins. Light brown unspotted underside. No conspicuous light fin margins. Ridges over eyes, first dorsal fin larger, over pelvic fin bases, second much smaller and over anal fin.

Distribution. Near Quilon, India, north-central Indian Ocean. A similar small swellshark occurs off Myanmar, Andaman Sea, but needs to be compared with this species.

Habitat. Bottom on the uppermost continental slope (collected at ~300 m).

Behaviour. Unknown.

Biology. Unknown.

Status. IUCN Red List: Not Evaluated. Reported to be relatively common in known small range.

Balloon Shark *Cephaloscyllium sufflans* Plate 36

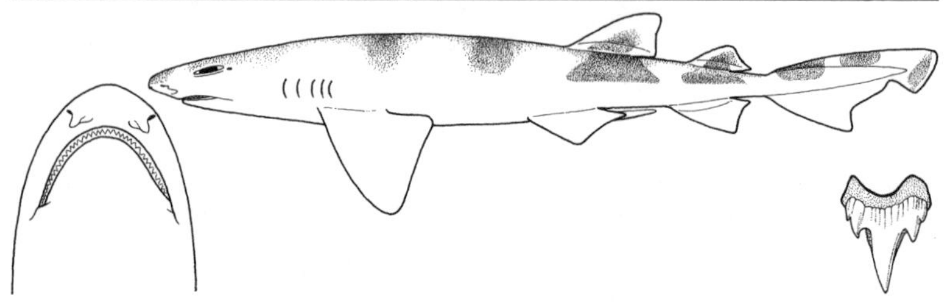

Measurements. Hatch: 20–22 cm TL. Mature: ~70–75 cm. Max: >106 cm TL.

Identification. Large shark with inflatable stomach. Seven light grey-brown saddles on lighter grey background are obscure or absent in adults. Pectoral fins dusky above, unspotted below. No obvious light fin margins. Ridges over eyes. First dorsal fin over pelvic fins, second much smaller and over anal fin.

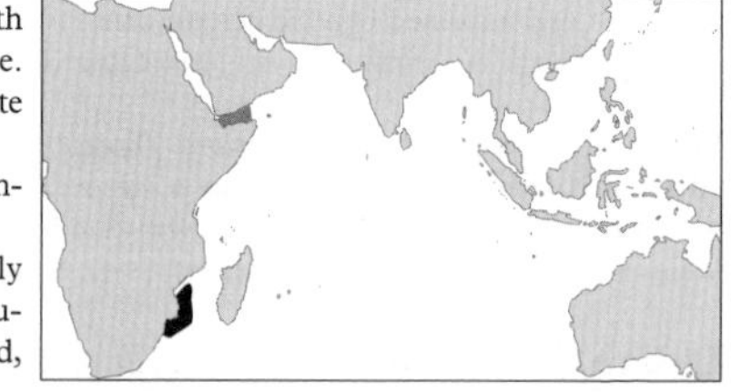

Distribution. Western Indian Ocean: South Africa (kwaZulu-Natal) and Mozambique. Dubious Gulf of Aden record may be a separate and smaller species.

Habitat. Sand and mud bottom, offshore continental shelf and uppermost slope, 40–440 m.

Behaviour. Juveniles and adults apparently segregated – immatures common off kwaZulu-Natal, not adults, and free eggcases not yet found, (may occur further north or in deeper water).

Biology. Oviparous. Feed on lobsters, shrimp and cephalopods, also teleosts and other elasmobranchs.

Status. IUCN Red List: Not Evaluated.

Japanese Swellshark *Cephaloscyllium umbratile* Plate 36

Measurements. Mature: ~98 cm ♂. Max: >98 cm TL.

Identification. Similar to and formerly synonymized with *Cephaloscyllium isabellum*, but longer snout and more mottled pattern. Pale brown dorsal surface with dense dark brown mottling and regular saddles separated by lighter reddish-brown areas. Scattered small white and dark brown spots. Underside lighter, mostly unspotted. Dorsal and caudal fins mottled and spotted, pectoral and pelvic fins mottled and spotted above, light below, anal fin mottled. Ridges over eyes, inflatable stomach, second dorsal fin much smaller than first. Anal fin.

Distribution. Western Pacific: Japan (wide-ranging), China and Taiwan.

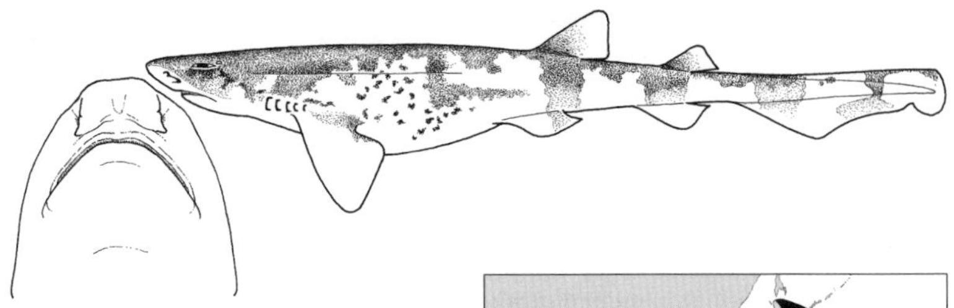

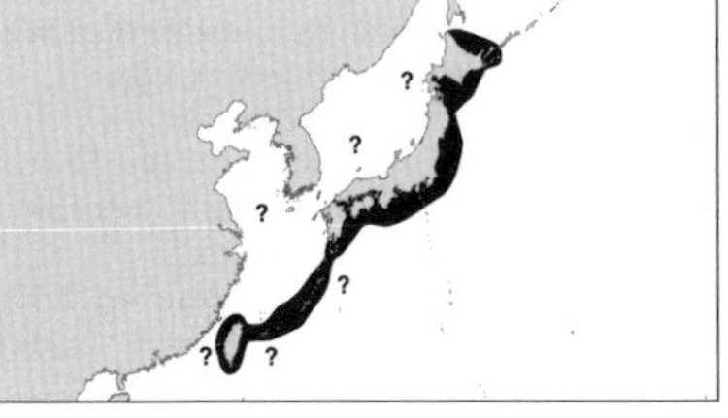

Habitat. Little-known, probably shallow water.
Behaviour. Little-known.
Biology. Oviparous.
Status. IUCN Red List: Not Evaluated. Records apparently include at least two mostly sympatric species, the large *C. umbratile* and a smaller, more simply patterned, undescribed species (*Cephaloscyllium* sp., false *umbratile*).

Swellshark *Cephaloscyllium ventriosum* Plate 36

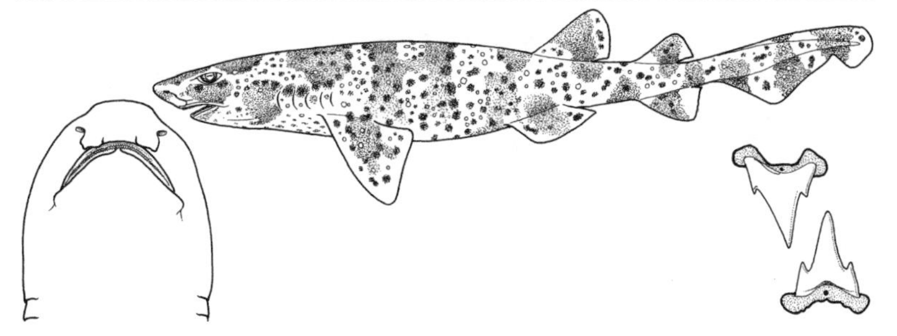

Measurements. Hatch: 13–15 cm TL. Adult: 82–85 cm ♂. Max: >100 cm TL.
Identification. Large strongly variegated catshark with close-set dark brown saddles and blotches, numerous dark spots and occasional light spots on a lighter yellow-brown background, undersides also heavily spotted. Ridges over eyes, two dorsal fins, second much smaller. Anal fin.
Distribution. Eastern Pacific. California to Mexico. Central Chile.
Habitat. Rocky bottom in kelp beds and other algae.
Behaviour. Relatively sluggish and mainly nocturnal. Lies motionless in rocky caves and crevices by day, often in small groups, and swims slowly at night. Inflates stomach when disturbed to wedge itself into crevices.
Biology. Oviparous. Eggs laid in large, unridged greenish-amber, purse-shaped eggcases, hatch in 7.5–10 months (depending on water temperature). Feeds on fish, probably crustacea.

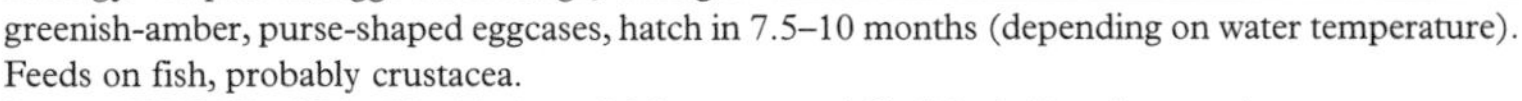

Status. IUCN Red List: Not Evaluated. Not commercially fished. Kept in aquaria.

Whitefin Swellshark *Cephaloscyllium* sp. A Plate 36

Measurements. Eggcase: ~11 x 5 cm. Mature: ~70 cm ♂. Max: at least 94 cm TL.
Identification. Large and stocky, broad head, rough skin, and pattern of broad dark blotches and

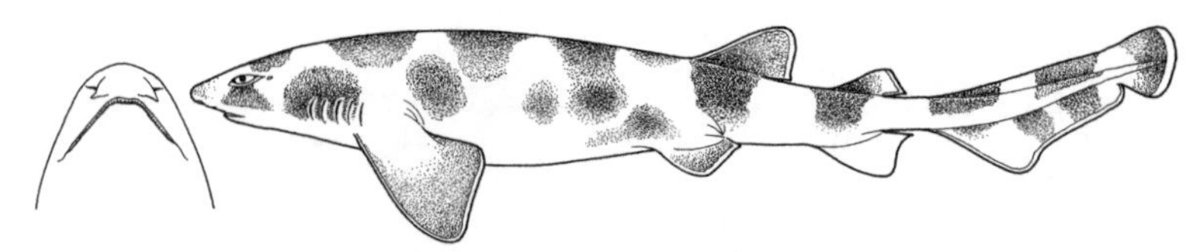

saddles on medium brownish or greyish background on sides and back (fainter in juveniles). Usually five pre-dorsal bars. Fins mostly dark above with pale margins, pale underneath. Ridges over eyes. Two dorsal fins, second much smaller than first. Anal fin. Flask-shaped, unridged eggcase.

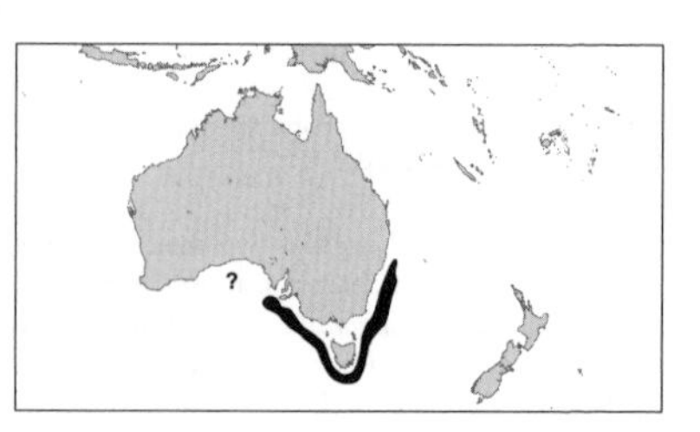

Distribution. Southeastern Australia.
Habitat. Upper continental slope, 240–550 m.
Behaviour. Inflates by swallowing water or air.
Biology. Oviparous, poorly known.
Status. IUCN Red List: Near Threatened. Taken in trawl bycatch and some population declines reported.

Saddled Swellshark *Cephaloscyllium* sp. B Plate 36

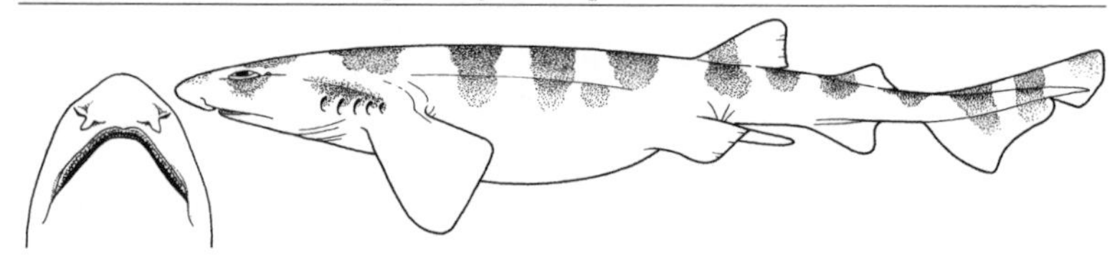

Measurements. Mature ~55 cm TL ♂. Max: >70 cm TL.
Identification. Fairly small catshark with slender head, ridges over eyes. Medium brownish or greyish with obvious dark saddles (usually five in front of dorsal fins). Blotches usually absent from sides. Fin margins sometimes pale, underside of body pale. Inflatable stomach. Two dorsal fins, close together, first behind pectorals, second much smaller than first and over anal fin.

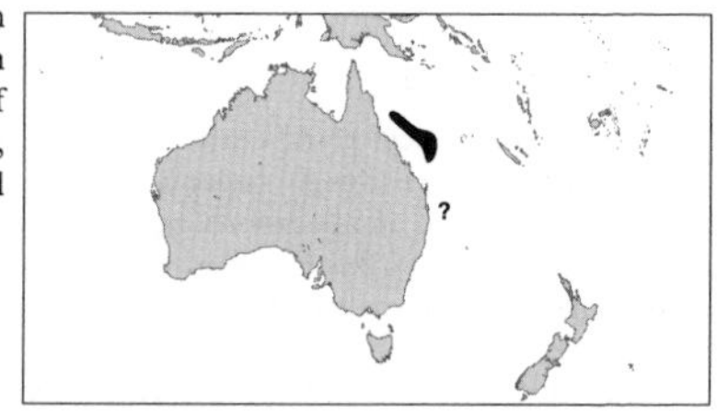

Distribution. Northeastern Australia.
Habitat. Continental slope, 380–590 m.
Behaviour. Unknown.
Biology. Oviparous, otherwise unknown.
Status. IUCN Red List: Data Deficient. Poorly known, distribution not heavily fished.

Northern Draughtboard Swellshark *Cephaloscyllium* sp. C Plate 36

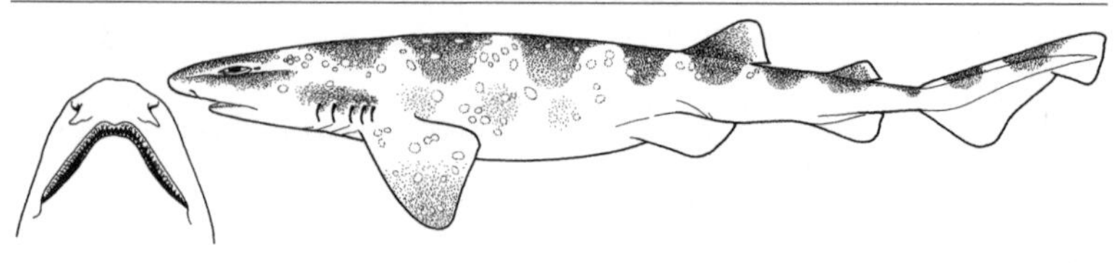

Measurements. Eggcase: ~7 x 4 cm TL. Reaches at least 65 cm TL.
Identification. Dark greyish-brown medium-sized slender catshark with indistinct darker saddles and blotches, covered with whitish blotches and flecks extending onto fins. No dark stripe along

belly. Juveniles paler with small widely-spaced spots. Ridges over eyes. Two dorsal fins, first over pelvic fins and larger than second, over anal fin. Inflatable stomach.

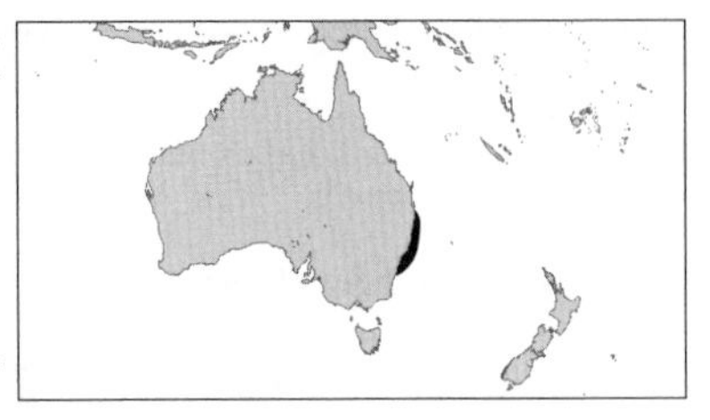

Distribution. Eastern Australia.
Habitat. Outer continental shelf, 90–140 m.
Behaviour. Unknown.
Biology. Oviparous, otherwise unknown.
Status. IUCN Red List: Near Threatened. Rarely recorded. Limited range is heavily trawled.

Narrowbar Swellshark *Cephaloscyllium* sp. D — Plate 36

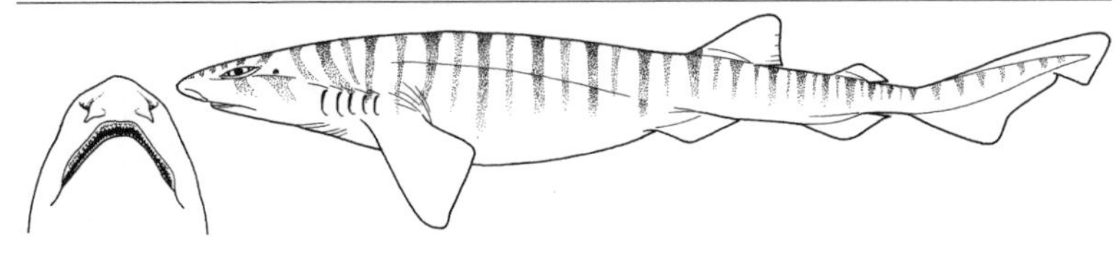

Measurements. Reaches at least 43 cm TL.
Identification. Very distinctive colour pattern of numerous narrow closely spaced dark bars on a dark-brownish to cream background. Bars do not join to form rings or saddles, 17–18 present predorsally. Irregular lines on snout, uniformly pale fins and underside. Ridge over eyes, inflatable stomach, two dorsal fins, first set behind pelvic fins, second very much smaller, over anal fin.

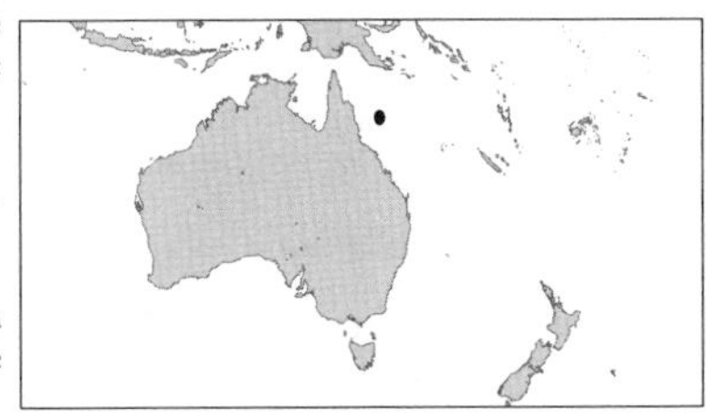

Distribution. Northeast Australia.
Habitat. Recorded from continental slope at 440 m.
Behaviour. Unknown.
Biology. Unknown.
Status. IUCN Red List: Data Deficient. Known from only a few specimens trawled from one location, subjected to minimal fishing pressure.

Speckled Swellshark *Cephaloscyllium* sp. E — Plate 36

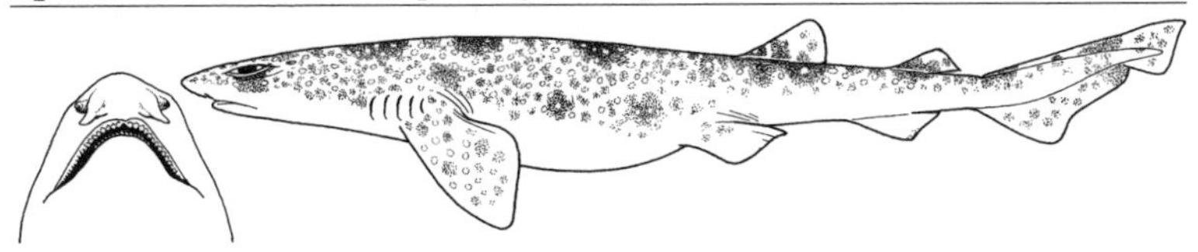

Measurements. Mature: ~64 cm TL ♂. Reaches at least 68 cm TL.
Identification. Pale greyish upper surface with intense mottling of small dark blotches, larger blotches and saddles with small white spots. Rounded blotches and white spots below eyes followed by three pre-dorsal saddles. Fins mostly pale with darker spots and blotches. Stocky, short-tailed, ridges over eyes, no labial furrows, an inflatable stomach, and second dorsal fin much smaller than the first. Long arched mouth reaches past the front end of the cat-like eyes. Two small dorsal fins and an anal fin. First dorsal base over or behind pelvic bases.

Distribution. Patchy records from tropical northeast and northwest Australia.

Habitat. Continental slope, 390–440 m (northwest) and 600–700 m (northeast).
Behaviour. Unknown.
Biology. Unknown.
Status. IUCN Red List: Data Deficient. Known from only a few specimens, may be rare, but subject to little fishing pressure.

Dwarf Oriental Swellshark *Cephaloscyllium* sp. Plate 36

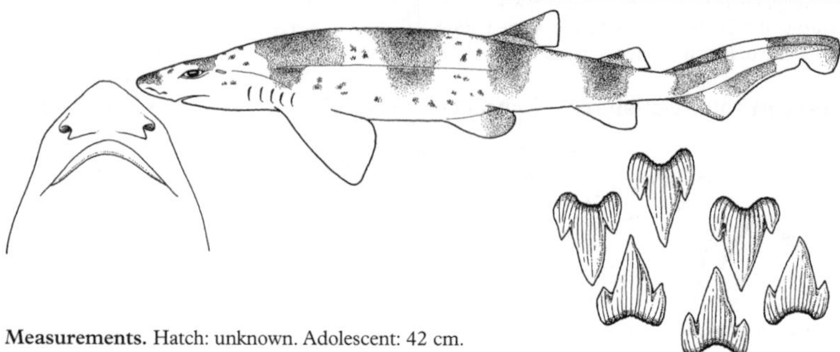

Measurements. Hatch: unknown. Adolescent: 42 cm.
Mature ~36–39 cm ♂. Adult: 42–44 cm ♀. Max: > 44 cm TL.
Identification. Similar to *Cephaloscyllium umbratile*, but with longer snout, longer anterior nasal flaps, simpler blotched color pattern with few dark spots, and small size. Ridges over eyes, an inflatable stomach, two dorsal fins, second much smaller than the first. Anal fin.

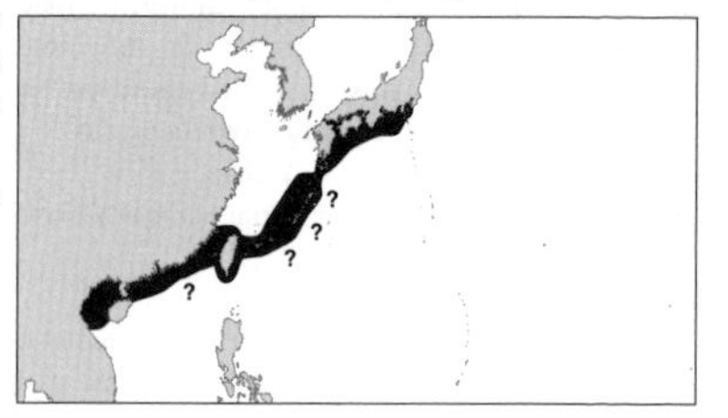

Distribution. Western Pacific: Vietnam, China, Taiwan, Japan.
Habitat. Poorly known, possibly inshore.
Behaviour. Unknown.
Biology. Poorly known. Oviparous.
Status. IUCN Red List: Not Evaluated.

Lollipop Catshark *Cephalurus cephalus* Plate 39

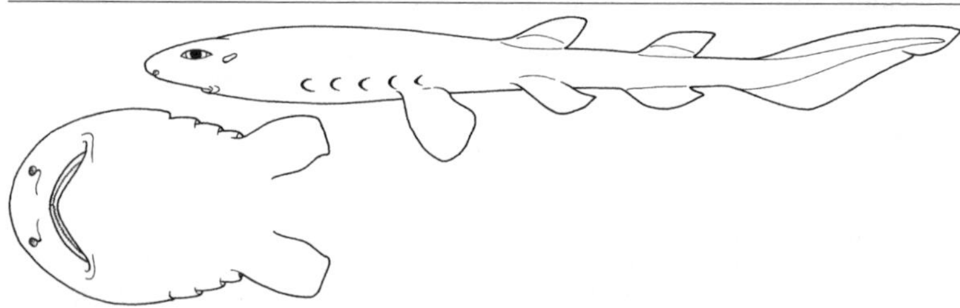

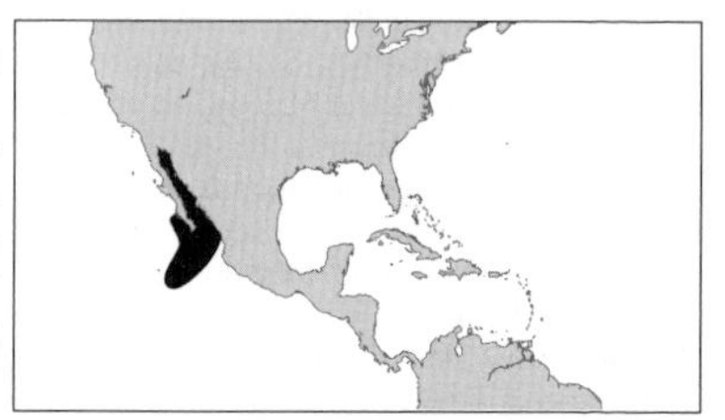

Measurements. Born: ~10 cm TL. Mature: ~19 cm. Max: ~28 cm TL.
Identification. Unmistakable very small, unpatterned tadpole-shaped catshark. Expanded, flattened rounded head and gill region; small, slender, very soft, thin skinned (almost gelatinous) body and tail. Two small dorsal fins and an anal fin. First dorsal origin slightly in front of pelvic fin origins.

Distribution. Southern Baja California and Gulf of California, Mexico.
Habitat. Upper continental slope and outermost shelf, on or near bottom, 155–927 m. The expanded gill area suggests that this species is adapted to areas of seabed with low dissolved oxygen levels.
Behaviour. Unknown.
Biology. Ovoviviparous, retaining its very thin-walled eggcases in the uterus until the two young per litters hatch.
Status. IUCN Red List: Not Evaluated.

Southern Lollipop Catshark *Cephalurus* sp. A — Not illustrated

Measurements. Adults reach 26–32 cm TL.
Identification. Very similar to but slightly larger than *Cephaloscyllium cephalus* and may have a slightly smaller eye. Possibly differs in coloration, with southern species being darker (dark brown to blackish, versus lighter brown in *C. cephalus*) and with a dusky mouth lining (white or yellow in *C. cephalus*).
Distribution. Peru and Chile.
Habitat. As for *C. cephalus.*
Behaviour. Unknown.
Biology. Probably ovoviviparous as in *C. cephalus.*
Status. IUCN Red List: Not Evaluated.

Antilles Catshark *Galeus antillensis* — Plate 37

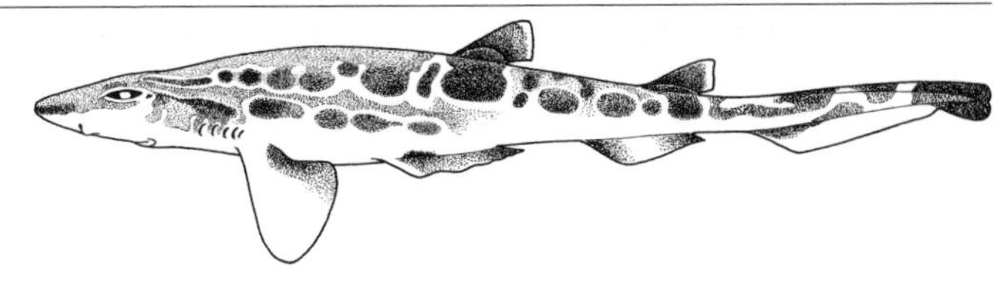

Measurements. Mature: 33–46 cm TL. Max: 46 cm TL.
Identification. Striking pattern of variegated dark brown saddle blotches (usually <eleven) and dark bands on tail. Saddles may be clearly outlined or obscure. Usually dark markings on flanks, no black marks on dorsal fin tips. Dark mouth lining. Distinct crest of enlarged dermal denticles along upper margin of elongated tail. No black marks on dorsal fin tips, caudal fin tip may have a dusky base and light web. Larger than *Galeus arae* and with more precaudal vertebrae.
Distribution. West Atlantic (Caribbean Sea). Straits of Florida and Caribbean from Hispaniola, Puerto Rico, Jamaica and Leeward Islands to Martinique.
Habitat. Deepwater upper insular slopes, on or near bottom, 293–658 m.
Behaviour. May school in large numbers.
Biology. Possibly oviparous. Feeds mainly on deepwater shrimp.
Status. IUCN Red List: Not Evaluated. Common to abundant where it occurs. Formerly considered an island subspecies of *G. arae.*

Roughtail Catshark *Galeus arae* — Plate 37

Measurements. Mature: ~27–33 cm. Max: ~33 cm TL.
Identification. Similar to *Galeus antillensis*, but smaller and with fewer precaudal vertebrae.
Distribution. Western Atlantic (including Gulf of Mexico and continental Caribbean Sea): two separate populations, one off USA (North Carolina to the Mississippi Delta), Mexico and Cuba, second off Belize, Honduras, Nicaragua, Costa Rica and adjacent islands.

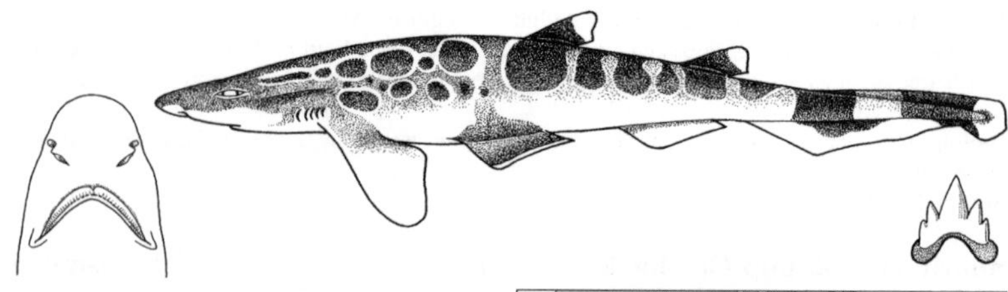

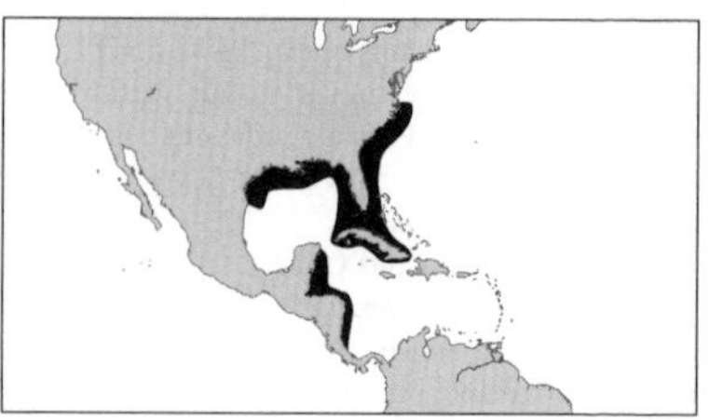

Habitat. Upper continental and insular slopes, on or near bottom, 292–732 m.
Behaviour. Partly segregated by depth: only adults in deep water, adults and juveniles above 450 m. May school in large numbers.
Biology. Reproduction possibly ovoviviparous, feeds mainly on deepwater shrimp.
Status. IUCN Red List: Not Evaluated. Common to abundant where it occurs.

Atlantic Sawtail Catshark *Galeus atlanticus* Plate 37

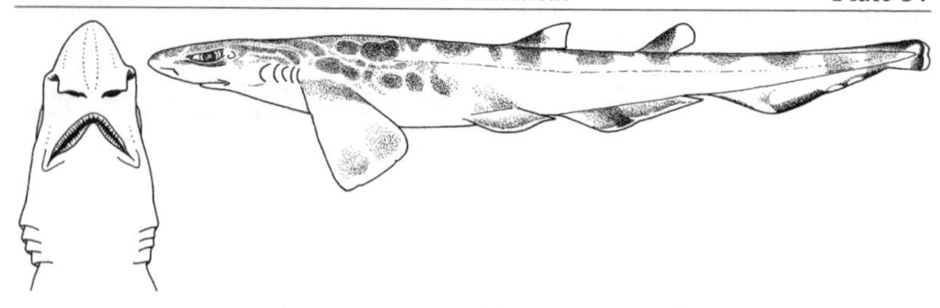

Measurements. Mature: ~38–42 cm ♂, 40–45 cm ♀. Max: ~45 cm TL.
Identification. Grey above, whitish below, dark grey blotches and saddles or narrow vertical bars along body and tail (rarely absent). Long angular snout, black mouth cavity. Eyes on lateral edges of head. Large pectoral fins, angular dorsal fins dusky with light rear webs, anal fin long and low. Distinct crest of enlarged dermal denticles along upper margin of elongated tail, caudal fin with dark margin, not black-tipped.
Distribution. Northeast Atlantic and Mediterranean: Morocco, Mediterranean Spain (around Straits of Gibraltar), possibly to Mauritania and Italy.
Habitat. Continental slopes, 400–600 m.
Behaviour. Essentially unknown.
Biology. Oviparous, apparently with multiple ovipary (nine eggcases in one female), suggesting a short hatching period outside the mother.
Status. IUCN Red List: Not Evaluated. Formerly synonymized with *Galeus melastomus*, possibly mistaken for *G. polli*, but apparently distinct.

Australian Sawtail Catshark *Galeus boardmani* Plate 37

Measurements. Mature: ~40 cm ♂, 43 cm ♀. Max: 61 cm TL.
Identification. Greyish with a variegated pattern of dark greyish-brown saddles and bars, three broad pale-edged pre-dorsal saddles with a narrower less distinct band between each, a band at and between each dorsal, and three broad bands post-dorsally. Bands and saddles sometimes with white flecks. Pale

underside. Distinct crests of enlarged dermal denticles along the upper margin of the elongated tail and along the ventral midline of the caudal peduncle.

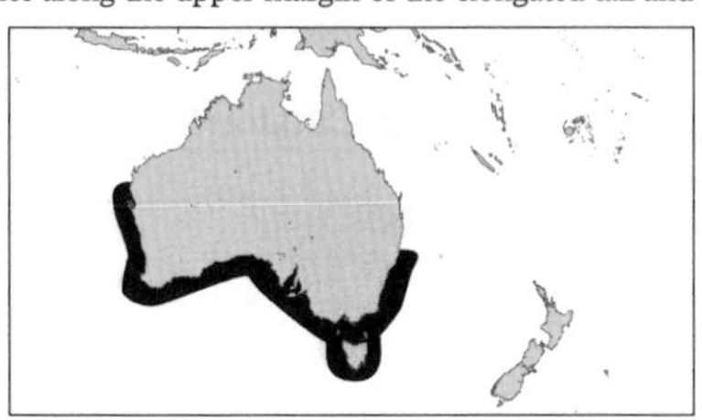

Distribution. Australia, southeast Queensland to Western Australia.
Habitat. Outer continental shelf and upper slope, at or near bottom at 150–640 m.
Behaviour. May sometimes aggregate by sex.
Biology. Oviparous. Feeds on fish, crustaceans and cephalopods, otherwise unknown.
Status. IUCN Red List: Least Concern. Widespread and apparently common.

Longfin Sawtail Catshark *Galeus cadenati* Plate 38

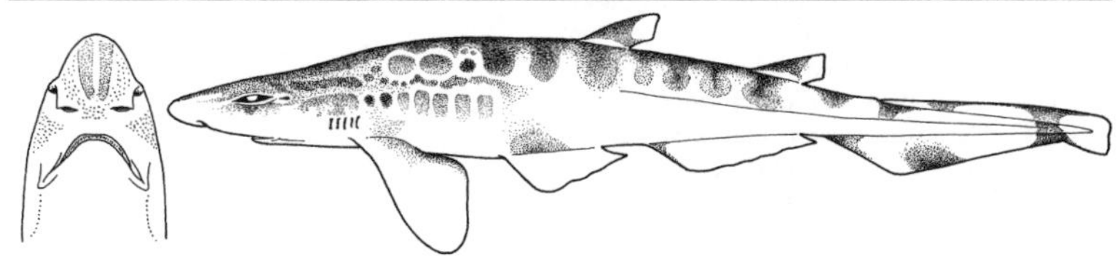

Measurements. Max: 35 cm TL.

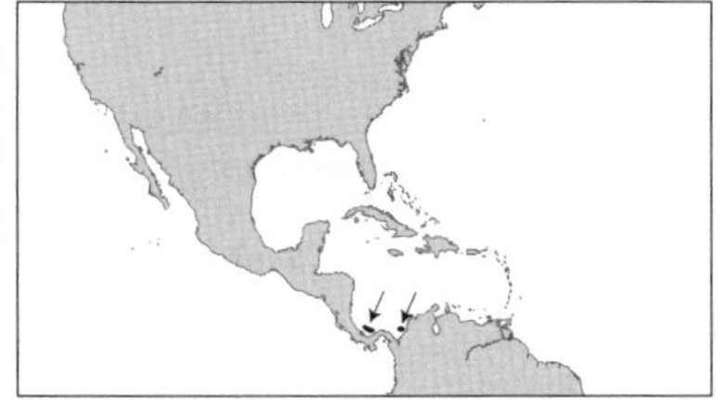

Identification. Similar to *G. arae* and *G. antillensis* and formerly considered to be a subspecies of *G. arae*. Has a noticeably longer lower anal fin than either species.
Distribution. Caribbean, off Panama and Colombia.
Habitat. Upper continental slopes at 439–548 m.
Behaviour. Unknown.
Biology. Reproduction oviparous.
Status. IUCN Red List: Not Evaluated. Rare and little-studied.

Gecko Catshark *Galeus eastmani* Plate 38

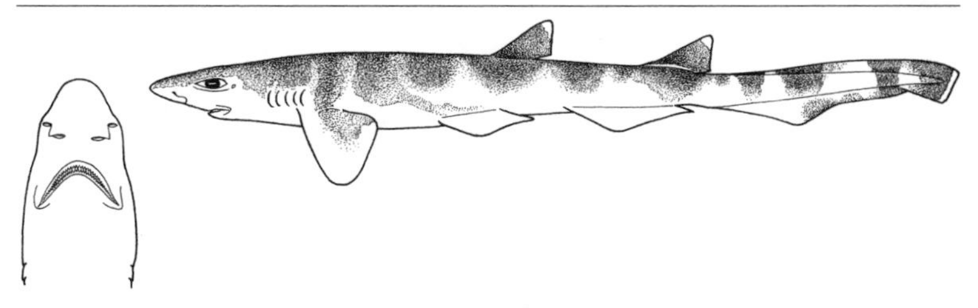

Measurements. Mature: ~31–32 cm ♂, 36–37 cm ♀. Max: possibly 50 cm TL.
Identification. Characterised by a white mouth lining and white-edged dorsal and caudal fins,

otherwise obscurely patterned slender body with a fairly small anal fin. Distinct crest of enlarged dermal denticles along the upper margin of the elongated tail.

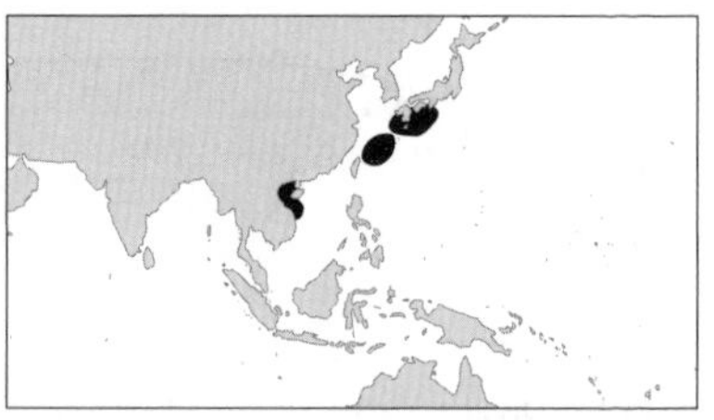

Distribution. Northwest Pacific: Japan, East China Sea, possibly Vietnam.
Habitat. Deepwater, on or near the bottom.
Behaviour. Unknown.
Biology. Oviparous, eggs laid in pairs (one per oviduct).
Status. IUCN Red List: Not Evaluated. Very common in Japanese waters, of no interest to fisheries.

Slender Sawtail Catshark *Galeus gracilis* Plate 38

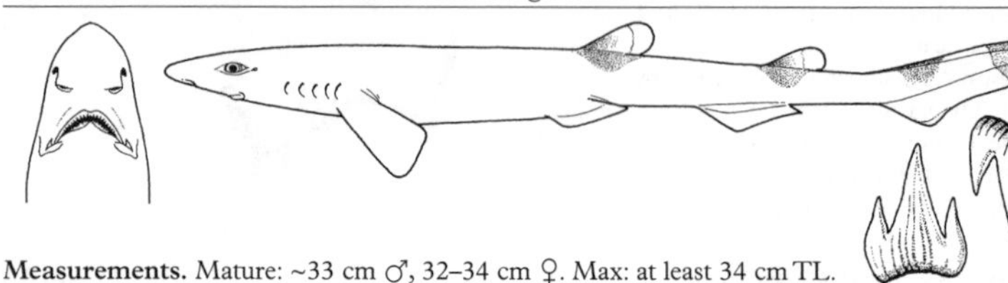

Measurements. Mature: ~33 cm ♂, 32–34 cm ♀. Max: at least 34 cm TL.
Identification. Pale grey, with four short dusky saddles (one beneath each dorsal fin and two on the caudal fin). No predorsal markings, pale under-side. Very small and slender with a distinct crest of enlarged dermal denticles along the upper margin of the elongated tail.

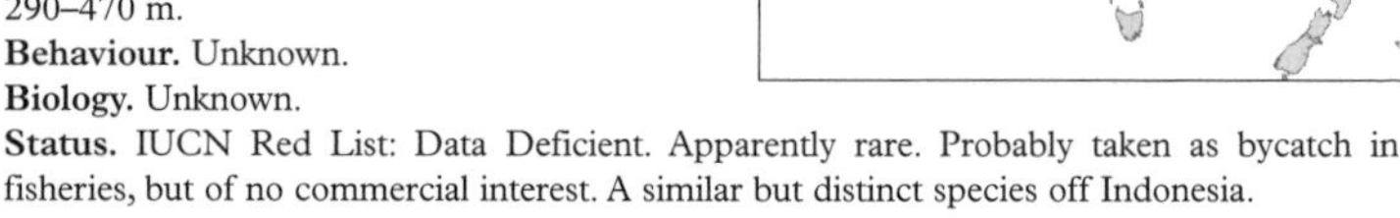

Distribution. Northern Australia.
Habitat. Continental slope, on or near bottom, 290–470 m.
Behaviour. Unknown.
Biology. Unknown.
Status. IUCN Red List: Data Deficient. Apparently rare. Probably taken as bycatch in trawl fisheries, but of no commercial interest. A similar but distinct species off Indonesia.

Longnose Sawtail Catshark *Galeus longirostris* Plate 38

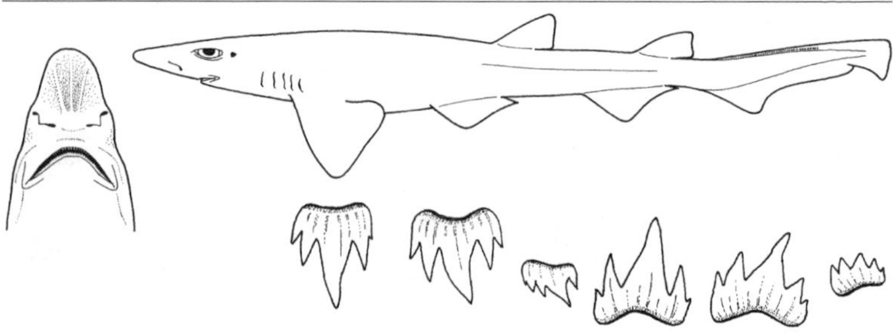

Measurements. Mature: ~66–71 cm ♂, 68–78+ cm ♀. Max: at least 80 cm TL.
Identification. Uniform grey above in adults, a few obscure dusky saddle blotches at dorsal fins and on caudal fin in young, whitish below. White-edged dorsal and pectoral fins, dorsal and caudal fins without black tips. Greyish-white mouth lining. Very long broadly rounded snout, distinct crest of enlarged dermal denticles along the upper margin of the elongated tail.
Distribution. Northwest Pacific: Japan (southern islands).

Habitat. On or near bottom, upper insular slopes at 350–550 m.
Behaviour. Unknown.
Biology. Unknown, fairly common where it occurs. No adult females yet recorded with egg-cases or young.
Status. IUCN Red List: Not Evaluated.

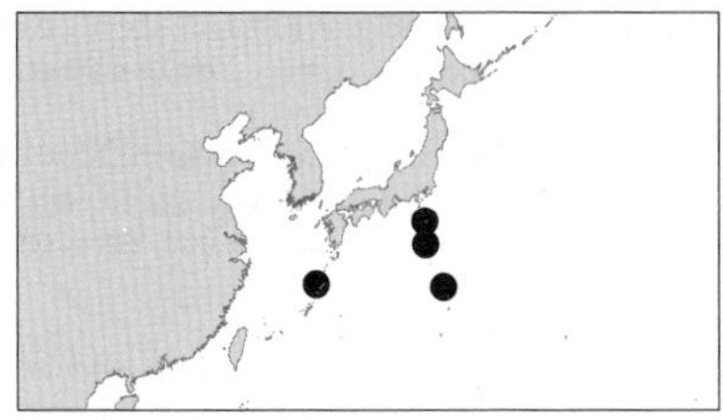

Blackmouth Catshark *Galeus melastomus* Plate 37

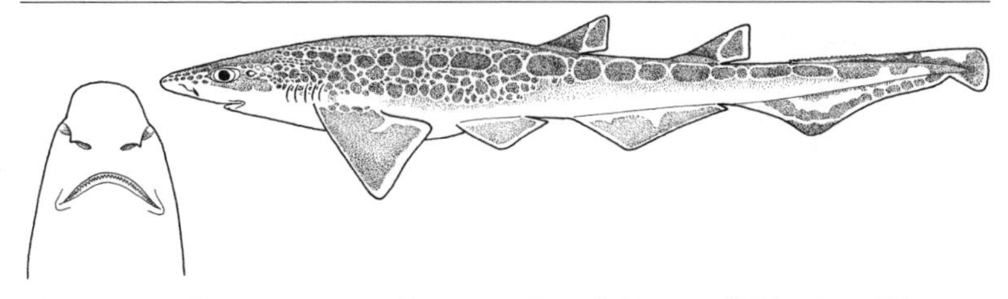

Measurements. Eggcase: ~6 x 3 cm. Mature: 34–42 cm ♂, 39–45 cm ♀. Max: 90 cm TL.
Identification. Striking pattern of 15–18 dark saddles, blotches and circular spots on back and tail, mouth cavity blackish, white edges to fins. Compressed precaudal tail, long anal fin reaching or extending past lower caudal origin, distinct crest of enlarged dermal denticles along upper margin of elongated tail.
Distribution. Northeast Atlantic and Mediterranean Sea: Faeroes and Norway, south to Senegal and Azores.

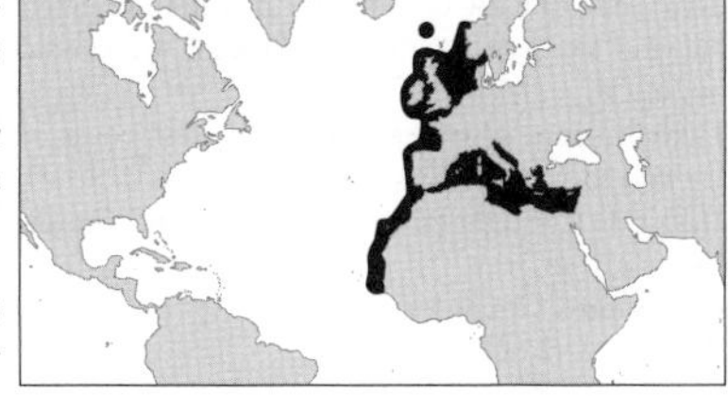

Habitat. Outer continental shelves and upper slopes, mainly 200–500 m, occasionally to 55 m and 1000 m.
Behaviour. Unknown.
Biology. Oviparous, up to 13 eggs/female. Feeds on bottom invertebrates (e.g. shrimp, cephalopods) and lanternfish.
Status. IUCN Red List: Not Evaluated. Common to abundant where it occurs. Taken as bycatch but low commercial value.

Southern Sawtail Catshark *Galeus mincaronei* Plate 37

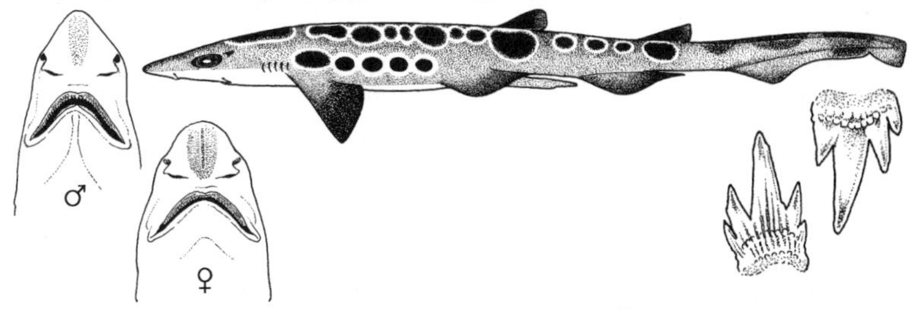

Measurements. Eggcase: 5–6 x 4 cm. Mature: ~40 cm ♂, 39 cm ♀. Max: >40 cm TL.
Identification. Reddish-brown above, whitish below. Striking pattern of eleven large oval or circular dark saddles and spots outlined in white on back and precaudal tail, fins dark but no black tips or

white edges. Long head, precaudal tail slightly compressed, moderately long anal fin nearly reaching past lower caudal origin, distinct crest of enlarged dermal denticles along upper (not lower) margin of elongated tail.

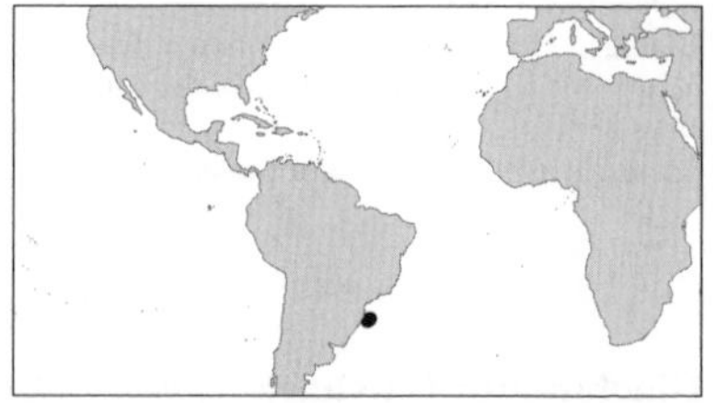

Distribution. Western Atlantic: southern Brazil.
Habitat. Deep-reef habitat with gorgonians, corals, sponges, crinoids and brittle stars on upper continental slope, 430 m.
Behaviour. Unknown.
Biology. Oviparous, single eggs in each oviduct, apparently laid in capsules. Often found with *Scyliorhinus haeckelii*.
Status. IUCN Red List: Not Evaluated.

Mouse Catshark *Galeus murinus* Plate 37

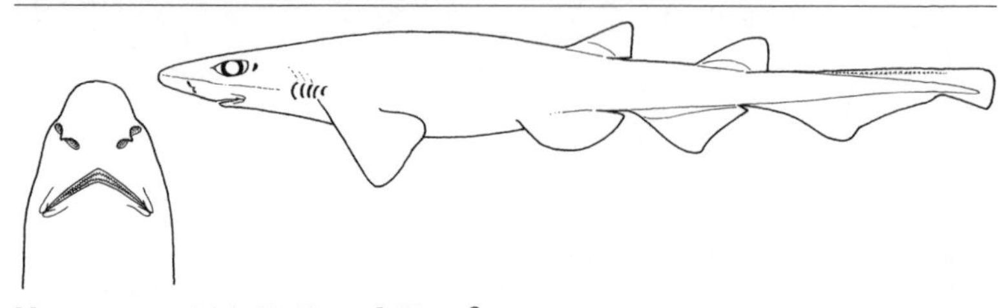

Measurements. Adult: 50–63 cm ♂, 53 cm ♀.
Identification. Uniform brown colour (slightly paler below), mouth blackish. Large, broadly rounded pelvic fins. Precaudal tail cylindrical, not flattened. Distinct crests of enlarged dermal denticles along the upper margin of the elongated tail and along the ventral midline of the caudal peduncle.

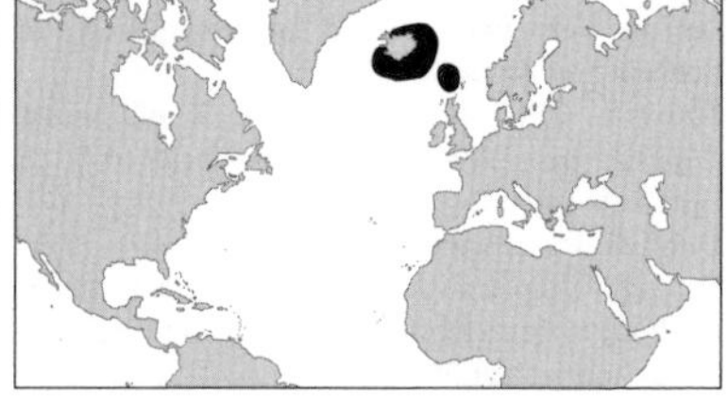

Distribution. Northeast Atlantic: west coast of Iceland to Faeroes Channel.
Habitat. Continental slopes, on or near bottom, 380–1250 m.
Behaviour. Unknown.
Biology. Unknown.
Status. IUCN Red List: Not Evaluated.

Broadfin Sawtail Catshark *Galeus nipponensis* Plate 37

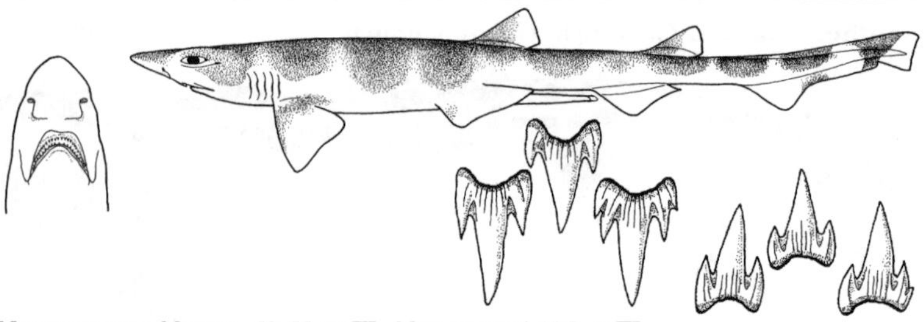

Measurements. Mature: ~53–55 cm TL. Max: reported 65.5 cm TL.
Identification. Greyish-white above with numerous obscure dusky saddles and blotches, white below. Black anterior margins to dorsal and anal fins, lower caudal lobe and rear tip. No black fin tips. White mouth lining. Long snout (tip to nostril > eye length). Distinct crest of enlarged dermal denticles along upper (not lower) margin of elongated tail. Mature males have slender, greatly

elongated claspers with bases covered by broad extensions of pelvic free rear tips (aprons), and a long space between pelvic fin bases and unusually short anal fin.

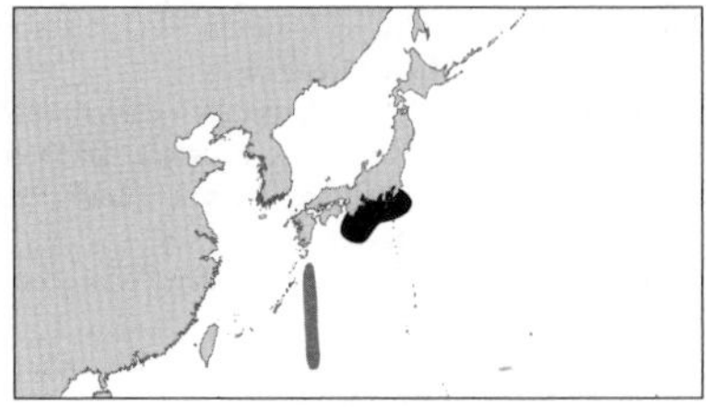

Distribution. Northwest Pacific: Common in deep water off Japan (southeast Honshu), also Kyushu-Palau Ridge.
Habitat. Deepwater on bottom, 362–540 m.
Behaviour. Unknown.
Biology. Oviparous, possibly lays pairs of eggs (one per oviduct).
Status. IUCN Red List: Not Evaluated. A similar smaller species in Philippines and off China is possibly distinct.

Peppered Catshark *Galeus piperatus* Plate 37

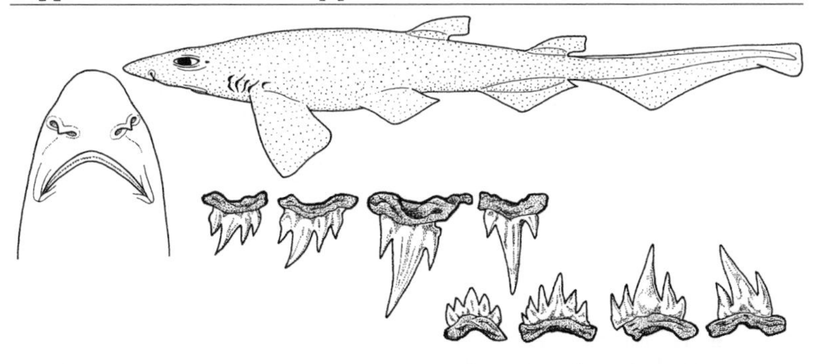

Measurements. Hatch: <7 cm TL. Mature: 28–29 cm ♂, 26–30 cm ♀. Max: 30 cm TL.
Identification. Very similar to *Galeus arae* and *G. antillensis* from the west Atlantic. Small, with a distinct crest of enlarged dermal denticles along upper (not lower) margin of elongated tail. May lack saddle markings or be patterned with variegated, white edged dark saddle blotches. White edged caudal fin, no black tips on dorsal or caudal fins. Mouth lining usually dark.

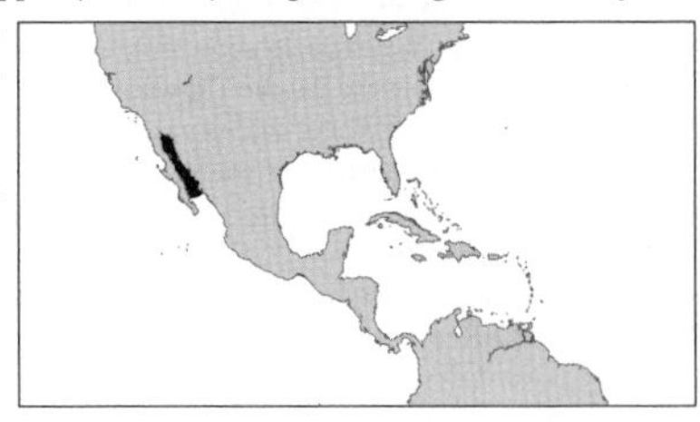

Distribution. Northeast Pacific: Mexico, northern Gulf of California.
Habitat. Bottom in deepwater, 275–1326 m.
Behaviour. Unknown.
Biology. Possibly oviparous.
Status. IUCN Red List: Not Evaluated.

African Sawtail Catshark *Galeus polli* Plate 37

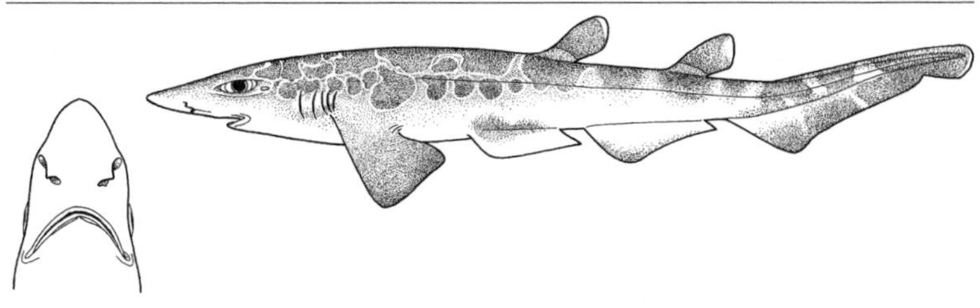

Measurements. Born: ~10–12 cm TL. Adolescent: 27–32 cm ♂, 29–38 cm ♀. Mature 30–46 cm ♂, 30–43 cm ♀. Max: 46 cm TL.

Identification. Fairly small, usually ~eleven or less well-defined dark grey or blackish-grey saddle blotches outlined in whitish on back and tail, sometimes uniform dark above. No black dorsal and caudal fin tips. Mouth lining dark. Distinct crest of enlarged dermal denticles along upper (not lower) margin of elongated tail.

Distribution. Eastern Atlantic: southern Morocco to South Africa (west coast).

Habitat. Upper continental slope and outer continental shelf, 159–720 m.

Behaviour. Unknown.

Biology. Ovoviviparous, 6–12 pups/litter. Eats small fish, squid and shrimp.

Status. IUCN Red List: Not Evaluated. Abundant off Namibia. Only the shallow part of its range is heavily fished, and this catshark is small enough to escape from trawls.

Blacktip Sawtail Catshark *Galeus sauteri* Plate 37

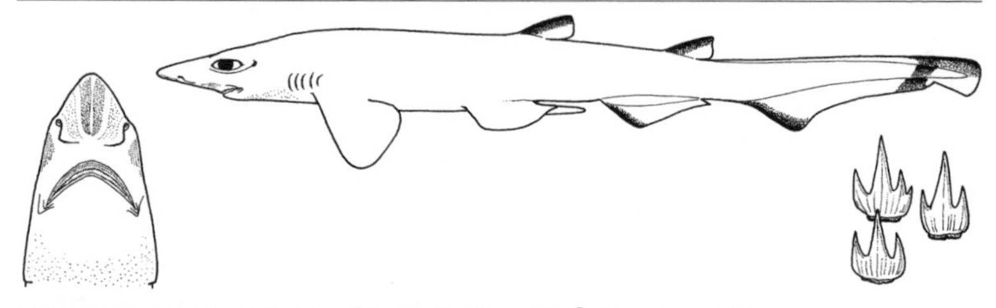

Measurements. Adult: 36–38 cm TL ♂, 42–45 cm TL ♀. Max: 45 cm TL.

Identification. No pattern of dark saddle blotches, dorsal fins and sometimes upper and lower caudal lobes with prominent black tips. Light mouth lining. Distinct crest of enlarged dermal denticles along upper (not lower) margin of elongated tail, caudal peduncle compressed.

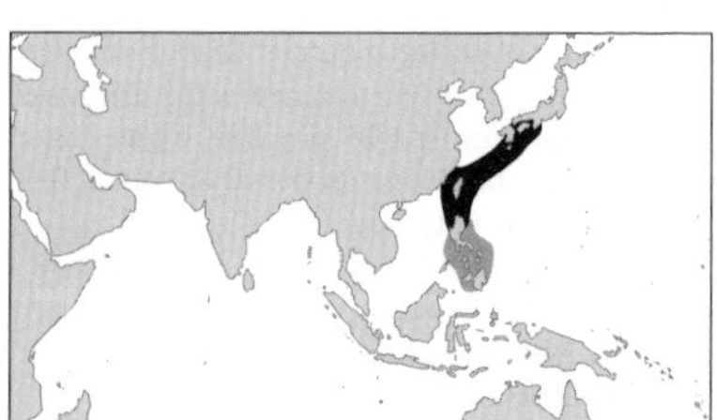

Distribution. Northwest Pacific: Taiwan Island, Philippines, Japan (including Okinawa).

Habitat. Continental shelves, offshore at 60–90 m in the Taiwan Strait, possibly deeper elsewhere.

Behaviour. Unknown.

Biology. Apparently oviparous.

Status. IUCN Red List: Not Evaluated.

Dwarf Sawtail Catshark *Galeus schultzi* Plate 37

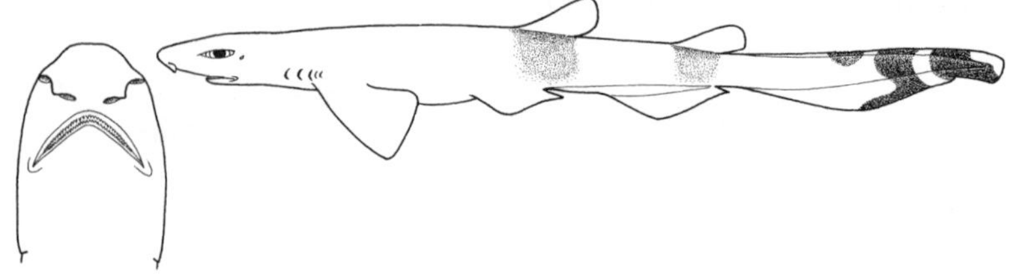

Measurements. Mature: ~25 cm ♂, 27–30 cm ♀. Max: 30 cm TL.

Identification. One of the smallest catsharks. Pattern of obscure dark saddle blotches below dorsal fins and two bands on the tail. No black tips to dorsal and caudal fin tips. Unusually short and rounded snout for a sawtail catshark and with very short labial furrows (confined to mouth corners). Mouth lining light or dusky. Distinct crest of enlarged dermal denticles along upper margin of elongated tail but not lower.

Distribution. Western Pacific: Philippines (off Luzon).
Habitat. Mainly on insular slopes, 329–431 m. One record on outer shelf, 50 m.
Behaviour. Unknown.
Biology. Poorly known.
Status. IUCN Red List: Not Evaluated.

Springer's Sawtail Catshark *Galeus springeri* Plate 37

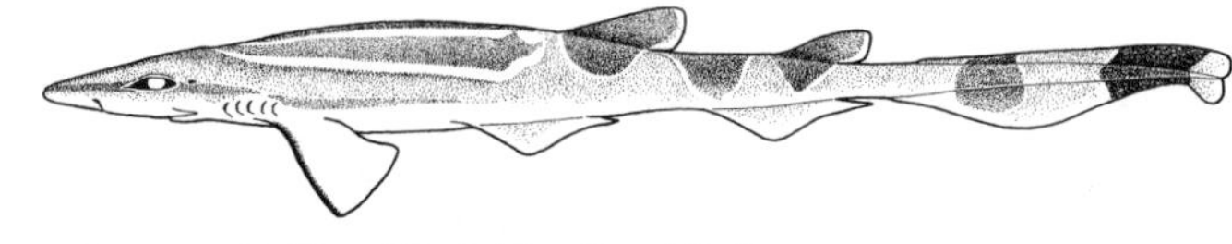

Measurements. Immature: 13–32 cm TL ♂. Max: 44 cm TL ♀.
Identification. Only sawtail catshark with a pre-dorsal pattern of dark longitudinal stripes outlined with white on a dark background; dark saddles on the tail, underside white. A distinct crest of enlarged dermal denticles along the upper and lower margin of the elongated tail.

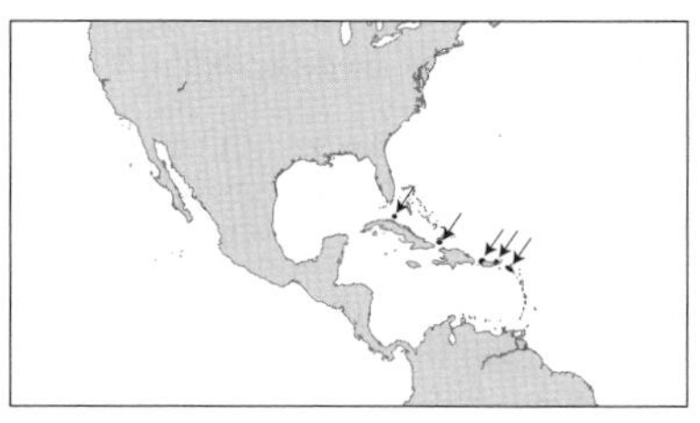

Distribution. Caribbean Sea: Cuba (north coast), Bahamas, Puerto Rico, Leeward Islands.
Habitat. Upper insular slopes, 457–699 m.
Behaviour. Unknown.
Biology. Possibly oviparous.
Status. IUCN Red List: Not Evaluated. Not important to fisheries.

Northern Sawtail Catshark *Galeus* sp. B Plate 38

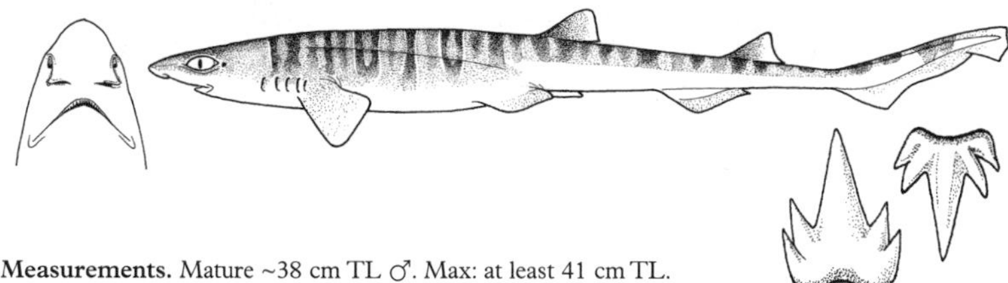

Measurements. Mature ~38 cm TL ♂. Max: at least 41 cm TL.
Identification. Pale greyish-brown with 10–16 darker, pale-edged saddles and bars in front of first dorsal fin. Bars below dorsal fins with pale centres. Underside and much of sides and lower fins uniformly pale. Dorsal and upper caudal fins dusky, with pale trailing edges. Distinct crest of enlarged dermal denticles along upper and lower margins of elongated tail.

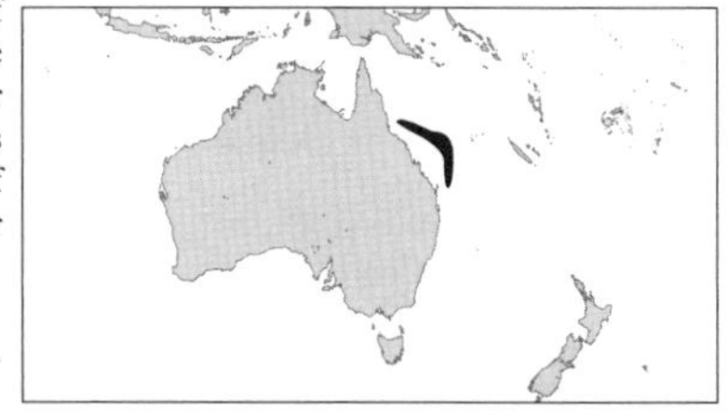

Distribution. Off northeastern Australia.
Habitat. Continental shelf, bottom at 310–420 m.
Behaviour. Unknown.

Biology. Unknown.
Status. IUCN Red List: Data Deficient. Little fishing activity within small range. Close to *Galeus boardmani*.

Speckled Catshark *Halaelurus boesemani* Plate 39

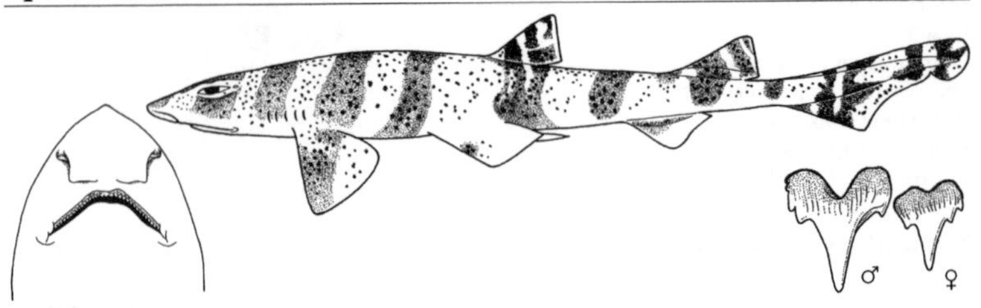

Measurements. Born/hatch: >7 cm TL. Adult: 42–48 cm ♂, 43–47 cm ♀. Max: 48 cm TL.
Identification. About eight dark saddles separated by narrower bars on yellow-brown back and tail. Dark blotches on dorsal and caudal fins, numerous small scattered dark spots on body, dorsal and caudal fins. Broad pale bands edge pectoral fins. Pale underside, pelvic and anal fins. Snout pointed, not upturned, eyes raised above head, gills on upper surface of head above mouth.

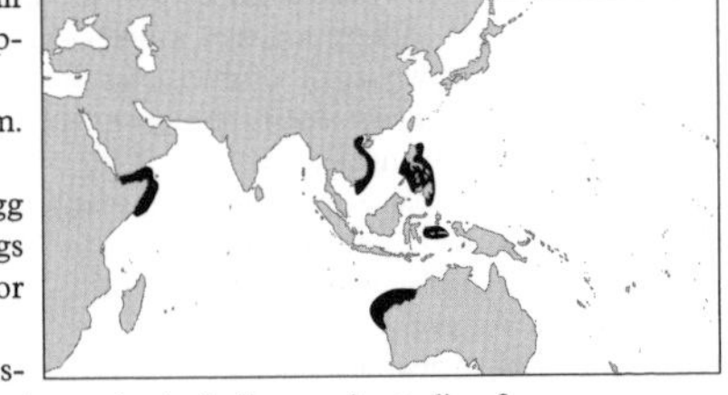

Distribution. Indo-west Pacific: Somalia, Gulf of Aden, western Australia, Indonesia, ?Philippines, Vietnam.
Habitat. Continental and insular shelves, 37–91 m.
Behaviour. Unknown.
Biology. Largely unknown. Up to four egg capsules per oviduct. May be ovoviviparous (eggs hatch inside the oviduct before birth), or oviparous.
Status. IUCN Red List: Least Concern (in Australia). This may be a complex of distinct endemic species including an Australian form.

Blackspotted or Darkspot Catshark *Halaelurus buergeri* Plate 39

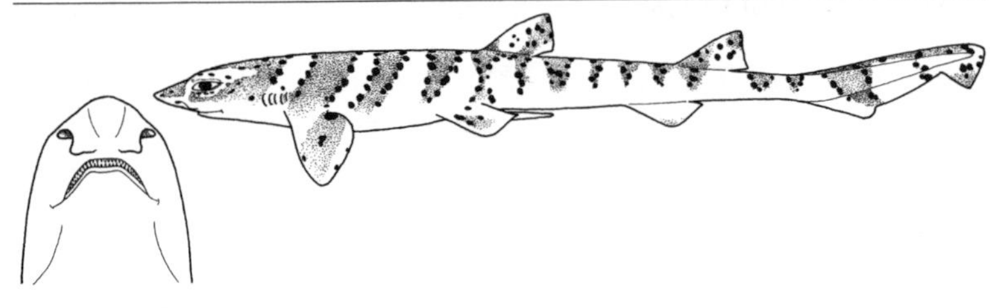

Measurements. Mature: 36 cm ♂, ~40 cm ♀. Max: 49 cm TL.

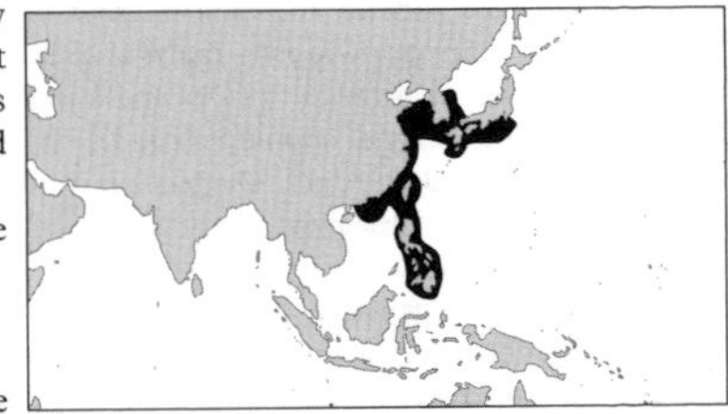

Identification. Variegated pattern with dusky bands outlined by large black spots on a light background. Snout pointed, not upturned, eyes raised above head, gills on upper surface of head above mouth.
Distribution. Northwest Pacific: Japan, the Koreas, Philippines, China.
Habitat. Continental shelf, 80–100 m.
Behaviour. Unknown.
Biology. Oviparous, several egg-capsules are

retained in the oviduct and not laid until embryos are advanced and close to hatching.
Status. IUCN Red List: Not Evaluated. Common offshore, not fished.

Lined Catshark *Halaelurus lineatus* Plate 39

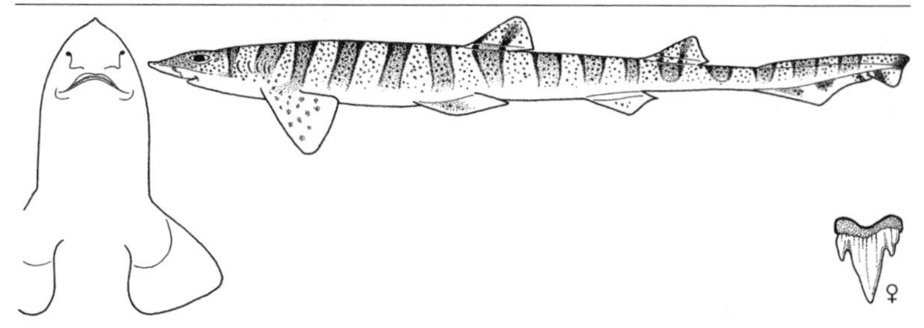

Measurements. Born: ~8 cm. Adult: 48–56 cm ♂, 46–52 cm ♀ TL. Max: 56 cm TL.
Identification. Pale brown above with ~13 pairs of narrow dark brown stripes outlining dusky saddles, many small spots and squiggles, cream below. Narrow head, upturned knob on snout, eyes raised above head, gills on upper surface of head above mouth.
Distribution. West Indian Ocean: South Africa (from East London) to Mozambique.
Habitat. Continental shelf, surf line to 290 m.
Behaviour. May segregate by depth or region; most captured off kwaZulu-Natal are pregnant females, adult males and young only rarely caught.
Biology. Possibly oviparous, up to eight egg-cases/oviduct retained until embryos close to hatching (eggs hatch in 23–26 days in captivity); possibly ovoviviparous in the wild. Eat mostly crustaceans, also bony fishes and cephalopods.
Status. IUCN Red List: Not Evaluated. Common in prawn trawler bycatch.

Tiger Catshark *Halaelurus natalensis* Plate 39

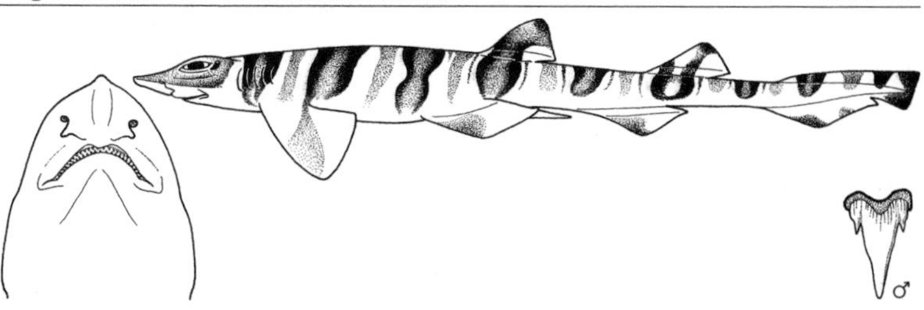

Measurements. Eggcase: ~4 x 1.5 cm. Adolescent: 29–35 cm ♂, 30–44 cm ♀. Adult: 35–45 cm ♂, 37–50 cm ♀. Max: 50 cm ♀ TL.
Identification. Yellow-brown above with ten pairs of broad dark brown bars enclosing lighter reddish areas, no spots, cream below. Pointed, upturned snout tip, eyes raised above broad head, gills on upper surface of head above mouth.
Distribution. Southeast Atlantic and western Indian Ocean: South Africa. kwaZulu-Natal and Mozambique records need confirmation, possibly based on *Halaelurus lineatus*.

Habitat. On or near bottom, continental shelf, close inshore to 114 m, most records <100 m, slope record at 355 m may be erroneous.
Behaviour. May segregate by size and depth. Offshore trawls mainly take adults.
Biology. Oviparous, 6–11 (mostly 6–9) eggcases per oviduct, laid when embryos close to hatching. Eats small bony fishes, fish offal and crustaceans, also polychaetes, cephalopods and small elasmobranchs.
Status. IUCN Red List: Not Evaluated. Common, bycaught in prawn trawls.

Quagga Catshark *Halaelurus quagga* Plate 39

Measurements. Born/hatch: ~8 cm TL. Adult: 28–35 cm TL ♂. Max: 35 cm TL.
Identification. Snout pointed but not upturned, eyes raised above head, gills on upper surface of head above mouth. Light brown above with >20 dark brown narrow vertical bars, pairs of bars forming saddles only under dorsal fins. Lighter below. No spots.

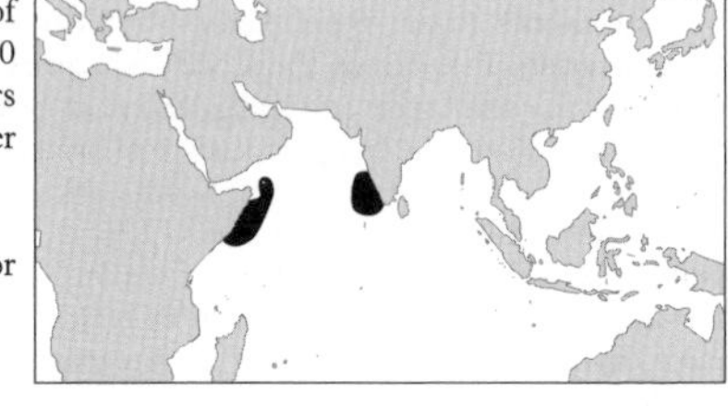

Distribution. Indian Ocean: Somalia, India.
Habitat. Offshore on continental shelf, on or near bottom, 54–186 m.
Behaviour. Unknown.
Biology. Unknown.
Status. IUCN Red List: Not Evaluated. Not important in fisheries.

Puffadder Shyshark or Happy Eddie *Haploblepharus edwardsii* Plate 40

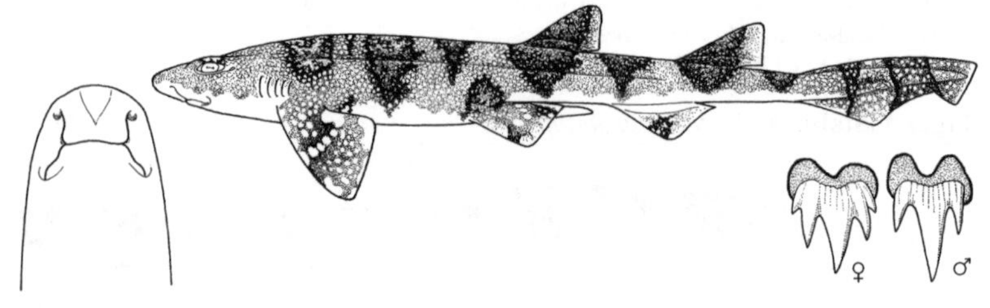

Measurements. Eggcase: ~3.5–5 x 1.5–3 cm. Hatch: ~10 cm TL. Adolescent: 36–45 cm ♂, 39–44 cm ♀. Mature: 37–60 cm ♂, 39–69 cm ♀. Max: 69 cm TL.
Identification. Pale to dark brown or grey-brown above with prominent golden-brown or reddish saddles with darker brown margins and many white spots on saddles or between them, mostly smaller or same size as spiracles. White below. Stocky, broad head, slender body. Very large nostrils. Greatly expanded anterior nasal flaps reach mouth. Gill slits on upper sides of body.

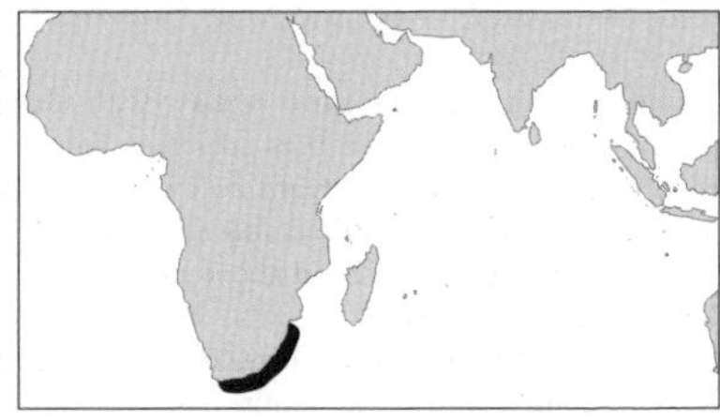

Distribution. Southeast Atlantic and western Indian Ocean: South Africa (eastern and western Cape).
Habitat. Continental shelf, on or near sandy and rocky bottom 0–133 m, possibly 288 m on slope; most common at 30–90 m, closer inshore in west.
Behaviour. Apparently social, resting in groups in captivity. Curls up with tail over eyes when captured.

Biology. Oviparous, eggcases laid in pairs (one/oviduct). Eat small bony fishes, fish offal, crustaceans, cephalopods and polychaetes.
Status. IUCN Red List: Near Threatened. Caught by surf anglers, discarded from bottom trawls. Kept in aquaria.

Brown Shyshark or Plain Happy *Haploblepharus fuscus* Plate 40

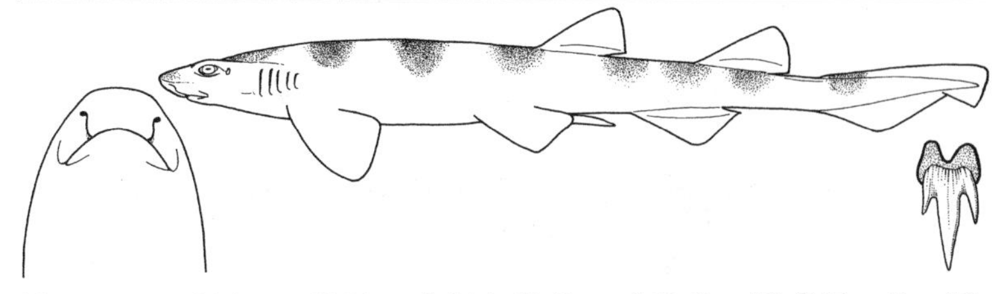

Measurements. Adolescent: 50-54 cm ♂. Adult: 55–69 cm ♂, 60–63 cm TL ♀. Max: 69 cm TL.
Identification. Brown colour above, sometimes with slightly darker, obscure saddles or small white or black spots, white below. Stocky, broad head, stocky body in large individuals, very large nostrils. Greatly expanded anterior nasal flaps reach mouth. Gill slits on upper sides of body.
Distribution. South Atlantic and western Indian Ocean: South Africa (<1000 km of coast).
Habitat. Inshore on continental shelf, often in shallow, sandy areas, rocky bottom, 0–35 m.
Behaviour. Curls up with tail over eyes when captured.
Biology. Oviparous (lays pairs of eggs). Eats lobsters and bony fishes.
Status. IUCN Red List: Near Threatened. Very limited geographic range. Considered a nuisance and sometimes killed by anglers.

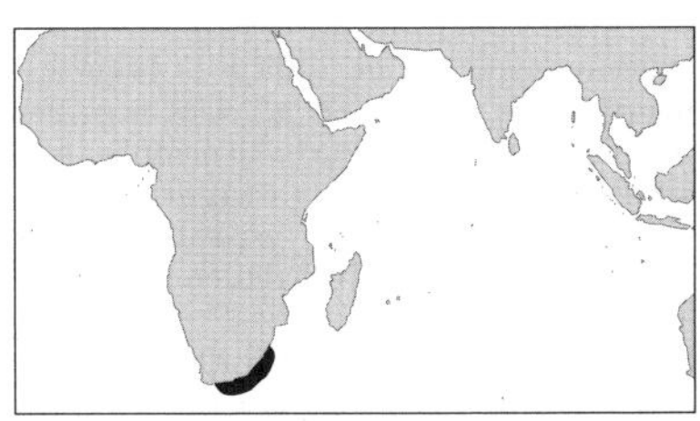

Dark Shyshark or Pretty Happy *Haploblepharus pictus* Plate 40

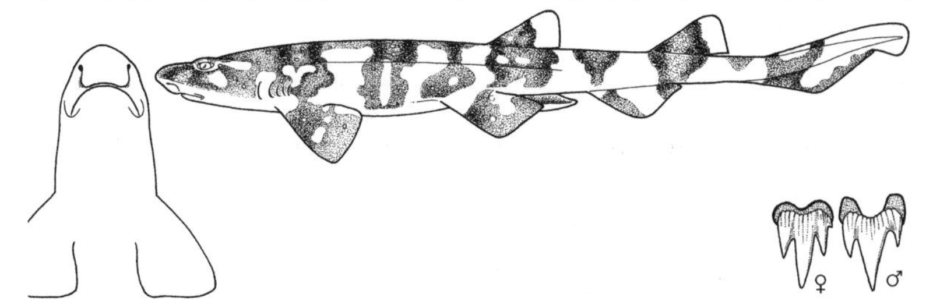

Measurements. Eggcase: ~6 x 3 cm. Hatch: ~11 cm TL. Adults: 40–57 cm ♂, 36–60 cm TL ♀. Max: 60 cm TL.
Identification. Dorsal saddle markings without obvious darker edges, sparsely dotted with large white spots, mostly larger than spiracles and absent between saddles. Broad head, very large nostrils. Greatly expanded anterior nasal flaps reach mouth. Gill slits on upper sides of stocky body.

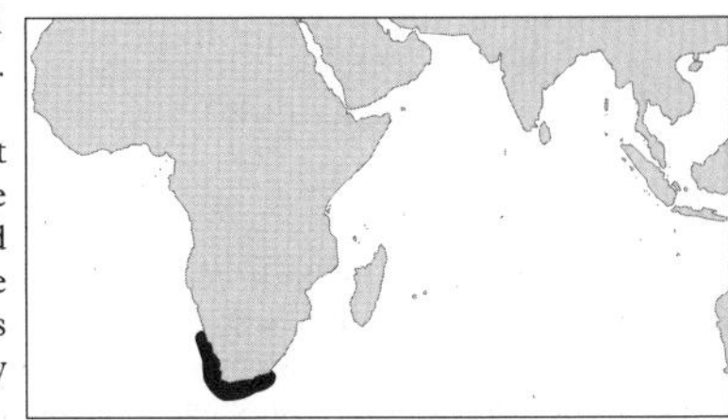

Distribution. Southeast Atlantic and southwest Indian Ocean: central Namibia to South Africa (East London).
Habitat. Continental shelf, kelp forests, rocky inshore reefs and sandy areas, close inshore to ~35 m.
Behaviour. Curls up with tail over eyes when captured.
Biology. Oviparous, one egg laid per oviduct, hatch in about 3.5 months in an aquarium. Eats bony fishes, sea snails, cephalopods, crustaceans, polychaetes and echinoderms, occasionally algae.
Status. IUCN Red List: Not Evaluated. Common within limited range. Killed by sports anglers, caught in lobster traps.

Eastern/Natal Shyshark/Happy Chappie *Haploblepharus* sp. A Plate 40

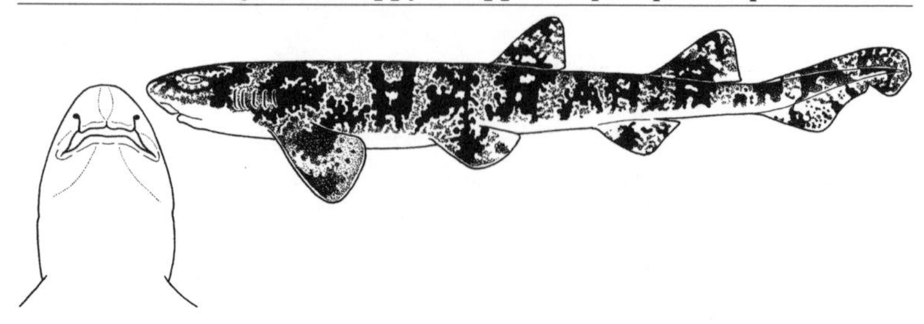

Measurements. Mature: ~50 cm ♂, 48 cm ♀. Max: >50 cm TL.
Identification. Body and tail with dark brown H-shaped dorsal saddle markings, margins conspicuous dark and dotted with numerous small white spots, interspaces between saddles and fins with dark mottling on brown background, underside white. Stocky, broad, flat head, slender body, very large nostrils. Greatly expanded anterior nasal flaps reach mouth. Gill slits on upper sides of body.
Distribution. Western Indian Ocean: South Africa (western Cape to eastern Cape and kwaZulu-Natal).
Habitat. Close inshore on continental shelf, sometimes in surf zone or on rocky reefs.
Behaviour. Unknown.
Biology. Unknown.
Status. IUCN Red List: Not Evaluated. Formerly considered identical to ***Haploblepharus edwardsii***.

Tropical or Crying Izak *Holohalaelurus melanostigma* Plate 40

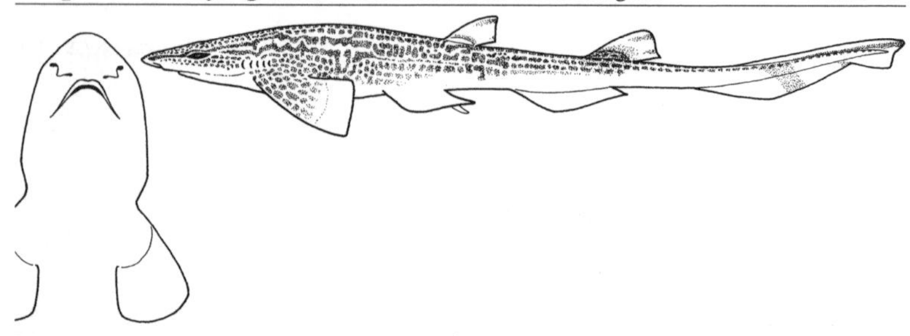

Measurements. Mature: ~38 cm ♂, unknown ♀. Max: >38 cm TL.
Identification. Scattered tiny black dots beneath very broad head. Grey-brown above, lighter below, upper surface with many large dark brown spots, some fused to form reticulations, blotches and horizontal stripes. Horizontal 'tear-marks' on snout in front of eyes. Dark lines and dark C-shaped mark on bases

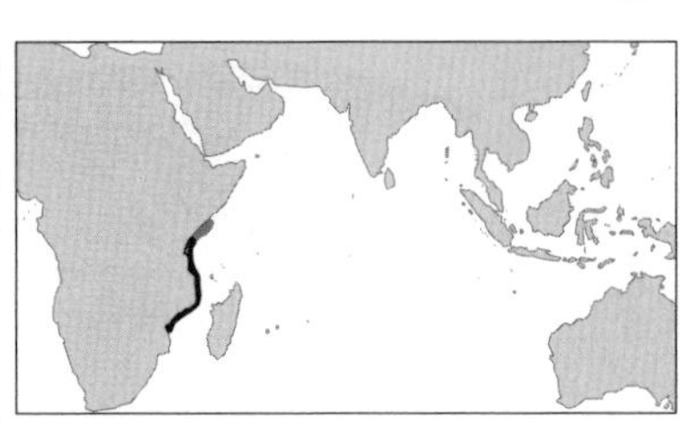

and webs of dorsal fins. Snout short, mouth long, dorsal fins short and angular, tail slender. Slightly enlarged rough denticles on middle of back.
Distribution. Western Indian Ocean: Mozambique, Tanzania including Zanzibar, possibly Kenya.
Habitat. Continental slope, 607–658 m.
Behaviour. Unknown.
Biology. Unknown.
Status. IUCN Red List: Not Evaluated. Apparently extremely rare, known only from a few museum specimens.

Whitespotted or African Spotted Izak *Holohalaelurus punctatus* Plate 40

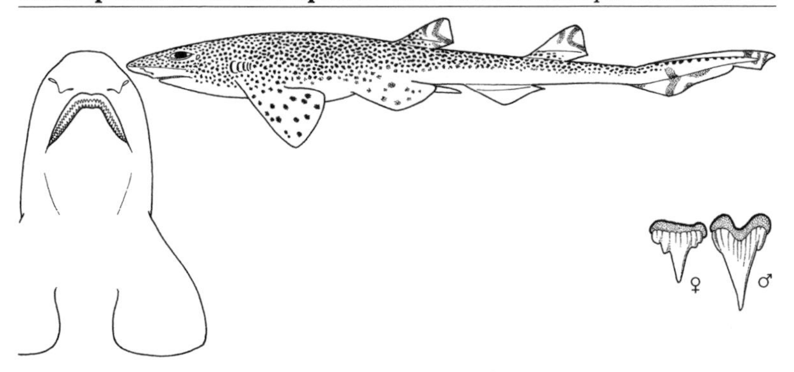

Measurements. Mature: ~ 24–33 ♂, ~22–24 cm ♀. Max: 34 cm.
Identification. Scattered tiny black dots beneath very broad head. Yellow-brown to dark brown upper surface densely covered with small close-spaced dark brown spots, faint saddles sometimes present but no reticulations, blotches, horizontal stripes, or tear-marks. Whitish below. White spots few and scattered on back and dorsal fin insertions. Highlighted C or V-shaped dark mark often on dorsal fin webs. Snout short, mouth long. Short angular dorsal fins, slender tail. No enlarged rough denticles on middle of back.
Distribution. Western Indian Ocean: South Africa (kwaZulu-Natal), southern Mozambique, (type specimens caught off Cape Point).
Habitat. Continental shelf and upper slope, 220–420 m.
Behaviour. Partial sexual segregation (♂ more numerous than ♀ off kwaZulu-Natal, in equal numbers off Mozambique).
Biology. Eggs laid in pairs. Eats small bony fishes, crustaceans and cephalopods.
Status. IUCN Red List: Not Evaluated. Vanished? Formerly common, but not caught in research trawls for ~15 years. Entire range heavily fished.

Izak *Holohalaelurus regani* Plate 40

Measurements. Typical subspecies – Eggcases: 3.5 x 1.5 cm. Hatch: <11 cm TL. Mature: 45–69 cm ♂, 40–52 cm ♀. Northeastern subspecies – Mature: 52–55 cm ♂, 38–44 cm ♀. Max: 69 cm TL.
Identification. Scattered tiny black dots beneath very broad head. Yellowish to yellow-brown, white below, upper surface covered with dark brown reticulations, bars and blotches, more spotted in young. Natal subspecies with smaller reticulations, more checkerboard-like or spotted. No horizontal

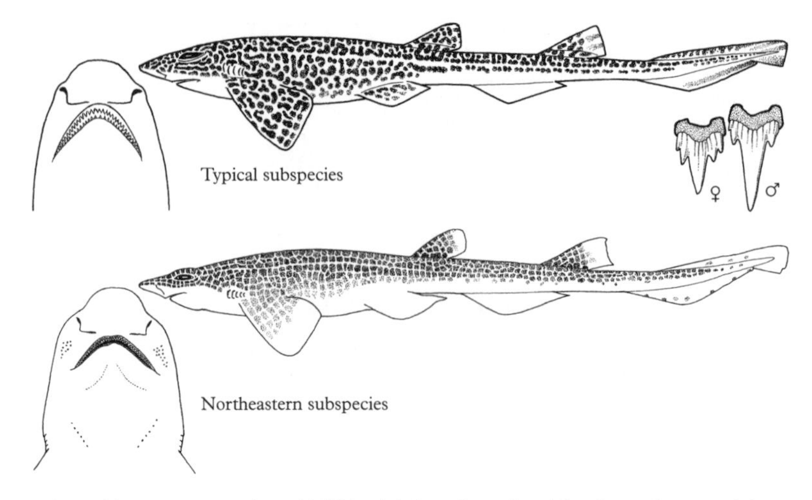

stripes, white spots, tear-marks, or highlighted dark marks on dorsal fins. Snout short, mouth long, dorsal fins short and angular, tail slender. Enlarged rough denticles on middle of back. Young dark and slender, with a line of white spots on sides, black bars on tiny fins and very long tail.

Distribution. Southeast Atlantic, southwest Indian Ocean: Typical subspecies: Namibia to South Africa (possibly to kwaZulu-Natal). Northeastern subspecies: South Africa (kwaZulu-Natal) and Mozambique (dark grey on map).

Habitat. Continental shelf and upper slopes. Typical subspecies 40–910 m, mainly ~100–300 m. Northeastern subspecies 200–740 m.

Behaviour. At least part of the population migrates inshore in autumn. Northeastern subspecies juveniles occur in deeper water than adults. Typical subspecies juveniles in shallower water than adults.

Biology. Oviparous, pairs of eggs laid year round. Eats small bony fishes, crustaceans, cephalopods, polychaetes, hydrozoans, occasionally kelp.

Status. IUCN Red List: Not Evaluated. Discarded bycatch. Typical subspecies population increasing, northeastern subspecies status uncertain.

East African Spotted or Grinning Izak *Holohalaelurus* sp. A Plate 40

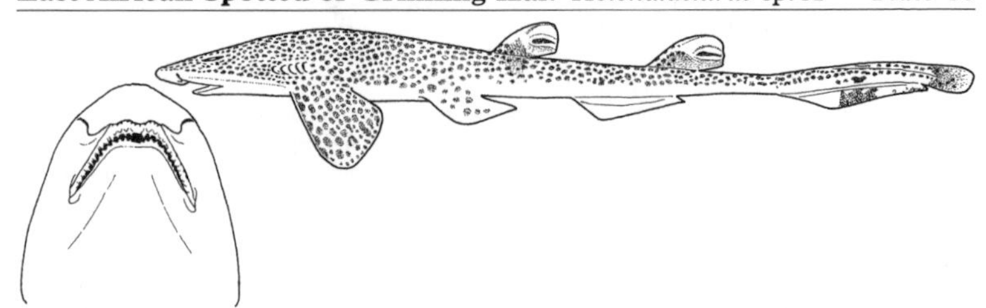

Measurements. Mature: ~27 cm ♂ TL, unknown ♀. Max: >27 cm TL.

Identification. Scattered tiny black dots beneath very broad head. Yellow-brown above, whitish below, upper surface densely covered with small close-spaced dark brown spots. No dark reticulations, blotches, horizontal stripes, or tear-marks. White spots few, large and conspicuous above pectoral fin insertions. Highlighted narrow dark bar on webs of dorsal fins. Snout short, mouth long and very wide, dorsal fins short and angular, tail slender. Enlarged rough denticles on middle of back.

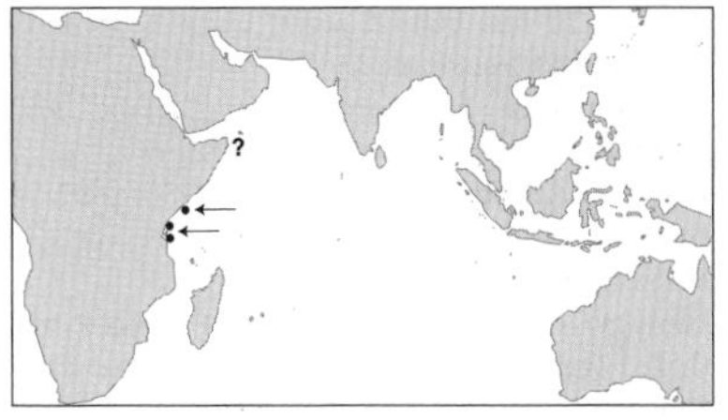

Distribution. Western Indian Ocean: Kenya, Tanzania, possibly Somalia.
Habitat. Upper continental slope, 238–300 m.
Behaviour. Essentially unknown.
Biology. Unknown.
Status. IUCN Red List: Not Evaluated.

Campeche Catshark *Parmaturus campechiensis* Plate 38

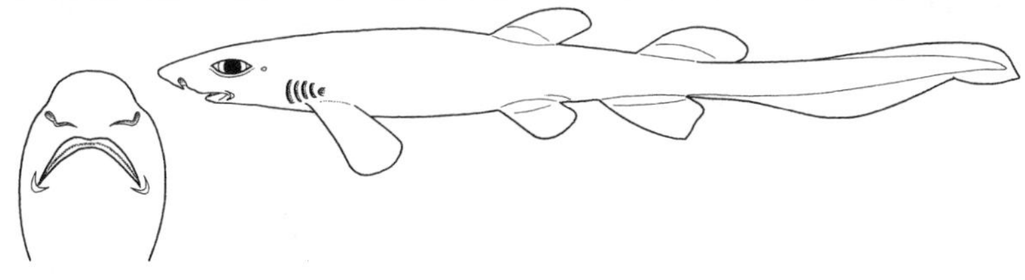

Measurements. Only known specimen is immature ♀, 16 cm TL.
Identification. Soft flabby greyish body, dusky on abdomen, around gills and on fin webs. Short snout, very small and low anterior nasal flaps, ridges under eyes, gills not greatly enlarged. Crest of saw-like denticles along top of tail but apparently not below. First dorsal fin origin slightly in front of pelvic origins, fin slightly smaller than second dorsal fin. Second dorsal about as large as anal fin and with insertion well behind that of anal fin.

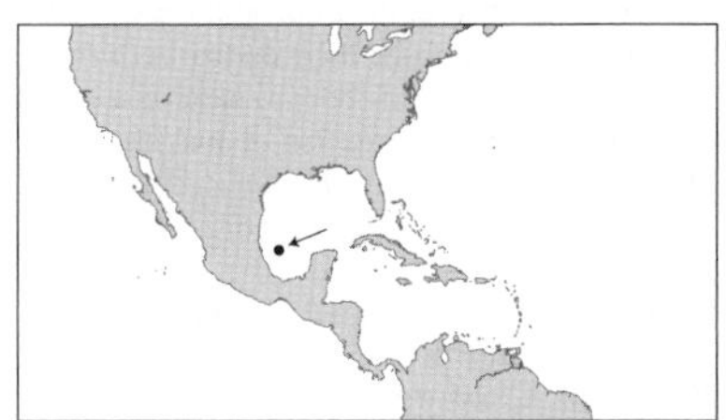

Distribution. Collected in Bay of Campeche, Gulf of Mexico.
Habitat. Continental slope, on or near bottom, 1097 m.
Behaviour. Unknown.
Biology. Unknown.
Status. IUCN Red List: Not Evaluated. Known only from the holotype.

New Zealand Filetail *Parmaturus macmillani* Plate 38

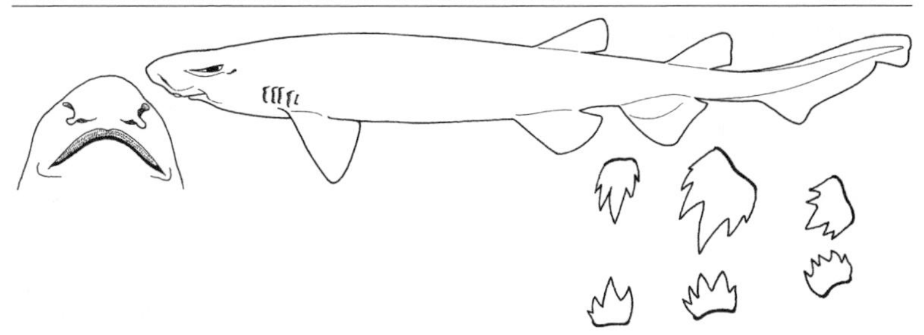

Measurements. Eggcase: 5.2 x 3.0 cm. Mature: unknown ♂, 52–53 cm ♀. Max: >53 cm TL.
Identification. Soft flabby grey body, lighter below, fins with dusky webs. Elongated lobate anterior nasal flaps, very short blunt snout, ridges under eyes, gills not greatly enlarged. Crest of saw-like denticles along top of caudal fin but not below. First dorsal fin about as large as second, origin

opposite or just behind pelvic origins. Second dorsal fin smaller than anal fin, insertion slightly behind that of anal fin.

Distribution. Northern New Zealand and submarine ridge south of Madagascar.
Habitat. Captured in deepwater, 950–1003 m (may extend deeper).
Behaviour. Unknown.
Biology. Oviparous, lays pairs of large stout-shelled eggcases (one per oviduct) without tendrils.
Status. IUCN Red List: Data Deficient. Known from five females (two New Zealand, three Madagascar).

Blackgill Catshark *Parmaturus melanobranchius* Plate 38

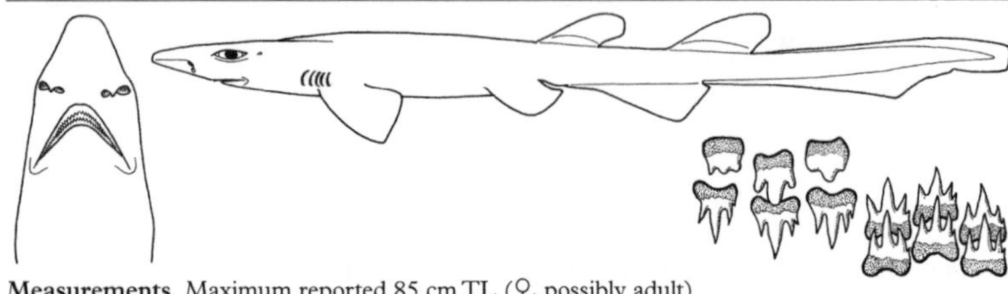

Measurements. Maximum reported 85 cm TL (♀, possibly adult).
Identification. Soft flabby plain body, grey to dark brown above, lighter below, blackish on gill septa. Moderately long blunt snout, anterior nasal flaps elongated and pointed, ridges under eyes, gills not greatly enlarged. Crests of saw-like denticles along top of caudal fin and on preventral caudal margin below. First dorsal fin noticeably smaller than second with origin well behind pelvic fin origins and opposite their midbases or insertions. Second dorsal fin about as large as anal fin, insertion well behind insertion of anal fin.

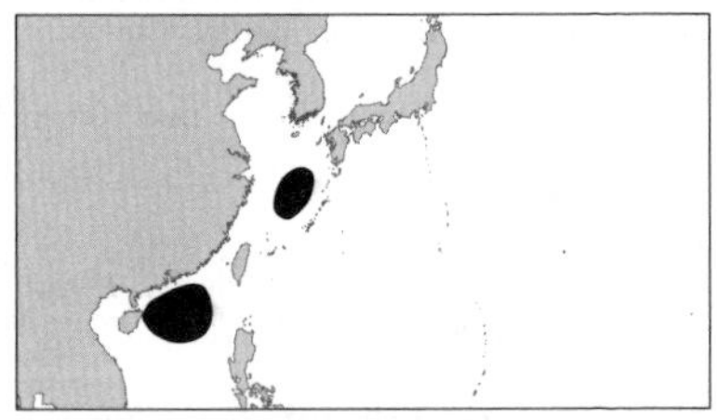

Distribution. Northwest Pacific. China (South China Sea off Hong Kong and near Taiwan), Japan (Ryu-Kyu Islands).
Habitat. Upper continental and insular slopes, on mud bottom at 540–835 m.
Behaviour. Unknown.
Biology. Unknown.
Status. IUCN Red List: Not Evaluated.

Salamander Shark *Parmaturus pilosus* Plate 38

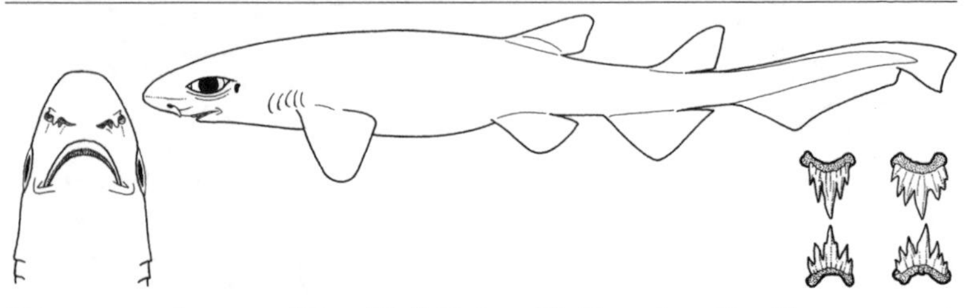

Measurements. Immature ♂ 56 cm TL. ♀ 59–64 cm TL. Max: at least 64 cm TL.
Identification. Soft flabby plain body, reddish above and white below, fin webs darker. Moderately long, blunt snout, anterior nasal flaps elongated and narrowly lobate, ridges under eyes, gills not greatly enlarged. Conspicuous crest of saw-like denticles along top of tail and short and

inconspicuous crest on preventral caudal margin. First dorsal fin about as large as second with origin about opposite pelvic origins. Second dorsal much smaller than anal fin, with insertion about opposite anal insertion.

Distribution. Northwest Pacific: Japan (southeastern Honshu and Riu-Kyu Islands), China near Taiwan.
Habitat. Upper continental and insular slopes. 358–895 m.
Behaviour. Unknown.
Biology. Unknown. Large quantities of squalene in liver may maintain neutral buoyancy.
Status. IUCN Red List: Not Evaluated.

Filetail Catshark *Parmaturus xaniurus* Plate 38

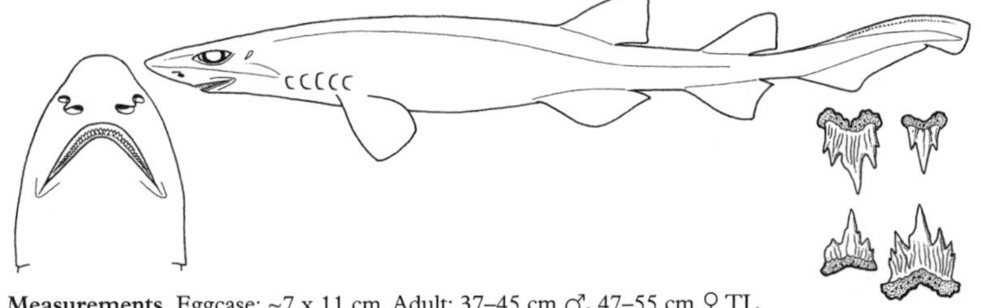

Measurements. Eggcase: ~7 x 11 cm. Adult: 37–45 cm ♂, 47–55 cm ♀ TL.
Identification. Soft flabby plain dark body, brownish-black above, lighter below, fins dark. Enlarged gill slits, moderately long blunt snout, large triangular anterior nasal flaps, ridges under eyes. Crests of saw-like denticles along top of tail, not below. Similar sized dorsal fins, origin of first slightly behind pelvic origins but well in front of midbases. Second dorsal much smaller than anal fin, insertion well in front of anal fin insertion.
Distribution. Northeast Pacific: USA (Oregon) to Mexico (Gulf of California). 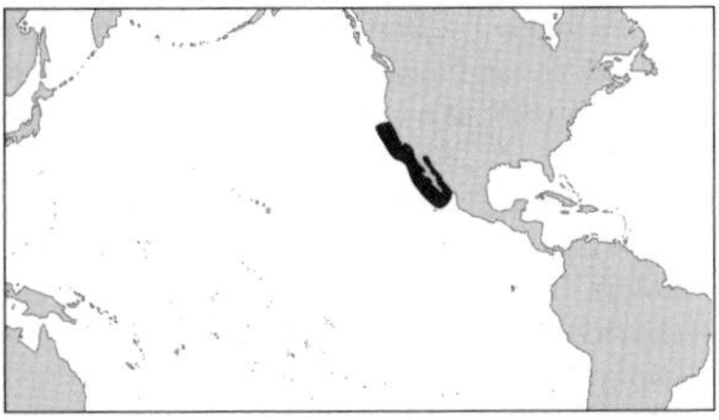
Habitat. Outer continental shelf and upper slope, often on or near bottom, 91–1251 m; Juveniles live in midwater up to 490 m above seabed in water over 1000 m deep.
Behaviour. Observed from submersible, feeding on moribund lanternfish in almost anoxic conditions.
Biology. Oviparous, eggcases with 'T'-shaped lateral flanges and short tendrils. Eats mainly pelagic crustaceans and small bony fishes. Squalene-filled liver may maintain neutral buoyancy. Enlarged gills enable it to thrive in low-oxygen habitats.
Status. IUCN Red List: Not Evaluated. Relatively common. Discarded from bottom trawls and sablefish traps.

Shorttail Catshark *Parmaturus* sp. A. Plate 38

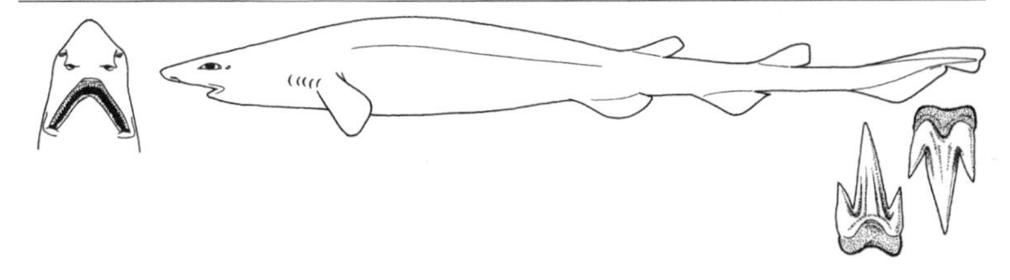

Measurements. Mature ♀, 72 cm TL.
Identification. Soft flabby pale yellowish-brown body with velvety skin, paler below, fins with paler margins. Moderately long pointed snout, small gill slits, large triangular anterior nasal flaps, ridges under eyes. Crests of saw-like denticles along top of tail and also below. First dorsal fin smaller than second, origin far behind pelvic origins and about over their insertions. Second dorsal fin slightly smaller than anal fin, insertion far behind anal fin insertion. Longer snout and dorsal fins set further back than *Parmaturus macmillani*.

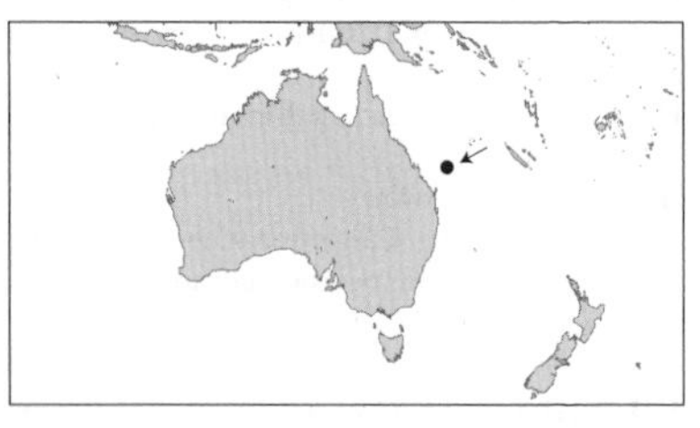

Distribution. Northeast Australia (Queensland). One offshore location.
Habitat. Offshore deepwater plateau, 590 m.
Behaviour. Unknown.
Biology. Oviparous (mature female contained one eggcase).
Status. IUCN Red List: Data Deficient. Known from only one specimen.

Onefin Catshark *Pentanchus profundicolus* Plate 39

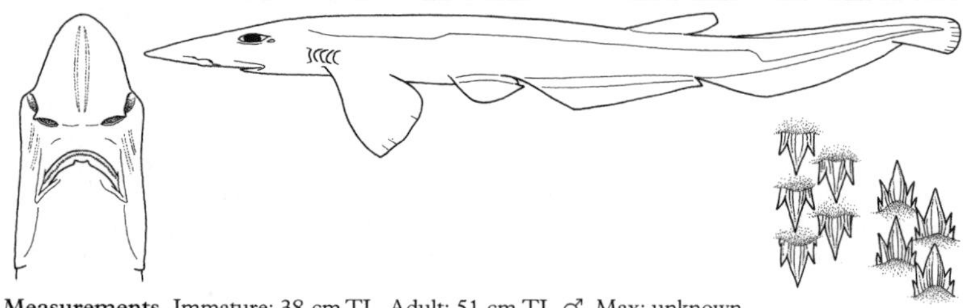

Measurements. Immature: 38 cm TL. Adult: 51 cm TL ♂. Max: unknown.
Identification. Plain brownish colour. Only species with five gill slits and one dorsal fin. Unusually short abdomen (pectoral and pelvic fins very close together). Broadly rounded snout, short mouth, short gill slits and incised gill septa. Long narrow caudal fin without a crest of denticles, very long low anal fin.

Distribution. Northwest Pacific: Philippines (Tablas Straits and Mindanao Sea)(east of Bohol).
Habitat. Insular slope, on bottom, 673–1070 m.
Behaviour. Unknown.
Biology. Unknown.
Status. IUCN Red List: Not Evaluated. Known from two specimens.

Pyjama Shark or Striped Catshark *Poroderma africanum* Plate 39

Measurements. Hatch: 14–15 cm TL. Mature: 58–76 ♂, 65–72 cm ♀. Max: 95 cm TL.
Identification. Unmistakable combination of striking longitudinal stripes, no spots, prominent but short nasal barbels and dorsal fins (second much smaller) set very far back on body.

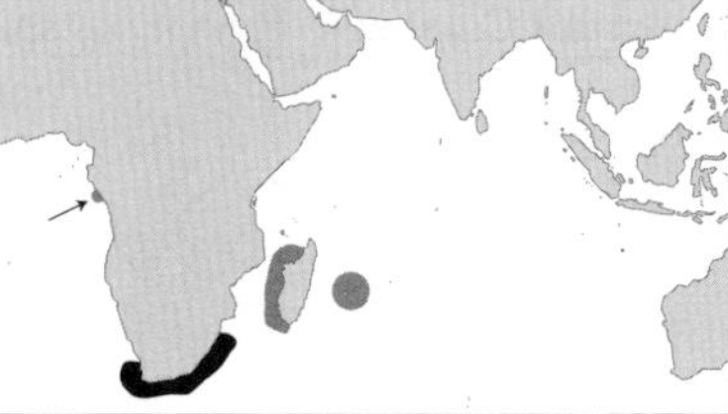

Distribution. Southeast Atlantic and western Indian Ocean: apparently endemic to South Africa (both Capes, rarely to kwaZulu-Natal). Records from Zaire, Madagascar and Mauritius require verification.
Habitat. Continental shelf to upper slope, from surfline and intertidal to 282 m. On or near bottom in rocky areas, often in caves.

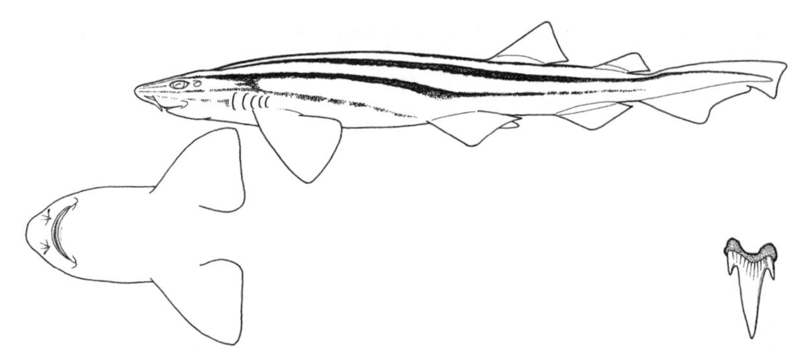

Behaviour. Nocturnal, sometimes day-active.
Biology. Oviparous, lays pairs of eggcases (one/oviduct); one hatched in an aquarium after about 5.5 months. Eats bony fish, hagfish, other small sharks, shark eggcases, and a wide range of invertebrates
Status. IUCN Red List: Not Evaluated. Hardy in captivity. Taken by trawlers and anglers.

Leopard Catshark *Poroderma pantherinum* Plate 39

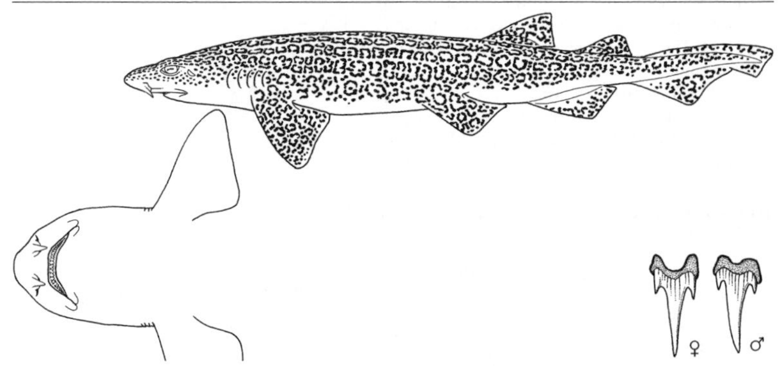

Measurements. Mature: 54–58 cm ♂, 58–61 cm ♀. Max: 74 cm TL.
Identification. Striking, leopard-like rosettes of dark spots and lines surrounding light centres, usually arranged in irregular longitudinal rows, variations include numerous small dense spots to very large dark spots and partial longitudinal stripes. Long nasal barbels reach mouth, which extends behind front of eyes. Dorsal fins set far back, first much larger than second.
Distribution. Southeast Atlantic and western Indian Ocean: apparently endemic to South Africa (both Capes, rarely to kwaZulu-Natal). Records from Mauritius and Madagascar require verification.
Habitat. Continental to upper slope from surf zone and intertidal to 256 m. On or near bottom.
Behaviour. Apparently nocturnal.
Biology. Oviparous, one egg/oviduct. Feeds on small bony fishes and invertebrates.
Status. IUCN Red List: Not Evaluated. Taken by trawlers and anglers. Hardy when kept in aquaria. Large-spotted variant formerly considered a separate species, *Poroderma marleyi*.

Narrowmouth Catshark *Schroederichthys bivius* Plate 41

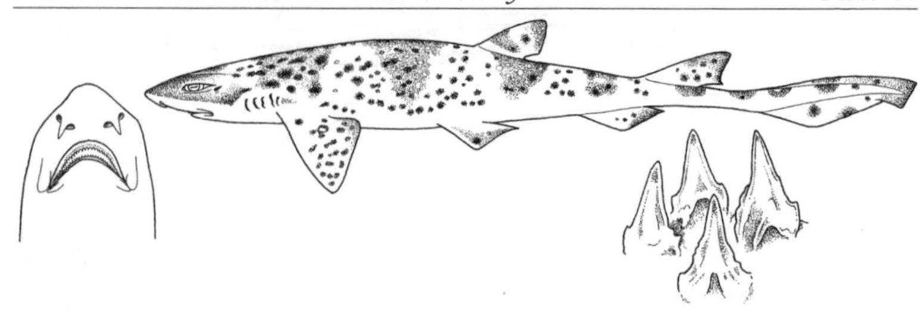

Measurements. Hatch: 14–20 cm TL. Mature: 53 cm ♂, 40 cm ♀. Max: 82 cm ♂, 70 cm TL ♀.
Identification. Fairly slim body, seven or eight dark brown saddles on grey-brown back (two conspicuous saddles in interdorsal space), scattered large dark spots not bordering the saddles, small white spots usually present. Short tail. Short, narrow rounded snout, anterior nasal flaps narrow and lobate. First dorsal origin slightly in front of pelvic insertions. Adult males longer and lighter, with much larger teeth and longer narrower mouths than females, young even longer and more slender.
Distribution. Southeast Pacific and southwest Atlantic: southern Chile to southern Brazil.
Habitat. Temperate continental shelf and upper slope, 14–359 m (mostly <130 m), in deeper water in south.
Behaviour. Largely unknown.
Biology. Oviparous, probably lays pairs of eggs (one/oviduct) in sheltered nursery grounds. Feeds mainly on squat lobsters.
Status. IUCN Red List: Data Deficient. May be declining.

Redspotted Catshark *Schroederichthys chilensis* Plate 41

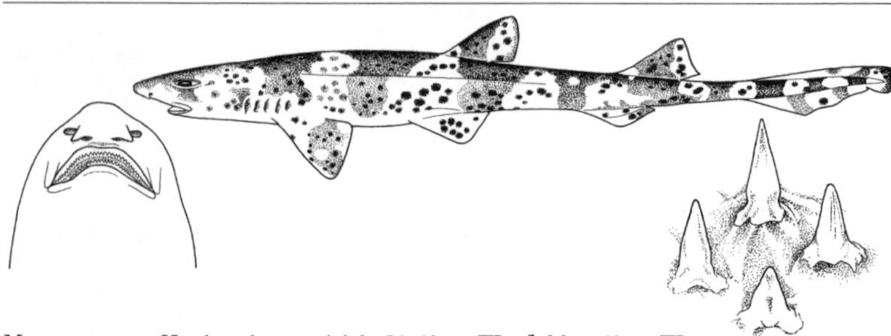

Measurements. Hatch: unknown. Adult: 56–62 cm TL ♂. Max: 62 cm TL.
Identification. Moderately slim body with conspicuous dark saddles, including two in the interdorsal space, numerous dark spots that do not border saddles, and with white spots few or absent. Short broad rounded snout, anterior nasal flaps broad and triangular, very broad wide mouth. Short tail, first dorsal origin slightly in front of pelvic insertions. Young more slender than adults.

Distribution. Southeast Pacific: South America off Peru and south-central Chile.
Habitat. Temperate inshore continental shelf, on

or near bottom, sometimes in very shallow inshore water, 1–50 m.
Behaviour. Unknown.
Biology. Oviparous, probably lays pairs of eggs (one egg per oviduct), with tendrils on cases.
Status. IUCN Red List: Not Evaluated. Common where it occurs.

Narrowtail Catshark *Schroederichthys maculatus* Plate 41

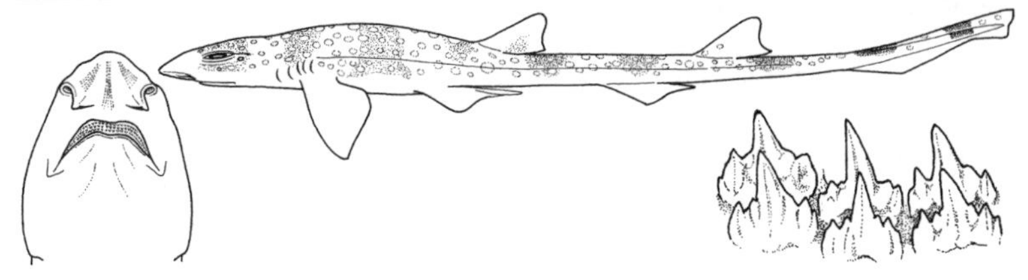

Measurements. Hatch: unknown. Mature: <28 cm TL ♂. Max: 35 cm TL.
Identification. Extremely slender and elongated trunk and tail in juveniles and adults. Juvenile pattern of six to nine light inconspicuous brown saddles on darker tan to grey background disappears in adults, three saddles in interdorsal space, numerous scattered white spots but no dark spots. Rounded snout, broad elongate-triangular anterior nasal flaps, broad wide mouth, first dorsal origin slightly behind pelvic insertions.

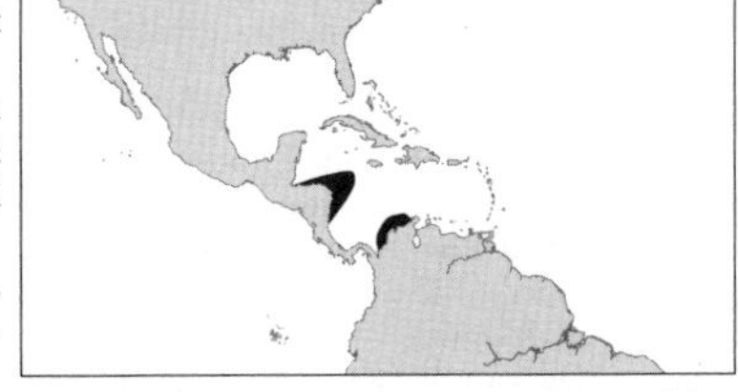

Distribution. West Atlantic: Central and South America off Honduras, Nicaragua, and Colombia, and between the Honduras bank and Jamaica.
Habitat. Shelly or sandy bottom in deepwater, 190–410 m, tropical outer shelf and upper slope.
Behaviour. Unknown.
Biology. Oviparous, probably lays pairs of eggs, with tendrils on cases. Feeds on small bony fishes and squids.
Status. IUCN Red List: Not Evaluated.

Lizard Catshark *Schroederichthys saurisqualus* Plate 41

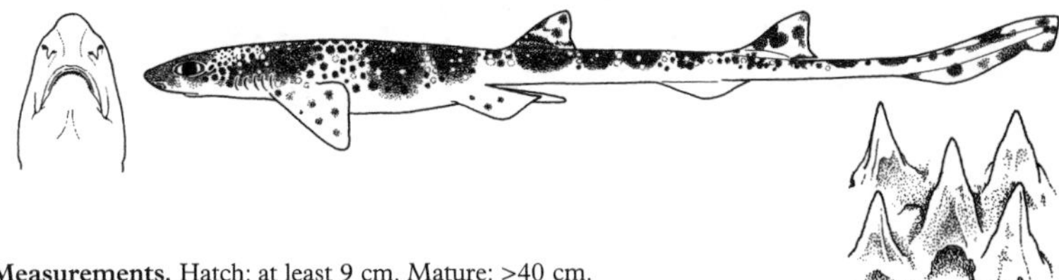

Measurements. Hatch: at least 9 cm. Mature: >40 cm.
Adult: 58–59 cm TL ♂. Mature: 69 cm ♀. Max: >69 cm TL.
Identification. Ten conspicuous dusky saddles on lighter grey or brown background in adults and subadults, four between dorsal fins, numerous white spots interspersed with dark. Slender elongated trunk and tail in adults and subadults, rounded snout, elongated narrow lobate anterior nasal flaps, moderately wide mouth. First dorsal origin slightly behind pelvic insertions.

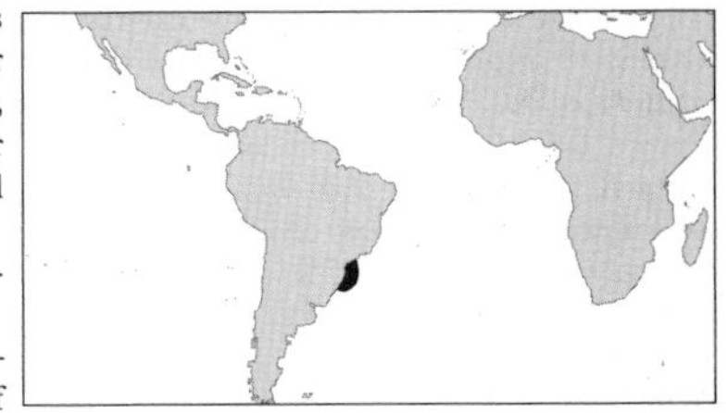

Distribution. Southwest Atlantic: South America off southern Brazil.
Habitat. Outer shelf and upper slope at 122–435 m. Mostly below 250 m on deep-reef

habitats. Occurs with deepwater gorgonians, hard corals, tube sponges, crinoids, brittle stars and Freckled Catshark *Scyliorhinus haeckelii*.
Behaviour. Unknown.
Biology. Oviparous, probably lays pairs of eggs, with tendrils on cases.
Status. IUCN Red List: Not Evaluated.

Slender Catshark *Schroederichthys tenuis* Plate 41

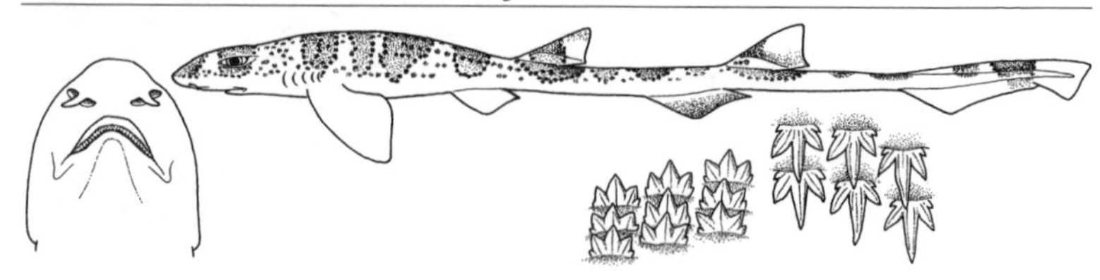

Measurements. Eggcase: 3.5 x 1.7 cm. Mature: 40–47 cm ♂, 37–46 cm ♀. Max: 47 cm TL.
Identification. Very slender light brown body with many small dark spots outlining and scattered between seven and eight conspicuous dark saddles, four between dorsal fins; no white spots. Broad snout, anterior nasal flaps narrow and lobate, mouth narrow, First dorsal origin slightly behind pelvic insertions. Males have much larger teeth and a longer more angular mouth than females (may reach a larger size).

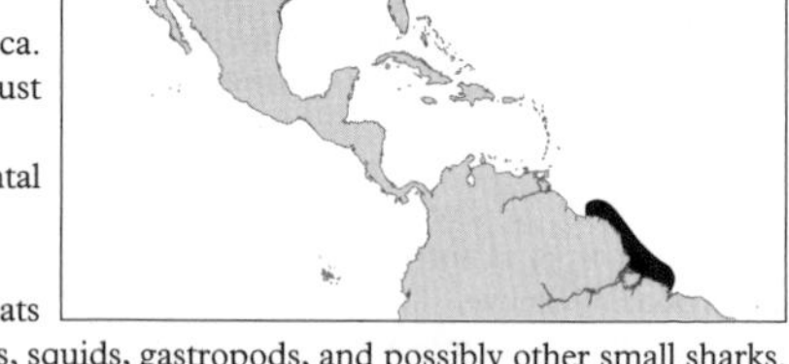

Distribution. Western Atlantic: South America. Surinam and northern Brazil (north and just south of Amazon River mouth).
Habitat. On or near bottom, outer continental shelf and upper continental slope, 72–450 m.
Behaviour. Unknown.
Biology. Oviparous, lays eggs in pairs. Eats foraminifera, small fishes, crustaceans, sponges, squids, gastropods, and possibly other small sharks.
Status. IUCN Red List: Not Evaluated.

Polkadot Catshark *Scyliorhinus besnardi* Plate 42

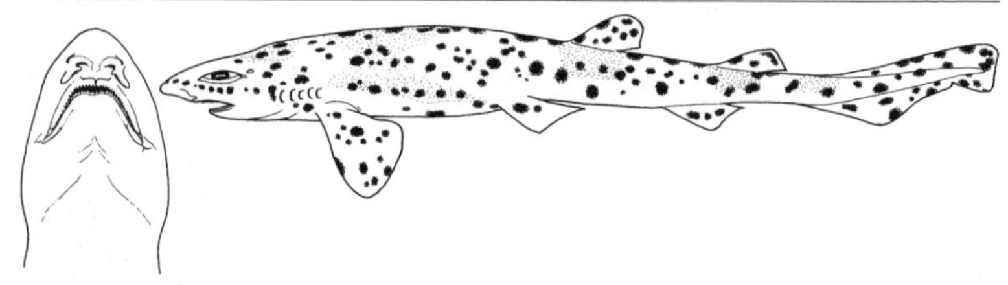

Measurements. Largest record: 47 cm TL (adult ♂).
Identification. Fairly small and slender, with sparsely scattered almost round black spots on fins, sides and back, sometimes with light centres. No white spots or prominent saddles. Small anterior nasal flaps end in front of mouth, no nasoral grooves, labial furrows only on lower jaw, second dorsal fin much smaller than first.

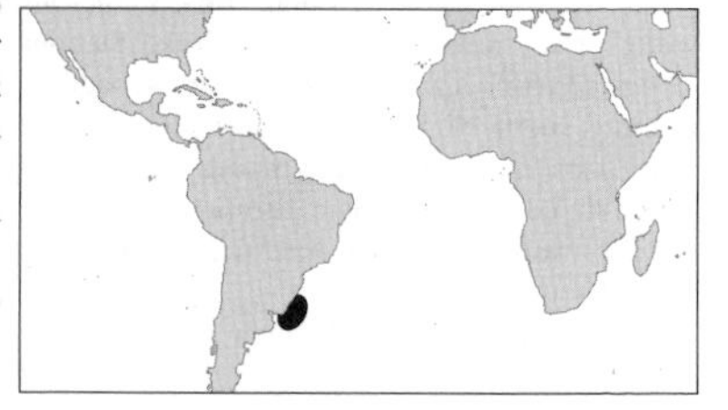

Distribution. Southwest Atlantic: northern Uruguay, Brazil.
Habitat. Outer continental shelf, on bottom, 140–190 m.
Behaviour. Unknown.

Biology. Presumably oviparous.
Status. IUCN Red List: Not Evaluated.

Boa Catshark *Scyliorhinus boa* Plate 42

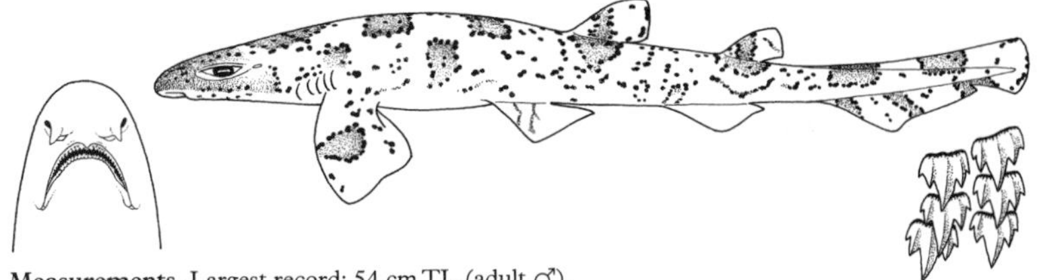

Measurements. Largest record: 54 cm TL (adult ♂).
Identification. Slender greyish catshark with inconspicuous saddles and flank markings outlined by many small black spots, sometimes in reticulating rows or broken black lines. Sometimes a few white spots. Few or no spots inside saddles. Small anterior nasal flaps do not reach mouth, no nasoral grooves, labial furrows on lower jaw only, second dorsal fin much smaller than first.

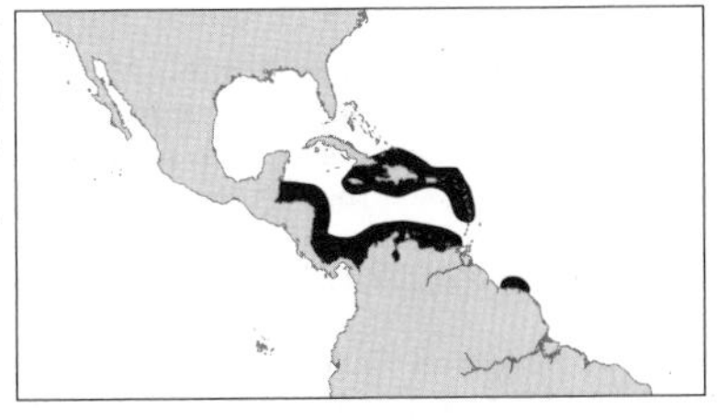

Distribution. Northwest Atlantic: Caribbean insular slope off Barbados, Hispanola, Jamaica, Leeward Islands, Windward Islands; continental slope off Nicaragua, Honduras, Panama, Colombia, Venezuela, and Surinam.
Habitat. Continental and insular slopes, on or near bottom 229–676 m.
Behaviour. Unknown.
Biology. Presumably oviparous.
Status. IUCN Red List: Not Evaluated.

Smallspotted Catshark *Scyliorhinus canicula* Plate 41

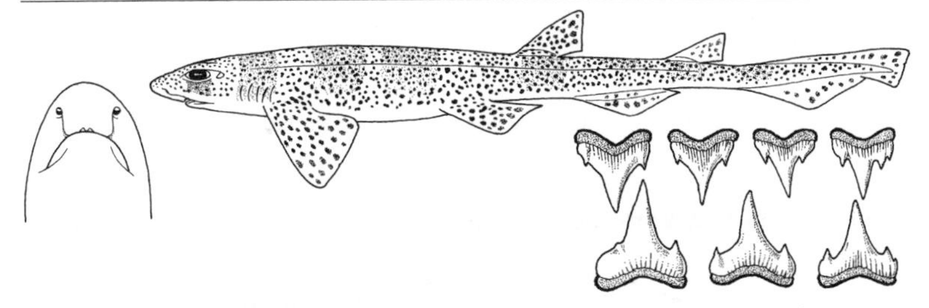

Measurements. Hatch: 9–10 cm TL. Mature: Mediterranean; 39 cm ♂, 44 cm ♀. Larger in Atlantic and North Sea. Max: Mediterranean; 60 cm TL. UK, North Sea; 100 cm TL.
Identification. Large slender body covered in numerous small dark spots on light background, scattered white spots sometimes present, eight or nine dusky saddles may be unclear. Greatly expanded anterior nasal flaps reach mouth and cover shallow nasoral grooves, labial furrows on lower jaw only. Second dorsal much smaller than first.

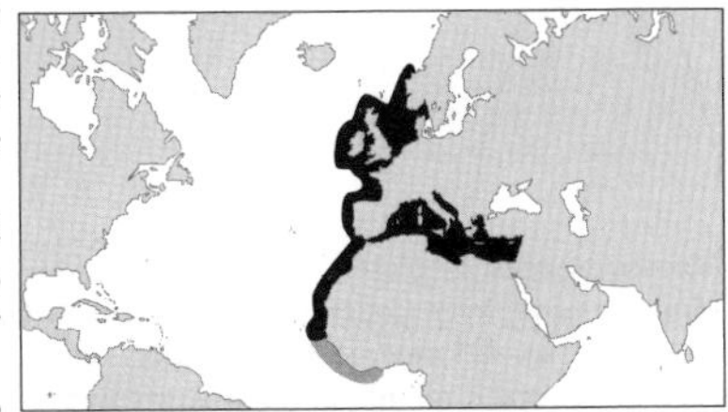

Distribution. Northeast Atlantic: Norway and British Isles to Mediterranean, Canary Islands, Azores, Morocco, Sahara Republic and Mauritania to Senegal, Ivory Coast.
Habitat. Continental shelves and upper slopes, on sediment from nearshore to 110 m (exceptionally 400 m).

Behaviour. Adults often found in single sex schools, young and hatchlings in shallower water.
Biology. Oviparous, eggcases deposited in pairs year-round (mostly November–July) on seaweed. Hatch in 5–11 (mostly 8–9) months. Egg size varies with female size (smaller in Mediterranean). Feed on small bottom invertebrates (crustaceans, gastropods, cephalopods, worms) and fishes.
Status. IUCN Red List: Not Evaluated. Taken in many fisheries, retained and discarded, but high discard survival and some populations are stable or increasing. Hardy aquarium species, breeds in captivity.

Yellowspotted Catshark *Scyliorhinus capensis* Plate 41

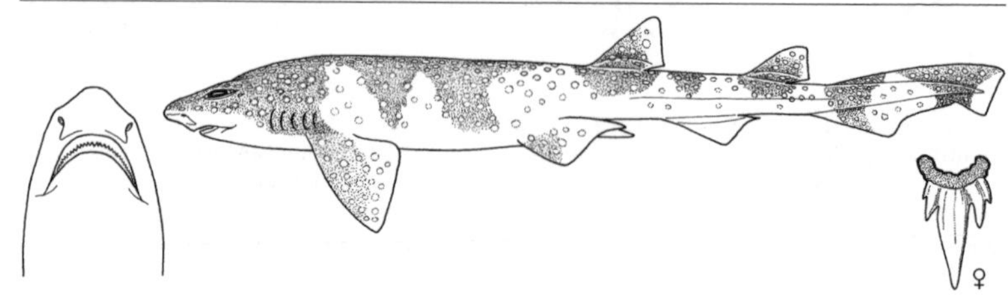

Measurements. Hatch: 25–27 cm TL. Mature: 72–83 cm ♂, 75–80 cm ♀. Max: 122 cm TL.
Identification. Fairly large grey catshark with eight or nine irregular darker grey saddles and numerous small bright yellow spots, no dark spots. Small anterior nasal flaps, no nasoral grooves, labial furrows on lower jaw only. Second dorsal fin much smaller than first.
Distribution. Southeast Atlantic and western Indian Ocean: Namibia and South Africa (both Capes).

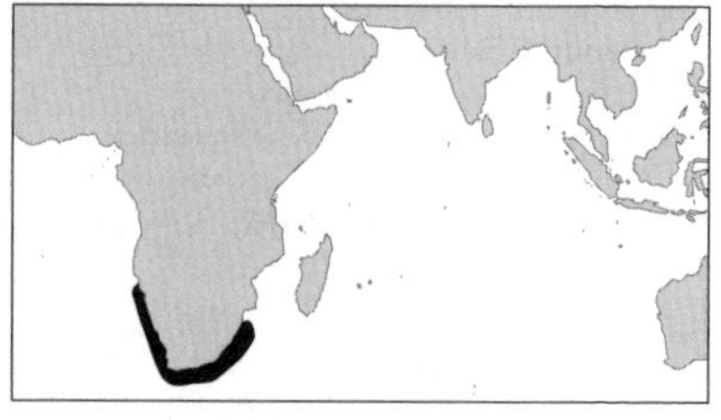

Habitat. Bottom on continental shelf and upper slope, including soft bottom where it is frequently trawled, 26–695 m, mostly 200–400 m, deeper in warmer water.
Behaviour. Coils up tightly when caught.
Biology. Oviparous, laying pairs of eggcases ~8 x 3 cm. Feeds on small fishes and many invertebrates.
Status. IUCN Red List: Not Evaluated. Moderately common on heavily fished offshore banks, discarded from bycatch.

West African Catshark *Scyliorhinus cervigoni* Plate 41

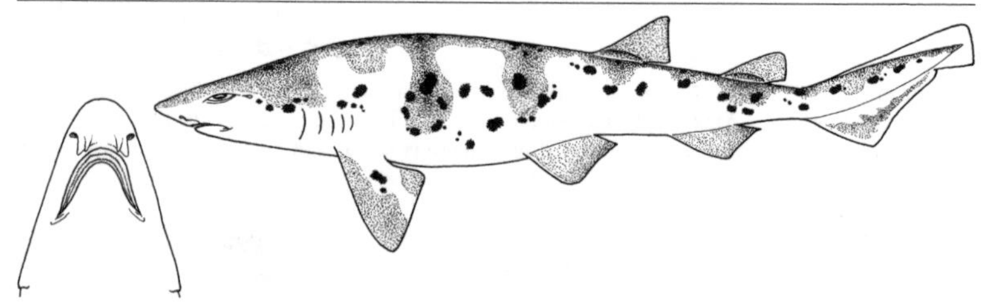

Measurements. Eggcase: 7–8 x 3 cm. Mature: ~65 cm ♂. Max: at least 76 cm TL ♀.
Identification. Very stout with a few relatively large and some small scattered dark spots, eight or nine dusky saddles centred on dark spots on midline of back, no white spots. Small anterior nasal flaps barely reach mouth, no nasoral grooves, labial furrows on lower jaw only. Second dorsal fin much smaller than first, interdorsal space slightly less than anal base.

Distribution. East Atlantic: tropical West Africa from Mauritania to Angola.
Habitat. Rocky and mud bottom on continental shelf and upper slope 45–500 m.
Behaviour. Unknown.
Biology. Probably oviparous. Eats bony fish.
Status. IUCN Red List: Not Evaluated. Probably taken in trawl fisheries. Has been recorded as *Scyliorhinus stellaris*.

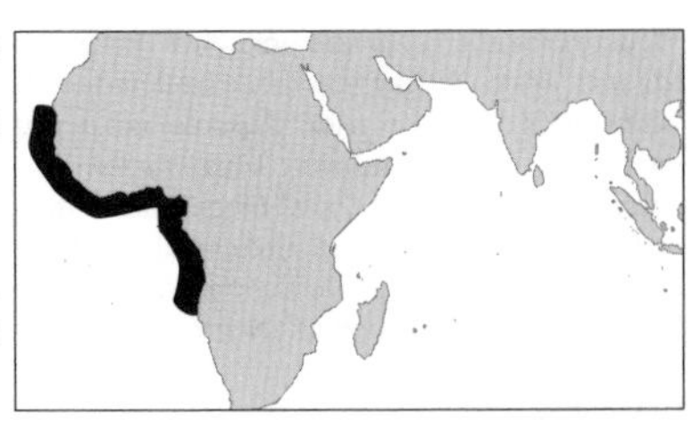

Comoro Catshark *Scyliorhinus comoroensis* Plate 41

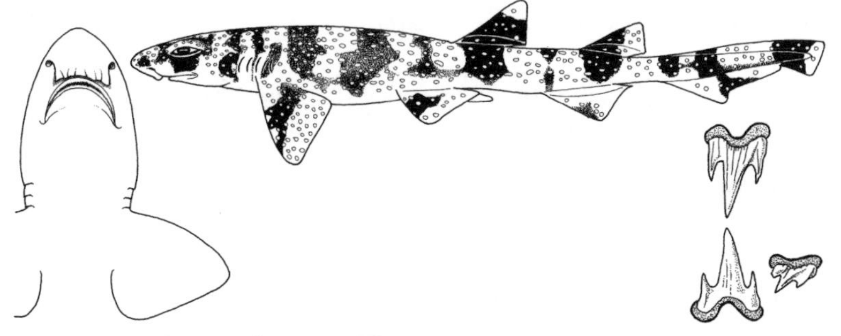

Measurements. One adult ♂ was 46 cm TL.
Identification. Small, patterned with bold, sharply defined dark grey-brown saddles centred on dark spots on the midline of the back and large blotches on a light grey-brown background. Numerous scattered small white spots in saddles and spaces between them, no small bold dark spots. Conspicuous dark bar under eye. Large anterior nasal flaps reach mouth but no nasoral grooves, labial furrows on lower jaw only. Second dorsal fin much smaller than first.
Distribution. Southwest Indian Ocean: Comoro Islands.
Habitat. Insular slopes, on bottom, 200–400 m.
Behaviour. Photographed in deep water by research submersible studying coelacanth *Latimeria chalumnae*.
Biology. Unknown.
Status. IUCN Red List: Not Evaluated. Apparently uncommon, known from only one specimen.

Brownspotted Catshark *Scyliorhinus garmani* Plate 42

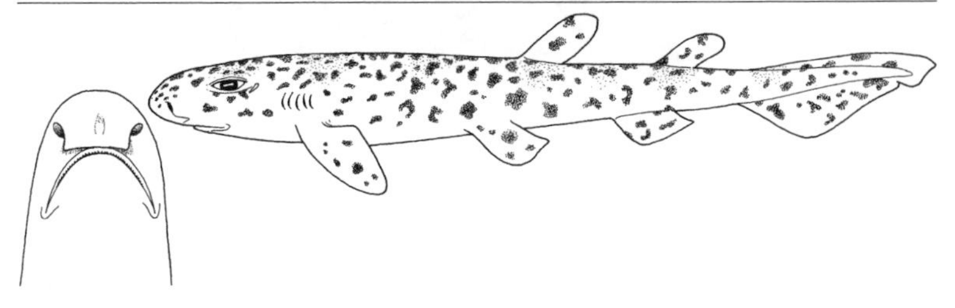

Measurements. Adult size unknown, specimens measure 24 or >36 cm TL.
Identification. Stocky, with large scattered round brown spots, seven indistinct saddle markings, no white spots. Anterior nasal flaps do not reach mouth, no nasoral grooves, labial furrows on lower jaw only. Second dorsal much smaller than first, anal base shorter than interdorsal space.
Distribution. Indo-west Pacific: collected in "East Indies", possibly Philippines (Negros Island).
Habitat. Unknown.
Behaviour. Unknown.
Biology. Unknown.
Status. IUCN Red List: Not Evaluated.

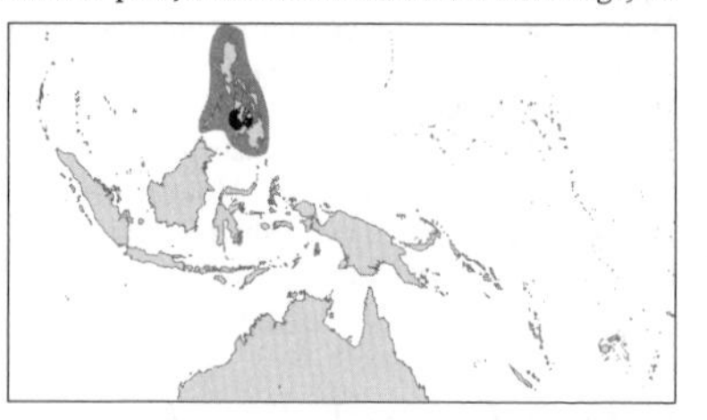

Freckled Catshark *Scyliorhinus haeckelii* Plate 42

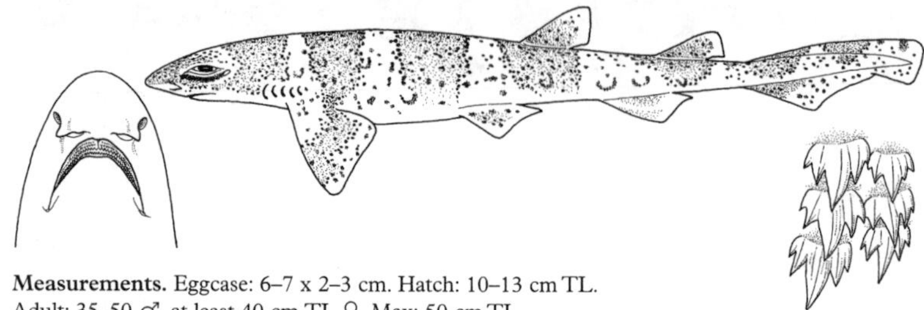

Measurements. Eggcase: 6–7 x 2–3 cm. Hatch: 10–13 cm TL.
Adult: 35–50 ♂, at least 40 cm TL ♀. Max: 50 cm TL.
Identification. Small, slender, with seven or eight dusky (sometimes faint) saddles. Conspicuous dark bar under eye. Very small black spots scattered over back and outlining saddles. No light spots. Small anterior nasal flaps do not reach mouth. Second dorsal fin much smaller than first. Adult males have larger teeth, longer mouths and larger lateral denticles than females.
Distribution. Western Atlantic: Venezuela to Uruguay.
Habitat. Continental shelf and upper slope, on or near seabed, 37–439 m. On deep-reef habitats (mostly below 250 m) off southern Brazil in association with deepwater gorgonians, hard corals, tube sponges, crinoids and brittle stars.
Behaviour. Unknown.
Biology. Oviparous, lays pairs of eggcases on corals and seafans.
Status. IUCN Red List: Not Evaluated.

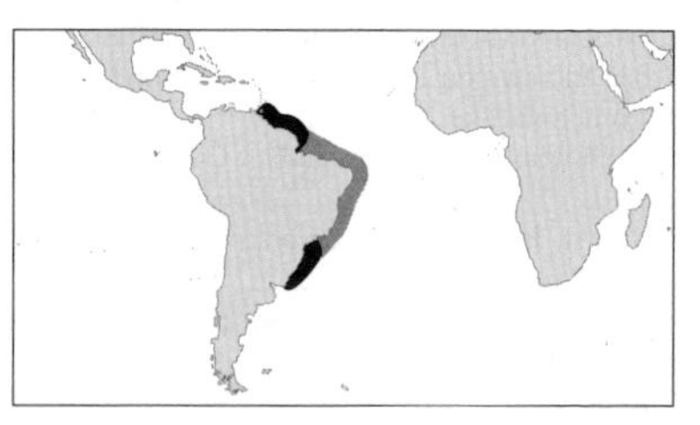

Whitesaddled Catshark *Scyliorhinus hesperius* Plate 42

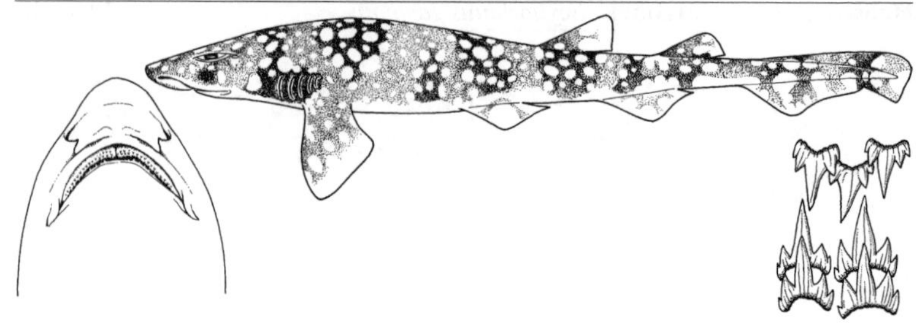

Measurements. Adult females reach at least 47 cm TL.
Identification. Fairly small and slender with conspicuous dark bar under eye, seven or eight well-defined dark saddles densely covered with large closely spaced white spots which sometimes extend to lighter spaces between saddles. No black spots. Small anterior nasal flaps end in front of mouth, no nasoral grooves, labial furrows on lower jaw only. Second dorsal fin much smaller than first.

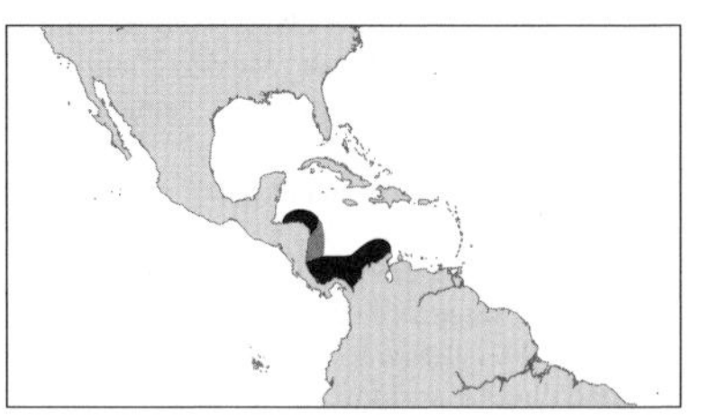

Distribution. West Atlantic (Caribbean): Honduras, Panama, Colombia.
Habitat. Upper continental slope, on or near bottom, 274–457 m.
Behaviour. Unknown.
Biology. Unknown.
Status. IUCN Red List: Not Evaluated.

Blotched Catshark *Scyliorhinus meadi* Plate 42

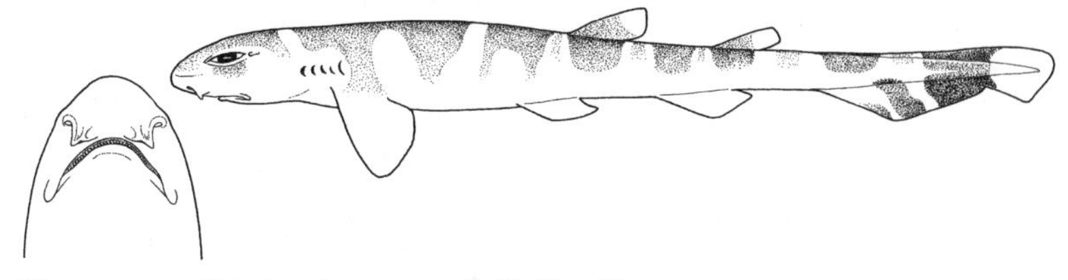

Measurements. Only immatures measured: 18–49 cm TL.
Identification. Broad-headed, stocky dark catshark with seven or eight darker and sometimes obscure saddles, no spots. Small anterior nasal flaps end just in front of mouth, no nasoral grooves, labial furrows on lower jaw only. Second dorsal fin much smaller than first.

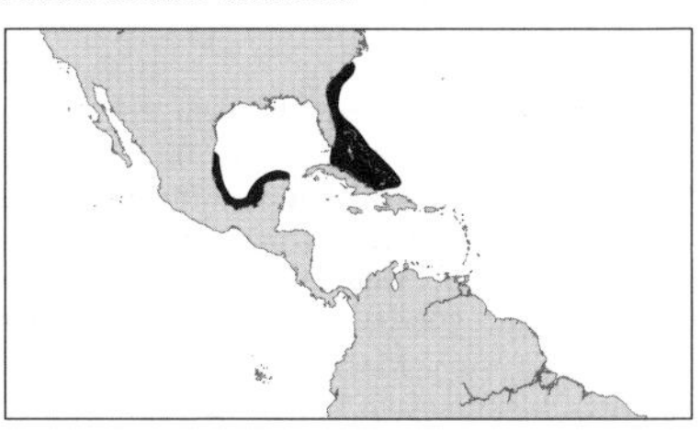

Distribution. Northwest Atlantic: North Carolina to Florida, USA; Santaren Channel between Cuba and Bahamas Bank; Mexico (Gulf of Mexico and northern Yucatan Peninsula).
Habitat. Continental slope, on or near bottom, 329–548 m.
Behaviour. Unknown.
Biology. Poorly known, presumably oviparous. Eats cephalopods, shrimp, and bony fishes.
Status. IUCN Red List: Not Evaluated. Relatively rare.

Chain Catshark *Scyliorhinus retifer* Plate 42

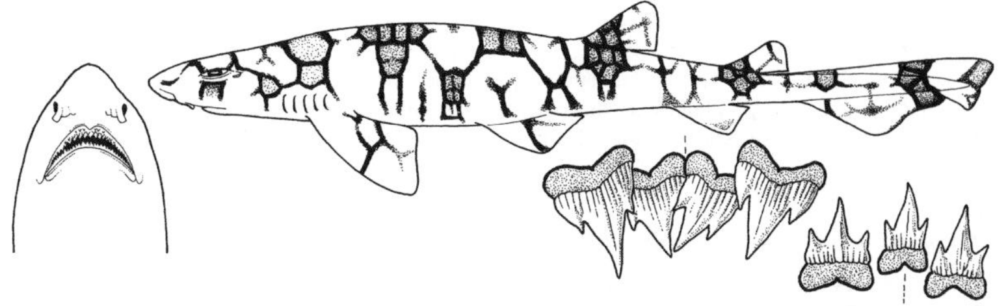

Measurements. Eggcase: ~2.7 x 6.5 cm. Hatch: ~10–11 cm TL. Mature: 38–50 cm ♂, 35–52 cm ♀. Max: 59 cm TL. Young reach 25–30 cm in two years in captivity.
Identification. Black chain patterning outlining faint dusky saddles, without spots, is unique to this

species and *Cephaloscyllium fasciatum* (Plate 36), but *Scyliorhinus retifer* has well-developed lower labial furrows and dorsal fins set well back.

Distribution. Northwest Atlantic, Gulf of Mexico, and Caribbean: USA (Georges Bank, Mass., to Florida and Texas), Mexico (Campeche Gulf), Barbados, area between Jamaica and Honduras, Nicaragua.

Habitat. Outer continental shelf and upper slope, on or near bottom, 73–754 m, deeper in the south. May be most common on very rough, rocky untrawlable areas suitable for egg laying.

Behaviour. Sluggish shark commonly found resting on bottom. May swallow small pebbles, perhaps for ballast. Females wrap eggcase tendrils up to 35 cm long around seabed projections such as corals by rapid circular swimming.

Biology. Oviparous, eggs laid in pairs every 8–15 days in captivity during spring and summer, ~44–52 eggs/year. Eggs hatch after about seven months in captivity at 11.7–12.8°C, longer in nursery areas at temperatures down to 7°. Feed on squid, bony fishes, polychaetes and crustaceans.

Status. IUCN Red List: Not Evaluated. Common where it occurs.

Nursehound *Scyliorhinus stellaris* Plate 41

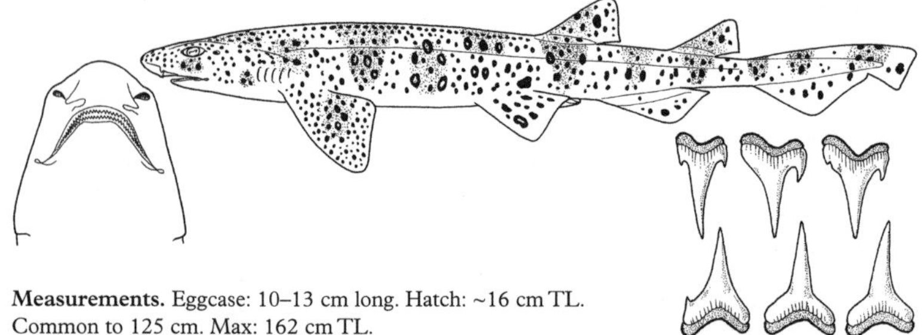

Measurements. Eggcase: 10–13 cm long. Hatch: ~16 cm TL. Common to 125 cm. Max: 162 cm TL.

Identification. Large and stocky with many large and small black spots, sometimes white spots, over pale background. Saddles very faint or absent. Large spots may be irregular, occasionally expand into large blotches totally covering the body.

Distribution. Northeast Atlantic: southern Scandinavia to Mediterranean, Morocco, Canaries, Mauritania to Senegal. Records further south to Gulf of Guinea and Congo River mouth may be *Scyliorhinus cervigoni*.

Habitat. Continental shelf, 1–125 m, commonly 20–63 m, on rocky or seaweed-covered bottom.

Behaviour. Deposit large thick-walled eggcases with strong tendrils at corners on algae.

Biology. Oviparous. Single egg/oviduct laid in spring and summer, may take nine months to hatch. Eats mostly crustaceans, cephalopods, other molluscs, bony fishes and other small sharks.

Status. IUCN Red List: Not Evaluated. Limited fisheries interest. Less common than *Scyliorhinus canicula*.

Izu Catshark *Scyliorhinus tokubee* Plate 42

Measurements. Eggcase: ~4.5 cm long. Hatch: uncertain. Mature: ~39–41 cm ♂, 38–39 cm ♀. Max: >41 cm TL.

Identification. Small, with bold, sharply defined reddish-brown saddles on midline of back and large lateral blotches on a light grey-brown background, yellow to white below. Very numerous

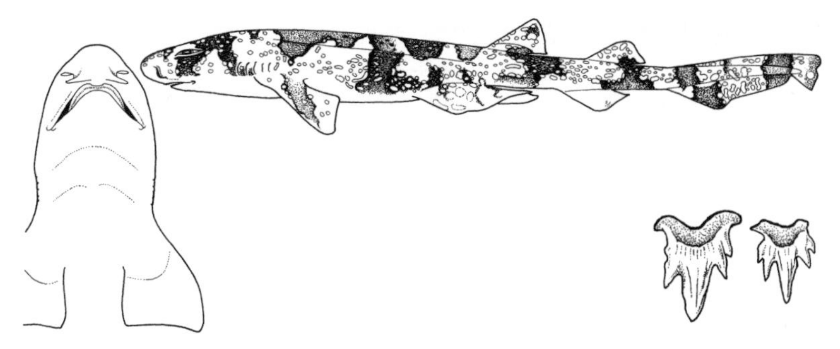

conspicuous densely set small yellowish spots in saddles and spaces between them as well as on fins, no dark spots. Conspicuous dark bar below eye. Small narrow anterior nasal flaps end in front of mouth, no nasoral grooves, labial furrows on lower jaw only. Second dorsal fin much smaller than first.

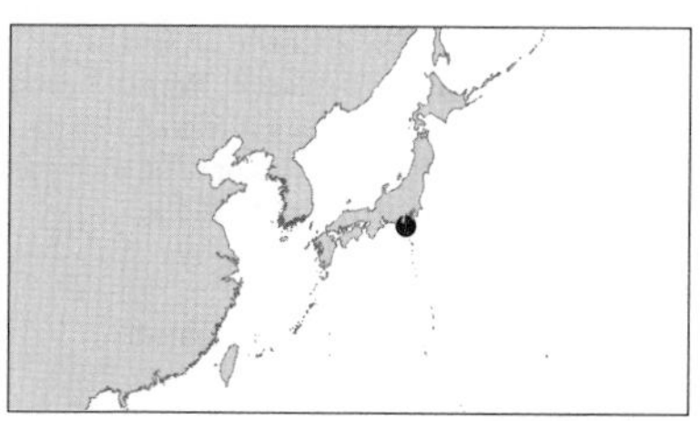

Distribution. Northwest Pacific, Japan (south-eastern Honshu, Izu Peninsula off Shirahama).
Habitat. Offshore on continental shelf, 100 m.
Behaviour. Little known.
Biology. Oviparous, one egg/oviduct. Hatchlings reach adulthood in five years or less.
Status. IUCN Red List: Not Evaluated. Readily breeds in captivity.

Cloudy Catshark *Scyliorhinus torazame* Plate 42

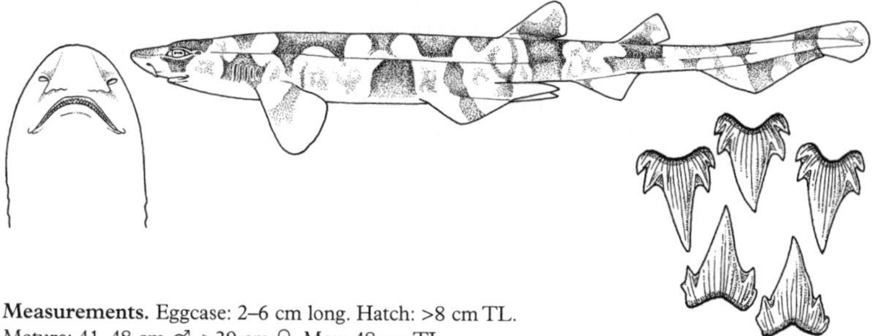

Measurements. Eggcase: 2–6 cm long. Hatch: >8 cm TL.
Mature: 41–48 cm ♂, >39 cm ♀. Max: 48 cm TL.
Identification. Fairly small, slender narrow-headed shark with six to nine darker saddles and, in larger specimens, many irregular large dark and light spots on very rough, dark skin. Small anterior nasal flaps do not reach mouth, no nasoral grooves, labial furrows on lower jaw only. Second dorsal fin much smaller than first.

Distribution. Northwest Pacific: Japan (Hokkaido and Honshu to Okinawa), Korea, China, possibly Philippines.
Habitat. Continental shelf and upper slope, close inshore to at least 320 m.
Behaviour. Eggcases deposited in a nursery or hatching ground.
Biology. Oviparous, one egg/oviduct.
Status. IUCN Red List: Not Evaluated.

Dwarf Catshark *Scyliorhinus torrei* Plate 42

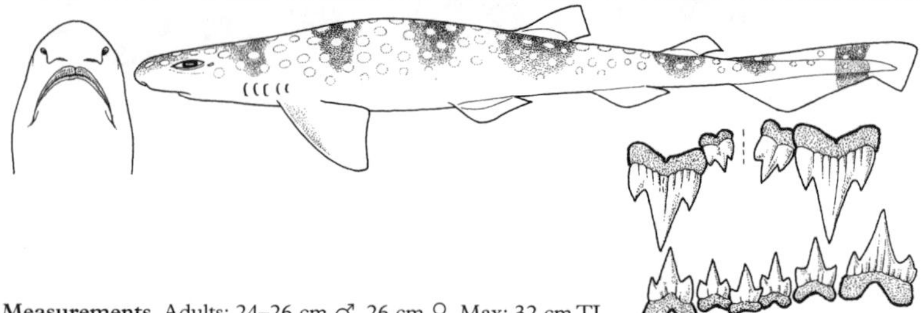

Measurements. Adults: 24–26 cm ♂, 26 cm ♀. Max: 32 cm TL.
Identification. Very small slender light brown catshark with seven or eight darker brown saddles (obscure in adults) and many large regularly scattered white spots on back, no black spots. Small anterior nasal flaps do not reach mouth, no nasoral grooves, labial furrows on lower jaw only, second dorsal fin much smaller than first.

Distribution. Northwest Atlantic: Florida Straits, Bahamas, northern Cuba, Virgin Islands.
Habitat. Upper continental slope on or near bottom, 229–550 m, most below 366 m.
Behaviour. Unknown.
Biology. Poorly known.
Status. IUCN Red List: Not Evaluated.

Proscylliidae: Finback Catsharks

Identification. Dwarf to small sharks (adults 15–65 cm) with narrowly rounded head and rounded or subangular snout, no deep groove in front of elongated, cat-like eyes. Rudimentary nictitating eyelids. No barbels or nasoral grooves, internarial space less than 1.3 times nostril width, long angular arched mouth reaching past anterior ends of eyes, small papillae on palate and edges of gill arches, labial furrows very short or absent. First dorsal fin base short and well ahead of pelvic fin bases, but closer to pelvic bases than pectoral bases, no precaudal pits, caudal fin without a strong ventral lobe or lateral undulations on its dorsal margin. Body and fin colour usually variegated, but *Eridacnis* species with plain bodies and striped caudal fins.
Biology. Most are ovoviviparous, except for oviparous *Proscyllium habereri*, and feed on small fishes and invertebrates.
Status. Poorly-known deepwater sharks of outer continental and insular shelves and upper slopes, on or near bottom, 50–766 m. Disjunct distribution, mostly in Indo-west Pacific but also in tropical northwest Atlantic.

Harlequin Catshark *Ctenacis fehlmanni* Plate 43

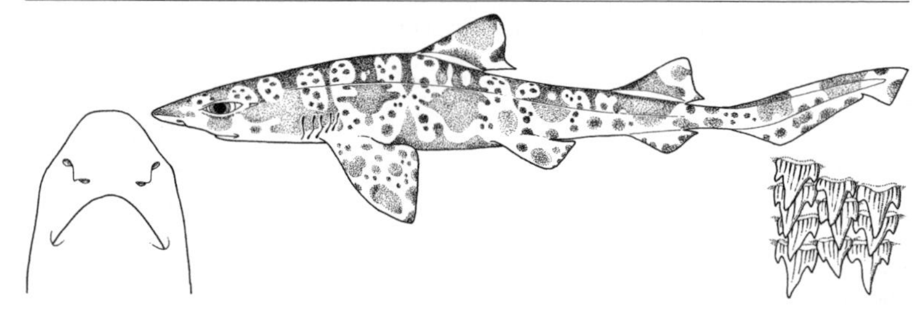

Measurements. Max: 46 cm TL (adult ♀).
Identification. Rather stout body and tail, with a unique colour pattern of large, reddish-brown, irregular dorsal saddle blotches, interspersed with smaller round spots and vertical bars, and spots on fins. Nictitating eyelids, large triangular mouth, very short labial furrows, short anterior nasal flaps not reaching mouth.

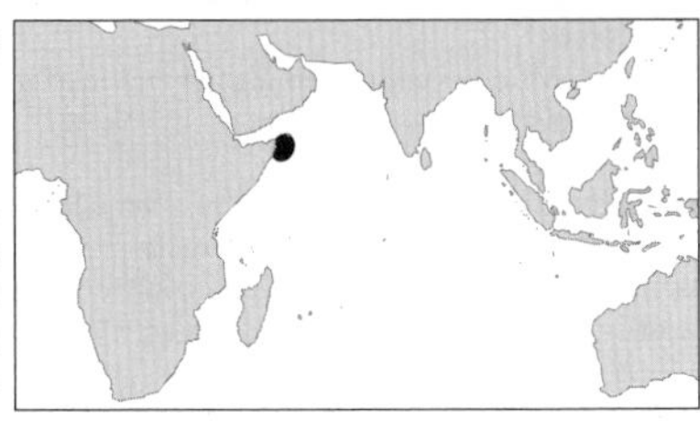

Distribution. Northwestern Indian Ocean: off Somalia.
Habitat. Outer continental shelf.
Behaviour. Unknown.
Biology. Possibly ovoviviparous (a very thin-walled large eggcase in each uterus of the holotype would probably have been retained until pair of young hatch). Large mouth, small teeth and large pharynx with gill rakers suggest a diet of very small invertebrates.
Status. IUCN Red List: Not Evaluated. Known from only a few specimens.

Cuban Ribbontail Catshark *Eridacnis barbouri* Plate 43

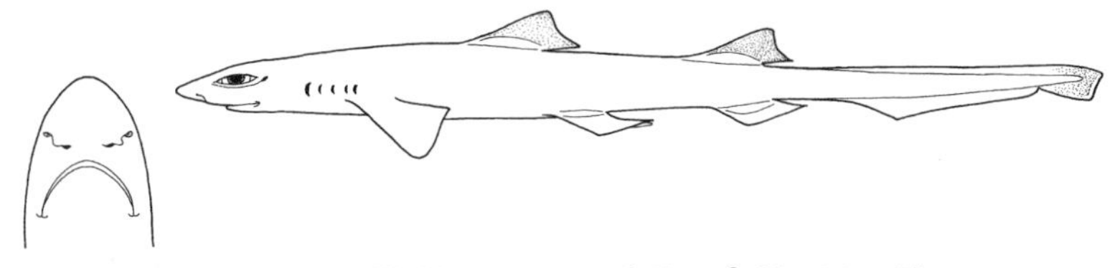

Measurements. Born: >10 cm TL. Mature: ~27 cm ♂, 28 cm ♀. Max: 34 cm TL.
Identification. Very small, slender, light greyish-brown with light edges on two dorsal fins and faint dark banding on long, narrow, ribbon-like caudal fin. Anal fin ~two-thirds of dorsal fin heights. Nictitating eyelids, triangular mouth, very short but well-developed labial furrows, short anterior nasal flaps do not reach mouth, no nasoral grooves or barbels.

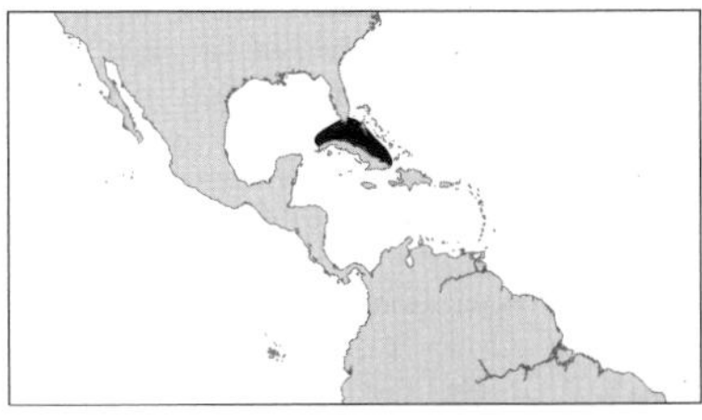

Distribution. Northwest Atlantic: Florida Straits and off north coast of Cuba.
Habitat. Upper continental and insular slopes, on the bottom, 430–613 m.
Behaviour. Unknown.
Biology. Ovoviviparous, two young/litter.
Status. IUCN Red List: Not Evaluated. Relatively common or formerly common in its limited range.

Pygmy Ribbontail Catshark *Eridacnis radcliffei* Plate 43

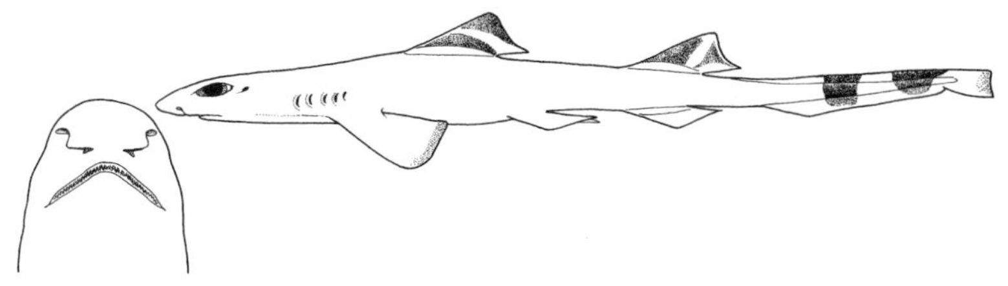

Measurements. Born: ~11 cm TL. Mature: 18–19 cm ♂, 15–16 cm ♀. Max: 24 cm TL.
Identification. Extremely small dark brown shark. Long, narrow, ribbon-like caudal fin with

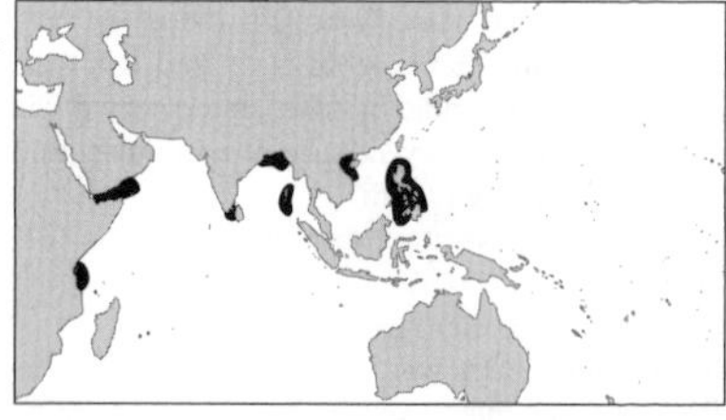

prominent dark banding, blackish markings on two spineless dorsal fins. Anal fin less than half height of dorsals.
Distribution. Patchy, Indo-west Pacific: Tanzania, Gulf of Aden, India, Andaman Islands, Vietnam, Philippines.
Habitat. Mud on upper continental and insular slopes, outer shelves, 71–766 m.
Behaviour. Unknown.
Biology. Ovoviviparous, one or two very large young/litter. Probably grow fast: females carry large eggs at 15 cm TL, small embryos at ~17 cm, large young at >180 cm TL. Feed mainly on small bony fishes and crustaceans, also squid.
Status. IUCN Red List: Not Evaluated. Common in southern India and Philippines.

African Ribbontail Catshark *Eridacnis sinuans* Plate 43

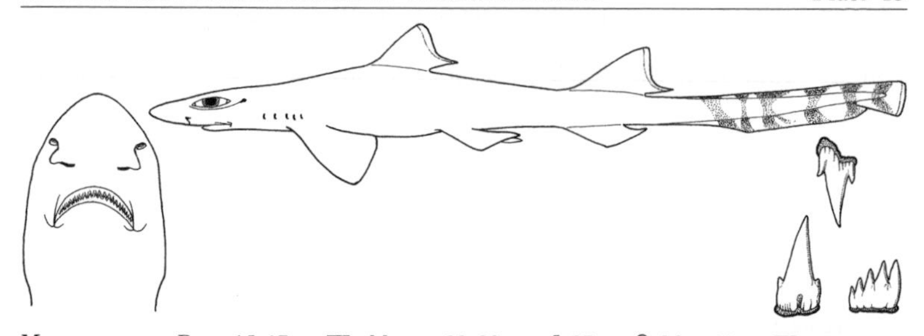

Measurements. Born: 15–17 cm TL. Mature: 29–30 cm ♂, 37 cm ♀. Max: 37 cm TL.
Identification. Dwarf, slender grey-brown sharklet with faint dark banding on long, narrow, ribbon-like caudal fin and light margins on dorsal fins. Fairly long snout, short anterior nasal flaps that do not reach triangular mouth, nictitating eyelids.
Distribution. Southwestern Indian Ocean, off South Africa, Mozambique and Tanzania.
Habitat. Upper continental slope and outer shelf, 180–4850 m.
Behaviour. Sexes apparently segregate by area or depth: mostly males are (or were) taken off Natal.
Biology. Ovoviviparous, two young/litter. Feeds on small bony fishes, crustaceans and cephalopods.
Status. IUCN Red List: Not Evaluated. Only part of its range is trawled, species is not utilised and other areas are unfished.

Graceful Catshark *Proscyllium habereri* Plate 43

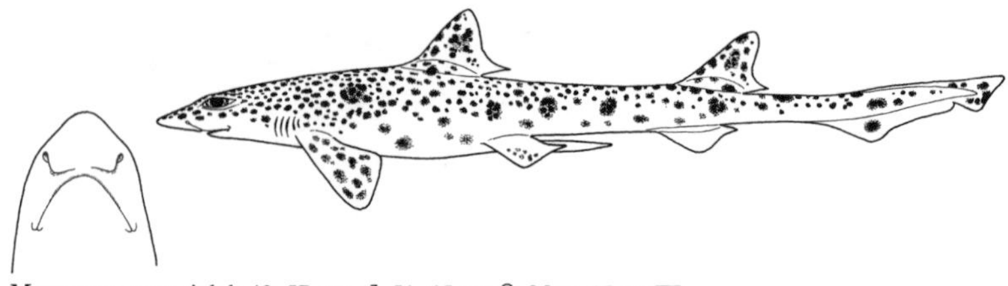

Measurements. Adult 42–57 cm ♂, 51–65 cm ♀. Max: 65 cm TL.
Identification. Slender body, rather long tail with broad caudal fin. Small to large dark brown spots,

sometimes small white spots and indistinct dusky saddle blotches on body and fins. Large eyes with nictitating eyelids. Large anterior nasal flaps nearly reaching triangular mouth, which extends past eyes.

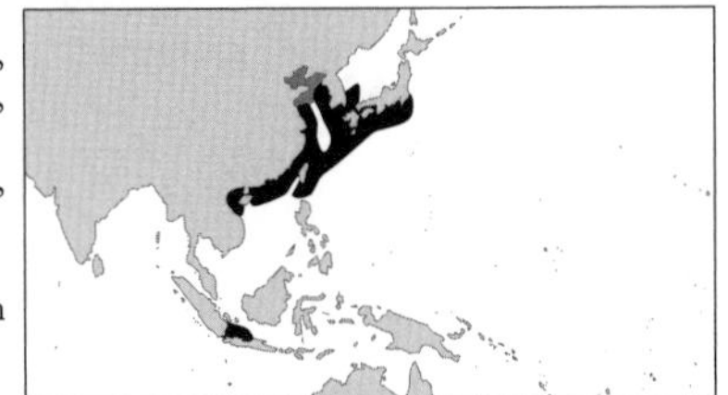

Distribution. Western Pacific: Northwest Java, Vietnam, China, Taiwan, Korea, Ryukyu Islands, southeast Japan.
Habitat. Continental and insular shelves, 50–100 m.
Behaviour. Unknown.
Biology. Oviparous, one egg/uterus. Feed on bony fishes, crustacea and cephalopods.
Status. IUCN Red List: Not Evaluated.

Clown or Magnificent Catshark *Proscyllium* sp. A — Plate 42

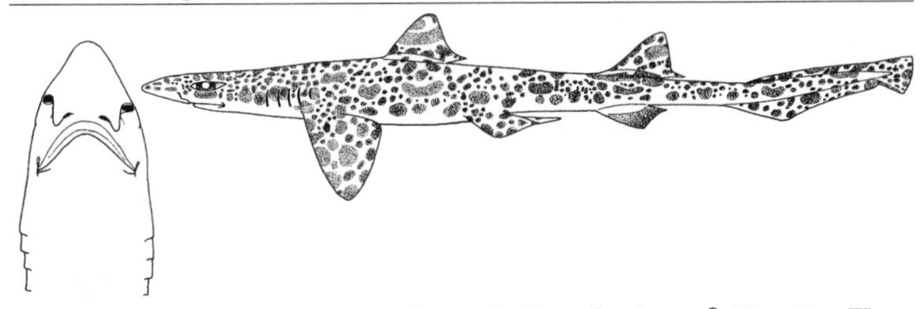

Measurements. Born or hatch: unknown. Mature: 47–49 cm ♂, unknown ♀. Max: 49 cm TL.
Identification. Similar to *P. habereri*, but with a more variegated pattern of small and large spots and dots, including clusters of two small round spots above a large upcurved spot and an intermediate small spot forming 'clown faces' below dorsal fins.

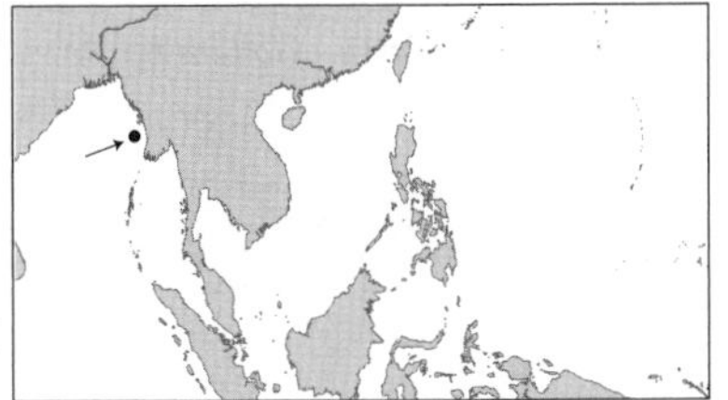

Distribution. Indian Ocean: Andaman Sea off Myanmar (Burma).
Habitat. Near edge of outer continental shelf, >200 m.
Behaviour. Unknown.
Biology. Unknown.
Status. IUCN Red List: Not Evaluated. Known from five specimens. This species has only recently been discovered and is awaiting scientific description.

Pseudotriakidae: False Catsharks and Gollumsharks

Identification. Small to large sharks (adults 56–295 cm) with narrowly rounded head and more or less elongated bell-shaped snout, a deep groove in front of elongated, cat-like eyes. Rudimentary nictitating eyelids. No barbels or nasoral grooves, internarial space over 1.5 times nostril width, long angular arched mouth reaching past anterior ends of eyes, no papillae inside mouth and none on edges of gill arches, labial furrows short but always present. First dorsal fin more or less elongated, base closer to pectoral fin bases than to pelvic fin bases, no precaudal pits, caudal fin with a weak ventral lobe or none, no lateral undulations on its dorsal margin. Colour usually plain grey to brown or blackish (white spots and fin margins on some *Gollum* species).
Biology. Ovoviviparous as far as known, at least two species are known to be oophagous, with foetuses eating nutritive eggs. Probably prey upon small fishes and invertebrates.
Status. Poorly-known deepwater sharks of outer continental and insular shelves and slopes, on or near the bottom, 129–1890 m. The large *Pseudotriakis microdon* is wide-ranging but small species have restricted distributions in western Indian Ocean and western Pacific. Absent from the South Atlantic and eastern Pacific.

Slender Smoothhound or Gollumshark *Gollum attenuatus* Plate 43

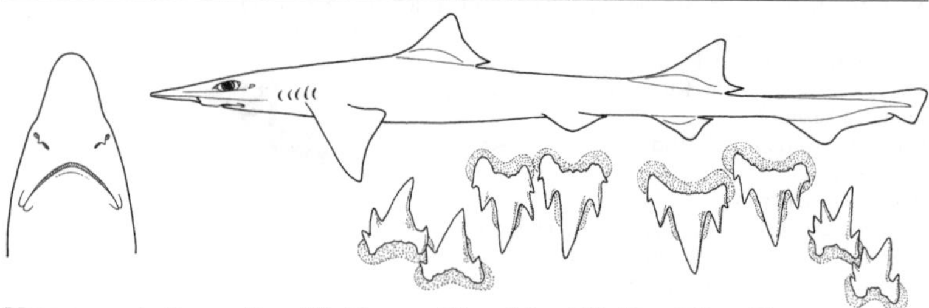

Measurements. Born: ~40 cm TL. Mature: ~70 cm (♂ and ♀). Max: 110 cm TL.
Identification. Unpatterned greyish small shark with slender body and tail, elongated cat-like eyes, nictitating eyelids, very long and angular snout (bell-shaped in dorsoventral view, wedge-shaped laterally), short anterior nasal flaps, angular mouth extending behind eyes, short labial furrows, small spiracles, numerous small teeth (<120 rows/jaw), relatively short subtriangular first dorsal fin, second as high or slightly higher.

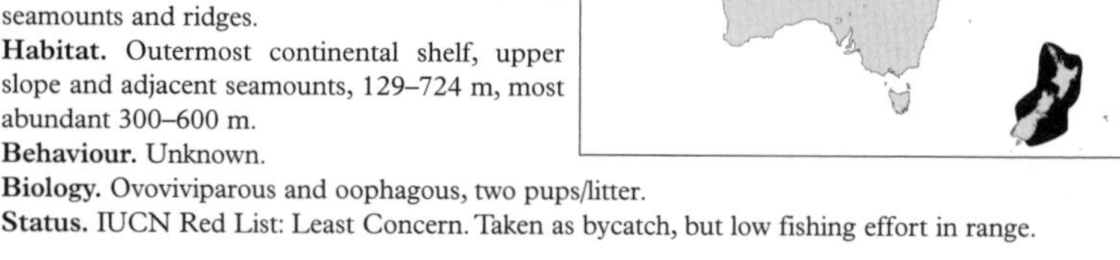

Distribution. New Zealand and surrounding seamounts and ridges.
Habitat. Outermost continental shelf, upper slope and adjacent seamounts, 129–724 m, most abundant 300–600 m.
Behaviour. Unknown.
Biology. Ovoviviparous and oophagous, two pups/litter.
Status. IUCN Red List: Least Concern. Taken as bycatch, but low fishing effort in range.

Sulu Gollumshark *Gollum* sp. A Plate 42

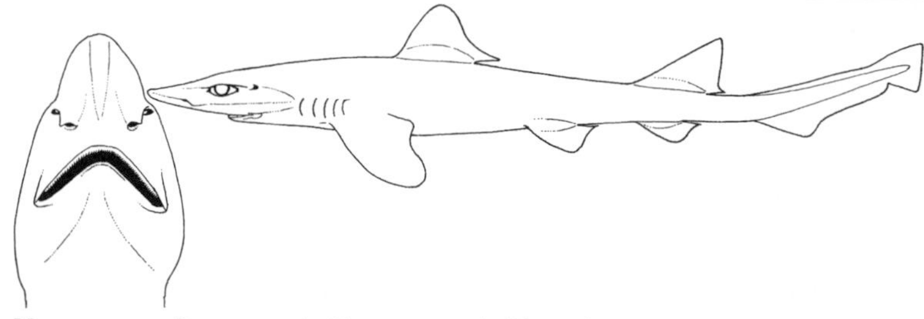

Measurements. Born: uncertain. Mature: uncertain. Max: at least 54 cm TL.
Identification. Distinguished from other species of gollumshark by shorter snout. Colour plain dark greyish-brown above, lighter grey below, no conspicuous white markings on body or fins. Narrow light posterior margins on dorsal, pectoral and pelvic fins. Web of terminal caudal lobe light, edged with dark grey or blackish.

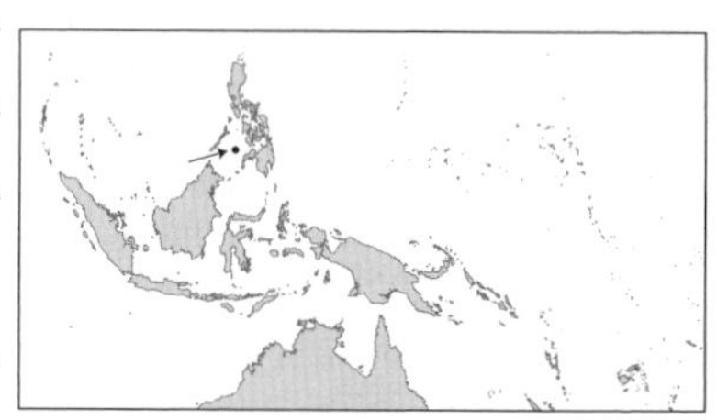

Distribution. Southwestern Pacific Ocean: Sulu Sea, Philippines.
Habitat. Unknown. Presumably deepwater on upper continental slope or upper shelf.
Behaviour. Unknown.
Biology. Unknown.
Status. IUCN Red List: Not Evaluated. An undescribed, recently discovered species.

Whitemarked Gollumshark *Gollum* sp. B — Plate 42

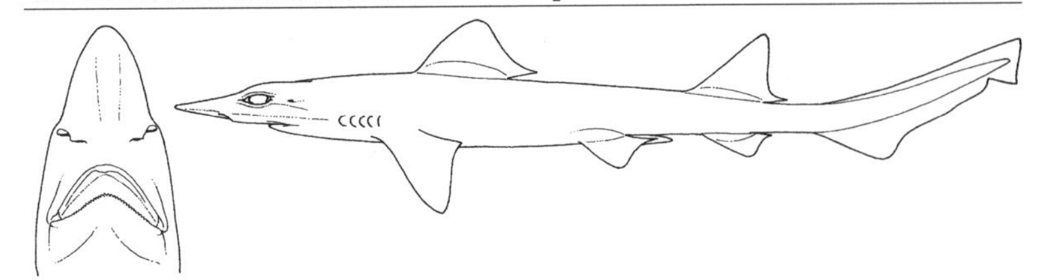

Measurements. Born: unknown. Mature: unknown. Max: c.60 cm TL.
Identification. Snout as long or longer than *Gollum attenuatus*. Grey-brown above, brownish-white below, with characteristic pattern of conspicuous row of white spots on the tail and bold white markings on the head and fins.

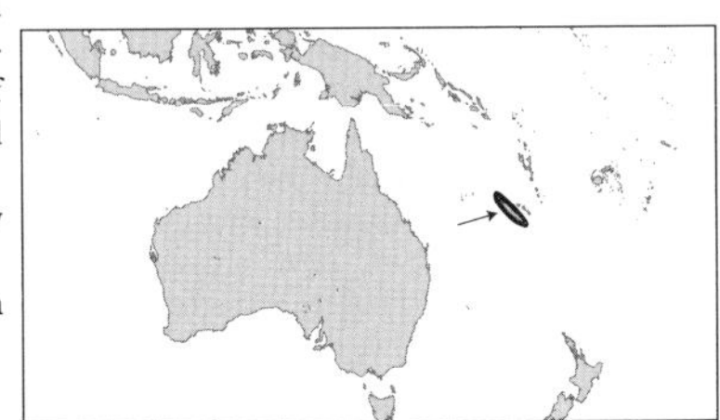

Distribution. Southwestern Pacific Ocean: New Caledonia.
Habitat. Unknown. Presumably deepwater on upper continental slope or upper shelf.
Behaviour. Unknown.
Biology. Unknown.
Status. IUCN Red List: Not Evaluated. An undescribed, recently discovered species.

False Catshark *Pseudotriakis microdon* — Plate 43

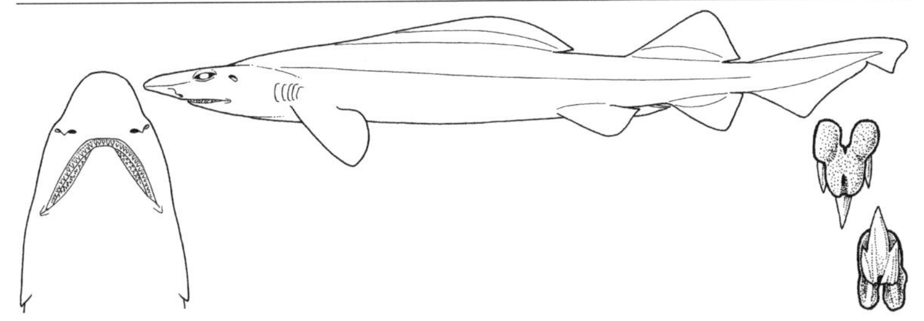

Measurements. Born: 70–85 cm TL. Mature: 200–269 cm ♂, 212–295 cm ♀. Max: 295 cm TL.
Identification. Unpatterned dark brown to blackish large shark with stocky, bulky soft body and tail. Elongated cat-like eyes, nictitating eyelids, short snout (bell-shaped in dorsoventral view, wedge-shaped laterally), short anterior nasal flaps, huge angular mouth extending behind eyes, short labial furrows, very large spiracles about as long as eyes, numerous small teeth (>200 rows/jaw). Long, low, keel-like first dorsal fin, second much higher.

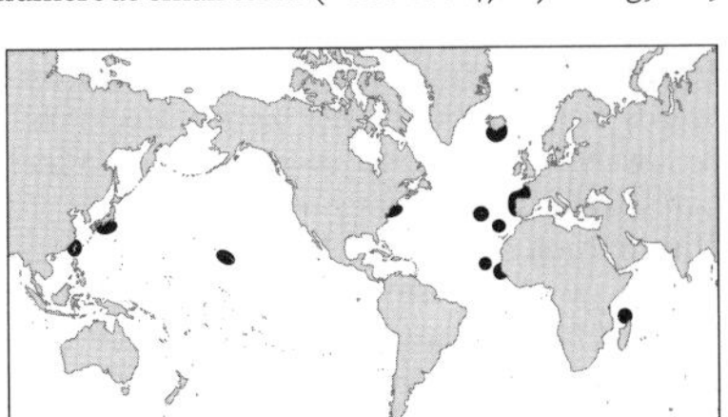

Distribution. Patchy world-wide, except (so far) South Atlantic or eastern Pacific.
Habitat. Deepwater seabed on continental and insular slopes 200–1890 m, occasionally continental shelves and shallower near submarine canyons.
Behaviour. Large body cavity, soft fins, skin and musculature indicates an inactive and sluggish shark of virtually neutral buoyancy. Photographed feeding on a fish in deepwater.
Biology. Ovoviviparous: 2, possibly 4 pups/litter. Apparently oophagous.
Status. IUCN Red List: Not Evaluated. Seemingly uncommon or rare wherever it occurs.

Pygmy False Catshark (new genus and species) Plate 43

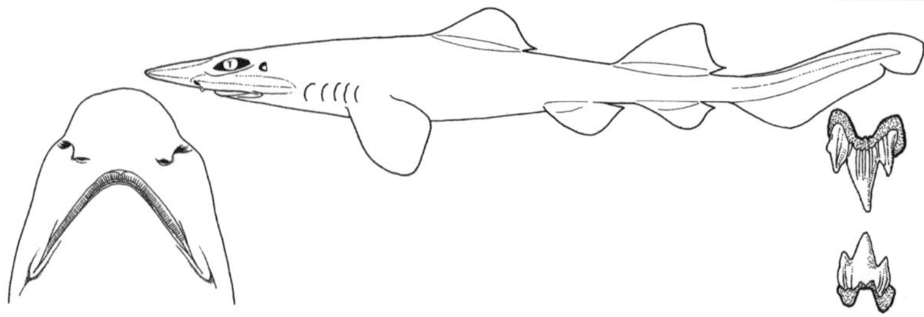

Measurements. Born: unknown. Mature: 49~56 cm ♂, unknown ♀. Max: >56 cm TL.
Identification. Unpatterned small, dark grey-brown shark with dark fins, moderately stocky soft body and tail, elongated cat-like eyes, nictitating eyelids, short snout (bell-shaped in dorsoventral view, wedge-shaped laterally), short anterior nasal flaps, angular mouth extending behind eyes, short labial furrows, small spiracles, numerous small teeth (<130 rows/jaw), relatively short sub-triangular first dorsal fin, second higher.

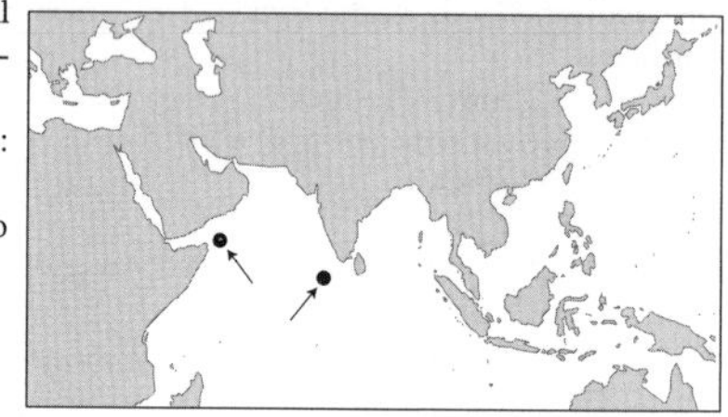

Distribution. Northwestern Indian Ocean: Arabian Sea (off Socotra Island) and Maldives.
Habitat. Continental and insular slopes down to 1120 m.
Behaviour. Unknown.
Biology. Unknown.
Status. IUCN Red List: Not Evaluated.

Leptochariidae: Barbeled Houndsharks

One species.

Barbeled Houndshark *Leptocharias smithii* Plate 43

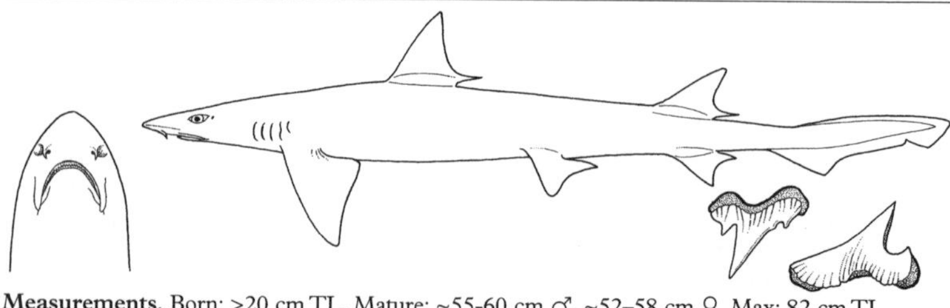

Measurements. Born: >20 cm TL. Mature: ~55-60 cm ♂, ~52–58 cm ♀. Max: 82 cm TL.
Identification. Small slender light grey-brown shark. Horizontally oval eyes, internal nictitating eyelids. Nostrils with slender barbels, long arched mouth reaching past anterior ends of eyes, labial furrows very long, small cuspidate teeth. Males have greatly enlarged anterior teeth.

Distribution. Eastern Atlantic: Mauritania to Angola, possibly to Morocco and Mediterranean.
Habitat. Continental shelf, near bottom, 10–75 m. Especially common on mud off river mouths.
Behaviour. Unknown. Males may use greatly enlarged front teeth in courtship and copulation.
Biology. Viviparous, with unique spherical placenta. Pregnant females occur July–October

off Senegal. Seven pups/litter, gestation period at least four months. Feeds on crustaceans, small bony fishes and skates.
Status. IUCN Red List: Near Threatened. Taken in many fisheries as utilized bycatch throughout its restricted habitat; status needs further study.

Triakidae: Houndsharks

One of the largest families of sharks, with over 40 species, distributed world-wide in warm and temperate coastal seas. Most species occur in continental and insular waters, from the shoreline and intertidal to the outermost shelf, often close to the bottom, with many in sandy, muddy and rocky inshore habitats, enclosed bays and near river mouths. A few deepwater species range down continental slopes to great depths, possibly >2000 m. Many are endemic with a very restricted distribution.
Identification. Small to medium-sized sharks with two medium to large spineless dorsal fins, the first dorsal base well ahead of pelvic bases, and an anal fin. Horizontally oval eyes with nictitating eyelids, no nasoral grooves, anterior nasal flaps not barbel-like (except in *Furgaleus*), a long, angular or arched mouth reaching past the front of the eyes, moderate to very long labial furrows. Caudal fin without a strong ventral lobe or lateral undulations on its dorsal margin. Some (e.g. *Mustelus*) can be very hard to identify without vertebral counts and many species are undescribed.
Biology. Some species are very active and swim almost continuously, others can rest on the bottom, and many swim close to the seabed. Some are most active by day, others at night. Houndsharks are either ovoviviparous or viviparous, the latter with a yolk sac placenta, bearing litters from 1 or 2 to 52 pups. They feed mainly on bottom and midwater invertebrates and bony fishes; some take largely crustaceans, some mainly fishes, and a few primarily cephalopods; none feed on birds and mammals.
Status. Houndsharks are generally fairly common and some are very abundant in coastal waters, fished extensively for their meat, liver oil and fins (e.g. *Galeorhinus* and *Mustelus*). Some of the smaller coastal species reproduce rapidly and can support well-managed fisheries. Others have a history of stock collapse where fisheries arc unregulated and require much more careful management. Some species are extremely rare. None are harmful to people.

Whiskery Shark *Furgaleus macki* Plate 44

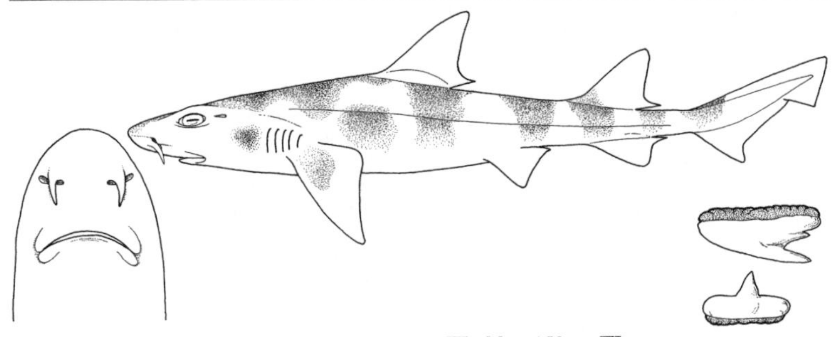

Measurements. Born: 25 cm TL. Mature ~110 cm TL. Max: 150 cm TL.
Identification. Only houndshark with anterior nasal flaps forming slender barbels. Stocky, almost humpbacked, obvious ridges below dorsolateral eyes, very short arched mouth. Grey above, with variegated dark blotches or saddles that fade with age, light below.

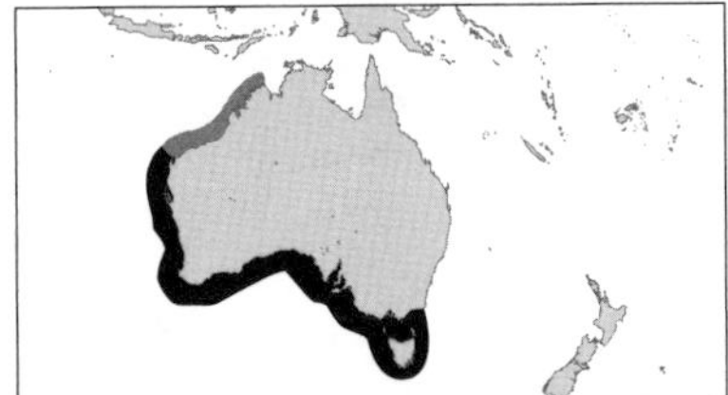

Distribution. Australia.
Habitat. Shallow temperate continental shelf, on or near bottom on rock, seagrass and in kelp.
Behaviour. Active-swimmer, specialist feeder on octopus, also takes squid, bony fishes and lobsters.

Biology. Ovoviviparous, no yolk-sac placenta, 4–29 pups/litter (average 19) every second year.
Status. IUCN Red List: Least Concern. Common. Stock depleted by target gillnet fishery to <30% of original size in 1960s–70s. Fishery now well managed, population fairly stable since mid-1980s and now increasing.

Tope, Soupfin or School Shark *Galeorhinus galeus* Plate 44

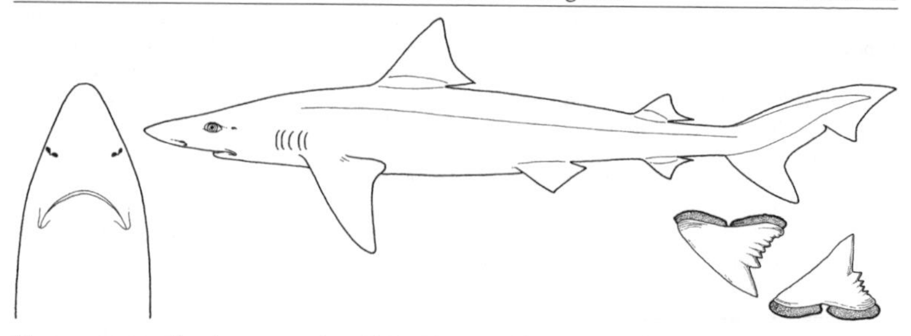

Measurements. Vary between regions. Born: 30–40 cm TL. Mature: ~120–170 cm ♂, 130–185 cm ♀. Max: 175 cm ♂, 195 cm ♀ TL.
Identification. Slender, long-nosed houndshark without obvious anterior nasal flaps or subocular ridges. Large arched mouth, small blade-like teeth. Second dorsal fin much smaller than first and about as large as anal fin. Extremely long terminal caudal lobe (~half the dorsal caudal margin). Greyish above, white below, black markings on fins of young.

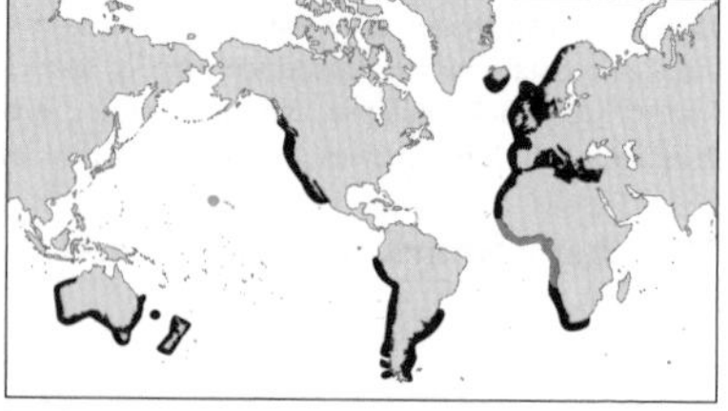

Distribution. World-wide in temperate waters.
Habitat. Most abundant in cold to warm temperate continental seas, from the surfline and very shallow water to well offshore (not oceanic), often near the bottom, 2–471 m.
Behaviour. Active, strong, long distance swimmers, occurring in small schools, partly segregated by size and sex, which are seasonally highly migratory in higher latitudes. Migrations of 1600 km recorded. Pregnant females move into shallow bays and estuaries to give birth, then return to offshore feeding grounds. Juveniles remain in nursery grounds for up to two years, then join schools of immatures in other locations.
Biology. Ovoviviparous, no yolk-sac placenta, 6–52 pups/litter, increasing with size of mother. Very low biological productivity; maximum age possibly 60 years, female maturity >10 years. Mainly an opportunistic feeder on bony fishes, also invertebrates. Predators include other large sharks and probably marine mammals.
Status. IUCN Red List: Vulnerable. Fished all over the world for meat, liver oil and fins. Also taken by sports anglers. Many populations are seriously depleted and most fisheries unmanaged.

Sailback Houndshark *Gogolia filewoodi* Plate 44

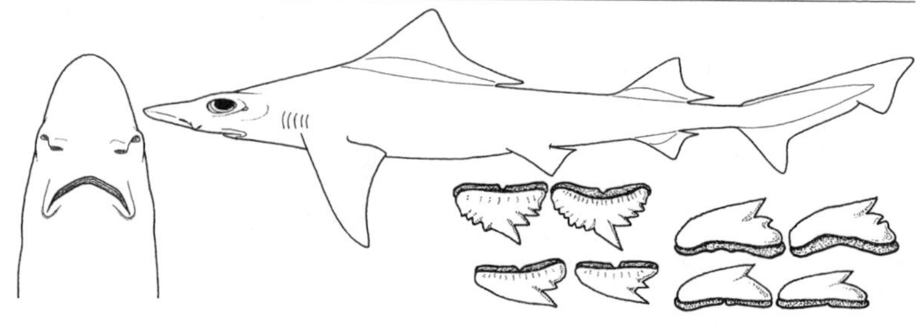

Measurements. Born: ~22 cm TL. Mature: ♀ measured 74 cm TL.
Identification. Small, grey-brown houndshark distinguished by huge triangular first dorsal fin (~length caudal fin) and long preoral snout (1.6–1.7 x mouth width).
Distribution. Southern Pacific: one site known, northern New Guinea.

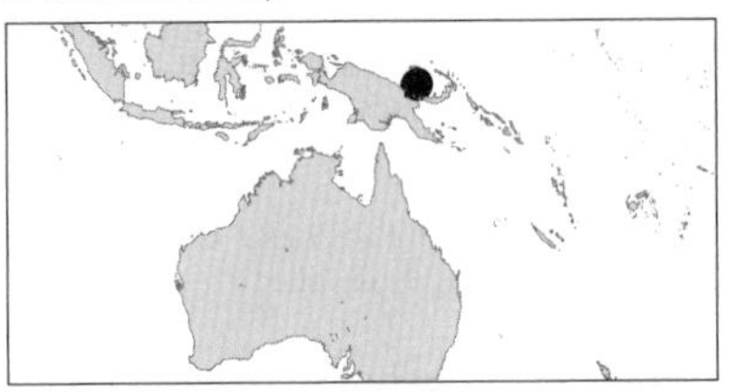

Habitat. Only known specimen from 73 m on continental shelf, probably near bottom.
Behaviour. Unknown.
Biology. Ovoviparous. Pregnant female was carrying two pups.
Status. IUCN Red List: Data Deficient. Presumably an uncommon endemic.

Darksnout or Deepwater Sicklefin Houndshark
Hemitriakis abdita Plate 44

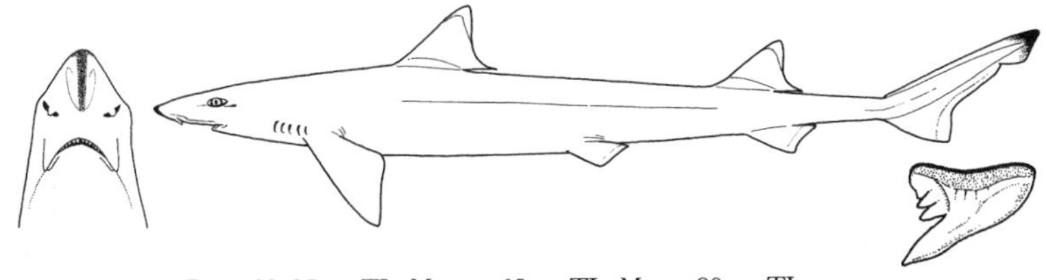

Measurements. Born: 20–25 cm TL. Mature: 65 cm TL. Max: >80 cm TL.
Identification. Slender greyish-brown houndshark with dark stripe underneath rather long, parabolic snout, small anterior nasal flaps, arched mouth, small blade-like teeth, distinct white tips to similar-sized dorsal fins, falcate dorsal, pectoral and anal fins. Juveniles have prominent dark bars and solid saddles on body and fins.

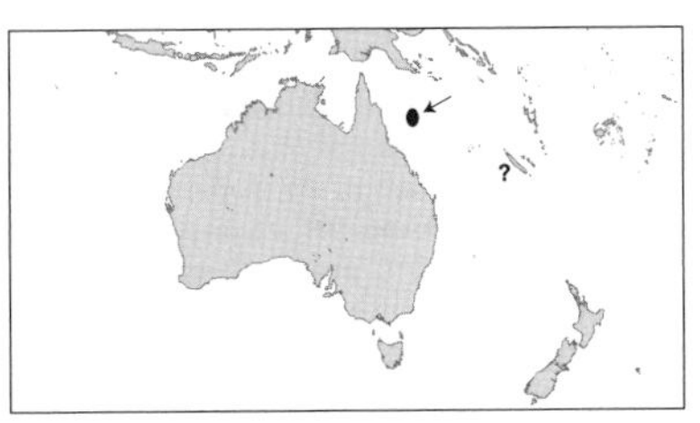

Distribution. Southwest Pacific: Australia, possibly New Caledonia.
Habitat. Deepwater on upper continental slope, 225–400 m.
Behaviour. Unknown.
Biology. Unknown.
Status. IUCN Red List: Data Deficient. Known from only a few specimens from a small area.

Sicklefin Houndshark *Hemitriakis falcata* Plate 44

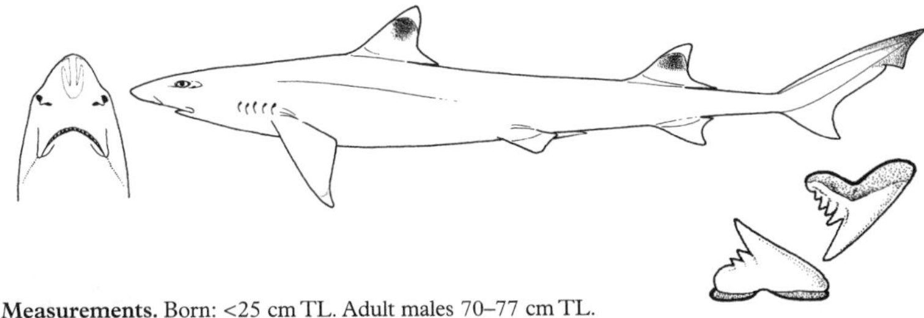

Measurements. Born: <25 cm TL. Adult males 70–77 cm TL.
Identification. Slender greyish-brown body, small blade-like teeth, small anterior nasal flaps, arched mouth, no stripe below snout, distinct white tips to similar-sized dorsal fins. Dorsal, pectoral and

anal fins strongly falcate in adults. First dorsal origin over pectoral insertions, in front of pectoral free rear tips. Juveniles have saddles and large spots with solid centres on body and fins.
Distribution. Northwestern Australia.
Habitat. Outer continental shelf, 146–197 m.
Behaviour. Unknown.
Biology. Unknown.
Status. IUCN Red List: Least Concern. Very limited range is largely unfished.

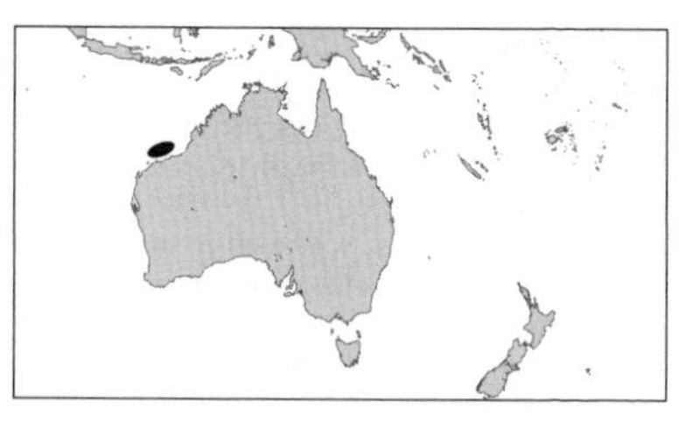

Japanese Topeshark *Hemitriakis japanica* Plate 44

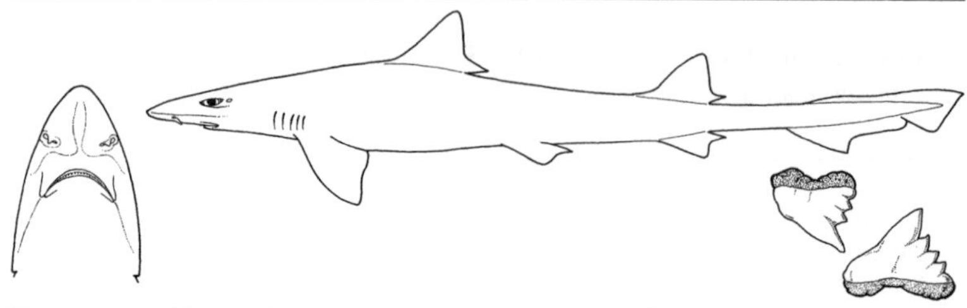

Measurements. Born: ~20 cm TL. Mature: ~85 ♂, 80–100 cm ♀. Max: 110 ♂, >120 cm TL ♀.
Identification. Fairly low slit-like eyes above ridges, small blade-like teeth. Conspicuous white-edged fins, not strongly falcate in adults. First dorsal origin over or behind pectoral free rear tips (except in newborns), anal fin much smaller than dorsals. Moderately long parabolic snout, broadly arched mouth, short anterior nasal flaps.
Distribution. West Pacific: China (incl. Taiwan), Korea, Japan.
Habitat. Temperate to subtropical continental shelf, close inshore to >100 m offshore.
Behaviour. Unknown.
Biology. Ovoviviparous, no yolk-sac placenta, 8–22 pups/litter (mean 10), increasing with size of female. In East China Sea, mating occurs June–September (mostly June–August) and pupping in June–August (mainly June).

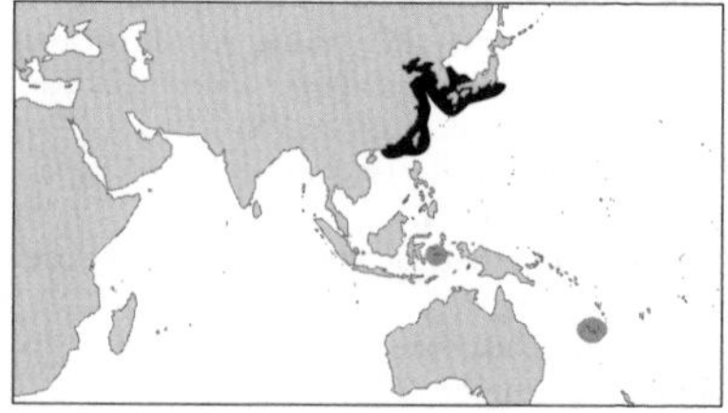

Status. IUCN Red List: Not Evaluated. A common utilised fisheries bycatch.

Whitefin Topeshark *Hemitriakis leucoperiptera* Plate 44

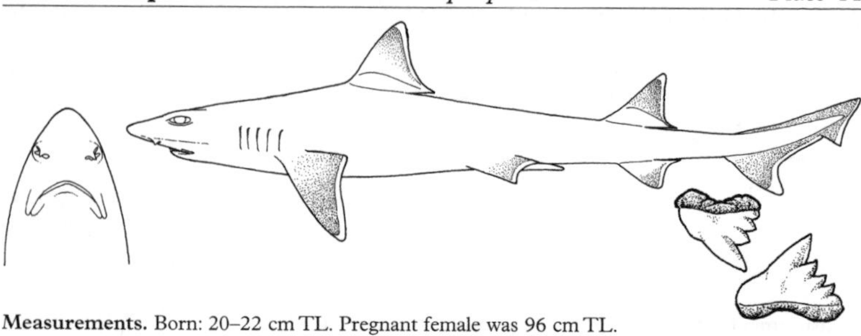

Measurements. Born: 20–22 cm TL. Pregnant female was 96 cm TL.
Identification. Fins strongly falcate with conspicuous white edges. First dorsal origin over pectoral inner margins, second smaller, anal fin much smaller than first dorsal. Moderately long parabolic

snout (no dark stripe below), broadly arched mouth, small blade-like teeth, moderately elongated dorsolateral eyes with prominent subocular ridges. Nostrils with short anterior nasal flaps.

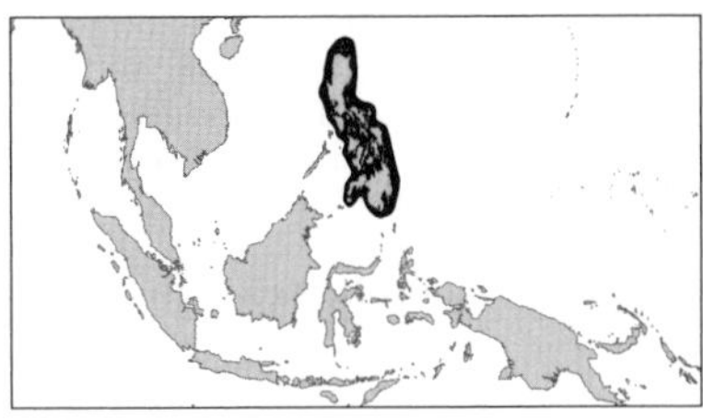

Distribution. Philippines.
Habitat. Coastal waters to 48 m.
Behaviour. Unknown.
Biology. Live-bearing 12 pups/litter.
Status. IUCN Red List: Endangered. Until recently known from just two specimens collected >50 years ago, a few additional specimens caught in last three years.

Ocellate Topeshark *Hemitriakis* sp. A — Plate 44

Measurements. Term foetuses were 16–18 cm TL. Adult size not known, presumably <100 cm TL.
Identification. Unborn (near term) young very brightly patterned with distinct '0'-shaped dark spots with light centres and dark bars and saddles.

Distribution. Philippines (Dumaguete, Negros Island).
Habitat. Probably inshore marine waters.
Behaviour. Unknown.
Biology. Bears at least four live young.
Status. IUCN Red List: Not Evaluated. Known only from a litter of near-term foetuses.

Blacktip Topeshark or Pencil Shark *Hypogaleus hyugaensis* — Plate 44

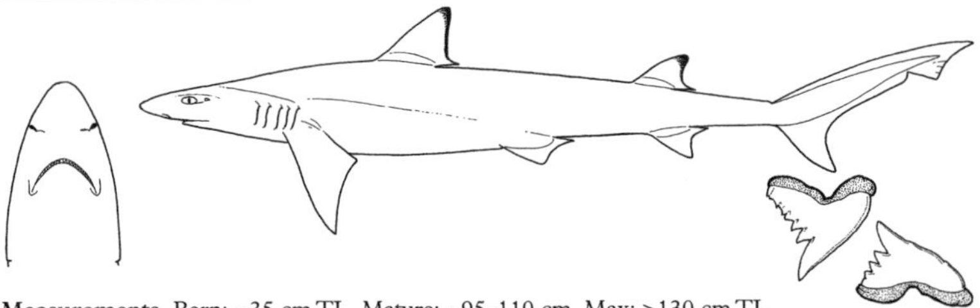

Measurements. Born: ~35 cm TL. Mature: ~95–110 cm. Max: >130 cm TL.
Identification. Fairly slender, medium-sized bronzy to grey-brown (lighter underside) with dusky dorsal and upper caudal fin tips, particularly in young. Long broadly pointed snout, large oval eyes, blade-like teeth. Second dorsal fin smaller than first but larger than anal fin, relatively short terminal caudal lobe (<half dorsal caudal margin).

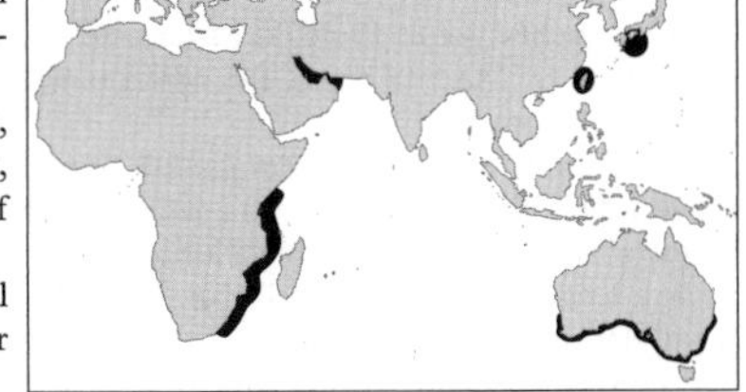

Distribution. West Indian Ocean: South Africa, Tanzania, Kenya. Northwest Pacific: Taiwan, Japan. Southwest Pacific: Australia. (Persian Gulf records may be based on *Paragaleus randalli.*)
Habitat. Tropical and warm-temperate continental shelves and uppermost slope, 40–230 m near bottom.

Behaviour. Unknown.
Biology. Viviparous, yolk-sac placenta, 8–11 pups/litter. Pups in December in South African waters and in February off Australia after estimated 15 month gestation. Eats bony fishes and cephalopods.
Status. IUCN Red List: Lower Risk (near threatened). Patchy distribution and generally low abundance, bycatch in fisheries.

Longnose Houndshark *Iago garricki* Plate 45

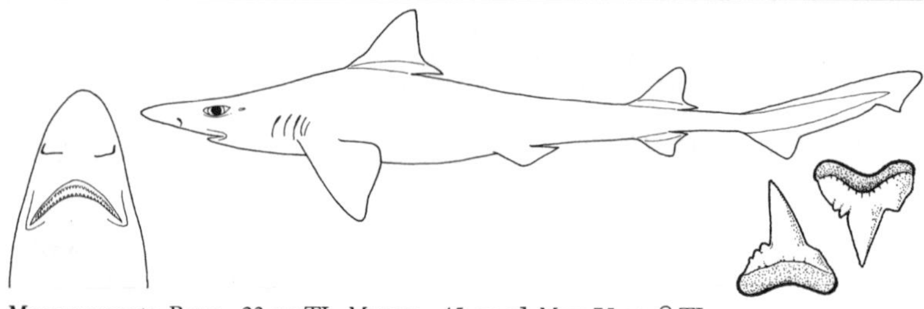

Measurements. Born: ~23 cm TL. Mature: ~45 cm ♂. Max: 75 cm ♀ TL.
Identification. Rather slender greyish-brown small houndshark with small blade-like teeth. Conspicuous black fin tips and upper rear edges, and pale free rear edges to dorsal fins. Tips and trailing edges of caudal and pectoral fins pale. First dorsal situated well forward (origin over pectoral fin inner margins), second nearly as large as first. Poorly developed ventral caudal lobe.

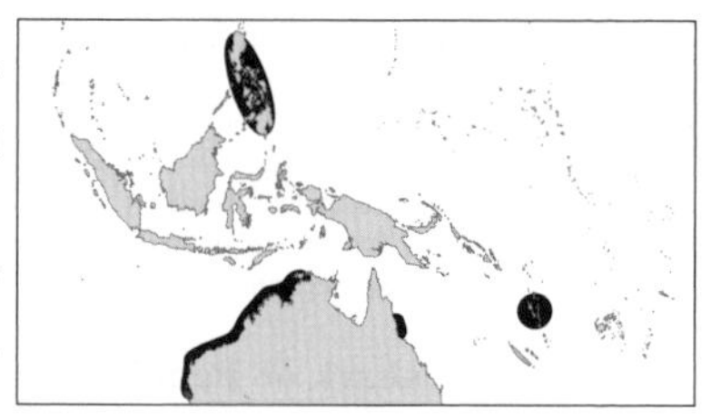

Distribution. Southwest Pacific: New Hebrides, Vanuatu, northern Australia, Philippines.
Habitat. Deepwater tropical upper continental and insular slopes, 250–475 m.
Behaviour. Unknown.
Biology. Viviparous, yolk-sac placenta, four or five pups/litter. Eats cephalopods.
Status. IUCN Red List: Least Concern. Only rarely taken by fisheries.

Bigeye Houndshark *Iago omanensis* Plate 45

Measurements. Born: >17 cm TL. Mature: <30 cm ♂, <40 cm ♀. Max: 58 cm ♀ TL. Males much smaller than females.
Identification. Slender, uniform greyish-brown above, lighter below, sometimes with darker dorsal fin margins. Large gill slits, width of longest nearly equal to length of large eyes, small blade-like teeth. Small dorsal fins, origin of first set far forward over pectoral fins bases. Small ventral caudal lobe.

Distribution. Indian Ocean: Red Sea, Gulf of Oman, Pakistan, southwest India, possibly Bay of Bengal and Myanmar (Burma). Sharks similar to *I. omanensis* in Bay of Bengal may be distinct.
Habitat. On or near bottom, continental shelf and slope, <110–1000+ m, possibly 2195 m in Red Sea. Often in warm, oxygen-poor conditions.
Behaviour. Unknown.
Biology. Viviparous, yolk-sac placenta, 2–10 pups/litter. Enlarged gills may permit survival in warm, low-oxygen, and probably hypersaline waters. Eat bony fishes and cephalopods.
Status. IUCN Red List: Not Evaluated.

Lowfin Houndshark *Iago* sp. A — Plate 45

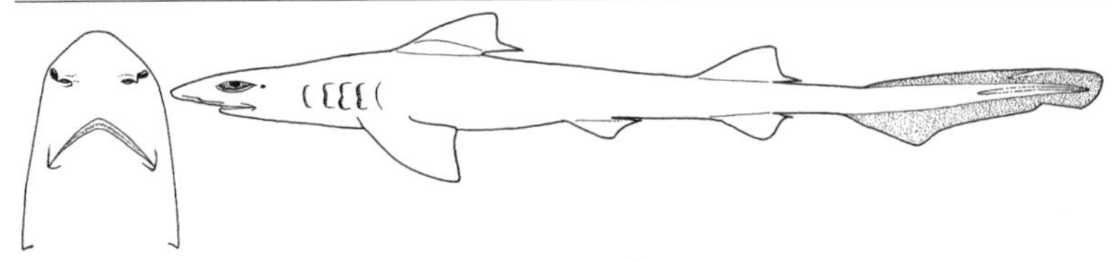

Measurements. Adolescent: 33 cm ♂. Mature: 41 cm ♀. Max: >41 cm TL.
Identification. Slender houndshark with large gill slits, similar to *Iago omanensis* but with shorter head, slimmer body, much lower dorsal fins, smaller pectoral fins, softer skin and muscles, and darker coloration.

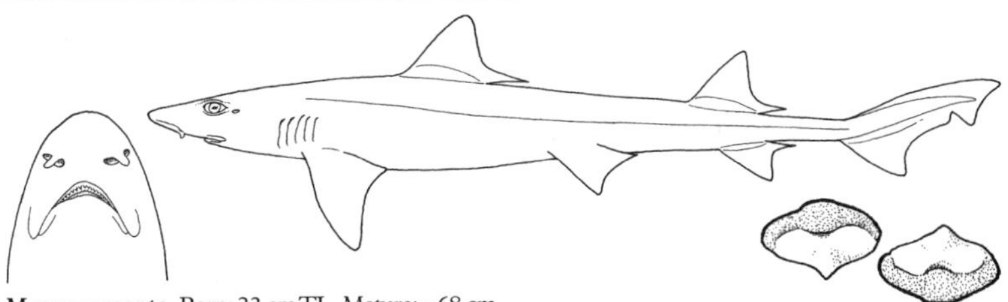

Distribution. Indian Ocean: Gulf of Aden, west coast of southern India.
Habitat. Outer continental shelf and upper slope below 183 m, also semipelagic. Collected at 0–330 m in pelagic trawls over bottom 500–2000 m deep. Large gills suggest that this species can tolerate water with low oxygen content.
Behaviour. Unknown.
Biology. Essentially unknown; confused with *I. omanensis* in Indian waters.
Status. IUCN Red List: Not Evaluated.

Gummy Shark *Mustelus antarcticus* — Plate 47

Measurements. Born: 33 cm TL. Mature: ~68 cm ♂, 80 cm ♀. Max: 148 ♂, 185 cm TL ♀.
Identification. Slender white-spotted bronze to greyish-brown houndshark (rarely with some black spots), pale underside. Angular mouth, flat 'pavement' of crushing teeth, long upper labial furrows, widely separated nostrils. Second dorsal fin nearly as large as first, fin margins not frayed, relatively small pectoral and pelvic fins.

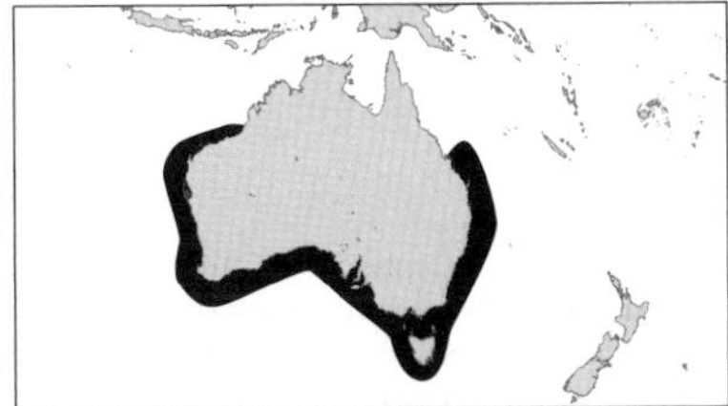

Distribution. Australia: the only *Mustelus* in temperate waters.

Habitat. On or near bottom, mainly on continental shelf, shore to ~80 m, also upper slope to 350 m.
Behaviour. No major migrations, some large females move long distances.
Biology. Aplacental viviparity, 1–38 pups/litter (depending on female size) in uterine compartments. Gestation 11 months. Females give birth annually or on alternate years in shallow coastal nurseries. Longevity 16 years. Eat crustaceans, marine worms and small fishes.
Status. IUCN Red List: Least Concern. Abundant, high productivity endemic harvested for meat throughout range. Management protects nursery grounds and large mature females; stocks have recovered from earlier depletion.

Starry Smoothhound *Mustelus asterias* Plate 46

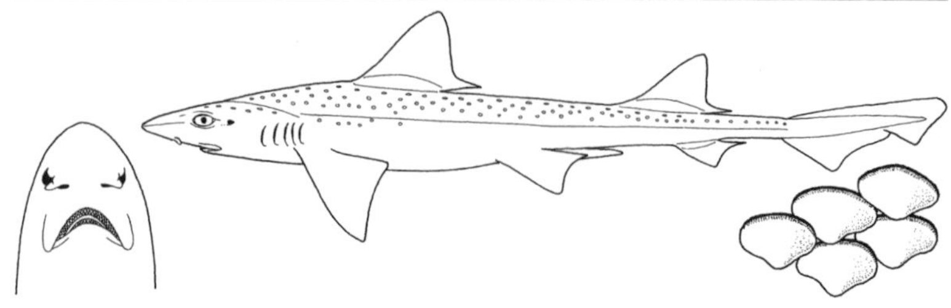

Measurements. Born: ~30 cm TL. Mature: 78–85 cm ♂, ~85 cm ♀. Max: 140 cm TL.
Identification. Only European smoothhound with many small white spots on grey or grey-brown sides and back (no dark spots or bars). Large, nostrils closer together than similar regional species. Pavement of flat teeth. Unfringed dorsal fins, pectorals and pelvics fairly small.
Distribution. Northeast Atlantic: North Sea to Canaries, Mediterranean, and Mauritania.

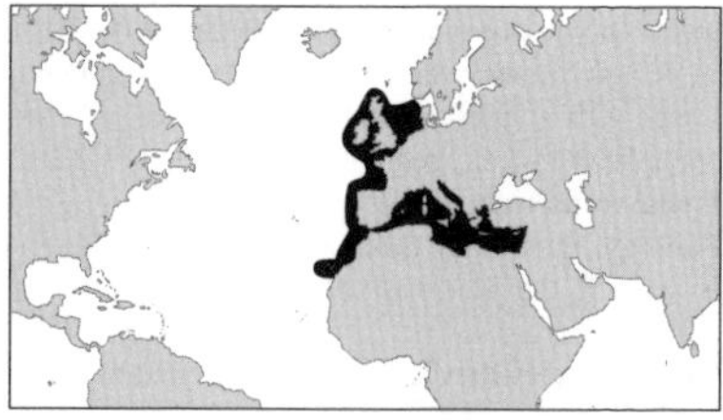

Habitat. Continental and insular shelves, on or near sand and gravel, intertidal to >100 m.
Behaviour. Swims actively in captivity. May migrate inshore in summer. Specialist feeder on crustaceans.
Biology. Ovoviviparous, no yolk-sac placenta, 7–15/litter, increasing with maternal size, born inshore in summer after ~one year gestation. Mature at two to three years.
Status. IUCN Red List: Least Concern. Fisheries bycatch with low market value. Taken by anglers, kept in aquaria.

Grey Smoothhound *Mustelus californicus* Plate 47

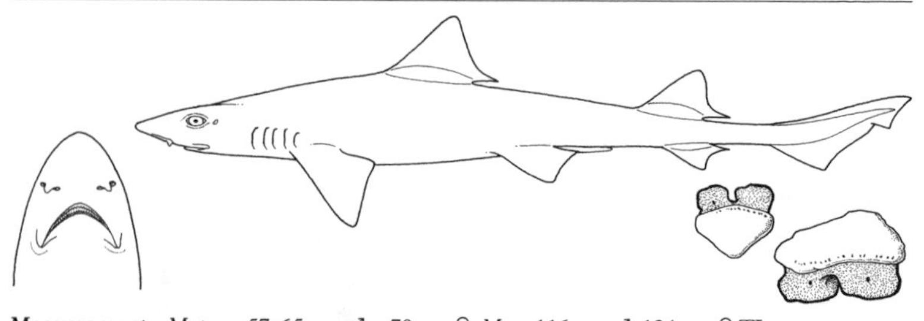

Measurements. Mature: 57–65 cm ♂, ~70 cm ♀. Max: 116 cm ♂, 124 cm ♀ TL.
Identification. Unpatterned, uniform grey above, light below. Short narrow head, broad internarial, fairly small eye, short mouth with pavement of flat teeth, equal-sized labial furrows. Triangular dorsal fins (first closer to pelvics than pectorals), poorly developed ventral caudal lobe.

Distribution. Northeast Pacific: northern California to Gulf of California.

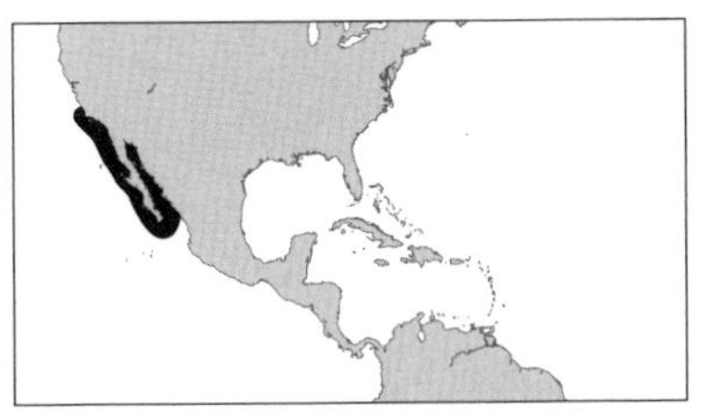

Habitat. Continental shelf and shallow muddy bays, inshore and offshore warm-temperate to tropical bottom of the continental shelves.

Behaviour. Summer visitor to north-central California waters, resident further south.

Biology. Viviparous, two to five pups/litter. Feeds mostly on crabs.

Status. IUCN Red List: Not Evaluated. Common where it occurs. Important in fisheries in south.

Dusky Smoothhound *Mustelus canis* Plate 46

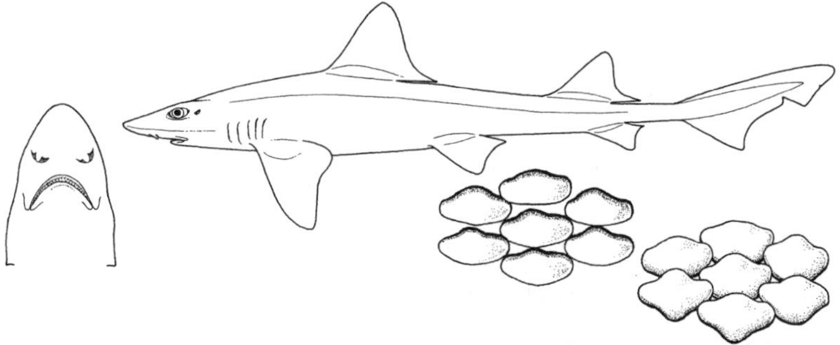

Measurements. Born: 34–39 cm TL. Mature: ~82 cm ♂, 90 cm ♀. Max: 150 cm TL.

Identification. Large, slender, usually unspotted grey above, white below. Short head and snout, low flat pavement teeth, broad internarial, large closely-spaced eyes, upper labial furrows longer than lowers, dorsal fins unfringed. Deeply notched caudal fin. Newborn young with dusky-tipped dorsal and caudal fins. Two subspecies (*Mustelus canis canis* on continental shelf and *M. c. insularis* in Caribbean) are very hard to distinguish. *M. c. canis* has slightly higher first dorsal fin and shorter terminal caudal lobe than latter, and indistinct juvenile white fin margins; latter are distinct in *M. c. insularis*.

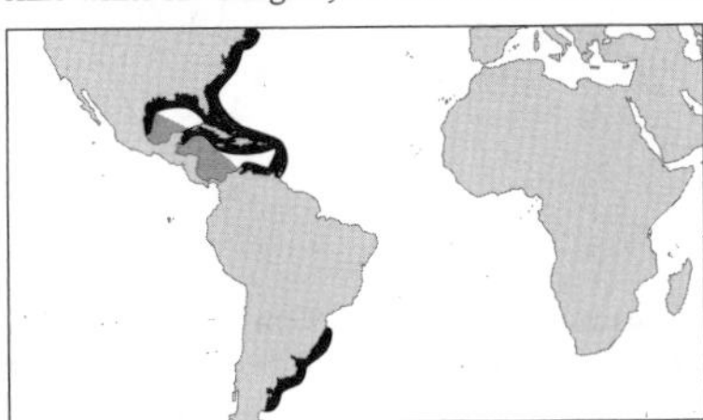

Distribution. Northwest Atlantic: Canada to Argentina, in several widely separated discrete populations.

Habitat. Continental shelf subspecies prefers mud and sand <18 m, but occurs to 200 m, rarely on the uppermost slopes to 360 m. Island subspecies prefers rough rocky bottoms on outer shelves and upper slopes at 137–808 m, most >200 m, and may occur in midwater off Cuba.

Behaviour. Very active shark, constantly patrolling for food that can be located even when hidden. Northern population migrates inshore and north in summer, south and offshore in winter. Not territorial in captivity, but larger animals dominant.

Biology. Viviparous, with yolk-sac placenta, bearing 4–20 pups/litter after ten month gestation. Matures in one to two years.

Status. IUCN Red List: Near Threatened. Abundant where it occurs. Largest females fished heavily by long-line and gill net, declines reported. Kept in public aquaria.

Sharpnose Smoothhound *Mustelus dorsalis* Plate 47

Measurements. Born: 21–23 cm TL. Mature: ~43 cm. Max: 64 cm TL.

Identification. Small fairly slender, plain grey or grey-brown (light below) with high-cusped teeth

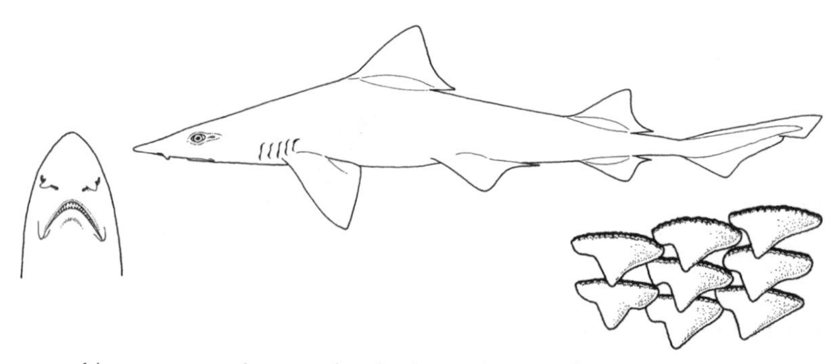

arranged in a pavement, a long, acutely pointed snout, large nostrils, very small widely-set eyes, lanceolate lateral trunk denticles and unfringed broadly triangular dorsal fins.

Distribution. East Pacific: Mexico to Ecuador.

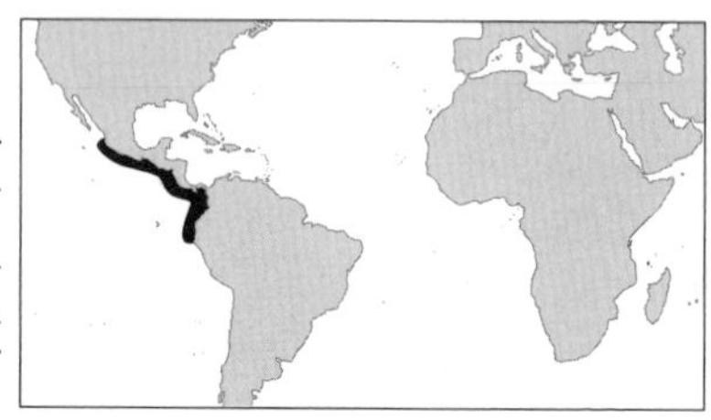

Habitat. Tropical inshore continental shelf.

Behaviour. Unknown.

Biology. Viviparous, yolk-sac placenta, four pups/litter. Eats manis shrimps and other crustaceans.

Status. IUCN Red List: Not Evaluated. Possibly uncommon compared with other houndsharks in region, but likely taken in fisheries with other species of 'tollo' (*Mustelus*) and used for meat.

Striped Smoothhound *Mustelus fasciatus* Plate 48

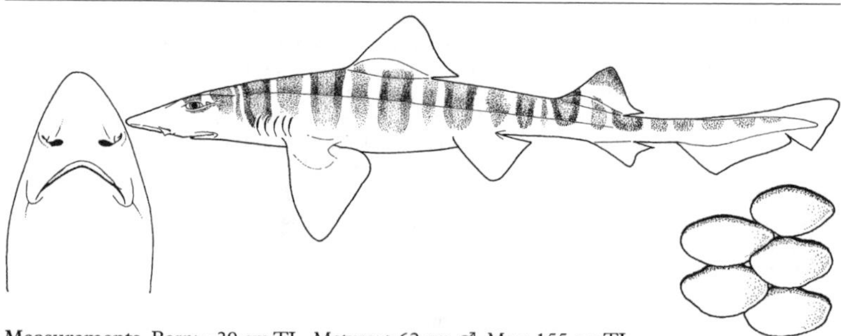

Measurements. Born: ~39 cm TL. Mature: >62 cm ♂. Max: 155 cm TL.

Identification. Vertical dark bars on unspotted body, at least in young. Fairly stocky with very long head and long, angular, acutely pointed snout, widely spaced nostrils. Eyes very small. Teeth arranged in pavement without cusps, crowns broadly rounded. Broadly triangular unfringed dorsal fins. Short caudal peduncle. Very distinctive, closest to *M. mento*, but has a longer and more angular head.

Distribution. Southwest Atlantic: southern Brazil, Uruguay, northern Argentina.

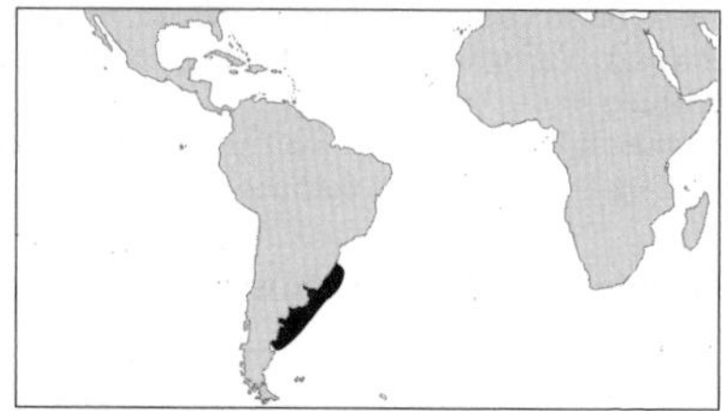

Habitat. Temperate continental shelf and uppermost slope, on bottom inshore and offshore, possibly from intertidal to 70 m, rarely at 10–500 m off southern Brazil.

Behaviour. Unknown.

Biology. Viviparous, with placenta.

Status. IUCN Red List: Not Evaluated. Uncommon to rare.

Spotless Smoothhound *Mustelus griseus* Plate 47

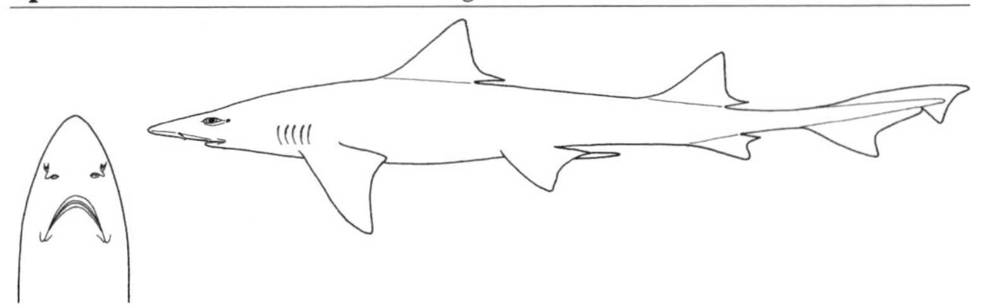

Measurements. Born: ~28 cm TL. Mature: 62–71 cm ♂, ~80 cm ♀. Max: 87 cm ♂, 101 cm TL ♀.
Identification. Plain grey or grey-brown, light below, moderate-sized with short head and snout, widely-spaced nostrils, small eyes, upper labial furrows equal or slightly shorter than lowers, low-crowned teeth with weak cusps. Unfringed dorsal fins, semifalcate ventral caudal lobe.
Distribution. Northwest Pacific: Japan, Koreas, China, Vietnam.

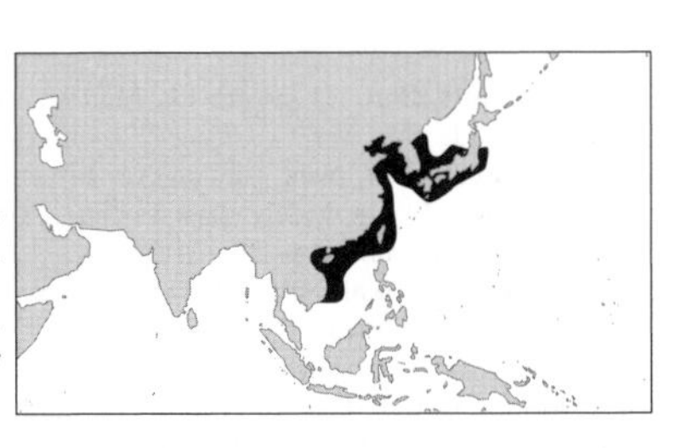

Habitat. Bottom inshore to at least 51 m.
Behaviour. Unknown.
Biology. Viviparous, yolk-sac placenta, 5–16 pups/litter, larger females bear more. Gestation ten months in Japanese waters; July mating, birth April–May.
Status. IUCN Red List: Not Evaluated. Common where it occurs. Important in fisheries off Japan and China (including Taiwan).

Brown Smoothhound *Mustelus henlei* Plate 47

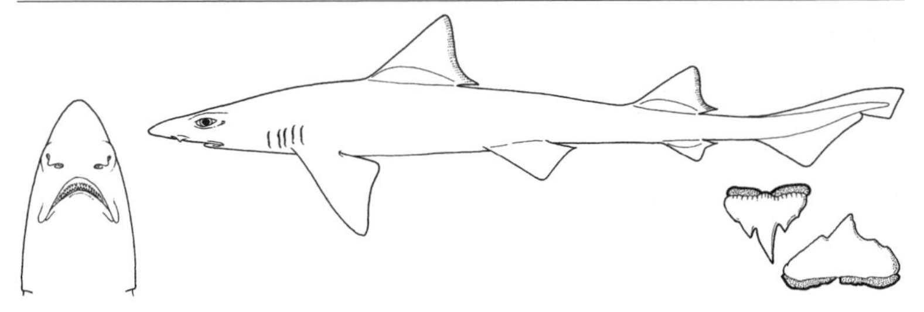

Measurements. Born: 19–21 cm TL. Mature: 52–66 cm ♂, 51–63 cm ♀. Max: 95 cm TL.
Identification. Unspotted, slender, usually iridescent bronzy-brown (occasionally greyish), white below. Short head, moderately long snout, large close-set eyes, nostrils widely spaced. High cusped 'pavement' teeth. Trailing dorsal fin edges appear broadly frayed with dark margins of bare ceratotrichia. Long caudal peduncle.
Distribution. East Pacific: northern California to Mexico; Ecuador and Peru.

Habitat. Continental shelf, intertidal to at least 200 m. Most abundant in enclosed, shallow, muddy bays. Most cold tolerant of three *Mustelus* spp. off California; resident in the north.
Behaviour. Active agile swimmers in captivity, mainly patrolling over bottom, sometimes in midwater and at surface. Often rest on seabed. One tagged individual migrated ~160 km in three months.

Biology. Viviparous, 3–5 pups/litter. Crustaceans most important prey, also polychaete worms and fish.
Status. IUCN Red List: Not Evaluated. Common to abundant where it occurs. Heavily fished. Kept in large aquaria.

Smalleye Smoothhound *Mustelus higmani* Plate 48

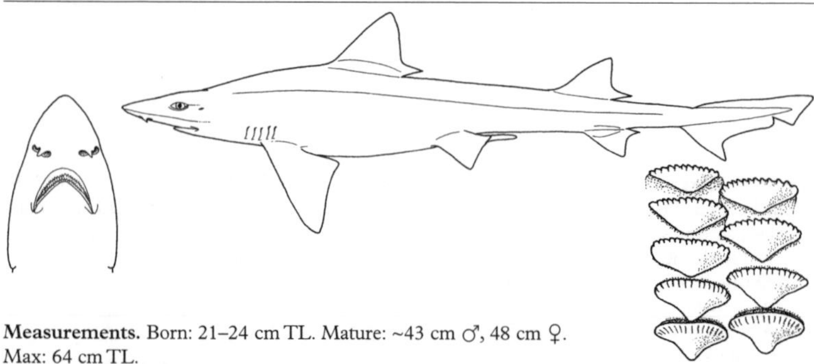

Measurements. Born: 21–24 cm TL. Mature: ~43 cm ♂, 48 cm ♀. Max: 64 cm TL.
Identification. Small, plain grey or grey-brown above, light below. Long, acutely pointed snout, fairly low-cusped teeth, widely spaced nostrils, very small widely separated eyes. Unfringed falcate dorsal fins, second almost as large as first.
Distribution. Tropical west Atlantic: northern Gulf of Mexico to Brazil.
Habitat. Continental shelf and upper slope. Close inshore to well offshore (130 m) on mud, sand and shell seabed off South America. Deep continental slope (at least 1281 m) in Gulf of Mexico.
Behaviour. Partial sexual segregation occurs.
Biology. Viviparous, yolk-sac placenta, one to seven pups/litter (usually three to five). Feeds mainly on crustacea.
Status. IUCN Red List: Not Evaluated. Locally common to abundant where it occurs. Fisheries unknown.

Spotted Estuary Smoothhound or Rig *Mustelus lenticulatus* Plate 47

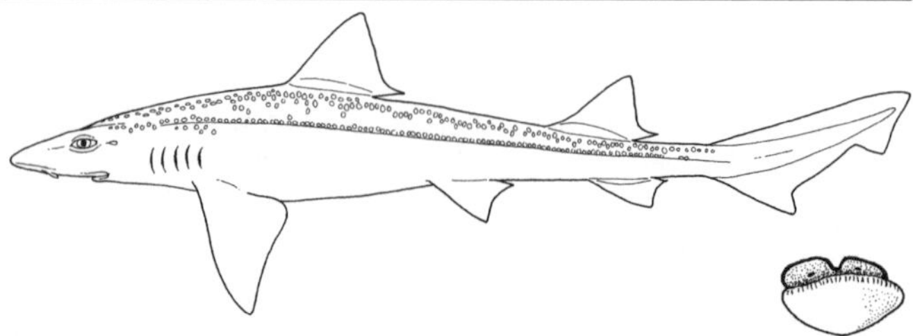

Measurements. Born: 20–32 cm TL. Mature: 78–89 ♂, 79–113 cm ♀. Max: 126 ♂, 151 cm ♀ TL.
Identification. Large, grey or grey-brown, lighter below, white-spotted (only *Mustelus* in range). Fairly large wide-set eyes, widely-spaced nostrils, long upper labial furrows. Dorsal fin margins unfrayed, relatively large pectoral and pelvic fins.
Distribution. New Zealand (five breeding stocks).
Habitat. Cold temperate insular shelf, close inshore to 220 m, deeper in winter.
Behaviour. Schooling species, move inshore to summer feeding and mating grounds. Segregated by

size and sex: immatures school separately from largely single-sex adult schools. Females migrate further than males.
Biology. Ovoviviparous, 2–23 pups/litter (increasing with female size), ~11 month gestation. Fast-growing, mature at five to eight years, longevity at least 15. Eat crustaceans, particularly crabs.
Status. IUCN Red List: Least Concern. Abundant, commercially-fished. Stocks rebuilt since fishery quotas introduced.

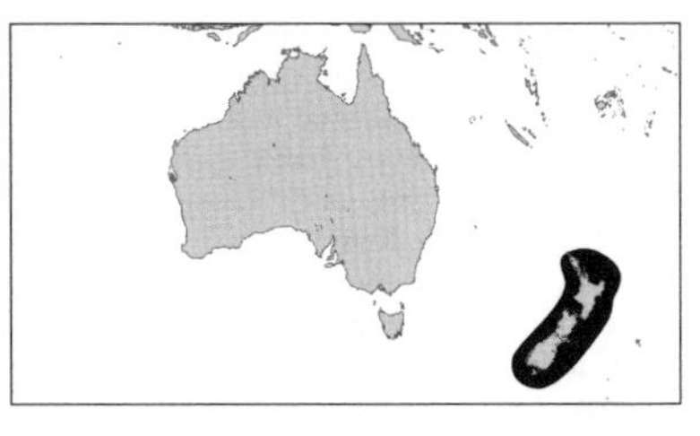

Sicklefin Smoothhound *Mustelus lunulatus* Plate 48

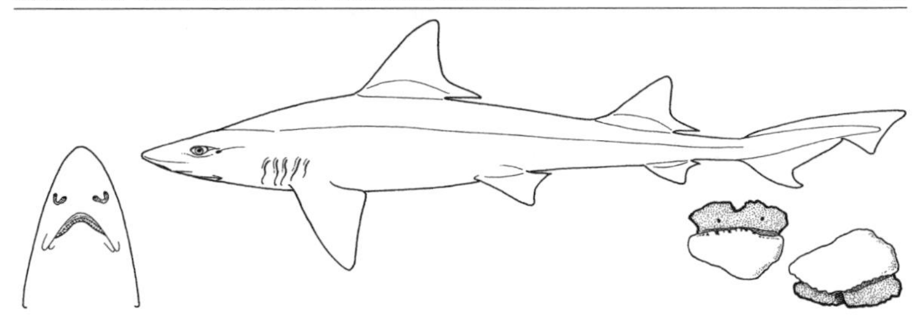

Measurements. Born: 32–35 cm TL. Mature: 70–83 cm ♂, reaching >110 cm; 97 cm ♀. Max: possibly 170 cm TL.
Identification. Large, unspotted grey or grey-brown above, light below. Distinguished from *Mustelus californicus* by more pointed snout, more widely separated eyes, shorter mouth, strongly falcate fins, and shorter upper labial furrow.
Distribution. Northeast Pacific: southern California (possibly only in warm summers) to Panama.
Habitat. Warm-temperate to tropical continental shelf, close inshore to well offshore.
Behaviour. Unknown.
Biology. Little-known. Presumably viviparous.
Status. IUCN Red List: Not Evaluated. Common to abundant where it occurs. Important in longline fisheries for meat in the Gulf of California.

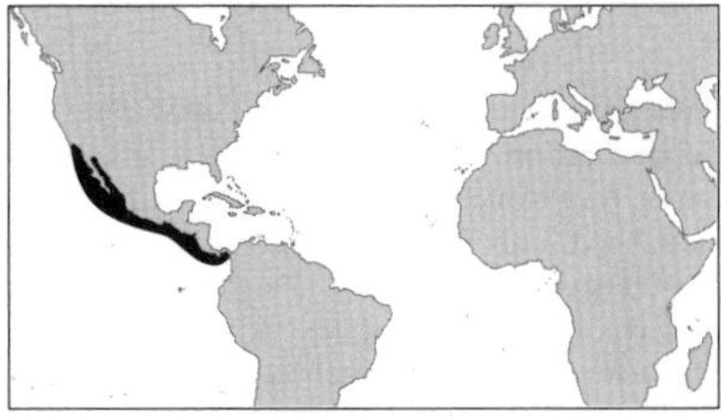

Starspotted Smoothhound *Mustelus manazo* Plate 47

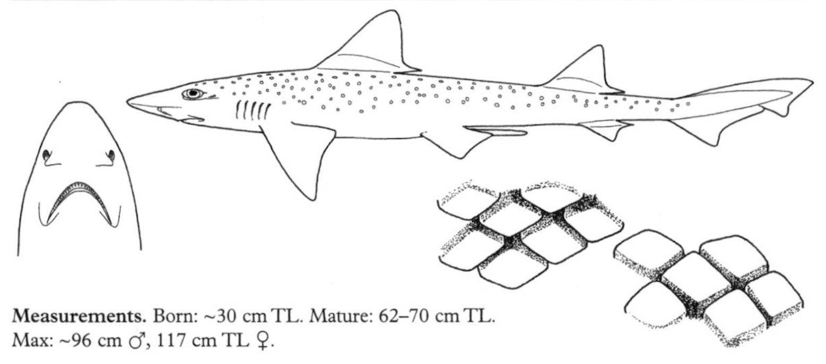

Measurements. Born: ~30 cm TL. Mature: 62–70 cm TL. Max: ~96 cm ♂, 117 cm TL ♀.

Identification. Medium-sized, grey to grey-brown, light below, many white spots. Fairly close-set nostrils, large close-set eyes, unfringed dorsal fins, relatively small pectoral and pelvic fins. Only white-spotted smoothhound in its range.

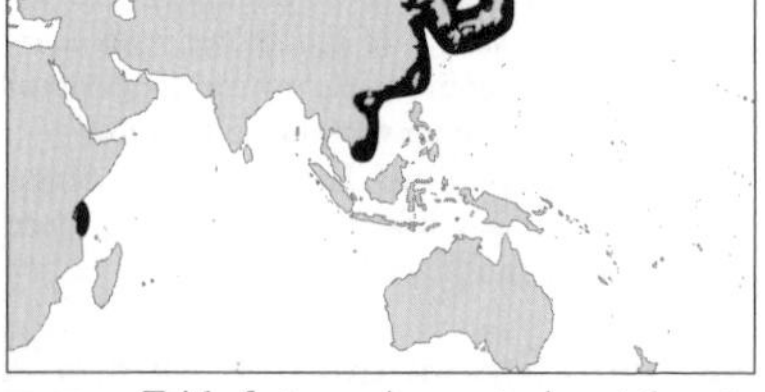

Distribution. Northwest Pacific: southern Siberia, Japan, Koreas, China, (incl. Taiwan), Vietnam. West Indian Ocean: Kenya.
Habitat. Temperate–tropical continental shelf, intertidal to close inshore, on mud and sand.
Behaviour. Not recorded.
Biology. Ovoviviparous, no yolk-sac placenta, 1–22 pups/litter (mostly two to six, average ~five), increasing with size of mother, born in spring after ~ten month gestation. Adults mate in summer. Fairly fast-growing, maturing at three to four years. Eat mostly bottom invertebrates, particularly crustacea.
Status. IUCN Red List: Not Evaluated. Generally abundant where it occurs. Important in longline fisheries for meat off Japan, also China and the Koreas.

Speckled Smoothhound *Mustelus mento* Plate 47

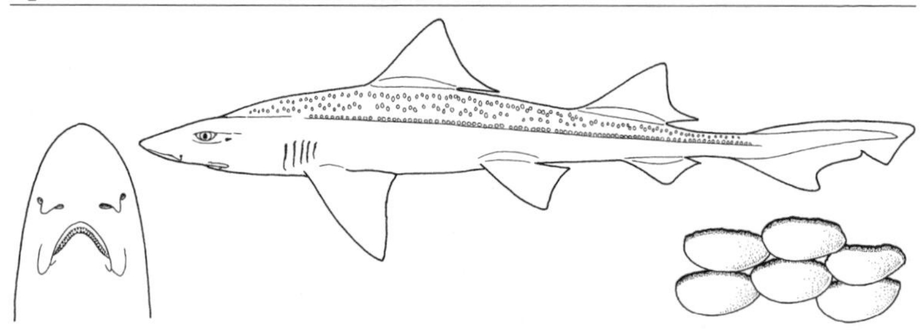

Measurements. Born: ~30 cm TL. Mature: 65–76 cm ♂, 86–90 cm ♀. Max: 130 cm TL.
Identification. White-spotted grey to grey-brown stocky body, light below. Bluntly angular, short snout, short caudal peduncle, and a crushing 'pavement' of broadly rounded uncusped teeth. Vertical dark bands only in young. Fairly similar to *Mustelus fasciatus* but with shorter more rounded head.
Distribution. Southeast Pacific: Galapagos Islands, Peru, Chile, Juan Fernandez Island. ?Southwest Atlantic: Argentina.
Habitat. Temperate continental and insular shelves, inshore and offshore 16–50 m.
Behaviour. Unknown.
Biology. Ovoviviparous, seven pups/litter.
Status. IUCN Red List: Not Evaluated. This and other species of 'tollo' (*Mustelus*) are important in fisheries and used for meat.

Venezuelan Dwarf Smoothhound *Mustelus minicanis* Plate 48

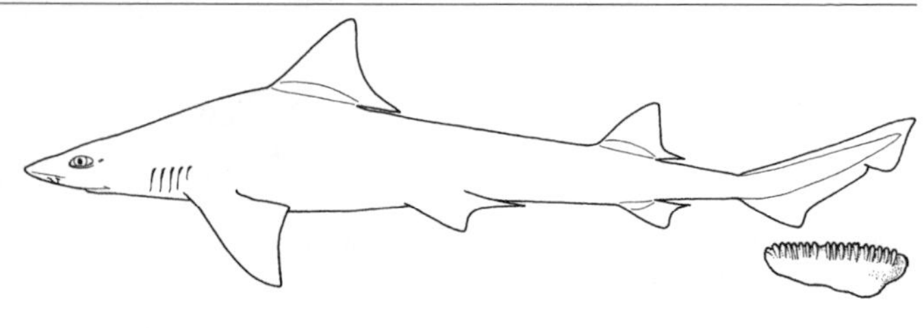

Measurements. Born: ~22 cm TL. Mature: 47 cm ♂. Max: 57 cm TL ♀.
Identification. Small, stout, uniform grey without spots. Short head and snout, widely-spaced nostrils, very large fairly close-set eyes, upper labial furrows somewhat longer than lowers, pavement of low-crowned teeth with weak cusps, unfringed dorsal fins, poorly-developed ventral caudal lobe. Newborns have dusky-tipped dorsal and caudal fins.

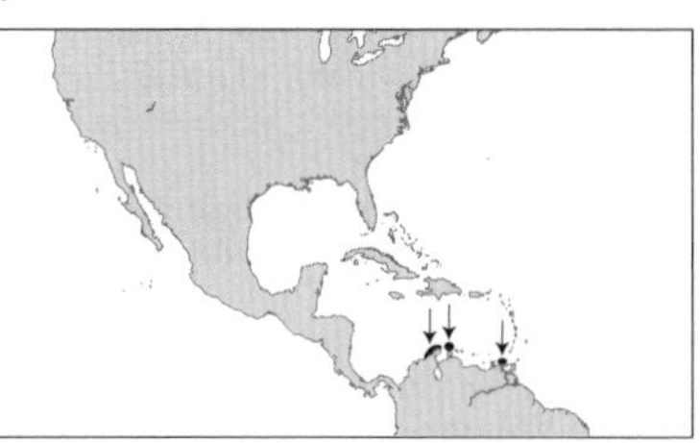

Distribution. West Atlantic: Colombia and Venezuela.
Habitat. Offshore outer continental shelf, 71–183 m.
Behaviour. Unknown.
Biology. Poorly known. Viviparous (placental), one known litter of five.
Status. IUCN Red List: Not Evaluated. Possibly rare or uncommon; only nine specimens known. Possibly caught in offshore trawl fisheries.

Arabian, Hardnose or Moses Smoothhound *Mustelus mosis* Plate 48

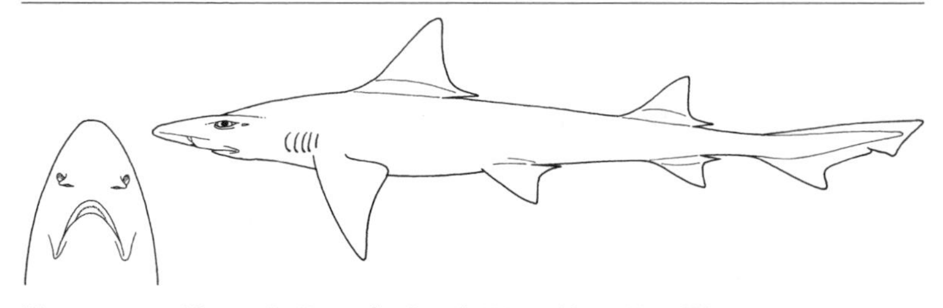

Measurements. Mature: 63–67 cm ♂ and reach 106 cm. Max: 150 cm TL.
Identification. Large, fairly slender unspotted grey or grey-brown, light below. Short head and snout, broad internarial, large fairly close-set eyes, labial furrows similar length, low-crowned teeth with weak cusps, unfringed dorsal fins, semifalcate ventral caudal lobe. White-tipped first dorsal and black-tipped second dorsal and caudal fins in South Africa.
Distribution. West Indian Ocean: Red Sea, Gulf, India, Pakistan and Sri Lanka; kwaZulu-Natal, South Africa. The only *Mustelus* in most of above.
Habitat. Continental shelf, bottom inshore and offshore, some on coral reefs.
Behaviour. Unrecorded.
Biology. Viviparous, six to ten pups/litter. Eats small bottom fish, molluscs, crustaceans.
Status. IUCN Red List: Not Evaluated. Common where present. Fished for food off Pakistan and India. Does well in captivity.

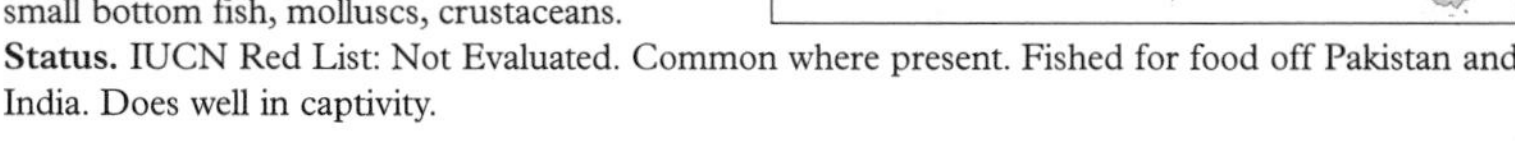

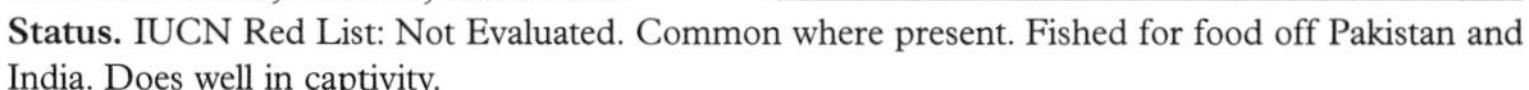

Smoothhound *Mustelus mustelus* Plate 46

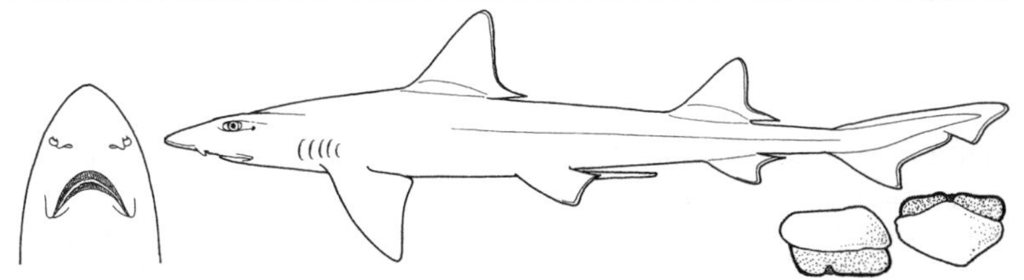

Measurements. Born: ~39 cm TL. Mature: 70–74 cm ♂, ~80 cm ♀. Max: at least 110 ♂, 164 cm TL ♀.
Identification. Large, fairly slender, usually unspotted grey to grey-brown, with short head and snout, broad internarial, large close-set eyes, upper labial furrows slightly longer than lowers, low-crowned teeth with weak cusps, unfringed dorsal fins, and semifalcate ventral caudal lobe. Occasionally dark spots.

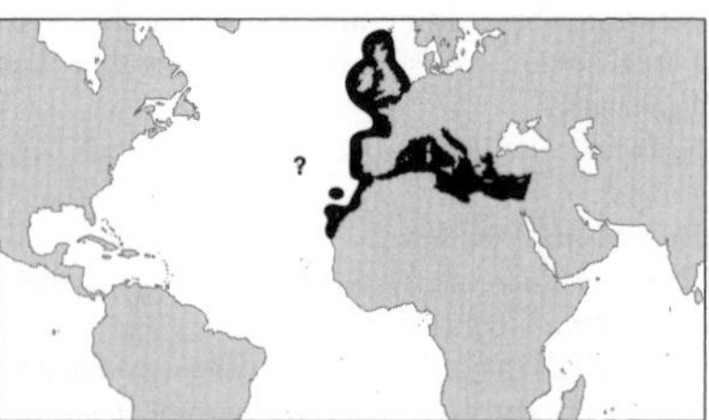

Distribution. Temperate East Atlantic: UK to Mediterranean, Morocco, Canaries, possibly Azores, Madeira. Angola to South Africa, including Indian Ocean coast.
Habitat. Continental shelves and upper slopes, usually 5–50 m, often in intertidal and to at least 350 m.
Behaviour. Prefers swimming near bottom, but sometimes found in midwater.

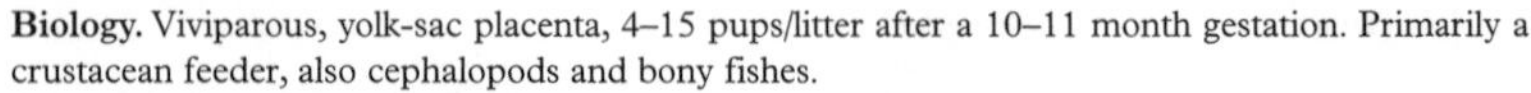

Biology. Viviparous, yolk-sac placenta, 4–15 pups/litter after a 10–11 month gestation. Primarily a crustacean feeder, also cephalopods and bony fishes.
Status. IUCN Red List: Not Evaluated. Common to abundant. Very important in fisheries in European, Mediterranean and West African waters. Taken in bottom trawls, fixed nets and line gear. Used fresh, froze, dried-salted and smoked for food; liver for oil, and for fishmeal. Also taken by sports fishers. Kept in aquaria.

Narrowfin or Florida Smoothhound *Mustelus norrisi* Plate 48

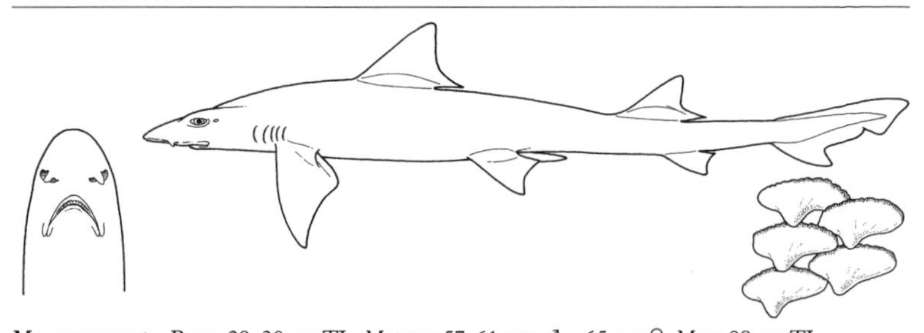

Measurements. Born: 29–30 cm TL. Mature: 57–61 cm ♂, ~65 cm ♀. Max: 98 cm TL.
Identification. Fairly large, grey and unspotted. Short narrow head, narrow internarial, relatively large eye, long mouth, upper labial furrows shorter than lowers, strongly falcate fins. Dorsal and caudal fin tips dusky in newborns. Smaller, slenderer and narrower-headed than *Mustelus canis*.
Distribution. West Atlantic: USA (Gulf of Mexico); southern Caribbean coast of Colombia and Venezuela; southern Brazil.

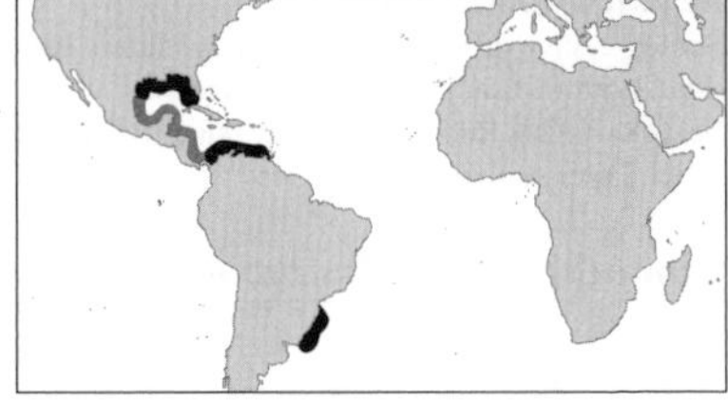

Habitat. Continental shelf, sandy and mud bottom, close inshore to >84 m, mostly <55 m.
Behaviour. Segregates by size and sex off Florida: adult males are inshore in winter. May be migratory in Gulf of Mexico, moving inshore to <55 m in winter, offshore in other seasons.
Biology. Viviparous, yolk-sac placenta, 7–14 pups/litter. Eats crabs and shrimp, also small fishes.
Status. IUCN Red List: Not Evaluated. Common where it occurs. No fisheries information.

Whitespot Smoothhound *Mustelus palumbes* Plate 46

Measurements. Mature: 76–88 cm ♂, 79–102 cm ♀. Max: 120 cm TL.
Identification. Large, uniform grey to grey-brown, usually white-spotted (only white-spotted smoothhound in southern African waters). Relatively broad internarial space, unfringed dorsal fins

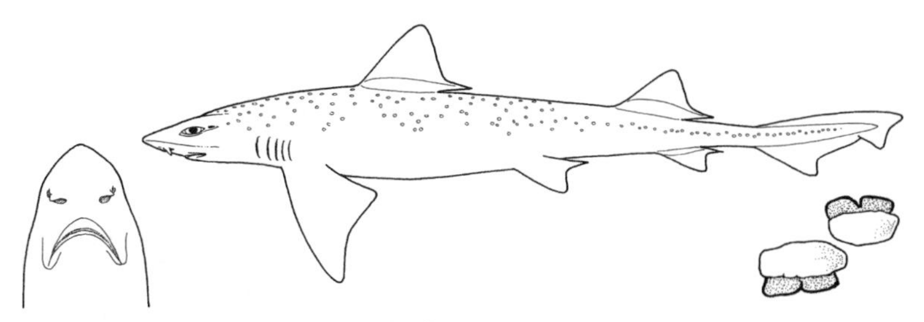

and relatively large pectoral and pelvic fins (larger than those of *Mustelus asterias*, *M. manazo* and *M. mustelus*).
Distribution. Southeast Atlantic, southwest Indian Ocean: Namibia, South Africa, southern Mozambique.
Habitat. Continental shelf and upper slope, intertidal to >360 m, on or near sand and gravel bottom.
Behaviour. Unknown.
Biology. Ovoviviparous, no yolk-sac placenta, three to eight pups/litter. Eats crabs and other crustacea.
Status. IUCN Red List: Not Evaluated. Common offshore. Taken by sports anglers and as bycatch, usually discarded.

Blackspot Smoothhound *Mustelus punctulatus* Plate 46

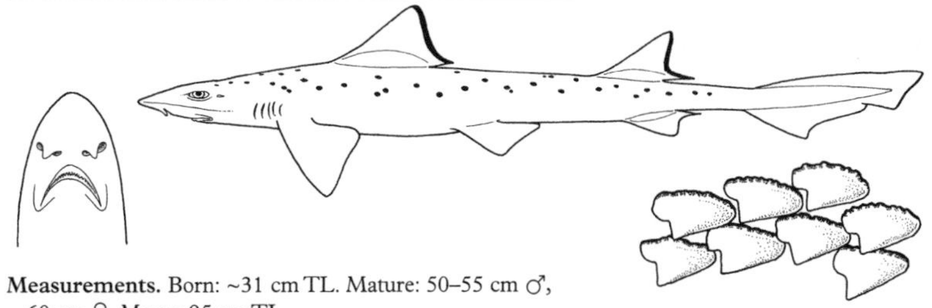

Measurements. Born: ~31 cm TL. Mature: 50–55 cm ♂, ~60 cm ♀. Max: >95 cm TL.
Identification. Smallish, slender, usually black-spotted on grey (light below) with short head and snout, narrow head and internarial, large eyes, upper labial furrows somewhat longer than lowers, pavement of low-crowned teeth with weak cusps, and prominently fringed dorsal fins. Often misidentified as more common *Mustelus mustelus*.

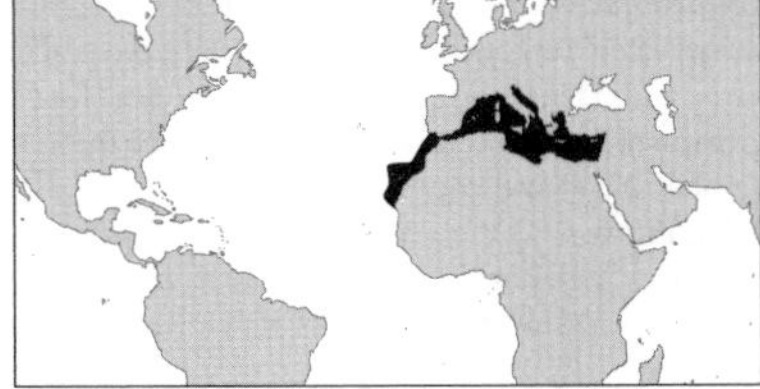

Distribution. East Atlantic: western Sahara, Mediterranean.
Habitat. Inshore continental shelf on bottom.
Behaviour. Unknown.
Biology. Little known, presumably viviparous. Probably a crustacean-feeder.
Status. IUCN Red List: Not Evaluated. Presumably fished for food, but recorded as *M. mustelus*.

Narrownose Smoothhound *Mustelus schmitti* Plate 46

Measurements. Born: ~36 cm TL. Mature: 62–67 cm ♂, 60–67 cm ♀. Max: 109 cm TL.
Identification. White-spotted with a very narrow internarial space and dorsal fins with naked

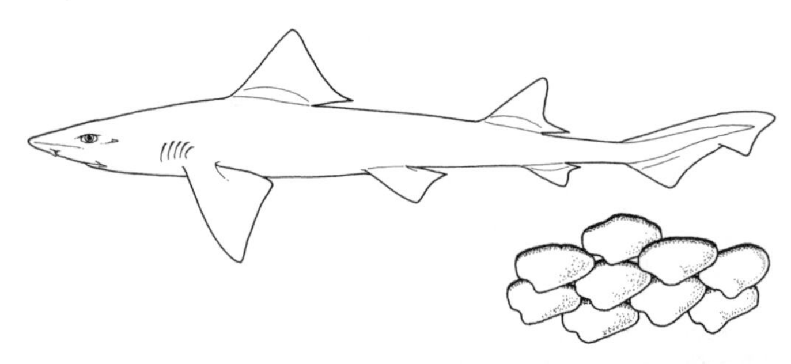

ceratotrichia on margins, giving a dark, frayed appearance. Easily distinguished from other white-spotted *Mustelus*.

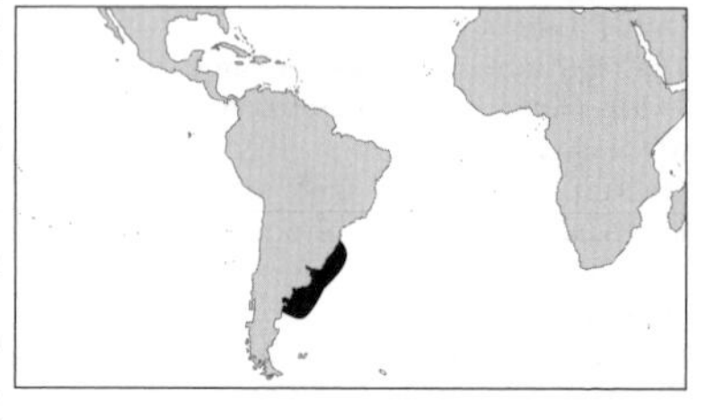

Distribution. Southwest Atlantic: southern Brazil to northern Argentina.
Habitat. Offshore continental shelf, 60–195 m.
Behaviour. Migrates seasonally from Brazil in winter to Uruguay and Argentina in summer.
Biology. Ovoviviparous, no placenta, two to seven pups/litter. Eats crabs, other crustacea, many other invertebrates and small benthic fishes.
Status. IUCN Red List: Not Evaluated. Important commercial fish species, exploited throughout range, including breeding and nursery grounds, seriously depleted at least in northern part of range.

Gulf of Mexico Smoothhound *Mustelus sinusmexicanus* Plate 48

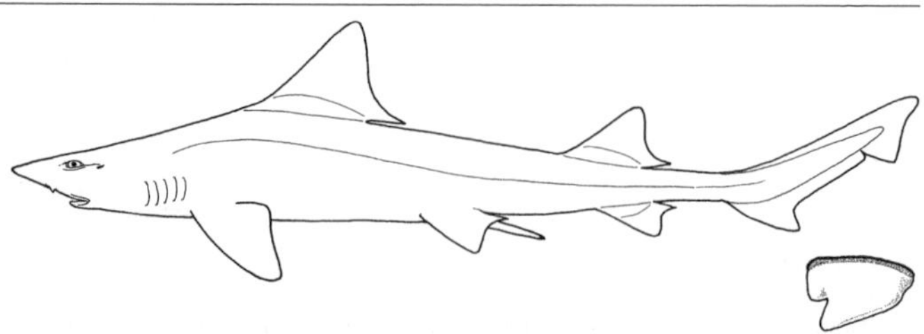

Measurements. Born: 39–43 cm TL. Mature: 70<80 cm ♂, <118 cm ♀. Max: ~140 cm TL.
Identification. Large, grey, unspotted, short head and snout, broad internarial, large close-set eyes, upper labial furrows longer than lowers, pavement of high-crowned teeth (unlike sympatric *Mustelus canis* and *M. norrisi*), unfringed dorsal fins, first large, and a non-falcate but strong and moderately expanded ventral caudal lobe. Young have dusky-tipped dorsal and caudal fins.

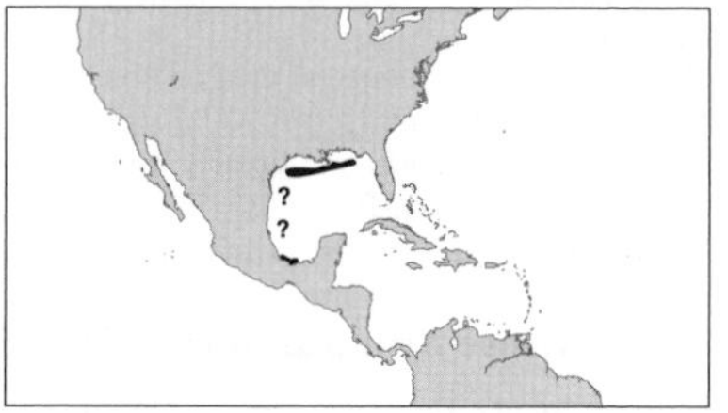

Distribution. Gulf of Mexico: USA and Mexico.
Habitat. Offshore continental shelf and upper slope, 36–229 m, mostly 42–91 m. Not in shallow water.
Behaviour. Unknown.
Biology. Poorly-known. Apparently placental viviparous, eight pups/litter.
Status. IUCN Red List: Not Evaluated. Recently described endemic, confused with *M. canis*.

Humpback Smoothhound *Mustelus whitneyi* Plate 48

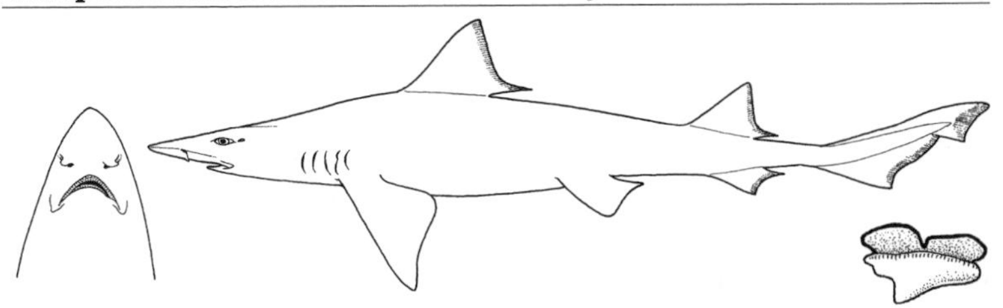

Measurements. Born: ~25 cm TL. Mature: <68 cm ♂. Max: at least 87 cm TL.
Identification. Grey, unspotted, rather stocky (almost hump-backed) with strongly cuspidate teeth, short caudal peduncle, and a dark margin of bare ceratotrichia on trailing edges of dorsal fins, giving them a broadly frayed appearance.

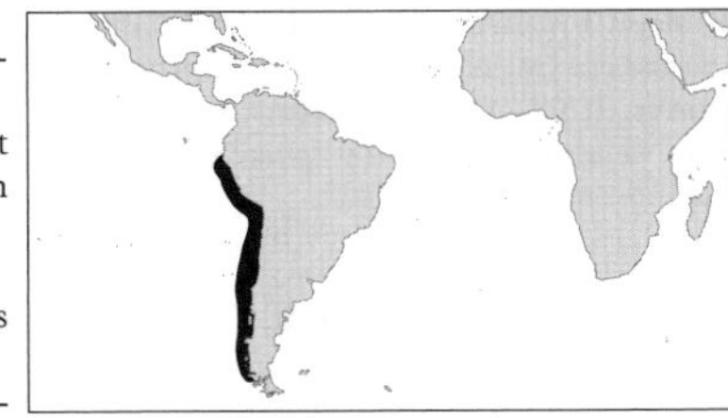

Distribution. Southeast Pacific: Peru to southern Chile.
Habitat. Continental shelf, 16–211 m, most common at 70–100 m. Prefers rocky bottom around islands.
Behaviour. Unknown.
Biology. Viviparous, five to ten pups/litter. Eats crabs, manis shrimp and small bony fishes.
Status. IUCN Red List: Not Evaluated. Common off Peru, probably fished with other 'tollo' (*Mustelus*).

Grey Gummy Shark *Mustelus* sp. A Plate 46

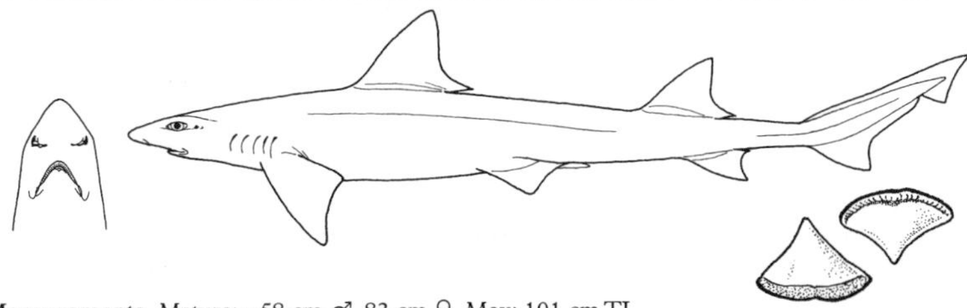

Measurements. Mature: ~58 cm ♂, 83 cm ♀. Max: 101 cm TL.
Identification. Slender, uniform unspotted bronzy houndshark with rather high-cusped crushing teeth arranged in a 'pavement', lower labial furrows longer than upper. Deeply notched caudal fin.
Distribution. Northern (tropical) Australia. 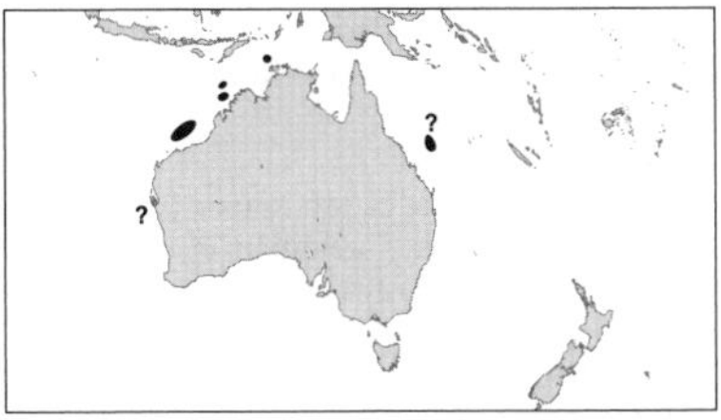
Habitat. Deep continental shelf, 100–300 m.
Behaviour. Unknown.
Biology. Largely unknown, bearing 6–24 pups/litter, presumably fairly productive.
Status. IUCN Red List: Least Concern. Taken as fisheries bycatch, but much of range unfished.

Whitespotted Gummy Shark *Mustelus* sp. B Plate 46

Measurements. Mature: ~60–70 cm ♀. Max: 117 cm TL.
Identification. Slender greyish-brown with numerous white spots. Flattened low-cusped crushing teeth arranged in 'pavement'. Dorsal fins sometimes have dark tips. Juveniles more strongly marked,

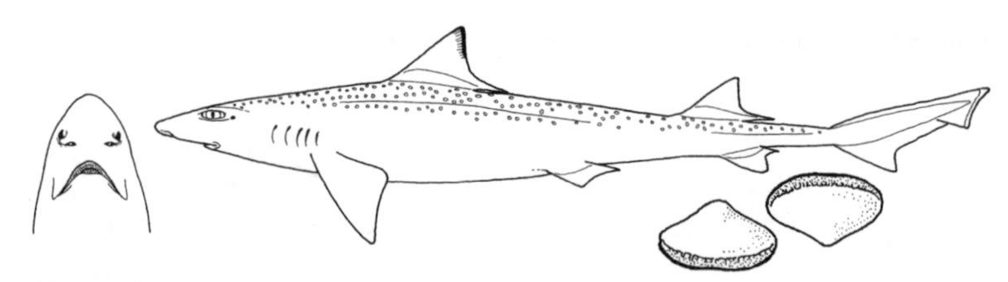

other fins also have pale tips and trailing fin edges are pale. Slight variation between eastern and western forms. Very similar to cold water *Mustelus antarcticus*, but range does not overlap.

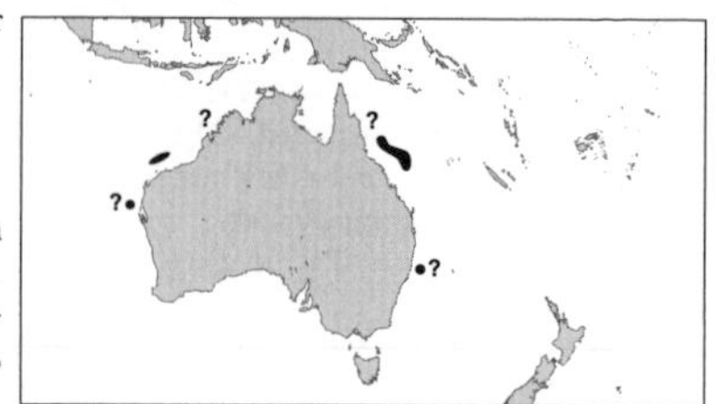

Distribution. Northern (tropical) Australia.
Habitat. Deep continental shelf, 120–400 m.
Behaviour. Unknown.
Biology. Viviparous, 4–17 pups/litter. Feeds on crustaceans (mainly crabs), fish and cephalopods.
Status. IUCN Red List: Least Concern. Probably widespread. Irregular bycatch in fisheries, but most of range unfished.

Flapnose Houndshark *Scylliogaleus quecketti* — Plate 45

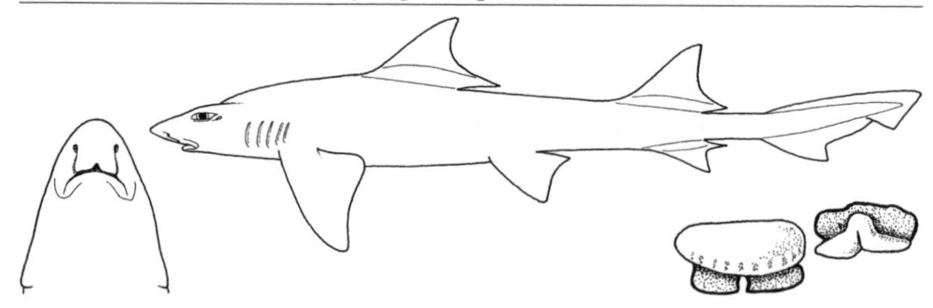

Measurements. Born: ~34 cm TL. Mature: <70 cm ♂, <80 cm ♀. Max: 102 cm TL.
Identification. Blunt short snout, large fused nasal flaps expanded to cover mouth, nasoral grooves, small, blunt pebble-like teeth. Second dorsal fin as large as first and much larger than anal fin. Grey above, cream below, newborns with white rear edges on dorsal, anal and caudal fins.
Distribution. West Indian Ocean: South Africa.
Habitat. Inshore continental shelf, at the surfline and close offshore.
Behaviour. Unknown.
Biology. Bears litters of two to four (usually two or three) pups after a gestation of nine to ten months. Feeds primarily on crustaceans (including lobsters), also squid.
Status. IUCN Red List: Vulnerable. Restricted distribution is heavily fished; targeted for overseas trade in shark meat.

Sharpfin Houndshark *Triakis acutipinna* — Plate 45

Measurements. Mature: 90 cm ♂, 102 cm TL ♀.
Identification. Short broadly rounded snout. Widely separated anterior nasal flaps do not reach the mouth. Long upper labial furrows reach the lower symphysis of the mouth. Teeth not blade-like.

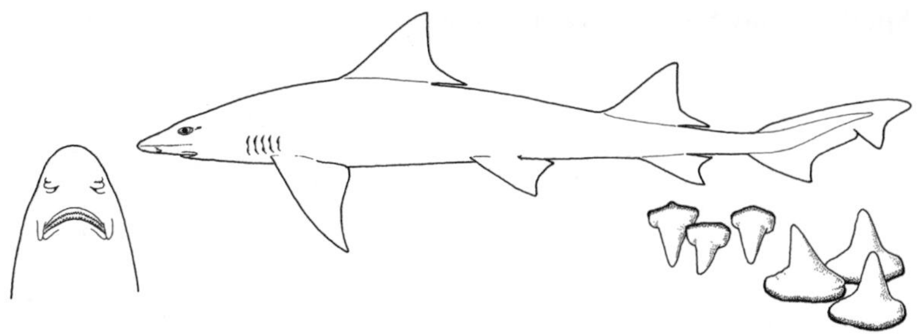

Narrow fins, pectorals narrowly falcate, first dorsal with an abruptly vertical posterior margin. No spots or bands.

Distribution. Southeast Pacific: Ecuador.
Habitat. Tropical continental waters
Behaviour. Unknown.
Biology. Largely unknown.
Status. IUCN Red List: Vulnerable. Rare, known from two specimens in heavily fished waters.

Spotted Houndshark *Triakis maculata* Plate 45

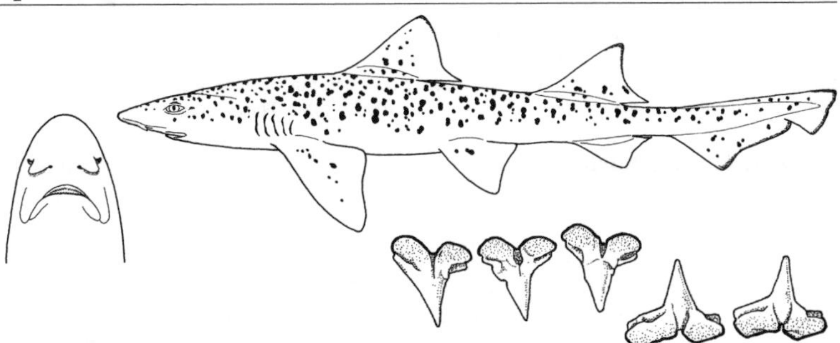

Measurements. Born: 40–43 cm TL. Max: 180, possibly 240 cm TL.
Identification. Very stout (for a houndshark) usually with many small black spots. Some are unspotted (plain females may have spotted young). Short broadly rounded snout, widely separated lobate anterior nasal flaps not reaching mouth, long upper labial furrows reaching lower symphysis of mouth, teeth not blade-like. Broad fins, pectorals broadly falcate and first dorsal fin with a backwards-sloping rear edge.

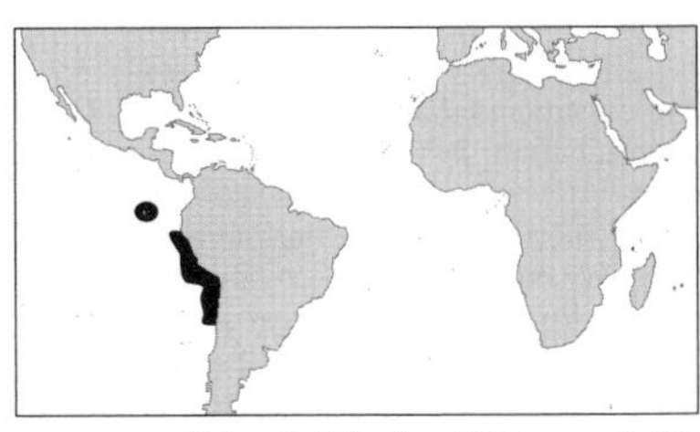

Distribution. East Pacific: Peru to northern Chile, Galapagos Islands.
Habitat. Inshore temperate continental shelf.
Behaviour. Unknown.
Biology. Poorly known. Probably ovoviviparous; one record of 14 pups/litter.
Status. IUCN Red List: Not Evaluated. Possibly uncommon. Taken in fisheries off Peru, probably Chile, used for food.

Spotted Gully Shark *Triakis megalopterus* Plate 45

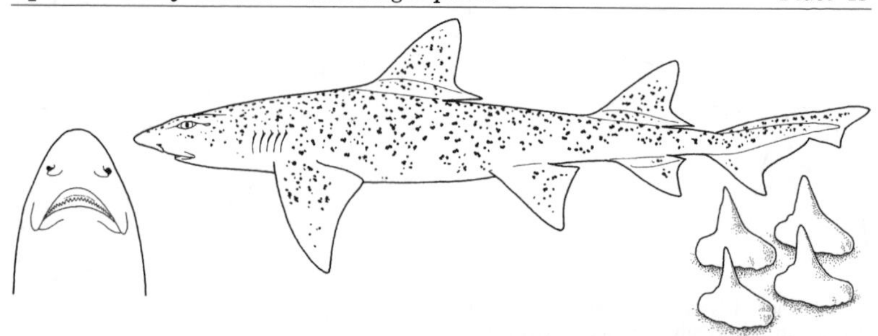

Measurements. Born: 30–32 cm TL. Mature: 130–140 cm ♂, 140–150 cm ♀. Max: poss. 170 cm TL.
Identification. Grey-bronzy, usually many black spots, white below, no or few spots in young (some adults also plain). Broad blunt snout, large mouth, pointed small teeth. Broad large fins. High interdorsal ridge. Short, heavy caudal peduncle.
Distribution. Southern Angola to South Africa.
Habitat. Shallow inshore to surfline, prefers sandy shores and rocks and crevices in shallow bays.
Behaviour. Schools in summer, often many pregnant females present. Actively patrol very close to bottom in captivity, sometimes in midwater but rarely in the open.
Biology. Ovoviviparous, no yolk-sac placenta, six to ten pups/litter. Eats crabs, bony fishes, small sharks.
Status. IUCN Red List: Near Threatened. Locally common but range is heavily fished. Caught by sports anglers and commercial fishers, but low commercial value. Hardy in captivity.

Banded Houndshark *Triakis scyllium* Plate 45

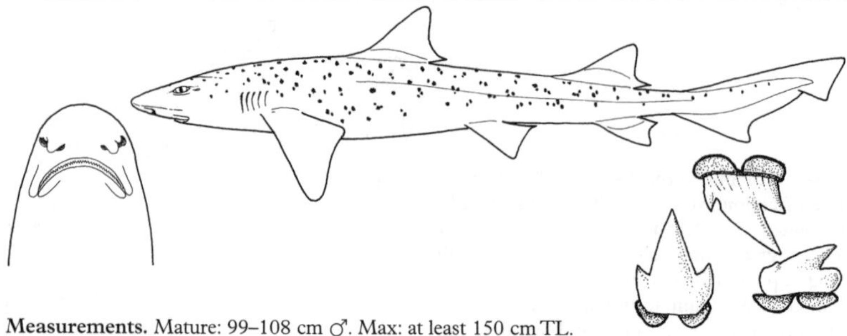

Measurements. Mature: 99–108 cm ♂. Max: at least 150 cm TL.
Identification. Fairly slender, sparsely scattered small black spots and broad dusky saddles in young, spots fading (sometimes absent) in adults. Relatively narrow fins, pectorals triangular in adults, first dorsal rear margin almost vertical. Short broadly rounded snout, partly blade-like teeth.

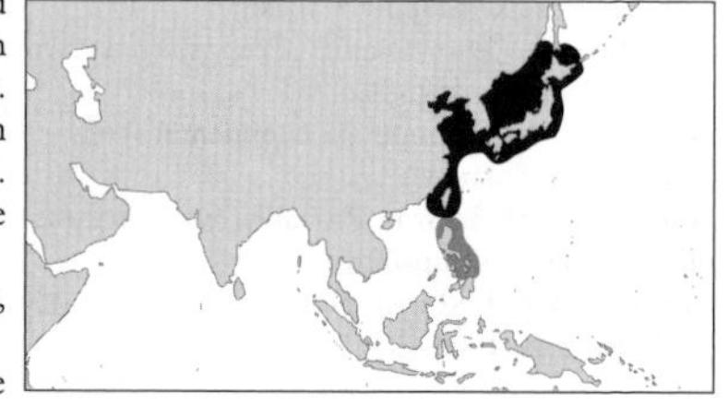

Distribution. Northwest Pacific: southern Siberia, Japan, Koreas, China (including Taiwan).
Habitat. Continental and insular shelves, close

inshore on or near bottom. Often in estuaries and shallow bays, on sand, seaweed and eel-grass flats.
Behaviour. Seldom gregarious, but some gather in seabed resting areas.
Biology. Ovoviviparous, no yolk-sac placenta, 10–20 pups/litter. Feeds on small fishes and probably invertebrates.
Status. IUCN Red List: Not Evaluated. Common to abundant where it occurs. Often fished.

Leopard Shark *Triakis semifasciata* Plate 45

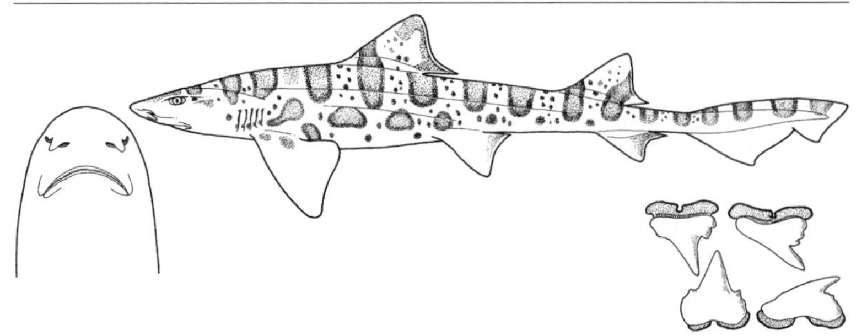

Measurements. Born: ~20 cm TL. Mature: 70–119 ♂, 110–129 cm ♀. Max: 150 ♂, 180 cm ♀ TL.
Identification. Unique pattern of striking black saddle marks and spots on a pale tan to greyish background, fading to whitish below. Saddle-marks become light-centred in adults.
Distribution. Northeast Pacific, Oregon (USA) to Gulf of California (Mexico).

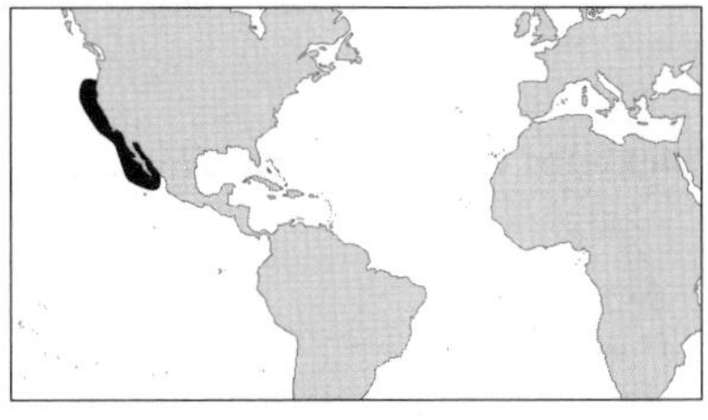

Habitat. Cool to warm temperate inshore and offshore continental shelf, commonest on or near bottom from intertidal to 4 m, also recorded to 91 m. Prefers shallow, enclosed, muddy bays, flat sandy areas, mud flats, and rock-strewn areas near rocky reefs and kelp beds.
Behaviour. Active, strong-swimming shark. Forms large nomadic schools (sometimes with smoothhounds *Mustelus californicus* and *M. henlei* and piked dogfish *Squalus acanthias*). Most have a small range, some travel up to 150 km. Sometimes rest on sand among rocks.
Biology. Ovoviviparous, no yolk-sack placenta, 4–29 pups/litter. Slow growing and late to mature. Feeds opportunistically on bottom animals, including burrowing invertebrates. Diet varies with size and season.
Status. IUCN Red List: Lower Risk (conservation dependent). Common to abundant where it occurs. Valuable flesh, intensive commercial and sports fishing led to declines, but USA population now well-managed. Status in Mexico unknown. Very hardy, readily adapts to captivity when young, maintained for >20 years in aquaria.

Hemigaleidae: Weasel Sharks

Identification. Small to moderate sized sharks with horizontal oval eyes with nicitating eyelids, small spiracles, long labial furrows, precaudal pits, spiral intestinal valves, and large second dorsal fins. Caudal fin with a strong ventral lobe and wavy dorsal edge.
Biology. Live-bearing (viviparous), with yolk sac placenta. Some are specialist feeders on cephalopods, others have a very varied diet.
Status. World-wide in fossil record, now restricted to shallow tropical continental and insular shallow shelf waters of east Atlantic and Indo-west Pacific. Common and important in inshore fisheries.

Hooktooth Shark *Chaenogaleus macrostoma* Plate 49

Measurements. Born: ≥20 cm TL. Mature: 68–97 cm ♂. Max: ~100 cm TL.
Identification. Small slender, light grey or bronzy shark, often without prominent markings, sometimes black second dorsal and terminal lobe of caudal fin. Extremely long, hooked smooth-edged lower teeth protrude from very long mouth. Fairly long angular snout, large lateral eyes with nictitating eyelids, small spiracles, gill slits at least twice eye length. Second dorsal ~two-thirds size of first, origin opposite or slightly ahead of smaller anal fin origin.

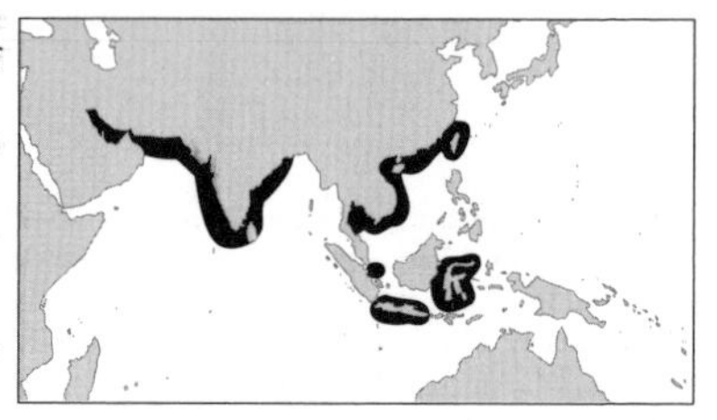

Distribution. Indo-west Pacific: The Gulf to Indonesia and China.
Habitat. Continental and insular shelves, to 59 m.
Behaviour. Unknown.
Biology. Poorly known; four pups/litter.
Status. IUCN Red List: Not Evaluated. Commonly caught in fisheries.

Sicklefin Weasel Shark *Hemigaleus microstoma* Plate 49

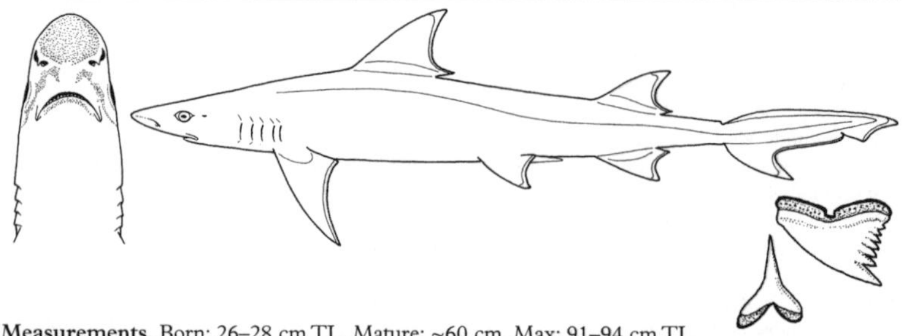

Measurements. Born: 26–28 cm TL. Mature: ~60 cm. Max: 91–94 cm TL.
Identification. Small, slender light grey or bronzy shark, dorsal fins with light margins and tips, sometimes white spots on sides. Fairly long rounded snout, short gill slits, very short arched mouth. Pelvic and dorsal fins, and ventral caudal lobe strongly falcate, second dorsal ~ two-thirds size of first, opposite origin of equal-sized anal fin.

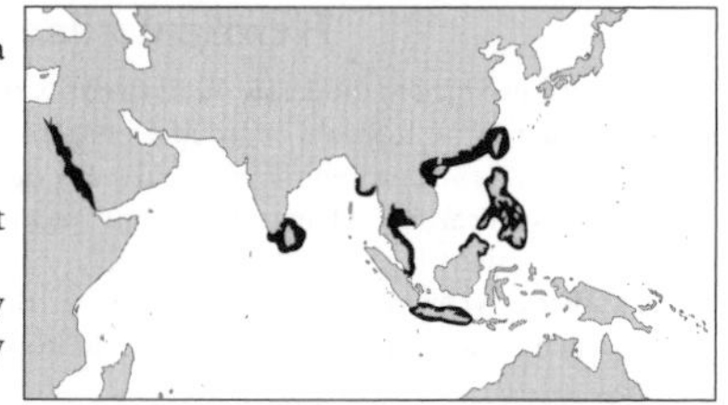

Distribution. Tropical Indo-west Pacific: India to Philippines and China.
Habitat. Continental shelf, on or near bottom.
Behaviour. Unknown.
Biology. Poorly known, probably a specialist feeder on cephalopods.
Status. IUCN Red List: Least Concern. Heavily fished, but apparently reproduces quite rapidly with fast population growth rates.

Australian Weasel Shark *Hemigaleus* sp. A Plate 49

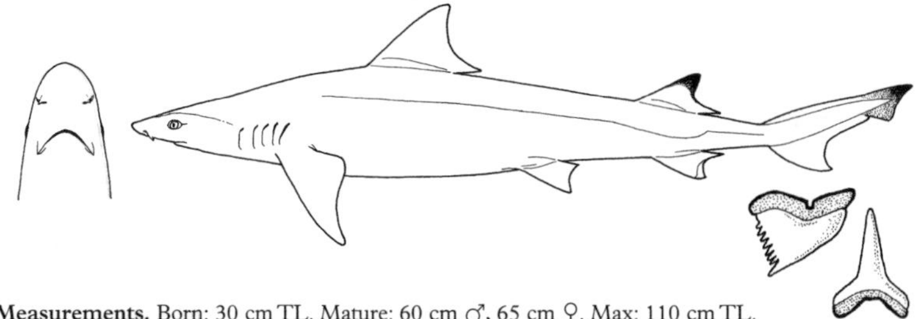

Measurements. Born: 30 cm TL. Mature: 60 cm ♂, 65 cm ♀. Max: 110 cm TL.
Identification. Similar to *Hemigaleus microstoma*, with pelvic and dorsal fins, and ventral caudal lobe strongly falcate. Has different coloration (second dorsal and caudal fins have dark margins and tips), vertebral counts and tooth counts.
Distribution. Northern Australia, Papua New Guinea.
Habitat. Continental shelf, on or near bottom, 12–167 m.
Behaviour. Essentially unknown
Biology. Litters of 4–19 pups after six month gestation. Specialist feeder on cephalopods, also takes crustaceans and echinoderms.
Status. IUCN Red List: Not Evaluated. Identified as *H. microstoma* in Australia, but differs from typical *H. microstoma* and is possibly a separate species.

Snaggletooth Shark *Hemipristis elongatus* Plate 49

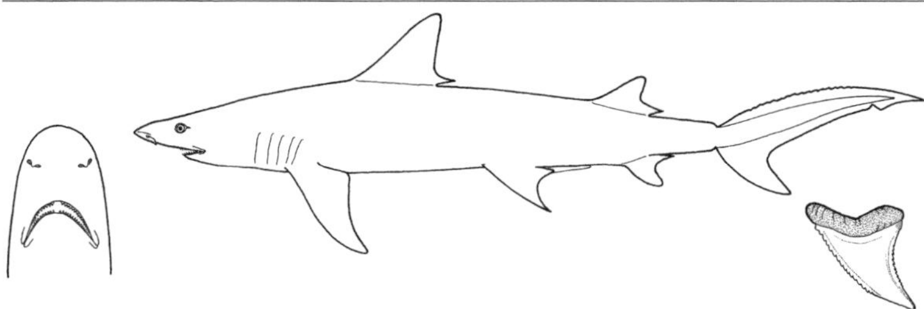

Measurements. Born: 45–52 cm TL. Mature: ~110–120 cm. Max: 230–240 cm TL.
Identification. A slender light grey or bronzy shark without prominent markings. Broadly rounded long snout, large curved saw-edged upper teeth, hooked lower teeth that protrude from mouth, long gill slits, strongly curved concave fins. Tip of second dorsal and terminal lobe of caudal fin sometimes with a dusky blotch, more prominent in juveniles than adults.
Distribution. Indo-west Pacific: South Africa to northern Australia, Philippines and China.
Habitat. Continental and insular shelves, 1–132 m.
Behaviour. Poorly known.
Biology. 2–11 pups/litter, increasing with female size, pregnancy 7–8 months, possibly breeds in alternate years. Feeds on cephalopods and fishes.
Status. IUCN Red List: Vulnerable. Important, intensively fished commercial species with declining stocks reported.

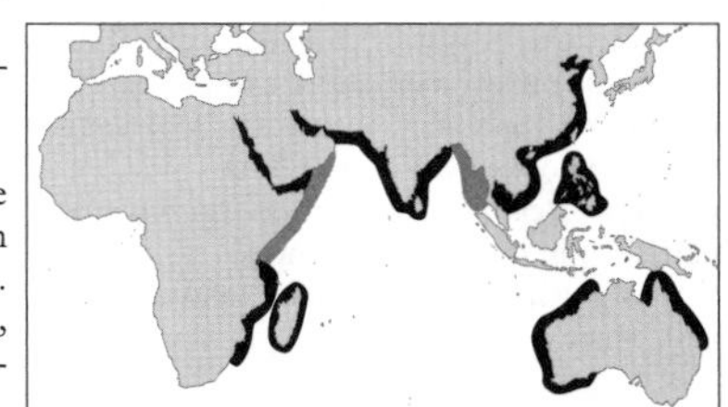

Whitetip Weasel Shark *Paragaleus leucolomatus* Plate 49

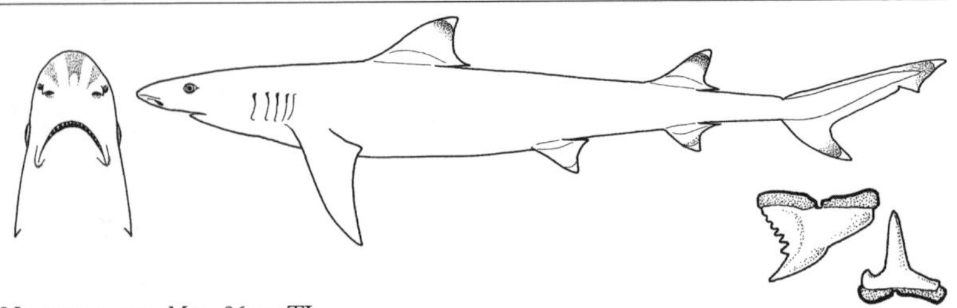

Measurements. Max: 96 cm TL.

Identification. Slender dark grey shark, with a long snout, large oval eyes, fairly long mouth with small serrated upper teeth, erect-cusped lower teeth. Prominent white tips and margins on most fins, large second dorsal black-tipped. Gills twice eye length. Pelvic and dorsal fins and ventral caudal lobe not falcate. Underside of snout with broad dusky patches, otherwise white below.

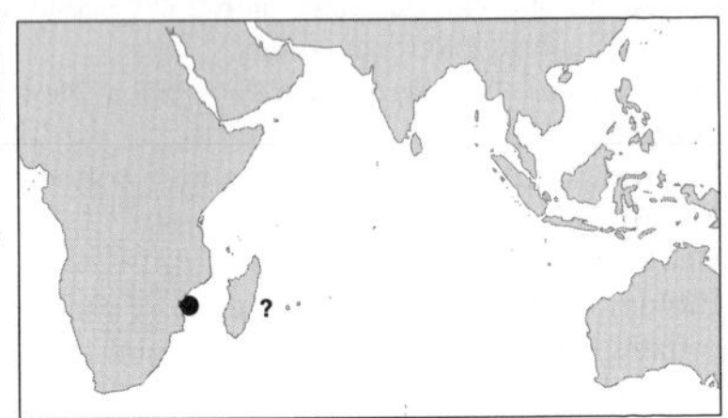

Distribution. Eastern South Africa.

Habitat. Coastal and tropical, in shallow water (to 20 m).

Behaviour. Unknown.

Biology. Two pups/litter.

Status. IUCN Red List: Not Evaluated.

Atlantic Weasel Shark *Paragaleus pectoralis* Plate 49

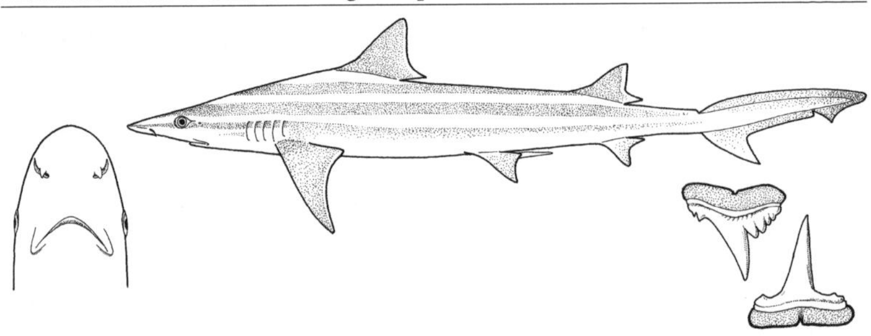

Measurements. Born: ~47 cm TL. Mature: ~80 cm ♂, 75–90 cm ♀. Max: 138 cm TL.

Identification. Slender shark with striking longitudinal yellow bands on light grey or bronze background, white below. Moderately long snout, large oval eyes, short small mouth, small serrated upper teeth, erect-cusped lower teeth, and plain fins.

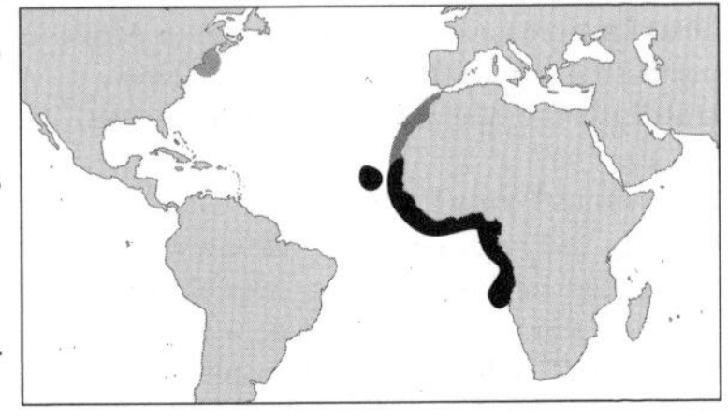

Distribution. East Atlantic: Cape Verde Islands and Mauritania to Angola, possibly north to Morocco. One northwest Atlantic record (1906).

Habitat. Tropical shelf, surfline to 100 m.

Behaviour. Specialist feeder on cephalopods (squid and octopi), also small fishes.

Biology. Bears one to four (mostly two) pups/litter in May–June in Senegal.

Status. IUCN Red List: Not Evaluated. Common where it occurs.

Slender Weasel Shark *Paragaleus randalli* Plate 49

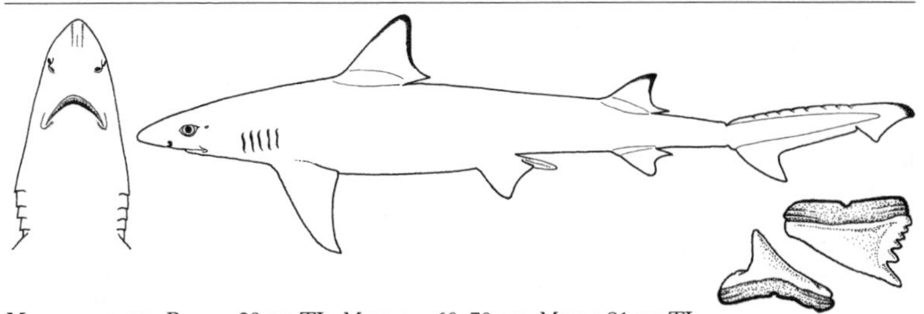

Measurements. Born: ~29 cm TL. Mature: ~60–70 cm. Max: >81 cm TL.
Identification. Snout with narrowly rounded tip with a pair of narrow black lines but no dark patches on underside. Gills equal to eye length. Long mouth and relatively deep lower jaw. Concave fins, mostly dark with inconspicuous light posterior margins, no obvious white or black tips. Second dorsal ~two-thirds area of first, origin ahead of smaller anal fin origin.

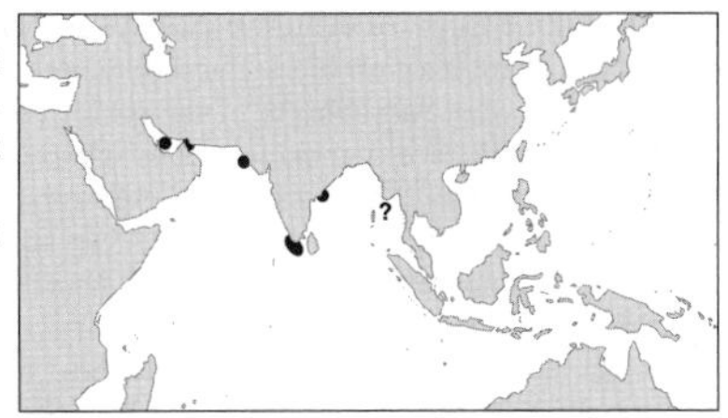

Distribution. Northern Indian Ocean: Arabian Gulf, Gulf of Oman, India, Sri Lanka. A somewhat similar *Paragaleus*, possibly a distinct species, occurs off Myanmar.
Habitat. Inshore, in shallow water (down to 18 m depth) on the continental shelf.
Behaviour. Unknown.
Biology. Two pups per litter (one/uterus).
Status. IUCN Red List: Not Evaluated.

Straighttooth Weasel Shark *Paragaleus tengi* Plate 49

Measurements. Mature: 78–88 cm TL ♂.
Identification. Small light grey slender shark with no prominent markings. Rounded, moderately long snout, large lateral eyes with nictitating eyelids, small spiracles, moderate-sized gill slits (~1.2–1.3 x eye length in adults; less in young), rather short arched mouth, lower teeth not prominently protruding.

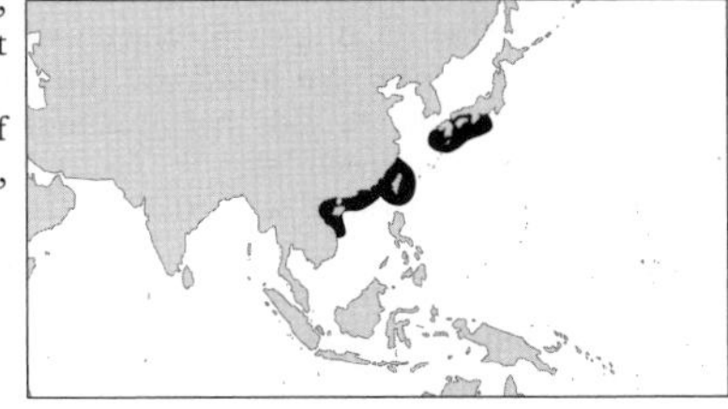

Distribution. West Pacific: Thailand (Gulf of Thailand), Vietnam, southern China, Taiwan, Japan.
Habitat. Inshore, depth range not reported.
Behaviour. Unknown.
Biology. Unknown.
Status. IUCN Red List: Not Evaluated.

Carcharhinidae: Requiem Sharks

Found in all warm and temperate seas. One of the largest, most important shark families with many common and wide-ranging species. The dominant sharks (often in biodiversity, abundance and biomass) in tropical waters on continental shelves and offshore. Several prefer coral reefs and oceanic islands; a few species are oceanic and wide-ranging. A few enter temperate waters or deep water. The family includes the only freshwater shark species.

Identification. Mostly medium to large (100–300 cm), some small (~100 cm TL). Long arched mouth with bladelike teeth (often broader in upper jaw), most with short labial furrows (except *Galeocerdo* and *Rhizoprionodon*), no nasoral grooves or barbels. Usually round (to horizontal) eyes with internal nictitating eyelids, usually no spiracles. Two dorsal fins, one anal fin, first dorsal fin medium to large with base well ahead of pelvic bases, second usually much smaller. Precaudal pits, caudal fin with a strong ventral lobe and lateral undulations on dorsal margin. Mostly unpatterned (particularly in genus *Carcharhinus*). The extremely rare river sharks (*Glyphis* species) are very difficult to distinguish without tooth and vertebral counts.

Biology. Active, strong swimmers. Some are 'ram-ventilators' needing to swim continually to oxygenate their gills, others can rest on the bottom. Many are more active at night or dawn and dusk than by day. Some are solitary or socialize in small groups, some are social schooling species. Some give specialized threat or defensive displays and may also interact aggressively when they encounter each other. There is a clear hierarchical dominance between certain species: Oceanic Whitetip Sharks are dominant over Silky Sharks of the same size, which in turn can dominate Grey Reef Sharks; Galapagos Sharks are dominant over Blacktip Sharks but subordinate to the Silvertip. Most are placental livebearers. These are major predators, taking a wide range of prey. Some larger species opportunistically eat carrion and rubbish, but none are obligate scavengers.

Status. Very important in tropical shark fisheries: commercial, subsistence and sports. Used for food, also for liver oil, fins and skin. Several species are potentially harmful, having bitten people and boats, but also important for dive ecotourism. Some are popular in captivity and have bred in large aquaria. A few (particularly Lemon and Blue Sharks) are the subject of intensive research activity.

Blacknose Shark *Carcharhinus acronotus* Plate 62

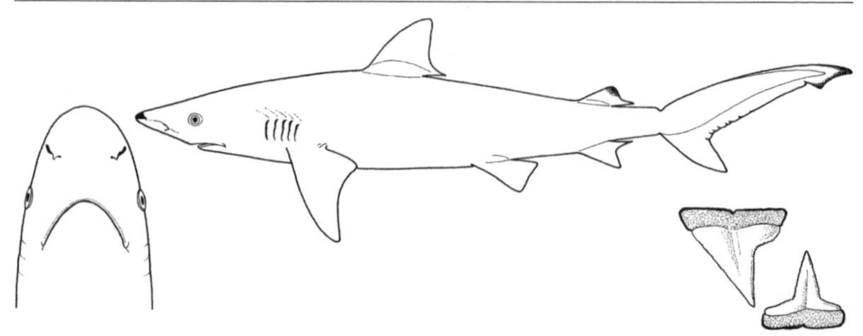

Measurements. Born: 38–50 cm TL. Mature: 97–106 cm ♂, ~103 cm ♀. Max: >137, poss. 200 cm TL.

Identification. Dark tip to snout. Second dorsal and upper caudal fin tips dark. Second dorsal fin origin ~over anal origin. No interdorsal ridge, small pectoral fins, dorsals with short rear tips, first small, second moderately large.

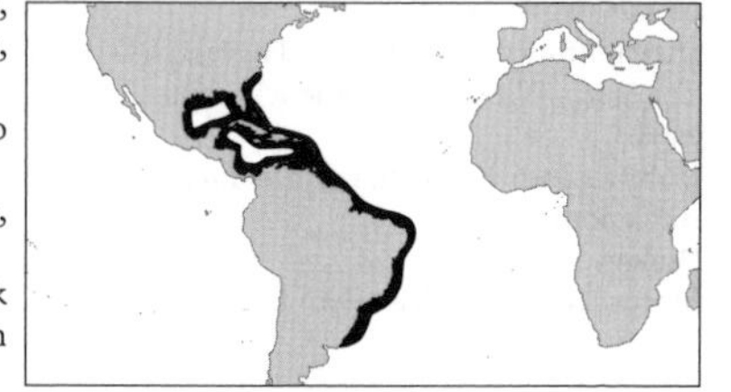

Distribution. West Atlantic: southern USA to south Brazil.

Habitat. Coastal continental and insular shelves, mainly over sand, shell and coral, 18–64 m.

Behaviour. Performs 'hunch' display (back arched, caudal lowered, head raised), when threatened. Migrates short distances seasonally.

Biology. Viviparous, yolk-sac placenta, three to six pups/litter after 10–11 month gestation. May mature aged two, breed in alternate years, and live for ~10 years.
Status. IUCN Red List: Not Evaluated. Fished for food in large numbers. Kept in public aquaria.

Silvertip Shark *Carcharhinus albimarginatus* Plate 52

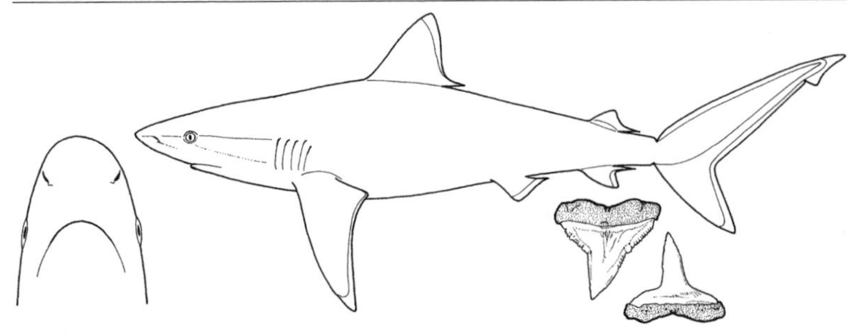

Measurements. Born: 63–68 cm TL. Mature: 160–180 cm ♂, 160–199 cm ♀. Max: ~300 cm TL.
Identification. Dark grey, sometimes bronze-tinged, white below, striking white tips and trailing edges on all fins. Pectorals narrow tipped, first dorsal apex narrowly rounded or pointed. Faint white band on sides. Upper teeth triangular.

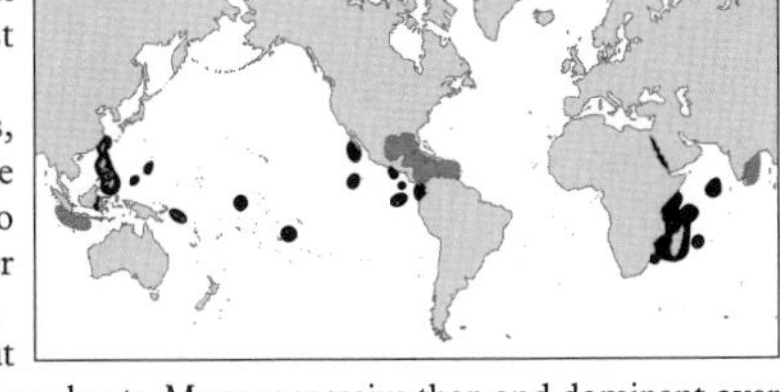

Distribution. Tropical Indo-Pacific, wide but patchy distribution. Unconfirmed in west Atlantic.
Habitat. Continental shelf, offshore islands, coral reefs and offshore banks; also inside lagoons, near drop-offs and offshore, surface to 600–800 m. Not oceanic. Young in shallower water closer to shore, adults more wide ranging.
Behaviour. Ranges from surface to bottom, but may not disperse widely between sites. Often follows boats. More aggressive than and dominant over *Carcharhinus galapagensis* and *C. limbatus*. Adults often scarred.
Biology. Viviparous, yolk-sac placenta, 1–11 pups/litter, often 5–6, after ~one year gestation. Feeds on a variety of midwater and bottom fishes, eagle rays, and octopi.
Status. IUCN Red List: Not Evaluated. Large and slow-growing. Even remote populations are likely highly vulnerable to target fisheries for meat or fins, particularly if limited dispersion between sites. Large, bold and potentially harmful, one confirmed shark-bite incident. Caution advised when encountering it underwater.

Bignose Shark *Carcharhinus altimus* Plate 57

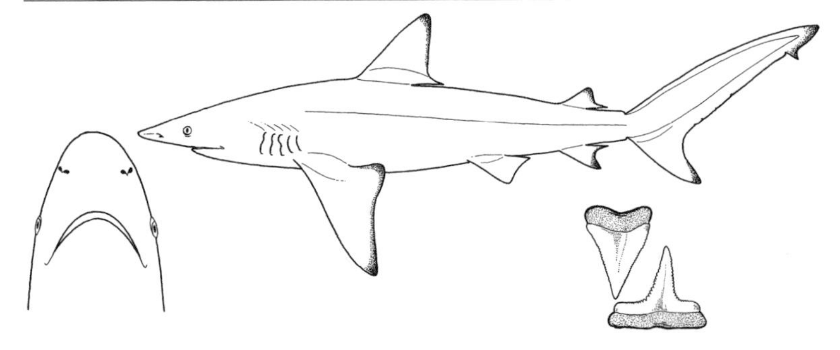

Measurements. Probably born 70–90 cm TL. Mature: 216 cm ♂, 226 cm ♀. Max: poss. ~300 cm TL.
Identification. Greyish (sometimes bronzy) above, white below, no conspicuous markings but dusky fin tips (except for pelvics), faint white flank marking. Cylindrical heavy body. Large, long, broad snout, long nasal flaps, high triangular saw-edged upper teeth. Prominent high interdorsal ridge, large straight pectoral and dorsal fins.

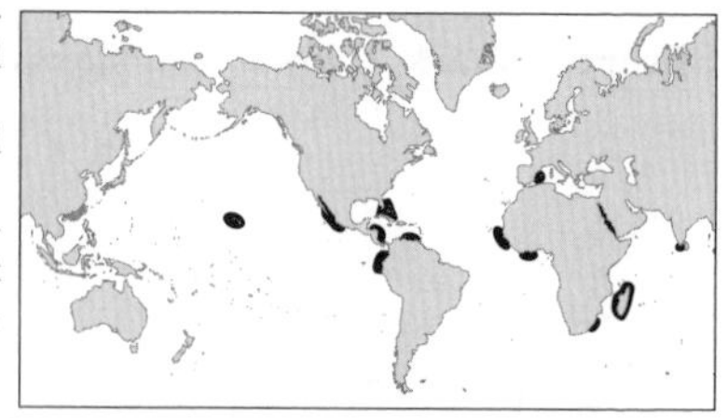

Distribution. Probably world-wide in tropical and warm seas, but records incomplete.
Habitat. Offshore, on deep continental and insular shelf edge and uppermost slopes, 90 m to at least 250–430 m. Young shallower, up to 25 m. Sometimes on surface.
Behaviour. Unknown.
Biology. Viviparous, 3–15 pups/litter. Eats bony fishes, other sharks, stingrays, cuttlefish.
Status. IUCN Red List: Not Evaluated. Bycatch of deep pelagic longlines, occasionally bottom trawls.

Graceful Shark *Carcharhinus amblyrhynchoides* Plate 53

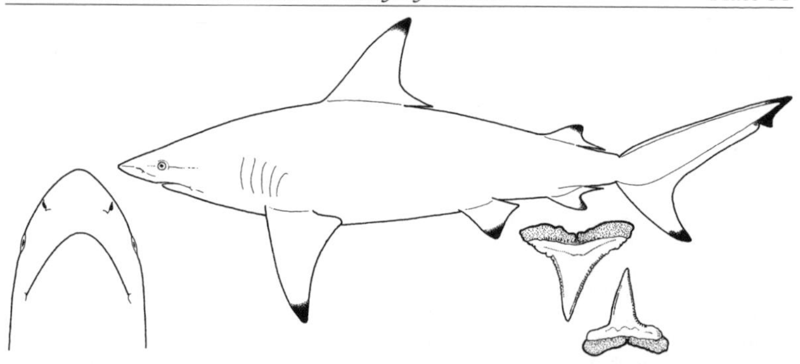

Measurements. Born: 52–55 cm TL. Mature: 140 cm ♂, 167 cm ♀. Max: >167 cm TL.
Identification. Largish tubby grey shark with conspicuous white flank mark, often black-tipped fins. Fairly short, wedge-shaped pointed snout, fairly large eyes, large gill slits, no interdorsal ridge, moderately large pectoral fins, large triangular first dorsal and moderately large second dorsal, both with short rear tip.

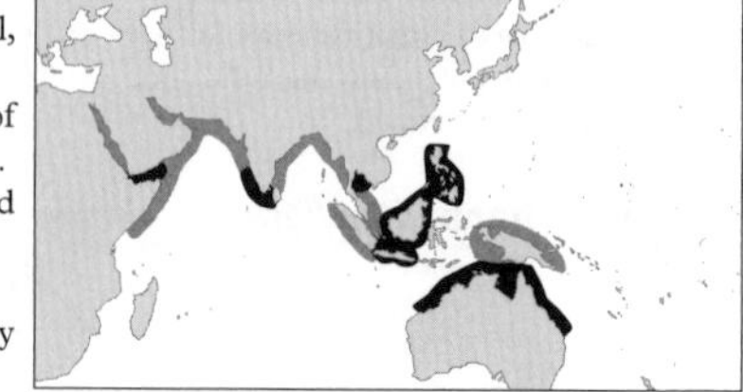

Distribution. Tropical Indo-west Pacific: Gulf of Aden to Philippines and northwestern Australia.
Habitat. Coastal-pelagic on continental and insular shelves.
Behaviour. Unknown.
Biology. Poorly known. Viviparous. Probably feeds mostly on fish.
Status. IUCN Red List: Near Threatened. Taken and landed as bycatch in commercial fisheries throughout range.

Grey Reef Shark *Carcharhinus amblyrhynchos* (inc. *C. wheeleri*) Plate 52

Measurements. Born: 45–75 cm TL. Mature: 110–145 cm ♂, 120–137 cm ♀. Max: poss. 233–255 cm TL.
Identification. Grey above, white below. First dorsal plain or irregularly to prominently white-edged, obvious broad black posterior margin to entire caudal fin, blackish tips to other fins. Moderately long, broadly rounded snout, eyes usually round, narrow serrated upper teeth, no interdorsal ridge, large second dorsal fin with short rear tip. Pectoral fins narrow and falcate.
Distribution. Indo-west to central Pacific.

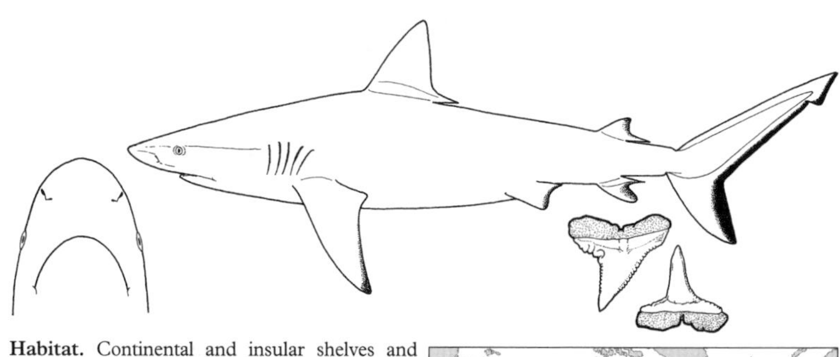

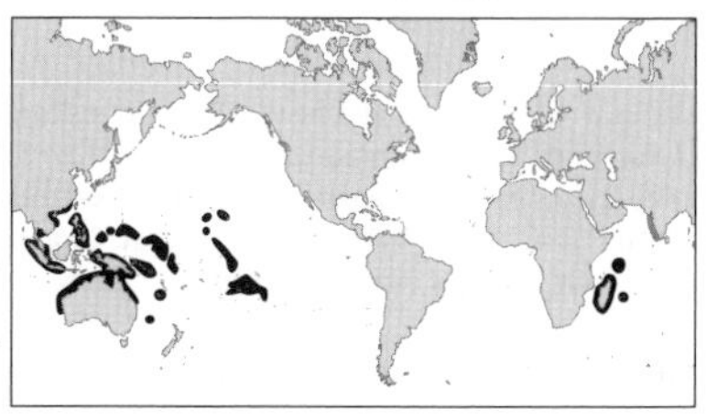

Habitat. Continental and insular shelves and adjacent oceanic waters, coastal-pelagic and inshore, 0–140 m. Common on coral reefs, often in deeper areas near drop-offs (including fringing reefs), in atoll passes, and shallow lagoons adjacent to strong currents. Prefers leeward sides of low, small coral islands. Found in deeper water than *Carcharhinus melanopterus* (which occupies shallow flats), but commonly found on the flats if *melanopterus* are absent. Young favour shallower water than adults.

Behaviour. Active, strong-swimming social species, aggregating by day in or near reef passes or lagoons. Groups of juveniles often seen on probable pupping and nursery grounds. Often cruises near bottom but will visit surface to investigate food sources and venture several km offshore before returning to home site. Even more active at night, when groups disperse. Inquisitive, several sharks may approach divers closely in seldom-dived areas, but soon disperse and seldom reappear except at a distance during repeated dives. Intimidating threat display (exaggerated swimming pattern with head and tail wagging in broad sweeps, back arching, head lifting, pectoral fin lowering and sometimes horizontal spiral swimming) performed if approached too closely or startled, unless food is present. Display varies from almost immediate flight, to a series of figure-eight loops, to biting then fleeing.

Biology. Viviparous, yolk-sac placenta, one to six pups/litter after ~year gestation. Individuals mature at ~7–7.5 years and longevity is at least 25. Feeds mostly off bottom on small bony reef fishes, also cephalopods (squids and octopi) and crustaceans.

Status. IUCN Red List: Near Threatened. One of the most abundant Indo-Pacific reef sharks (with *C. melanopterus* and *Triaenodon obesus*). Formerly common in clear tropical coastal waters and oceanic atolls, now under threat because of restricted inshore habitat, site fidelity, small litter size, relatively late age at maturity and increasing unmanaged fishing pressure. More valuable protected for dive tourism than for fisheries, but can be aggressive (particularly when spearfishing is occurring) and may bite if cornered, harassed, or when food stimuli are present. It should be treated with respect, although most dive encounters with it are without incident.

Pigeye or Java Shark *Carcharhinus amboinensis* Plate 54

Measurements. Born: ~71–72 cm TL. Mature: ~195 ♂, 198–223 cm ♀. Max: 280 cm TL.

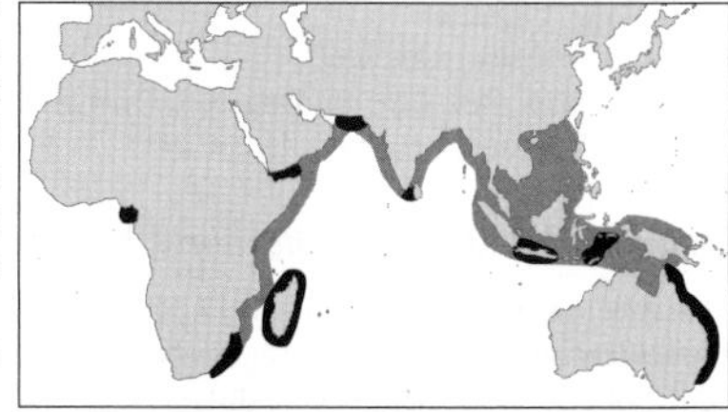

Identification. Greyish, underside white, fin tips dusky, but not strikingly marked. Large with massive thick head, very short, broad, blunt snout, small eyes, large triangular saw-edged upper teeth. Large angular pectorals, high erect triangular first dorsal (over three times second dorsal height), dorsals with short rear tips, no interdorsal ridge.

Distribution. Indo-west Pacific: South Africa to Australia. East Atlantic: Nigeria.

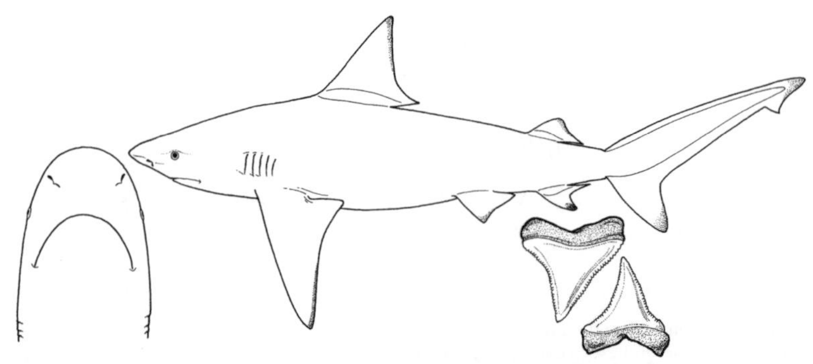

Habitat. Continental and insular shelves, inshore near surf line and along beaches, 0–60 m.
Behaviour. Less common where *Carcharhinus leucas* are common; possible competitive exclusion?
Biology. Apparently viviparous. Feeds on bottom fishes, crustaceans and molluscs.
Status. IUCN Red List: Data Deficient. Not common? Shark fisheries intensifying in its range. Potentially harmful (size, large jaws and teeth), but no records of it biting people.

Borneo Shark *Carcharhinus borneensis* Plate 60

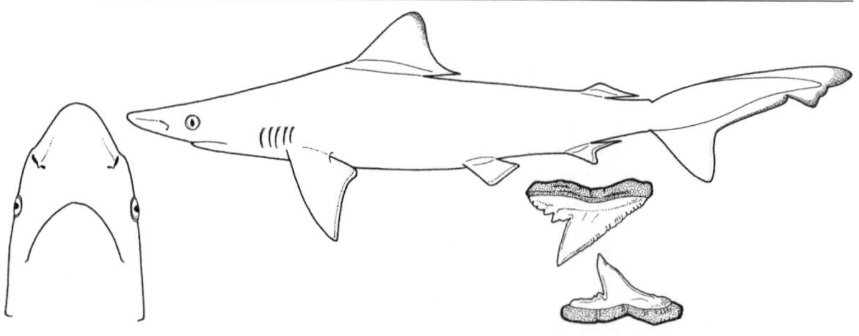

Measurements. Born: 24-28 cm TL. Max: estimated ~70 cm TL.
Identification. Brown above, white below, dusky to blackish markings on the second dorsal and upper caudal tip, paired fins and anal fin with inconspicuous light edges. Small, with long pointed snout, enlarged pores alongside mouth corners, large eyes, black spot beneath snout tip, small dorsal fins with a short rear tips, origin of second ~over anal midbase, no interdorsal ridge, small pectoral fins.
Distribution. Indo-west Pacific: Java(?), Borneo, China, Philippines(?).
Habitat. Tropical inshore/coastal.
Behaviour. Unknown.
Biology. Virtually unknown.
Status. IUCN Red List: Endangered. Only five known specimens, taken in inshore fisheries in a heavily fished area. Last recorded in 1937.

Bronze Whaler *Carcharhinus brachyurus* Plate 61

Measurements. Born: 59–70 cm TL. Mature: 200–229 cm ♂, <240 cm ♀. Max: 294 cm TL.
Identification. Olive-grey to bronzy above, white below. Most fins with inconspicuous darker edges and dusky to black tips, not boldly marked. Fairly prominent white band on flank. Large, with a

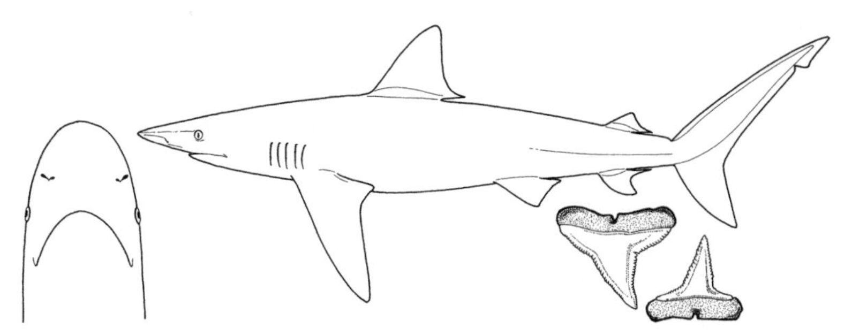

bluntly pointed broad snout, narrow bent cusps on upper teeth and usually no interdorsal ridge. Long pectoral fins, small dorsals with short rear tips.

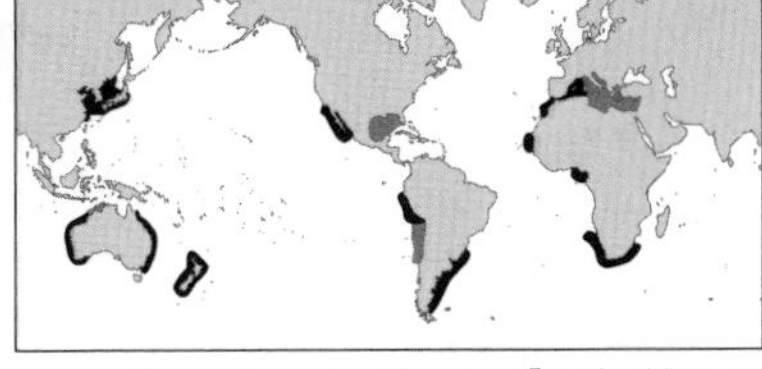

Distribution. Most warm temperate waters in Indo-Pacific, Atlantic and Mediterranean.
Habitat. Close inshore to at least 100 m offshore.
Behaviour. Active, seasonally migratory in at least part of its range but very little exchange between adjacent regional populations. Large numbers follow winter sardine run off kwaZulu-Natal, South Africa. Nursery grounds in inshore bays and coasts.
Biology. Viviparous, yolk-sac placenta. Gestation probably ~one year, litter of 13–24 pups every other year. May mature at ~13 years ♂ and ~20 years ♀. Feeds on bony fishes, elasmobranchs and cephalopods.
Status. IUCN Red List: Near Threatened. Exceptionally slow growth rate, large size, inshore habitat vulnerable to damage, and easy to capture. Caught for food and by sports anglers. Potentially harmful to swimmers and divers and sometimes aggressive during encounters with divers.

Spinner Shark *Carcharhinus brevipinna* Plate 57

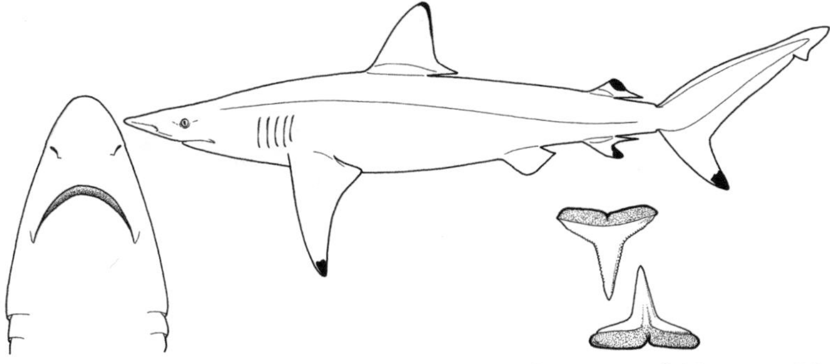

Measurements. Born: ~60–75 cm TL. Mature: 159–203 cm ♂, 170–200 cm ♀. Max: 278 cm TL.
Identification. Most fins of adults and juveniles (not young) have obvious black tips, inconspicuous white band on flanks, underside white. Slender, long narrow pointed snout, small circular eyes, small narrow-cusped teeth, prom-inent labial furrows (longer than any other *Carcharhinus*), long gill slits. Short relatively inconspicuous anterior nasal flaps. Small pectoral and first dorsal fins, second dorsal moderately large, both with short rear tips, no interdorsal ridge.
Distribution. Warm-temperate and tropical Atlantic, Mediterranean and Indo-west Pacific.
Habitat. Coastal-pelagic on continental and insular shelves, common close inshore (<30 m) to at least 75 m, from the surface to the bottom.

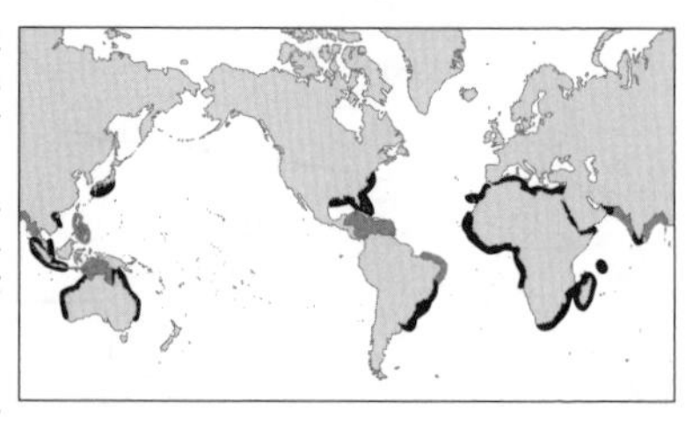

Behaviour. Active, schooling shark. Vertical feeding runs through fish schools end with a spinning leap out of the water. Highly migratory in Gulf of Mexico (and possibly elsewhere): move inshore in spring-summer to feed and breed, possibly south and into deeper water in winter. Some segregation by age and sex; young prefer lower water temperatures.
Biology. Viviparous, yolk-sac placenta, 3–15 pups/litter (increasing with female size) born every other year. Primarily a fish-eater, also stingrays and cephalopods.
Status. IUCN Red List: Near Threatened. Common, but inshore distribution vulnerable to fishing pressure and habitat change. Valuable meat and fins, important in multi-species fisheries. Has bitten swimmers, but probably of limited concern to people because of its feeding habits.

Nervous Shark *Carcharhinus cautus* Plate 59

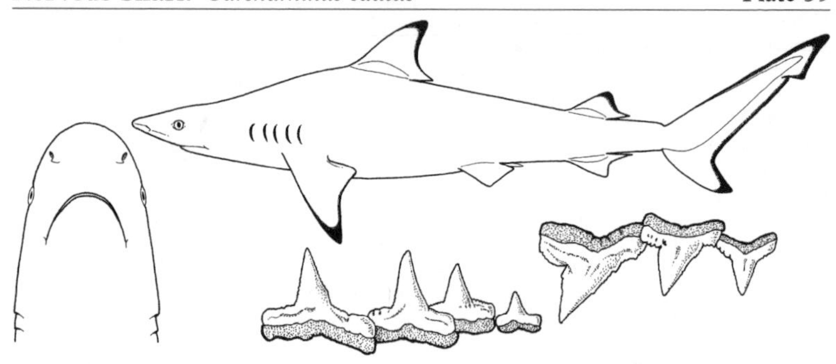

Measurements. Born: 35–40 cm TL. Mature: ~80 cm ♂, ~85 cm ♀. Max: 150 cm TL.
Identification. Grey to light brownish, conspicuous white band on flank, white below. Black dorsal and caudal fin edges (no obvious black blotch on first dorsal fin), black tips on caudal lobes and pectoral fins. Medium-sized, short bluntly rounded snout, horizontally oval eyes. Nipple-like anterior nasal flaps, short labial furrows. No interdorsal ridge, fairly large second dorsal with a short rear tip.

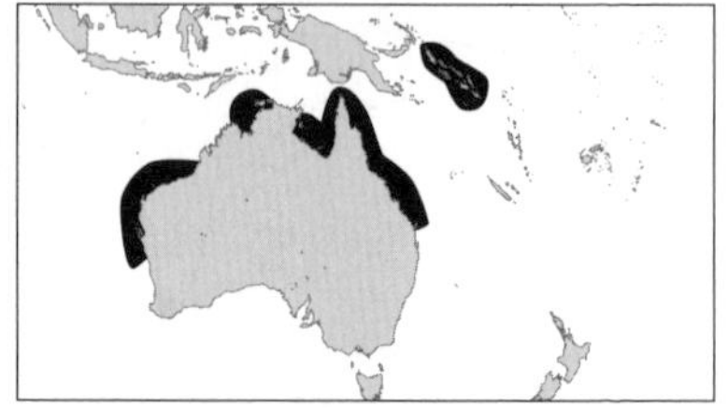

Distribution. East Indian Ocean, southwest Pacific.
Habitat. Shallow inshore water on continental and insular shelves, coral reefs and estuaries; may range into deeper water.
Behaviour. Reported skittish and timid when accosted by people (hence name).
Biology. Viviparous, one to five pups/litter after eight to nine month gestation. Eats small fishes.
Status. IUCN Red List: Data Deficient. Relatively common in northern Australia.

Whitecheek Shark *Carcharhinus dussumieri* Plate 56

Measurements. Born: 37–38 cm TL. Mature: 65–70 cm ♂, 70–75 cm ♀. Max: ~100 cm TL.
Identification. Small grey to grey-brown shark with black or dusky spot only on second dorsal, other fins with pale trailing edges, inconspicuous light flank stripe. Moderately long rounded snout, small widely spaced nostrils, short inconspicuous upper labial furrows, fairly large horizontally-oval eyes, oblique-cusped serrated teeth, small semifalcate pectoral fins, small triangular first dorsal, both dorsals with short rear tips. (Often confused with *C. sealei.*)

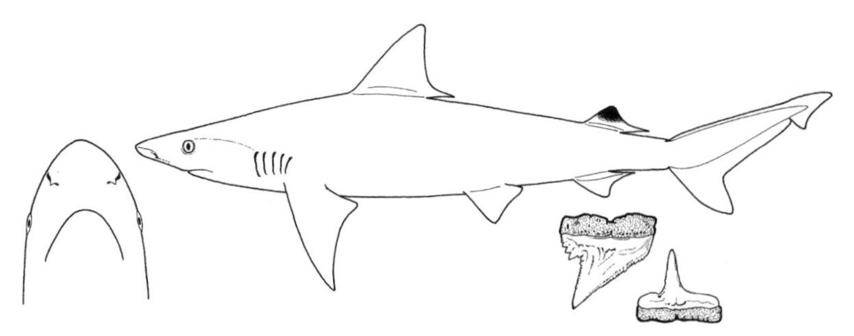

Distribution. Indo-west Pacific, Arabian Sea to Japan.
Habitat. Tropical inshore continental and insular shelves.
Behaviour. Poorly known.
Biology. Viviparous, yolk-sac placenta, normally two pups/litter, exceptionally up to four
Status. IUCN Red List: Near Threatened. Common but heavily fished for meat and sometimes fins in most of range. Harmless to people.

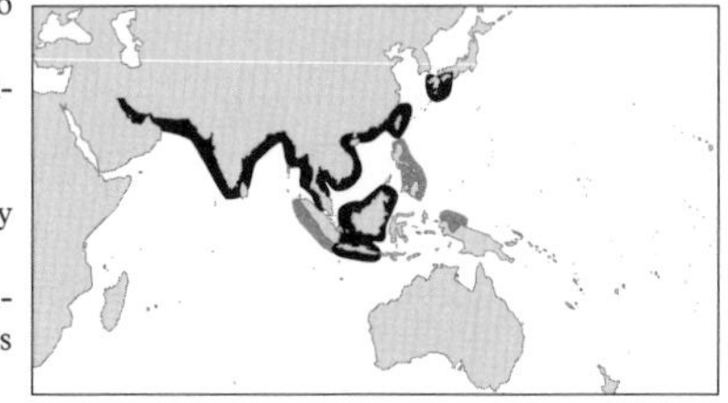

Silky Shark *Carcharhinus falciformis* Plate 51

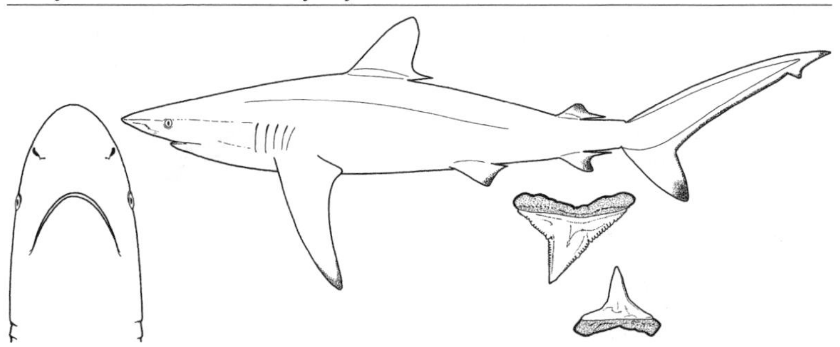

Measurements. Born: ~70–87 cm TL. Mature: ~187–217 cm ♂, 213–230 cm ♀. Max: ~330 cm TL.
Identification. Large slim dark grey to grey-brown or nearly blackish above, inconspicuous pale flank band, white below. Fin markings inconspicuous, tips dusky except for first dorsal. Fairly long, flat rounded snout, small jaws, large eyes, oblique-cusped serrated upper teeth. First dorsal behind pectorals, second low with greatly elongated inner margin and rear tip, narrow interdorsal ridge, long narrow pectoral fins. No caudal keels.
Distribution. World-wide in tropical seas.
Habitat. Oceanic and epipelagic, surface to at least 500 m. Commonest in water >200 m near continental and insular shelf edge and over deepwater reefs, also in open sea, occasionally inshore to 18 m. Longline catches much more abundant offshore near land than in the open ocean.

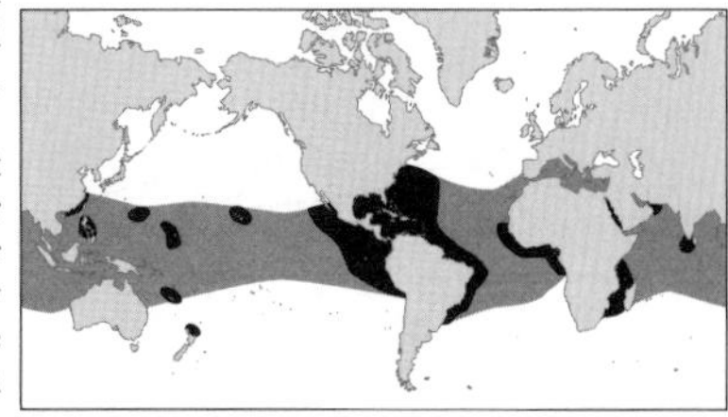

Behaviour. Active, swift, bold, inquisitive and sometimes aggressive shark. 'Hunch' display (back arched, head raised, caudal fin lowered) reported by divers. Segregate by size (young in offshore nursery areas on shelf edge and oceanic banks, adults further out to sea). Found with tuna schools in east Pacific.

Biology. Viviparous, yolk-sac placenta, 2–14 pups/litter, possibly on alternate years. Primarily eats fish, also cephalopods and pelagic crabs.

Status. IUCN Red List: Least Concern (under review). One of the three most common oceanic sharks, huge bycatch in oceanic fisheries but often misidentified. Landed for meat and fins. Susceptible to overfishing; serious declines reported in some areas. Potentially harmful to people, but rarely encountered except at offshore reefs near deep water. Important for ecotourism in the Red Sea.

Creek Whaler *Carcharhinus fitzroyensis* Plate 59

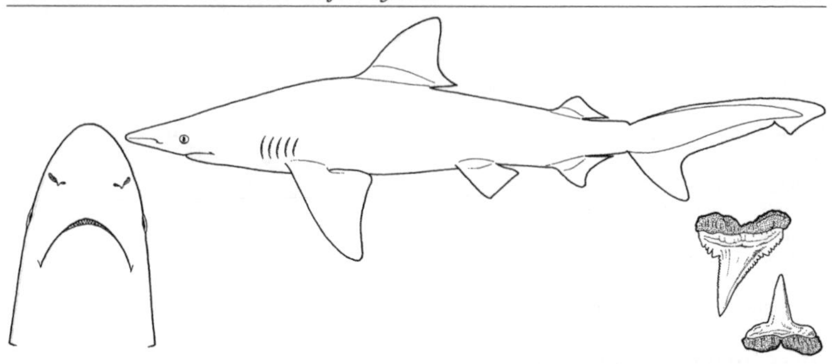

Measurements. Born: ~50 cm TL. Mature ~80 cm ♂, 90 cm ♀. Max: 135 cm TL.

Identification. Fairly large, bronze to greyish-brown shark, light below, lacking any conspicuous fin or body markings. Triangular broad fins (first dorsal origin over pectoral fin rear tips, second ~over anal origin), long parabolic snout, short labial furrows, lobate anterior nasal flaps, short gill slits, narrow teeth, no interdorsal ridge.

Distribution. Northern Australia.

Habitat. Mainly inshore, from intertidal to at least 40 m.

Behaviour. Unrecorded.

Biology. Viviparous, one to seven pups/litter born annually. Young use embayments as nursery grounds. Eats mainly small fishes, some crustaceans.

Status. IUCN Red List: Least Concern. Small numbers caught in inshore gillnet fisheries.

Galapagos Shark *Carcharhinus galapagensis* Plate 58

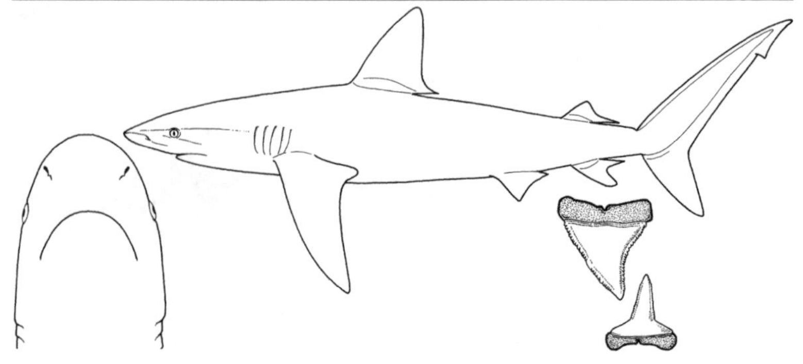

Measurements. Born: 57–80 cm TL. Mature 170–236 ♂, ~235 cm ♀. Max: possibly 370 cm TL.
Identification. Large, brownish-grey above, white below, no conspicuous fin markings but most tips fins dusky (not black or white), inconspicuous white band on flank. Fairly long broadly rounded snout, low anterior nasal flaps, fairly large eyes, large erect teeth, low interdorsal ridge, large semifalcate pectoral fins, moderately large first dorsal with short rear tip originating over pectoral fins inner margin.

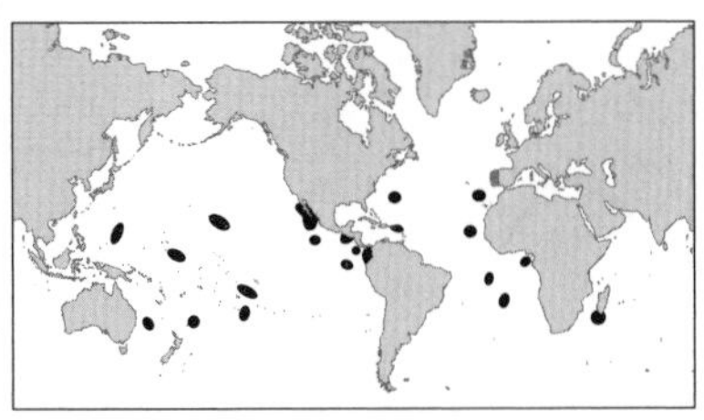

Distribution. World-wide, patchy, mainly around oceanic islands.
Habitat. Coastal pelagic, not oceanic, from shallow inshore (2 m) to well offshore (0–180 m) on or near insular and (occasionally) continental shelves, in clear water over rugged coral or rocky bottom. Juvenile nursery grounds >25 m, adults further offshore.
Behaviour. Found in aggregations, but not coordinated schools. Often swim close to bottom, coming to surface to feed or investigate disturbances. Inquisitive and sometimes aggressive, performs 'hunch' display (arched back, raised head, lowered caudal and pectoral fins, while twisting and rolling) near divers that may be followed by biting.
Biology. Viviparous, yolk-sac placenta, 6–16 pups/litter. Feeds mainly on bottom fishes.
Status. IUCN Red List: Near Threatened. Common or abundant in restricted habitat, but often heavily fished with reports of extirpations of some populations around Central America. Can be a nuisance to divers because of its inquisitiveness, occasionally bites people (one fatality reported).

Pondicherry Shark *Carcharhinus hemiodon* Plate 56

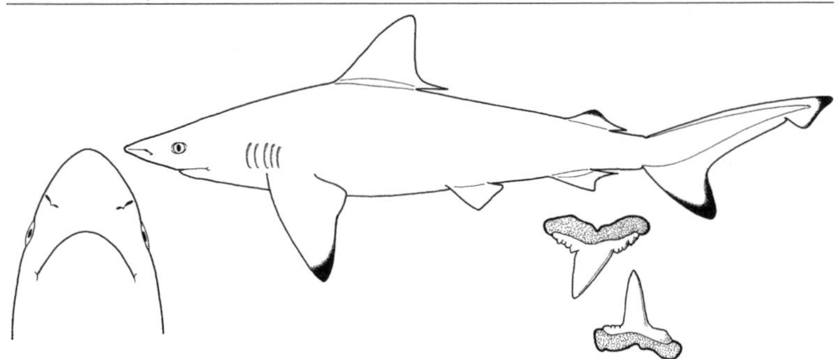

Measurements. Born: <32 cm TL. Mature: ? (♂ & ♀). Max: about 102 cm TL.
Identification. Grey above, white below, black tips to pectorals, second dorsal, dorsal and ventral caudal lobes and a conspicuous white band on flank. Probably small with fairly long pointed snout, fairly large eyes, an interdorsal ridge, small pectoral fins, fairly large first dorsal and moderately large second dorsal, both with short rear tips.
Distribution. Indo-west Pacific.

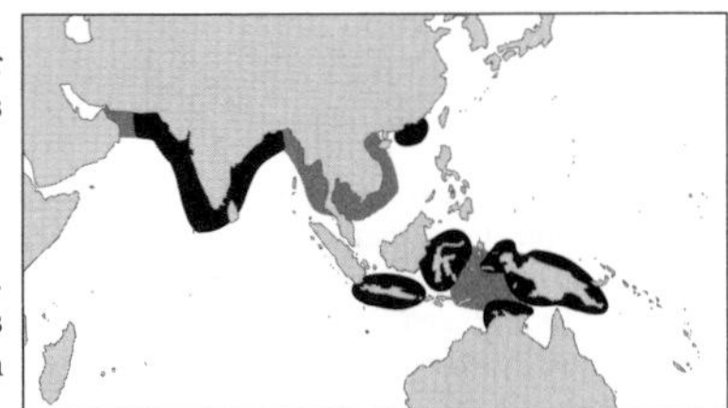

Habitat. Coastal, on continental and insular shelves. Unconfirmed reports from river mouths and fresh water upstream.
Behaviour. Unknown.
Biology. Virtually unknown.
Status. IUCN Red List: Critically Endangered. Very rare, known from ~20 museum specimens collected from heavily fished sites. Last seen in 1979.

Finetooth Shark *Carcharhinus isodon* Plate 62

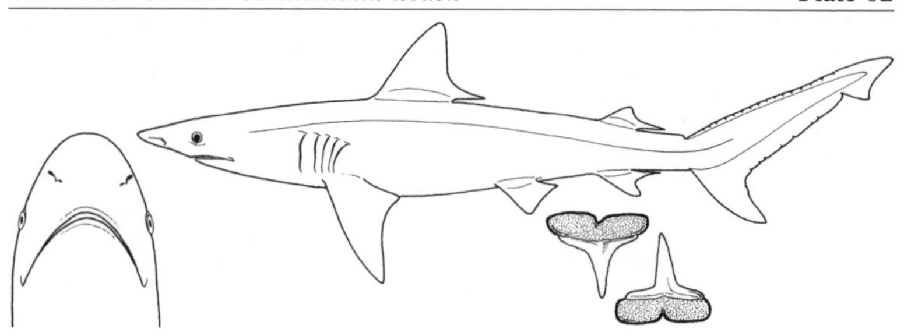

Measurements. Born: 55–58 cm TL. Mature: ~133 cm ♂, 125–135 cm ♀. Max: poss. 189–200 cm TL.
Identification. Small dark bluish-grey, white below, inconspicuous white band on flank, no prominent fin markings. Moderately long pointed snout, fairly large eyes, very long gill slits, erect smooth or irregularly serrated teeth, no interdorsal ridge, small pectoral fins, small first dorsal, moderately large second dorsal, both with short rear tips.
Distribution. West Atlantic: United States to Brazil.

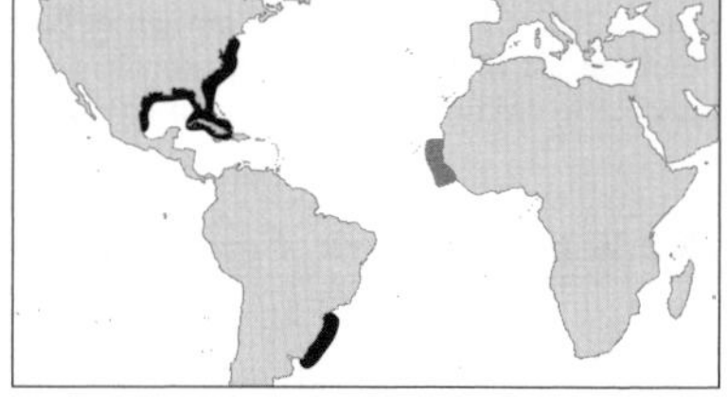

Habitat. Warm temperate to tropical inner continental shelf, intertidal to ~20 m.
Behaviour. Very active. Large schools or aggregations migrate seasonally with changing water temperature.
Biology. Viviparous, yolk-sac placenta, two to six pups/litter in May–June, two-year reproductive cycle. Feeds on small fishes, also shrimp.
Status. IUCN Red List: Not Evaluated. Locally common, bycaught throughout range, also targeted on migration. Inshore habitat vulnerable to degradation.

Smoothtooth Blacktip *Carcharhinus leiodon* Plate 62

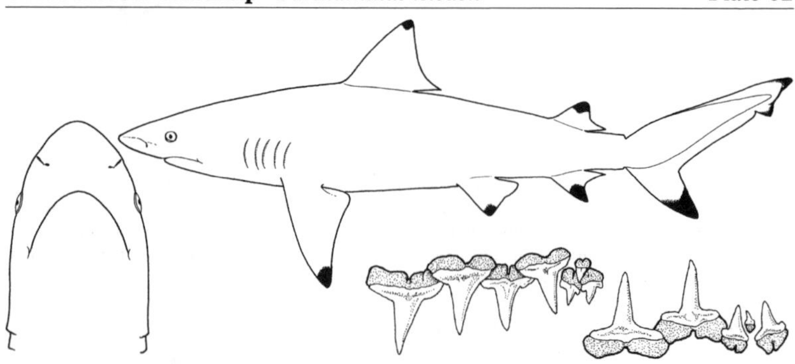

Measurements. Only known specimen was an immature male, 75 cm TL.
Identification. Conspicuous black tips on all fins. Short bluntly-pointed snout, fairly large eyes, smooth and erect-cusped upper teeth, long gill slits, no interdorsal ridge, small pectoral fins, fairly large first dorsal and moderately large second dorsal, both with short rear tips. Similar to *Carcharhinus amblyrhynchoides*, which has serrated upper tooth cusps and less prominently marked fin tips.

Distribution. Gulf of Aden.
Habitat. Probably inshore.
Behaviour. Unknown.
Biology. Unknown.
Status. IUCN Red List: Vulnerable. Known from one specimen collected in 1980s from a heavily fished region.

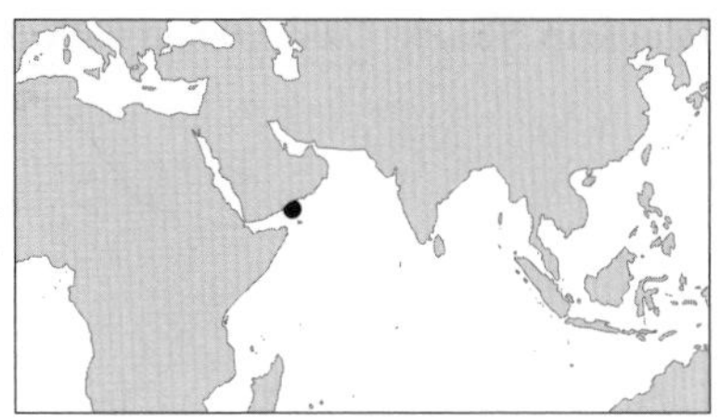

Bull Shark *Carcharhinus leucas* Plate 50

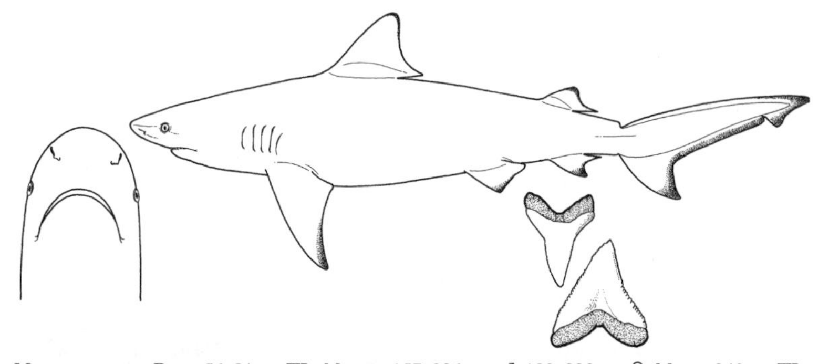

Measurements. Born: 56–81 cm TL. Mature: 157–226 cm ♂, 180–230 cm ♀. Max: ~340 cm TL.
Identification. Underside white, fin tips dusky but not conspicuously marked, except in juveniles. Large, massive, thick-headed greyish shark. Very short, broad, bluntly-rounded snout, small eyes, triangular saw-edged upper teeth, large angular pectoral fins, broad triangular first dorsal fin (<3.2 x second dorsal height), both dorsals with short rear tips, no interdorsal ridge, inconspicuous caudal keels.
Distribution. World-wide, tropical and subtropical seas.

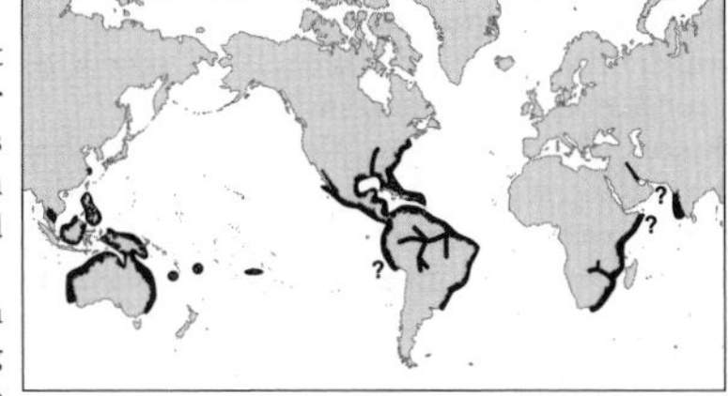

Habitat. Usually close inshore (1–30 m, to at least 152 m) in hypersaline lagoons, bays, river mouths, passages between islands, near wharves and along surf line; also many 100 km upstream in warm rivers (often in very turbid water) and freshwater lakes.
Behaviour. Usually cruises slowly near bottom but agile and quick when chasing and attacking prey. Young sharks often seen spinning out of the water. Migrate seasonally as sea temperature changes.
Biology. Viviparous, yolk-sac placenta, 1–13 pups/litter usually born in estuaries and rivers after estimated 10–11 month gestation. Mature at 15–20 years, longevity >25 years. Takes a very broad range of food, from bony fishes, invertebrates and elasmobranchs to sea turtles, birds, dolphins, whale offal, and terrestrial mammals.
Status. IUCN Red List: Near Threatened. Bycatch throughout its range; populations depleted rapidly if targeted. Nearshore and estuarine habitat vulnerable to human impacts. Large size, massive jaws and teeth, indiscriminate appetite, abundance in coastal areas, lakes and rivers, and proximity to human activities makes this the most potentially dangerous of tropical sharks; well known for several fatal and nonfatal bites on people, particularly swimmers and bathers. Very popular for dive ecotourism, divers often encounter it without incident although it should be treated with respect. Popular sports angling trophy. Hardy in captivity (kept in aquaria for over 15 years).

Blacktip Shark *Carcharhinus limbatus* Plate 57

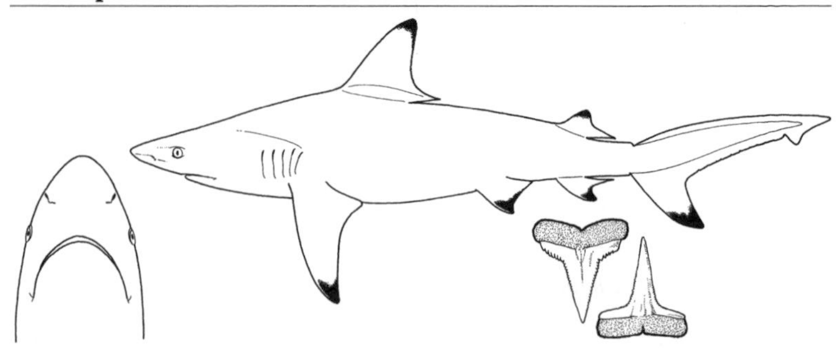

Measurements. Born: 38–72 cm TL. Mature: 135–180 cm ♂, 120–190 cm ♀. Max: 255 cm TL.
Identification. Fairly large stout grey to grey-brown shark, white below, conspicuous white band on flanks. Fin tips usually black on pectorals, second dorsal, and ventral caudal lobe, sometimes on pelvic and anal fins (anal usually plain); usually black edges on first dorsal apex and dorsal caudal lobe (some adults lack black tips). Long narrow pointed snout, small eyes, narrow-cusped erect upper teeth, long gill slits, high first dorsal fin, no interdorsal ridge.
Distribution. Widespread, tropical and subtropical seas.

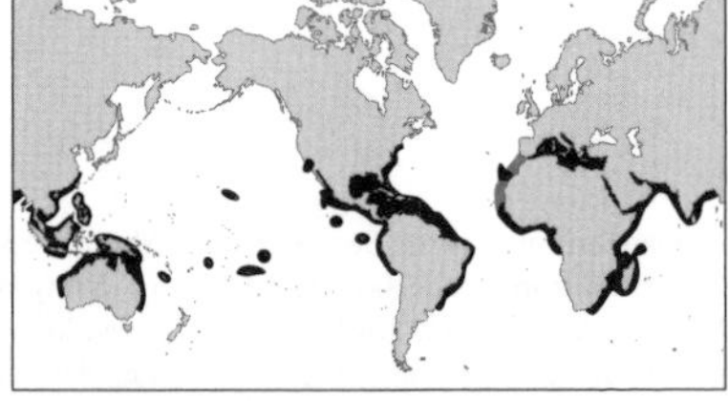

Habitat. Continental and insular shelves. Usually close inshore (off river mouths, in estuaries, shallow muddy bays, saline mangrove swamps, island lagoons and coral reefs dropoffs), sometimes offshore but rarely >30 m. Tolerates reduced salinities, not fresh water.
Behaviour. Often segregated by age and sex; pregnant females seasonally migratory. Very active, fast swimmer, often in large surface schools. May leap out of the water and rotate up to three times around its axis before dropping back into the sea at the end of a feeding run on small schooling fishes, but less often than Spinner Sharks. May enter feeding frenzy on highly concentrated food source.
Biology. Viviparous, yolk-sac placenta, one to ten pups/litter (commonly four to seven) born in inshore nursery grounds after 10–12 month gestation in alternate years. Mainly eats fishes, also cephalopods and crustaceans.
Status. IUCN Red List: Near Threatened. Inshore distribution heavily fished and susceptible to habitat degradation. An important commercial and sports fishery species (meat and fins valuable). Very few serious bites on people reported – not potentially harmful unless stimulated by food (may harass spearfishers when fish speared). Important for dive tourism.

Oceanic Whitetip Shark *Carcharhinus longimanus* Plate 50

Measurements. Born: 60–65 cm TL. Mature: 175–198 cm ♂, ~180–200 cm ♀. Max: possibly 350– 395 cm TL.

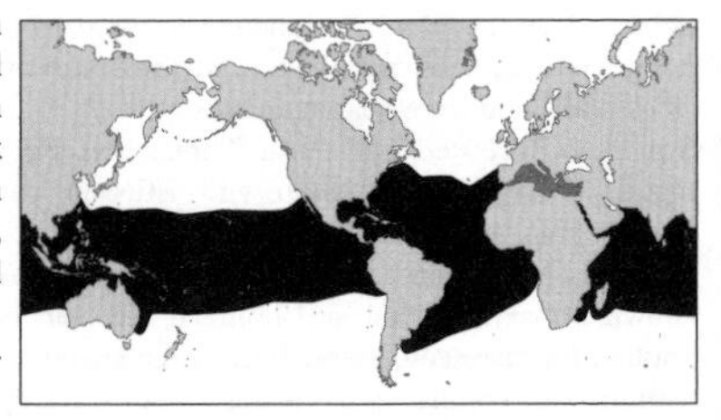

Identification. Large stocky grey or brownish shark, underside white, huge rounded first dorsal fin and long paddle-like pectoral fins with obvious white-mottled tips. Juveniles have black tips on some fins and black patches or saddles on the caudal peduncle. Snout bluntly rounded, small eyes, upper teeth triangular. Interdorsal ridge present, inconspicuous caudal keels.
Distribution. World-wide in tropical and warm-temperate waters.

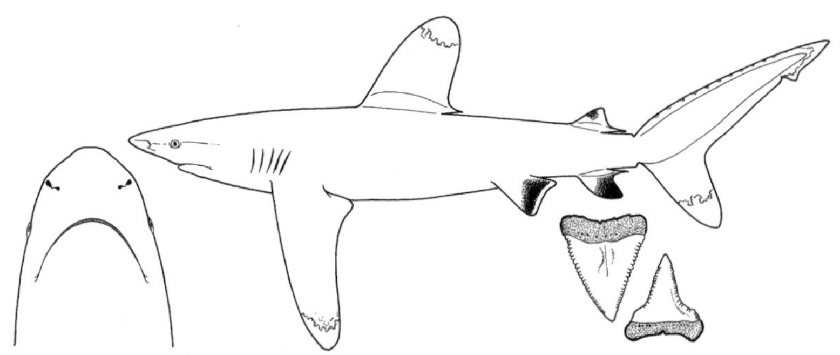

Habitat. Oceanic-epipelagic (occasionally coastal), usually found far offshore in the open sea in temperatures of 18–28°C. Occasionally in shallow water (37 m) off oceanic islands or where the continental shelf is very narrow, but is usually from surface to at least 152 m in water >184 m deep.
Behaviour. Slow-moving but active by day and at night, cruising slowly at or near the surface with huge pectoral fins outspread. Very inquisitive, aggressive and persistent, especially when competing for food with Silky Sharks and sometimes when investigating divers. Some size and sexual segregation reported.
Biology. Viviparous, yolk-sac placenta, 1–15 pups/litter (increasing with female size) after ~one year gestation. Mainly feeds on oceanic bony fishes and cephalopods, also stingrays, sea birds, turtles, marine gastropods, crustaceans, marine mammal carrion and garbage.
Status. IUCN Red List: Near Threatened. Originally widespread and common, but low reproductive capacity and has been taken in huge numbers in bycatch and directed fisheries, with some massive population decreases recently reported. Huge fins are very high value in trade. Rarely encountered in the water, but has been responsible for biting swimmers and boats.

Hardnose Shark *Carcharhinus macloti* Plate 60

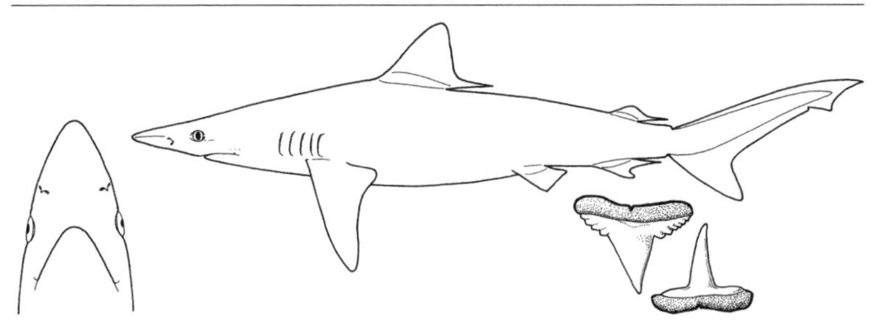

Measurements. Born: 40–50 cm TL. Mature: ~69–74 cm ♂, 70–75 cm ♀. Max: 110 cm TL.
Identification. Small slender shark, grey or grey-brown above, white below, fins light edged but not conspicuously marked, inconspicuous light flank marks. Long pointed snout, fairly large eyes, oblique-cusped smooth-edged upper teeth, no interdorsal ridge, dorsal fins small (second very low) with extremely long rear tips. Only *Carcharhinus* with a hypercalcified rostrum (felt if snout pinched).

Distribution. Indo-west Pacific.
Habitat. Continental and insular shelves, from close inshore to 170 m offshore.
Behaviour. Forms large sexually segregated aggregations.

Biology. Viviparous, yolk-sac placenta, one or two (usually two) pups/litter after ~one year gestation and one year resting before next breeding cycle. Eats mostly small fishes, also cephalopods and crustaceans.
Status. IUCN Red List: Near Threatened. Fished throughout range, population likely reduced.

Blacktip Reef Shark *Carcharhinus melanopterus* Plate 52

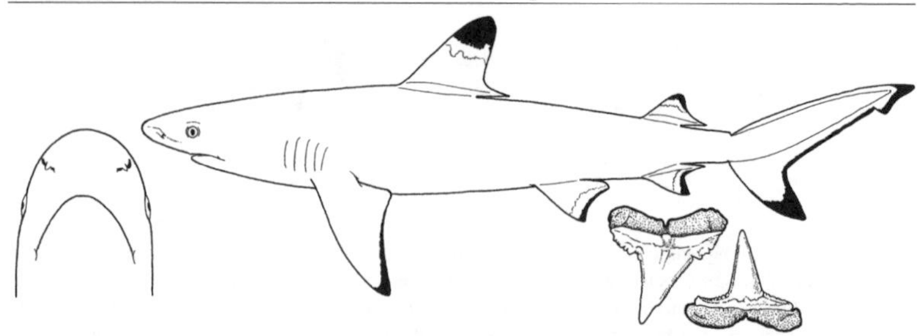

Measurements. Born: 33–52 cm TL. Mature: 91–100 cm ♂, 96–112 cm ♀. Max: <200 cm TL.
Identification. Medium-sized brownish-grey, underside white, brilliant black fin tips, highlighted by white. Short bluntly rounded snout, horizontally oval eyes, narrow-cusped teeth, no interdorsal ridge, largish second dorsal fin with short rear tip.

Distribution. West Pacific, Indian Ocean and eastern Mediterranean (through Suez Canal).
Habitat. Very shallow water on coral reefs and reef flats, also near reef dropoffs, rarely close offshore and in brackish water.
Behaviour. Active strong swimmer, dorsal fins above surface in very shallow water, alone or in small groups (not strongly schooling).
Biology. Viviparous, yolk sac placenta, two to four pups/litter after possible 16-month gestation. Eats small fishes and invertebrates.
Status. IUCN Red List: Near Threatened. One of most common Indo-Pacific reef sharks. Regularly caught by inshore fisheries and likely depleted. Probably adversely affected by reef destruction. Important for aquarium display and for dive tourism, very occasionally bites people that are swimming or wading on reefs but more circumspect when encountering divers.

Dusky Shark *Carcharhinus obscurus* Plate 58

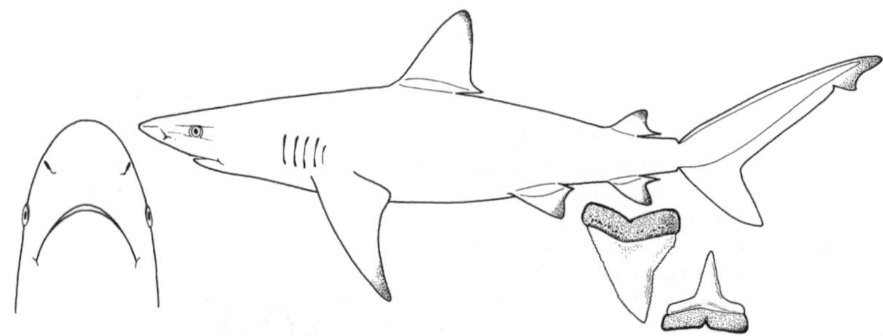

Measurements. Born: 69–100 cm TL. Mature: ~280 cm ♂, 257–300 cm ♀. Max: 360–400 cm TL.
Identification. Large grey to bronzy shark with most fin tips dusky, but not boldly marked. Underside white, inconspicuous white band on flank. Broadly rounded snout, triangular saw-edged upper teeth, curved moderate-sized pectoral fins, large falcate first dorsal fin, low interdorsal ridge.
Distribution. Possibly world-wide in tropical and warm temperate waters.

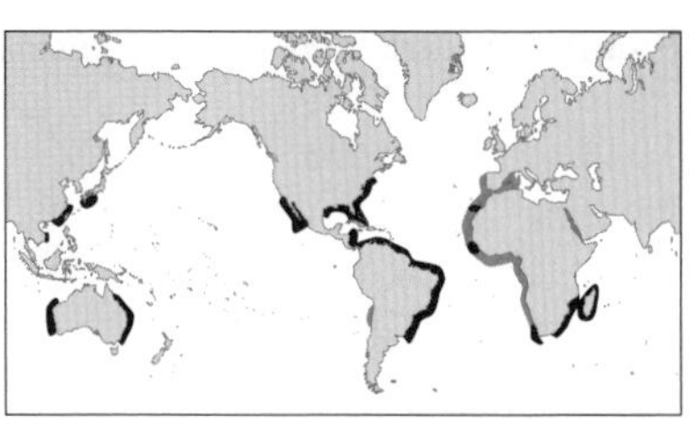

Habitat. Continental and insular shelves, from shoreline to adjacent oceanic waters, 0–400 m. Avoids estuaries. Often seen offshore following ships.
Behaviour. Strongly migratory with changing temperatures in temperate and subtropical areas. Newborns segregate in nursery areas where they are born (mothers depart immediately after birth). Young form large feeding schools or aggregations. Juveniles and adults may undertake different, partially sexually-segregated movements north-south and inshore-offshore.
Biology. Viviparous, yolk-sac placenta, 3–14 pups/litter (varies between regions but not correlated with size of mother). Females mature at ~17 years old, mate every other year, move inshore to give birth after an estimated 16 month gestation. Bony fishes are primary prey, followed by elasmobranchs and crustaceans, with other species also taken. Other big sharks eat young *Carcharhinus obscurus*.
Status. IUCN Red List: Near Threatened. One of the most vulnerable of vertebrates to exploitation because it reproduces so slowly. Difficult to manage or protect because taken in mixed species fisheries, and has a high mortality rate when taken as bycatch. Very occasionally bites people. Young are kept in aquaria.

Caribbean Reef Shark *Carcharhinus perezi* Plate 61

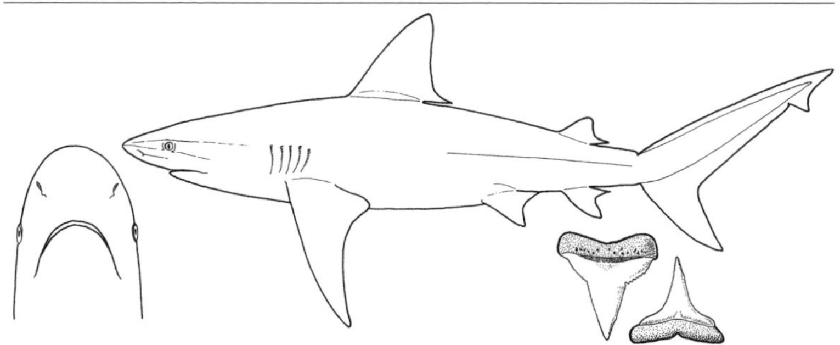

Measurements. Born: 70–73 cm TL. Mature: ~152–168 cm. Max: 295 cm TL.
Identification. Large dark grey or grey-brown reef shark, white below; undersides of paired fins, anal and ventral caudal lobe dusky but not prominently marked, inconspicuous white band on flanks. Short, bluntly rounded snout, narrow teeth, large narrow pectoral fins, a small first dorsal and moderately large second dorsal with short rear tips; interdorsal ridge present.

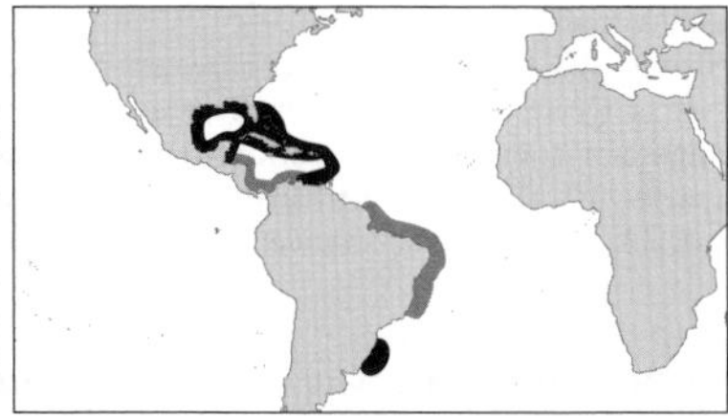

Distribution. West Atlantic and Caribbean: North Carolina (USA) to Brazil.
Habitat. Commonest Caribbean coral reef shark, near bottom and near dropoffs on outer reef to at least 30 m. On hard bottom (including calcareous algae) and on mud off river deltas in Brazil,
Behaviour. Poorly studied. Can lie motionless on the bottom, pumping water over gills with its pharynx. Can be closely approached while lying 'sleeping' in caves or in the open.
Biology. Viviparous, yolk-sac placenta, three to six pups / litter born after ~one year gestation every two years. There may be a pupping ground off the northern coast of Brazil. Eats bony fishes.
Status. IUCN Red List: Not Evaluated. Common and quite heavily fished for human consumption, hides, oil and fins but far more valuable for dive tourism. Often encountered by divers without incident; only rarely bites people (some during feeding to attract sharks to divers). Kept in a few large aquaria for public viewing, and has given birth in captivity.

Sandbar Shark *Carcharhinus plumbeus* Plate 58

Measurements. Varies between stocks. Born: 56–75 cm TL. Mature: ~140–180 cm. Max: ~240, possibly 300 cm TL.

Identification. Stout grey-brown or bronzy shark, underside white; tips and posterior edges of fins often dusky, but no obvious markings; white band on flank inconspicuous. Moderately long rounded snout, high triangular saw-edged upper teeth. Very large erect first dorsal fin with origin over/slightly ahead of large pectoral fins insert, interdorsal ridge present.

Distribution. World-wide, tropical and warm temperate waters.

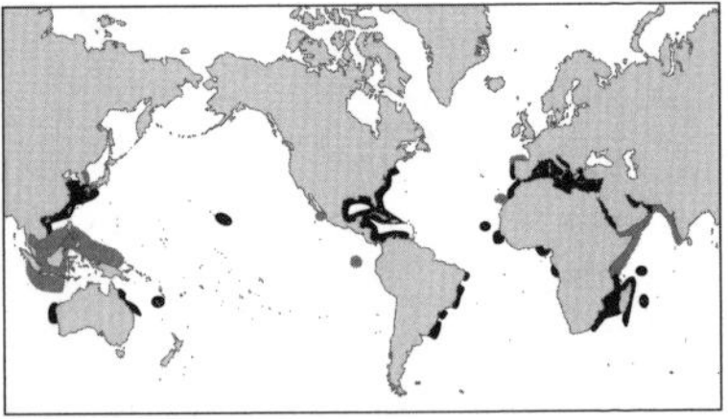

Habitat. Common in bays, harbours and at river mouths, also offshore in adjacent deep water and on oceanic banks. Usually near bottom, 20–55 m, ranging 1–280 m.

Behaviour. Some stocks migrate seasonally, often in large schools, as water temperatures change. Young form mixed-sex schools on shallow coastal nursery grounds, moving into deeper, warmer water in winter. Adults are segregated from juveniles (to reduce cannibalism?) sexes usually also separate except when mating in spring and summer. Males apparently follow and bite the female in the back until they swim upside down, then mate with both claspers. Mating wounds found on females during mating season. Feed slightly more actively at night.

Biology. Viviparous, yolk-sac placenta, 1–14 pups/litter (5–12 common) increasing with size of female, after a gestation of 8–12 months. Females give birth every second or third year. One of the slowest growing, latest maturing of known sharks. Feeds mainly on small bottom fishes, also some molluscs and crustaceans.

Status. IUCN Red List: Near Threatened. Large, slow-growing, late-maturing and low-fecundity coastal species, common and widespread in subtropical and warm temperate waters world-wide. An important component of shark fisheries in most areas where it occurs, although catch data are sparse. Severely overfished in the western North Atlantic where meat and fins are extremely valuable. Hardy and spectacular in public aquaria.

Smalltail Shark *Carcharhinus porosus* Plate 54

Measurements. Born: 31–40 cm TL. Mature: 72–78 cm ♂, ~84 cm ♀. Max: <150 cm TL.

Identification. Small grey shark, light below, with tips of pectoral, dorsal and caudal fins frequently dusky or blackish, not conspicuously so, and inconspicuous white flank band. Long pointed snout, short labial furrows, large circular eyes. Large falcate first dorsal fin, small second dorsal (origin over anal midbase), small pectoral fins, deeply notched anal fin, no interdorsal ridge.

Distribution. West Atlantic and east Pacific (continental coast, not offshore islands).

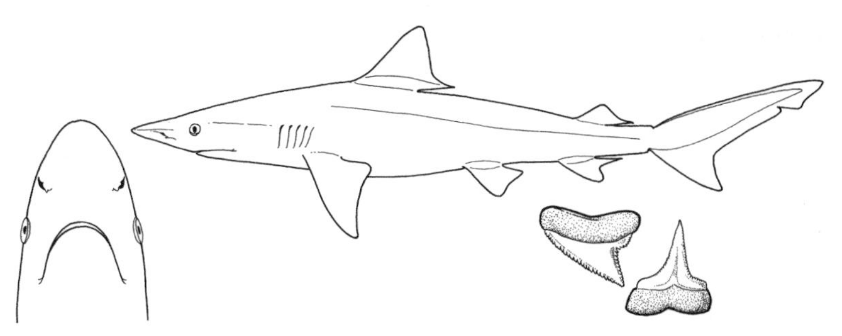

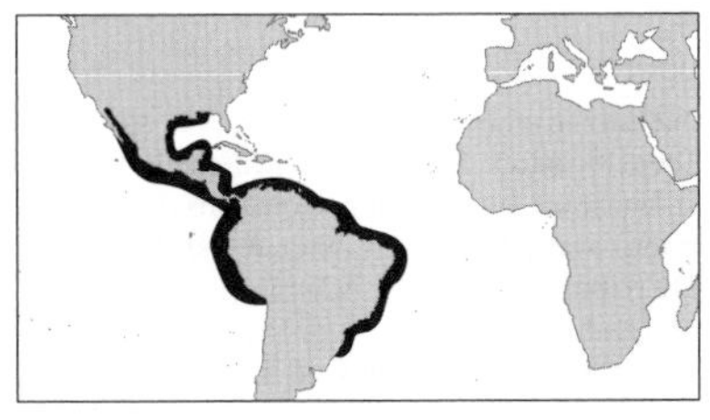

Habitat. Shallow continental shelf and estuaries, near muddy bottom from close inshore to at least 36 m.
Behaviour. Unknown.
Biology. Viviparous, yolk sac placenta, two to seven pups/ litter, probably annually. Eats bony fishes, small sharks, crabs and shrimp.
Status. IUCN Red List: Not Evaluated. Important in fisheries throughout range, but landings not recorded. Harmless to people.

Blackspot Shark *Carcharhinus sealei* Plate 60

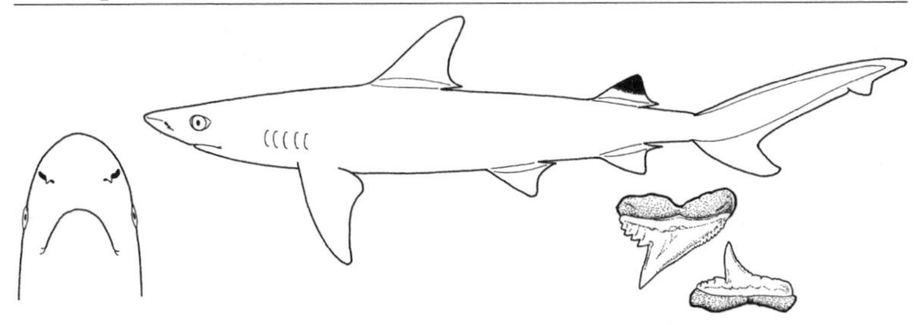

Measurements. Born: 33-36 cm TL. Mature: 70–80 cm ♂, 68–75 cm ♀. Max: 95 cm TL.
Identification. Small slender shark, grey or tan above, lighter below. Conspicuous black or dusky tip on second dorsal fin. Other fins with pale trailing edges, no dark marks. Inconspicuous light stripes on flanks. Long rounded snout, large oval eyes, oblique-cusped teeth. No or very low interdorsal ridge.

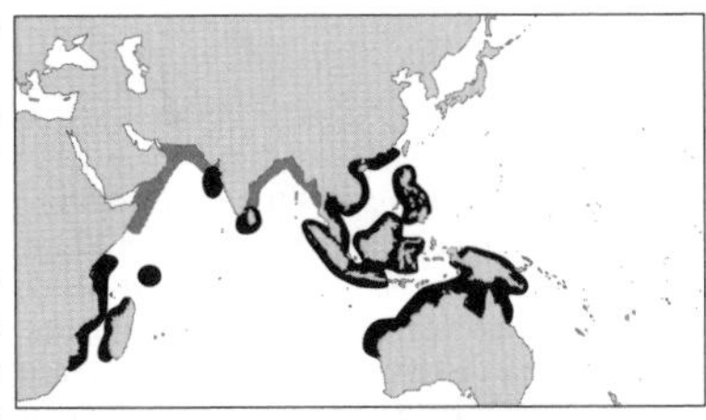

Distribution. Indo-west Pacific: South Africa to China.
Habitat. Coastal, continental and insular shelves, from surf line and intertidal to 40 m. Not off river mouths.
Behaviour. Not migratory off South Africa.
Biology. Viviparous, yolk sac placenta, one to two pups/ litter born in spring after ~nine month gestation. Fast-growing, mature at ~one year, maximum age at least five years. Eats small fishes, squid and prawns.
Status. IUCN Red List: Near Threatened. Occurs in intensively-fished coastal areas, declining in some areas.

Night Shark *Carcharhinus signatus* Plate 62

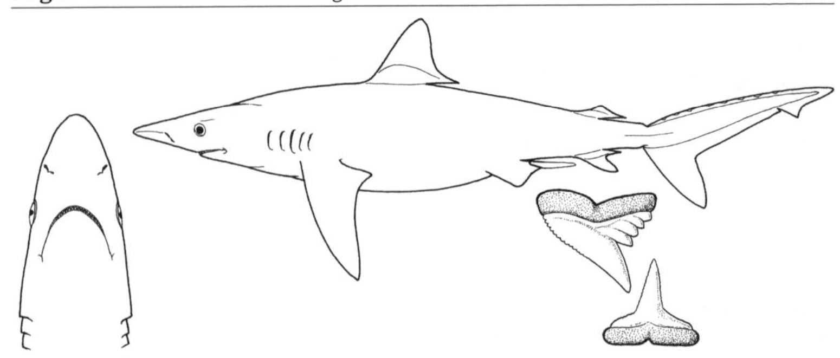

Measurements. Born: 60–72 cm TL. Mature: ~190–200 cm. Max: 280 cm TL.
Identification. Slim grey-brown shark, white below, no conspicuous fin markings. Long, pointed snout, small jaws, small pectoral fins, large eyes, oblique-cusped serrated upper teeth, front of first dorsal over pectoral fins, both dorsals low with elongated rear tips, interdorsal ridge.
Distribution. Tropical Atlantic: USA to Argentina and west Africa.

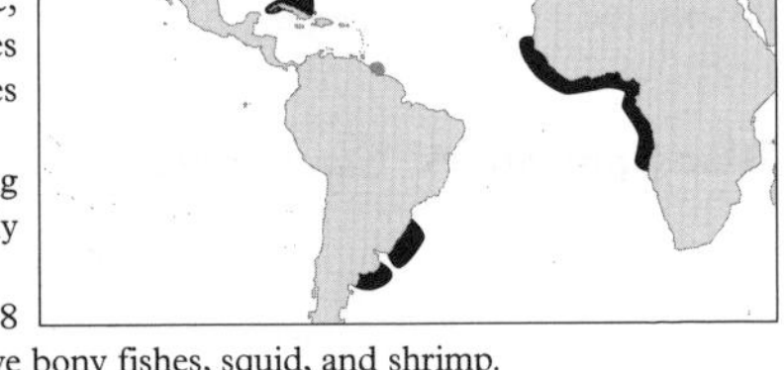

Habitat. Deepwater coastal and semi-oceanic, on or along outer continental and insular shelves and off upper slopes. Prefers 50–100 m, ranges 0–600 m.
Behaviour. Active schooling shark, migrating vertically into shallower water at night. Possibly seasonal geographic migrations.
Biology. Viviparous, yolk-sac placenta, 4–18 (usually 12–18) pups/litter. Feeds on small active bony fishes, squid, and shrimp.
Status. IUCN Red List: Not Evaluated. Formerly very common in Caribbean fisheries, now apparently rare. Harmless to people.

Spottail Shark *Carcharhinus sorrah* Plate 59

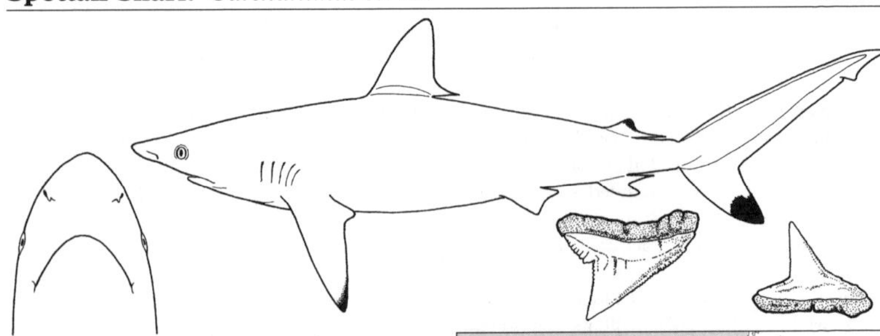

Measurements. Born: 45–60 cm TL. Mature: 106 cm ♂, 110–118 cm ♀. Max: 160 cm TL.
Identification. Small, spindle-shaped shark, medium-grey above, white below, with large conspicuous black tip on pectorals; conspicuous white band on flank. Very low elongated second dorsal, and ventral caudal lobe (first dorsal plain or black edged). Long rounded snout, large circular eyes, oblique-cusped serrated teeth, interdorsal ridge.

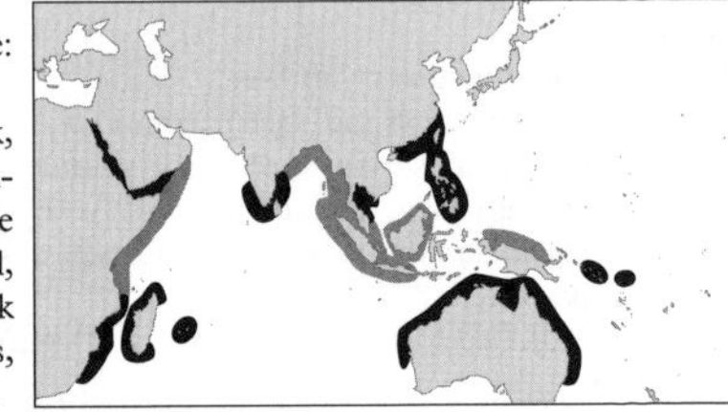

Distribution. Tropical Indo-west Pacific: South Africa to China and Solomon Islands.
Habitat. Continental and insular shelves, shallow water and around coral reefs, usually 20–50 m, ranging 0–140 m.
Behaviour. Young occur in quiet, shallow water, segregated from adults.
Biology. Viviparous, yolk-sac placenta, one to eight pups/litter, commonly six, after ten month gestation. Feeds on bony fishes and octopi.
Status. IUCN Red List: Data deficient. Fished heavily in part of range but not monitored.

Australian Blacktip Shark *Carcharhinus tilsoni* Plate 62

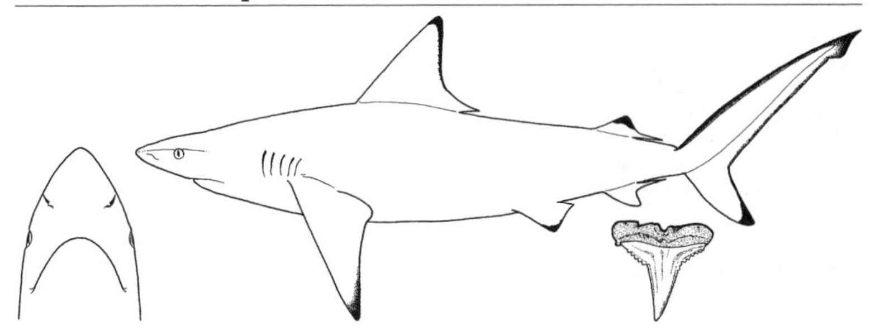

Measurements. Born: 60 cm TL. Mature: ~110 cm ♂, ~115 cm ♀. Max: 200 cm TL.
Identification. Medium-sized bronzy to grey above, pale below, pale flank stripe, and black-tipped fins (pelvic and anal fins may be plain). Long snout, slender erect serrated teeth, first dorsal origin ~over pectoral fin insertions, no interdorsal ridge. Distinguished from *Carcharhinus limbatus* by vertebral counts.
Distribution. Tropical Australia.
Habitat. Continental shelf waters, close inshore to ~150 m, midwater or near surface.
Behaviour. Often in large aggregations.
Biology. Viviparous, one to six pups/litter born in January after ~ten month gestation. Matures in three to four years. Feeds on bony fishes and cephalopods.
Status. IUCN Red List: Least Concern. Fast growth rates, early maturity and relatively high fecundity make it fairly resilient to fisheries, catches have been reduced under management.

False Smalltail Shark *Carcharhinus* sp. A Plate 62

Measurements. Born: < ~ 34 cm TL. Mature: unknown. Max. < ~ 100 cm TL.
Identification. Small slender grey shark, light below, inconspicuous white flank band. Pectoral, dorsal and caudal fin webs often dusky but not conspicuously marked. Moderately long bluntly pointed snout, large eyes, short labial furrows, pores alongside mouth corners not greatly enlarged. No interdorsal ridge, large first dorsal (origin over rear thirds of pectoral bases), small second dorsal (origin over first third of anal base), moderately large pectoral fins, broadly notched anal fin.

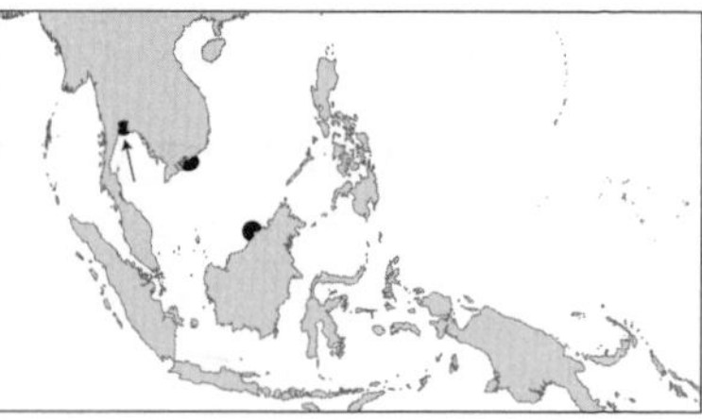

Distribution. South China Sea: Borneo (Sarawak), Vietnam (Ho Chi Minh City) and Thailand (Bangkok).
Habitat. Unknown, presumably inshore.
Behaviour. Unknown.
Biology. Essentially unknown.
Status. IUCN Red List: Not Evaluated. Very rare, only three specimens known, all collected in fish markets, last in 1934. Formerly placed in American species *Carcharhinus porosus*, but has shorter blunter snout, larger more anterior first dorsal fin, larger pectoral fins, and differences in the structure of its skull.

Tiger Shark *Galeocerdo cuvier* Plate 50

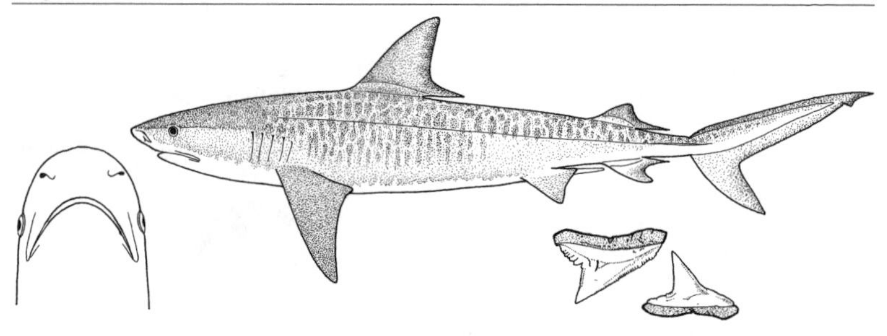

Measurements. Born: 51–76 cm TL. Mature: 226–290 cm ♂, 250–350 cm ♀. Max: >550 cm TL (one record of 740 cm).
Identification. Colour grey above with vertical black to dark grey bars and spots, bold in young but fading in adults, white below. Huge striped shark with a broad, bluntly rounded snout, very long upper labial furrows, big mouth with large, saw-edged, cockscomb-shaped teeth, spiracles. Prominent interdorsal ridge and low caudal keels.
Distribution. World-wide, temperate and tropical seas.

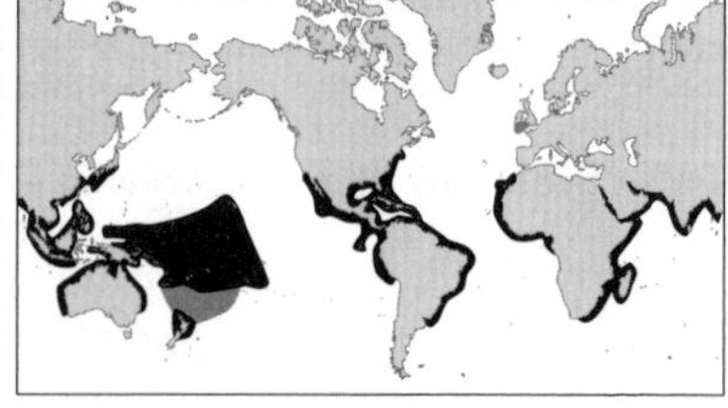

Habitat. On or near continental and insular shelves, surface and intertidal to possibly 140 m. May prefer turbid areas with high runoff of fresh water, including estuaries and harbours, also coral atolls and lagoons. Can travel long distances between islands.
Behaviour. Apparently nocturnal, moving in-shore at night to very shallow water, deeper areas by day. Smaller Tiger Sharks may be more active by day. Usually solitary but may aggregate when feeding. Strong swimmers; often moving through a very large semi-resident home range.
Biology. The only ovoviviparous carcharhinid. Very large litters (10–82 pups) born in spring-early summer. Gestation may be slightly over a year. Fast growing: mature at four to six years, longevity at least 12 years. Sometimes called 'a garbage can with fins': eats bony fishes, elasmobranchs (including Tiger Shark pups), sea turtles, sea snakes, marine iguanas, seabirds, marine mammals, carrion and rubbish.
Status. IUCN Red List: Near Threatened. Common or formerly common large shark but caught

regularly in target and non-target fisheries (fins are high value). Some declines reported, but juvenile survival increases when adults are depleted. Potentially harmful; responsible for fatal attacks on people, but often unaggressive when encountered underwater. Valuable for dive tourism.

Identification of *Glyphis* species

The Indo-west Pacific river sharks (Genus *Glyphis*) are rare, poorly known and difficult to identify. Specimens and good photographs of whole animals, measurements, jaws and teeth, and ideally vertebral counts (see table) are essential for confirmation.

SPECIES	TOOTH ROWS	VERTEBRAL COUNTS			
		total	precaudal	monospondylous precaudal	caudal
Glyphis gangeticus	30–37/31–34	169	80	50	89
Glyphis glyphis (possibly = *Glyphis* sp. A)	probably 26–29/27–29 (Papua New Guinea)				90 (type specimen)
Glyphis siamensis	29/29	209	117	66	92
Glyphis sp. A	26–29/27–29	213–217	124	70–71	93
Glyphis sp. B	28–31/29–32	196–205	108–114	63–67	82–92
Glyphis sp. C	32–34/32–34	147–148	79–83	46–50	65–68

Ganges Shark *Glyphis gangeticus* Plate 55

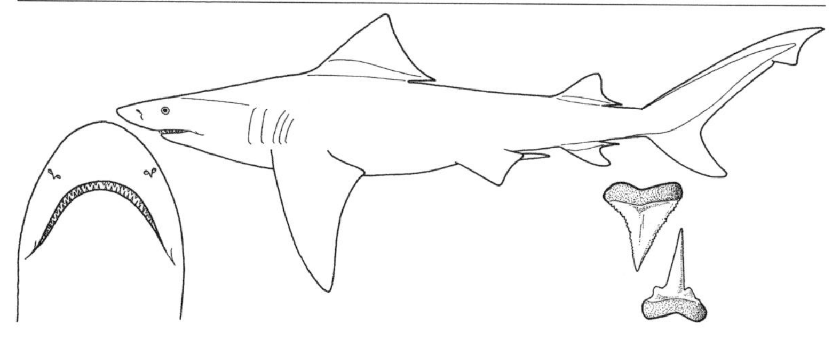

Measurements. Born: 56–61 cm TL. Mature: ~178 cm ♂. Max: possibly at least 204 cm TL and probably larger.

Identification. Large stocky shark, grey above, white below, without conspicuous markings. Broadly rounded short snout, tiny eyes, upper teeth with high, broad, serrated triangular cusps, first few lower front teeth with weakly serrated cutting edges and low cusplets on crown foot. No interdorsal ridge, longitudinal upper precaudal pit, first dorsal origin over rear thirds of pectoral bases, second dorsal about half height of first, anal fin with deeply notched posterior margin.

Distribution. Indo-west Pacific: Ganges-Hooghly river system, India. Unconfirmed old record from near Karachi, Pakistan. Other (nominal) records over a wide range in the Indo-west Pacific were often based on *Carcharhinus leucas* or other species and cannot be confirmed.

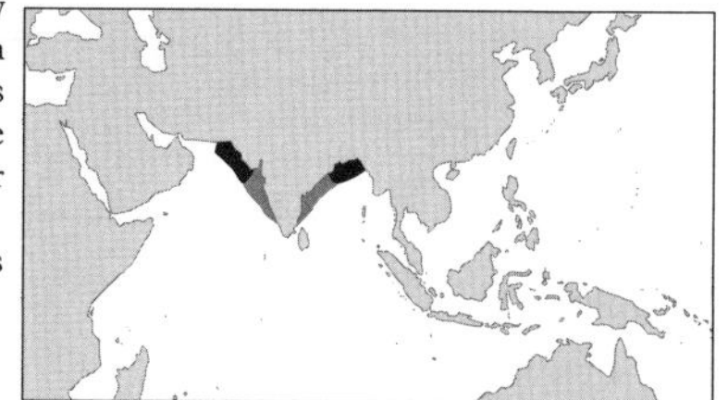

Habitat. Freshwater in rivers, possibly estuaries and inshore marine.

Behaviour. Unknown.

Biology. Probably viviparous.

Status. IUCN Red List: Critically Endangered. Originally known from three 19th century museum specimens. Recent records need confirmation; a set of jaws was recently collected from a specimen of what seems to be this species in the mouth of the Hooghly River. Habitat heavily fished and degraded. Man-eating reputation may be caused by *C. leucas*.

Speartooth Shark *Glyphis glyphis* Plate 55

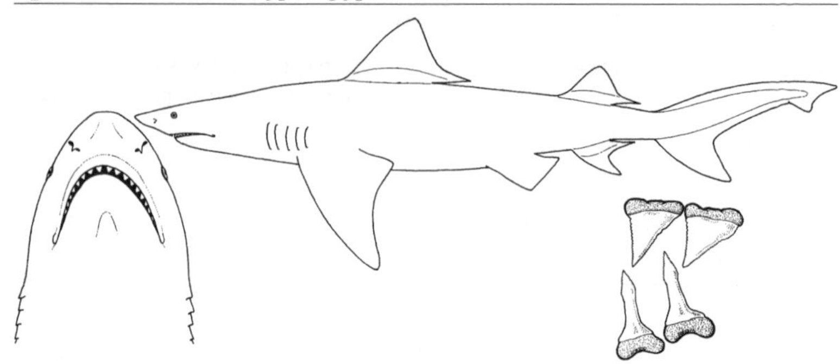

Measurements. Max: at least 100 cm, possibly 200–300 cm TL.
Identification. Large stocky shark, grey above, white below, without conspicuous markings. Broadly rounded short snout, tiny eyes, upper teeth with high, broad, serrated triangular cusps, first few lower front teeth teeth with long, hooked, protruding cusps, with unserrated or serrated cutting edges confined to spearlike tips and no cusplets. No interdorsal ridge, longitudinal upper precaudal pit, first dorsal origin over rear ends of pectoral bases, second dorsal rather large (~3/5 height of first), anal fin with deeply notched posterior margin.
Distribution. Indo-Pacific. Originally known from a single specimen without locality. Jaws of large river sharks with similar teeth and tooth counts have been collected in Papua New Guinea and in the Bay of Bengal.
Habitat. Presumably inshore, possibly rivers, estuaries and coastal waters.

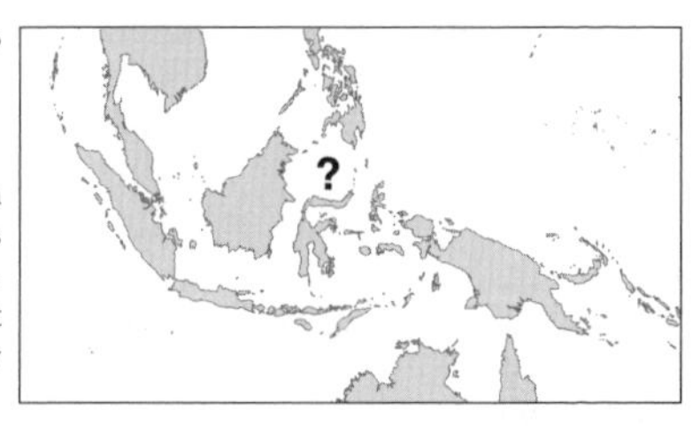

Behaviour. Unknown.
Biology. Unknown.
Status. IUCN Red List: Endangered. Known from a single museum specimen and a few jaws of possibly the same species. Extremely rare. Probably confined to habitats under significant development and exploitation pressure. Possibly the same as *Glyphis* sp. A.

Irrawaddy River Shark *Glyphis siamensis* Plate 55

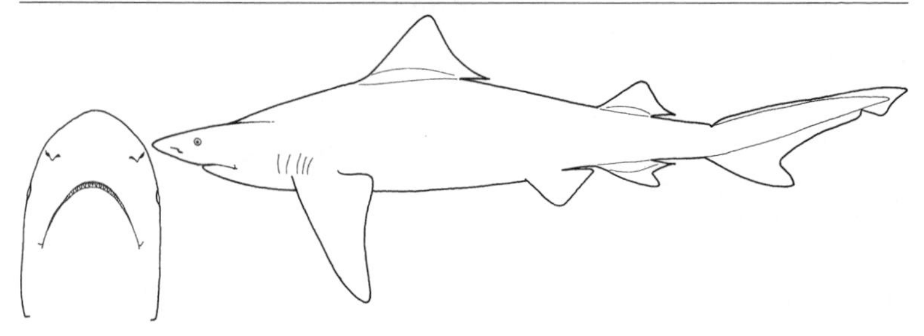

Measurements. Birth and maturity: unknown. Single known specimen: 60 cm immature male. Max: unknown, probably 100–300 cm TL.

Identification. Stocky shark, grey-brown above, white below, with no conspicuous markings, broadly rounded short snout, tiny eyes. Upper teeth with high, broad, serrated triangular cusps, first few lower front teeth with weakly serrated cutting edges and low cusplets on crown foot. No interdorsal ridge, longitudinal upper precaudal pit, first dorsal origin over rear ends of pectoral bases, second dorsal about half height of first, anal fin with deeply notched posterior margin.

Distribution. Indian Ocean: Myanmar (Irawaddy River mouth).

Habitat. Estuarine or riverine?

Behaviour. Unknown.

Biology. Unknown.

Status. IUCN Red List: Not Evaluated. Extremely rare. Known from single 19th century museum specimen. Habitat heavily fished and possibly degraded. Close to *Glyphis gangeticus* but with higher vertebral counts and fewer teeth.

Bizant River Shark *Glyphis* sp. A [? = *G. glyphis*] Plate 55

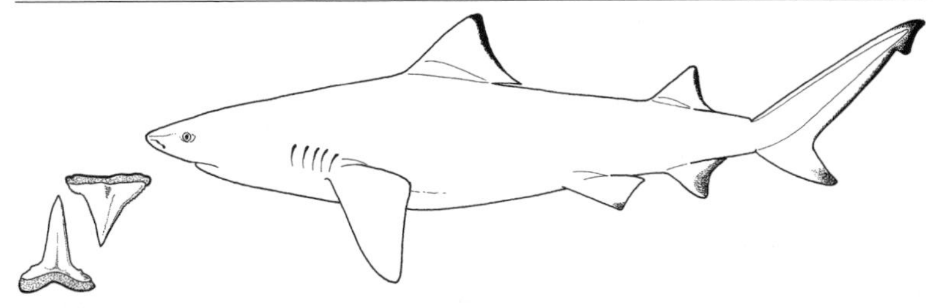

Measurements. Specimens range 70–131 cm TL. May reach 200–300 cm TL.

Identification. Stocky greyish shark with inconspicuous pale flank stripe and dusky flank patch, blackish fin margins in young, otherwise no conspicuous markings. Broadly rounded short snout, tiny eyes, upper teeth with high, broad, serrated triangular cusps, first few lower front teeth with long, hooked, protruding cusps with entire, unserrated cutting edges confined to spearlike tips and small cusplets. No interdorsal ridge, longitudinal upper precaudal pit, first dorsal origin over rear thirds of pectoral bases, second dorsal rather large (> 3/5 height of first), anal fin with deeply notched posterior margin.

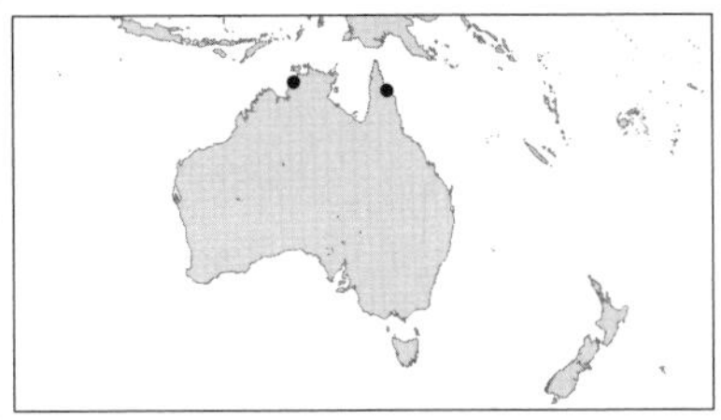

Distribution. Australia: Bizant and Adelaide Rivers.

Habitat. Presumably inshore and often in freshwater.

Behaviour. Unknown.

Biology. Unknown, presumably viviparous.

Status. IUCN Red List: Critically Endangered. Known from a few specimens collected in 1982 and 1999. Not recorded during 2002 River Shark survey. Very rare, population extremely small and fragmented. Possibly the same as *Glyphis glyphis*; differences in lower anterior teeth shape between the two species could be size-related or definitive and need further study.

Borneo River Shark *Glyphis* sp. B Plate 55

Measurements. Born: ~50 cm TL. Mature: unknown, specimens immature at 50–78 cm. Max: possibly > 200 cm TL.

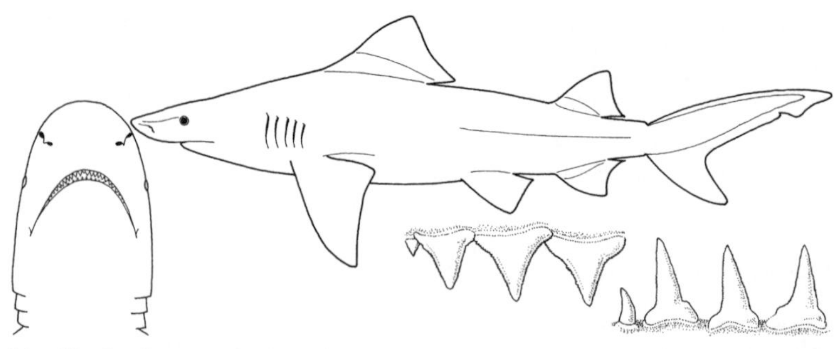

Identification. Large, stocky shark with dusky to blackish fin margins and tips, conspicuous dusky blotch on each flank and pectoral base. Broadly rounded short snout, tiny eyes, upper teeth with high, broad, serrated triangular cusps, first few lower front teeth with weakly serrated cutting edges and low cusplets on crown foot. No interdorsal ridge, longitudinal upper precaudal pit, first dorsal origin over rear thirds of pectoral bases, second dorsal slightly more than half to two-thirds of first, anal fin with deeply notched posterior margin.

Distribution. Borneo.

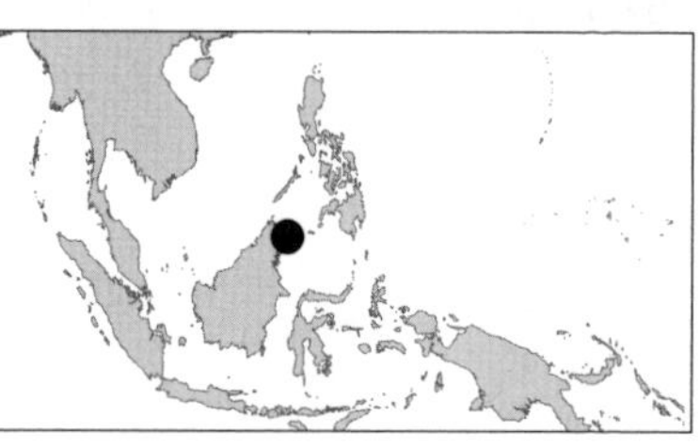

Habitat. Turbid brackish to freshwater in rivers.

Behaviour. Unknown.

Biology. Presumably viviparous.

Status. IUCN Red List: Not Evaluated. Very rare, population extremely small and fragmented. Habitat heavily fished and possibly degraded by logging and other land-based activities. Close to *Glyphis siamensis* but differs in vertebral counts, coloration and higher second dorsal fin.

New Guinea River Shark *Glyphis* sp. C — Plate 55

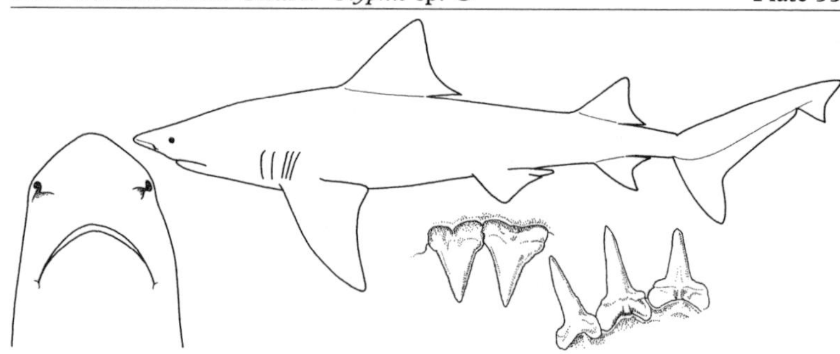

Measurements. Born: below 70 cm TL. Mature: ~200 cm ♂. Max: >200 cm TL.

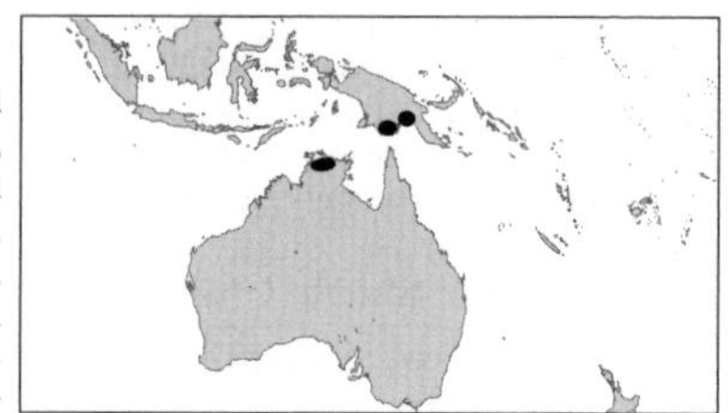

Identification. Large, slender, rather flat-headed grey shark without conspicuous body markings, dusky fin margins, white below. Broadly rounded short snout, tiny eyes, upper teeth with high, broad, serrated triangular cusps, first few lower front teeth with long, hooked, protruding cusps, with serrated cutting edges confined to spearlike tips and no cusplets. No interdorsal ridge, longitudinal upper

precaudal pit, first dorsal origin over rear thirds of pectoral bases, second dorsal about two-thirds height of first, anal fin with deeply notched posterior margin.
Distribution. Northern Australia, Papua New Guinea.
Habitat. Turbid and brackish freshwater in rivers and adjacent marine waters.
Behaviour. Unknown.
Biology. Unknown, presumably viviparous.
Status. IUCN Red List: Critically Endangered. Very rare, population extremely small and fragmented.

Daggernose Shark *Isogomphodon oxyrhynchus* Plate 54

Measurements. Born: 38–43 cm TL. Mature: 90–110 cm ♂, 105–112 cm ♀. Max: 152, possibly 200–244 cm TL.

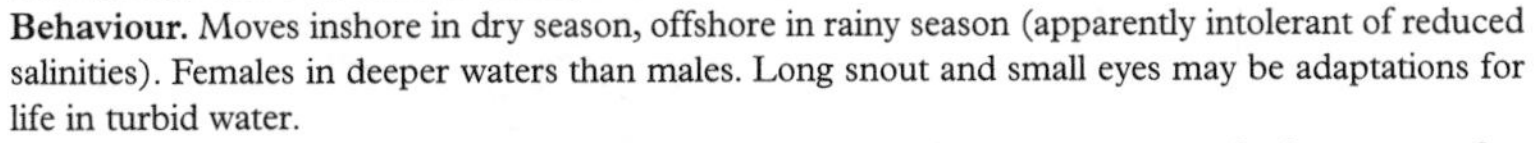

Identification. Unpatterned grey or yellow-grey shark with unmistakable extremely long, flat, pointed snout. Tiny circular eyes, large paddle-shaped pectoral fins, >45 rows of small spike-like teeth in both jaws, upper teeth serrated.
Distribution. Tropical west Atlantic, northern South America.
Habitat. Turbid water in estuaries, mangroves, river mouths and over shallow banks, 4–40 m.
Behaviour. Moves inshore in dry season, offshore in rainy season (apparently intolerant of reduced salinities). Females in deeper waters than males. Long snout and small eyes may be adaptations for life in turbid water.
Biology. Viviparous, yolk-sac placenta, three to eight pups/litter born at start of rainy season after ~one year gestation. Possible two year birth cycle. Feeds on small schooling fishes.
Status. IUCN Red List: Not Evaluated. Bycatch in fisheries. Populations declining steeply.

Broadfin Shark *Lamiopsis temmincki* Plate 59

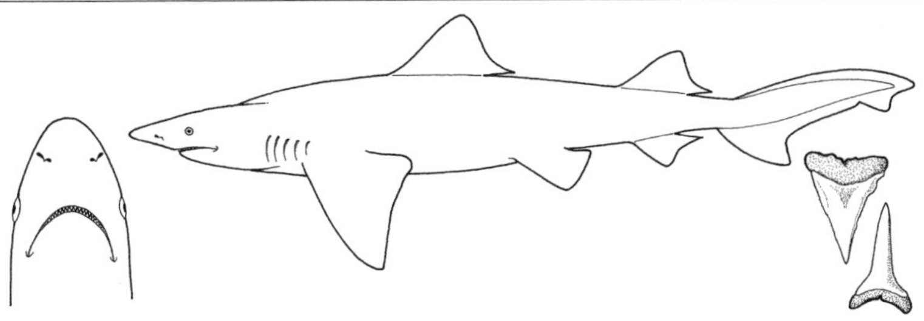

Measurements. Born: 40–60 cm TL. Mature: ~ 114 cm ♂, <130 cm ♀. Max: 168 cm TL.
Identification. Small, rather stocky light grey or tan above, light below, no prominent markings.

Moderately long snout nearly equal to mouth width, upper teeth serrated with broad triangular cusps, lower teeth smooth, hooked, narrow-cusped, small round eye. Fifth gill ~half height of first, second dorsal fin nearly as large as first, longitudinal upper precaudal pit, broad triangular pectoral fins, anal fin posterior margin nearly straight.

Distribution. Scattered in Indian Ocean and west Pacific. Common (formerly?) near Bombay, west India.

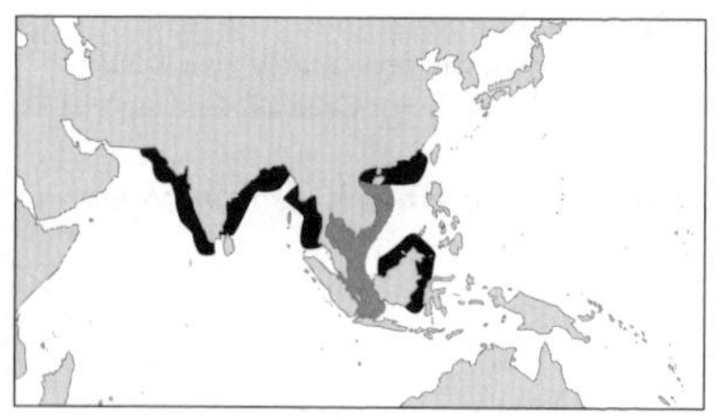

Habitat. Inshore continental shelf.

Behaviour. Unknown.

Biology. Viviparous, four to eight pups/litter (usually eight) born before the monsoon. Possibly eight month gestation.

Status. IUCN Red List: Not Evaluated. Rare. Taken in fisheries, trends unknown.

Sliteye Shark *Loxodon macrorhinus* Plate 60

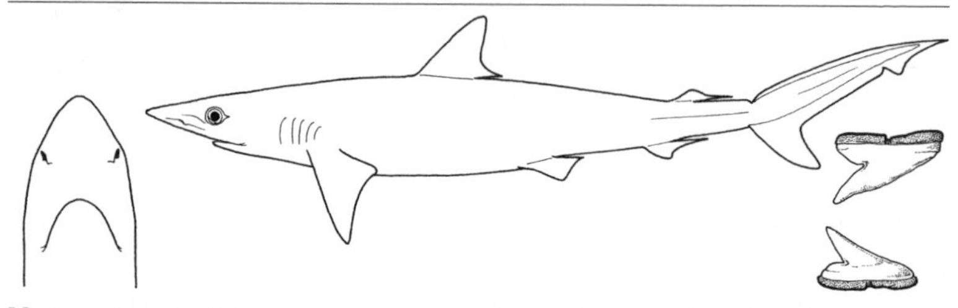

Measurements. Born: 40–43 cm TL. Mature: 62–66 cm ♂, 79 cm ♀. Max: 91 cm TL.

Identification. Small, very slim, grey to brownish, underside white, fins with inconspicuous light rear edges and black margin on first dorsal and caudal fins. Long narrow snout, big eyes with posterior notches, short labial furrows, small smooth-edged oblique-cusped teeth, small low second dorsal fin with very large free posterior margin behind larger anal fin, interdorsal ridge absent or rudimentary.

Distribution. Indo-west Pacific.

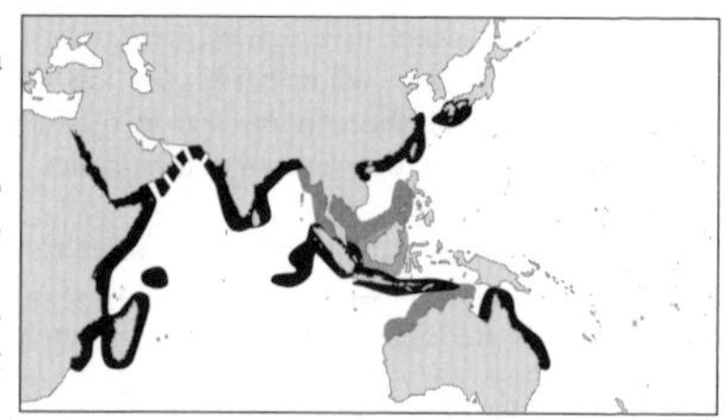

Habitat. Continental and insular shelves in shallow, clear water, 7–80 m.

Behaviour. Unreported.

Biology. Viviparous, yolk-sac placenta, two to four pups/litter. Eats small bony fishes, shrimp and cuttlefish.

Status. IUCN Red List: Not Evaluated. Commonly caught in fisheries. Presumed fast growing and able to sustain reasonable fishing pressure.

Whitenose Shark *Nasolamia velox* Plate 59

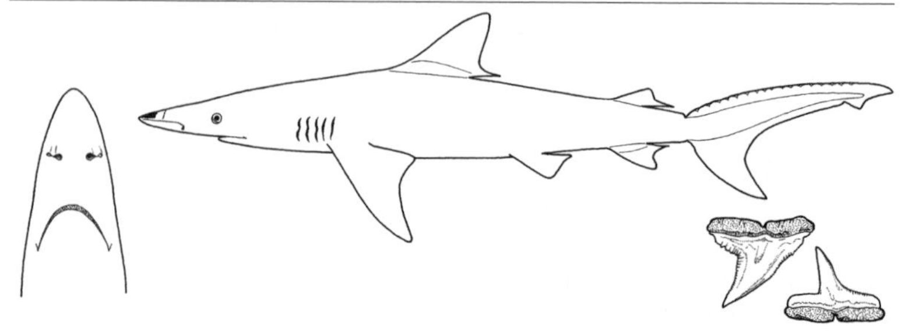

Measurements. Born: ~53 cm TL. Mature: >106 cm ♂. Max: at least 150 cm TL.
Identification. Slender grey-brown to light brown shark with very long conical snout, large round eyes, very large close-set nostrils separated by a space only slightly greater than the nostril width, and a prominent black spot outlined with white on upper snout tip.
Distribution. Tropical east Pacific, Central America: Baja California, Mexico, to Peru.

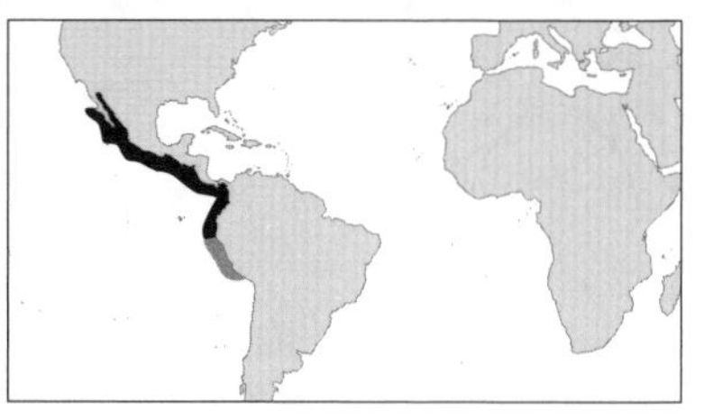

Habitat. Continental shelf, inshore and offshore, usually 15–24 m or less, occasionally to 192 m.
Behaviour. Unknown.
Biology. Viviparous, yolk-sac placenta, five pups/litter. Feeds on small bony fishes and crabs.
Status. IUCN Red List: Not Evaluated. Taken with longlines for food and fishmeal. Formerly uncommon to rare where it occurs.

Sharptooth Lemon Shark *Negaprion acutidens* Plate 53

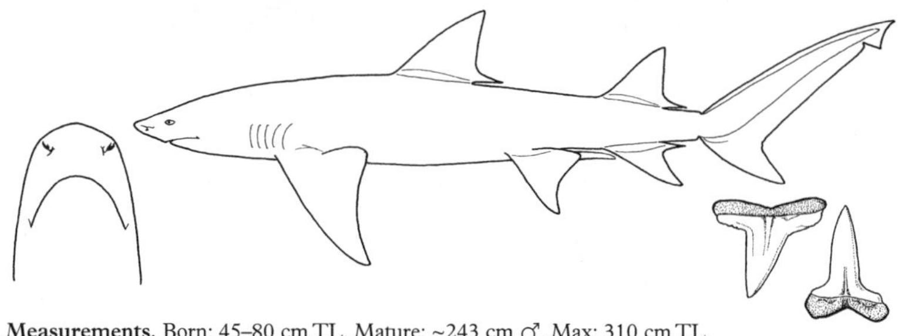

Measurements. Born: 45–80 cm TL. Mature: ~243 cm ♂. Max: 310 cm TL.
Identification. Big, stocky, yellowish shark, second dorsal fin aboutas large as first. Broad, blunt snout. Narrow, smooth-cusped teeth in both jaws.
Distribution. Tropical Indo-west and central Pacific.
Habitat. Inshore, on or near bottom, 0–30 m. Prefers turbid, still water in bays, estuaries, sandy plateaus, outer reef shelves and reef lagoons. Young sharks found on very shallow reef flats, dorsal fins exposed.
Behaviour. Sluggish, swimming slowly near or resting on the bottom. May surface when stimulated by food, but shy and reluctant to approach divers. Young sharks are more inquisitive.

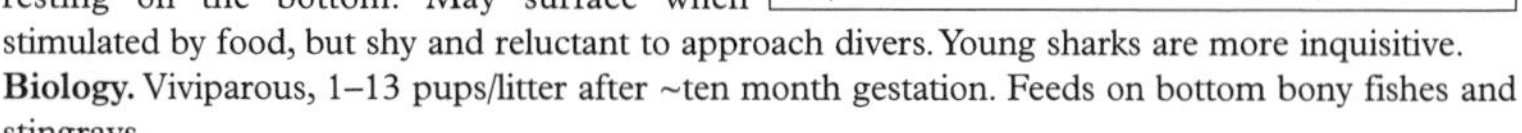

Biology. Viviparous, 1–13 pups/litter after ~ten month gestation. Feeds on bottom bony fishes and stingrays.
Status. IUCN Red List: Vulnerable. Heavily fished outside Australia throughout range. Hardy in captivity. Valuable for dive tourism. Potentially harmful, can be aggressive when provoked.

Lemon Shark *Negaprion brevirostris* Plate 61

Measurements. Born: 60–65 cm TL. Mature: ~224 cm ♂, ~239 cm ♀. Max: 340 cm TL.
Identification. Big, stocky, short-nosed, pale yellow-brown shark. Second dorsal about as large as first, narrow, smooth-cusped teeth in both jaws. Usually with less falcate fins than *Negaprion acutidens.*
Distribution. Tropical west Atlantic, northeast Atlantic (west Africa), east Pacific (Mexico to Ecuador).
Habitat. Inshore and coastal, from surface and intertidal to at least 92 m. Usually around coral keys, mangrove fringes, docks, on sand or coral mud, in saline creeks, enclosed sounds or bays, and river mouths. Adapted to low oxygen shallow water environments. May travel short distances upriver.

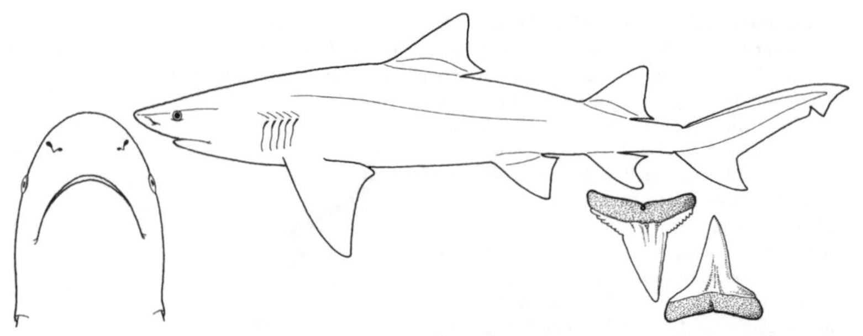

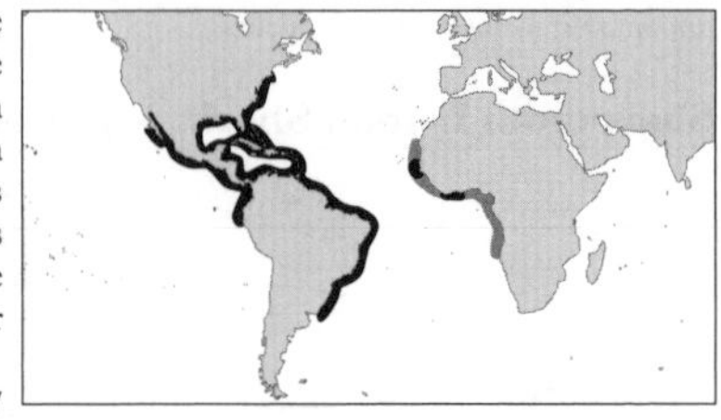

Behaviour. Occurs singly or forms loose aggregations of up to 20 individuals, with some segregation by size and sex. Most active at dawn and dusk. Pups have small home ranges within nursery grounds (that are only visited by adults when females give birth). Their range expands as sharks grow. Can rest on the bottom. Some populations migrate seasonally, sometimes at or near surface in open ocean.

Biology. Viviparous, yolk-sac placenta, 4–17 pups/litter after a 10–12 month gestation. Slow-growing, maturing at 6.5 years old, longevity ~27 years. Mainly eats fishes, also crustaceans and molluscs.

Status. IUCN Red List: Near Threatened. Some coastal nursery grounds are subject to habitat deterioration. Evidence of some population depletion by largely unmanaged fisheries in east Pacific and west Atlantic. Valuable for dive tourism. Potentially harmful, can be aggressive when provoked.

Blue Shark *Prionace glauca* Plate 51

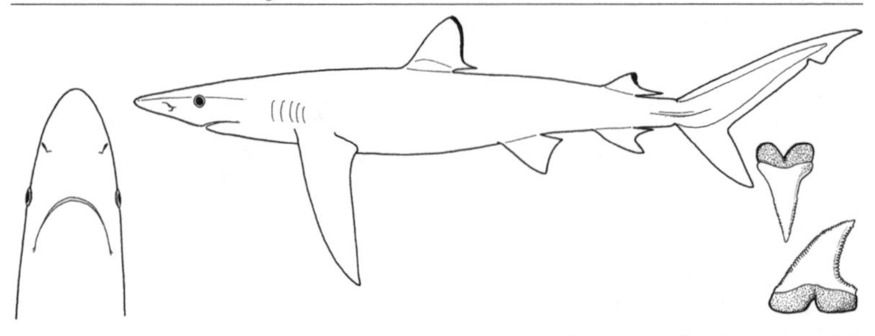

Measurements. Born: 35–44 cm TL. Mature: 182–281 cm ♂, ~220 cm ♀. Max: 380 cm TL (larger records unconfirmed).

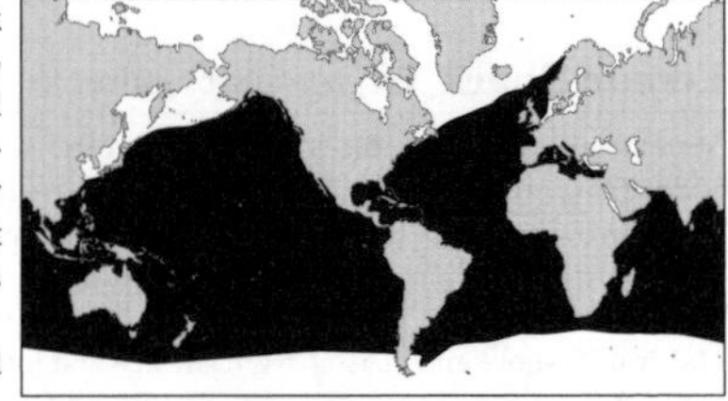

Identification. Slim, graceful shark with dark blue back, bright blue flanks and sharp demarcation to white underside. Long conical snout, large eyes (no spiracles), curved, saw-edged, triangular upper teeth. Long narrow scythe-shaped pectoral fins well in front of first dorsal fin, no interdorsal ridge. Mature females often bear bite wounds (mating scars).

Distribution. World-wide in temperate and tropical oceanic waters (temperature 7–25°C, preferably 12–20°C, latitude 60°N to 50°S). Possibly the most wide-ranging of sharks.

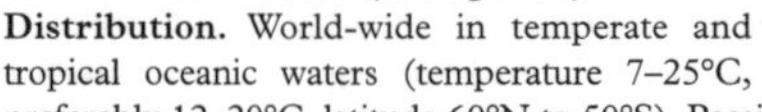

Habitat. Oceanic and pelagic, usually off the edge of the continental shelf, 0–350 m (deeper in warmer waters). Migrations often follow major trans-oceanic currents. Occasionally venture inshore at night, particularly around oceanic islands or where the continental shelf is narrow. Nursery areas offshore.
Behaviour. Cruise slowly at the surface, tips of dorsal and tail fins out of the water and long pectoral fins extended. Most active in early evening and at night, when they may move inshore. Forms large aggregations (where still sufficiently abundant) to feed on shoals of prey or carrion. Have been seen biting at floating objects. Known to harass spearfishers. May circle swimmers, boats and divers for some time before approaching and biting. Highly migratory, with complex movements related to prey availability and reproductive cycles. Segregate by age, sex and reproductive phase: juveniles, sub-adults, mature sharks and pregnant females are usually found in separate areas, with adult males and females meeting only briefly to mate. Move seasonally to higher latitudes where prey is more abundant in productive oceanic convergence or boundary zones. Frequent vertical excursions made into deep water or to the thermocline, returning regularly to the surface (possibly to prevent body cooling). Tagging studies have demonstrated that Atlantic Blue Sharks undertake numerous trans-Atlantic migrations, swimming slowly with the major current systems. Pacific Blue Sharks may migrate up to 9200 km.
Biology. Viviparous, yolk-sac placenta, 4–135 pups/litter (usually 15–30) born in spring and summer after 9–12 month gestation. In European waters, pups remain in offshore nursery area until ~130 cm TL, when they begin to migrate with other sharks of the same age and sex. Males mature at four to six years, females at five to seven years old. Mature females may breed annually, or on alternate years. Longevity ~20 years. Feed on relatively small prey: usually squid and pelagic fish, also invertebrates, bottom-dwelling fish and small sharks, sometimes seabirds taken at the surface.
Status. IUCN Red List: Near Threatened. The most heavily fished shark in the world, many millions taken annually, mainly as bycatch. Meat is low value but large valuable fins are kept and enter the international shark fin trade (often after carcasses are discarded at sea). Recent reports of ~60–80% decline in catch rates and reductions in sightings frequency, but data are inadequate to assess global population decline. Valuable for offshore dive tourism. Potentially harmful, responsible for a few fatal bite incidents on people, but often timid.

Milk Shark *Rhizoprionodon acutus* Plate 56

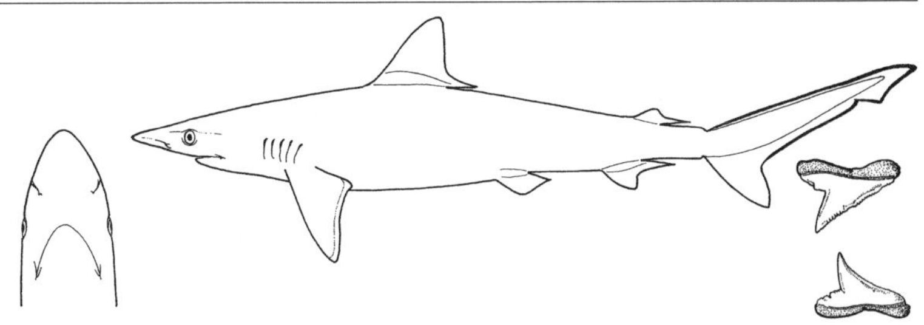

Measurements. Born: 25–39 cm TL. Mature: ~68–72 cm ♂, ~70–81 cm ♀. Max: 178 cm TL (usually <110 cm).
Identification. Small bronze to greyish shark, white below, most fin tips slightly pale; juvenile dorsal and upper caudal fin tips dark (sometimes in adults). Long narrow snout, big eyes, oblique narrowly triangular smooth-edged teeth, small low second dorsal fin behind larger anal fin. Only requiem shark in its range with long upper and lower labial furrows, second dorsal origin well behind larger anal fin origin, and long preanal ridges.

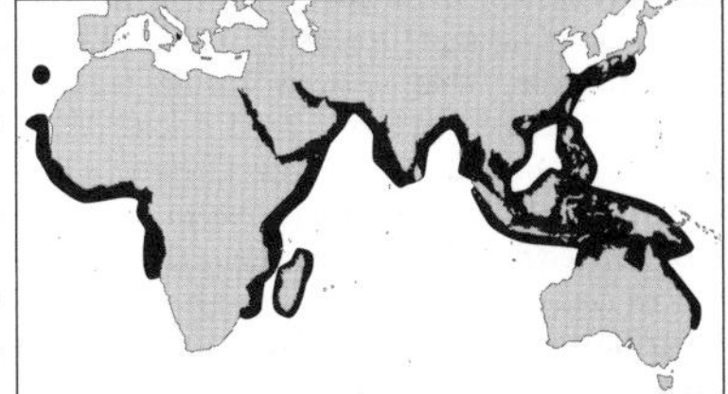

Distribution. East Atlantic, Indo-west Pacific, also Mediterranean Sea (Gulf of Taranto off Italy).
Habitat. Continental shelf, midwater to near bottom, 1–200 m. Often off sandy beaches, sometimes in estuaries (not very low salinity).

Behaviour. Unknown.
Biology. Viviparous, yolk-sac placenta, one to eight pups/litter (usually two to five), ~one year gestation. Matures at two years. Feeds mainly on bony fishes. Eaten by larger sharks.
Status. IUCN Red List: Least Concern. Heavily fished, but productive, common and widespread.

Brazilian Sharpnose Shark *Rhizoprionodon lalandei* Plate 60

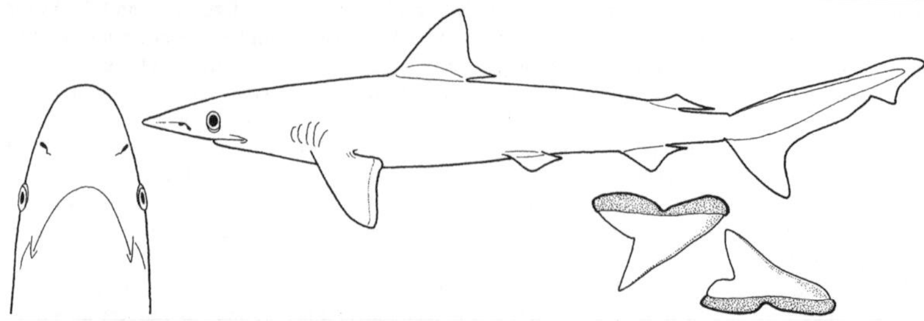

Measurements. Born: 33–34 cm TL. Mature: 45–50 cm ♂, <54 cm ♀. Max: ~77 cm TL.
Identification. Only requiem shark in its range with long upper and lower labial furrows, small second dorsal fin with origin far behind anal origin, long anal ridges, and apex of pectoral fin when fin is laid back against body falling in front of first dorsal midbase. Dark grey or grey-brown above, light below, light margins to pectorals and dusky dorsal fins. Long snout, small, wide-spaced nostrils, no spiracles, large eyes.
Distribution. West Atlantic: Panama to southern Brazil.

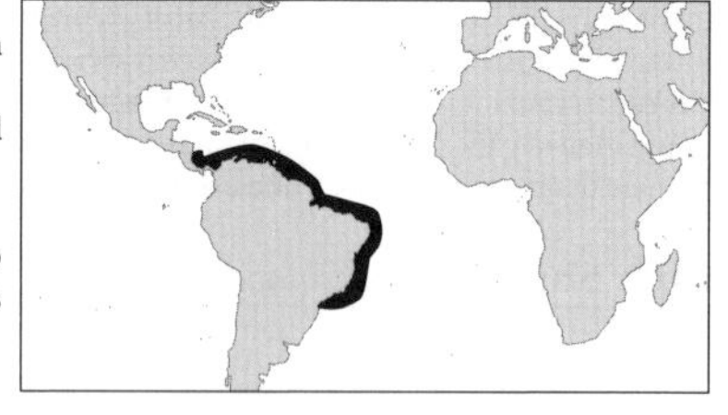

Habitat. Littoral, continental shelf, 3–70 m, on sand and mud, not normally in lagoons and estuaries.
Behaviour. Unknown.
Biology. Viviparous, yolk-sac placenta, one to four pups/litter. Summer mating season. Eats small bony fishes, shrimp and squid.
Status. IUCN Red List: Not Evaluated.

Pacific Sharpnose Shark *Rhizoprionodon longurio* Plate 56

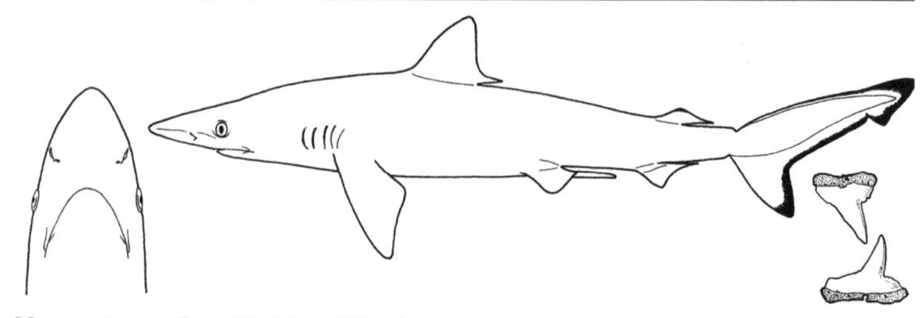

Measurements. Born: 33–34 cm TL. Mature: 58–69 cm ♂. Max: 110, possibly 154 cm TL.
Identification. Only eastern Pacific requiem shark with long labial furrows and second dorsal origin well behind anal origin. Grey or grey-brown above, white below, light edged pectorals, dusky tipped dorsal fins. Long snout, small, wide-spaced nostrils, no spiracles, large eyes.
Distribution. Tropical east Pacific: California to Peru.

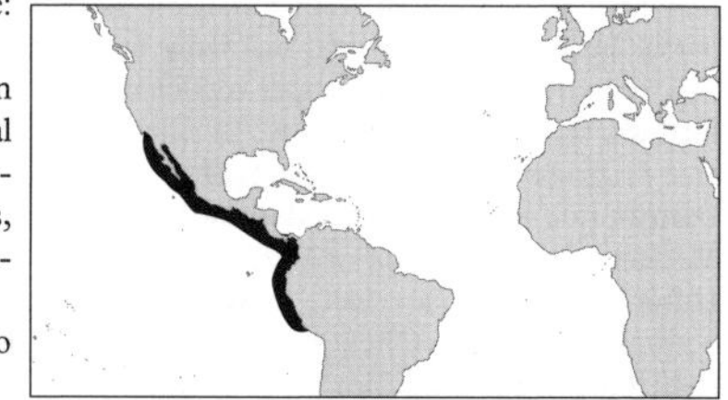

Habitat. Littoral on continental shelf, from intertidal to at least 27 m.
Behaviour. Unknown.
Biology. Poorly known. Probably viviparous.
Status. IUCN Red List: Not Evaluated. Locally very abundant.

Grey Sharpnose Shark *Rhizoprionodon oligolinx* Plate 60

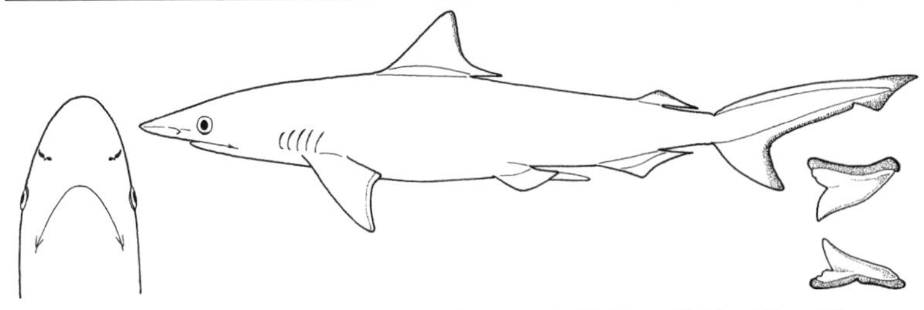

Measurements. Born: 21–26 cm TL. Mature: 29–38 cm ♂, 32–41 cm ♀. Max: 70 cm TL.
Identification. Very small shark, grey or brownish-grey to bronzy above, pale below, inconspicuous dusky fin edges. Long snout, short labial furrows, small wide-spaced nostrils, no spiracles, large eyes, oblique and narrow-cusped small teeth in both jaws. Second dorsal origin far behind anal origin.

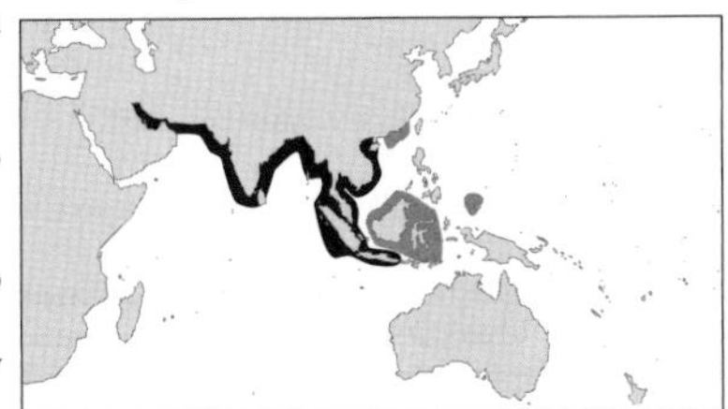

Distribution. Tropical Indo-west Pacific.
Habitat. Littoral, continental and insular shelves, inshore and offshore.
Behaviour. Unknown.
Biology. Viviparous, yolk-sac placenta, three to five pups/litter.
Status. IUCN Red List: Least Concern. Heavily fished, but abundant and productive.

Caribbean Sharpnose Shark *Rhizoprionodon porosus* Plate 54

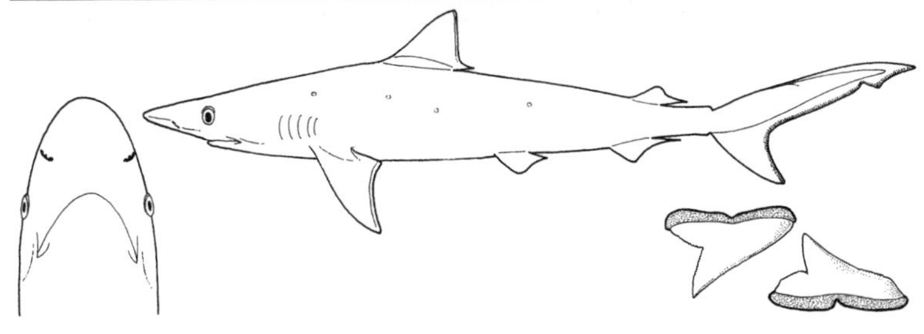

Measurements. Born: 31–39 cm TL. Mature: ~60 cm ♂, ~80 cm ♀. Max: ~110 cm TL.
Identification. Small brown or grey-brown shark, sometimes with white spots on sides and white-edged fins, white below. Long snout, small wide-spaced nostrils, no spiracles, long labial furrows, serrated teeth, fairly large eyes.

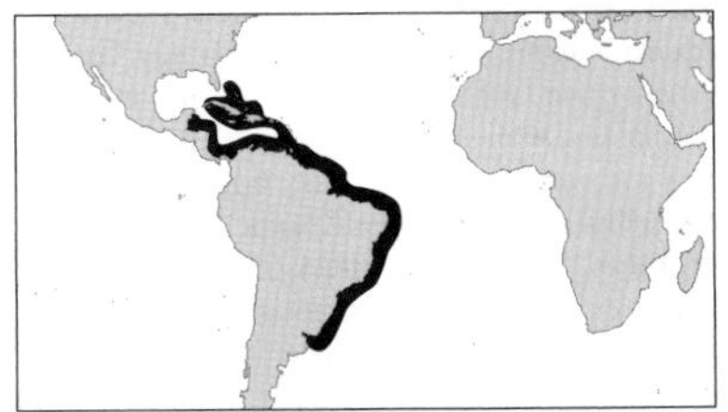

Distribution. Tropical west Atlantic: Caribbean and South America.
Habitat. Usually close inshore on continental and insular shelves, also offshore to 500 m.
Behaviour. Unknown.

Biology. Viviparous, yolk-sac placenta, two to six pups/litter, 10–11 month gestation, born in spring or early summer off southern Brazil. Mostly eats small bony fishes, also invertebrates.
Status. IUCN Red List: Not Evaluated. Very common.

Australian Sharpnose Shark *Rhizoprionodon taylori* Plate 60

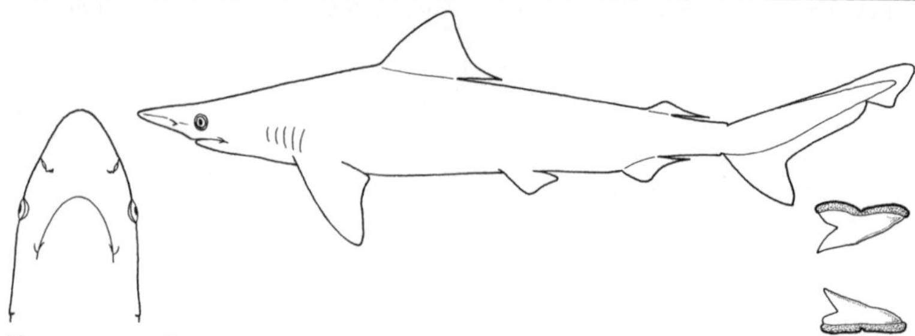

Measurements. Born: 25–30 cm TL. Mature: ~40 cm ♂, ~45 cm ♀. Max: 67 cm TL.
Identification. Small shark, very similar to *R. oligolinx*. Bronze to greyish above, pale below. Not conspicuously marked, but dorsal and caudal fin margins and upper caudal fin tip dark, other fins light-edged.

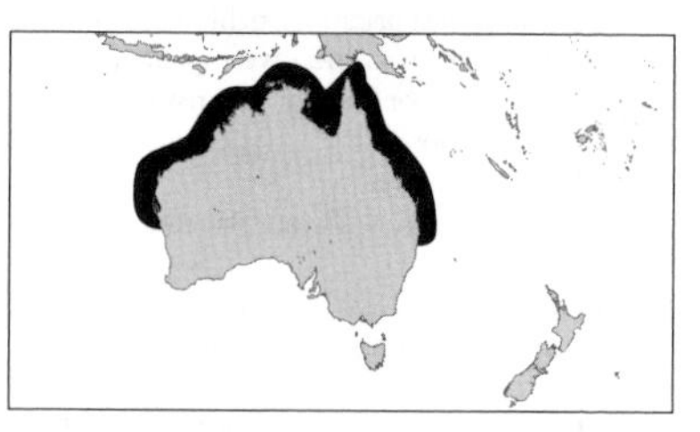

Distribution. Australia.
Habitat. Tropical inshore continental shelf.
Behaviour. Unknown.
Biology. Viviparous, yolk-sac placenta, one to ten pups/litter.
Status. IUCN Red List: Least Concern. Taken in bycatch but abundant and one of the most productive species of shark known, growing very rapidly, maturing after one year with females producing litters each year.

Atlantic Sharpnose Shark *Rhizoprionodon terraenovae* Plate 54

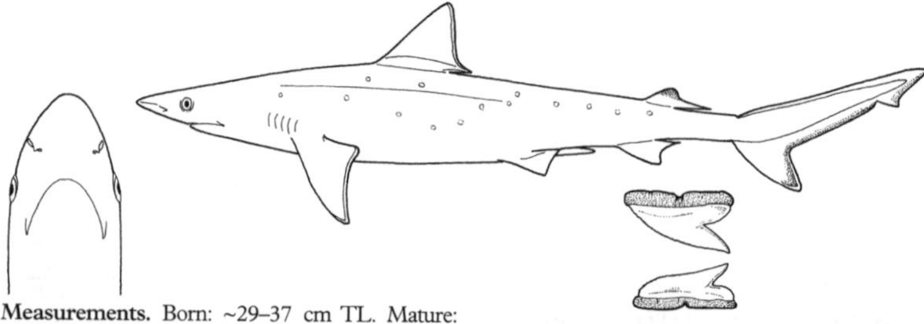

Measurements. Born: ~29–37 cm TL. Mature: 65–80 cm ♂, 85–90 cm ♀. Max: at least 110 cm TL.
Identification. Very similar to *R. porosus*. Grey to grey-brown with small light spots in large specimens, white below, light-margined pectoral fins with white margins, dusky tipped dorsals.

Distribution. Northwest Atlantic.
Habitat. Coastal in enclosed bays, sounds, harbours, and marine to brackish estuaries, from intertidal to ~280 m, usually <10 m. Often close to surf zone off sandy beaches.

Behaviour. Migrates seasonally in Gulf of Mexico, offshore in winter, inshore in spring.
Biology. Viviparous, yolk-sac placenta, one to seven pups/litter, (usually four to six, increasing with female size) born inshore in spring-summer after 10–11 month gestation. Feeds mainly on small bony fishes.
Status. IUCN Red List: Least Concern. Abundant and able to sustain fairly intensive fishing pressure.

Spadenose Shark *Scoliodon laticaudus* Plate 60

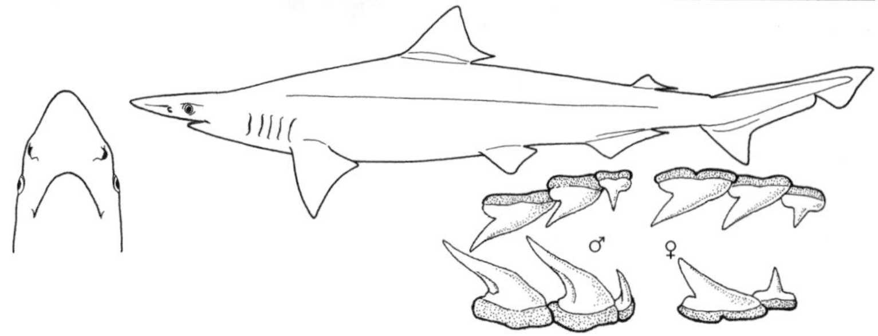

Measurements. Born: 12–15 cm TL. Mature: 24–36 cm ♂, 33–35 cm ♀. Max: ~74 cm TL.
Identification. Small stocky bronzy-grey shark, no conspicuous markings. Unmistakable very long, flattened, spade-like snout, small eyes, small smooth-edged blade-like teeth, short broad triangular pectoral fins, rear tip of first dorsal over pelvic midbases, second much smaller and originating well behind origin of larger anal fin. No interdorsal ridge.
Distribution. Indo-west Pacific.

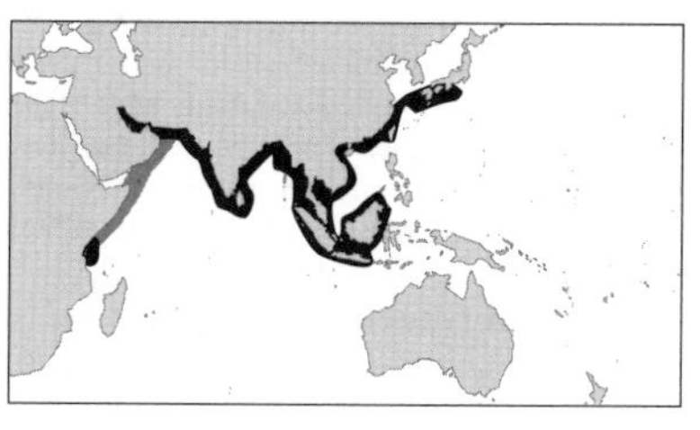

Habitat. Close inshore, often in rocky areas and lower reaches of large tropical rivers.
Behaviour. Occurs in large schools.
Biology. Viviparous, unusual columnar placenta and numerous long appendiculae on umbilical cord (egg yolks far too small to nourish pups). Breed year-round, 1–14 pups/litter. Mature at one to two years, longevity five to six years. Eat small pelagic schooling and bottom-living bony fishes, also shrimp and cuttlefish.
Status. IUCN Red List: Near Threatened. Abundant but very heavily fished.

Whitetip Reef Shark *Triaenodon obesus* Plate 53

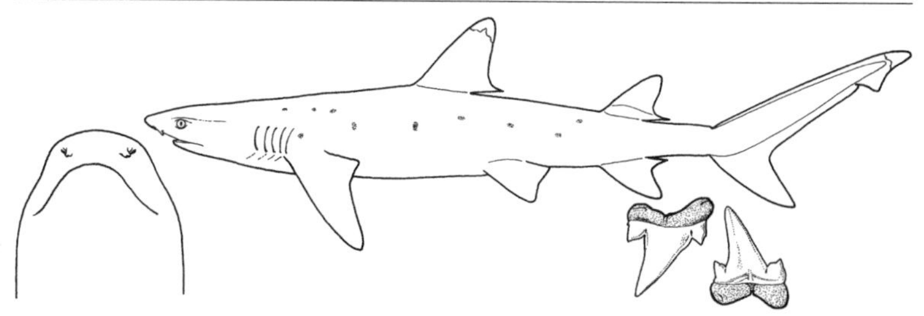

Measurements. Born: 52–60 cm TL. Mature: 104–105 cm ♂, 105–109+ cm ♀. Max: ~200 cm TL.
Identification. Small, slender greyish-brown shark with brilliant, very conspicuous white tips on first dorsal and upper caudal fin. Underside lighter, sometimes scattered dark spots on sides. Very short broad snout, oval eyes, first dorsal fin behind pectorals, second dorsal nearly as large as first, no interdorsal ridge.
Distribution. Indo-Pacific Ocean.

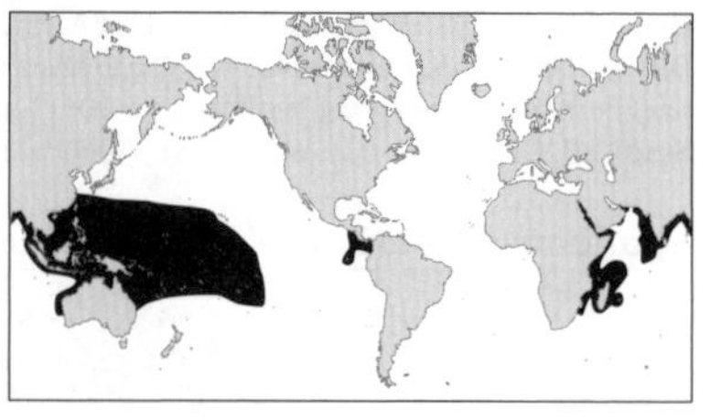

Habitat. Continental shelves and island terraces. Usually on or near bottom in crevices or caves in coral reefs, and in coral lagoons in shallow clear water, 8–40 m but ranging from 1–330 m.
Behaviour. Often seen resting on the bottom, in caves and under ledges in coral and on sand by day. More active during slack tide and at night. Small home range occupied for months or years. Social but not territorial; share home range without conflict. Specialise in capturing bottom prey in crevices, holes and caves in coral heads and ledges, located by scent and sound, sometimes in packs. Readily attracted to bait and may be handfed by divers; rarely aggressive.
Biology. Viviparous, one to five pups/litter (commonly two to three) after at least five month gestation. May mature at five years, longevity at least 25 years.
Status. IUCN Red List: Near Threatened. Often very common, but restricted depth range and habitat and small litter size suggest that increasing fishing pressure may be a threat. Popular for dive tourism, kept in captivity.

Sphyrnidae: Hammerheads

Identification. Unmistakable hammer-shaped heads. These function as a submarine-like bow plane to improve manoeuvrability and increase sensory capacity by enhancing stereoscopic vision and ability to triangulate sources of scent and electromagnetic signals.
Biology. Live bearing (viviparous), with yolk sac placenta. Feed on bony fishes, smaller sharks, rays, cephalopods and invertebrates, but not on marine mammals or other large vertebrates.
Status. Found worldwide in tropical and warm temperate seas, on or adjacent to continental and insular shelves and seamounts, from surface to at least 275 m, sometimes in large schools. Target and bycatch fisheries have depleted many populations; the fins are very valuable and hammerheads die very quickly when hooked or entangled, thus live release of bycatch is unusual. Large species (Great Hammerhead and possibly Scalloped and Smooth Hammerheads) have occasionally bitten divers and swimmers, but most are shy and very difficult to approach, not aggressive.

Winghead Shark *Eusphyra blochii* Plate 63

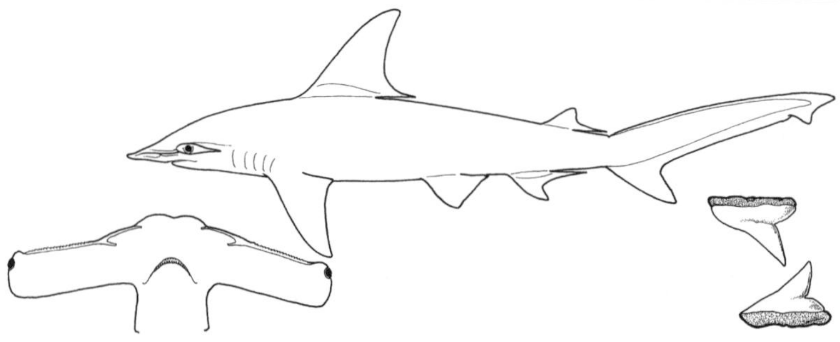

Measurements. Born: 32–45 cm TL. Mature: ~100 cm TL. Max: 186 cm TL.
Identification. Unmistakable immense wing-shaped head with narrow blades, width between eyes about half total length. First dorsal fin origin over pectoral fin bases, further forward than other hammerheads. Upper precaudal pit longitudinal, not crescent-shaped.
Distribution. Indo-west Pacific: northern Indian

Ocean to Australia and China.
Habitat. Shallow water, continental and insular shelves.
Behaviour. Pregnant females reportedly fight each other.
Biology. Six (usually) to nine pups/litter after 8–11 month pregnancy.
Status. IUCN Red List: Near Threatened. Heavily fished in much of range. Not known to bite people.

Mallethead Shark *Sphyrna corona* Plate 63

Measurements. Born: ≥23 cm TL. Mature: ~60 cm. Max: 92 cm TL.
Identification. Grey above, white below extending to back of head. Small, with moderately broad, anteriorly arched, mallet-shaped head with medial and lateral indentations on its anterior edge and transverse posterior margins, no prenarial grooves, snout rather long and about 2/5 of head width, small, strongly arched mouth, free rear tip of first dorsal fin over pelvic insertions, posterior margin of anal fin nearly straight. Transverse, crescentric upper pre-caudal pit.

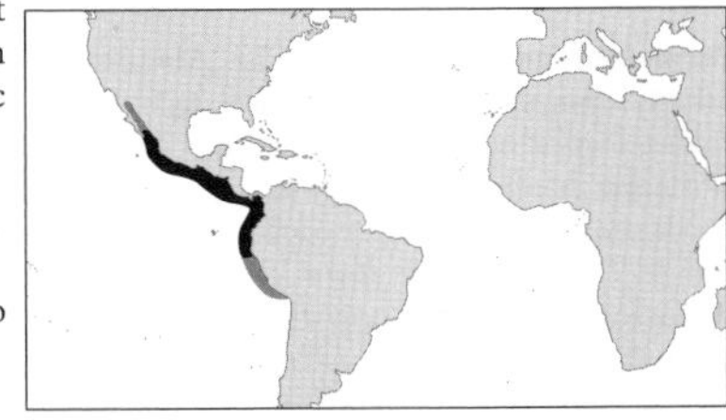

Distribution. East Pacific: Mexico to Peru.
Habitat. Continental shelf, presumably inshore.
Behaviour. Unknown.
Biology. Virtually unknown. Possibly two pups/litter.
Status. IUCN Red List: Not Evaluated.

Scalloped Hammerhead *Sphyrna lewini* Plate 64

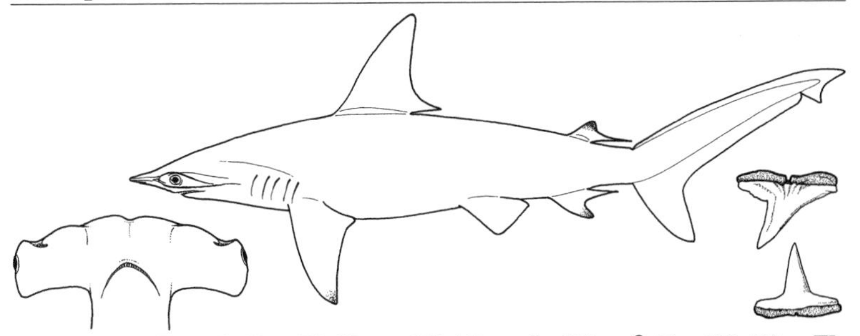

Measurements. Born: 42–55 cm TL. Mature: 140–165 cm ♂, ~212 cm ♀. Max: 370–420 cm TL.
Identification. Large with broadly arched, narrow-bladed head, central notch and two smaller lateral indentations. First dorsal moderately high, second dorsal and pelvic fins low. Light grey or bronzy above, white below, dusky or black-tipped pectoral fins, dark blotch on lower caudal fin lobe.
Distribution. World-wide, warm temperate and tropical seas.
Habitat. Over continental and insular shelves and adjacent deep water, surface to >275 m, often close inshore and in enclosed bays and estuaries. Juveniles mainly inshore.

Behaviour. Seasonally migratory, schooling, coastal-pelagic semi-oceanic shark.
Biology. 13–31 pups/litter after 9–10 month pregnancy. Eats bony fishes, sharks, rays, invertebrates.
Status. IUCN Red List: Near Threatened. Common and widespread, but extremely heavily fished in most areas.

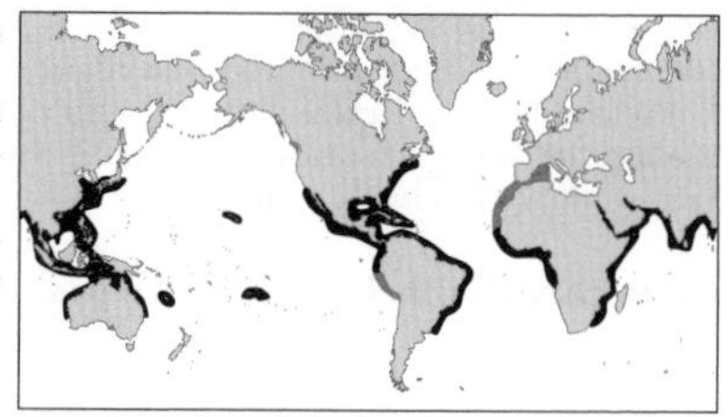

Scoophead Shark *Sphyrna media* Plate 63

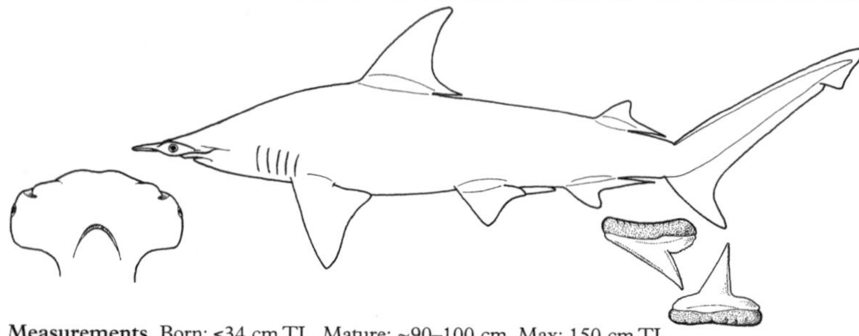

Measurements. Born: ≤34 cm TL. Mature: ~90–100 cm. Max: 150 cm TL.
Identification. Grey-brown above, light below. Small, with moderately broad, anteriorly arched, mallet-shaped head with weak medial and lateral indentations on its anterior edge and transverse posterior margins, no prenarial grooves, snout rather short and about one-third of head width, moderately large, broadly arched mouth, free rear tip of first dorsal fin over pelvic insertions. Transverse, crescentric upper pre-caudal pit.
Distribution. West Atlantic: Panama to southern Brazil. East Pacific: Gulf of California to Ecuador, probably northern Peru.
Habitat. Continental shelves.
Behaviour. Unknown.
Biology. Little known.
Status. IUCN Red List: Not Evaluated.

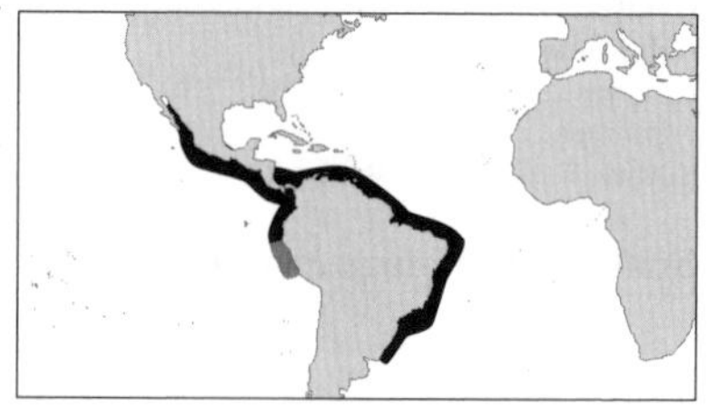

Great Hammerhead *Sphyrna mokarran* Plate 64

Measurements. Born: 50–70 cm TL. Mature: 234–269 ♂, 250–300 cm ♀. Max: >550–610 cm TL.
Identification. Light grey or grey-brown above, white below, fins unmarked. Very large hammerhead with a notch at the centre of head. First dorsal very high and curved, second dorsal and pelvic fins high, with deeply concave rear margins.

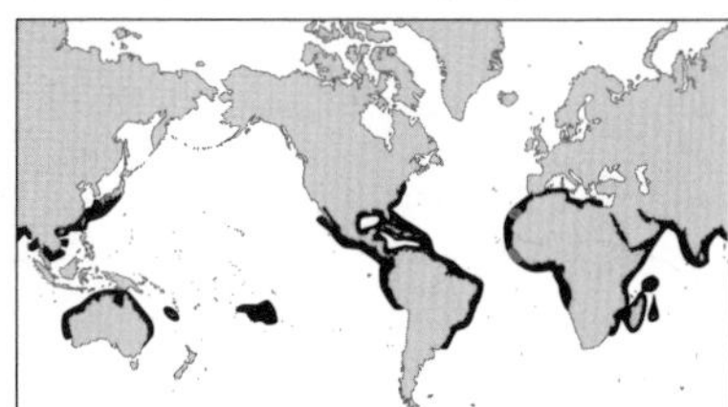

Distribution. World-wide, tropical seas.
Habitat. Coastal-pelagic and semi-oceanic, over continental shelves, island terraces, in passes and lagoons of coral atolls and on coral reefs, close inshore to well offshore, 1–80+ m.
Behaviour. Nomadic and seasonally migratory.
Biology. 6–42 pups/litter after ~11 month pregnancy. Varied prey; apparently prefers stingrays and other batoids, groupers and sea catfishes.
Status. IUCN Red List: Data Deficient. Not abundant. May occasionally bite people.

Bonnethead Shark *Sphyrna tiburo* Plate 63

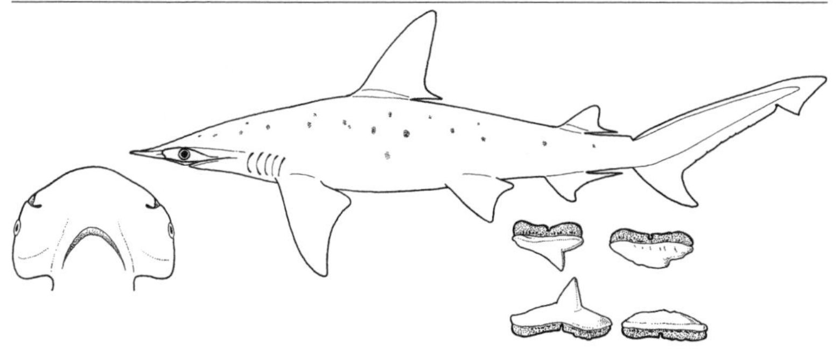

Measurements. Born: 35–40 cm TL. Mature: 52–75 cm ♂, ~84 cm ♀. Max: 150 cm TL.
Identification. Grey or grey-brown above, often with small dark spots, light below. Small. Unique, very narrow, smooth shovel-shaped head. First dorsal rear tip in front of pelvic fin origins, shallowly concave posterior anal fin margin.
Distribution. West Atlantic and east Pacific tropical seas.
Habitat. Continental and insular shelves, over mud and sand, on coral reefs, in estuaries, shallow bays and channels, 0–80 m (mainly 10–25 m). Pupping females most common in shallow water.
Behaviour. Migratory and social. Often in groups of 3–15, rarely alone. Behaviour well studied and complex.
Biology. 4–16 pups/litter. Mainly eats crustacea, also bivalves, octopi, small fishes.
Status. IUCN Red List: Least Concern. Abundant, minor fisheries importance. Kept in public aquaria.

Smalleye Hammerhead *Sphyrna tudes* Plate 63

Measurements. Born: ~30 cm TL. Mature: 80 cm ♂, 98 cm ♀. Max: 122–150 cm TL.
Identification. Grey-brown to golden above, light below. Broad, arched mallet-shaped head, deeply indented in front, straight behind. Fairly large, broadly arched mouth, short snout. Free rear tip of first dorsal over pelvic insertions. Transverse, crescentric upper pre-caudal pit.
Distribution. West Atlantic: Venezuela to Uruguay. Old records from Mediterranean Sea and West Africa require confirmation.

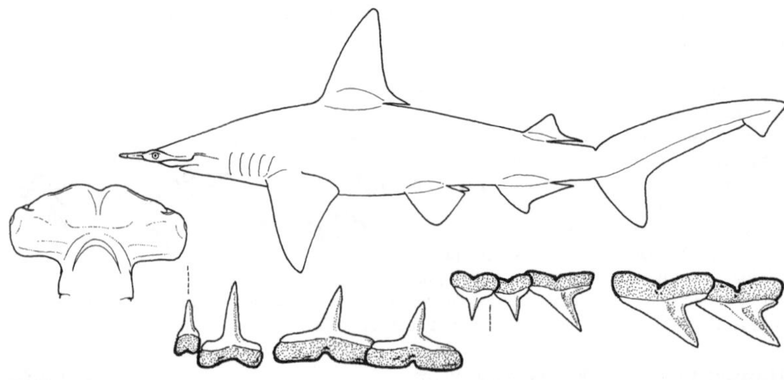

Habitat. Continental shelf, to ≥12 m.
Behaviour. Nursery grounds on shallow muddy beaches.
Biology. 5–12 pups/litter, ten month pregnancy. Eats small bony fishes, newborn scalloped hammerheads, small crustaceans, squid.
Status. IUCN Red List: Not Evaluated.

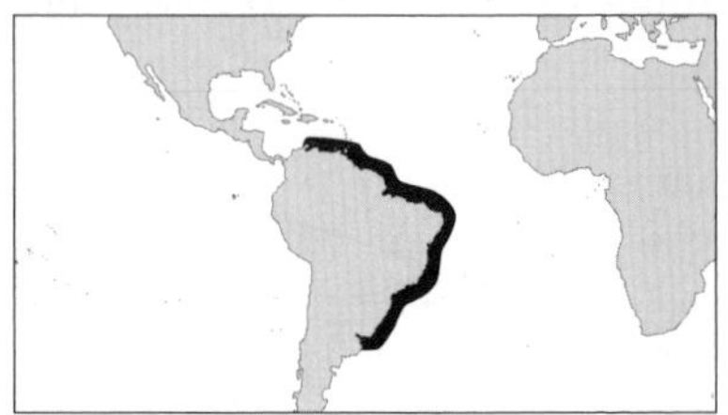

Smooth Hammerhead *Sphyrna zygaena* Plate 64

Measurements. Born: 50–61 cm TL. Mature: ~210–240 cm. Max: 370–400 cm TL.
Identification. Large hammerhead without a notch at the centre of curved head. First dorsal moderately high, second dorsal and pelvic fins low. Olive-grey or dark grey-brown above, white below, undersides of pectoral fin tips dusky.
Distribution. World-wide in tropical and temperate waters.
Habitat. Continental and insular shelves, from close inshore to well offshore, 0–20+ m.
Behaviour. Young sharks (≤1.5 m) sometimes occur in huge migrating schools.
Biology. 29–37 pups/ litter. Feeds on bony fishes, small sharks, skates and stingrays.
Status. IUCN Red List: Near Threatened. May be confused with *Sphyrna lewini* in tropics.

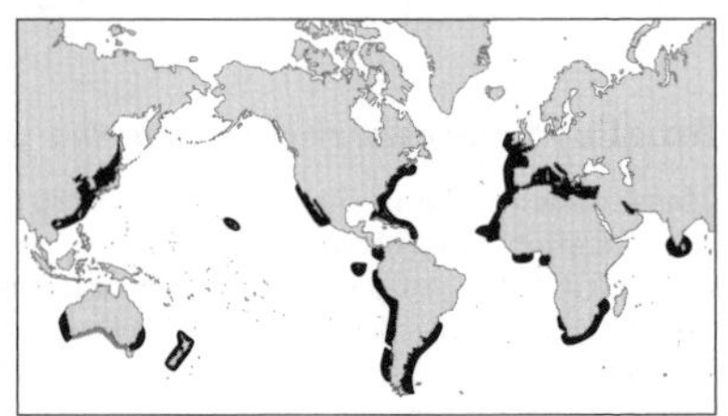

Glossary

Abdominal ridges or keels. In some sharks, paired longitudinal dermal ridges that extend from the bases of the pectoral fins to the pelvic fin bases.

Abyss. The deep sea bottom, ocean basins or abyssal plain descending from 2000 m to about 6000 m. (See p. 336).

Abyssal plain. The extensive, flat, gently sloping or nearly level region of the ocean floor from about 2000 m to 6000m depth.

Anal fin. A single fin on the ventral surface of the tail between the pelvic fins and caudal fin of some sharks, absent in batoids, dogfish, sawsharks, angel sharks, and some chimaeras.

Annular rings or annuli. In a vertebral centrum in cross section, rings of calcified cartilage separated by uncalcified cartilage that occupy the intermedialia only, or concentric rings that cross both the intermedialia and basalia.

Anterior. Forward, in the [longitudinal] direction of the snout tip.

Anterior margin. In precaudal fins, the margin from the fin origin to its apex.

Anterior nasal flap. A flap on the front edges of the nostrils, that serves to partially divide the nostril into incurrent and excurrent apertures or openings.

Apex (plural, Apices). In precaudal fins, the distal tip, which can be acutely pointed to broadly rounded.

Apical. In oral teeth, towards the tip of the crown or cusp. Can also be used as indicating direction towards the apex or tip of a fin, fin-spine, etc.

Aplacental yolksac viviparity. A reproductive mode where the maternal adult gives birth to live young which are primarily nourished by the yolk in their yolksac. The yolk is gradually depleted and the yolksac reabsorbed until the young are ready to be born. Often referred to as ovoviviparity.

Aplacental viviparity. A reproductive mode where the maternal adult gives birth to live young which do not have a yolksac placenta.

Barbels. Long conical paired dermal lobes on the snouts of sharks, which may serve to locate prey. Sawsharks have barbels on the underside of the snout in front of the nostrils (as in sturgeon), but most barbelled sharks have them associated with the nostrils.

Base. In precaudal fins, the proximal part of the fin between the origin and insertion, extending distally, and supported by the cartilaginous fin skeleton. In the caudal fin, that thickened longitudinal part of the fin enclosing the vertebral column and between the epaxial and hypaxial lobes or webs of the fin. In oral teeth, the proximal root and crown foot, in apposition to the distal cusp. In denticles, the proximal anchoring structures, often with four or more lobes, holding the denticles in the skin.

Bathyal. Benthic habitats from 200 m to 2000 m depth. (See page 336).

Bathypelagic zone. That part of the oceans beyond the continental and insular shelves, from about 1000 m to over 4000 m and above the middle and lower continental rises and the abyssal plain, the sunless zone. Some oceanic sharks may transit the epipelagic, mesopelagic and bathypelagic zones to the bottom while migrating vertically. (See page 336).

Batoid. A ray or flat, winged, or pancake shark, a neoselachian of the superorder Squalomorphii, order Rajiformes. a sawfish, sharkray, wedgefish, guitarfish, thornray, panray, electric ray, skate, stingray, butterfly ray, eagle ray, cownose ray, devilray or manta. Rays are closely allied to the sawsharks (Pristiophoriformes) and angel sharks (Squatiniformes), but differ from them in having the pectoral fins fused to the sides of the head over the gill openings, which are ventral rather than laterally or ventrolaterally directed.

Benthic or Demersal. Referring to organisms that are bottom-dwelling.

Bioluminscence. Light produced by biochemical means in some organisms.

Boreal. Cold northern regions.

Blade. In oral teeth, an arcuate, convex-edged section of the cutting edge of the crown foot, without cusplets.

Body ridges. Elongated longitudinal dermal ridges on the sides of the trunk and precaudal tail in certain carpet sharks (Orectolobiformes), in the whale, zebra, and some bamboo sharks.

Body. Can refer to an entire shark, sometimes restricted to the trunk and precaudal tail.

Branchial arches. The paired cartilaginous arches that support the gills.

Breach. A complete or almost complete leap above the water's surface followed by a splash on re-entry.

Bycatch. The part of a catch taken incidentally in addition to the target species towards which fishing effort is directed. In a broad context, this includes all non-targeted catch including byproduct, discards and other interactions with gear.

Calcified cartilage. Shark skeletons are formed of hyaline cartilage or gristle, but this is often reinforced with layers of calcified cartilage, cartilage impregnated with a mineral, hydroxyapatite, similar to that of bone.

Calcification. The process by which calcium and carbonate ions are combined to form calcerous skeletal materials.

Cannibal vivipary. See Uterine cannibalism.

Carcharhinoid. A ground shark, a member of the order Carcharhiniformes, and including the catsharks, false catsharks, finbacked catsharks, barbeled houndsharks, houndsharks, weasel sharks, requiem sharks, and hammerheads.

Cartilaginous fishes. Members of the class Chondrichthyes.

Caudal crest. A prominent sawlike row of enlarged pointed denticles along the dorsal caudal margin and sometimes along the ventral caudal margin of the caudal fin. Found in certain sharks including hexanchoids and some carcharhinoids.

Caudal fin. The fin on the end of the tail in sharklike fishes, lost in some batoids.

Caudal keels. A dermal keel on each side of the caudal peduncle that may extend onto the base of the caudal fin, and may, in a few sharks, extend forward as a body keel to the side of the trunk.

Caudal peduncle. That part of the precaudal tail extending from the insertions of the dorsal and anal fins to the front of the caudal fin.

Centrum (plural, Centra). Spool-shaped, partially or usually fully calcified structures articulated together to form the vertebral column.

Cephalopod. Referring to Cephalopoda.

Cephalopoda. A class of marine molluscs that have prehensile tentacles large eyes, which include nautiloids, cuttlefishes, squids and octopuses; also extinct groups such as the ammonites and belemnites.

Ceratotrichia. Slender soft or stiff filaments of an elastic protein, superficially resembling keratin or horn (from the Greek keratos, horn, and trichos, hair). Ceratotrichia run in parallel and radial to the fin base and support the fin webs. The prime ingredient of shark-fin soup.

Cetacean. Referring to Cetacea.

Cetacea. An order of marine mammals with modified forelimbs, no visible hind limbs and horizontal tail fin, which comprised of whales, dolphins and porpoises.

Chimaera or Chimaeroid. A chimaera, ratfish, silver shark, ghost shark, spookfish, or elephant fish, a member of the order Chimaeriformes, subclass Holocephali.

Chondrichthyan. Referring to the class Chondrichthyes.

Chondrichthyes. The class Chondrichthyes (from Greek chondros, cartilage, and ichthos, fish), a major taxonomic group of aquatic, gill-breathing, jawed, finned vertebrates with primarily cartilaginous skeletons, one to seven external gill openings, oral teeth in transverse rows on their jaws, and mostly small, toothlike scales or dermal denticles. Chondrichthyes include the living elasmobranchs and holocephalans and their numerous fossil relatives, and also can be termed sharklike fishes or simply sharks.

Chondrocranium. See neurocranium.

Circumglobal. Occurring around the world.

Circumnarial fold. A raised semicircular, lateral flap of skin around the incurrent aperture of a nostril, in heterodontoids, orectoloboids, and a few batoids, defined by a circumnarial groove.

Circumnarial groove. A shallow groove defining the lateral bases of the circumnarial folds.

Circumtropical. Occurring around the tropical regions of the world.

CITES. Convention on International Trade in Endangered Species of Fauna and Flora. An international agreement which aims to ensure that international trade in specimens of wild fauna and flora does not threaten the survival of species. Appendix II of CITES includes "species not necessarily threatened with extinction, but in which trade must be controlled in order to avoid utilisation incompatible with their survival" (http.//www.cites.org).

Claspers. Paired copulatory organs present on the pelvic fins of male cartilaginous fishes, for internal fertilization of eggs.

Class. One of the taxonomic groups of organisms, containing related orders; related classes are grouped into phlya.

Classification. The ordering of organisms into groups on the basis of their relationships, which may be by similarity or common ancestry.

Cloaca. The common chamber at the rear of the body cavity of elasmobranchs through which body wastes and reproductive products including sperm, eggs, and young pass, to be expelled to the outside through a common opening or vent.

Common name. The informal vernacular name for an organism, which may vary from location to location.

Continental rise. The gently sloping base of the continental shelf made up of sediment deposits; see also Rise. (See page 336).

Continental shelf. The gently sloping, shelf-like part of the seabed adjacent to the coast extending to a depth of about 200 m. (See page 336).

Continental slope. The often steep, slope-like part of the seabed extending from the edge of the continental shelf to a depth of about 2000 m. (See page 336).

Crustacean. Referring to Crustacea.

Crustacea. Diverse subphylum of arthropods, with jointed appendages, chitinous exoskeleton and two pairs of antennae; most are marine, many are filter-feeders, some are parasitic; representatives include crabs, lobsters, shrimps, copepods, isopods, amphipods and barnacles.

Cusp. A usually pointed large distal projection of the crown. A primary cusp is situated on the midline of the crown foot. Multicuspid refers to oral teeth or denticles with more than one cusp. In lateral trunk denticles, the posterior ends of the crown may have medial and lateral cusps, sharp or blunt projections associated with the medial and lateral ridges.

Cusplet. As with a cusp, but a small projection in association with a cusp, and usually mesial and distal but not medial on the crown foot.

Cutting edge. In oral teeth, the compressed sharp longitudinal ridge on the mesodistal edges of the crown.

DELASS. 'Development of Elasmobranch Assessments'. A project funded by the European Union to develop elasmobranch assessments to improve the scientific basis for the regulation of fisheries.

Demersal. Occurring or living near or on the bottom of the ocean (cf. pelagic).

Dermal denticle or Placoid scale. A small tooth-like scale found in cartilaginous fishes, covered with enameloid, with a core and base of dentine and usually small and often close-set to one another and covering the body. A few non-batoid sharks, many batoids, and chimaeroids generally have them enlarged and sparse or reduced in numbers.

Dermal lobes. In wobbegongs, family Orectolobidae, narrow or broad-based, simple or branched projections of skin along the horizontal head rim and on the chin.

Diphycercal. A caudal fin with the vertebral axis running horizontally into the fin base, which is not elevated.

Distal. In any direction, at the far end of a structure. In oral teeth, used in a special sense for structures on the teeth towards the posterolateral mouth corners or rictuses. See apical and basal.

Dorsal fin spine. A small to large enameloid-covered, dentine-cored spine located on the anterior margins of one or both of the dorsal fins, found on bullhead sharks (Heterodontiformes), many dogfish sharks, fossil (but not living) batoids, chimaeroids, but lost entirely or buried in the fin bases of other sharklike fishes.

Dorsal fin. A fin located on the trunk or precaudal tail or both, and between the head and caudal fin. Most sharks have two dorsal fins, some batoids one or none.

Dorsal lobe. In the caudal fin, the entire fin including its base, epaxial and hypaxial webs but excepting the ventral lobe.

Dorsal margin. In the caudal fin, the margin from the upper origin to its posterior tip. Usually continuous, but in angel sharks (Squatiniformes) with their hypocercal, superficially inverted caudal fins, it is subdivided. See Squatinoid caudal fin.

Dorsal. Upwards, in the vertical direction of the back. See ventral.

Drop-off. Steep or sheer underwater cliff or precipice.

Echinoderms. Referring to members of the Echinodermata.

Echinodermata. An entirely marine invertebrate phylum with a five-part symmetry, water-based vascular system and tube feet; the diverse group includes brittlestars, starfish, sea urchins, sea cucumbers and sea lilies

Ecosystem. The living community of different species, interdependent on each other, together with their non-living environment.

EEZ. Exclusive Economic Zone. A zone under national jurisdiction (up to 200-nautical miles wide) declared in line with the provisions of 1982 United Nations Convention of the Law of the Sea, within which the coastal State has the right to explore and exploit, and the responsibility to conserve and manage, the living and nonliving resources.

Eggcase. A stiff-walled elongate-oval, rounded rectangular, conical, or dart-shaped capsule that surrounds the eggs of oviparous sharks, and is deposited by the female shark on the substrate. It is analogous to the shell of a bird's egg and is made of protein, which is a type of collagen that superficially resembles horn or keratin. Egg cases often have pairs of tendrils or hornlike structures on their ends, or flat flanges on their sides or spiral flanges around their lengths, which anchor the cases to the bottom. As the egg travels from the ovaries into the oviducts and through the nidamental glands, the egg case is secreted around it and the egg is fertilized. Live-bearing sharks may retain egg cases, and these vary from being rigid and similar to those of oviparous sharks to soft, bag like, degenerate and membranous. Soft egg cases may disintegrate during the birth cycle.

Elasmobranch. Referring to the subclass Elasmobranchii.

Elasmobranchii. The subclass Elasmobranchii, (from Greek elasmos, plate, and branchos, gills, in allusion to their platelike gill septa), the sharklike fishes other than the Holocephali or chimaeras, and including the living non-batoid sharks, batoids, and a host of fossil species. They differ from holocephalans in having 5–7 pairs of gill openings open to the exterior and not covered by a soft gill cover, oral teeth separate and not formed as tooth plates, a fixed first dorsal fin with or without a fin spine, and a short spined or spineless second dorsal.

Embryo. An earlier development stage of the young of a live-bearing shark, ranging from nearly microscopic to moderate-sized but not like a miniature adult. See foetus.

Enameloid. The shiny hard external coating of the crowns of shark oral teeth, superficially similar to enamel in land vertebrates.

Endemic. A species or higher taxonomic group of organisms that is only found in a given area. It can include national endemics found in a river system or along part or all of the coast of a given country, but also regional endemics, found off or in adjacent countries with similar habitat, but not elsewhere.

Epibenthic zone. The area of the ocean just above and including the sea bottom, from shallow seas to deep abysses, epibenthic sharks live on or near the bottom. (See page 336).

Epipelagic zone. That part of the oceans beyond the continental and insular shelves, in oceanic waters, from the surface to the limits of where most sunlight penetrates, about 200 m. Also known as the sunlit sea or 'blue water'. Most epipelagic sharks are found in the epipelagic zone, but may penetrate the mesopelagic zone. (See page 336).

Euryhaline. An organism that is able to tolerate a large variance in salinity found mainly in and around estuaries.

Euselachian. Referring to the Euselachii.

Euselachii. The cohort Euselachii (Greek Eu, true, good or original, and selachos, shark or cartilaginous fish), the spined or 'phalacanthous' sharks, including the modern sharks or Neoselachii, and fossil shark groups including the hybodonts, the ctenacanths, and the xenacanths, all primitively with anal fins and having two dorsal fins with fin spines.

Excretion. The separation and removal of waste material from an organism usually applied to nitrogen waste normally in the form of urea or ammonia.

Excurrent apertures. The posterior and ventrally facing openings of the nostrils, which direct water out of the nasal cavities and which are often partially covered by the anterior nasal flaps. These are usually medial on the nostrils and posteromedial to the incurrent apertures, but may be posterior to the incurrent apertures only.

Extant. Applied to a group that has all or some living representatives.

Extinct. Applied to a group that has no living representatives.

Eye notch. A sharp anterior or posterior indentation in the eyelid, where present cleanly dividing the upper and lower eyelids.

Eyespots or ocelli. Large eye-like pigment spots located on the dorsal surface of the pectoral fins or bodies of some sharks including rays, angel sharks, and some bamboo sharks, possibly serving to frighten potential enemies.

Falcate. Sickle-shaped.

Family. One of the taxonomic groups of organisms, containing related genera; related families are grouped into orders.

FAO. United Nations Food and Agricultural Organisation.

Fauna. The community of animals peculiar to a region, area, specified environment or period.

Fecundity. A measure of the capacity of the maternal adult to produce young.

Filter-feeding. A form of feeding whereby suspended food particles are extracted from the water using the gill rakers.

Fin base. See base.

Fin insertion. See insertion.

Fin origin. See origin

Fin web. The usually thin, compressed part of the fin, distal to the base, that is supported by ceratotrichia alone (in aplesodic fins) or by ceratotrichia surrounding expanded fin radials or by radials only (plesodic fins).

Finning. The practice of slicing off a shark's valuable fins and discarding the body at sea.

First dorsal fin. The anteriormost dorsal fin of two, ranging in position from over the pectoral fin bases to far posterior on the precaudal tail.

FL. fork length. A standard morphometric measurement used for sharks, from the tip of the snout to the fork of the caudal fin.

Flank. Sides of the shark excluding the head and tail regions.

Foetus. A later development stage of the unborn young of a live-bearing shark, that essentially resembles a small adult. Term foetuses are ready to be born, and generally have oral teeth and denticles erupting, have a colour pattern (often more striking than adults) and, in ovoviviparous sharks, have their yolk-sacs resorbed.

Free rear tips. The pectoral, pelvic, dorsal, and anal fins all have a movable rear corner or flap, the free rear tip, that is separated from the trunk or tail by a notch and an inner margin. In some sharks the rear tips of some fins are very elongated.

Fusiform. Spindle- or torpedo-shaped, elongated with tapering ends.

Galeomorph. Referring to the Galeomorphii.

Galeomorphii. The neoselachian superorder Galeomorphii, including the heterodontoid, lamnoid, orectoloboid, and carcharhinoid sharks.

Generation. Measured as the average age of parents of newborn individuals within the population. Where generation length varies under threat, the more natural, i.e. pre-disturbance, generation length should be used.

Genus (plural, Genera). One of the taxonomic groups of organisms, containing related species; related genera are grouped into families. The first of two scientific names assigned to each species.

Gestation period. The period between conception and birth in live-bearing animals.

Gill openings or slits. In elasmobranchs, the paired rows of five to seven transverse openings on the sides or underside of the head for the discharge of water through the gills. Chimaeras have their four gill openings hidden by a soft gill cover and discharge water through a single external gill opening.

Gill raker denticles. In the basking shark (Cetorhinidae), elongated denticles with hairlike cusps arranged in rows on the internal gill openings, which filter out planktonic organisms.

Habitat. The locality or environment in which an animal lives.

Hadal. The benthic zone of the deep trenches, 6000 to about 11,000 m, from which no cartilaginous fishes have been observed or recorded. (See page 336).

Hadopelagic zone. The pelagic zone inside the deep trenches, 6000 to about 11,000 m, from which no chondrichthyans have been observed or recorded. (See page 336).

Head. That part of a cartilaginous fish from its snout tip to the last (or in chimaeras only) gill slits.

Heterocercal. A caudal fin with the vertebral axis slanted dorsally into the fin base, which is also dorsally elevated.

Heterodontoid. A bullhead shark, horn shark, or Port Jackson shark, a member of the order Heterodontiformes, family Heterodontidae.

Heterodonty. In oral teeth, structural differences between teeth in various positions on the jaws, between teeth in the same position during different life stages, or between teeth in the same positions in the two sexes.

Hexanchoid. A cowshark or frilled shark, members of the order Hexanchiformes, and including the sixgill sharks, sevengill sharks, and frilled sharks.

Holocephalan. Referring to the Holocephali.

Holocephali. The subclass Holocephali (from Greek holos, entire, and kephalos, head), the living chimaeras and their numerous fossil relatives, a major subdivision of the class Chondrichthyes. The name is in reference to the fusion of the upper jaws or palatoquadrates to the skull in all living species and in many but not all fossils. The living holocephalans include three families in the order Chimaeriformes. The living species differ from elasmobranchs in having four pairs of gill openings covered by a soft gill cover and with a single pair of external gill openings, oral teeth fused and reduced to three pairs of ever-growing tooth plates, an erectile first dorsal fin with a spine and a long, low spineless second dorsal.

Holotype. Either the only specimen used and mentioned in an original description of a species, with or without a designation of such, or one of two or more specimens used and mentioned in an original description of a species and designated as such. This becomes the 'name-bearer' of the species, and is used to validate the species or scientific name by anchoring it to a single specimen.

Homodonty. In oral teeth, structural similarity between teeth in various positions on the jaws, between teeth in the same position during different life stages, or between teeth in the same positions in the two sexes.

Hypaxial web. The entire fin web below the vertebral column (vertebral axis) and the caudal base.

Hypercalcified structures. Parts of the skeleton that have developed extremely dense calcified cartilage, primarily during growth and maturation, which sometimes swell to knobs that distort and engulf existing cartilaginous structures. The rostrum of the Salmon Shark (*Lamna ditropis*) is a particularly impressive hypercalcified structure.

Hypocercal. A caudal fin with the vertebral axis slanted ventrally into the fin base, which is also ventrally depressed. Found only in angelsharks (Squatiniformes) among living sharks.

ICES. International Council for the Exploration of the Seas.

Incurrent apertures. The anterior and ventrally facing openings of the nostrils, which direct water into the nasal cavities. These are usually lateral on the nostrils and anterolateral to the excurrent apertures, but may be anterior to the excurrent apertures only.

Indo-Pacific. An area covering the Indian Ocean and the western Pacific Ocean.

Inner margin. In precaudal fins including the pectoral, pelvic, dorsal and anal fins, the margin from the fin insertion to the rear tip.

Insertion. The posterior or rear end of the fin base in precaudal fins. See Origin.

Inshore. Shallow waters seaward side of the surf zone.

Insular shelf. See Shelf.

Interdorsal ridge. A ridge of skin on the midback of sharks, in a line between the first and second dorsal fins; particularly important in identifying grey sharks (genus *Carcharhinus*, family Carcharhinidae).

Intermediate teeth. Small oral teeth between the laterals and anteriors of the upper jaw, found in most lamnoids.

Intertidal zone. Shoreline between high and low tide marks that is diurnally exposed to the air by tidal movement. (See page 336).

Intestinal valve. A dermal flap inside the intestine, protruding into its cavity or lumen, and of various forms in different cartilaginous fishes. Often formed like a corkscrew or augur. See spiral, ring and scroll valves.

Invertebrate. Animals without a backbone.

IPOA-Sharks. International Plan of Action for the Conservation and Management of Sharks.

Island slope. See Slope.

ITQ. Individual Transferable Quota. a catch limit or quota (a part of the Total Allowable Catch) allocated to an individual fisher or vessel owner which can either be harvested or sold to others.

IUCN. The World Conservation Union. A union of sovereign states, government agencies and non-governmental organisations. www.iucn.org.

Jaws. See mandibular arch.

Juvenile. The life stage between hatching from egg or birth to sexual maturity and adulthood.

Keel. See Caudal keel.

Labial cartilages. Paired internal cartilages that support the labial folds at the lateral angles of the mouth. Living neoselachians typically have two pairs of upper labial cartilages, the anterodorsal and posterodorsal labial cartilages, and one pair of ventral labial cartilages, but these are variably reduced and sometimes absent in many sharks.

Labial folds. Lobes of skin at the lateral angles of the mouth, usually with labial cartilages inside them, separated from the sides of the jaws by pockets of skin (labial grooves or furrows).

Labial furrows or Labial grooves. Grooves around the mouth angles on the outer surface of the jaws of many cartilaginous fishes, isolating the labial folds. Primitively there is a distinct upper labial furrow above the mouth corner and a lower labial furrow below it.

Lagoon. Area of relatively shallow water that is partially or completely separated from the open water by a land barrier.

Lamnoid. A mackerel shark, a member of the order Lamniformes, and including the sandtiger sharks, Goblin Shark, Crocodile Shark, Megamouth Shark, thresher sharks, Basking Shark, and the makos, Porbeagle, Salmon Shark, and White Shark.

Lateral. Outwards, in the transverse direction towards the periphery of the body. See medial.

Lateral keel. See Caudal keel.

Lateral line. A sensory canal system of pressure sensitive cells, that run along the sides of the body, often branching at the head, which detect water movements, disturbances and vibrations.

Lateral ridges. Reinforced ridges along the side of the body one of which is often an extension of the caudal keel.

Lateral teeth. Large broad-rooted, compressed, high crowned oral teeth on the sides of the jaws between the anteriors and laterals.

Lateral trunk denticle. A dermal denticle from the dorsolateral surface of the back below the first dorsal fin base.

Littoral zone. That part of the oceans over the continental and insular shelves, from the intertidal to 200 m. (See page 336).

Live-bearing. A mode of reproduction in which female sharks give birth to young sharks, which are miniatures of the adults. See Vivipary.

Longevity. The maximum expected age, on average, for a species or population in the absence of human-induced or fishing mortality.

Lower origin. In the caudal fin, the anteroventral beginning of the hypaxial or lower web of the caudal fin, at the posterior end of the anal-caudal or pelvic-caudal space.

Lower postventral margin. In the caudal fin, the lower part of the postventral margin of the hypaxial web, from the ventral tip to the posterior notch.

Mandibular arch. The paired primary jaw cartilages of sharks, including the dorsal palatoquadrates and the ventral Meckel's cartilages.

Medial. Inwards, in the transverse direction towards the middle of the body. See lateral.

Medial teeth. Small oral teeth, generally symmetrical and with narrow roots, in one row at the symphysis and often in additional paired rows on either side of the symphysial one.

Mesopelagic zone. That part of the oceans beyond the continental and insular shelves, in oceanic waters, from about 200–1000 m, the twilight zone where little light penetrates. (See page 336).

Migratory. The systematic (as opposed to random) movement of individuals from one place to another, often related to season and breeding or feeding. Knowledge of migratory patterns helps to manage shared stocks and to target aggregations of fish.

Molariform. In oral teeth, referring to a tooth with a broad flat crown with low cusps or none, for crushing hard-shelled invertebrate prey.

Mono-specific. Genus containing only one known extant species.

MPA. Marine Protected Area. Any area of the intertidal or subtidal terrain, together with its overlying water and associated flora, fauna, historical and cultural features, which has been reserved by law or other effective means to protect part or all of the enclosed environment.

MSY. Maximum sustainable yield. The largest theoretical average catch or yield that can continuously be taken from a stock under existing environmental conditions without causing the stock to become depleted (assuming that removals and natural mortality are balanced by stable recruitment and growth).

Multiple ovipary. A mode of egg-laying or ovipary in which female sharks retain several pairs of cased eggs in the oviducts, in which embryos grow to advanced developmental stages. When deposited on the bottom (in captivity) the eggs may take less than a month to hatch. Found only in the scyliorhinid genus *Halaelurus*, with some uncertainty as to whether the eggs are normally retained in the oviducts until hatching. Eggs laid by these sharks may be abnormal, unusual, or an alternate to ovovivipary. The whale shark (*Rhincodon typus*) may have multiple retention of egg cases; near-term foetuses have been found in their uteri and eggcases with developing foetuses have been collected on the bottom.

Nares. see Nostrils.

Nasal aperture. On the neurocranium, an aperture in the anteroventral surface or floor of each nasal capsule, through which the nostril directs water into and out of the nasal organ.

Nasal flap. One of a set of dermal flaps associated with the nostrils, and serving to direct water into and out of them, including the anterior, posterior, and mesonarial flaps.

Nasoral grooves. Many bottom-dwelling, relatively inactive sharks have nasoral grooves, shallow or deep grooves on the ventral surface of the snout between the excurrent apertures and the mouth. The nasoral grooves are covered by expanded anterior nasal flaps that reach the mouth, and form water channels that allow the respiratory current to pull water by partial pressure into and out of the nostrils and into the mouth. This allows the shark to actively irrigate its nasal cavities while sitting still or when slowly moving. Nasoral grooves occur in heterodontoids, orectoloboids, chimaeroids, some carcharhinoids, and most batoids.

Neoselachian. Referring to the Neoselachii.

Neoselachii. From Greek neos, new, and selachos, shark. The modern sharks, the subcohort Neoselachii, consisting of the living elasmobranchs and their immediate fossil relatives. See Euselachii.

Neotype. A specimen, not part of the original type series for a species, which is designated by a subsequent author, particularly if the holotype or other types have been destroyed, were never designated in the original description, or are presently useless.

Neritic province. A body of water from the shoreline to 200 m deep that includes the continental shelf. (See page 336).

Neurocranium. In sharks, a box-shaped complex cartilaginous structure at the anterior end of the vertebral column, containing the brain, housing and supporting the nasal organs, eyes, ears, and other sense organs, and supporting the visceral arches or splanchnocranium. Also termed chondrocranium, chondroneurocranium, or endocranium.

Nictitating lower eyelid. In the ground sharks (order Carcharhiniformes), a movable lower eyelid that has special posterior eyelid muscles that lift it and, in some species, completely close the eye opening.

Nictitating upper eyelid. In parascylliid orectoloboids, the upper eyelid has anterior eyelid muscles that pull it down and close the eye opening, analogous to the nictitating lower eyelids of carcharhinoids.

Nomenclature. In biology, the application of distinctive names to groups of organisms.

Non-target species. Species which are not the subject of directed fishing effort (cf. target catch), including the bycatch and byproduct.

Nostrils. The external openings of the cavities of the nasal organs, or organs of smell.

Nursery ground. Area (often inshore and in sheltered waters and with abundant food organisms) where newborn sharks live and feed and grow to a certain size in their life cycle, then move elsewhere.

Oceanic. Referring to organisms inhabiting those parts of the oceans beyond the continental and insular shelves, over the continental slopes, ocean floor, sea mounts and abyssal trenches. The open ocean.

Oceanic ridge. Part of the ocean floor where new material is added to plate margins by magma eruptions forming an ever expanding parallel line of deepwater ridges.

Oceanic seamount. See seamount.

Oceanic trench. The deepest part of the ocean floor, typically formed when one tectonic plate slides under another. (See page 336).

Ocellus (plural Ocelli). Eye-like marking in which the central colour is bordered in a full or broken ring of another colour.

Ocular. Of or associated with the eye.

Offshore. Waters that are over continental shelves and slope edges. (See page 336).

Olfactory. Parts of the body that are associated with the sense of smell.

Oophagy. Egg-eating, a mode of live-bearing reproduction employing uterine cannibalism; early foetuses deplete their yolk-sacks early and subsist by eating nutritive eggs produced by the mother. Known in several lamnoid sharks, the carcharhinoid family Pseudotriakidae, and in the orectoloboid family Ginglymostomatidae (*Nebrius ferrugineus*).

Order. One of the taxonomic groups of organisms, containing related families; related orders are grouped into classes.

Orectoloboid. A carpet shark, a member of the order Orectolobiformes, including barbelthroat carpet sharks, blind sharks, wobbegong sharks, bamboo sharks, epaulette sharks, nurse sharks, Zebra Shars, and Whale Shark.

Origin. The anterior or front end of the fin base in all fins. The caudal fin has upper and lower origins but no insertion. See insertion.

Ovipary or Oviparity. A mode of reproduction in which female sharks deposit eggs enclosed in oblong or conical egg-cases on the bottom, which hatch in less than a month to more than a year, producing young sharks which are miniatures of the adults.

Ovoviviparity. See aplacental yolksac viviparity.

Ovovivipary. Generally equivalent to yolk-sac vivipary, live-bearing in which the young are nourished primarily by the yolk in the yolk-sac, which is gradually depleted and the yolk-sac resorbed until the young are ready to be born. Sometimes used to cover all forms of aplacental vivipary, including cannibal vivipary.

Paired fins. The pectoral and pelvic fins.

Parasite. An organism that lives in or on another organism, the host, and obtains nutrients from the host at it's expense.

Paratype. Each specimen of a type series other than the holotype. Specimens other than the holotype automatically become paratypes unless the author designates them as referred specimens that are not part of the type series.

Pectoral fins. A symmetrical pair of fins on each side of the trunk just behind the head and in front of the abdomen. These are present in all cartilaginous fishes and correspond to the forelimbs of a land vertebrate (a tetrapod or four-footed vertebrate).

Pelagic. Referring to organisms that live in the water column, not on the sea bottom.

Pelvic fin. A symmetrical pair of fins on the sides of the body between the abdomen and precaudal tail which correspond to the hindlimbs of land vertebrate (a tetrapod or four-footed vertebrate). Also, ventral fins.

Pharynx. Part of the gut that links the mouth to the oesophagus.

Photophores. Conspicuously pigmented small spots on the bodies of most lantern sharks (family Etmopteridae) and some kitefin sharks (family Dalatiidae). These are tiny round organs that are covered with a conspicuous dark pigment (melanin) and produce light by a low-temperature chemical reaction.

Phylum. In animal taxonomy it is one of the major groupings dividing organisms into general similarities and comprises of superclasses, classes and all other lower groupings.

Placenta. See Yolk-sac placenta.

Placental vivipary. Live-bearing in which the young develop a yolk-sac placenta, which is apparently confined to the carcharhinoid sharks.

Poikilothermic. An organism whose body temperature varies in accordance with the temperature of its surroundings.

Population. A group of individuals of a species living in a particular area. (This is defined by IUCN (2001) as the total number of mature individuals of the taxon, with subpopulations defined as geographically or otherwise distinct groups in the population between which there is little demographic or genetic exchange (typically one successful migrant individual or gamete per year or less).)

Pores, pigmented. In a few sharks and skates, the pores for the lateral line and ampullae of Lorenzini are conspicuously black-pigmented, and look like little black specks.

Posterior. Rearwards, in the longitudinal direction of the caudal fin tip or tail filament. Also caudad.

Posterior margin. In precaudal fins, the margin from the fin apex to either the free rear tip (in sharks with distinct inner margins) or the fin insertion (for those without inner margins).

Posterior nasal flaps. Low flaps or ridges arising on the posterior edges of the excurrent apertures of the nostrils.

Posterior notch. In the caudal fin, the notch in the postventral margin dividing it into upper and lower parts.

Posterior tip. The posteriormost corner or end of the terminal lobe of the caudal fin.

Postventral margin. In the caudal fin, the margin from the ventral tip to the subterminal notch of the caudal fin. See lower and upper postventral margins.

Preanal ridges. A pair of low, short to long, narrow ridges on the midline of the caudal peduncle entending anteriorly from the anal fin base.

Precaudal fins. All fins in front of the caudal fin.

Precaudal pit. A depression at the upper and sometimes lower origin of the caudal fin where it joins the caudal peduncle.

Precaudal tail. That part of the tail from its base at the vent to the origins of the caudal fin

Predorsal ridge. A low narrow ridge of skin on the midline of the back anterior to the first dorsal fin base.

Preventral margin. In the caudal fin, the margin from the lower origin to the ventral tip of the caudal fin.

Pristiophoroid. A saw shark, order Pristiophoriformes, family Pristiophoridae.

Productivity. Relates to the birth, growth and mortality rates of a fish stock. Highly productive stocks are characterised by high birth, growth and mortality rates and can usually sustain higher exploitation rates and, if depleted, could recover more rapidly than comparatively less productive stocks.

Protrusible. Capable of being thrust forward, term normally applied to jaws.

Proximal. In any direction, at the near end of a structure.

Pupping ground. Area favoured for giving birth and depositing young.

Ray. See Batoid.

Red List of Threatened Species. Listing of the conservation status of the world's flora and fauna administered by IUCN. www.redlist.org

Reef channel. Channels that develop in the reef linking the water in the reef flat and lagoons to the open sea.

Reef face. Seaward face to the coral reef, often known as 'drop-off' and the most productive area of the reef. (See page 336).

Reef flat (Reef terrace). Horizontal part of the reef extending from the shore to, the reef crest, usually only a few metres down from the seas surface. (See page 336).

Ring valve. A type of spiral intestinal valve in which the valve turns are very numerous and short resembling a stack of washers.

Rise. The transitional and less steep bottom zone from the lower slope to the abyss or ocean floor, between 2250 m and 4500 m. The rise can be divided into upper (2250–3000 m), middle (3000–3750 m) and lower (3750–4500 m) rises. Few sharks are known from the rise, and those mostly from the upper rise. See Abyss, Hadal, shelf and slope. (See page 336).

Rostral keel. In squaloids, a large vertical plate on the underside of the rostrum and internasal septum, sometimes reduced, and with the cavities of the subnasal fenestrae on either side of the keel.

Rostrum. The cartilaginous anteriormost structure that supports the prenasal snout including lateral line canals and masses of ampullae. It is absent in a few nonbatoid sharks and in many batoids.

Row. In oral teeth, a single replicating line of teeth, approximately transverse to the longitudinal jaw axis, which includes functional teeth and their replacements, derived from one tooth-producing area on the jaw.

Saddle. Darker dorsal marking that extends downwards either side of the shark but does not meet on the ventral surface.

Sargasso Sea. Central body of water surrounded by the North Atlantic circulation gyre.

Scientific name. The formal binomial name of a particular organism, consisting of the genus and specific names; a species only has one valid scientific name.

Scroll valve. A type of spiral intestinal valve in requiem and hammerhead sharks in which the valve has uncoiled and resembles a rolled-up bib or scroll.

Seagrass. The collective name for marine flowering plants that are found in shallow waters in mud and sand.

Seamount. A large isolated elavation in the open ocean, characteristically of conical form, that rises at least 1000 m from the ocean floor; often a productive area for deepwater fisheries. (See page 336).

Second dorsal fin. The posteriormost dorsal fin of two in cartilaginous fishes, not always present, ranging in position from over the pelvic fin bases to far posterior on the precaudal tail.

Secondary caudal keels. Low horizontal dermal keels on the ventral base of the caudal fin in mackerel sharks (Lamnidae) and some somniosids.

Secondary lower eyelid. The eyelid below or lateral to the nictitating lower eyelid, separated from it by a subocular groove or pocket, and, in many carcharhinoids with internal nictitating lower eyelids, functionally replacing them as lower eyelids.

Seine netting. A fishing method using nets to surround an area of water where the ends of the nets are drawn together to encircle the fish (includes purse seine and Danish seine netting).

Serrations. In oral teeth, minute teeth formed by the cutting edge of the crown that enhance the slicing abilities of the teeth.

Shark. Generally used for cylindrical or flattened cartilaginous fishes with five to seven external gill openings on the sides of their heads, pectoral fins that are not attached to the head above the gill openings, and a large, stout tail with a large caudal fin; that is, all living elasmobranchs except the rays or batoids. Living sharks in this sense are all members of the Neoselachii, the modern sharks and rays. Rays are essentially flattened sharks with the pectoral fins attached to their heads, while living chimaeras are also called ghost sharks or silver sharks. Hence shark is used here in an alternate and broader sense to include the rays and chimaeras.

Shelf, continental and insular. The sloping plateau-like area along the continents and islands between the shoreline and approximately 200 m depth. It is roughly divided into inshore (intertidal to 100 m), and offshore (100–200 m) zones. The shelves have the greatest diversity of cartilaginous fishes. See Abyss, rise and slope. (See page 336).

Single ovipary. A mode of egg-laying or ovipary in which female sharks produce encased eggs in pairs, which are not retained in the oviducts and are deposited on the bottom. Embryos in the egg cases are at an early developmental stage, and take a few months to over a year to hatch. Found in almost all oviparous cartilaginous fishes.

Slope, continental and insular. The precipitous bottom zone from the edge of the outer shelf down to the submarine rise, between 200 m–2250 m. The slope can be divided into upper (200–750 m), middle (750–1500 m) and lower (1500–2250 m) slopes, of which the upper and middle slope has the highest diversity of deepwater benthic sharks. See Abyss, rise and shelf.

Snout. That part of a cartilaginous fish in front of its eyes and mouth, and including the nostrils.

Species. A group of interbreeding individuals with common characteristics that produce fertile (capable of reproducing) offspring and which are not able to interbreed with other such groups, that is, a population that is reproductively isolated from others; related species are grouped into genera.

Spiracle. A small to large opening between the eye and first gill opening of most sharks and rays, representing the modified gill opening between the jaws and hyoid (tongue) arch. Lost in chimaeras and some sharks.

Squalene. A long-chain oily hydrocarbon present in the liver oil of deepwater cartilaginous fishes. It is highly valued for industrial and medicinal use.

Squaloid. A dogfish shark, a member of the order Squaliformes, including bramble sharks, spiny dogfish, gulper sharks, lantern sharks, viper sharks, rough sharks, sleeper sharks, kitefin sharks, and cookiecutter sharks.

Squalomorph. Referring to the Squalomorphii.

Squalomorphii. The neoselachian superorder Squalomorphii, including the hexanchoid, squaloid, squatinoid, pristiophoroid, and batoid sharks.

Squatinoid caudal fin. Angelsharks (Squatiniformes) are unique among living sharks in having hypocercal caudal fins that resemble inverted caudal fins of ordinary sharks. The dorsal margin is subdivided into a predorsal margin from the upper origin to its dorsal tip (analogous to the preventral margin and ventral tips in ordinary sharks), a postdorsal margin (like the postventral margin) from the dorsal tip to its supraterminal notch (similar to the subterminal notch), and a short supraterminal margin and large ventral terminal margin (similar to the subterminal and terminal margins) between the supraterminal notch and the ventral tip of the caudal. The ventral margin has a preventral margin forming a ventral lobe with the ventral tip and the ventral terminal margin.

Squatinoid. An angelshark, order Squatiniformes, family Squatinidae.

SSC. Species Survival Commission. One of six volunteer commissions of IUCN.

SSG. Shark Specialist Group (part of the IUCN Species Survival Commission network).

Stock. A group of individuals in a species, which are under consideration from the point of view of actual or potential utilisation, and which occupy a well defined geographical range independent of other stocks of the same species. A stock is often regarded as an entity for management and assessment purposes.

Subcaudal keel. In a few dogfish sharks (family Centrophoridae), a single longitudinal dermal keel on the underside of the caudal peduncle.

Sublittoral zone (subtidal zone). The benthic region from low water mark to the outer limit of the continetal shelf, usually about 200 m deep. (See page 336).

Subterminal margin. In the caudal fin, the margin from the subterminal notch to the ventral beginning of the terminal margin.

Subterminal mouth or ventral mouth. Mouth located on the underside of the head, behind the snout.

Subterminal notch. On the caudal fin of most sharks, the notch in the lower distal end of the caudal fin, between the postventral and subterminal margins, and defining the anterior end of the terminal lobe.

Subtropical region. The intermediate region between the tropical and temperate zones.

Supraorbital crest. On the neurocranium, an arched horizontal plate of cartilage forming the dorsal edge of the orbit on each side; it arises from the medial orbital wall and the cranial roof and extends horizontally from the preorbital process to the postorbital process. It is apparently primitive for sharklike fishes but is variably reduced or absent in some living elasmobranchs.

Supraorbital or brow ridge. A dermal ridge above each eye, particularly well-developed in heterodontoids and some orectoloboids.

Sympatric. Different species which inhabit the same or overlapping geographic areas.

Symphyseal or symphysial groove. A longitudinal groove on the ventral surface of the lower jaw of some orectoloboid sharks, extending posteriorly from the lower symphysis.

Symphysial teeth. Larger oral teeth in one row on either side of the symphysis, distal to medials or alternates where present. Symphysials are broader than medials and usually have asymmetrical roots.

Symphysis. The midline of the upper and lower jaws, where the paired jaw cartilages articulate with each other.

Syntype. Two or more specimens used and mentioned in an original description of a species, where there was no designation of a holotype or a holotype and paratype(s) by the describer of the species.

Systematics. Scientific study of the kinds and diversity of organisms, including relationships between them.

Tail. That part of a cartilaginous fish from the cloacal opening or vent (anus in chimaeroids, which lack a cloaca) to the tip of the caudal fin or caudal filament, and including the anal fin, usually the second dorsal fin when present, and caudal fin.

Target catch. The catch which is the subject of directed fishing effort within a fishery; the catch consisting of the species primarily sought by fishers.

Taxon, plural taxa. A taxonomic group at any level in a classification. Thus the taxon Chondrichthyes is a class with two taxa as subclasses, Elasmobranchii and Holocephali, and the taxon *Galeorhinus*, a genus, has one taxon as a species, *G. galeus*.

Taxonomy. Often used as a synonym of systematics or classification, but narrowed by some researchers to the theoretical study of the principles of classification.

Temperate. Two circumglobal bands of moderate ocean temperatures usually ranging between 10° and 22°C at the surface, but highly variable due to currents and upwelling. including the north temperate zone between the Tropic of Cancer, 23°27' N latitude, to the Arctic Circle, 66°30' N; and the south temperate zone between the Tropic of Capricorn, 23°27' S latitude, to the Antarctic Circle, 66°30' N. (See page 338).

Terminal lobe. In the caudal fin of most non-batoid sharks and at least one batoid, the free rear wedge-shaped lobe at the tip of the caudal fin, extending from the subterminal notch to the posterior tip.

Terminal margin. In the caudal fin, the margin from the ventral end of the subterminal margin to the posterior tip.

Terminal mouth. Mouth located at the very front of the animal. Most cartilaginous fishes have subterminal mouths, but some species (Viper Shark, wobbegongs, angelsharks, frilled sharks, Whale Shark, Megamouth Shark, and mantas) have it terminal or nearly so.

Thorn. In many batoids, most angel sharks and the Bramble Shark (*Echinorhinus brucus*), enlarged, flat conical denticles with a sharp, erect crown and a flattened base (which may grow as the shark grows).

TL. Total length. A standard morphometric measurement for sharks and some batoids, from the tip of snout or rostrum to the end of the upper lobe of the caudal fin.

Transverse. Across the long access of the body.

Tropical. Circumglobal band of warm coastal and oceanic water, usually above 22°C at the surface (but varying because of currents and upwelling), between the latitudes of 23°27' N (Tropic of Cancer) and 23°27' S (Tropic of Capricorn) and including the Equator. (See page 338).

Truncate. Blunt, abbreviated.

Trunk. That part of a cartilaginous fish between its head and tail, from the last gill openings to the vent, including the abdomen, back, pectoral and pelvic fins, and often the first dorsal fin.

Umbilical cord. A modified yolk stalk in placental viviparous sharks, carrying nutrients from the placenta to the foetus.

Undescribed species. An organism not yet formally described by science and so does not yet a have a formal binomial scientific name. Usually assigned a letter or number designation after the generic name, for example, *Squatina* sp. A is an undescribed species of angel shark belonging to the genus *Squatina*.

Unpaired fins. The dorsal, anal, and caudal fins.

Upper eyelid. The dorsal half of the eyelid, separated by a deep pocket (conjunctival fornix) from the eyeball. The upper eyelid fuses with the eyeball and the pocket is lost in all batoids.

Upper origin. In the caudal fin, the anterodorsal beginning of the epaxial or upper web of the caudal fin, at the posterior end of the dorso-caudal space.

Upper postventral margin. In the caudal fin, the upper part of the postventral margin of the hypaxial web, from the posterior notch to the subterminal notch.

Uterine cannibalism or cannibal vivipary. A mode of reproduction in which foetuses deplete their yolk-sacks early and subsist by eating nutritive eggs produced by the mother (see oophagy) or first eat smaller siblings and then nutritive eggs (see adelphophagy).

Vent. The opening of the cloaca on the ventral surface of the body between the inner margins and at the level of the pelvic fin insertions.

Ventral fin. See pelvic fin.

Ventral lobe. In the caudal fin, the expanded distal end of the preventral and lower postventral margins, defined by the posterior notch of the caudal fin.

Ventral margin. In the caudal fin, the entire ventral margin from lower origin to posterior tip, either a continuous margin or variably subdivided into preventral, postventral, subterminal and terminal margins.

Ventral tip. In the caudal fin, the ventral apex of the caudal fin.

Ventral. Downwards, in the vertical direction of the abdomen. See dorsal.

Vertebra, plural vertebrae. A single unit of the vertebral column, including a vertebral centrum and associated cartilages that form arches and ribs.

Vertebral axis. That part of the vertebral column inside the base of the caudal fin.

Vertebral column. The entire set or string of vertebrae, or 'backbone', of a shark, from the rear of the chondrocranium to the end of the caudal base. Living elasmobranchs range from having as few as 60 vertebrae (some squaloids of the family Dalatiidae) to as many as 477 vertebrae (thresher sharks).

Viviparity. A reproductive mode where the maternal adult gives birth to live young. Encompasses aplacental viviparity and placental viviparity.

Vivipary. Used in two ways in recent literature, as being equivalent to placental vivipary only, that is for carcharhinoid sharks with a yolk-sac placenta; or for all forms of live-bearing or aplacental vivipary.

Yolk sac or yolk sack. Almost all sharks start embryonic development somewhat like a chicken, as a large spherical yolky egg inside an elongated shell, the egg case. A small disk of dividing cells represents the pre-embryo or blastula atop the huge yolk mass. The blastula expands around the sides and ventral surface of the yolk mass, and differentiates into an increasingly sharklike embryo, the yolk sac or baglike structure containing the yolk, and a narrow tubular yolk stalk, between the abdomen of the embryo and the yolk sac.

Yolk-sac placenta. An organ in the uterus of some ground sharks (order Carcharhiniformes), formed from the embryonic yolk-sac of the embryo and maternal uterine lining, through which maternal nutriment is passed to the embryo. It is analogous to the placenta of live-bearing mammals. There are several forms of yolk-sac placentas in carcharhinoid sharks, including entire, discoidal, globular, and columnar placentas (see Compagno, 1988).

Yolk-sac vivipary. Live-bearing in which the young are nourished primarily by the yolk in the yolk-sacs, which is gradually depleted and the yolk-sacs resorbed until the young are ready to be born

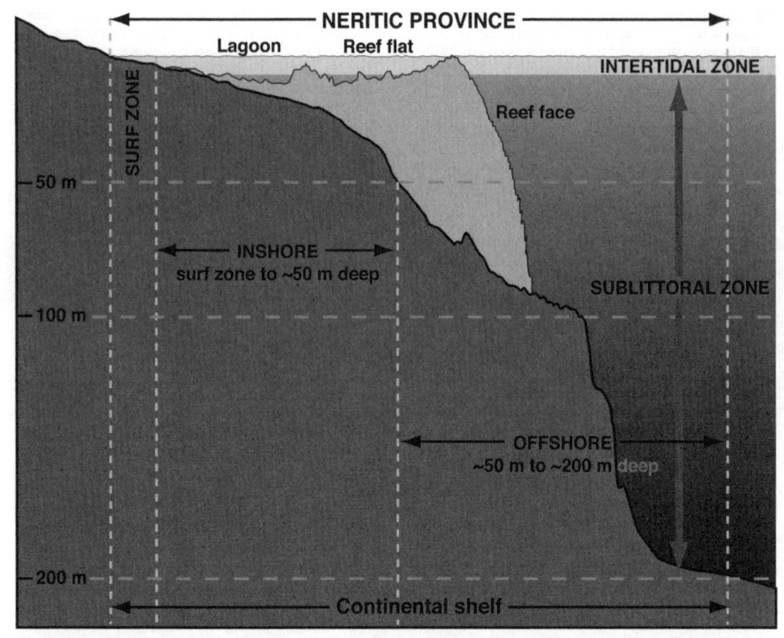

Figure 53a: (above) continental shelf zoes, including reef zones; figure 53b: (below) oceanic zones.

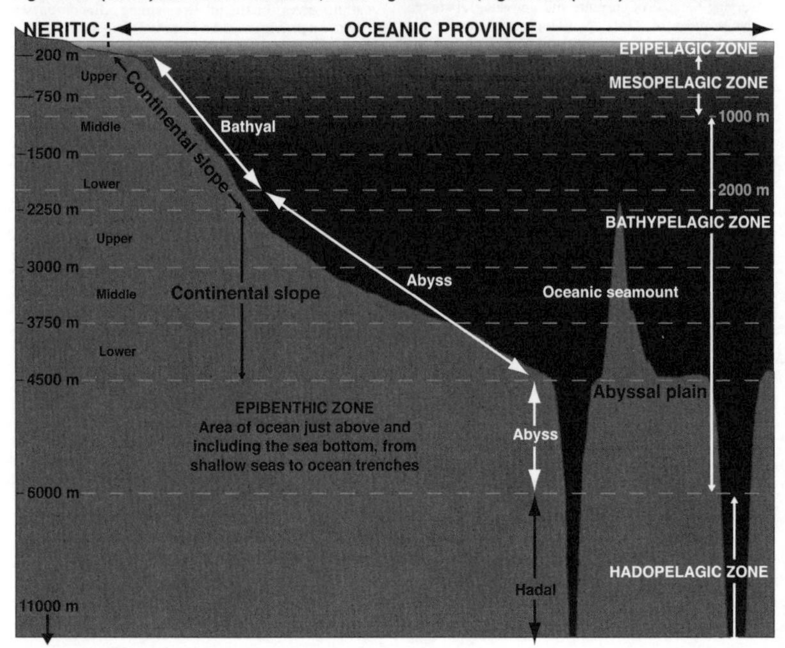

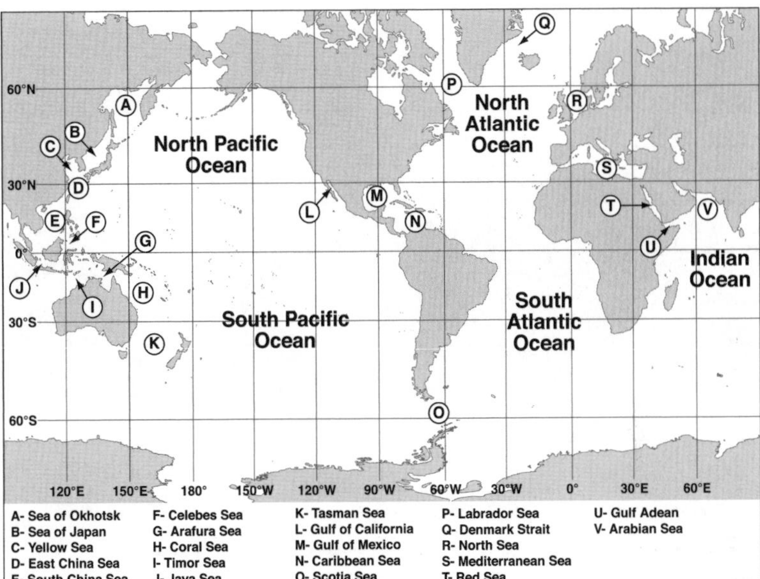

Figure 54: Large bodies of water referred to inthis book.

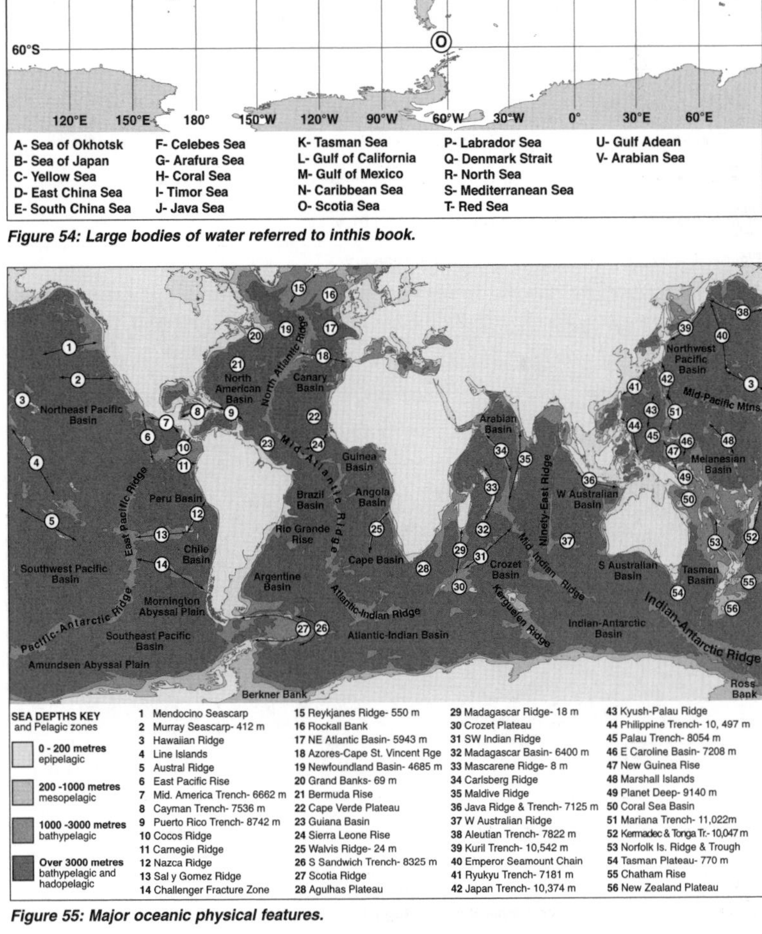

Figure 55: Major oceanic physical features.

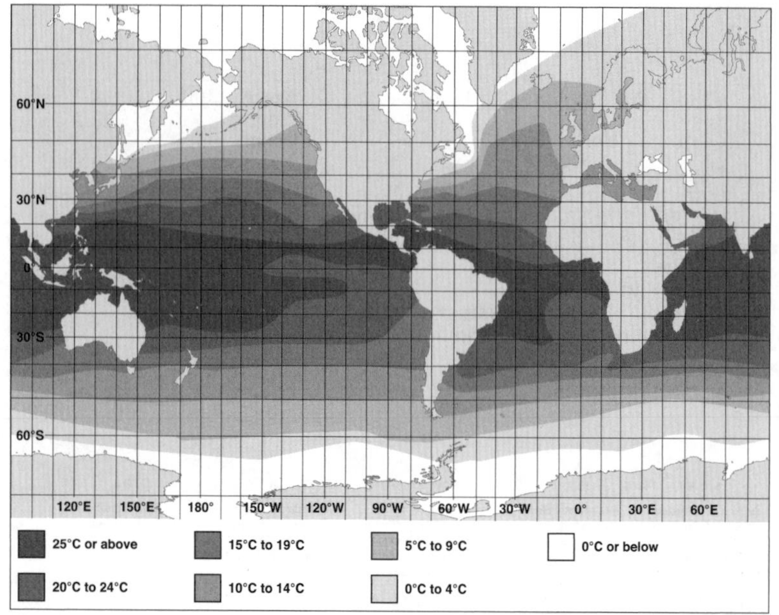

Figure 56: Average sea surface temperatures in February.

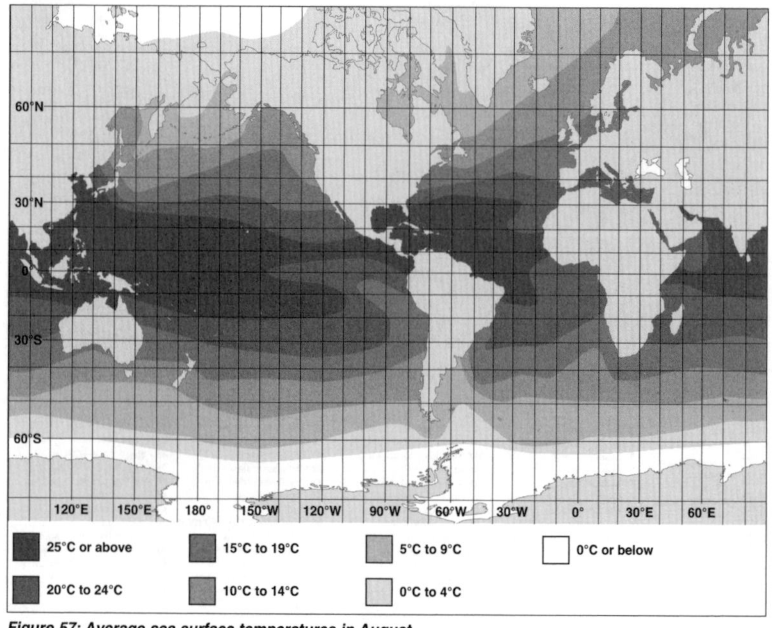

Figure 57: Average sea surface temperatures in August.

Field observations

Many readers will use this book for identifying sharks that they see while diving or fishing and will probably mainly come across fairly common and widely distributed species. Even so, it is worth keeping good records of your observations because scientific knowledge of even some of the most widespread species is lacking. Anyone may be able to contribute important new information, particularly if they are in a part of the world where little research is underway.

Before handling a live shark, please remember that these animals live supported by water. Their internal organs (gut, liver etc.) do not have the surrounding protection and support provided by the ribs and abdominal muscles of land animals. They can, therefore, be extremely easily damaged if lifted out of the water without good abdominal support (i.e. in a sling). Please take as many measurements as possible in the water alongside the boat, don't take your shark out to hug it, and try to ensure that it is released in good condition! You may need to spend some time helping the shark to get oxygenated water flowing over the gills again, if it has been motionless for a while. Pushing or pulling it gently through the water (headfirst, naturally) should help.

Fish markets, particularly (but not exclusively) in tropical countries, are some of the best hunting grounds for poorly known or even completely unknown sharks. One of the authors collected an undescribed (but relatively common) species of houndshark on her very first visit to a small fish market in Borneo! You may have to get up extremely early in the morning to see the fish as they are landed and before they are chopped up and sold and are advised to wear old clothes and shoes that can be washed (or thrown away) afterwards, but the experience can be really memorable. With luck, stallholders will be enthusiastic and helpful if you show a keen interest in their stock.

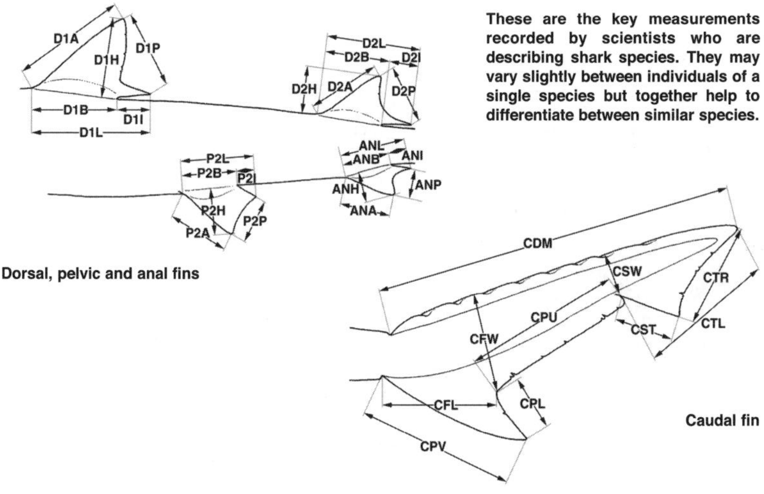

Figure 58: Fin measurements

D1A 1st dorsal fin anterior margin
D1P 1st dorsal fin posterior margin
D1H 1st dorsal fin height
D1B 1st dorsal fin base
D1I 1st dorsal fin inner margin
D1L 1st dorsal fin length
D2A 2nd dorsal fin anterior margin
D2P 2nd dorsal fin posterior margin
D2H 2nd dorsal fin height
D2B 2nd dorsal fin base
D2I 2nd dorsal fin inner margin
D2L 2nd dorsal fin length

P2A Pelvic fin anterior margin
P2P Pelvic fin posterior margin
P2H Pelvic fin height
P2B Pelvic fin base
P2I Pelvic fin inner margin
P2L Pelvic fin length
ANA Anal fin anterior margin
ANP Anal fin posterior margin
ANH Anal fin height
ANB Anal fin base
ANI Anal fin inner margin
ANL Anal fin length

CDM Caudal fin dorsal margin
CPU Caudal fin upper postventral margin
CFW Caudal fin fork width
CSW Caudal fin subterminal width
CFL Caudal fin fork length
CPV Caudal fin preventral margin
CPL Caudal fin lower postventral margin
CST Caudal fin subterminal margin
CTR Caudal fin terminal margin
CTL Caudal fin terminal lobe

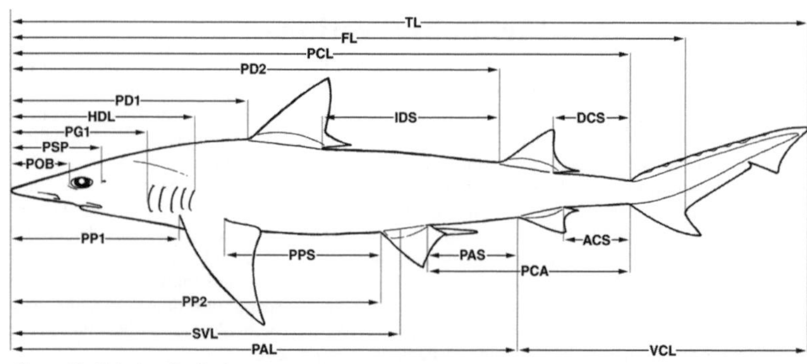

Figure 59: Main longitudinal measurements

TL	Total length	**PSP**	Prespiracular length	**DCS**	Dorsal fin–caudal fin space
FL	Fork length	**POB**	Preorbital length	**PPS**	Pectoral fin–pelvic fin space
PCL	Precaudal fin length	**PP1**	Pre-pectoral fin length	**PAS**	Pelvic fin–anal fin space
PD1	Pre-first dorsal fin length	**PP2**	Pre-pelvic fin length	**ACS**	Anal fin–caudal fin space
PD2	Pre-second dorsal fin length	**SVL**	Snout–vent length	**PCA**	Pelvic fin–caudal fin space
HDL	Head length	**PAL**	Pre-anal fin length	**VCL**	Vent–caudal fin length
PG1	Prebranchial length	**IDS**	Interdorsal space		

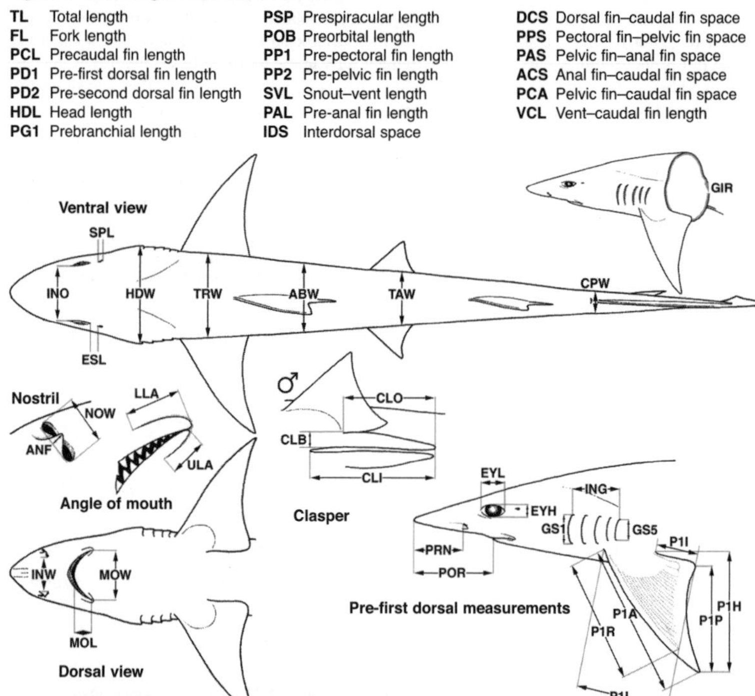

Figure 60: Other measurements

INO	Interorbital space	**LLA**	Lower labial furrow length	**EYH**	Eye height
ESL	Eye–spiracle space	**CLO**	Clasper outer length	**GS1**	First gill slit height
SPL	Spiracle length	**CLB**	Clasper base width	**GS5**	Fifth gill slit height
HDW	Head width	**CLI**	Clasper inner length	**ING**	Intergill length
TRW	Trunk width	**GIR**	Girth	**P1A**	Pectoral fin anterior margin
ABW	Abdomen width	**INW**	Internarial space	**P1R**	Pectoral fin radial length
TAW	Tail width	**MOL**	Mouth length	**P1I**	Pectoral fin inner margin
CPW	Caudal fin–peduncle width	**MOW**	Mouth width	**P1P**	Pectoral fin posterior margin
ANF	Anterior nasal flap length	**PRN**	Prenarial length	**P1H**	Pectoral fin height
NOW	Nostril width	**POR**	Preoral length	**P1L**	Pectoral fin length
ULA	Upper labial furrow length	**EYL**	Eye length		

Where possible the following information should be recorded, whether from live sharks in the water, on hook and line alongside a boat, onboard or onshore, or from dead specimens.

i. Name and address of observer or collector

ii. Date, time, location, habitat and water depth (where available).

iii. Other relevant observations (e.g. behaviour).

iv. Photographs of the whole specimen, particularly if it is an uncommon record our outside its usual geographic range. These should be taken from the side (standard scientific procedure is to illustrate the left side of the shark), from above, and from below, with close ups of the underside of the head and pectoral fins. Put something in the photograph to provide a known measurement scale.

v. Measurements. Anglers like to record weights, but since these are very variable, depending upon time of year, point in the reproductive cycle etc., scientists prefer to record length as well, if not instead of weight. Total length should be recorded as a point-to-point distance, not over the curve of the body (which produces an overestimate of length). Precaudal length, from tip of snout to the tail fork, is another important measurement and can be easier to record than total length. Other useful measurements are illustrated opposite.

vi. Record whether male (a photograph of the claspers may be useful in judging maturity) or female, and whether there are any signs of pregnancy.

vii. If the specimen is dead, remove and dry a strip of teeth from each of the upper and lower jaws (label them!) or keep the entire jaws. The latter should be pinned to a board to dry in order to prevent distortion. If it is also possible to keep and dry the fins and the vertebral column, these can be very useful in confirming identification of species that are difficult to identify. DNA can also be extracted from dried tissue for scientific studies.

viii. If the specimen is dead, small and apparently unusual, it can be useful to keep the whole animal. Freezing is the easiest way to do this in the short term. In the longer term, it will be necessary to fix and preserve the specimen, using formalin and alcohol, but these procedures are usually undertaken in museums because of the difficulties of storing safely these toxic and flammable chemicals. For more information, see Compagno 2001.

Remember, that many of the species described in this field guide may well turn out to be two or more very similar species. They may be distinguished by careful examination and measurements, by differences in size (they may mature at quite different lengths), by their use of different habitats and prey species, or because of different (but sometimes overlapping) distribution. Confusion between such similar species makes it very hard to understand their distribution, habitat, life history and other biological characters. Good records can help to overcome these problems.

Records of sharks, particularly of unusual species or species recorded outside their usual range, should be sent to the relevant national or state museum, or fishery department, of the country in whose waters they were recorded. County biological records centres may be interested in receiving UK records. At minimum, a good photograph, ideally the specimen itself, should accompany unusual records, particularly if it is the first record from that country or region, but please contact the curator with details before sending in any specimens (particularly if these are large). While every country needs a national fish collection for reference purposes, to help train its fisheries staff and researchers and for the use of visiting scientists, and should be offered new specimens, some institutes may not have the necessary facilities to curate and keep them in good condition. In these cases, or if more than one important specimen has been collected, it may be necessary to send them to one of the major international fish collections as well, or instead (see Museums and research organisations on page 344).

Suggested Further Reading

Allen, T.B. 1996. *Shadows in the Sea: the sharks, skates and rays.* Lyons and Burford, New York.

Bonfil, R. and M. Abdallah. 2004. *Field identification guide to the sharks and rays of the Red Sea and Gulf of Aden. FAO Species Identification Guide for Fishery Purposes.* FAO, Rome, Italy.

Camhi, M., S. Fowler, J. Musick, A. Brautigam, and S. Fordham. 1998. *Sharks and Their Relatives: Ecology and Conservation.* IUCN Occasional Paper No. 20. IUCN Cambridge UK and Gland Switzerland.

Carrier, J.C., J.A. Musick, and M.R. Heithaus. 2004. *Biology of Sharks and Their Relatives.* CRC Press.

Cawardine, M. 2004. *Shark.* BBC Books, BBC, UK.

Carwardine, M., and K. Watterson. 2002. *The Shark-Watcher's Handbook: A Guide to Sharks and Where to See Them.* Princeton University Press, Princeton.

Castro, J.I. 1983. *The Sharks of North American Waters.* Texas A and M University Press, College Station.

Compagno, L.J.V., C. Simpfendorfer, J.E. McCosker, K. Holland, C. Lowe, B. Wetherbee, A. Bush, and C. Meyer. 1998. *Sharks (Reader's Digest explores).* Reader's Digest.

Compagno, L.J.V., D.A. Ebert, and M.J. Smale. 1989. *Guide to the sharks and rays of southern Africa.* Struik Pub., Cape Town.

Cook, L. and J. Simonetti. 2003. *Why I care about sharks.* Big Fish Press.

Daley, R., J. Stevens, P. Last and G. Yearsley. 2002. *Field Guide to Australian Sharks and Rays.* CSIRO Marine Research, Australia.

Ebert, D. A. 2003. *Sharks, rays, and chimaeras of California.* University of California Press, Berkley.

Fowler, S.L., Camhi, M., Burgess, G.H., Cailliet, G., Fordham, S.V., Cavanagh, R.D., Simpfendorfer, C.A. and Musick, J.A. (In Press) *Sharks, rays and chimaeras: the status of the chondrichthyan fishes.* IUCN SSC Shark Specialist Group. IUCN, Gland, Switzerland and Cambridge, UK

Gilbert, P.W., ed. 1963. *Sharks and survival.* D. C. Heath, Boston.

Gruber, S.H. ed. 1990. Discovering sharks. *Underwater Naturalist., Bull. Amer. Littor. Soc.* 19(4)–20(1).

Hennemann, R.M. 2001. *Sharks and Rays: Elasmobranch Guide of the World.* Ikan - Unterwasserarchiv. Frankfurt, Germany.

Klimley, A.P. 2003. *The Secret Life of Sharks.* Simon & Schuster.

MacQuitty, M. 2000. *Eyewitness: Shark.* Dorling Kindersley Publishing, London.

Musick, J.A. (editor). 1999. *Life in the Slow Lane: ecology and conservation of long-lived marine animals.* American Fisheries Society Symposium 23. AFS, Bethesda, Maryland, USA.

Musick, J.A. and B. McMillan. 2002. *The Shark Chronicles.* Times Books, Henry Holt & Co. New York.

Rose, D. 1996. *An overview of world trade in sharks and other cartilaginous fishes.* TRAFFIC International, Cambridge, UK.

Springer, V.G. and J.P. Gold. 1989. *Sharks in Question: the Smithsonian answer book.* Smithsonian Institution, Washington, USA.

Stafford-Deitsch, J. 1988. *Shark: a photographer's story.* Sierra Club Books.

Stafford-Deitsch, J. 1999. *Red Sea Sharks.* Trident Press, UK.

Stafford-Deitsch, J. 2000. *Sharks of Florida, the Bahamas, the Caribbean, the Gulf of Mexico.* Trident Press, UK.

Stevens, J.D., (editor) 1999. *Sharks.* Revised edition. Weldon Owen publishers, Australia.

Scientific societies and conservation organisations

Shark Trust. UK-based wildlife conservation charity that promotes the study, management and conservation of sharks, skates and rays. Colour newsletter. UK Conference. Shark photo-identification projects. EEA member. www.sharktrust.org.

European Elasmobranch Association (EEA). A network of European organisations that promote elasmobranch research, management and conservation. Members include:

ElasmoFrance www.fredshark.net

German Elasmobranch Society (DEG) www.elasmo.de

Italian Research Group for the study and conservation of sharks (GRIS) http://digilander.libero.it/infogris/

Portuguese Association for the Study & Conservation of Sharks (APECE) www.apece.org

Shark Trust (see above)

Spanish Association for Sharks and Rays (AITYR) http://idd00e7k.eresmas.net/

Uno squalo per amico (San Marino) www.unosqualoperamico.org

American Elasmobranch Society. Scientific non-profit organization, USA. Annual scientific meetings and newsletter. www.elasmo.org

Sociedade Brasileira para Estudos de Elasmobranquios (SBEEL Brazilian Society for Elasmobranch Study). www.sbeel.hpg.ig.com.br

IUCN Shark Specialist Group. International expert network. Detailed shark conservation and management information on line. Newsletter and technical publications. www.flmnh.ufl.edu/fish/Organizations/SSG/SSG.htm

Japanese Elasmobranch Society

Shark Foundation. Swiss-based non-profit organisation. www.shark.ch

Shark Research Institute. US-based non-profit research and conservation organisation. www.sharks.org

Swedish Elasmobranch Society http://hem2.passagen.se/frogboy/

Other useful Websites

Elasmo.com. Information on extinct and fossil sharks. www.elasmo.com

Elasmoworld. Scientific information and literature database. www.elasmoworld.com

Fishbase. Global information system and searchable database. www.fishbase.org

Florida Museum of Natural History Ichthyology Department. Detailed online information and news. Links to many other sites. www.flmnh.ufl.edu/fish

International Shark Attack File. www.flmnh.ufl.edu/fish/ /ISAF/ISAF.htm

Mediterranean Shark Site. Information and news. www.zoo.co.uk/~z9015043

Red List of Threatened Species. Searchable website. Includes annual updates of shark status information presented in this volume. www.redlist.org

ReefQuest Centre for Shark Research. www.elasmo-research.org

Museums and other research organisations

Most major European capital cities have an important fish collection in their state museums. Many also have detailed online information (search for 'shark'). There is insufficient space to list them all here; this is just a selection:

Natural History Museum, London www.nhm.ac.uk

National Museums & Galleries of Wales www.nmgw.ac.uk

National Museums of Scotland www.nms.ac.uk/royal

American Museum of Natural History www.amnh.org

Australian Museum www.austmus.gov.au

Muséum National d'histoire Naturelle, Paris www.mnhn.fr/museum

National Museum of Natural History, Smithsonian Institution www.mnh.si.edu

South African Museum Shark Research Centre www.museums.org.za/sam/src/sharks.htm

Australian National Fish Collection www.marine.csiro.au/facilities/fish

California Academy of Sciences, Department of Ichthyology. The CAS Catalogue of Fishes resides here, an invauable research tool for shark systematics, also the on-line catalogue fro the CAS fish collection which has one of the largest shark collections in the world. www.calacademy.org/research/ichthyology

Canadian Shark Research Laboratory www.mar.dfo-mpo.gc.ca/science/shark

Centre for Environment Fisheries and Aquaculture Science (CEFAS, UK). Handles UK shark and ray tagging data, undertaking basking shark research. www.cefas.co.uk

CSIRO Marine Laboratory, Australia www.marine.csiro.au

Food and Agricultural Organisation of the United Nations, Fisheries Department, Species Identification and Data Program (SIDP). For free PDF versions of the 1984 and 2001 shark catalogues, plus much else that is shark-related. www.fao.org/fi/sidp

Natal Sharks Board. Protects beach users against shark attack in South Africa, undertakes research and supports the conservation of sharks. www.shark.co.za

National Shark Research Consortium www.flmnh.ufl.edu/fish/sharks/nsrc/nsrc.htm Comprised of four US shark research organisations:

> **Pacific Shark Research Centre**, Moss Landing Marine Laboratories, California, www.mlml.calstate.edu
>
> **Centre for Shark Research**, Mote Marine Laboratory, Florida, www.mote.org
>
> **Florida Program for Shark Research**, University of Florida, Florida Museum of Natural History, www.flmnh.ufl.edu/fish
>
> **Shark Research Program**, Virginia Institute of Marine Science, www.vims.edu

Bibliography

Bass, A.J., J.D. D'Aubrey, and N. Kistnasamy. 1973–1976. *Sharks of the east coast of southern Africa.* Volumes I–VI. S. Afr. Ass. Mar. Biol. Res., Oceanogr. Res. Inst., Invest. Reports 33, 37-39, 40, 45.

Bigelow, H.B., and W.C. Schroeder. 1948. Chapter Three, Sharks. In: *Fishes of the Western North Atlantic.* Mem. Sears Fnd. Mar. Res. 1.

Bright, M. 1999. *The Private Life of Sharks.* Robson Books, London.

Compagno, L.J.V. 1984. *Sharks of the World. Volume 4. FAO Species Catalogue for Fisheries Purposes.* FAO Fisheries Synopsis No. 125. vol. 4, pt. 1-2. FAO, Rome, Italy.

Compagno, L.J.V. 1988. *Sharks of the Order Carcharhiniformes.* Princeton University Press, Princeton, New Jersey. Reprinted by Blackburn Press, 2003.

Compagno, L.J.V. 1990. *Alternate life history styles of cartilaginous fishes in time and space.* Environm. Biol. Fishes, 28: 33-75, figs. 1-36.

Compagno, L.J.V. 2001. *Sharks of the World. Volume 2. Bullhead, mackerel and carpet sharks (Heterodontiformes, Lamniformes and Orectolobiformes). An annotated and illustrated catalogue of the shark species known to date. FAO Species Catalogue for Fisheries Purposes.* FAO, Rome, Italy.

Compagno, L.J.V. (in prep. a). *Sharks of the World. Volume 1. Cow, frilled, dogfish, angel and saw sharks (Hexanchiformes, Squaliformes, Squatiniformes, and Pristiophoriformes). An annotated and illustrated catalogue of the shark species known to date. FAO Species Catalogue for Fisheries Purposes.* FAO, Rome, Italy.

Compagno, L.J.V. (in prep. b). *Sharks of the World. Volume 3. Ground sharks (Carcharhiniformes). An annotated and illustrated catalogue of the shark species known to date. FAO Species Catalogue for Fisheries Purposes.* FAO, Rome, Italy.

Dando, M., Burchett, M., and Waller, G. 2000. *Sealife: A guide to the marine environment.* Pica Press, Mountfield, UK.

Garman, S. 1913. *The Plagiostomia. Memoirs of the Museum of Comparative Zoology, Harvard 36.* Reprinted by Benthic Press, Los Angeles, California, 1997.

Hamlett, W.C. (editor). 1999. *Sharks, Skates, and Rays: The Biology of Elasmobranch Fishes.* Johns Hopkins University Press.

Klimley, A.P. and D.G. Ainley. 1998. *Great White Sharks: The Biology of* Carcharodon carcharias. Academic Press, London.

Last, P.R., & J.D. Stevens. 1994. *Sharks and rays of Australia.* CSIRO, Australia.

Perrine, D. 1999. *Sharks and Rays of the World.* Voyageur Press, Stillwater.

Pratt, H.L. Jr., S.H. Gruber, & T. Taniuchi, eds. 1990. *Elasmobranchs as living resources: Advances in the biology, ecology, systematics, and the status of the fisheries.* NOAA Tech. Rept. (90).

Schaeffer, B. 1967. Comments on elasmobranch evolution. *In* P.W. Gilbert, R.F. Mathewson, and D.P. Rall (editors), *Sharks, Skates, and Rays,* p. 3-35. Johns Hopkins Press, Baltimore.

Taylor, L.R., Jr. (editor) 1997. *The Nature Company Guides. Sharks and Rays.* The Nature Company, Time/Life books, Weldon-Owen, Sydney.

Tricas, T.C. and S.H. Gruber. 2001. *The behavior and sensory biology of elasmobranch fishes.* Kluwer Academic Publications, Netherlands.

Walker, T.I. (editor). 1998. *Shark Fisheries Management and Biology. Marine and Freshwater Research Special Issue.* Volume 49(7). CSIRO Publishing, Australia.

Index of scientific names

Index of english names

Checklist

Order Hexanchiformes: Cow and Frilled Sharks

Family Chlamydoselachidae: Frilled Sharks

- ☐ *Chlamydoselachus anguineus* Frilled Shark
- ☐ *Chlamydoselachus* sp. A Southern African Frilled Shark

Family Hexanchidae: Sixgill and Sevengill Sharks

- ☐ *Heptranchias perlo* Sharpnose Sevengill Shark
- ☐ *Hexanchus griseus* Bluntnose Sixgill Shark
- ☐ *Hexanchus nakamurai* Bigeye Sixgill Shark
- ☐ *Notorynchus cepedianus* Broadnose Sevengill Shark

Order Squaliformes: Dogfish Sharks

Family Echinorhinidae: Bramble Sharks

- ☐ *Echinorhinus brucus* Bramble Shark
- ☐ *Echinorhinus cookei* Prickly Shark

Family Squalidae: Dogfish Sharks

- ☐ *Cirrhigaleus asper* Roughskin Spurdog
- ☐ *Cirrhigaleus barbifer* Mandarin Dogfish
- ☐ *Squalus acanthias* Piked Dogfish
- ☐ *Squalus blainvillei* Longnose Spurdog
- ☐ *Squalus cubensis* Cuban Dogfish
- ☐ *Squalus japonicus* Japanese Spurdog
- ☐ *Squalus megalops* Shortnose Spurdog
- ☐ *Squalus melanurus* Blacktail Spurdog
- ☐ *Squalus mitsukurii* Shortspine Spurdog
- ☐ *Squalus rancureli* Cyrano Spurdog
- ☐ *Squalus* sp. A Bartail Spurdog
- ☐ *Squalus* sp. B Eastern Highfin Spurdog
- ☐ *Squalus* sp. C Western Highfin Spurdog
- ☐ *Squalus* sp. D Fatspine Spurdog
- ☐ *Squalus* sp. E Western Longnose Spurdog
- ☐ *Squalus* sp. F Eastern Longnose Spurdog

Family Centrophoridae: Gulper Sharks

- ☐ *Centrophorus acus* Needle Dogfish
- ☐ *Centrophorus atromarginatus* Dwarf Gulper Shark
- ☐ *Centrophorus granulosus* Gulper Shark
- ☐ *Centrophorus harrissoni* Longnose Gulper Shark
- ☐ *Centrophorus isodon* Blackfin Gulper Shark
- ☐ *Centrophorus lusitanicus* Lowfin Gulper Shark
- ☐ *Centrophorus moluccensis* Smallfin Gulper Shark
- ☐ *Centrophorus niaukang* Taiwan Gulper Shark
- ☐ *Centrophorus squamosus* Leafscale Gulper Shark
- ☐ *Centrophorus tessellatus* Mosaic Gulper Shark
- ☐ *Deania calcea* Birdbeak Dogfish
- ☐ *Deania hystricosum* Rough Longnose Dogfish

- ☐ *Deania profundorum* Arrowhead Dogfish
- ☐ *Deania quadrispinosum* Longsnout Dogfish

Family Etmopteridae: Lantern Sharks

- ☐ *Aculeola nigra* Hooktooth Dogfish
- ☐ *Centroscyllium* excelsum Highfin Dogfish
- ☐ *Centroscyllium fabricii* Black Dogfish
- ☐ *Centroscyllium granulatum* Granular Dogfish
- ☐ *Centroscyllium kamoharai* Bareskin Dogfish
- ☐ *Centroscyllium nigrum* Combtooth Dogfish
- ☐ *Centroscyllium ornatum* Ornate Dogfish
- ☐ *Centroscyllium ritteri* Whitefin Dogfish
- ☐ *Etmopterus baxteri* Giant Lanternshark
- ☐ *Etmopterus bigelowi* Blurred Smooth Lanternshark
- ☐ *Etmopterus brachyurus* Shorttail Lanternshark
- ☐ *Etmopterus bullisi* Lined Lanternshark
- ☐ *Etmopterus carteri* Cylindrical Lanternshark
- ☐ *Etmopterus caudistigmus* Tailspot Lanternshark
- ☐ *Etmopterus decacuspidatus* Combtooth Lanternshark
- ☐ *Etmopterus dianthus* Pink Lanternshark
- ☐ *Etmopterus dislineatus* Lined Lanternshark
- ☐ *Etmopterus evansi* Blackmouth Lanternshark
- ☐ *Etmopterus fusus* Pygmy Lanternshark
- ☐ *Etmopterus gracilispinis* Broadband Lanternshark
- ☐ *Etmopterus granulosus* Southern Lanternshark
- ☐ *Etmopterus hillianus* Caribbean Lanternshark
- ☐ *Etmopterus litvinovi* Smalleye Lanternshark
- ☐ *Etmopterus lucifer* Blackbelly Lanternshark
- ☐ *Etmopterus molleri* Slendertail Lanternshark
- ☐ *Etmopterus perryi* Dwarf Lanternshark
- ☐ *Etmopterus polli* African Lanternshark
- ☐ *Etmopterus princeps* Great Lanternshark
- ☐ *Etmopterus pseudosqualiolus* False Pygmy Lanternshark
- ☐ *Etmopterus pusillus* Smooth Lanternshark
- ☐ *Etmopterus pycnolepis* Densescale Lanternshark
- ☐ *Etmopterus robinsi* West Indian Lanternshark
- ☐ *Etmopterus schultzi* Fringefin Lanternshark
- ☐ *Etmopterus sentosus* Thorny Lanternshark
- ☐ *Etmopterus spinax* Velvet Belly
- ☐ *Etmopterus splendidus* Splendid Lanternshark
- ☐ *Etmopterus unicolor* Brown Lanternshark
- ☐ *Etmopterus villosus* Hawaiian Lanternshark
- ☐ *Etmopterus virens* Green Lanternshark
- ☐ *Miroscyllium sheikoi* Rasptooth Dogfish
- ☐ *Trigonognathus kabeyai* Viper Dogfish

Family Somniosidae: Sleeper Sharks

- ☐ *Centroscymnus coelolepis* Portugese Dogfish
- ☐ *Centroscymnus owstoni* Roughskin Dogfish
- ☐ *Centroselachus crepidater* Longnose Velvet Dogfish
- ☐ *Proscymnodon macracanthus* Largespine Velvet Dogfish
- ☐ *Proscymnodon plunketi* Plunket Shark
- ☐ *Scymnodalatias albicauda* Whitetail Dogfish
- ☐ *Scymnodalatias garricki* Azores Dogfish
- ☐ *Scymnodalatias oligodon* Sparsetooth Dogfish
- ☐ *Scymnodalatias sherwoodi* Sherwood Dogfish
- ☐ *Scymnodon ringens* Knifetooth Dogfish
- ☐ *Somniosus antarcticus* Southern Sleeper Shark
- ☐ *Somniosus longus* Frog Shark
- ☐ *Somniosus microcephalus* Greenland Shark
- ☐ *Somniosus pacificus* Pacific Sleeper Shark
- ☐ *Somniosus rostratus* Little Sleeper Shark
- ☐ *Somniosus* sp. A Longnose Sleeper Shark
- ☐ *Zameus ichiharai* Japanese Velvet Dogfish
- ☐ *Zameus squamulosus* Velvet Dogfish

Family Oxynotidae: Roughsharks

- ☐ *Oxynotus bruniensis* Prickly Dogfish
- ☐ *Oxynotus caribbaeus* Caribbean Roughshark
- ☐ *Oxynotus centrina* Angular Roughshark
- ☐ *Oxynotus japonicus* Japanese Roughshark
- ☐ *Oxynotus paradoxus* Sailfin Roughshark

Family Dalatiidae: Kitefin Sharks

- ☐ *Dalatias licha* Kitefin Shark
- ☐ *Euprotomicroides zantedeschia* Taillight Shark
- ☐ *Euprotomicrus bispinatus* Pygmy Shark
- ☐ *Heteroscymnoides marleyi* Longnose Pygmy Shark
- ☐ *Isistius brasiliensis* Cookiecutter or Cigar Shark
- ☐ *Isistius labialis* South China Cookiecutter Shark
- ☐ *Isistius plutodus* Largetooth Cookiecutter Shark
- ☐ *Mollisquama parini* Pocket Shark
- ☐ *Squaliolus aliae* Smalleye Pygmy Shark
- ☐ *Squaliolus laticaudus* Spined Pygmy Shark

Order Pristiophoriformes: Sawsharks

Family Pristiophoridae: Sawsharks

- ☐ *Pliotrema warreni* Sixgill Sawshark
- ☐ *Pristiophorus cirratus* Longnose Sawshark
- ☐ *Pristiophorus japonicus* Japanese Sawshark
- ☐ *Pristiophorus nudipinnis* Shortnose Sawshark
- ☐ *Pristiophorus schroederi* Bahamas Sawshark
- ☐ *Pristiophorus* sp. A Eastern Sawshark
- ☐ *Pristiophorus* sp. B Tropical Sawshark

- ☐ *Pristiophorus* sp. C Philippine Sawshark
- ☐ *Pristiophorus* sp. D Dwarf Sawshark (Western Indian Ocean)

Order Squatiniformes: Angel Sharks

Family Squatinidae: Angel Sharks

- ☐ *Squatina aculeata* Sawback Angelshark
- ☐ *Squatina africana* African Angelshark
- ☐ *Squatina argentina* Argentine Angelshark
- ☐ *Squatina armata* Chilean Angelshark
- ☐ *Squatina australis* Australian Angelshark
- ☐ *Squatina californica* Pacific Angelshark
- ☐ *Squatina dumeril* Sand Devil
- ☐ *Squatina formosa* Taiwan Angelshark
- ☐ *Squatina guggenheim* Hidden Angelshark
- ☐ *Squatina japonica* Japanese Angelshark
- ☐ *Squatina nebulosa* Clouded Angelshark
- ☐ *Squatina oculata* Smoothback Angelshark
- ☐ *Squatina punctata* Angular Angelshark
- ☐ *Squatina squatina* Angelshark
- ☐ *Squatina tergocellata* Ornate Angelshark
- ☐ *Squatina tergocellatoides* Ocellated Angelshark
- ☐ *Squatina* sp. A Eastern Angelshark
- ☐ *Squatina* sp. B Western Angelshark

Order Heterodontiformes: Bullhead Sharks

Family Heterodontidae: Bullhead Sharks

- ☐ *Heterodontus francisci* Horn Shark
- ☐ *Heterodontus galeatus* Crested Bullhead Shark
- ☐ *Heterodontus japonicus* Japanese Bullhead Shark
- ☐ *Heterodontus mexicanus* Mexican Hornshark
- ☐ *Heterodontus portusjacksoni* Port Jackson Shark
- ☐ *Heterodontus quoyi* Galapagos Bullhead Shark
- ☐ *Heterodontus ramalheira* Whitespotted Bullhead Shark
- ☐ *Heterodontus zebra* Zebra Bullhead Shark
- ☐ *Heterodontus* sp. A Oman Bullhead Shark

Order Orectolobiformes: Carpet Sharks

Family Parascylliidae: Collared Carpetsharks

- ☐ *Cirrhoscyllium expolitum* Barbelthroat Carpetshark
- ☐ *Cirrhoscyllium formosanum* Taiwan Saddled Carpetshark
- ☐ *Cirrhoscyllium japonicum* Saddled Carpetshark
- ☐ *Parascyllium collare* Collared Carpetshark
- ☐ *Parascyllium ferrugineum* Rusty Carpetshark
- ☐ *Parascyllium variolatum* Necklace Carpetshark
- ☐ *Parascyllium sparsimaculatum* Ginger Carpetshark

Family Brachaeluridae: Blind Sharks

- ☐ *Brachaelurus waddi* Blind Shark

- ☐ *Heteroscyllium colcloughi* Bluegray Carpetshark

Family Orectolobidae: Wobbegongs

- ☐ *Eucrossorhinus dasypogon* Tasselled Wobbegong
- ☐ *Orectolobus japonicus* Japanese Wobbegong
- ☐ *Orectolobus maculatus* Spotted Wobbegong
- ☐ *Orectolobus ornatus* Ornate Wobbegong
- ☐ *Orectolobus wardi* Northern Wobbegong
- ☐ *Orectolobus* sp. A Western Wobbegong
- ☐ *Sutorectus tentaculatus* Cobbler Wobbegong

Family Hemiscylliidae: Longtailed Carpetsharks

- ☐ *Chiloscyllium arabicum* Arabian Carpetshark
- ☐ *Chiloscyllium burmensis* Burmese Bambooshark
- ☐ *Chiloscyllium griseum* Gray Bambooshark
- ☐ *Chiloscyllium hasselti* Indonesian Bambooshark
- ☐ *Chiloscyllium indicum* Slender Bambooshark
- ☐ *Chiloscyllium plagiosum* Whitespotted Bambooshark
- ☐ *Chiloscyllium punctatum* Brownbanded Bambooshark
- ☐ *Hemiscyllium freycineti* Indonesian Speckled Carpetshark
- ☐ *Hemiscyllium hallstromi* Papuan Epaulette Shark
- ☐ *Hemiscyllium ocellatum* Epaulette Shark
- ☐ *Hemiscyllium strahani* Hooded Carpetshark
- ☐ *Hemiscyllium trispeculare* Speckeled Carpetshark

Family Ginglymostomatidae: Nurse Sharks

- ☐ *Ginglymostoma cirratum* Nurse Shark
- ☐ *Nebrius ferrugineus* Tawny Nurse Shark
- ☐ *Pseudoginglymostoma brevicaudatum* Shorttail Nurse Shark

Family Stegostomatidae: Zebra Sharks

- ☐ *Stegostoma fasciatum* Zebra Shark

Family Rhincodontidae: Whale Sharks

- ☐ *Rhincodon typus* Whale Shark

Order Lamniformes: Mackerel Sharks

Family Odontaspididae: Sandtiger Sharks

- ☐ *Carcharias taurus* Sandtiger or Spotted Raggedtooth
- ☐ *Odontaspis ferox* Smalltooth Sandtiger or Bumpytail Raggedtooth
- ☐ *Odontaspis noronhai* Bigeye Sandtiger

Family Pseudocarchariidae: Crocodile Sharks

- ☐ *Pseudocarcharias kamoharai* Crocodile Shark

Family Mitsukurinidae: Goblin Sharks

- ☐ *Mitsukurina owstoni* Goblin Shark

Family Megachasmidae; Megamouth Sharks

- ☐ *Megachasma pelagios* Megamouth Shark

Family Alopiidae: Thresher Sharks

- ☐ *Alopias pelagicus* Pelagic Thresher
- ☐ *Alopias superciliosus* Bigeye Thresher
- ☐ *Alopias vulpinus* Thresher Shark

Family Cetorhinidae: Basking Sharks

- ☐ *Cetorhinus maximus* Basking Shark

Family Lamnidae: Mackerel Sharks

- ☐ *Carcharodon carcharias* White Shark
- ☐ *Isurus oxyrinchus* Shortfin Mako
- ☐ *Isurus paucus* Longfin Mako
- ☐ *Lamna ditropis* Salmon Shark
- ☐ *Lamna nasus* Porbeagle Shark

Order Carcharhiniformes: Ground Sharks

Family Scyliorhinidae: Catsharks

- ☐ *Apristurus albisoma* White-bodied Catshark
- ☐ *Apristurus aphyodes* White Ghost Catshark
- ☐ *Apristurus brunneus* Brown Catshark
- ☐ *Apristurus canutus* Hoary Catshark
- ☐ *Apristurus exsanguis* Flaccid Catshark
- ☐ *Apristurus fedorovi* Stout Catshark
- ☐ *Apristurus gibbosus* Humpback Catshark
- ☐ *Apristurus herklotsi* Longfin Catshark
- ☐ *Apristurus indicus* Smallbelly Catshark
- ☐ *Apristurus internatus* Shortnose Demon Catshark
- ☐ *Apristurus investigatoris* Broadnose Catshark
- ☐ *Apristurus japonicus* Japanese Catshark
- ☐ *Apristurus kampae* Longnose Catshark
- ☐ *Apristurus laurussoni* Iceland Catshark
- ☐ *Apristurus longicephalus* Longhead Catshark
- ☐ *Apristurus macrorhynchus* Flathead Catshark
- ☐ *Apristurus macrostomus* Broadmouth Catshark
- ☐ *Apristurus manis* Ghost Catshark
- ☐ *Apristurus microps* Smalleye Catshark
- ☐ *Apristurus micropterygeus* Smalldorsal Catshark
- ☐ *Apristurus nasutus* Largenose Catshark
- ☐ *Apristurus parvipinnis* Smallfin Catshark
- ☐ *Apristurus pinguis* Fat Catshark
- ☐ *Apristurus platyrhynchus* Spatulasnout Catshark
- ☐ *Apristurus profundorum* Deepwater Catshark
- ☐ *Apristurus riveri* Broadgill Catshark
- ☐ *Apristurus saldanha* Saldanha Catshark
- ☐ *Apristurus sibogae* Pale Catshark
- ☐ *Apristurus sinensis* South China Catshark
- ☐ *Apristurus spongiceps* Spongehead Catshark
- ☐ *Apristurus stenseni* Panama Ghost Catshark

- ☐ *Apristurus* sp. A Freckled Catshark
- ☐ *Apristurus* sp. B Bigfin Catshark
- ☐ *Apristurus* sp. C Fleshynose Catshark
- ☐ *Apristurus* sp. D Roughskin Catshark
- ☐ *Apristurus* sp. E Bulldog Catshark
- ☐ *Apristurus* sp. F. Bighead Catshark
- ☐ *Apristurus* sp. G Pinocchio Catshark
- ☐ *Asymbolus analis* Grey Spotted Catshark
- ☐ *Asymbolus funebris* Blotched Catshark
- ☐ *Asymbolus occiduus* Western Spotted Catshark
- ☐ *Asymbolus pallidus* Pale Spotted Catshark
- ☐ *Asymbolus parvus* Dwarf Catshark
- ☐ *Asymbolus rubiginosus* Orange Spotted Catshark
- ☐ *Asymbolus submaculatus* Variegated Catshark
- ☐ *Asymbolus vincenti* Gulf Catshark
- ☐ *Atelomycterus fasciatus* Banded Sand Catshark
- ☐ *Atelomycterus macleayi* Australian Marbled Catshark
- ☐ *Atelomycterus marmoratus* Coral Catshark
- ☐ *Aulohalaelurus kanakorum* New Caledonia Catshark
- ☐ *Aulohalaelurus labiosus* Blackspotted Catshark
- ☐ *Bythaelurus canescens* Dusky Catshark
- ☐ *Bythaelurus clevai* Broadhead Catshark
- ☐ *Bythaelurus dawsoni* New Zealand Catshark
- ☐ *Bythaelurus hispidus* Bristly Catshark
- ☐ *Bythaelurus immaculatus* Spotless Catshark
- ☐ *Bythaelurus lutarius* Mud Catshark
- ☐ *Bythaelurus* sp. A Dusky Catshark
- ☐ *Bythaelurus* sp. B Galapagos catshark
- ☐ *Cephaloscyllium fasciatum* Reticulated Swellshark
- ☐ *Cephaloscyllium isabellum* Draughtsboard Shark
- ☐ Cephaloscyllium laticeps Australian Swellshark
- ☐ *Cephaloscyllium silasi* Indian Swellshark
- ☐ *Cephaloscyllium sufflans* Balloon Shark
- ☐ *Cephaloscyllium umbratile* Japanese Swellshark
- ☐ *Cephaloscyllium ventriosum* Swellshark
- ☐ *Cephaloscyllium* sp. A Whitefin Swellshark
- ☐ *Cephaloscyllium* sp. B Saddled Swellshark
- ☐ *Cephaloscyllium* sp. C Northern Draughtboard Shark
- ☐ *Cephaloscyllium* sp. D Narrowbar Swellshark
- ☐ *Cephaloscyllium* sp. E Speckled Swellshark
- ☐ *Cephaloscyllium* sp. Dwarf Oriental Swellshark
- ☐ *Cephalurus cephalus* Lollipop Catshark
- ☐ *Cephalurus* sp. A Southern Lollipop Catshark
- ☐ *Galeus antillensis* Antilles Catshark
- ☐ *Galeus arae* Roughtail Catshark
- ☐ *Galeus atlanticus* Atlantic Sawtail Catshark

- ☐ *Galeus boardmani* Australian Sawtail Catshark
- ☐ *Galeus cadenati* Longfin Sawtail Catshark
- ☐ *Galeus eastmani* Gecko Catshark
- ☐ *Galeus gracilis* Slender Sawtail Catshark
- ☐ *Galeus longirostris* Longnose Sawtail Catshark
- ☐ *Galeus melastomus* Blackmouth Catshark
- ☐ *Galeus mincaronei* Southern Sawtail Catshark
- ☐ *Galeus murinus* Mouse Catshark
- ☐ *Galeus nipponensis* Broadfin Sawtail Catshark
- ☐ *Galeus piperatus* Peppered Catshark
- ☐ *Galeus polli* African Sawtail Catshark
- ☐ *Galeus sauteri* Blacktip Sawtail Catshark
- ☐ *Galeus schultzi* Dwarf Sawtail Catshark
- ☐ *Galeus springeri* Springer's Sawtail Catshark
- ☐ *Galeus* sp. B Northern Sawtail Catshark
- ☐ *Halaelurus boesemani* Speckled Catshark
- ☐ *Halaelurus buergeri* Darkspot/Blackspotted Catshark
- ☐ *Halaelurus lineatus* Lined Catshark
- ☐ *Halaelurus natalensis* Tiger Catshark
- ☐ *Halaelurus quagga* Quagga Catshark
- ☐ *Haploblepharus edwardsii* Puffadder Shyshark or Happy Eddie
- ☐ *Haploblepharus fuscus* Brown Shyshark or Plain Happy
- ☐ *Haploblepharus pictus* Dark Shyshark or Pretty Happy
- ☐ *Haploblepharus* sp. A Eastern or Natal Shyshark or Happy Chappie
- ☐ *Holohalaelurus melanostigma* Tropical or Crying Izak
- ☐ *Holohalaelurus punctatus* Whitespotted or African Spotted Catshark
- ☐ *Holohalaelurus regani* Izak
- ☐ *Holohalaelurus polystigma* East African Spotted Catshark or Grinning Izak
- ☐ *Parmaturus campechiensis* Campeche Catshark
- ☐ *Parmaturus macmillani* New Zealand Filetail
- ☐ *Parmaturus melanobranchius* Blackgill Catshark
- ☐ *Parmaturus pilosus* Salamander Shark
- ☐ *Parmaturus xaniurus* Filetail Catshark
- ☐ *Parmaturus* sp. A Shorttail Catshark
- ☐ *Pentanchus profundicolus* Onefin Catshark
- ☐ *Poroderma africanum* Striped Catshark or Pyjama Shark
- ☐ *Poroderma pantherinum* Leopard Catshark
- ☐ *Schroederichthys bivius* Narrowmouth Catshark
- ☐ *Schroederichthys chilensis* Redspotted Catshark
- ☐ *Schroederichthys maculatus* Narrowtail Catshark
- ☐ *Schroederichthys saurisqualus* Southern Sawtail
- ☐ *Schroederichthys tenuis* Slender Catshark
- ☐ *Scyliorhinus besnardi* Polkadot Catshark
- ☐ *Scyliorhinus boa* Boa Catshark
- ☐ *Scyliorhinus canicula* Smallspotted Catshark
- ☐ *Scyliorhinus capensis* Yellowspotted Catshark

- ☐ *Scyliorhinus cervigoni* West African Catshark
- ☐ *Scyliorhinus comoroensis* Comoro Catshark
- ☐ *Scyliorhinus garmani* Brownspotted Catshark
- ☐ *Scyliorhinus haeckelii* Freckled Catshark
- ☐ *Scyliorhinus hesperius* Whitesaddled Catshark
- ☐ *Scyliorhinus meadi* Blotched Catshark
- ☐ *Scyliorhinus retifer* Chain Catshark
- ☐ *Scyliorhinus stellaris* Nursehound
- ☐ *Scyliorhinus tokubee* Izu Catshark
- ☐ *Scyliorhinus torazame* Cloudy Catshark
- ☐ *Scyliorhinus torrei* Dwarf Catshark

Family **Proscylliidae: Finback Catsharks**

- ☐ *Ctenacis fehlmanni* Harlequin Catshark
- ☐ *Eridacnis barbouri* Cuban Ribbontail Catshark
- ☐ *Eridacnis radcliffei* Pygmy Ribbontail Catshark
- ☐ *Eridacnis sinuans* African Ribbontail Catshark
- ☐ *Proscyllium habereri* Graceful Catshark
- ☐ *Proscyllium* sp. A Clown or Magnificent Catshark

Family **Pseudotriakidae: False Catsharks**

- ☐ *Gollum attenuatus* Slender Smoothhound or Gollumshark
- ☐ *Gollum* sp. A Sulu Gollumshark
- ☐ *Gollum* sp. B Whitemarked Gollumshark
- ☐ *Pseudotriakis microdon* False Catshark
- ☐ New genus/new species Pygmy False Catshark

Family **Leptochariidae: Barbeled Houndsharks**

- ☐ *Leptocharias smithii* Barbeled Houndshark

Family **Triakidae: Houndsharks**

- ☐ *Furgaleus macki* Whiskery Shark
- ☐ *Galeorhinus galeus* Tope, Soupfin or School Shark
- ☐ *Gogolia filewoodi* Sailback Houndshark
- ☐ *Hemitriakis abdita* Darksnout or Deepwater Sicklefin Houndshark
- ☐ *Hemitriakis falcata* Sicklefin Houndshark
- ☐ *Hemitriakis japanica* Japanese Topeshark
- ☐ *Hemitriakis leucoperiptera* Whitefin Topeshark
- ☐ *Hemitriakis* sp. A Ocellate Topeshark
- ☐ *Hypogaleus hyugaensis* Blacktip Topeshark or Pencil Shark
- ☐ *Iago garricki* Longnose Houndshark
- ☐ *Iago omanensis* Bigeye Houndshark
- ☐ *Iago* sp. A Lowfin Houndshark
- ☐ *Mustelus antarcticus* Gummy Shark
- ☐ *Mustelus asterias* Starry Smoothhound
- ☐ *Mustelus californicus* Grey Smoothhound
- ☐ *Mustelus canis* Dusky Smoothhound
- ☐ *Mustelus dorsalis* Sharpnose Smoothhound
- ☐ *Mustelus fasciatus* Striped Smoothhound

- ☐ *Mustelus griseus* Spotless Smoothhound
- ☐ *Mustelus henlei* Brown Smoothhound
- ☐ *Mustelus higmani* Smalleye Smoothhound
- ☐ *Mustelus lenticulatus* Spotted Estuary Smoothhound or Rig
- ☐ *Mustelus lunulatus* Sicklefin Smoothhound
- ☐ *Mustelus manazo* Starspotted Smoothhound
- ☐ *Mustelus mento* Speckled Smoothhound
- ☐ *Mustelus minicanis* Venezuelan Dwarf Smoothhound
- ☐ *Mustelus mosis* Arabian, Hardnose or Moses Smoothhound
- ☐ *Mustelus mustelus* Smoothhound
- ☐ *Mustelus norrisi* Narrowfin or Florida Smoothhound
- ☐ *Mustelus palumbes* Whitespot Smoothhound
- ☐ *Mustelus punctulatus* Blackspot Smoothhound
- ☐ *Mustelus schmitti* Narrownose Smoothhound
- ☐ *Mustelus sinusmexicanus* Gulf of Mexico Smoothhound
- ☐ *Mustelus whitneyi* Humpback Smoothhound
- ☐ *Mustelus* sp. A Grey Gummy Shark
- ☐ *Mustelus* sp. B Whitespotted Gummy Shark
- ☐ *Scylliogaleus quecketti* Flapnose Houndshark
- ☐ *Triakis acutipinna* Sharpfin Houndshark
- ☐ *Triakis maculata* Spotted Houndshark
- ☐ *Triakis megalopterus* Spotted Gully Shark
- ☐ *Triakis scyllium* Banded Houndshark
- ☐ *Triakis semifasciata* Leopard Shark

Family **Hemigaleidae: Weasel Sharks**

- ☐ *Chaenogaleus macrostoma* Hooktooth Shark
- ☐ *Hemigaleus microstoma* Sicklefin Weasel Shark
- ☐ *Hemigaleus* sp. A Australian Weasel Shark
- ☐ *Hemipristis elongatus* Snaggletooth Shark
- ☐ *Paragaleus leucolomatus* Whitetip Weasel Shark
- ☐ *Paragaleus pectoralis* Atlantic Weasel Shark
- ☐ *Paragaleus randalli* Slender Weasel Shark
- ☐ *Paragaleus tengi* Straighttooth Weasel Shark

Family **Carcharhinidae: Requiem Sharks**

- ☐ *Carcharhinus acronotus* Blacknose Shark
- ☐ *Carcharhinus albimarginatus* Silvertip Shark
- ☐ *Carcharhinus altimus* Bignose Shark
- ☐ *Carcharhinus amblyrhynchoides* Graceful Shark
- ☐ *Carcharhinus amblyrhynchos* Grey Reef Shark
- ☐ *Carcharhinus amboinensis* Pigeye or Java Shark
- ☐ *Carcharhinus borneensis* Borneo Shark
- ☐ *Carcharhinus brachyurus* Bronze Whaler or Copper Shark
- ☐ *Carcharhinus brevipinna* Spinner Shark
- ☐ *Carcharhinus cautus* Nervous Shark
- ☐ *Carcharhinus dussumieri* Whitecheek Shark

- ☐ *Carcharhinus falciformis* Silky Shark
- ☐ *Carcharhinus fitzroyensis* Creek Whaler
- ☐ *Carcharhinus galapagensis* Galapagos Shark
- ☐ *Carcharhinus hemiodon* Pondicherry Shark
- ☐ *Carcharhinus isodon* Finetooth Shark
- ☐ *Carcharhinus leiodon* Smoothtooth Blacktip
- ☐ *Carcharhinus leucas* Bull Shark
- ☐ *Carcharhinus limbatus* Blacktip Shark
- ☐ *Carcharhinus longimanus* Oceanic Whitetip Shark
- ☐ *Carcharhinus macloti* Hardnose Shark
- ☐ *Carcharhinus melanopterus* Blacktip Reef Shark
- ☐ *Carcharhinus obscurus* Dusky Shark
- ☐ *Carcharhinus perezi* Caribbean Reef Shark
- ☐ *Carcharhinus plumbeus* Sandbar Shark
- ☐ *Carcharhinus porosus* Smalltail Shark
- ☐ *Carcharhinus sealei* Blackspot Shark
- ☐ *Carcharhinus signatus* Night Shark
- ☐ *Carcharhinus sorrah* Spottail Shark
- ☐ *Carcharhinus tilsoni* Australian Blacktip Shark
- ☐ *Carcharhinus* sp. A False Smalltail Shark
- ☐ *Galeocerdo cuvier* Tiger Shark
- ☐ *Glyphis gangeticus* Ganges Shark
- ☐ *Glyphis glyphis* Speartooth Shark
- ☐ *Glyphis siamensis* Irrawaddy River Shark
- ☐ *Glyphis* sp. B Borneo River Shark
- ☐ *Glyphis* sp. C New Guinea River Shark
- ☐ *Isogomphodon oxyrhynchus* Daggernose Shark
- ☐ *Lamiopsis temmincki* Broadfin Shark
- ☐ *Loxodon macrorhinus* Sliteye Shark
- ☐ *Nasolamia velox* Whitenose Shark
- ☐ *Negaprion acutidens* Sharptooth Lemon Shark
- ☐ *Negaprion brevirostris* Lemon Shark
- ☐ *Prionace glauca* Blue Shark
- ☐ *Rhizoprionodon acutus* Milk Shark
- ☐ *Rhizoprionodon lalandei* Brazilian Sharpnose Shark
- ☐ *Rhizoprionodon longurio* Pacific Sharpnose Shark
- ☐ *Rhizoprionodon oligolinx* Grey Sharpnose Shark
- ☐ *Rhizoprionodon porosus* Caribbean Sharpnose Shark
- ☐ *Rhizoprionodon taylori* Australian Sharpnose Shark
- ☐ *Rhizoprionodon terraenovae* Atlantic Sharpnose Shark
- ☐ *Scoliodon laticaudus* Spadenose Shark
- ☐ *Triaenodon obesus* Whitetip Reef Shark

Family Sphyrnidae: Hammerhead Sharks

- ☐ *Eusphyra blochii* Winghead Shark
- ☐ *Sphyrna corona* Mallethead Shark
- ☐ *Sphyrna lewini* Scalloped Hammerhead

- ☐ *Sphyrna media* Scoophead Shark
- ☐ *Sphyrna mokarran* Great Hammerhead
- ☐ *Sphyrna tiburo* Bonnethead Shark
- ☐ *Sphyrna tudes* Smalleye Hammerhead
- ☐ *Sphyrna zygaena* Smooth Hammerhead